Student Solutions Manual

Principles of Physical Chemistry

LIONEL M. RAFF

OKLAHOMA STATE UNIVERSITY

Prentice Hall

PRENTICE HALL
Upper Saddle River, NJ 07458

Executive Editor: *John Challice*
Associate Editor: *Kristen Kaiser*
Editorial Assistants: *Gillian Buonanno and Eliana Ortiz*
Special Projects Manager: *Barbara A. Murray*
Production Editor: *Barbara A. Till*
Text Composition and Electronic Page Makeup: *WestWords, Inc.*
Supplemental Cover Manager: *Paul Gourhan*
Supplemental Cover Designer: *PMWorkshop, Inc.*
Manufacturing Buyer: *Lisa McDowell*

Prentice Hall

Upper Saddle River, New Jersey 07458

Printed in the United States of America
10 9 8 7 6 5 4 3 2 1

ISBN 0-13-040664-3

Prentice-Hall International (UK) Limited, *London*
Prentice-Hall of Australia Pty. Limited, *Sydney*
Prentice-Hall Canada Inc., *Toronto*
Prentice-Hall Hispanoamericana, S.A., *Mexico*
Prentice-Hall of India Private Limited, *New Delhi*
Prentice-Hall of Japan, Inc., *Tokyo*
Pearson Education Asia Pte. Ltd.
Editora Prentice-Hall do Brasil, Ltda., *Rio de Janeiro*

CONTENTS

CHAPTER 1

Properties of Gases

1.1 One mole of Ar atoms are confined in a vessel whose volume is 1,000 cm^3. If all of the atoms are moving with identical speeds of 230 m s^{-1} in the x, y, and z directions, compute the expected pressure, in kPa, inside the vessel. What is the pressure in atm?

Solution

Eq. 1.4 shows that

$$P = \frac{mv^2}{V} = \frac{[39.95 \text{ g mol}^{-1}][230]^2 \text{ m}^2 \text{ s}^{-2}[10^{-3} \text{ kg g}^{-1}]}{[10^3 \text{ cm}^3][10^{-2} \text{ m cm}^{-1}]^3}$$

$$= 2.113 \times 10^6 \text{ kg m s}^{-2}/\text{m}^2$$

$$= 2{,}113 \times 10^3 \text{ N m}^{-2} = \underline{2{,}113 \text{ kPa}}, \quad (1)$$

or in atm, we have

$$P = [2{,}113 \text{ kPa}]\left[\frac{1 \text{ atm}}{101.325 \text{ kPa}}\right] = \underline{20.85 \text{ atm}}. \quad (2)$$

1.3 It is found that the lengths of two metal rods, denoted A and B, are linear functions of the absolute temperature T. That is,

$$L_A = a_o + a_1 T,$$

where a_o is the length of rod A at $T = 0$ K and a_1 is a positive constant, and

$$L_B = b_o + b_1 T,$$

with similar definitions for b_o and b_1. An investigator now defines two temperature scales based on the lengths of rods A and B, respectively. The temperatures t_A and t_B are defined by

$$t_A = \frac{100\,(L_A - L_o)}{(L_{100} - L_o)}$$

and

$$t_B = \frac{100\,(L_B - L_o)}{(L_{100} - L_o)},$$

where L_o and L_{100} are the lengths of the rod at the normal freezing and boiling points of water, respectively.

(A) Determine the relationship between t_A and T.

(B) Show that $t_A = t_B$ at all values of T, even if $a_o \neq b_o$ and $a_1 \neq b_1$.

Solution

(A) Direct substitution of the dependence of L_A upon T into the definition of t_A gives

$$t_A = \frac{100[(a_o + a_1 T) - (a_o + 273.15a_1)]}{[(a_o + 373.15a_1) - (a_o + 273.15a_1)]}$$

$$= \frac{100a_1(T - 273.15)}{a_1(373.15 - 273.15)} = \boxed{T - 273.15}, \quad (1)$$

which is the required relationship between t_A and T. In deriving the result in Eq. 1, we have made use of the fact that T_o = 273. 15 K and T_{100} = 373.15 K for the normal freezing and boiling points of H_2O, respectively.

(B) Proceeding in the same manner for the temperature scale t_B, we obtain

$$t_B = \frac{100[(b_o + b_1T) - (b_o + 273.15b_1)]}{[(b_o + 373.15b_1) - (b_o + 273.15b_1)]}$$

$$= \frac{100b_1(T - 273.15)}{b_1(373.15 - 273.15)} = T - 273.15, \qquad (2)$$

and we have $t_A = t_B = T - 273.15$ for all values of T no matter what the values of a_o, a_1, b_o and b_1 are. Thus, if the rod length is a linear function of T for all materials, all thermometers will give the same temperature reading in all systems.

1.5 Here is a problem for hot-air balloon enthusiasts. An investigator decides to construct a thermometer using the volume of a balloon filled with Ar as a measuring device. She defines her temperature scale by $t = 75(V - V_{25})/(V_{100} - V_{25})$, where V_{25} and V_{100} are the volumes of the balloon at 25°C and 100°C, respectively, when the pressure is 1 atm. The thermometer is placed outside on the ground and slowly heated with a laser beam. The air temperature is 25°C, and the outside pressure is 1 atm. Heating is continued until the investigator notes that her thermometer has risen off the ground and is floating in the air. The balloon itself weighs 4 grams and contains 15 grams of Ar gas. What is the temperature of the thermometer, t, at the point it rises off the ground? Assume that all gases are ideal and that the air is 20% O_2 and 80% N_2 by mass. Ignore the effect of the balloon's elasticity on the pressure and volume of the balloon.

Solution

First, we need to compute the constants appearing in the definition for t, V_{25} and V_{100}. Since the gas is ideal,

$$V_{25} = \frac{nRT}{P} = \frac{(15\text{ g}/39.95\text{ g mol}^{-1})(0.08206\text{ L atm mol}^{-1}\text{ K}^{-1})(298.15\text{ K})}{1\text{ atm}}$$

$$= 9.186\text{ L} \qquad (1)$$

and

$$V_{100} = \frac{(15\text{ g}/39.95\text{ g mol}^{-1})(0.08206\text{ L atm mol}^{-1}\text{ K}^{-1})(373.15\text{ K})}{1\text{ atm}}$$

$$= 11.50\text{ L}. \qquad (2)$$

The balloon will rise off the ground when the density of the Ar + balloon equals that of the air. Therefore, we need to compute the air density. The average molar mass of the air is

$$\langle M \rangle = f_{O_2}M_{O_2} + f_{N_2}M_{N_2}, \qquad (3)$$

where f and M are the fraction by mass and the molar mass, respectively. The subscripts denote oxygen (O_2) and nitrogen (N_2).

$$\langle M \rangle = 0.2000(32.00) + 0.8000(28.00) = 28.80\text{ g mol}^{-1}. \qquad (4)$$

Using Eq. 1.39,

$$d_{air} = \frac{MP}{RT} = \frac{(28.80\text{ g mol}^{-1})(1\text{ atm})}{(0.08206\text{ L atm mol}^{-1}\text{ K}^{-1})(298.15\text{ K})} = 1.177\text{ g L}^{-1}. \qquad (5)$$

We now need to determine the volume of the balloon when the total density of the Ar and the balloon will equal d_{air}.

$$d_{\text{thermometer}} = \frac{(m_{\text{Ar}} + m_{\text{balloon}})}{V} = d_{\text{air}} = 1.177 \text{ g L}^{-1}. \quad (6)$$

Solving for V,

$$V = \frac{(m_{\text{Ar}} + m_{\text{balloon}})}{1.177 \text{ g L}^{-1}} = \frac{19 \text{ g}}{1.177 \text{ g L}^{-1}} = 16.14 \text{ L}. \quad (7)$$

Substituting into the definition for t, we compute

$$t = \frac{75(V - V_{25})}{(V_{100} - V_{25})} = \frac{75(16.14 - 9.186)}{(11.50 - 9.186)} = \underline{225.4} \quad (8)$$

at the point the balloon rises off the ground.

1.7 The coefficient of thermal expansion is defined to be $\alpha = V^{-1}(\partial V/\partial T)_P$.
(A) Obtain an expression for α for an ideal gas.
(B) Show that, for a Dieterici gas, α may be written in the form

$$\alpha = \frac{RV + a/T}{PV^2 \exp\{a/VRT\} - a}$$

where the notation $\exp\{x\}$means e^x and a is a parameter in the Dieterici equation of state, which is given in Table 1.2.

Solution

(A) Solving the ideal-gas equation for V, we obtain

$$V = \frac{nRT}{P}. \quad (1)$$

Taking the partial derivative of both sides with respect to T at constant P yields

$$\left(\frac{\partial V}{\partial T}\right)_P = \frac{nR}{P}, \quad (2)$$

so that

$$\boxed{\alpha = V^{-1}\left(\frac{\partial V}{\partial T}\right)_P = \frac{nR}{PV} = T^{-1}}, \quad (3)$$

which is the required expression.

(B) The Dieterici equation of state is

$$P(V - b)\exp\left[\frac{a}{RTV}\right] = RT, \quad (4)$$

where we have written V for the molar volume for simplicity. Differentiating both sides of Eq. 4 with respect to T at constant P gives

$$P\exp\left[\frac{a}{RTV}\right]\left(\frac{\partial V}{\partial T}\right)_P - P(V - b)\exp\left[\frac{a}{RTV}\right]\left[\frac{a}{RT^2V} + \frac{a}{RTV^2}\left(\frac{\partial V}{\partial T}\right)_P\right] = R. \quad (5)$$

Substituting Eq. 4 into the second term on the left side of Eq. 5 produces

$$P\exp\left[\frac{a}{RTV}\right]\left(\frac{\partial V}{\partial T}\right)_P - RT\left[\frac{a}{RT^2V} + \frac{a}{RTV^2}\left(\frac{\partial V}{\partial T}\right)_P\right] = R. \quad (6)$$

Collecting terms, we obtain

$$\left(\frac{\partial V}{\partial T}\right)_P\left[P\exp\left[\frac{a}{RTV}\right]-\frac{a}{V^2}\right]=R+\frac{a}{TV}. \tag{7}$$

Multiplying both sides by V^2 gives

$$\left(\frac{\partial V}{\partial T}\right)_P\left[PV^2\exp\left[\frac{a}{RTV}\right]-a\right]=RV^2+\frac{aV}{T}. \tag{8}$$

Therefore, we have

$$\left(\frac{\partial V}{\partial T}\right)_P=\frac{RV^2+\dfrac{aV}{T}}{\left[PV^2\exp\left[\dfrac{a}{RTV}\right]-a\right]}. \tag{9}$$

Consequently, α is given by

$$\boxed{\alpha=\frac{RV+\dfrac{a}{T}}{\left[PV^2\exp\left[\dfrac{a}{RTV}\right]-a\right]}}, \tag{10}$$

which is the required expression.

1.9 A container is divided into two compartments. Compartment A holds ideal gas A at 400 K and 5 atm of pressure. Compartment B is filled with ideal gas B at 400 K and 8 atm. The partition between the compartments is removed and the gases are allowed to mix. (It will be shown in later chapters that this mixing produces no change in temperature if the gases are ideal.) The mole fraction of A in the mixture is found to be $25/43 = 0.581395\ldots$. The total volume of both compartments is 29 liters. Determine the original volumes of compartments A and B.

Solution

We know P and T for both compartments. If we also knew n, the original number of moles of gas in each compartment, we could compute the volumes from the ideal-gas law. Consequently, let us determine n_A and n_B. We know that

$$X_A=\frac{n_A}{n_A+n_B}=\frac{25}{43}. \tag{1}$$

Both n_A and n_B can be written in terms of the ideal-gas law:

$$n_A=\frac{PV_A}{RT}=\frac{5V_A}{400R}$$

and

$$n_B=\frac{PV_B}{RT}=\frac{8V_B}{400R}. \tag{2}$$

Substitution of Eq. 2 into Eq. 1 gives

$$X_A=\frac{5V_A}{5V_A+8V_B}=\frac{25}{43}. \tag{3}$$

We also know that

$$V_A + V_B = 29 \text{ liters}. \quad \textbf{(4)}$$

Thus,

$$V_A = 29 - V_B. \quad \textbf{(5)}$$

Substituting this result into Eq. 3 gives

$$\frac{5(29 - V_B)}{5(29 - V_B) + 8V_B} = \frac{145 - 5V_B}{145 + 3V_B} = \frac{25}{43}. \quad \textbf{(6)}$$

Cross multiplying the fractions yields

$$6{,}235 - 215V_B = 3{,}625 + 75V_B. \quad \textbf{(7)}$$

Solving for V_B, we obtain

$$V_B = \frac{2{,}610}{290} = \underline{9 \text{ liters}}. \quad \textbf{(8)}$$

Using Eq. 4, we obtain V_A

$$V_A = 29 - V_B = \underline{20 \text{ liters}}. \quad \textbf{(9)}$$

1.11 A container is known to hold a pure rare gas. A 1-liter sample of the gas at 298 K and 1 atm pressure is found to weigh 3.427 grams. What gas is inside the container?

Solution

From Eq. 1.39, we have

$$M = \frac{dRT}{P} = \frac{(3.427 \text{ g L}^{-1})(0.08206 \text{ L atm mol}^{-1}\text{ K}^{-1})(298 \text{ K})}{1 \text{ atm}} = 83.80 \text{ g mol}^{-1}. \quad \textbf{(1)}$$

An inspection of the periodic table shows that this is the molar mass of krypton.

1.13 Consider the equation $Z = A[y^2 + x^2]\exp[-xy/a]$, where Z is a function of x and y, while a and A are constants. The notation $\exp[w]$ represents e^w. Z has units of joules, while x and y each have units of meters.

(A) What are the units on the constant A?

(B) What are the units on the constant a?

(C) Is it possible for Z to be given by the function

$$Z = B[y + x^2]\exp[-xy/a],$$

where B is another constant? Explain.

Solution

(A) Since the exponential is unitless, A must have units such that Ax^2 will be in joules. This means the units on $\boxed{A \text{ must be kg s}^{-2}}$ so that Ax^2 will have units of kg m^2 s^{-2}, which is a joule.

(B) The argument of the exponential must be unitless. Since xy has units of m^2, the units on a must also be $\boxed{\text{the units on } a \text{ must also be m}^2}$.

(C) No, it is not possible, since we can never add y to x^2, as y and x^2 have different units. All the terms in a sum must have the same units.

1.15* The ideal-gas constant is obtained by measuring P–V data for a real gas at a fixed temperature. The ratio PV/T is then computed at each measured pressure. The result is fitted by an appropriate least-squares procedure and extrapolated to zero pressure, at which point the gas will behave ideally. This problem illustrates the procedure. An investigator measures the pressure of 1 mole of a real gas at various volumes at a fixed temperature of 300 K. Her data are as follows:

Volume (liters)	Pressure (atm)
20	1.223046
21	1.165159
22	1.112504
23	1.064403
24	1.020288
25	0.979685
26	0.942189
27	0.907458
28	0.875196
29	0.845150
30	0.817097
35	0.700794
40	0.613473

(A) Compute the apparent value of R at each of the data points.

(B) Use the last six data points (at volumes $V = 27$ liters to $V = 40$ liters) to execute a least-squares fit of the computed value of R to a linear function of the pressure. That is, fit the function

$$R = a_o + a_1 P$$

to the computed values of R at the six lowest pressures.

(C) Using the fitted function, obtain the limit of R as $P \longrightarrow 0$.

(D) Plot the fitted function and compare your curve with the measured data points.

Solution

(A) Since we have

$$R = \frac{PV}{T}, \tag{1}$$

the calculation is straightforward. The results are shown in the table on the top of the next page.

(B) To execute the least-squares fit to a straight line, we use Eqs. 1.108 and 1.109. For this, we need four sums:

$$\sum_{i=1}^{6} R_i = 0.490319, \tag{2}$$

Volume (liters)	Pressure (atm)	$R = \frac{PV}{T}$ (L atm mol^{-1} K^{-1})
20	1.223046	0.0815364
21	1.165159	0.08156113
22	1.112504	0.08158363
23	1.064403	0.08160423
24	1.020288	0.08162304
25	0.979685	0.0816404
26	0.942189	0.0816564
27	0.907458	0.0816712
28	0.875196	0.0816850
29	0.845150	0.0816978
30	0.817097	0.0817097
35	0.700794	0.0817593
40	0.613473	0.0817964

where the last six points are labeled as $i = 1$ to $i = 6$;

$$\sum_{i=1}^{6} P_i = 4.75917; \tag{3}$$

$$\sum_{i=1}^{6} P_i^2 = 3.83884; \tag{4}$$

and

$$\sum_{i=1}^{6} P_i R_i = 0.388892. \tag{5}$$

We may now compute a_o from Eq. 1.108:

$$a_0 = \frac{(0.490319)(3.83884) - (4.75917)(0.388892)}{6(3.83884) - (4.75917)^2} = \underline{0.0820498}. \tag{6}$$

a_1 can be obtained using Eq. 1.109:

$$a_1 = \frac{6(0.388892) - (4.75917)(0.490319)}{6(3.83884) - (4.75917)^2} = \underline{-0.000416014}. \tag{7}$$

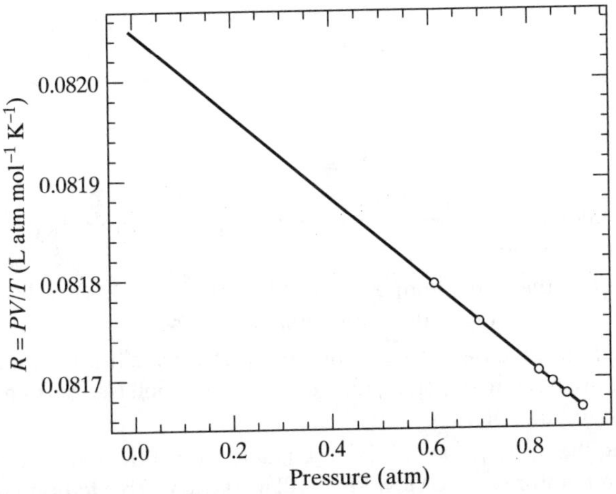

(C) The limiting value of R as $P \longrightarrow 0$ is a_o. Thus, we have

$$\lim_{P \longrightarrow 0} R = \underline{0.08205\ \text{L atm mol}^{-1}\ \text{K}^{-1}}. \quad (8)$$

This result is almost identical to best experimental value for R, 0.0820578 L atm $\text{mol}^{-1}\ \text{K}^{-1}$.

(D) The comparison is given in the plot on the previous page. The plot shows a comparison of the data (solid points) and the least-squares fit

$$R = 0.0820498 - 0.000416014P,$$

which is presented as a solid line, illustrated in the graph on the previous page.

1.17 All those concerned about the temperature of hell will find this problem interesting. Because the earth's population is increasing in a near-exponential fashion, it is reasonable to assume that the number of souls in hell is also increasing exponentially with time. That is,

$$n = \text{number of moles of souls in hell} = A \exp\{at\},$$

where A and a are positive constants. Since it is likely that at $t = 0$ we had only one soul (the Devil) in hell, we know that $A = 1/6.022 \times 10^{23}\ \text{mol}^{-1} = 1.66 \times 10^{-24}$ mol. It is also reasonable to assume that souls entering hell do not leave. Let us further assume that we may treat a collection of souls as an ideal gas. Under these conditions, the temperature of hell will be given by

$$T = \frac{PV}{nR} = \frac{PV}{AR \exp\{at\}}.$$

If the pressure of hell is constant at 1 atm, as it is on earth (this assumption is reasonable, since there are no suggestions that hell is a place of very high or very low pressure), the temperature will be dependent on how fast hell is expanding as souls enter it.

(A) There have been approximately 10^{10} people on earth since the Devil entered hell. If we assume that 10% of these people have entered hell and that the Devil was thrown into hell 5,000 years ago, determine the value of the constant a. What are the units on a?

(B) Because at $t = 0$ there was only one occupant of hell, the initial volume must have been rather small. The volume of an average house is probably a good estimate. If this house had 2,000 ft^2 of floor space and 8-ft ceilings, we would have $V_o = 16{,}000\ \text{ft}^3 = 4.5307 \times 10^5$ liters. Given that the number of souls in hell is rising exponentially, the volume must also be increasing in exponential fashion; that is

$$V = V_o \exp\{bt\}.$$

Use the foregoing assumptions to obtain the temperature of hell as a function of the parameter b and time.

(C) Compute the initial temperature of hell at $t = 0$. This result shows the origin and the inaccuracy of the popular slang phrase "hot as hell."

(D) Note that if we have $b < a$, hell will eventually freeze over. Let $b = a/2$, and compute how many years must elapse before hell freezes over (reaches the freezing point of water).

(E) Using the same value of b as in (D), determine how long it will take hell to reach a temperature of 1 K. How long will it take to reach a temperature of 0.01 K? How long will it take to reach a temperature of 10^{-10} K? How long will it take to reach a temperature of 0 K? The answer to this last question shows that $T = 0$ K is not attainable. We will later show that the second law of thermodynamics prohibits reaching absolute zero in hell or anywhere else.

Solution

(A) This is straightforward substitution of values:

$$n = A\exp[at]. \tag{1}$$

If at $t = 5{,}000$ years, we have

$$n = \frac{10^9 + 1(\text{for the Devil})}{6.022 \times 10^{23}} = 1.66 \times 10^{-15}\ \text{mol}. \tag{2}$$

Since $A = 1.66 \times 10^{-24}$ mol, we have, from Eq. 1,

$$1.66 \times 10^{-15}\ \text{mol} = (1.66 \times 10^{-24}\ \text{mol})\exp[5{,}000\, a] \tag{3}$$

when $t = 5{,}000$ years. Dividing by 1.66×10^{-24} mol and taking logarithms of both sides produces

$$(5{,}000\ \text{years})a = \ln\left[\frac{1.66 \times 10^{-15}}{1.66 \times 10^{-24}}\right] = \ln[10^9] = 20.72, \tag{4}$$

which gives

$$\underline{a = 0.004144\ \text{years}^{-1}} \tag{5}$$

(B) Substituting into the equation given in the problem yields

$$T = \frac{PV}{nR} = \frac{PV}{AR\exp\{at\}} = \frac{PV_o\exp\{bt\}}{AR\exp\{at\}}. \tag{6}$$

Substituting values produces

$$T = \frac{(4.5307 \times 10^5\ \text{L})(1\ \text{atm})\exp\{(b - 0.004144)t\}}{(1.66 \times 10^{-24}\ \text{mol})(0.08206\ \text{L atm mol}^{-1}\,\text{K}^{-1})}$$

$$= \boxed{3.33 \times 10^{30}\exp[(b - 0.004144)t]\ \text{K}} \tag{7}$$

provided that we measure t in years. This is the desired function for the temperature of hell.

(C) At $t = 0$, Eq. 7 shows the temperature to be

$$T(t = 0) = 3.33 \times 10^{30}\ \text{K}. \qquad \text{(PRETTY HOT!)}$$

(D) If $b = a/2 = 0.002072$ years^{-1}, then we have

$$T = 3.33 \times 10^{30}\exp[-0.002072t]. \tag{8}$$

Equating T to the freezing point of H_2O, we obtain

$$273.15 = 3.33 \times 10^{30}\exp[-0.002072t], \tag{9}$$

which must be solved for t. Taking logarithms,

$$-0.002072t = \ln\left[\frac{273.15}{3.33 \times 10^{30}}\right] = -64.67, \tag{10}$$

from which we compute

$$t = \underline{3.12 \times 10^4} \tag{11}$$

years until hell freezes over.

(E) Using Eq. 8, we may repeat the calculation to obtain the time required for hell to reach any specified temperature. The results are as follows:

T (K) of hell	time (years)
1	3.39×10^4
0.01	3.61×10^4
10^{-10}	4.50×10^4
0	∞

Obviously, hell is not going to reach absolute zero.

1.19 The intermolecular forces between two gas molecules are described by the $LJ(12, 6)$ potential given in Eq. 1.55. Show that the potential minimum occurs at $r = \sigma$ and that the well depth, defined as $[V_{LJ}(r = \infty) - V_{LJ}(r = \sigma)]$, is equal to ε.

Solution

The $LJ(12, 6)$ potential is given by

$$V_{LJ}(r) = \varepsilon[(\sigma/r)^{12} - 2(\sigma/r)^6]. \quad (1)$$

At the minimum, we have $\partial V_{LJ}(r)/\partial r = 0$. The first derivative is given by

$$\frac{\partial V_{LJ}(r)}{\partial r} = \varepsilon[-12\sigma^{12}r^{-13} + 12\sigma^6 r^{-7}]\Big|_{r = r_{min}} = 0. \quad (2)$$

Thus,

$$12\varepsilon[-\sigma^{12}r_{min}^{-13} + \sigma^6 r_{min}^{-7}] = 0. \quad (3)$$

Solving for r_{min}, we obtain

$$r_{min}^6 = \sigma^6 \text{ or } \boxed{r_{min} = \sigma}, \quad (4)$$

as required.

Substituting the result of Eq. 4 into Eq. 1 gives

$$V_{LJ}(r = \sigma) = \varepsilon[(\sigma/\sigma)^{12} - 2(\sigma/\sigma)^6] = -\varepsilon. \quad (5)$$

We also have

$$V_{LJ}(r = \infty) = \varepsilon[(\sigma/\infty)^{12} - 2(\sigma/\infty)^6] = 0. \quad (6)$$

Therefore, the well depth is

$$\text{Well depth} = [V_{LJ}(r = \infty) - V_{LJ}(r = \sigma)] = 0 - (-\varepsilon) = \varepsilon, \quad (7)$$

as required by the problem.

1.21 A helium nucleus is separated from an electron by a distance of 5 Å (5×10^{-10} m). If both particles are treated as point charges, compute the potential energy between them in units of joules and kcal mol^{-1}.

Solution

From Eq. 1.49, we have

$$V(r) = \left[\frac{q_1 q_2}{4\pi\varepsilon_0 r}\right]. \quad (1)$$

The charge on the proton is 1.6022×10^{-19} C. The electron charge is the negative of this value. The He nucleus contains 2 protons. Thus, direct substitution gives

$$V(r = 5\text{Å}) = -\frac{(2)(1.6022 \times 10^{-19}\ \text{C})^2}{(4)(3.14159)(8.854 \times 10^{-12}\ \text{J}^{-1}\ \text{C}^2\ \text{m}^{-1})(5 \times 10^{-10}\ \text{m})}$$

$$= \underline{-9.229 \times 10^{-19}\ \text{J}} \qquad (2)$$

Conversion to kcal mol^{-1} is as follows:

$$-9.229 \times 10^{-19}\ \text{J}[6.022 \times 10^{23}\ \text{mol}^{-1}]\left[\frac{1\ \text{cal}}{4.184\ \text{J}}\right]\left[\frac{1\ \text{kcal}}{1{,}000\ \text{cal}}\right] = \underline{-132.8\ \text{kcal mol}^{-1}}. \qquad (3)$$

1.23 What mass of N_2 gas is present in a 50-liter container at 400 K under 20 atm of N_2 pressure if

(A) the gas is ideal and

(B) the gas obeys the van der Waals equation of state?

Solution

(A) Expressed in terms of the mass, the ideal-gas equation is

$$PV = \frac{mRT}{M}, \qquad (1)$$

where m is the gaseous mass and M is the molar mass. Solving for m, we obtain

$$m = \frac{PVM}{RT} = \frac{(20\ \text{atm})(50\ \text{L})(28.00\ \text{g mol}^{-1})}{(0.08206\ \text{L atm mol}^{-1}\ \text{K}^{-1})(400\ \text{K})} = \underline{853.0\ \text{g of}\ N_2}. \qquad (2)$$

(B) The van der Waals equation of state for n moles of gas is

$$\left(P + \frac{n^2a}{V^2}\right)(V - nb) = nRT. \qquad (3)$$

We wish to solve for n. Expanding the equation gives

$$PV - Pnb + \frac{n^2a}{V} - \frac{n^3ab}{V^2} - nRT = 0. \qquad (4)$$

Collecting terms in n, we have

$$\frac{n^3ab}{V^2} + \frac{n^2a}{V} - n(Pb + RT) + PV = 0. \qquad (5)$$

The equation is, therefore, cubic in n, and we must effect a numerical solution. Inserting the values of the quantities from the problem and Table 1.1 gives

$$\frac{ab}{V^2} = \frac{(1.37\ \text{L}^2\ \text{bar mol}^{-2})(0.0387\ \text{L mol}^{-1})}{(50\ \text{L})^2} = 2.12 \times 10^{-5}\ \text{L bar mol}^{-3} \qquad (6)$$

$$\frac{a}{V} = \frac{1.37\ \text{L}^2\ \text{bar mol}^{-2}}{50\ \text{L}} = 0.0274\ \text{L bar mol}^{-2}. \qquad (7)$$

$$Pb = (20\ \text{atm})\frac{1.01325\ \text{bar}}{\text{atm}}(0.0387\ \text{L mol}^{-1}) = 0.784\ \text{L bar mol}^{-1} \qquad (8)$$

$$RT = (0.083145\ \text{L bar mol}^{-1}\ \text{K}^{-1})(400\ \text{K}) = 33.258\ \text{L bar mol}^{-1} \qquad (9)$$

and

$$PV = (20\ \text{atm})\frac{1.01325\ \text{bar}}{\text{atm}}(50\ \text{L}) = 1{,}013.25\ \text{L bar}. \qquad (10)$$

Note how the pressure in Eqs. 8 and 10 had to be converted to maintain a consistent set of units. Thus, Eq. 5 becomes

$$2.12 \times 10^{-5} n^3 + 0.0274 n^2 - 34.042 n + 1{,}013.25 = 0. \qquad \textbf{(11)}$$

To solve this cubic equation, we search for the roots of the function

$$F(n) = 2.12 \times 10^{-5} n^3 + 0274 n^2 - 34.042 n + 1{,}013.25 = 0. \qquad \textbf{(12)}$$

Since the molar mass of N_2 is 28.00 g mol^{-1}, the ideal-gas result suggests that the answer lies in the region around 30 to 31 moles if N_2 is close to ideal. A grid search shows that the sign on $F(n)$ changes between $n = 30.5325$ mol and $n = 30.5335$ mol. A second search in this range shows that the root is 30.533 mol. The mass of N_2 is, therefore,

$$m = nM = 30.533 \text{ mol} \times 28.00 \text{ g mol}^{-1} = \underline{854.9 \text{ g of } N_2} \qquad \textbf{(13)}$$

if it is a van der Waals gas.

Note: To obtain the same 20-atm pressure at 400 K for a van der Waals gas in a 50-L container requires more $N_2(g)$ than is the case if the gas is ideal. The physical reason underlying this result lies in the attractive forces between N_2 molecules that are present in a van der Waals gas, but not an ideal gas. These forces tend to hold the N_2 molecules together and thereby reduce the force they exert when they collide with the container walls. To compensate for this reduced impact force, we must have more wall collisions per second for the van der Waals gas, which means that more $N_2(g)$ molecules are required. Hence, more N_2 mass is present in the van der Waals gas.

1.25 Using the data in Table 1.1, estimate the second and third virial coefficients for CO_2 at 300 K. At what temperature would we expect the second virial coefficient for CO_2 to be zero?

Solution

If CO_2 behaves as a van der Waals gas, the second virial coefficient will be

$$C_2(T) = b - \frac{a}{RT} = 0.04286 \text{ L mol}^{-1} - \frac{3.658 \text{ L}^2 \text{ bar mol}^{-2}}{(0.083145 \text{ L bar mol}^{-1} \text{ K}^{-1})(300 \text{ K})}$$

$$= \underline{-0.1038 \text{ L mol}^{-1}}. \qquad \textbf{(1)}$$

For $C_2(T)$ to be zero, we must have

$$b = \frac{a}{RT}. \qquad \textbf{(2)}$$

This will occur when T is given by

$$T = \frac{a}{bR} = \frac{(3.658 \text{ L}^2 \text{ bar mol}^{-2})}{(0.04286 \text{ L mol}^{-1})(0.083145 \text{ L bar mol}^{-1} \text{ K}^{-1})} = \underline{1026 \text{ K}}. \qquad \textbf{(3)}$$

1.27 Consider a gas that obeys the equation of state

$$P = \frac{RT}{V} \exp\left[-\frac{a}{VRT}\right],$$

where a is a constant and the notation $\exp[x]$ means e^x.

(A) Determine the second and third virial coefficients for this gas as a function of a, R, and T.

(B) Determine the residual volume of the gas as a function of a, R, and T.

Solution

(A) To expand in a virial-type equation, we first change variables. Let $y = V^{-1}$. Substitution into the equation of state yields

$$P = RTy \exp\left[-\frac{ay}{RT}\right] = F(y). \tag{1}$$

We now expand $F(y)$ in a power series in y:

$$P = F(y) = \sum_{i=0}^{\infty} a_i y^i. \tag{2}$$

The coefficients, a_i, are given by

$$a_0 = F(0) = 0, \tag{3}$$

$$a_1 = \left.\frac{\partial F(y)}{\partial y}\right|_{y=0} = F^1(0) = RT \exp\left[-\frac{ay}{RT}\right] + RTy \exp\left[-\frac{ay}{RT}\right]\left[-\frac{a}{RT}\right]$$

$$= RT \exp\left[-\frac{ay}{RT}\right] - ay \exp\left[-\frac{ay}{RT}\right]\Bigg|_{y=0} = RT, \tag{4}$$

$$a_2 = (2!)^{-1}F^2(0)$$

$$= (2!)^{-1}\left[-a \exp\left[-\frac{ay}{RT}\right] - a \exp\left[-\frac{ay}{RT}\right] + \frac{a^2y}{RT}\exp\left[-\frac{ay}{RT}\right]\right]\Bigg|_{y=0}$$

$$= (2!)^{-1}\left[-2a \exp\left[-\frac{ay}{RT}\right] + \frac{a^2y}{RT}\exp\left[-\frac{ay}{RT}\right]\right]\Bigg|_{y=0} = -a, \tag{5}$$

and

$$a_3 = (3!)^{-1}\left[\frac{2a^2}{RT}\exp\left[-\frac{ay}{RT}\right] + \frac{a^2}{RT}\exp\left[-\frac{ay}{RT}\right] - \frac{a^3y}{R^2T^2}\exp\left[-\frac{ay}{RT}\right]\right]\Bigg|_{y=0}$$

$$= (3!)^{-1}\left[\frac{3a^2}{RT}\exp\left[-\frac{ay}{RT}\right] - \frac{a^3y}{R^2T^2}\exp\left[-\frac{ay}{RT}\right]\right]\Bigg|_{y=0} = \frac{a^2}{2RT}. \tag{6}$$

The virial expansion is, therefore,

$$P = F(y) = RT\left[y - \frac{a}{RT}y^2 + \frac{a^2}{2R^2T^2}y^3 + \cdots\right]. \tag{7}$$

The second virial coefficient is

$$\boxed{C_2(T) = -\frac{a}{RT}}, \tag{8}$$

and the third virial coefficient is

$$\boxed{C_3(T) = \frac{a^2}{2R^2T^2}}. \tag{9}$$

(B) Multiplying Eq. 7 by V and replacing y with V^{-1}, we obtain

$$PV = RT - \frac{a}{V} + \frac{a^2}{2RTV^2} + \ldots. \tag{10}$$

Dividing both sides of Eq. 10 by P gives

$$V = \frac{RT}{P} - \frac{a}{PV} + \frac{a^2}{2RTPV^2} + \cdots. \tag{11}$$

Rearranging Eq. 11 and taking the limit of both sides as $P \longrightarrow 0$ and $V \longrightarrow \infty$, we obtain

$$\lim_{P \longrightarrow 0}\left[V - \frac{RT}{P}\right] = V_{\text{res}} = \lim_{P \longrightarrow 0}\left[-\frac{a}{PV} + \frac{a^2}{2RTPV^2} + \cdots\right] \tag{12}$$

Since $PV \longrightarrow RT$ at low pressure, we have

$$\boxed{V_{\text{res}} = \lim_{P \longrightarrow 0}\left[-\frac{a}{RT} + \frac{a^2}{2R^2T^2V} + \cdots\right] = -\frac{a}{RT}}. \tag{13}$$

We could also do Part (A) by using the series expansion for the exponential. We then obtain

$$P = F(y) = RTy\left[1 - \frac{ay}{RT} + \frac{a^2}{2R^2T^2}y^2 + \cdots\right]$$

$$= RT\left[y - \frac{a}{RT}y^2 + \frac{a^2}{2R^2T^2}y^3 + \cdots\right], \tag{14}$$

which is identical to Eq. 7. This method is faster, but it is less instructive.

1.29 A gas is represented by the equation of state

$$P = RT\left[\frac{1}{V} + \frac{a}{V^2} + \frac{b}{V^3}\right],$$

where a and b are constants. By requiring that this equation of state satisfy the three constraints at the critical point, express R, a, and b in terms of the critical variables, and put the equation of state in reduced form.

Solution

The three constraints required to fit the critical point are

$$P_c = RT_c\left[\frac{1}{V_c^2} + \frac{2a}{V_c^2} + \frac{b}{V_c^3}\right], \tag{1}$$

$$\left.\frac{\partial P}{\partial V}\right|_c = -RT_c\left[\frac{1}{V_c^2} + \frac{2a}{V_c^3} + \frac{3b}{V_c^4}\right] = 0, \tag{2}$$

and

$$\left.\frac{\partial^2 P}{\partial V^2}\right|_c = RT_c\left[\frac{2}{V_c^3} + \frac{6a}{V_c^4} + \frac{12b}{V_c^5}\right] = 0. \tag{3}$$

The factor RT_c in Eqs. 2 and 3 may be divided out. If we use Eq. 2 to solve for a in terms of b, we obtain

$$\frac{2a}{V_c^3} = -\frac{1}{V_c^2} - \frac{3b}{V_c^4}, \tag{4}$$

which may be rearranged to give

$$a = \frac{V_c^3}{2}\left[-\frac{1}{V_c^2} - \frac{3b}{V_c^4}\right] = -\frac{V_c}{2} - \frac{3b}{2V_c}. \tag{5}$$

Substituting this result into Eq. 3 permits us to obtain b in terms of the critical variables:

$$\frac{2}{V_c^3} + 6\left[-\frac{V_c}{2} - \frac{3b}{2V_c}\right]\frac{1}{V_c^4} + \frac{12b}{V_c^5} = 0$$

$$= \frac{2}{V_c^3} - \frac{3}{V_c^3} - \frac{9b}{V_c^5} + \frac{12b}{V_c^5} = -\frac{1}{V_c^3} + \frac{3b}{V_c^5} \quad (6)$$

Solving for b, we obtain

$$\boxed{b = \frac{V_c^2}{3}}. \quad (7)$$

Combining Eqs. 5 and 7 gives

$$\boxed{a = -\frac{V_c}{2} - \frac{3}{2V_c}\left[\frac{V_c^2}{3}\right] = -\frac{V_c}{2} - \frac{V_c}{2} = -V_c}. \quad (8)$$

Substituting Eqs. 7 and 8 into Eq. 1 yields

$$P_c = RT_c\left[\frac{1}{V_c} - \frac{V_c}{V_c^2} + \frac{V_c^2}{3V_c^3}\right] = RT_c\left[\frac{1}{V_c} - \frac{1}{V_c} + \frac{1}{3V_c}\right] = \frac{RT_c}{3V_c}, \quad (9)$$

which gives

$$\boxed{R = \frac{3P_cV_c}{T_c}}. \quad (10)$$

Substituting Eqs. 7, 8, and 10 into the equation of state yields

$$P = \frac{3P_cV_c}{T_c}T\left[V^{-1} - V_cV^{-2} + \frac{V_c^2V^{-3}}{3}\right] = 3P_cT_R\left[V_R^{-1} - V_R^{-2} + \frac{V_R^{-3}}{3}\right]. \quad (11)$$

Thus, the reduced form of the equation of state is

$$\boxed{P_R = 3T_R\left[\frac{1}{V_R} - \frac{1}{V_R^2} + \frac{1}{3V_R^3}\right]}. \quad (12)$$

This form of the equation of state will not be as accurate as the form containing a and b where a and b are adjusted by a least-squares method to fit P–V–T data.

1.31 For a gas represented by a virial equation of state, determine the residual volume in terms of the virial coefficients.

Solution

The virial equation of state is

$$P = RT\left[\frac{1}{V_m} + \sum_{n=2}^{\infty} C_n(T)V_m^{-n}\right], \quad (1)$$

where V_m is the molar volume and the $C_n(T)$ are the virial expansion coefficients. To obtain the residual volume, we need to formulate an analytic expression for $V_m - RT/P$. This is easily done by multiplying both sides by V_m and then dividing by P. These operations produce

$$V_m = \frac{RT}{P}\left[1 + \sum_{n=2}^{\infty} C_n(T)V_m^{-n+1}\right]. \quad (2)$$

We now subtract $[RT/P]$ from both sides to obtain

$$\left[V_m - \frac{RT}{P}\right] = \frac{RT}{P}\sum_{n=2}^{\infty} C_n(T)V_m^{-n+1}, \quad \textbf{(3)}$$

which we can write in the form

$$\left[V_m - \frac{RT}{P}\right] = RT\sum_{n=2}^{\infty}(PV_m)^{-1}C_n(T)V_m^{-n+2}. \quad \textbf{(4)}$$

We now take the limit of both sides as P approaches zero and use the fact that $PV_m \longrightarrow RT$ in this limit. This gives

$$\lim_{P\longrightarrow 0}\left[V_m - \frac{RT}{P}\right] = \lim_{P\longrightarrow 0}\left[RT\sum_{n=2}^{\infty}(PV_m)^{-1}C_n(T)V_m^{-n+2}\right]$$

$$= \lim_{P\longrightarrow 0}\left[\sum_{n=2}^{\infty} C_n(T)V_m^{-n+2}\right], \quad \textbf{(5)}$$

where the limit $P \longrightarrow 0$ corresponds to having $V_m \longrightarrow \infty$.

Clearly the only terms that will remain on the right-hand side are those that do not contain factors of $(1/V_m)$ to some power other than zero. The only such term is $n = 2$. Therefore, we have

$$\boxed{V_{\text{res}} = \lim_{P\longrightarrow 0}\left[V_m - \frac{RT}{P}\right] = C_2(T)}. \quad \textbf{(6)}$$

The residual volume of a virial gas is given by the second virial coefficient.

V (liters)	P (atm)
2.000	11.297
2.590	8.903
3.179	7.342
3.769	6.246
4.359	5.435
4.949	4.809
5.538	4.313
6.128	3.909
6.718	3.575
7.308	3.293
7.897	3.052
8.487	2.844
9.077	2.663
9.667	2.503
10.256	2.362
10.846	2.235
11.436	2.122
12.026	2.019
12.615	1.926
13.205	1.841
13.795	1.763
14.385	1.692
14.974	1.626
15.564	1.565
16.154	1.508
16.744	1.456
17.333	1.407
17.923	1.361
18.513	1.318
19.103	1.278
19.692	1.240
20.282	1.204
20.872	1.170
21.462	1.138
22.051	1.108
22.641	1.079
23.231	1.052
23.821	1.026
24.410	1.002

1.33* The set of pressure and volume data in the table to the left is obtained for 1 mole of a nonideal gas at 300 K.

(A) Using a least-squares method, fit the data to a truncated virial equation of state of the form $P = a_o + a_1 y + a_2 y^2$, where $y = V^{-1}$, and obtain the best values of a_o, a_1, and a_2. Be certain to show the equations that are being solved, and give the values of all required sums.

(B) Plot the data and the fit obtained in (A) on the same graph.

(C) Is the expression used in (A) an appropriate equation of state at low pressures and large volumes? Explain.

(D) Set $a_o = 0$ and $a_1 = RT$ in the equation in (A), and use a least-squares procedure to obtain a_2.

(E) Estimate the residual volume for this nonideal gas.

Solution

(A) The required equations are those given in Eq. 1.106 in the text. There are 39 data points. Therefore, we need the followings sums:

$$\sum_{i=1}^{39} y_i = 4.5247, \quad \textbf{(1)}$$

$$\sum_{i=1}^{39} y_i^2 = 0.91626, \quad \textbf{(2)}$$

$$\sum_{i=1}^{39} y_i^3 = 0.28203, \quad \textbf{(3)}$$

$$\sum_{i=1}^{39} y_i^4 = 0.10776, \quad \textbf{(4)}$$

$$\sum_{i=1}^{39} P_i = 107.682, \quad \textbf{(5)}$$

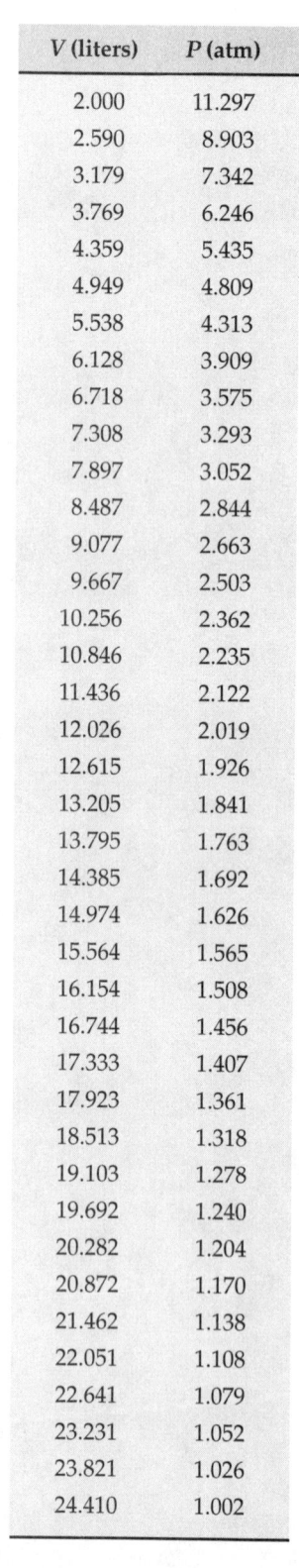

$$\sum_{i=1}^{39} P_i y_i = 21.415, \tag{6}$$

and

$$\sum_{i=1}^{39} P_i y_i^2 = 6.5069. \tag{7}$$

The equations that must be solved are

$$39a_0 + 4.5247a_1 + 0.91626a_2 = 107.682;$$

$$4.5247a_0 + 0.91626a_1 + 0.28203a_2 = 21.415;$$

$$0.91626a_0 + 0.28203a_1 + 0.10776a_2 = 6.5069. \tag{8}$$

The determinant of the coefficients is

$$\begin{vmatrix} 39 & 4.5247 & 0.91626 \\ 4.5247 & 0.91626 & 0.28203 \\ 0.91626 & 0.28203 & 0.10776 \end{vmatrix} = 0.11171 = D. \tag{9}$$

The determinants required for each solution are

$$N_1 = \begin{vmatrix} 107.682 & 4.5247 & 0.91626 \\ 21.415 & 0.91626 & 0.28203 \\ 6.5069 & 0.28203 & 0.10776 \end{vmatrix} = 0.000038, \tag{10}$$

$$N_2 = \begin{vmatrix} 39 & 107.682 & 0.91626 \\ 4.5247 & 21.415 & 0.28203 \\ 0.91626 & 6.5069 & 0.10776 \end{vmatrix} = 2.7494, \tag{11}$$

and

$$N_3 = \begin{vmatrix} 39 & 4.5247 & 107.682 \\ 4.5247 & 0.91626 & 21.415 \\ 0.91626 & 0.28203 & 6.5069 \end{vmatrix} = -0.45099. \tag{12}$$

The least-squares coefficients are, therefore,

$$\boxed{a_o = \frac{N_1}{D} = 0.00034}, \tag{13}$$

$$\boxed{a_1 = \frac{N_2}{D} = 24.612}, \tag{14}$$

and

$$\boxed{a_2 = \frac{N_3}{D} = -4.0372}. \tag{15}$$

As a partial check on the solution, we may compare the results with the ideal-gas law. We expect the leading term, a_o, to be near zero, and it is. We also expect a_1 to be approximately RT. The value of RT at 300 K is 24.618 L atm mol^{-1} K^{-1}, which we see is very nearly a_1.

(B) The requested plot is shown at the top of the next page.

(C) If we take the equation of state at face value, we have

$$P = a_o + a_1 y + a_2 y^2 = a_o + \frac{a_1}{V_m} + \frac{a_2}{V_m^2}. \tag{16}$$

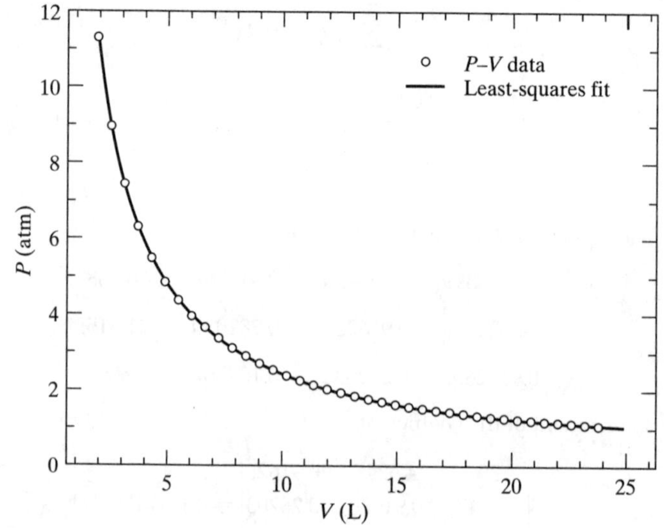

Multiplying by V_m and then dividing by RT gives

$$\frac{PV_m}{RT} = \frac{a_o V_m}{RT} + \frac{a_1}{RT} + \frac{a_2}{RTV_m}. \tag{17}$$

If we now take the limit of both sides as $P \longrightarrow 0$ and $V_m \longrightarrow \infty$, we obtain

$$\lim_{P \longrightarrow 0}\left[\frac{PV_m}{RT}\right] = \lim_{P \longrightarrow 0}\left[\frac{a_o V_m}{RT} + \frac{a_1}{RT} + \frac{a_2}{RTV_m}\right] = \infty + 0.99977\cdots = \infty, \tag{18}$$

which is an unacceptable result since the ratio $(PV_m)/(RT)$ must approach unity as $P \longrightarrow 0$. Consequently, although the fit is good over the range of volumes considered, the equation cannot be used as $P \longrightarrow 0$ and $V_m \longrightarrow \infty$.

(D) The equation $P = a_o + a_1 y + a_2 y^2$ cannot be used in the low-pressure, large-volume regime because of the decision to fit the P–V–T data to a form that includes a nonzero a_o term and to permit a_1 to be a fitting parameter. The normal way of analyzing the data is to fit the data to a virial expansion of the form

$$P = RT[V_m^{-1} + C_2(T)V_m^{-2}]. \tag{19}$$

Such an expansion means we are requiring that $a_o = 0$ and that $a_1 = RT$. Consequently, we have only one parameter available with which to fit the data, $C_2(T)$, which plays the role of a_2 in our problem. With these restrictions, the equation that must be solved to obtain the best value of a_2 is simply the third equation in Eq. 1.106 with $a_o = 0$ and $a_1 = RT$. This gives

$$a_2 = \frac{\sum_{n=1}^{39} P_i y_i^2 - RT\sum_{n-1}^{39} y_i^3}{\sum_{n=1}^{39} y_i^4} = \frac{6.5069 - (0.08206)(300)(0.28203)}{0.10776}$$
$$= -4.0471 \text{ L}^2 \text{ atm mol}^{-2}. \tag{20}$$

As can be seen, this result differs from that in Eq. 15 only in the third significant digit.

(E) With $a_o = 0$ and $a_1 = RT$, we have

$$\left[V_m - \frac{RT}{P}\right] = \frac{a_2}{PV_m}. \tag{21}$$

Taking the limit of both sides as $P \longrightarrow 0$, we obtain the following after substituting RT for PV_m in the limit:

$$V_{\text{res}} = \frac{a_2}{RT} = \frac{-4.0471 \text{ L}^2 \text{ atm mol}^{-2}}{(0.08206 \text{ L atm mol}^{-1} \text{ K}^{-1})(300 \text{ K})} = \underline{-0.1642 \text{ L mol}^{-1}}. \quad (22)$$

1.35 (A) Develop an expression for the compression factor for a gas described by a virial equation of state.
(B) Show that $Z(T, P) \to 1$ as $P \to 0$.

Solution

(A) The virial equation of state for 1 mole of gas is

$$P = RT\left[\frac{1}{V_m} + \frac{C_2(T)}{V_m^2} + \frac{C_3(T)}{V_m^3} + \cdots\right]. \quad (1)$$

Multiplying by V_m and then dividing by RT gives

$$\boxed{Z(T, P) = \frac{PV_m}{RT} = \left[1 + \frac{C_2(T)}{V_m} + \frac{C_3(T)}{V_m^2} + \cdots\right]}, \quad (2)$$

which is the required expression.

(B) As the pressure approaches zero, V_m goes to infinity, so that the limit we seek is

$$\lim_{V_m \longrightarrow \infty} Z(T, P) = \lim_{V_m \longrightarrow \infty}\left[1 + \frac{C_2(T)}{V_m} + \frac{C_3(T)}{V_m^2} + \cdots\right] = 1, \quad (3)$$

since the terms containing V_m in the denominator vanish in the limit.

1.37 A young scientist has recently broken up with her boyfriend. The former boyfriend is angry and decides to sabotage her research for revenge. It seems that the young lady is doing pressure, volume, and temperature measurements on gases. Late at night, her former boyfriend enters her laboratory and places a spring inside the cylinder she is using in the experiments. (See accompanying figure.) The force on the piston face produced by the spring is

$$F = -k\,[L^{-1} - L_o^{-1}] + C\,[L^{-2} - L_o^{-2}],$$

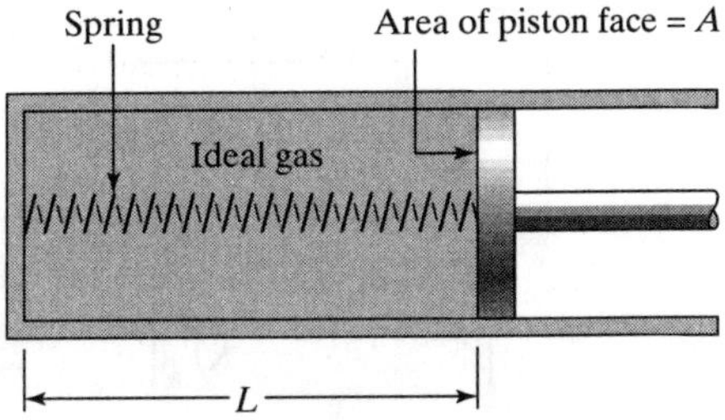

where k, C, and L_o are constants. At $L = L_o$, the spring is at equilibrium. The former boyfriend chooses the spring such that when 1 mole of ideal gas is placed in the cylinder at 298 K and 1 atm pressure, the cylinder length is precisely L_o.

The next day, the young scientist enters her laboratory and places 1 mole of an ideal gas in the apparatus at 298 K and 1 atm pressure. She then moves the piston to the right, so as to increase the volume of the gas at constant temperature, and measures the resulting pressure. This procedure is repeated until she has an extensive set of pressure and volume data at 298 K. Being unaware of the presence of the spring, she believes that her gas is behaving in a very nonideal fashion. Therefore, she fits her pressure, volume, and temperature data to a virial equation of state.

(A) Obtain an expression for the apparent "pressure" measured by the investigator as a function of the cylinder volume, the temperature, k, C, $V_o = AL_o$, and A.

(B) When the pressure and volume data are fitted to a virial equation of state, what values, in terms of T, k, C, and V_o, will be obtained for the virial coefficients?

(C) What features of the results will tell the investigator that the data cannot possibly represent the behavior of a nonideal gas?

Solution

(A) The pressure the investigator measures will be sum of two components: (1) the pressure due to the ideal gas itself, P_g, and (2) the pressure exerted by the spring, P_s. The pressure exerted by the spring is given by the force per unit area:

$$P_s = \frac{F}{A} = \frac{-k[L^{-1} - L_o^{-1}] + C[L^{-2} - L_o^{-2}]}{A}. \quad (1)$$

We can put this equation in terms of the volume by noting that $V = LA$ and $V_o = L_oA$. Thus,

$$P_s = \frac{-kA[V^{-1} - V_o^{-1}] + CA^2[V^{-2} - V_o^{-2}]}{A}$$

$$= -k[V^{-1} - V_o^{-1}] + CA[V^{-2} - V_o^{-2}]. \quad (2)$$

The total pressure is, therefore, given by

$$\boxed{P = P_g + P_s = \frac{RT}{V} - k[V^{-1} - V_o^{-1}] + CA[V^{-2} - V_o^{-2}]}, \quad (3)$$

which is the desired expression.

(B) If we collect the coefficients of $1/V_o$, $1/V$, and $1/V_2$ in Eq. 3 and write the equation in virial form, we obtain

$$P = RT\left[\left\{\frac{k}{RTV_o} - \frac{CA}{RTV_o^2}\right\} + \left\{1 - \frac{k}{RT}\right\}\frac{1}{V} + \left\{\frac{CA}{RT}\right\}\frac{1}{V^2}\right]. \quad (4)$$

Hence, the virial coefficients that the investigator will obtain when she fits her data to a virial expansion are

$$\boxed{C_o(T) = \frac{k}{RTV_o} - \frac{CA}{RTV_o^2}}, \quad (5)$$

$$\boxed{C_1(T) = 1 - \frac{k}{RT}}, \quad (6)$$

$$\boxed{C_2(T) = \frac{CA}{RT}}, \quad (7)$$

and

$$\boxed{C_n(T) = 0 \text{ for all } n > 2}.$$

(C) Since all ideal gases have $C_o(T) = 0$, $C_1(T) = 1$, and $C_2(T) = 0$, the investigator will know that the gas she has placed in the apparatus is not ideal. Furthermore, all real gases have $C_o(T)$ values that are near zero and $C_1(T)$ values that are close to unity. Equations 5 and 6 show that the values of $C_o(T)$ and $C_1(T)$ that the investigator will obtain upon fitting her data to a virial expansion will be far removed from zero and unity, respectively. Consequently, she will know that her data have been corrupted for some reason that will remain a mystery to her until she opens her apparatus and discovers the hidden spring.

1.39 Don't compute or derive anything. Prepare a brief outline of important points you would discuss if you were presenting a lecture on the van der Waals equation of state. When you're done, relax, have a snack, and get ready to begin your study of the first law of thermodynamics.

Solution

There is no unique solution to this problem. Points that might be raised are as follows:

1. The van der Waals' equation of state is a two-parameter equation of state.
2. The two parameters, a and b, give an approximate account of the intermolecular forces and the molecular volume, respectively.
3. The equation may be seen to approach $PV = nRT$ as $p \longrightarrow 0$ and $V \longrightarrow \infty$.
4. The residual volume predicted by the equation is $V_{res} = b - a/RT$, which is also the value predicted for the second virial coefficient.
5. The third viral coefficient predicted by the equation is b. No other virial coefficient is dependent upon a. Consequently, all of the information concerning intermolecular forces for a van der Waals gas is contained in the second virial coefficient.
6. The van der Waals equation can be put in reduced form by requiring that the three constraining conditions at the inflection point of the critical isotherm be satisfied.
7. The reduced form of the van der Waals equation is generally not as accurate as the form containing the a and b parameters adjusted to measured (P, V, T) data.

Now I think I'll have that snack before we start the first law.

CHAPTER 2

The First Law of Thermodynamics

2.1 Determine whether the expressions that follow are exact differentials. If so, determine the function F whose total differential is equal to the expression given.

(A) $[\cos(3y) + x\exp(-y^2)]\,dx + [2xy\exp(-y^2) - 3x\sin(3y)]\,dy$

(B) $[2x^3 - xy^2 - 2y + 3]\,dx - [x^2y + 2x]\,dy$

Solution

(A) If we define $f(x,y) = [\cos(3y) + x\exp(-y^2)]$ and $g(x,y) = [2xy\exp(-y^2) - 3x\sin(3y)]$, the Euler criterion for exactness requires that

$$\left(\frac{\partial f(x,y)}{\partial y}\right)_x = \left(\frac{\partial g(x,y)}{\partial x}\right)_y \tag{1}$$

for the differential to be exact. Evaluating these partial derivatives gives

$$\left(\frac{\partial f(x,y)}{\partial y}\right)_x = -3\sin(3y) - 2xy\exp(-y^2) \tag{2}$$

and

$$\left(\frac{\partial g(x,y)}{\partial x}\right)_y = 2y\exp(-y^2) - 3\sin(3y). \tag{3}$$

Inspection of Eqs. 2 and 3 shows that the right-hand sides are not equal. Thus, the differential is inexact, and there exists no function $z(x,y)$ whose total differential is equal to the differential in Part (A).

(B) We now define $f(x,y) = [2x^3 - xy^2 - 2y + 3]$ and $g(x,y) = -[x^2y + 2x]$. The partial derivatives needed to apply the Euler criterion are

$$\left(\frac{\partial f(x,y)}{\partial y}\right)_x = -2xy - 2 \tag{4}$$

and

$$\left(\frac{\partial g(x,y)}{\partial x}\right)_y = -2xy - 2. \tag{5}$$

The Euler criterion is satisfied and the differential is exact. Therefore, we have a function $z(x,y)$ such that

$$dz = \left(\frac{\partial z}{dx}\right)_y dx + \left(\frac{\partial z}{dy}\right)_x dy = f(x,y)dx + g(x,y)dy. \tag{6}$$

For Eq. 6 to hold, we must have

$$f(x,y) = \left(\frac{\partial z}{dx}\right)_y \tag{7}$$

and

$$g(x,y) = \left(\frac{\partial z}{dy}\right)_x. \tag{8}$$

Integrating both sides of Eq. 7 gives

$$z(x,y) = \int f(x,y)dx = \int [2x^3 - xy^2 - 2y + 3]dx$$

$$= \frac{x^4}{2} - \frac{x^2y^2}{2} - 2xy + 3x + F(y). \tag{9}$$

Similarly, integrating both sides of Eq. 8 yields

$$z(x,y) = \int g(x,y)dy = -\int [x^2y + 2x]dy = -\frac{x^2y^2}{2} + 2xy + G(x). \quad (10)$$

The right-hand sides of Eqs. 9 and 10 must be equal, since both are equal to $z(x,y)$. This can be true only if we have

$$F(y) = C \text{ and } G(x) = \frac{x^4}{2} + 3x + C, \quad (11)$$

where C is a constant. The required function is, therefore,

$$\boxed{z(x,y) = \frac{x^4}{2} - \frac{x^2y^2}{2} - 2xy + 3x + C}. \quad (12)$$

2.3 An automobile is moving down a straight highway, which we denote as the X-axis. A reasonable person suggests that the distance the car has moved should be measured along that axis, using the starting point as the reference point $X = 0$. Thus, if the car were at the point $X = X_1$, we would simply measure the distance of the car from the starting point, set that distance equal to X_1, and state the location of the car as $X = X_1$. However, a less intelligent person suggests that the position of the car be measured relative to an observer off to one side at point P. He suggests that instead of measuring the value of X_1 directly, we measure the distance d of the car from point P, as shown in the diagram on the left.

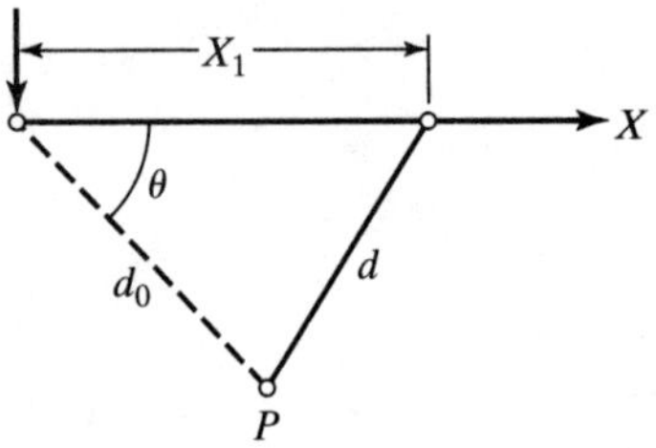

The distance between the observer at point P and the starting point is d_o, and the angle between a vector drawn between point P and the X-axis is given by θ.

(A) Obtain an expression giving the value of X_1, the car's position along the X-axis, in terms of d, d_o, and θ.

(B) Does a measurement of d uniquely determine the position of the car? Under what conditions will it do so?

(C) Comment on the choice of this reference point for the measurement.

Solution

(A) Using the law of cosines, we obtain

$$d^2 = d_o^2 + X_1^2 - 2d_oX_1 \cos\theta. \quad (1)$$

Rearranging Eq. 1 gives

$$X_1^2 - 2d_oX_1\cos\theta + (d_o^2 - d^2) = 0. \quad (2)$$

This is a quadratic whose solutions are

$$\boxed{X_1 = d_o\cos\theta \pm 0.5[4d_o^2\cos^2\theta - 4(d_o^2 - d^2)]^{1/2}}, \quad (3)$$

which is the required result.

(B) No, not in the general case, since there are two possible solutions to Eq. 3. If, however, the radical in Eq. 3 has a magnitude greater than $2d_o\cos\theta$, then the solution for X_1 is unique, since we must have $X_1 > 0$ and only the positive root will have physical significance.

(C) The reference point is very poorly chosen. First, a measurement of d does not uniquely determine X_1, except under the conditions given in the answer to Part (B). If the magnitude of the radical in Eq. 3 is less than $2d_o\cos\theta$, we must execute a second measurement to determine the value of X_1. We could, for example, measure the angle formed by d_o and d. If it is larger than $\pi/2 - \theta$, we would use the plus sign in Eq. 3. If it is less than this angle, the minus sign would be used. In addition, it is necessary to measure both d_o and θ before the reference point can be employed. We have, therefore, made things very difficult and inconvenient by choosing this reference frame.

2.5 Two moles of an ideal gas at 300 K and 10 atm pressure are expanded isothermally against a constant external pressure of 5 atm until the internal pressure reaches a value of 7 atm. At this point, the external pressure is reduced to zero and the gas is further expanded into a vacuum until a final state with $P = 1$ atm and $T = 300$ K is reached. Compute, if possible, $\Delta U, \Delta H, q$, and w for the process. You may ignore acceleration effects. (See explanatory note in Example 2.3.)

Solution

Since we have $dT = 0$ for the process in the first step and the gas is ideal,

$$dU = C_V dT = 0 \tag{1}$$

and

$$dH = C_P dT = 0. \tag{2}$$

Therefore, in this step, we have $\Delta U_1 = \Delta H_1 = 0$. The work is given by

$$w_1 = -\int_{V_1}^{V_2} P_{ext} dV = -P_{ext}\int_{V_1}^{V_2} dV = -P_{ext}[V_2 - V_1] = -P_{ext}\left[\frac{nRT}{P_2} - \frac{nRT}{P_1}\right]$$

$$= -nRTP_{ext}[P_2^{-1} - P_1^{-1}], \tag{3}$$

or

$$w_1 = -(2\text{ mol})(0.08206\text{ L atm mol}^{-1}\text{ K}^{-1})(300\text{ K})(5\text{ atm})\left[\frac{1}{7} - \frac{1}{10}\right]\text{atm}^{-1}$$

$$= -10.55\text{ L atm} = -1{,}069\text{ J}. \tag{4}$$

Since $\Delta U_1 = 0$, we must have $q_1 = -w_1$. Thus,

$$q_1 = 1{,}069\text{ J}. \tag{5}$$

In the second step, we also have $dT = 0$; thus,

$$\Delta H_2 = \Delta U_2 = 0. \tag{6}$$

We further have $P_{ext} = 0$; therefore,

$$w_2 = -\int_{V_1}^{V_2} P_{ext} dV = 0, \tag{7}$$

because $P_{ext} = 0$ at all points during the expansion.

Since $\Delta U_2 = 0$, we must have $q_2 = -w_2$. Therefore, $q_2 = 0$. The total results are

$$\boxed{\Delta U = \Delta U_1 + \Delta U_2 = 0}; \tag{8}$$

$$\boxed{\Delta H = \Delta H_1 + \Delta H_2 = 0}; \tag{9}$$

$$w = w_1 + w_2 = -1{,}069\text{ J} + 0 = \underline{-1{,}069\text{ J}}. \tag{10}$$

$$q = q_1 + q_2 = 1{,}069\text{ J} + 0 = \underline{1{,}069\text{ J}}. \tag{11}$$

2.7 Two moles of an ideal gas at 300 K and 10 atm are expanded isothermally to a final pressure of 2 atm with $T = 300$ K.

(A) What is the minimum magnitude of the work for this expansion?

(B) What is the maximum magnitude of the work for the expansion?

Solution

(A) The minimum magnitude for the work is zero. Since we have

$$w = -\int_{V_1}^{V_2} P_{ext} dV, \quad \textbf{(1)}$$

if P_{ext} is zero, the work will be zero.

(B) The maximum magnitude is obtained for a reversible process with $P_{ext} = P$. This gives

$$w_{max} = -\int_{V_1}^{V_2} P\,dV = -nRT \int_{V_1}^{V_2} \frac{dV}{V} = -nRT \ln\left[\frac{V_2}{V_1}\right] = -nRT \ln\left[\frac{P_1}{P_2}\right], \quad \textbf{(2)}$$

or

$$w_{max} = -(2 \text{ mol})(8.314 \text{ J mol}^{-1} \text{ K}^{-1})(300 \text{ K}) \ln\left[\frac{10}{2}\right] = \underline{-8{,}029 \text{ J}}. \quad \textbf{(3)}$$

Therefore, the maximum magnitude of the work is

$$|w_{max}| = \underline{8{,}029 \text{ J}}. \quad \textbf{(4)}$$

2.9 (A) Four moles of O_2 gas at 400 K and an internal pressure of 10 atm are expanded isothermally and irreversibly to a pressure of 4 atm against an external pressure that is equal to $0.5P_{int}$ at all points during the expansion. When P_{ext} reaches a pressure of 2 atm, it is held constant at that pressure, and isothermal expansion continues until the internal pressure reaches 2 atm. Compute q, w, ΔU, and ΔH for the process if O_2 is an ideal gas. You may ignore acceleration effects. (See explanatory note in Example 2.3.)

(B) Does the work computed in (A) have the maximum magnitude possible for the given expansion? If so, prove the result to be a maximum. If not, compute the maximum magnitude the work can have for the expansion.

Solution

(A) We need to divide the process into two parts. Part 1 is an isothermal, irreversible expansion against an external pressure equal to 0.5 P up to the point where $P_{ext} = 2$ atm $= P/2$, so that $P = 4$ atm at the end of this part. Part 2 is an isothermal, irreversible expansion from $P = 4$ atm to $P = 2$ atm against a constant external pressure of 2 atm. Since the entire process is isothermal and the gas is ideal, we have

$$\boxed{\Delta U = \Delta H = 0}. \quad \textbf{(1)}$$

For Part 1 of the expansion,

$$w_1 = -\int_{V_1}^{V_2} P_{ext} dV = -0.5 \int_{V_1}^{V_2} P\,dV = -0.5\,nRT \int_{V_1}^{V_2} \frac{dV}{V}$$

$$= -0.5\,nRT \ln\left[\frac{V_2}{V_1}\right] = -0.5\,nRT \ln\left[\frac{P_1}{P_2}\right]. \quad \textbf{(2)}$$

Substituting values gives

$$w_1 = -0.5(4\text{ mol})(8.314\text{ J mol}^{-1}\text{ K}^{-1})(400\text{ K})\ln\left[\frac{10}{4}\right] = -6{,}094\text{ J}. \quad \textbf{(3)}$$

The work in the second part is

$$w_2 = -\int_{V_1}^{V_2} P_{\text{ext}}dV = -P_{\text{ext}}\int_{V_1}^{V_2} dV = -P_{\text{ext}}[V_2 - V_1] = -nRTP_{\text{ext}}[P_2^{-1} - P_1^{-1}]$$

$$= -(4\text{ mol})(8.314\text{ J mol}^{-1}\text{ K}^{-1})(400\text{ K})(2\text{ atm})\left[\frac{1}{2} - \frac{1}{4}\right]\text{atm}^{-1} = -6{,}651\text{ J}. \quad \textbf{(4)}$$

The total work in the process is, therefore,

$$w_{\text{total}} = w_1 + w_2 = \underline{-12{,}745\text{ J}}. \quad \textbf{(5)}$$

Since $\Delta U = 0$, we have $q = -w$,
so that

$$q_{\text{total}} = \underline{12{,}745\text{ J}}. \quad \textbf{(6)}$$

(B) No, it is not a maximum. The maximum magnitude is obtained for the reversible process between the same limits. Thus,

$$w_{\max} = -\int_{V_1}^{V_2} PdV = -nRT\int_{V_1}^{V_2}\frac{dV}{V} = -nRT\ln\left[\frac{V_2}{V_1}\right] = -nRT\ln\left[\frac{P_1}{P_2}\right]. \quad \textbf{(7)}$$

Substituting values gives

$$w_{\max} = -(4\text{ mol})(8.314\text{ J mol}^{-1}\text{ K}^{-1})(400\text{ K})\ln\left[\frac{10}{2}\right] = \underline{-2.141\times10^4\text{ J}}, \quad \textbf{(8)}$$

which is obviously larger in magnitude than the work obtained in Part (A).

2.11 (A) 100 g of N_2 are heated from 300 to 500 K at a constant pressure of 1 atm. Using the data in Table 2.1, compute q, ΔH, w, and ΔU.

(B) The system in (A) is heated from 300 K to 500 K at constant volume. Compute q, w, ΔH, and ΔU.

(C) Why is $|q|_{\text{Part A}} > |q|_{\text{Part B}}$?

Solution

(A) For an ideal gas, we have

$$dH = C_p dT = nC_p^m dT. \quad \textbf{(1)}$$

The number of moles of N_2 is

$$n = \frac{100\text{ g}}{28.0134\text{ g mol}^{-1}} = 3.5697. \quad \textbf{(2)}$$

The data in Table 2.1 give $C_p^m = 29.125\text{ J mol}^{-1}\text{ K}^{-1}$. Therefore,

$$\Delta H = \int dH = nC_p^m\int_{T_1}^{T_2} dT = nC_p^m[T_2 - T_1]$$

$$= (3.5697\text{ mol})(29.125\text{ J mol}^{-1}\text{ K}^{-1})(500 - 300)\text{K} = \underline{20{,}794\text{ J}}. \quad \textbf{(3)}$$

Since the process is carried out at constant pressure, we have

$$q = q_p = \Delta H = \underline{20{,}794\text{ J}}. \quad \textbf{(4)}$$

For an ideal gas, we have

$$dU = C_v dT = nC_v^m dT, \tag{5}$$

where we note that we may compute dU along any path we choose, since dU is exact. Consequently, even though the actual path is one of constant pressure, dU may be computed along a constant-volume path, and since dU depends only upon T for an ideal gas, we need only ensure that we have the correct initial and final temperatures. Therefore,

$$\Delta U = \int dU = nC_v^m \int_{T_1}^{T_2} dT = nC_v^m[T_2 - T_1] = n[C_p^m - R][T_2 - T_1], \tag{6}$$

because the difference between C_p^m and C_v^m for an ideal gas is R. Substituting values gives

$$\Delta U = (3.5697\text{ mol})(29.125 - 8.314)\text{ J mol}^{-1}\text{ K}^{-1}(500 - 300)\text{ K} = \underline{14{,}858\text{ J}}. \tag{7}$$

From the first law, we have

$$\Delta U = q + w. \tag{8}$$

Solving for w yields

$$w = \Delta U - q = 14{,}858\text{ J} - 20{,}794\text{ J} = \underline{-5{,}936\text{ J}}. \tag{9}$$

(B) If the heating is done at constant volume, the values for ΔH and ΔU will not change, since the initial and final temperatures are still 300 and 500 K, respectively, and dU and dH depend only upon T for an ideal gas. Therefore, we have

$$\Delta U = \underline{14{,}858\text{ J}} \tag{10}$$

and

$$\Delta H = \underline{20{,}794\text{ J}}. \tag{11}$$

However, the path is now different, so q and w change. For a constant-volume path, we have

$$q = q_v = \Delta U = \underline{14{,}858\text{ J}}. \tag{12}$$

The volume is constant, so the amount of work done must be zero. This is also clear from the first law:

$$\Delta U = q + w. \tag{13}$$

But $\Delta U = q$; so $\underline{w = 0}$.

(C) Clearly, we have $|q|_{\text{Part A}} > |q|_{\text{Part B}}$, since 20,794 J > 14,858 J. The reason for this inequality is that work is done in Part A, but not in Part B. Thus, to heat the N_2 gas to 500 K in Part A, we must provide enough energy not only to heat the gas, but also to do the 5,936 joules of work that are done on the surroundings. When the heating is done at constant volume, there is no work, so we need only provide sufficient energy to heat the gas. This result shows that the reason C_p is larger than C_v is the work done in the constant-pressure process that is not present in the constant-volume heating.

2.13 The oxygen in a high-pressure cylinder is at 300 K and 100 atm pressure. The valve of the cylinder is opened, and the gas is allowed to expand to an atmospheric pressure of 1 atm. Assuming that the expansion is adiabatic and conducted under conditions of constant enthalpy, estimate the final temperature of O_2 after expansion. Do not assume O_2 to be an ideal gas.

Solution

Under the stated conditions, we have

$$\left(\frac{\partial T}{\partial P}\right)_H = \mu. \tag{1}$$

Therefore,

$$\partial T = \mu \, \partial P. \tag{2}$$

Integrating both sides between corresponding limits gives

$$\int_{T_1}^{T_2} dT = \int_{P_1}^{P_2} \mu dP = T_2 - T_1. \tag{3}$$

If we may regard the Joule–Thomson coefficient as being nearly independent of pressure, we may estimate

$$T_2 = T_1 + \mu \int_{P_1}^{P_2} dP = T_1 + \mu(P_2 - P_1). \tag{4}$$

Using the value of μ for O_2 given in Table 2.3, we obtain

$$T_2 = 300 \text{ K} + 0.31 \text{ K atm}^{-1} (1 - 100) \text{ atm} = \underline{269 \text{ K}}. \tag{5}$$

As expected, the expansion of a gas with a positive Joule–Thomson coefficient results in a cooling of the gas.

2.15 One mole of an ideal gas with $C_p^m = (5/2)R$ is heated reversibly at a constant pressure of 1 atm from 273.15 K to 373.15 K.

(A) Compute the work involved in the process.

(B) If the gas were expanded isothermally and reversibly at 273.15 K from an initial pressure of 1 atm, what would the final pressure need to be in order for the work to be equal to that calculated in (A)?

Solution

(A) The heat and ΔH are identical, since we have $dP = 0$. Thus,

$$\begin{aligned}\Delta H = q_p &= \int_{T_1}^{T_2} C_p dT = nC_p^m \int_{T_1}^{T_2} dT \\ &= nC_p^m[T_2 - T_1] = (1)(2.5R)(373.15 - 273.15) \\ &= (1 \text{ mol})(2.5)(8.314 \text{ J mol}^{-1} \text{ K}^{-1})(100 \text{ K}) = \underline{2{,}078 \text{ J}}.\end{aligned} \tag{1}$$

The change in internal energy for the process is

$$\begin{aligned}\Delta U &= \int_{T_1}^{T_2} C_v dT = nC_v^m \int_{T_1}^{T_2} dT = nC_v^m[T_2 - T_1] \\ &= (1 \text{ mol})(2.5R - R) \text{ J mol}^{-1} \text{ K}^{-1}(100 \text{ K}) = 1{,}247 \text{ J}.\end{aligned} \tag{2}$$

The work is, therefore,

$$w = \Delta U - q = 1{,}247 - 2{,}078 \text{ joules} = \underline{-831 \text{ J}}. \tag{3}$$

(B) We need to have

$$\begin{aligned}w &= -\int_{V_1}^{V_2} P dV = -nRT \int_{V_1}^{V_2} \frac{dV}{V} \\ &= -nRT \ln\left[\frac{V_2}{V_1}\right] = -nRT \ln\left[\frac{P_1}{P_2}\right] = nRT \ln\left[\frac{P_2}{P_1}\right].\end{aligned} \tag{4}$$

Solving for P_2, we get

$$P_2 = P_1 \exp\left[\frac{w}{nRT}\right] = (1\text{ atm}) \exp\left[\frac{-831}{(8.314)(273.15)}\right] = \underline{0.694\text{ atm}}. \quad (5)$$

2.17 Suppose that N_2 gas may be described by a van der Waals equation of state. One mole of N_2 is isothermally and reversibly expanded from a volume of 1 liter to 10 liters at 300 K. Compute the work done in the process. Is the magnitude of the result larger or smaller than that for an ideal gas undergoing the same process? Explain.

Solution

The van der Waals equation of state for 1 mole of gas is

$$\left(P + \frac{a}{V_m^2}\right)(V_m - b) = RT. \quad (1)$$

Therefore, the pressure is given by

$$P = \frac{RT}{V_m - b} - \frac{a}{V_m^2}. \quad (2)$$

The reversible, isothermal work of expansion for such a gas is thus

$$w = -\int_{V_1}^{V_2} P\,dV = -\int_{V_1}^{V_2} \left\{\frac{RT}{V_m - b} - \frac{a}{V_m^2}\right\} dV_m$$

$$= -RT \ln\left\{\frac{V_2 - b}{V_1 - b}\right\} - a\{V_2^{-1} - V_1^{-1}\}. \quad (3)$$

For N_2, the data in Table 1.1 give $b = 0.0387\text{ L mol}^{-1}$ and $a = 1.37\text{ L}^2\text{ bar mol}^{-2} = 1.35\text{ L}^2\text{ atm mol}^{-2}$. Substituting these values and V_1 and V_2 into Eq. 3 yields

$$w = -(8.314\text{ J mol}^{-1}\text{ K}^{-1})(300\text{ K}) \ln\left\{\frac{10 - 0.0387}{1 - 0.0387}\right\}$$

$$- 1.35\text{ L}^2\text{ atm mol}^{-2}\left\{\frac{1}{10} - \frac{1}{1}\right\}\text{mol L}^{-1} \times \frac{8.314\text{ J mol}^{-1}\text{ K}^{-1}}{0.08206\text{ L atm mol}^{-1}\text{ K}^{-1}}$$

$$= -5{,}832\text{ J mol}^{-1} - (-123)\text{ J mol}^{-1} = -5{,}709\text{ J mol}^{-1}. \quad (4)$$

Since we have one mole present, $w = \underline{-5{,}709\text{ J}}$.
(*Note*: Care must be taken to put both terms in the same units.) For an ideal gas,

$$w_{\text{ideal}} = -nRT \ln\left[\frac{V_2}{V_1}\right] = -(1\text{ mol})(8.314\text{ J mol}^{-1}\text{ K}^{-1})(300\text{ K}) \ln\left(\frac{10}{1}\right)$$

$$= -5{,}743\text{ J}. \quad (5)$$

Thus, $|W|_{\text{van der Waals}} < |W|_{\text{ideal}}$ because the attractive forces reduce the pressure and the work.

2.19 Two moles of an ideal gas at 500 K and $P = 10$ atm are contained in a cylinder–piston arrangement. As this gas is expanded isothermally and reversibly, more gas at 500 K is continuously added to the cylinder in a quantity sufficient to maintain the gas pressure constant at 10 atm. The process is continued until the total work done equals -500 L atm. Calculate the number of moles of gas contained in the cylinder at this point.

Solution

The reversible, isothermal work is given by

$$w = -\int_{V_1}^{V_2} P\,dV = -P[V_2 - V_1], \tag{1}$$

since the pressure is constant during the expansion. The final volume of the system is, therefore,

$$V_2 = V_1 - \frac{w}{P}. \tag{2}$$

The final volume may also be written in terms of the ideal-gas equation of state as

$$V_2 = \frac{n_2 RT}{P}, \tag{3}$$

where n_2 is the number of moles present after the expansion. This is the quantity we seek to compute. Combining Eqs. 2 and 3 gives

$$\frac{n_2 RT}{P} = V_1 - \frac{w}{P}. \tag{4}$$

Solving Eq. 4 for n_2, we obtain

$$n_2 = \frac{PV_1}{RT} - \frac{w}{RT} = \frac{PV_1 - w}{RT}. \tag{5}$$

The initial volume may be obtained from the ideal-gas equation of state:

$$V_1 = \frac{n_1 RT}{P} = \frac{(2\text{ mol})(0.08206\text{ L atm mol}^{-1}\text{ K}^{-1})(500\text{ K})}{10\text{ atm}} = 8.206\text{ L}. \tag{6}$$

Substituting of values into Eq. 5 gives the solution:

$$n_2 = \frac{(10\text{ atm})(8.206\text{ L}) - (-500\text{ L atm})}{(0.08206\text{ L atm mol}^{-1}\text{ K}^{-1})(500\text{ K})} = \underline{14.19\text{ moles of gas.}} \tag{7}$$

2.21 More about the thermodynamics of hell. (*Note:* This problem is dependent upon the analysis and assumptions contained in Problems 1.17 and 1.18). In Problems 1.17 and 1.18, we found that if hell is expanding at a slower rate than the rate at which additional souls enter the place, the temperature of hell will decrease with time. With certain assumptions we made in those problems, it could be determined that the temperature of hell is given by

$$T = 3.326 \times 10^{30} \exp\left[\left(b - \frac{\ln[10^{10} f + 1]}{5{,}000}\right)t\right]$$

in degrees K,

where b is a constant that determines the rate of expansion of the volume of hell and f is the fraction of people whose souls will enter hell. In Chapter 1, we employed the value $b = 0.0020723$ years^{-1}. We also found that the number of moles of souls in hell is given by

$$n = A \exp[at],$$

where $A = 1.66 \times 10^{-24}$ mol and $a = \ln(10^{10} f + 1)/5{,}000$ years^{-1}.

(A) If the average molar heat capacity of hell at constant pressure is 3.5R, determine the total heat change q of hell since the Devil entered it 5,000 years ago (by assumption).

(B) Given that hell is expanding at constant pressure, compute the total amount of pressure–volume work done by the expansion of hell since the Devil entered it.
(C) Compute ΔH and ΔU for hell over the last 5,000 years.

Solution

(A) Since we have assumed that hell is at a constant pressure of 1 atm, we have

$$C = nC_p^m = \frac{\delta q}{dT}. \tag{1}$$

Consequently, the total heat change is

$$q = q_p = \int \delta q = \int_{T_1}^{T_2} nC_p^m dT = C_p^m \int_{T_1}^{T_2} n dT. \tag{2}$$

Care must be exercised in integrating because n varies with T. Since we know n and T as functions of time, let us change the variables in Eq. 2 to time. From the time dependence of T, we obtain

$$dT = 3.326 \times 10^{30} \exp\left[\left(b - \frac{\ln[10^{10}f + 1]}{5{,}000}\right)t\right]\left[b - \frac{\ln[10^{10}f + 1]}{5{,}000}\right]dt. \tag{3}$$

The product ndT is, therefore,

$$ndT = 1.66 \times 10^{-24} \exp\left[\left\{\frac{\ln(10^{10}f + 1)}{5{,}000}\right\}t\right]$$

$$\times\, 3.326 \times 10^{30} \exp\left[\left(b - \frac{\ln[10^{10}f + 1]}{5{,}000}\right)t\right]\left[b - \frac{\ln[10^{10}f + 1]}{5{,}000}\right]dt$$

$$= 5.521 \times 10^{6}\left[b - \frac{\ln[10^{10}f + 1]}{5{,}000}\right]\exp(bt)dt. \tag{4}$$

If we have $f = 0.1$ and $b = 0.0020723$ years^{-1}, Eq. 4 becomes

$$ndT = -1.1441 \times 10^{4} \exp[0.0020723\, t]. \tag{5}$$

Substituting Eq. 5 into Eq. 2 gives

$$q_p = 3.5R(-1.1441 \times 10^4)\int_0^{5{,}000 \text{ years}} e^{0.0020723\, t}dt$$

$$= -\frac{3.329 \times 10^5}{0.0020723}[\exp(0.0020723\, t)]_0^{5{,}000} = -1.606 \times 10^8[e^{10.361} - 1]\,\mathrm{J}$$

$$= -1.606 \times 10^8(3.162 \times 10^4)\,\mathrm{J} = \underline{-5.078 \times 10^{12}\,\mathrm{J}}. \tag{6}$$

(B) The pressure–volume work is given by

$$w = -\int_{V_1}^{V_2} P dV = -P\int_{V_1}^{V_2} dV = -P(V_2 - V_1), \tag{7}$$

since the pressure is constant. In Problem 1.17, we found that

$$V = 4.5307 \times 10^5 \exp[bt]\ \mathrm{L}. \tag{8}$$

Thus, at $t_1 = 0$ years,

$$V_1 = 4.5307 \times 10^5 \exp[0]\ \mathrm{L} = 4.5307 \times 10^5\ \mathrm{L}. \tag{9}$$

At $t = 5{,}000$ years,

$$V_2 = 4.5307 \times 10^5 \exp[0.0020723 * 5{,}000] = 1.433 \times 10^{10}\ \mathrm{L}. \tag{10}$$

Substituting into Eq. 7 gives

$$w = -(1\text{ atm})(1.433 \times 10^{10} - 4.5307 \times 10^{5})\text{ L} = -1.432 \times 10^{10}\text{ L atm}$$

$$= \underline{-1.451 \times 10^{12}\text{ J}}. \quad \textbf{(11)}$$

(C) Since the heating is done at constant pressure, we have

$$q_p = \Delta H = \underline{-5.078 \times 10^{12}\text{ J}}. \quad \textbf{(12)}$$

ΔU may now be computed from

$$\Delta H = \Delta U + \Delta(PV) = \Delta U + P\Delta V, \quad \textbf{(13)}$$

because P is constant. Therefore, we have

$$\Delta U = \Delta H - P\Delta V = \Delta H + \text{work} = -5.078 \times 10^{12}\text{ J} + (-1.451 \times 10^{12}\text{ J})$$

$$= \underline{-6.529 \times 10^{12}\text{ J}}. \quad \textbf{(14)}$$

2.23 Two moles of an ideal gas at 500 K and 10 atm pressure are contained in the insulated piston–cylinder arrangement shown in Problem 2.22. The gas inlet valve is closed and the gas is heated at constant pressure to 800 K. During the heating, the piston expands to allow reversible work to be done on the surroundings. If $C_v^m = 1.5R$, calculate q, w, and ΔU for the process.

Solution

We first compute ΔU. For an ideal gas, we have

$$dU = C_v dT = nC_v^m dT. \quad \textbf{(1)}$$

Therefore,

$$\Delta U = \int_{T_1}^{T_2} nC_v^m dT = nC_v^m(T_2 - T_1)$$

$$= (2\text{ mol})(1.5)(8.314\text{ J mol}^{-1}\text{ K}^{-1})(800 - 500)\text{ K} = \underline{7{,}483\text{ J}}. \quad \textbf{(2)}$$

The work for a constant-pressure path is given by

$$w = -\int_{V_1}^{V_2} P\,dV = -P(V_2 - V_1). \quad \textbf{(3)}$$

The volumes can be computed using the ideal-gas law:

$$V_2 = \frac{nRT_2}{P} \text{ and } V_1 = \frac{nRT_1}{P}. \quad \textbf{(4)}$$

Substituting the results in Eq. 4 into Eq. 3 gives

$$w = -nR(T_2 - T_1) = -(2\text{ mol})(8.314\text{ J mol}^{-1}\text{ K}^{-1})(800 - 500)\text{ K} = \underline{-4{,}988\text{ J}}. \quad \textbf{(5)}$$

The first law now gives q:

$$\Delta U = q + w; \quad \textbf{(6)}$$

$$q = \Delta U - w = 7{,}483\text{ J} - (-4{,}988\text{ J}) = \underline{12{,}471\text{ J}}. \quad \textbf{(7)}$$

2.25 One mole of an ideal gas is contained in an insulated piston–cylinder arrangement ($dT = 0$) in an initial state (T_1, P_1, V_1). The gas is allowed to expand adiabatically and irreversibly against a constant external pressure P_o until a point is reached where the internal pressure becomes equal to P_o. If C_v^m for the gas is constant and equal to $1.5R$, derive an expression giving the final temperature of the gas in terms of P_o, V_1, T_1, and R. Ignore acceleration effects. (See explanatory note in Example 2.3.)

Solution

The general equation for an adiabatic process is

$$C_v dT + \left\{\left(\frac{\partial U}{\partial V}\right)_{T,n} + P_{\text{ext}}\right\} dV = 0. \tag{1}$$

Since we have an ideal gas, $(\partial U/\partial V)_{T,n} = 0$, and since the problem states that $P_{\text{ext}} = P_o$ = a constant, Eq. 1 becomes

$$C_v dT + P_o dV = nC_v^m dT + P_o dV = 0. \tag{2}$$

Rearranging terms and then integrating between corresponding limits gives

$$nC_v^m \int_{T_1}^{T_2} dT = -P_o \int_{V_1}^{V_2} dV, \tag{3}$$

so that

$$C_v^m(T_2 - T_1) = P_o(V_1 - V_2), \tag{4}$$

since $n = 1$. We know that the final internal pressure is P_o. Therefore, the final volume is

$$V_2 = \frac{nRT_2}{P_o} = \frac{RT_2}{P_o}. \tag{5}$$

Substituting Eq. 5 into Eq. 4 produces

$$C_v^m(T_2 - T_1) = P_oV_1 - RT_2. \tag{6}$$

Substituting the value of C_v^m and rearranging Eq. 6 gives

$$T_2(1.5R + R) = 2.5RT_2 = 1.5RT_1 + P_oV_1. \tag{7}$$

The final temperature is given by

$$\boxed{T_2 = \frac{(1.5RT_1 + P_oV_1)}{2.5R}}, \tag{8}$$

which is the required expression.

2.27 One mole of a monatomic ideal gas is heated along a reversible path such that $C = R$. What is the functional relationship between T and V along this path? (For a monatomic ideal gas, $C_v^m = 1.5R$.)

Solution

Equation 2.50 in the text gives the general equation for the heat capacity:

$$C = \left(\frac{\partial U}{\partial T}\right)_{V,n} + \left\{\left(\frac{\partial U}{\partial V}\right)_{T,n} + P_{\text{ext}}\right\}\frac{dV}{dT} = C_v^m + P\frac{dV}{dT}, \tag{1}$$

since we have 1 mole of an ideal gas for which $(\partial U/\partial V)_{T,n} = 0$ and the process is reversible, so that $P_{\text{ext}} = P$. Substituting of the value of C_v^m and the ideal-gas equation of state for P produces

$$C = 1.5R + \frac{RT}{V}\frac{dV}{dT} = R. \tag{2}$$

Canceling R, we obtain

$$1.5 + \frac{T}{V}\frac{dV}{dT} = 1. \tag{3}$$

Separating the variables gives

$$\frac{dV}{V} = -0.5\frac{dT}{T}. \tag{4}$$

Taking indefinite integrals of both sides produces

$$\int \frac{dV}{V} = \ln(V) = -0.5\int \frac{dT}{T} = -0.5\ln T + \text{Constant} = \ln T^{-0.5} + K. \tag{5}$$

Exponentiation of both sides results in

$$\boxed{V = \exp(K)\,T^{-1/2} = aT^{-1/2}}, \tag{6}$$

where a is a constant. Thus, for C to be equal to R, the path must have V inversely proportional to the square root of T.

2.29 Consider a gas whose equation of state over a certain temperature range can be represented by

$$PV = RT + aT^2$$

for 1 mole of gas, where a is a constant. Using the second law, we can show that, for such a gas,

$$\left(\frac{\partial U}{\partial V}\right)_T = \frac{aT^2}{V}.$$

(A) The gas is expanded isothermally and reversibly from V_1 to V_2. Derive an expression for the work associated with this process.
(B) Obtain an expression giving the change in the internal energy for the process in (A).
(C) Obtain an expression giving q for the process in (A).

Solution

(A) The isothermal, reversible work is

$$w = -\int_{V_1}^{V_2} P\,dV = -\int_{V_1}^{V_2}\left[\frac{RT}{V} + \frac{aT^2}{V}\right]dV = \boxed{-[RT + aT^2]\ln\left[\frac{V_2}{V_1}\right]}. \tag{1}$$

(B) The total differential for U is

$$dU = C_v dT + \left(\frac{\partial U}{\partial V}\right)_T dV. \tag{2}$$

Since the process is isothermal, $dT = 0$. Hence,

$$dU = \left(\frac{\partial U}{\partial V}\right)_T dV = \frac{aT^2}{V}dV. \tag{3}$$

Integrating both sides gives ΔU:

$$\boxed{\Delta U = \int_{V_1}^{V_2} \frac{aT^2}{V}dV = aT^2\ln\left[\frac{V_2}{V_1}\right]}. \tag{4}$$

(C) The quantity q can now be obtained from the first law:

$$q = \Delta U - w = aT^2 \ln\left\{\frac{V_2}{V_1}\right] + [RT + aT^2] \ln\left[\frac{V_2}{V_1}\right] = \boxed{[RT + 2aT^2] \ln\left[\frac{V_2}{V_1}\right]}. \quad (5)$$

The right-most expression is the required one.

2.31 Suppose the house of Problem 2.30 were cooled from temperature T_1 to temperature T_2 with the temperature of the outside air at T_o.

(A) How much heat must be extracted from the air within the house to execute this cooling if you take into account the entry of additional air from the outside upon cooling?

(B) If $T_1 = 25°C$, $T_2 = 15°C$, $T_o = 35°C$, $C_p^m = 2.5R$, $P = 1$ atm, and $V = 2,000\ m^3$, how many joules must be extracted? The amount of thermal energy required to heat the house over a similar temperature range obtained in Problem 2.30 is 1.728×10^7 J. Is it more difficult to cool or to heat a house?

Solution

(A) We need to break this problem into two parts. First, we will obtain an expression for the amount of heat we need to extract to cool the initial amount of air in the house from T_1 to T_2. We will then determine the number of additional moles of gas that must enter the house to maintain the pressure constant at P atm. Finally, we will derive an expression for the heat that must be extracted from these additional moles of gas to reduce their temperature to T_2.

Step 1: The amount of heat that must be extracted from gas initially in the house at $T_1 = q_1$ is given by

$$\delta q_1 = C_p dT = n_1 C_p^m dT. \quad (1)$$

Integrating gives

$$q_1 = \int_{T_1}^{T_2} n_1 C_p^m dT = n_1 C_p^m \int_{T_1}^{T_2} dT = n_1 C_p^m (T_2 - T_1) = \frac{PV}{RT_1} C_p^m (T_2 - T_1), \quad (2)$$

since initially we have $n_1 = PV/RT_1$.

Step 2: Let n_2 = number of moles of gas in house at $T = T_2$. We have

$$n_2 = \frac{PV}{RT_2}. \quad (3)$$

Thus, the additional gas that must enter is given by

$$\Delta n = n_2 - n_1 = \frac{PV}{RT_2} - \frac{PV}{RT_1} = \frac{PV}{R}[T_2^{-1} - T_1^{-1}]. \quad (4)$$

This number of moles of gas must be cooled from temperature T_o to temperature T_2. From the result obtained in Eq. 2, the amount of heat that must be removed to effect this cooling is

$$q_2 = \frac{PV}{R}[T_2^{-1} - T_1^{-1}] C_p^m (T_2 - T_o). \quad (5)$$

The total amount of heat that must be removed is, therefore,

$$q_{\text{total}} = q_1 + q_2 = \frac{PV}{RT_1} C_p^m (T_2 - T_1) + \frac{PV}{R}[T_2^{-1} - T_1^{-1}] C_p^m (T_2 - T_o)$$

$$= \boxed{\frac{PVC_p^m}{R}\left\{\frac{(T_2 - T_1)}{T_1} + (T_2 - T_o)(T_2^{-1} - T_1^{-1})\right\}}. \quad (6)$$

Equation 6 is the required expression.

(B) Direct substitution into Eq. 6 yields the desired answer:

$$q_{\text{total}} = (2.5)(1 \text{ atm})(2{,}000 \text{ m}^3)$$
$$\left\{\frac{(288.15 - 298.15)}{298.15} + (288.15 - 308.15)\left(\frac{1}{288.15} - \frac{1}{298.15}\right)\right\}$$
$$= 5{,}000 \text{ atm m}^3[-0.033540 - 0.0023280] = -179.34 \text{ atm m}^3. \quad (7)$$

Converting this number to L atm, we obtain

$$-179.34 \text{ atm m}^3 \times \frac{1 \text{ L}}{1{,}000 \text{ cm}^3} \times \left[\frac{100 \text{ cm}}{1 \text{ m}}\right]^3 = -1.793 \times 10^5 \text{ L atm}$$
$$= \underline{-1.817 \times 10^7 \text{ J}}. \quad (8)$$

Thus, more heat must be extracted to cool the house from 25°C to 15°C than need be added to heat the house over a similar interval. This is one reason that electric bills for cooling in the summer are usually higher than electric bills for heating in the winter.

2.33 One mole of an ideal gas is subjected to the following sequence of steps:
Step 1. The gas is heated from 25°C to 100°C at constant volume.
Step 2. The gas is then expanded freely into a vacuum to double its volume.
Step 3. The gas is cooled reversibly to 25°C at constant pressure.
Calculate, if possible, ΔU, ΔH, q, and w for the overall process (Step 1 + Step 2 + Step 3). Do you need to know the heat capacities of the gas?

Solution

Since we have an ideal gas, dH and dU depend only upon T. As we see, $dT = 0$ for the entire process. Therefore, we have $\boxed{\Delta H = \Delta U = 0}$.

In Step 1, the heat is

$$q_1 = \int_{298.15}^{373.15} C_v^m \, dT. \quad (1)$$

For Step 2, the work and heat are both zero, since there is no external pressure. For Step 3, the heat is

$$q_3 = \int_{373.15}^{298.15} C_p^m \, dT. \quad (2)$$

Therefore, the total heat for the process is

$$q_{\text{total}} = q_1 + q_3 = \int_{298.15}^{373.15} C_v^m \, dT + \int_{373.15}^{298.15} C_p^m \, dT = \int_{298.15}^{373.15} (C_v^m - C_p^m) dT. \quad (3)$$

The difference $C_v^m - C_p^m = -R$ Therefore, we have

$$q_{\text{total}} = -R\,(373.15 - 298.15) = -75R = \underline{-623.6 \text{ J}}. \quad (4)$$

The total work is now given by the first law:

$$w_{\text{total}} = \Delta U - q_{\text{total}} = 0 - (-623.6 \text{ joules}) = \underline{623.6 \text{ J}}. \quad (5)$$

We see that the heat capacities, C_v^m and C_p^m, are not required.

CHAPTER 3

Thermochemistry

3.1 At 298.15 K, the molar heat of combustion in oxygen of dipropyl ketone, $H_7C_3{-}CO{-}C_3H_7$, is −4,395.3 kJ when the products are $CO_2(g)$ and $H_2O(l)$. The standard partial molar enthalpies of $CO_2(g)$ and $H_2O(l)$ are −393.51 kJ and −285.83 kJ, respectively.

(A) Compute the standard partial molar enthalpy of dipropyl ketone.

(B) What is the standard molar heat of formation of dipropyl ketone?

(C) How much heat is liberated if 10 grams of dipropyl ketone are burned in excess oxygen in a constant-volume calorimeter?

Solution

(A) The first step is to write the balanced chemical equation to obtain the stoichiometric coefficients. The result is

$$H_7C_3{-}CO{-}C_3H_7(l) + 10\,O_2(g) \longrightarrow 7\,CO_2(g) + 7\,H_2O(l).$$

Eq. 3.15 may now be used to obtain an equation giving ΔH^o for the foregoing reaction:

$$\Delta H^o = 7\overline{H}^o_{CO_2(g)} + 7\overline{H}^o_{H_2O(l)} - \overline{H}^o_{C_7H_{14}O(l)}. \qquad (1)$$

Solving for the standard partial molar enthalpy of dipropyl ketone, we obtain

$$\begin{aligned}\overline{H}^o_{C_7H_{14}O(l)} &= 7\overline{H}^o_{CO_2(g)} + 7\overline{H}^o_{H_2O(l)} - \Delta H^o.\\ &= (7)(-393.51\text{ kJ mol}^{-1}) + (7)(-285.83\text{ kJ mol}^{-1}) - (-4{,}395.3\text{ kJ})\\ &= \underline{-360.1\text{ kJ mol}^{-1}}. \qquad (2)\end{aligned}$$

(B) The standard molar heat of formation of dipropyl ketone and its standard partial-molar enthalpy are numerically identical if we choose the reference point for standard partial-molar enthalpies to be the elements in their most stable state under 1 bar of pressure and then assign these values to be zero. Thus,

$$\overline{H}^o_{C_7H_{14}O(l)} = \Delta H^o_f(C_7H_{14}O(l)) = \underline{-360.1\text{ kJ mol}^{-1}}. \qquad (3)$$

(C) The number of moles of dipropyl ketone burned is given by

$$\Pi n = \frac{m}{M_w} \qquad (4)$$

where m is the mass of dipropyl ketone and M_w is its molar mass. Thus,

$$n = \frac{10\text{ g}}{114.17\text{ g mol}^{-1}} = 0.08759\text{ mol}. \qquad (5)$$

At constant pressure, the heat liberated would be equal to ΔH^o. Therefore, at conditions for which $dP = 0$,

$$q_p = \Delta H^o = n\overline{H}^o_{C_7H_{14}O(l)} = 0.08759\text{ mol}(-360.1\text{ kJ mol}^{-1}) = -31.54\text{ kJ}. \qquad (6)$$

However, in a constant-volume calorimeter, $q = q_v = \Delta U$, so that we need to compute

$$\begin{aligned}\Delta U^o &= \Delta H^o_f(C_7H_{14}O(l)) - RT\,\Delta n\\ &= -360.1\text{ kJ mol}^{-1} - (8.314\text{ J mol}^{-1}\text{ K}^{-1})(298.15\text{ K})(-3)\left(\frac{1\text{ kJ}}{1{,}000\text{ J}}\right)\\ &= -352.7\text{ kJ mol}^{-1}. \qquad (7)\end{aligned}$$

In Eq. 7, Δn is -3, since the balanced chemical equation has 7 moles of gaseous products and 10 moles of gaseous reactants. Note also that the units on the two terms must be converted so that they are the same.

When we burn 0.08759 mol of dipropyl ketone under constant-volume conditions, the heat released is

$$q = q_v = n\,\Delta U^o = (0.08759 \text{ mol})(-352.7 \text{ kJ mol}^{-1}) = \underline{-30.89 \text{ kJ}}. \quad \textbf{(8)}$$

3.3 Using the data in Problems 3.1 and 3.2 and in Appendix A, compute the heat of combustion of hexadecane $[C_{16}H_{34}]$ at 298.15 K and 1 bar of pressure when the products are CO(g) and H_2O(l).

Solution

The reaction of interest is

$$C_{16}H_{34}(s) + 16.5\,O_2(g) \longrightarrow 16\,CO(g) + 17\,H_2O(l).$$

Using Eq. 3.15, we have

$$\Delta H^o = 16\overline{H}^o_{CO\,(g)} + 17\overline{H}^o_{H_2O\,(l)} - \overline{H}^o_{C_{16}H_{34}(l)}. \quad \textbf{(1)}$$

The standard partial-molar enthalpies in Eq. 1 are numerically equal to the corresponding standard molar enthalpies of formation. Therefore,

$$\Delta H^o = (16)(-110.53 \text{ kJ mol}^{-1}) + (17)(-285.83 \text{ kJ mol}^{-1})$$

$$- (1)(-447.97 \text{ kJ mol}^{-1}) = -6{,}179.6 \text{ kJ mol}^{-1}. \quad \textbf{(2)}$$

Since the reaction is conducted under conditions of constant pressure, we have

$$q = q_p = \Delta H^o = \underline{-6{,}179.6 \text{ kJ mol}^{-1}}. \quad \textbf{(3)}$$

3.5 At 298.15 K and 1 bar of pressure, the molar heat of combustion of benzoic acid, $C_6H_5COOH(s)$, required to form $CO_2(g)$ and $H_2O(l)$ is $-3{,}227.5$ kJ mol^{-1}.
(A) Use this fact and the data in Appendix A to determine the standard molar heat of formation of benzoic acid.
(B) If the heat capacity of a constant-volume calorimeter plus the products of combustion is 2,150 J K^{-1}, determine the temperature change that would be observed for the calorimeter if 1.00 gram of benzoic acid were reacted to form $CO_2(g)$ and $H_2O(l)$ at 298.15 K and 1 bar of pressure.

Solution

(A) The combustion reaction is

$$C_6H_5COOH(s) + 7.5\,O_2(g) \longrightarrow 7\,CO_2(g) + 3\,H_2O(l).$$

Eq. 3.15 may now be used to obtain an equation giving ΔH^o for the preceding reaction:

$$\Delta H^o = 7\overline{H}^o_{CO_2(g)} + 3\overline{H}^o_{H_2O(l)} - \overline{H}^o_{C_7H_6O_2(s)}. \quad \textbf{(1)}$$

Solving for the standard partial molar enthalpy of benzoic acid, we obtain

$$\overline{H}^o_{C_7H_6O_2(s)} = 7\overline{H}^o_{CO_2(g)} + 3\overline{H}^o_{H_2O\,(l)} - \Delta H^o$$

$$= (7)(-393.51 \text{ kJ mol}^{-1}) + (3)(-285.83 \text{ kJ mol}^{-1}) - (-3227.5 \text{ kJ})$$

$$= -384.56 \text{ kJ mol}^{-1}. \quad \textbf{(2)}$$

The standard molar heat of formation of benzoic acid and its standard partial-molar enthalpy are numerically identical if we choose the reference point for standard partial-molar enthalpies to be the elements in their most stable state under 1 bar of pressure and then assign these values to be zero. Thus,

$$\overline{H}^o_{C_7H_6O_2(s)} = \Delta H_f^o(C_7H_6O_2(s)) = \underline{-384.56 \text{ kJ mol}^{-1}}. \quad (3)$$

(B) The heat released at constant volume is given by ΔU^o. Therefore, we need to first obtain ΔU^o from ΔH^o. We have

$$\Delta U^o = \Delta H^o - \Delta(PV) = \Delta H^o - RT\,\Delta n \quad (4)$$

if all gases are treated as being ideal. Hence,

$$\Delta U^o = -3{,}227.5 \text{ kJ mol}^{-1} - (8.314 \text{ J mol K}^{-1})(298.15 \text{ K})(-0.5 \text{ mol})\frac{1 \text{ kJ}}{1{,}000 \text{ J}}$$

$$= -3{,}226.3 \text{ kJ mol}^{-1}. \quad (5)$$

If we burn 1.00 grams of benzoic acid, the number of moles is

$$n = \frac{m}{\text{molar mass}} = \frac{1.00 \text{ g}}{122.11 \text{ g mol}^{-1}} = 0.008189 \text{ mol benzoic acid.} \quad (6)$$

The total heat released from burning 0.008189 mol of benzoic acid at constant volume at 298.15 K is

$$q = n\,\Delta U^o = (0.008189 \text{ mol})(-3{,}226.3 \text{ kJ mol}^{-1}) = -26.42 \text{ kJ}. \quad (7)$$

The temperature change of the calorimeter is related to the heat released by

$$q = C\,dT \quad (8)$$

so that

$$\Delta T = \frac{q}{C} = \frac{(-26.42 \text{ kJ})(1{,}000 \text{ J kJ}^{-1})}{2{,}150 \text{ J K}^{-1}} = \underline{12.29 \text{ K}}. \quad (9)$$

3.7 Xenon forms a hexafluoride, XeF_6. The standard molar enthalpy of formation of solid $XeF_6(s)$ is -368.2 kJ mol^{-1}. The enthalpy of sublimation of $XeF_6(s)$ at 298.15 K is 62.34 kJ mol^{-1}. [The enthalpy of sublimation is ΔH for the process $XeF_6(s) \longrightarrow XeF_6(g)$.] The standard molar enthalpy of formation of F(g) is 78.99 kJ mol^{-1}. Estimate the Xe–F bond enthalpy.

Solution

Consider the formation of $XeF_6(s)$ by two different paths as shown in the following figure:

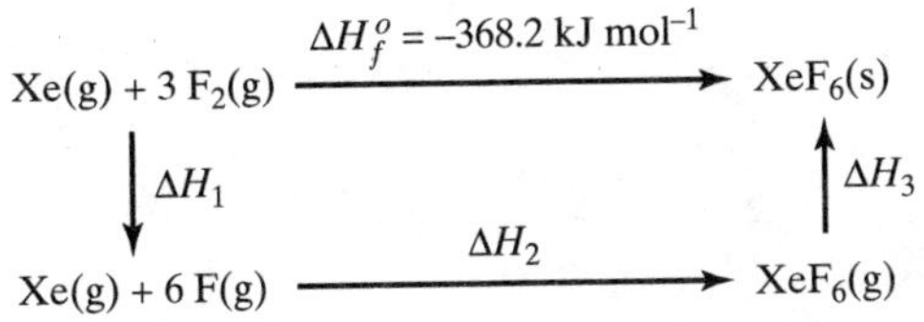

Because dH is exact and independent of the path, we must have

$$\Delta H_f^o(XeF_6(s)) = -368.2 \text{ kJ mol}^{-1} = \Delta H_1 + \Delta H_2 + \Delta H_3. \quad (1)$$

ΔH_1 is six times the standard molar enthalpy of formation of F(g):

$$\Delta H_1 = 6\,\Delta H_f^\circ(\mathrm{F(g)}). \tag{2}$$

ΔH_3 is the negative of the enthalpy of sublimation of $XeF_6(s)$:

$$\Delta H_3 = -\Delta H_{\mathrm{sub}}(\mathrm{XeF_6(s)}). \tag{3}$$

Finally, ΔH_2 is the formation of 6 Xe–F bonds. Therefore, we have

$$\Delta H_2 = -6\,H_{\mathrm{Xe-F}}. \tag{4}$$

Combining Eqs. 1–4, we obtain

$$-368.2\ \mathrm{kJ\ mol^{-1}} = 6\,\Delta H_f^\circ(\mathrm{F(g)}) - 6\,H_{\mathrm{Xe-F}} - \Delta H_{\mathrm{sub}}(\mathrm{XeF_6(s)}). \tag{5}$$

We may now solve for $H_{\mathrm{Xe-F}}$ to obtain

$$H_{\mathrm{Xe-F}} = \frac{6\,\Delta H_f^\circ(\mathrm{F(g)}) - \Delta H_{\mathrm{sub}}(\mathrm{XeF_6(s)}) + 368.2}{6}$$

$$= \frac{(6(78.99) - 62.34 + 368.2)}{6}\ \mathrm{kJ\ mol^{-1}} = \underline{130.0\ \mathrm{kJ\ mol^{-1}}}. \tag{6}$$

3.9 Use the average bond enthalpies given in Table 3.2 to estimate ΔH for the reaction

$$C_2H_2(g) + 2.5\,O_2(g) \longrightarrow 2\,CO_2(g) + H_2O(g)$$

at 298.15 K. Use standard partial molar enthalpies for these compounds to determine the percent error present in the bond enthalpy calculation. The percent error in Problem 3.8 was 3.71%. Why is the error in this calculation so much worse?

Solution

The reaction may be written in the form $HC{\equiv}CH(g) + 2.5\,O_2(g) \longrightarrow 2\,CO_2(g) + H_2O(g)$. In this form, it is clear that a C≡C, 2.5 O=O and two C—H bonds are broken. Four C=O bonds and two O—H bonds are formed. Using Table 3.2, we obtain

$$\sum_{\substack{\text{bonds}\\\text{broken}}} H_i = \mathrm{H_{C\equiv C}} + 2.5\,\mathrm{H_{O=O}} + 2\,\mathrm{H_{C-H}}$$

$$= [838 + 2.5(497) + 2(415)]\ \mathrm{kJ\ mol^{-1}} = 2{,}911\ \mathrm{kJ\ mol^{-1}} \tag{1}$$

and

$$\sum_{\substack{\text{bonds}\\\text{formed}}} H_i = 4\,\mathrm{H_{C=O}} + 2\,\mathrm{H_{O-H}} = [4(734) + 2(463)]\ \mathrm{kJ\ mol^{-1}}$$

$$= 3{,}862\ \mathrm{kJ\ mol^{-1}}. \tag{2}$$

Equation 3.56 gives

$$\Delta H \approx\ = \sum_{\substack{\text{bonds}\\\text{broken}}} H_i - \sum_{\substack{\text{bonds}\\\text{formed}}} H_j = [2{,}911 - 3{,}862]\ \mathrm{kJ\ mol^{-1}} = \underline{-951\ \mathrm{kJ\ mol^{-1}}}. \tag{3}$$

The correct ΔH for the reaction can be computed with the use of standard partial-molar enthalpies. This gives

$$\Delta H = 2\overline{H}^\circ_{\mathrm{CO_2(g)}} + \overline{H}^\circ_{\mathrm{H_2O(g)}} - \overline{H}^\circ_{\mathrm{C_2H_2(g)}}$$

$$= [2(-393.51) - 241.82 - 226.73]\ \mathrm{kJ\ mol^{-1}} = -1{,}255.6\ \mathrm{kJ\ mol^{-1}}. \tag{4}$$

The percent error in the bond enthalpy calculation is therefore

$$\% \text{ error} = \frac{100(-951 - (-1{,}255.6))}{1{,}255.6} = 24.3\ \% . \quad \textbf{(5)}$$

The percent error here is much larger than for the hydrogenation reaction because the C=O bond in $CO_2(g)$ is unusual in that there are two C=O bonds on the same carbon. Therefore, the average C=O bond energy is not a good approximation for the C=O bond in $CO_2(g)$.

3.11 N moles of liquid water are supercooled to 270.15 K in a container insulated from the surroundings at a pressure of 1 bar. Some of the water is then allowed to freeze, and the entire system is brought to a final equilibrium temperature of 273.15 K. The process is conducted under constant-pressure, adiabatic conditions. At this point, what fraction of the water is in the form of liquid H_2O? The enthalpy change upon melting is termed the molar enthalpy of fusion, ΔH_{fusion}. For water, $\Delta H^o_{\text{fusion}} = 6{,}004.4\ \text{J mol}^{-1}$ at 273.15 K and 1 bar of pressure. The constant-pressure heat capacities of ice and water are 38.73 J mol^{-1} K^{-1} and 75.94 J mol^{-1} K^{-1}, respectively. Assume that these values are constant from 270 to 273.15 K.

Solution

Since $\delta q_{\text{total}} = 0$ in the adiabatic process, the heat liberated when the water freezes must equal the thermal energy required to heat the system to $T = 273.15$ K. When liquid water freezes, the process is $H_2O(l) \longrightarrow H_2O(s)$. For this process,

$$\Delta H^o = \overline{H}^o_{H_2O(s)} - \overline{H}^o_{H_2O(l)} = -6{,}004.4\ \text{J mol}^{-1}. \quad \textbf{(1)}$$

Since $dP = 0$ for the process, we have $q = q_p = \Delta H^o = -6{,}004.4\ \text{J mol}^{-1}$. Let f equal the fraction of the N moles of liquid water that freeze. The heat liberated by freezing this much water is

$$q_{\text{total}} = q_1 = fN\Delta H^o = (-6{,}004.4\ \text{J mol}^{-1})f(\text{N mol}) = -6{,}004.4\ fN\text{J}. \quad \textbf{(2)}$$

This energy must be utilized to heat the solid and liquid water from 270.15 to 273.15 K. The differential heat required for this process is

$$\delta q_2 = [C_p]_{H_2O(l)}dT + [C_p]_{H_2O(s)}dT = \{[C_p]_{H_2O(l)} + [C_p]_{H_2O(s)}\}dT. \quad \textbf{(3)}$$

The total heat capacities of liquid and solid water are

$$[C_p]_{H_2O(l)} = (1-f)N[C_p^m]_{H_2O(l)} \quad \textbf{(4)}$$

and

$$[C_p]_{H_2O(s)} = fN[C_p^m]_{H_2O(s)}. \quad \textbf{(5)}$$

Combining Eqs. 3 through 5 yields

$$\delta q_2 = N[[C_p^m]_{H_2O(l)} + f\{[C_p^m]_{H_2O(s)} - [C_p^m]_{H_2O(l)}\}]dT. \quad \textbf{(6)}$$

Integrating both sides from 270.15 to 273.15 K gives

$$q_2 = (3\ \text{K})N[[C_p^m]_{H_2O(l)} + f\{[C_p^m]_{H_2O(s)} - [C_p^m]_{H_2O(l)}\}]. \quad \textbf{(7)}$$

Since $\delta q_{\text{total}} = 0$, we must have $q_1 + q_2 = 0$. This gives

$$q_1 + q_2 = -6{,}004.4fN + (3\ \text{K})N[[C_p^m]_{H_2O(l)} + f\{[C_p^m]_{H_2O(s)} - [C_p^m]_{H_2O(l)}\}] = 0. \quad \textbf{(8)}$$

Notice that N may be divided out, showing that the solution is independent of the quantity of water originally in the container. Solving Eq. 8 for f, we obtain

$$f = \frac{(3\text{ K})[C_p^m]_{H_2O(l)}}{6{,}004.4 - (3\text{ K})\{[C_p^m]_{H_2O(s)} - [C_p^m]_{H_2O(l)}\}}. \quad (9)$$

Inserting the heat capacity data gives

$$f = \frac{(3\text{ K})(75.94\text{ J mol}^{-1}\text{ K}^{-1})}{6{,}004.4\text{ J mol}^{-1} - (3\text{ K})(38.73 - 75.94)\text{ J mol}^{-1}\text{ K}^{-1}} = 0.03725. \quad (10)$$

Consequently, the fraction of the N moles that are liquid after the process is completed is 0.9628.

3.13 Consider a gas-phase dimerization reaction, $2A(g) \longrightarrow B(g)$. An experimental thermodynamicist determines that the enthalpy change in the reaction can be accurately represented by a quadratic function of temperature; that is,

$$\Delta H = aT^2 + bT + c,$$

where a, b, and c are constants.

(A) Obtain an expression for ΔC_p for this reaction as a function of a, b, c, and T.
(B) If C_p for molecule $B(g)$ is a constant equal to 8.368 J mol^{-1} K^{-1}, determine C_p for molecule $A(g)$ as a function of a, b, and c.

Solution

(A) Equation 3.23 shows that

$$\left(\frac{\partial \Delta H^o}{\partial T}\right)_P = \Delta C_p. \quad (1)$$

Therefore, we need only take the partial derivative of ΔH with respect to T at constant pressure to obtain ΔC_p. Using the data given in the problem, we obtain

$$\left(\frac{\partial \Delta H^o}{\partial T}\right)_P = \boxed{2aT + b = \Delta C_p}. \quad (2)$$

(B) For the dimerization reaction, we have

$$\Delta C_p = [C_p]_B - 2[C_p]_A = 8.368\text{ J mol}^{-1}\text{ K}^{-1} - 2[C_p]_A = 2aT + b. \quad (3)$$

Solving Eq. 3 for $[C_p]_A$ gives

$$\boxed{[C_p]_A = 0.5[8.368 - 2aT - b] = \left(4.184 - aT - \frac{b}{2}\right)\text{J mol}^{-1}\text{ K}^{-1}}, \quad (4)$$

which is the desired function.

3.15* Gaseous compounds A and B react to give gaseous compound C. The balanced chemical equation is

$$3\,A(g) + B(g) \longrightarrow 2\,C(g).$$

The following standard partial molar enthalpies are found:

$$\overline{H}^o_{A(g)} = 0.0 \text{ kJ mol}^{-1};$$
$$\overline{H}^o_{B(g)} = 0.0 \text{ kJ mol}^{-1};$$
$$\overline{H}^o_{C(g)} = -46.11 \text{ kJ mol}^{-1}.$$

Heat capacities at constant pressure, C_p, have also been measured over the temperature range 300 K $\leq T \leq$ 540 K for $A(g)$, $B(g)$, and $C(g)$. The results of these measurements are given in the table below.

	C_p (J mol^{-1} K^{-1})		
Temp. (K)	**A(g)**	**B (g)**	**C(g)**
300.0	28.7080	29.2610	35.8850
310.0	28.7711	29.2682	36.0414
320.0	28.8352	29.2744	36.1948
330.0	28.9003	29.2796	36.3451
340.0	28.9664	29.2838	36.4922
350.0	29.0335	29.2870	36.6363
360.0	29.1016	29.2892	36.7772
370.0	29.1708	29.2904	36.9150
380.0	29.2408	29.2906	37.0498
390.0	29.3119	29.2898	37.1814
400.0	29.3840	29.2880	37.3100
410.0	29.4751	29.2852	37.4353
420.0	29.5312	29.2814	37.5578
430.0	29.6063	29.2766	37.6771
440.0	29.6824	29.2708	37.7932
450.0	29.7595	29.2640	37.9063
460.0	29.8376	29.2562	38.0162
470.0	29.9167	29.2474	38.1231
480.0	29.9968	29.2376	38.2258
490.0	30.0779	29.2268	38.3275
500.0	30.1600	29.2150	38.4250
510.0	30.2431	29.2022	38.5194
520.0	30.3272	29.1884	38.6108
530.0	30.4123	29.1736	38.6990
540.0	30.4984	29.1578	38.7842

(A) Compute ΔH^o for the given reaction at 298.15 K.

(B) By fitting the heat capacity data with a power series expansion in T of the form

$$C_p(T) = \sum_{n=0}^{n=2} A_n T^n$$

in which the expansion coefficients A_o, A_1, and A_2 are determined using a least-squares procedure, obtain ΔH as an analytic function of temperature. On the same graph, plot the data and the analytic curves obtained from the least-squares fits.

(C) Compute the value of the heat of reaction at 500 K. How much error is made in this value by assuming that ΔH^o is independent of temperature?

Solution

(A) The standard enthalpy change at 298.15 K is given by

$$\Delta H^o = 2\overline{H}^o_{C(g)} - \overline{H}^o_{B(g)} - 3\overline{H}^o_{A(g)} = 2(-46.11 \text{ kJ mol}^{-1}) - 0 - 0$$

$$= \underline{-92.22 \text{ kJ mol}^{-1}}. \tag{1}$$

(B) We need to execute three least-squares fits, one each for the heat capacities of $A(g)$, $B(g)$, and $C(g)$. With a quadratic fit, the three linear equations that must be solved are given by Eq. 1.106. Here, we have used the polynomial fitting package contained in "Passage." The results are

$$[C_p]_{A(g)} = 27.2520 + 3.402 \times 10^{-3}T + 4.8289 \times 10^{-6}T^2, \tag{2}$$

$$[C_p]_{B(g)} = 28.5798 + 3.771 \times 10^{-3}T - 5.0013 \times 10^{-6}T^2, \tag{3}$$

and

$$[C_p]_{C(g)} = 29.7499 + 2.510 \times 10^{-2}T - 1.5500 \times 10^{-5}T^2. \tag{4}$$

The data and least-square fits are shown in the accompanying figure. If a different fitting routine is employed, the fit may differ slightly due to differences in rounding errors produced by the different software packages. These fits to the data are shown graphically in the figure.

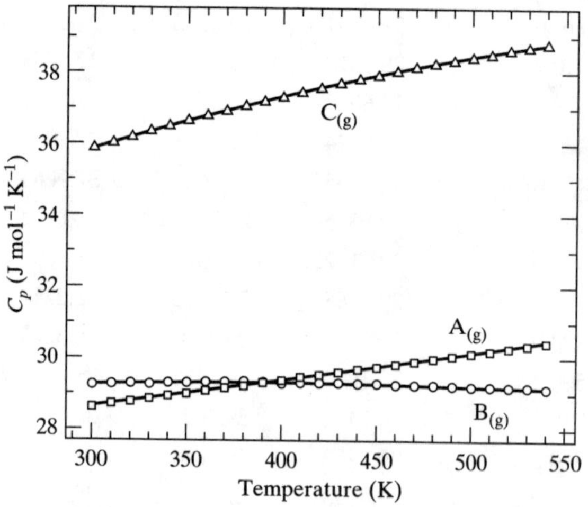

(C) The temperature dependence of ΔH is given by Eq. 3.25:

$$\Delta H^o(T) = \Delta H^o(298.15) + \int_{298.15 \text{ K}}^{T} \Delta C_p dT. \tag{5}$$

ΔC_p is given by

$$\Delta C_p = 2[C_p]_{C(g)} - [C_p]_{B(g)} - 3[C_p]_{A(g)}. \tag{6}$$

Using Eqs. 2, 3, and 4, we compute

$$\Delta C_p = -50.836 + 0.03622T - 4.0485 \times 10^{-5}T^2. \tag{7}$$

Substituting into Eq. 5 gives

$$\Delta H(500\text{ K}) = -9.222 \times 10^4 \text{ J mol}^{-1} + \int_{298.15}^{500} [-50.836 + 0.03622\,T - 4.0485 \times 10^{-5} T^2\} dT$$

$$= -9.222 \times 10^4 \text{ J mol}^{-1} - 50.836(500 - 298.15) + 0.01811(500^2 - 298.15^2) - 1.3495 \times 10^{-5}(500^3 - 298.15^3)$$

$$= -1.009 \times 10^5 \text{ J mol}^{-1} = \underline{-100.9 \text{ kJ mol}^{-1}}. \quad (8)$$

If we had taken ΔH to be independent of temperature, we would have assumed that $\Delta H(500\text{ K}) = \Delta H(298.15\text{ K}) = -92.22$ kJ mol^{-1}. The percent error would then have been

$$\%\text{ error} = \frac{100 \times [-92.22 - (-100.9)]}{100.9} = \underline{8.60\%}. \quad (9)$$

3.17 Consider the reaction $CO(g) + 0.5\,O_2(g) \longrightarrow CO_2(g)$. The heat capacities and Joule–Thomson coefficients are as follows:

Compound	C_p (cal mol^{-1} K^{-1})	μ (K bar^{-1})
$CO(g)$	6. 3423 + 0.0018363 T	1.20
$O_2(g)$	6.148 + 0.003102 T	1.15
$CO_2(g)$	6.369 + 0.0101 T	1.10

Using these data and the data in Appendix A, compute ΔH for the given reaction at 298.15 K with all gases at a pressure of 30 bar. Assume that μ and C_p are independent of pressure.

Solution

Using standard partial-molar enthalpies from Appendix A, we compute

$$\Delta H^o = \overline{H}^o_{CO_2(g)} - 0.5\overline{H}^o_{O_2(g)} - \overline{H}^o_{CO(g)}$$

$$= [-393.51 - 0.5(0) - (-110.53)] \text{ kJ mol}^{-1} = -282.98 \text{ kJ mol}^{-1}. \quad (1)$$

The rate of change of ΔH with pressure is given by Eq. 3.32:

$$\left(\frac{\partial \Delta H^o}{\partial P}\right)_T = \sum_{s=1}^{s=J} \nu_s[-\mu_s[C_p^m]_s] - \sum_{r=1}^{r=K} \nu_r[-\mu_r[C_p^m]_r] \equiv \Delta[-\mu C_p^m]. \quad (2)$$

Using the data given in the problem, we obtain, at $T = 298.15$ K,

$$\Delta(-\mu C_p^m) = -(1.10 \text{ K bar}^{-1})(6.369 + 0.0101(298.15)) \text{ cal mol}^{-1} \text{ K}^{-1}$$
$$+ 0.5(1.15 \text{ K bar}^{-1})(6.148 + 0.003102(298.15) \text{ cal mol}^{-1} \text{ K}^{-1}$$
$$+ (1.20 \text{ K bar}^{-1})(6.3424 + 0.0018363(298.15)) \text{ cal mol}^{-1} \text{ K}^{-1}$$
$$= +2.0164 \text{ cal mol}^{-1} \text{ bar}^{-1}. \quad (3)$$

Converting all energy units to joules, we have $\Delta H^o = -2.8298 \times 10^5$ J mol^{-1} and $\Delta(-\mu C_p^m) = 8.437$ J mol^{-1} bar^{-1}. Separating variables in Eq. 2 and then integrating between corresponding limits gives

$$\int_{\Delta H_{1\text{ bar}}}^{\Delta H_{30\text{ bar}}} d\Delta H = \int_{1\text{ bar}}^{30\text{ bar}} \Delta(-\mu C_p^m) dP. \quad (4)$$

This yields

$$\Delta H_{30\text{ bar}} = -2.8298 \times 10^5\text{ J mol}^{-1} + 8.437\text{ J mol}^{-1}\text{ bar}^{-1}(29\text{ bar})$$
$$= -2.8274 \times 10^5\text{ J mol}^{-1} = \underline{-282.74\text{ kJ mol}^{-1}}. \quad \textbf{(5)}$$

3.19* A differential scanning calorimetric measurement is made to determine ΔH for the unimolecular decomposition of a compound. In the experiment, the voltage is maintained at a constant value V_o, while the current is varied so as to maintain a zero temperature differential between the reference and sample cells. The data obtained are as follows:

Time (minutes)	$I(t)$ (amperes)
0	0.00
20	0.0050
22.5	0.0140
25.0	0.0280
27.5	0.0450
28.75	0.0490
30.0	0.0460
32.5	0.0230
35.0	0.0070
40.0	0.0000

A nonlinear least-squares fit of the data is made, and it is found that $I(t)$ is accurately represented by the function

$$I(t) = 0.024457\exp[-0.0809327\,(t - 28.75)^2]$$
$$+ 0.0247461\exp[-0.0240641(t - 28.75)^2],$$

where t is expressed in minutes. Also, 0.0007929 mole of the compound were used in the experiment, and it is known that $\Delta H = -65.56$ kJ mol^{-1}.

(A) Plot the fitted function $I(t)$ versus t between 0 and 40 minutes. On the same graph, show the measured data.

(B) Using a suitable method, determine the area beneath the curve $I(t)$ versus t.

(C) Determine the constant voltage that was used in the DSC experiment.

Solution

(A) As can be seen in the graph on the next page, the fit is reasonably good.

(B) The area beneath the graph can be obtained by a variety of techniques. The most simpleminded is to cut out the rectangular area of the graph and weigh it on an analytical balance. Then cut out the area beneath the graph and weigh that portion of the graph. Assuming the graph paper to have uniform density, the ratio of the weights will be ratio of the areas. Alternatively, we may proceed analytically. The desired integral is

$$\text{Area} = \int_{t=0}^{t=40} \text{I(t)dt} = 0.024457\int_0^{40} \exp[-0.0809327(t - 28.75)^2 dt$$
$$+ 0.0247461\int_0^{40} \exp[-0.0240641(t - 28.75)^2]dt. \quad \textbf{(1)}$$

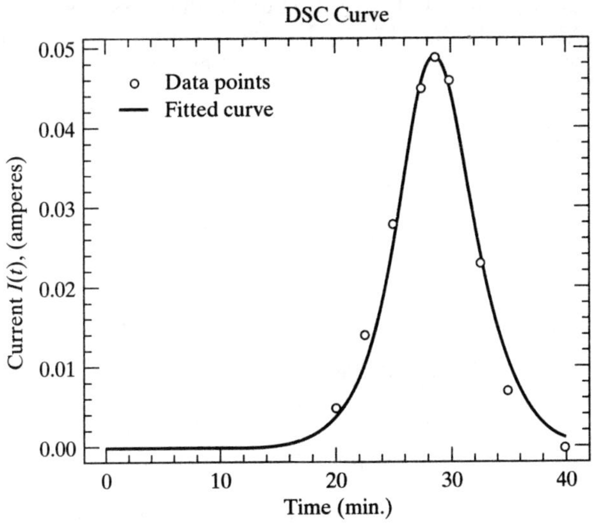

Neither of these integrals has a closed-form analytic solution. We may, however, use a numerical method such as Simpson's rule. To use this procedure, the interval from $t = 0$ to $t = 40$ minutes is divided into $N + 1$ equal intervals (with N an odd integer) whose spacing is Δt. The integral $\int_0^{40} I(t)\,dt$ is then given by

$$\int_0^{40} I(t)\,dt \approx [I(0) + 4I(\Delta t) + 2I(2\Delta t) + 4I(3\Delta t) + 2I(4\Delta t) + \cdots + 4I((N-1)\Delta t) + I(40)]\frac{\Delta t}{3}. \quad (2)$$

Using this method, we obtain

$$\underline{\text{Area} \approx 0.433 \text{ amp min}}. \quad (3)$$

(C) From Eq. 3.54, we have

$$\Delta H_{\text{total}} = -\int_0^{\infty} \omega(t)\,dt \approx -V_o \int_0^{40} I(t)\,dt, \quad (4)$$

since $I(t)$ is essentially zero after $t = 40$ minutes. The result in Part (B) gives

$$\Delta H_{\text{total}} = -(0.433 \text{ ampere minutes})V_o \quad (5)$$

If we express time in seconds and V_o in volts, ΔH_{total} will be in joules. Therefore,

$$\Delta H_{\text{total}} = -(0.433 \text{ amp min})\left[\frac{60 \text{ s}}{\text{min}}\right]V_o = -25.98V_o \text{ amp s}. \quad (6)$$

The value of ΔH_{total} is given by

$$\Delta H_{\text{total}} = n\,\Delta H^o = 0.0007929 \text{ mol}\,(-6.556 \times 10^4 \text{ J mol}^{-1}) = -51.98 \text{ J} = -51.98 \text{ volts amp s}, \quad (7)$$

since an ampere-second is a coulomb and a volt-coulomb is a joule. Combining Eqs. 6 and 7 gives

$$V_o = \frac{\Delta H_{\text{total}}}{-25.98 \text{ amp s}} = \frac{-51.98 \text{ volts amp s}}{-25.98 \text{ amp s}} = \underline{2.000 \text{ volts}}. \quad (8)$$

3.21 N moles of an alkane, C_nH_{2n+2}, are burned in excess oxygen to give $CO_2(g)$ and $H_2O(l)$ at 298.15 K. Obtain a general equation for the total enthalpy change for the process as a function of N, n, and $\overline{H}^o_{C_nH_{2n+2}}$.

Solution

If 1 mole of the alkane is burned, the reaction will be

$$C_nH_{2n+2} + \left[\frac{3n+1}{2}\right] O_2(g) \longrightarrow n\,CO_2(g) + (n+1)\,H_2O(l).$$

For this reaction,

$$\Delta H^o = n\overline{H}^o_{CO_2(g)} + (n+1)\overline{H}^o_{H_2O(l)} - \overline{H}^o_{C_nH_{2n+2}}. \quad \textbf{(1)}$$

Substituting values from Appendix A gives

$$\Delta H^o = n(-393.51 \text{ kJ mol}^{-1}) + (n+1)(-285.83 \text{ kJ mol}^{-1}) - \overline{H}^o_{C_nH_{2n+2}}$$
$$= [-679.34n - 285.83 - \overline{H}^o_{C_nH_{2n+2}}] \text{ kJ mol}^{-1}. \quad \textbf{(2)}$$

If N moles of the alkane are burned, the total change in enthalpy will be

$$\boxed{\Delta H_{\text{total}} = N\Delta H^o = N[-679.34n - 285.83 - \overline{H}^o_{C_nH_{2n+2}}] \text{ kJ mol}^{-1}}, \quad \textbf{(3)}$$

which is the desired function.

3.23 One mole of an alkane, C_nH_{2n+2}, is burned in excess oxygen to form $CO_2(g)$ and $H_2O(l)$ at 298.15 K and 1 bar of pressure. At constant pressure, the heat released in the process is found to be 4,816.8 kJ. If the products of the combustion at 298.15 K and 1 bar of pressure are $CO_2(g)$ and $H_2O(g)$, the heat released is 4,464.7 kJ. Determine the molecular formula of the alkane and its standard partial molar enthalpy.

Solution

The general equation for the combustion of the alkane is

$$C_nH_{2n+2} + \left[\frac{3n+1}{2}\right] O_2(g) \longrightarrow n\,CO_2(g) + (n+1)\,H_2O(l).$$

ΔH^o for this reaction is

$$\Delta H^o_{H_2O(l)} = n\overline{H}^o_{CO_2(g)} + (n+1)\overline{H}^o_{H_2O(l)} - \overline{H}^o_{C_nH_{2n+2}}. \quad \textbf{(1)}$$

Substituting values from Appendix A gives

$$\Delta H^o_{H_2O(l)} = n(-393.51 \text{ kJ mol}^{-1}) + (n+1)(-285.83 \text{ kJ mol}^{-1}) - \overline{H}^o_{C_nH_{2n+2}}$$
$$= [-679.34n - 285.83 - \overline{H}^o_{C_nH_{2n+2}}] \text{ kJ mol}^{-1} = -4{,}816.8 \text{ kJ mol}^{-1}. \quad \textbf{(2)}$$

If $H_2O(g)$ is formed, we have

$$\Delta H^o_{H_2O(g)} = n\overline{H}^o_{CO_2(g)} + (n+1)\overline{H}^o_{H_2O(g)} - \overline{H}^o_{C_nH_{2n+2}}. \quad \textbf{(3)}$$

Substituting values from Appendix A gives

$$\Delta H^o_{H_2O(g)} = n(-393.51 \text{ kJ mol}^{-1}) + (n+1)(-241.82 \text{ kJ mol}^{-1}) - \overline{H}^o_{C_nH_{2n+2}}$$
$$= [-635.33n - 241.82 - \overline{H}^o_{C_nH_{2n+2}}] \text{ kJ mol}^{-1} = -4{,}464.7 \text{ kJ mol}^{-1}. \quad \textbf{(4)}$$

Subtracting Eq. 4 from Eq. 2 yields

$$\Delta H^o_{H_2O(l)} - \Delta H^o_{H_2O(g)} = -352.1 = -44.01n - 44.01. \quad \textbf{(5)}$$

Solving Eq. 5 for n, we obtain

$$n = \frac{-352.1 + 44.01}{-44.01} = 7. \tag{6}$$

Therefore, the molecular formula is $\boxed{C_7H_{16}}$.

Substituting $n = 7$ into Eq. 2 and then solving for $\overline{H}^o_{C_7H_{16}}$, we obtain

$$\overline{H}^o_{C_7H_{16}} = 4{,}816.8 - 679.34(7) - 285.83 \text{ kJ mol}^{-1} = \underline{-224.4 \text{ kJ mol}^{-1}}. \tag{7}$$

3.25 Derive a general equation giving the rate of change of the reaction enthalpy with temperature at constant pressure for the combustion of an alkane, C_nH_{2n+2}, to form $CO_2(g)$ and $H_2O(g)$ as a function of n and C_p^m for the alkane. Assume that the heat capacities are independent of temperature, and use the data given in Table 2.1 as needed.

Solution

The general equation for the combustion reaction is

$$C_nH_{2n+2} + \left[\frac{3n+1}{2}\right] O_2(g) \longrightarrow n\, CO_2(g) + (n+1)\, H_2O(g).$$

The rate of change of the reaction enthalpy with temperature at $dP = 0$ is given by Eq. 3.22:

$$\left(\frac{\partial \Delta H}{\partial T}\right)_P = \Delta C_p = \sum_{s=1}^{s=J} \nu_s[C_p^m]_s - \sum_{r=1}^{r=K} \nu_r[C_p^m]_r. \tag{1}$$

For the combustion reaction, the result is

$$\left(\frac{\partial \Delta H}{\partial T}\right)_P = n[C_p^m]_{CO_2(g)} + (n+1)[C_p^m]_{H_2O\,(g)} - \left[\frac{3n+1}{2}\right][C_p^m]_{O_2(g)} - [C_p^m]_{C_nH_{2n+2}}. \tag{2}$$

Using the data in Table 2.1, we have

$$\left(\frac{\partial \Delta H}{\partial T}\right)_P = n(37.11) + (n+1)(33.58) - \left[\frac{3n+1}{2}\right](29.355) - [C_p^m]_{C_nH_{2n+2}}. \tag{3}$$

Collecting terms in Eq. 3 gives

$$\boxed{\left(\frac{\partial \Delta H}{\partial T}\right)_P = 26.66n + 18.90 - [C_p^m]_{C_nH_{2n+2}} \text{ J mol}^{-1}\text{ K}^{-1}}, \tag{4}$$

which is the desired function.

3.27 Let us assume that a car completely oxidizes its fuel to $CO_2(g)$ and $H_2O(g)$. Let us further assume that gasoline can be accurately represented as *n*-octane. (Actually, it is a complex mixture of branched-chain octanes and other hydrocarbons, as well as a variety of additives.)

(A) What is the total energy content of a tank of gasoline if the tank holds 20 U.S. gallons?

(B) Ethanol produced by fermentation of grain has been proposed as an alternative automotive fuel. If the ethanol is also completely oxidized to $CO_2(g)$ and $H_2O(g)$ in the car's engine, what is the total energy content of a tank of ethanol?

(C) In order to be cost competitive, what is the maximum price that can be charged for ethanol if gasoline is \$1.20 per U.S. gallon? In view of the approximation made concerning the composition of gasoline, it is appropriate to ignore the temperature and pressure dependence of the reaction enthalpy. Useful data are as follows: Density of $C_2H_5OH(l) = 0.7893$ g ml^{-1}; density of n-$C_8H_{18}(l) = 0.7036$ g ml^{-1}; 1 U.S. gallon = 3,785.20 ml.

Solution

(A) The oxidation reaction for *n*-octane is

$$C_8H_{18}(l) + 12.5\,O_2(g) \longrightarrow 8\,CO_2(g) + 9\,H_2O(g).$$

If we assume that ΔH is independent of temperature and pressure, the energy released by this reaction is given by

$$\Delta H^o = 8\overline{H}^o_{CO_2(g)} + 9\overline{H}^o_{H_2O(g)} - \overline{H}^o_{C_8H_{18}(l)}$$
$$= 8(-393.51) + 9(-241.82) + 249.9 = -5{,}074.56 \text{ kJ mol}^{-1}. \quad \textbf{(1)}$$

The mass of *n*-octane in a 20-gallon tank is

$$m_{\text{octane}} = 20 \text{ U.S. gal}\left[\frac{3{,}785.20 \text{ ml}}{1 \text{ U.S. gallon}}\right]\left[\frac{0.7036 \text{ g}}{\text{ml}}\right] = 5.327 \times 10^4 \text{ g of } n\text{-octane.} \quad \textbf{(2)}$$

The number of moles of *n*-octane in the tank is

$$n_{\text{octane}} = 5.327 \times 10^4 \text{ g}\left[\frac{1 \text{ mol}}{114.23 \text{ g mol}^{-1}}\right] = 466.4 \text{ mol of } n\text{-octane.} \quad \textbf{(3)}$$

The total energy content of the tank of *n*-octane is

$$q_{\text{octane}} = -5{,}074.56 \text{ kJ mol}^{-1})(466.4 \text{ mol}) = \underline{-2.367 \times 10^6 \text{ kJ}}\,. \quad \textbf{(4)}$$

As usual, the negative sign indicates that combustion will release the energy.

(B) The oxidation reaction for ethanol is

$$C_2H_5OH(l) + 3\,O_2(g) \longrightarrow 2\,CO_2(g) + 3\,H_2O(g).$$

If we assume that ΔH is independent of temperature and pressure, the energy released by this reaction is given by

$$\Delta H^o = 2\overline{H}^o_{CO_2(g)} + 3\overline{H}^o_{H_2O(g)} - \overline{H}^o_{C_2H_5OH(l)}$$
$$= 2(-393.51) + 3(-241.82) + 277.69 = -1{,}234.79 \text{ kJ mol}^{-1}. \quad \textbf{(5)}$$

The mass of ethanol in a 20-gallon tank is

$$m_{\text{octane}} = 20 \text{ U.S. gal}\left[\frac{3{,}785.20 \text{ ml}}{1 \text{ U.S. gallon}}\right]\left[\frac{0.7893 \text{ g}}{\text{ml}}\right] = 5.975 \times 10^4 \text{ g.} \quad \textbf{(6)}$$

The number of moles of ethanol in the tank is

$$n_{\text{octane}} = 5.975 \times 10^4 \text{ g}\left[\frac{1 \text{ mol}}{46.07 \text{ g mol}^{-1}}\right] = 1{,}297 \text{ mol.} \quad \textbf{(7)}$$

The total energy content of the tank of ethanol is

$$q_{\text{ethanol}} = -1{,}234.79 \text{ kJ mol}^{-1})(1{,}297 \text{ mol}) = \underline{-1.602 \times 10^6 \text{ kJ}}\,. \quad \textbf{(8)}$$

(C) To be cost competitive on an energy basis, a U.S. gallon of ethanol should cost no more than

$$\text{ethanol price per U.S. gallon} = \left[\frac{q_{\text{ethanol}}}{q_{\text{octane}}}\right][\text{octane price per U.S. gallon}]$$

$$= \frac{-1.602 \times 10^6}{-2.367 \times 10^6} \times \$1.20 = \underline{\$0.812 \text{ per U.S. gallon}}. \quad \textbf{(9)}$$

3.29 Let us examine a little of the economics of running hell. Natural gas, which is primarily methane [$CH_4(g)$], is sold by the hundred cubic feet (Cef). The retail price is generally around \$0.30 per Cef.

(A) Compute the energy available from the complete combustion of 1 Cef of $CH_4(g)$ to $CO_2(g)$ and $H_2O(l)$ at 298.15 K and 1 bar of pressure, assuming that the volume of the methane is measured at 1 atm pressure and 298.15 K. (1 ft^3 = 28.3171 liters).

(B) In Problem 1.17, we estimated the volume of hell to be $V = 4.5307 \times 10^5 \exp[0.0020723t]$ liters, where t is in years. The factor 4.5307×10^5 liters is the volume of a 2,000 ft^2 house with 8-ft ceilings. The Devil feels that, to keep up appearances and protect his image, he must keep one roaring fire going in each section of hell whose volume is equal to that of a 2,000 ft^2 house with 8-ft ceilings. A typical home in the winter might use 400 Cef per month of natural gas to provide heat. The Devil, however, wants more intense fires than that found in a normal household. Therefore, he opts to use furnaces that consume 2,000 Cef each month. Compute the total Cef consumption per month needed to keep the fires of hell burning. (*Note:* In Problem 1.17, it was assumed that the Devil entered hell 5,000 years ago; consequently, t = 5,000 years at the present time.)

(C) Compute the Devil's gas bill per month at the aforementioned retail price of natural gas. (*Note:* The Devil is far behind in his payments. The gas company has threatened to shut off his gas and thus quench the fires of hell. Congress, however, has decided that the Devil is an endangered species and, therefore, that his gas cannot be turned off. The gas company has appealed to the Supreme Court, but the Devil has an army of lawyers at his disposal, and the hopes of the gas company appear to be dim at the present time.)

Solution

(A) The combustion reaction is

$$CH_4(g) + 2\,O_2(g) \longrightarrow CO_2(g) + 2\,H_2O(l).$$

Therefore,

$$\Delta H^o = \overline{H}^o_{CO_2(g)} + 2\overline{H}^o_{H_2O\,(l)} - \overline{H}^o_{CH_4(g)}$$
$$= -393.51 + 2(-285.83) + 74.81 \text{ kJ mol}^{-1} = -890.36 \text{ kJ mol}^{-1}. \quad \textbf{(1)}$$

The number of moles of methane in 1 Cef is

$$n = \frac{PV}{RT} = \frac{(1 \text{ atm})(100 \text{ ft}^3 \text{ Cef}^{-1})}{(0.08206 \text{ L atm mol}^{-1} \text{ K}^{-1})(298.15 \text{ K})} \times \left[\frac{28.3171 \text{ L}}{1 \text{ ft}^3}\right]$$
$$= 115.7 \text{ mol } CH_4(g) \text{ Cef}^{-1}. \quad \textbf{(2)}$$

Thus, the energy available per Cef is

$$q = -890.36 \text{ kJ mol}^{-1}(115.7 \text{ mol Cef}^{-1}) = \underline{-1.0301 \times 10^5 \text{ kJ Cef}^{-1}}. \quad \textbf{(3)}$$

(B) The current volume of hell is

$$V = 4.5307 \times 10^5 \exp[0.0020723(5{,}000)] \text{ L} = 4.5037 \times 10^5(3.1619 \times 10^4) \text{ L}. \quad \textbf{(4)}$$

This volume is the equivalent of 31,619 houses, each with a volume of 4.5037×10^5 liters. Therefore, each month, the furnaces of hell require a natural-gas consumption of

$$\text{Total Cef} = 2{,}000 \text{ Cef month}^{-1} \text{ house}^{-1}(31{,}619 \text{ equivalent house volumes})$$
$$= \underline{6.3238 \times 10^7 \text{ Cef month}^{-1}}. \quad \textbf{(5)}$$

(C) The gas bill per month will be

$$\$ \text{ per month} = 6.3238 \times 10^7 \text{ Cef month}^{-1}(\$0.30 \text{ Cef}^{-1})$$
$$= \underline{\$1.8971 \times 10^7 \text{ month}^{-1}}, \quad (6)$$

or about $19 million per month!

3.31 Benzene $[C_6H_6(g)]$, is a six-membered ring of carbon atoms with alternating double and single bonds. Such structures are more stable than might be expected, due to a delocalization of electrons around the ring. The enhanced stability is termed the *resonance energy* of the molecule. The reasons underlying this effect will be treated in greater detail in Chapters 12 and 14. Here, we will obtain an estimate of the resonance energy in gaseous benzene. Consider the reaction

$$6\,C(g) + 6\,H(g) \longrightarrow C_6H_6(g).$$

(A) Using the bond enthalpies given in Table 3.2, obtain an estimate of ΔH for this reaction at 298.15 K.

(B) Using standard partial molar enthalpies from Appendix A, compute the correct ΔH for this reaction at 298.15 K.

(C) Compute the difference in the answers obtained in (A) and (B). This difference is the resonance energy of benzene.

Solution

(A) Benzene contains six C—H bonds, three C=C double bonds, and three C—C single bonds. Therefore,

$$\Delta H = -6\,H_{C-H} - 3\,H_{C=C} - 3\,H_{C-C} = -6(415) - 3(612) - 3(348) \text{ kJ mol}^{-1}$$
$$= \underline{-5{,}370 \text{ kJ mol}^{-1}}. \quad (1)$$

(B) The correct value of ΔH for the reaction is

$$\Delta H = \overline{H}^o_{C_6H_6(g)} - 6\overline{H}^o_{C(g)} - 6\overline{H}^o_{H(g)} = [82.93 - 6(716.68) - 6(217.97)] \text{ kJ mol}^{-1}$$
$$= \underline{-5{,}524.97 \text{ kJ mol}^{-1}}. \quad (2)$$

(C) The resonance energy of benzene is approximately $-5{,}370 - (-5{,}524.97) \approx \underline{155 \text{ kJ mol}^{-1}}$.

Note that the value of ΔH computed using standard partial-molar enthalpies is lower (more stable) than that predicted from a summation of three double and three single carbon–carbon bond enthalpies.

3.33 One mole of silver nitrate and 1 mole of NaCl are added to a large quantity of water at 298.15 K. The result is the precipitation of 1 mole of AgCl(s).

(A) Write ionic reactions, showing all processes that occur.

(B) Using Table 3.2 and the data in Appendix A, compute the change in enthalpy for the entire process. Assume that the solutions are sufficiently dilute that standard partial molar enthalpies at infinite dilution may be used for the ions present in the solution.

Solution

(A) The reactions are

$$NaCl(s) \xrightarrow{\infty H_2O(l)} Na^+_{(sol,\,\infty)} + Cl^-_{(sol,\,\infty)}, \quad (A)$$
$$AgNO_3(s) \xrightarrow{\infty H_2O(l)} Ag^+_{(sol,\,\infty)} + NO^-_{3(sol,\,\infty)}, \quad (B)$$

and

$$Ag^{+}_{(sol,\infty)} + Cl^{-}_{(sol,\infty)} \longrightarrow AgCl(s). \quad \textbf{(C)}$$

(B) The enthalpy change for the total process is the sum of the enthalpy changes in each of the preceding reactions—that is, $\Delta H = \Delta H_A + \Delta H_B + \Delta H_C$. We have

$$\Delta H_A = \overline{H}^o_{Na^+_{(sol,\infty)}} + \overline{H}^o_{Cl^-_{(sol,\infty)}} - \overline{H}^o_{NaCl(s)}$$
$$= -240.1 - 167.2 - (-411.15)\text{ kJ mol}^{-1} = 3.85\text{ kJ mol}^{-1}, \quad \textbf{(1)}$$

$$\Delta H_B = \overline{H}^o_{Ag^+_{(sol,\infty)}} + \overline{H}^o_{NO^-_{3(sol,\infty)}} - \overline{H}^o_{AgNO_3(s)}$$
$$= 105.6 - 207.4 - (-124.39) = +22.59\text{ kJ mol}^{-1}, \quad \textbf{(2)}$$

and

$$\Delta H_C = \overline{H}^o_{AgCl(s)} - \overline{H}^o_{Ag^+_{(sol,\infty)}} - \overline{H}^o_{Cl^-_{(sol,\infty)}}$$
$$= -127.07 - 105.6 - (-167.2)\text{ kJ mol}^{-1} = -65.47\text{ kJ mol}^{-1}. \quad \textbf{(3)}$$

Thus, the total enthalpy change for all three reactions is

$$\Delta H = \Delta H_A + \Delta H_B + \Delta H_C = 3.85 + 22.59 - 65.47\text{ kJ mol}^{-1}$$
$$= \underline{-39.03\text{ kJ mol}^{-1}}. \quad \textbf{(4)}$$

3.35 A hydrocarbon is known to be either an alkene with one C=C double bond or a saturated alkane. When 1 mole of the hydrocarbon is burned in excess $O_2(g)$ at 298.15 K, the enthalpy change in the reaction is found to be 176.04 kJ mol^{-1} larger if the products of the reaction are $CO_2(g)$ and $H_2O(g)$ than is the case if the products are $CO_2(g)$ and $H_2O(l)$. If the amount of $O_2(g)$ is limited so that the products of combustion at 298.15 K are $CO(g)$ and $H_2O(l)$, ΔH^o is 1,132.92 kJ mol^{-1} larger than the enthalpy change when the products are $CO_2(g)$ and $H_2O(l)$. Using these data and the facts that $\overline{H}^o_{CO_2(g)} = -393.51$ kJ mol^{-1}, $\overline{H}^o_{CO_2(g)} = -110.53$ kJ mol^{-1}, $\overline{H}^o_{H_2O(l)} = -285.83$ kJ mol^{-1}, and $\overline{H}^o_{H_2O(g)} = -241.82$ kJ mol^{-1} *alone*, determine the molecular formula of the hydrocarbon.

Solution

If the hydrocarbon is an alkane, the combustion reaction is

$$C_nH_{2n+2} + \left[\frac{3n+1}{2}\right] O_2(g) \longrightarrow n\,CO_2(g) + (n+1)\,H_2O(l\text{ or }g).$$

Let $\Delta H^o(H_2O(g))$ and $\Delta H^o(H_2O(l))$ be the standard enthalpy changes for $H_2O(g)$ and $H_2O(l)$ products, respectively. We will then have

$$\Delta H^o(H_2O(l)) = n\overline{H}^o_{CO_2(g)} + (n+1)\overline{H}^o_{H_2O(l)} - \overline{H}^o_{C_nH_{2n+2}} \quad \textbf{(1)}$$

and

$$\Delta H^o(H_2O(g)) = n\overline{H}^o_{CO_2(g)} + (n+1)\overline{H}^o_{H_2O(g)} - \overline{H}^o_{C_nH_{2n+2}}. \quad \textbf{(2)}$$

Subtracting Eq. 1 from Eq. 2 gives

$$\Delta H^o(H_2O(g)) - \Delta H^o(H_2O(l)) = (n+1)[\overline{H}^o_{H_2O(g)} - \overline{H}^o_{H_2O(l)}]$$
$$= (n+1)[-241.82 - (-285.83)] = 44.01(n+1)\text{ kJ mol}^{-1}. \quad \textbf{(3)}$$

Since the difference $\Delta H^o(H_2O(g)) - \Delta H^o(H_2O(l))$ is stated in the problem to be 176.04 kJ mol^{-1}, we have

$$44.01(n + 1) = 176.01 \tag{4}$$

so that

$$n = \frac{176.01 - 44.01}{44.01} = 3.000 \tag{5}$$

if the hydrocarbon is an alkane.

On the other hand, if the hydrocarbon is an alkene with one C=C double bond, we have

$$C_nH_{2n} + \left[\frac{3n}{2}\right] O_2(g) \longrightarrow n\, CO_2(g) + n\, H_2O(\text{l or g}).$$

The same analysis as that given for the alkane leads to the result

$$\Delta H^o(H_2O(g)) - \Delta H^o(H_2O(l)) = n[\overline{H}^o_{H_2O(g)} - \overline{H}^o_{H_2O(l)}]$$
$$= n[-241.82 - (-285.83)] = 44.01\, n \text{ kJ mol}^{-1}. \tag{6}$$

This gives

$$n = \frac{176.01}{44.01} = 4.000. \tag{7}$$

Consequently, if the hydrocarbon is an alkene with one C=C double bond, n must be 4.

The combustion reactions for an alkane to yield either CO(g) or $CO_2(g)$ and $H_2O(l)$ are

$$C_nH_{2n+2} + \left[\frac{2n + 1}{2}\right] O_2(g) \longrightarrow n\, CO(g) + (n + 1)\, H_2O(l)$$

and

$$C_nH_{2n+2} + \left[\frac{3n + 1}{2}\right] O_2(g) \longrightarrow n\, CO_2(g) + (n + 1)\, H_2O(l).$$

For an alkene, the reactions are

$$C_nH_{2n} + n\, O_2(g) \longrightarrow n\, CO(g) + n\, H_2O(l)$$

and

$$C_nH_{2n} + \left[\frac{3n}{2}\right] O_2(g) \longrightarrow n\, CO_2(g) + n\, H_2O(l).$$

The standard enthalpy changes in these reactions for the alkane are

$$\Delta H^o(CO) = n\overline{H}^o_{CO(g)} + (n + 1)\overline{H}^o_{H_2O(l)} - \overline{H}^o_{C_nH_{2n+2}} \tag{8}$$

and

$$\Delta H^o(CO_2) = n\overline{H}^o_{CO_2(g)} + (n + 1)\overline{H}^o_{H_2O(l)} - \overline{H}^o_{C_nH_{2n+2}}. \tag{9}$$

The corresponding changes for the alkene are

$$\Delta H^o(CO) = n\overline{H}^o_{CO(g)} + n\overline{H}^o_{H_2O(l)} - \overline{H}^o_{C_nH_{2n}} \tag{10}$$

and

$$\Delta H^o(CO_2) = n\overline{H}^o_{CO_2(g)} + n\overline{H}^o_{H_2O(l)} - \overline{H}^o_{C_nH_{2n}}. \tag{11}$$

Subtracting Eq. 9 from Eq. 8 or Eq. 11 from Eq. 10 gives

$$\Delta H°(CO) - \Delta H°(CO_2) = n[\overline{H}^o_{CO(g)} - \overline{H}^o_{CO_2(g)}]$$
$$= n[-110.53 - (-393.51)]\ \text{kJ mol}^{-1}$$
$$= 282.98n\ \text{kJ mol}^{-1}. \qquad \textbf{(10)}$$

The data given in the problem tells us that $\Delta H°(CO) - \Delta H°(CO_2) =$ 1,132.92 kJ mol^{-1}. Therefore, regardless of whether the hydrocarbon is an alkane or alkene, we must have

$$n = \frac{1{,}132.92}{282.98} = 4. \qquad \textbf{(11)}$$

Since $n = 4$, the data related to the enthalpy differences when the product is $H_2O(g)$ or $H_2O(l)$ show that the compound has to be an alkene. The molecular formula is, therefore, $\boxed{C_4H_8}$.

3.37 Which do you consider to be the most efficient way to use fodder to heat water?

(A) Burn the fodder under a kettle of water.

(B) Feed the fodder to a horse attached to a friction machine immersed in a kettle of water.

(C) Feed the fodder to a horse immersed in a kettle of water. Explain your answer [This problem is taken from Henry Bent's *The Second Law* (Oxford University Press, New York, 1965), Library of Congress Number 65-15608.]

Solution

The most efficient method is (a). In methods (b) and (c), some of the energy provided by the fodder will be used by the horse to build muscle, provide heat for its body, etc.

CHAPTER 4

The Second Law of Thermodynamics

4.1 A Carnot engine is being used to provide the work needed to compress a large cylinder of gas. It is determined that 2,000 kJ of work will be required to do the job.

(A) If the high-temperature heat reservoir of the engine is at 500 K and heat is discharged at 300 K, how much heat must be added to the engine to obtain the 2,000 kJ of work?

(B) How much methane must be burned in $O_2(g)$ to produce $CO_2(g)$ and $H_2O(g)$ in order to obtain the equivalent quantity of heat? Ignore the variation of the heat of reaction with temperature.

Solution

(A) The efficiency of the engine is given by

$$\mathcal{E} = \frac{T_1 - T_2}{T_1} = \frac{500 - 300}{500} = 0.4 = \frac{|w|}{|q_1|}. \quad (1)$$

The amount of heat needed is, therefore,

$$|q_1| = \frac{|w|}{0.4} = \frac{2{,}000 \text{ kJ}}{0.4} = \underline{5{,}000 \text{ kJ}}. \quad (2)$$

(B) The combustion of methane to $CO_2(g)$ and $H_2O(g)$ is

$$CH_4(g) + 2\,O_2(g) \longrightarrow CO_2(g) + 2\,H_2O(g).$$

The heat of reaction at constant pressure is given by

$$\begin{aligned} q = q_p = \Delta H \approx \Delta H^o &= \overline{H}^o_{CO_2(g)} + 2\overline{H}^o_{H_2O(g)} - \overline{H}^o_{CH_4(g)} \\ &= -393.51 \text{ kJ mol}^{-1} + (2)(-241.82 \text{ kJ mol}^{-1}) - (-74.81 \text{ kJ mol}^{-1}) \\ &= -802.34 \text{ kJ mol}^{-1}. \end{aligned} \quad (3)$$

To obtain 5,000 kJ of heat, we must burn

$$n = \frac{5{,}000 \text{ kJ}}{802.34 \text{ kJ mol}^{-1}} = \underline{6.232 \text{ mol } CH_4(g)}, \quad (4)$$

or

$$n = 6.232 \text{ mol } CH_4 = \underline{99.98 \text{ g of } CH_4(g)}. \quad (5)$$

4.3 We have shown that $(\partial H/\partial P)_T = V - T(\partial V/\partial T)_P$ and that $(\partial U/\partial V)_T = T(\partial P/\partial T)_V - P$. Analyze the units on the left and right sides of these two equations, and show that the units are in agreement.

Solution

The units on $(\partial H/\partial P)_T$ are energy pressure^{-1}, which we could express as L atm atm^{-1}. We see that this expression has units of volume. The units on $V - T(\partial V/\partial T)_P$ are clearly units of volume. Thus, the units agree.

The units on $(\partial U/\partial V)_T$ are energy volume^{-1}, which we could express in L atm L^{-1}, which has the units of pressure. The right-hand side, $T(\partial P/\partial T)_V - P$, clearly has units of pressure. The units, therefore, agree.

This type of unit analysis can often be utilized to check the validity of an equation. If the units agree, the expression may be correct. If the units do not agree, the expression is always wrong.

4.5 The coefficient of performance of a Carnot refrigerator is $C = T_2/(T_1 - T_2)$.

(A) Derive an expression for the total differential of C that shows the rate at which the coefficient of performance changes as we vary T_1 and T_2.

(B) C increases if T_2 is raised or if T_1 is lowered. Using the result obtained in (A), predict whether C increases faster as T_2 is raised or as T_1 is lowered.

Solution

(A) The total differential of C is given by

$$dC = \left(\frac{\partial C}{\partial T_1}\right)_{T_2} dT_1 + \left(\frac{\partial C}{\partial T_2}\right)_{T_1} dT_2. \qquad (1)$$

The required partial derivatives are

$$\left(\frac{\partial C}{\partial T_1}\right)_{T_2} = -\frac{T_2}{(T_1 - T_2)^2} \qquad (2)$$

and

$$\left(\frac{\partial C}{\partial T_2}\right)_{T_1} = \frac{(T_1 - T_2) - T_2(-1)}{(T_1 - T_2)^2} = \frac{T_1}{(T_1 - T_2)^2}. \qquad (3)$$

Consequently, the total differential of C is

$$\boxed{dC = -\frac{T_2}{(T_1 - T_2)^2} dT_1 + \frac{T_1}{(T_1 - T_2)^2} dT_2}, \qquad (4)$$

which is the desired expression.

(B) The absolute value of the ratio of rates is

$$\left|\frac{\left(\frac{\partial C}{\partial T_1}\right)_{T_2}}{\left(\frac{\partial C}{\partial T_2}\right)_{T_1}}\right| = \frac{\frac{T_2}{(T_1 - T_2)^2}}{\frac{T_1}{(T_1 - T_2)^2}} = \frac{T_2}{T_1}. \qquad (5)$$

Since $T_2 < T_1$, we have $(\partial C/\partial T_1)_{T_2} < (\partial C/\partial T_2)_{T_1}$, and C increases faster if T_2 is raised than if T_1 is lowered.

4.7 Refrigeration capacity is usually given in units of tons. One ton of refrigeration capacity is defined to be a rate of energy flow equal to the rate of extraction of the heat of fusion when 1 short ton (2,000 pounds) of ice is produced from water at the same temperature in 24 hours. (isn't the ton a wonderful unit!) The ton is equal to 200 British thermal units (BTUs) per minute. It is also equal to 3516.85 watts. A family purchases a 5-ton air-conditioning unit for its home.

(A) If the unit has a coefficient of performance of 3.75 under the conditions in which it will be used, how much heat will the family be able to remove from the house each day with this air conditioner?

(B) If the cost of electrical energy is \$0.0714 per kilowatt-hour (kWh) for the first 600 kWh and \$0.0410 for each kWh thereafter, figured on a monthly basis, what will it cost the family to operate the 5-ton unit 12 hours per day for a 31-day month? (See Problem 4.6 for a definition of kWh.).

Solution

(A) If the unit is run for 24 hours, the total energy input (work) to the air conditioner will be

$$|w| = [5 \text{ ton}] \times \left[\frac{3{,}516.85 \text{ J s}^{-1}}{\text{ton}}\right] \times \left[\frac{3{,}600 \text{ s}}{\text{hour}}\right] \times \left[\frac{24 \text{ hours}}{\text{day}}\right]$$

$$= 1.51928 \times 10^9 \text{ J day}^{-1} = 1.51928 \times 10^6 \text{ kJ day}^{-1}. \quad (1)$$

Equation 4.23 defines the coefficient of performance as

$$\mathcal{C} = \frac{|q|}{|w|}. \quad (2)$$

Therefore, the heat that the family will be able to remove is

$$|q| = \mathcal{C}|w| = 3.75(1.51928 \times 10^6 \text{ kJ day}^{-1}) = \underline{5.70 \times 10^6 \text{ kJ day}^{-1}}. \quad (3)$$

(B) The conversion factor between kWh and joules is

$$1 \text{ kWh} = 1 \text{ kWh} \times \frac{10^3 \text{ J s}^{-1}}{\text{kW}} \times \frac{1 \text{ kJ}}{10^3 \text{ J}} \times \frac{3{,}600 \text{ s}}{\text{hour}} = 3.6 \times 10^3 \text{ kJ}. \quad (4)$$

The total number of kWh used for 12-hour-per-day operation for 31 days is

$$\# \text{kWh} = 1.51928 \times 10^6 \text{ kJ day}^{-1}\left[\frac{31 \text{ days}}{2}\right] \times \left[\frac{1 \text{ kWh}}{3.6 \times 10^3 \text{ kJ}}\right] = 6{,}541 \text{ kWh}. \quad (5)$$

The operational cost will be

$$\text{Cost} = 600 \text{ kWh} \times \left[\frac{\$0.0714}{\text{kWh}}\right] + (6{,}541 - 600) \times \left[\frac{\$0.041}{\text{kWh}}\right]$$

$$= \underline{\$286 \text{ month}^{-1}}. \quad (6)$$

4.9 A Carnot engine's efficiency is found to be numerically equal to its coefficient of performance when the engine is reversed and operated as a refrigerator. What is the engine's efficiency?

Solution

The efficiency and coefficient of performance are given by Eqs. 4.22 and 4.25, respectively:

$$\mathcal{E} = \frac{T_1 - T_2}{T_1} = 1 - \frac{T_2}{T_1}. \quad (1)$$

$$\mathcal{C} = \frac{T_2}{T_1 - T_2}. \quad (2)$$

Equation 1 may easily be solved for $T_1 - T_2$:

$$T_1 - T_2 = \mathcal{E}T_1. \quad (3)$$

Substituting Eq. 3 into Eq. 2 and replacing $\mathcal{C}$ with $\mathcal{E}$ (since we are told that $\mathcal{C} = \mathcal{E}$) gives

$$\mathcal{E} = \frac{T_2}{\mathcal{E}T_1}. \quad (4)$$

Rearranging terms in Eq. 4 yields

$$\mathcal{E}^2 = \frac{T_2}{T_1}. \tag{5}$$

Substituting Eq. 5 into Eq. 1 gives

$$\mathcal{E} = 1 - \mathcal{E}^2, \tag{6}$$

which may be written in quadratic form as

$$\mathcal{E}^2 + \mathcal{E} - 1 = 0. \tag{7}$$

Solving for $\mathcal{E}$, we obtain

$$\mathcal{E} = \frac{[-1 + (1 + 4)^{1/2}]}{2} = \frac{-1 + 5^{1/2}}{2} = \underline{0.61803\ldots}. \tag{8}$$

4.11 Consider a Carnot heat engine operating between finite energy reservoirs at initial temperatures T_1° and T_2° with $T_1^\circ > T_2^\circ$. As the engine runs, T_1 decreases and T_2 increases until both reservoirs reach some final common temperature T_f.

(A) If reservoirs 1 and 2 have total heat capacities C_1 and C_2, respectively, which are constant, obtain expressions in terms of C_1, C_2, T_1°, T_2°, and T_f giving the total amount of work produced by the engine.

(B) If $C_1 = C_2$, $T_1^\circ = 650$ K, and $T_2^\circ = 298$ K, compute T_f.

Solution

(A) Let an infinitesimal amount of heat dq_1 be added to the engine in Stroke 1 and an amount of heat dq_3 be discharged to the surroundings in Stroke 3. Integration of dS around the Carnot cycle gives

$$\int_{\text{cycle}} dS = \frac{dq_1}{T_1} + \frac{dq_3}{T_2} = 0, \tag{1}$$

since the integration of any exact differential about a closed cycle is zero. We may now express both dq_1 and dq_3 in terms of the heat capacities and associated temperature changes. We have

$$dq_1 = -C_1 dT_1 \tag{2}$$

and

$$dq_3 = -C_2 dT_2, \tag{3}$$

where the minus signs are present because $dq_1 > 0$ but $dT_1 < 0$ and because $dq_3 < 0$ while $dT_2 > 0$. Substituting Eqs. 2 and 3 into Eq. 1 gives

$$\frac{-C_1 dT_1}{T_1} + \frac{-C_2 dT_2}{T_2} = 0. \tag{4}$$

Rearranging Eq. 4 gives

$$\frac{-C_1 dT_1}{T_1} = \frac{C_2 dT_2}{T_2}. \tag{5}$$

We may now sum all the differential heat changes for repeated cycles of the engine by integrating both sides of Eq. 5 between the corresponding limits of T_1° and T_f for T_1 and T_2° and T_f for T_2:

$$-C_1 \int_{T_1^\circ}^{T_f} \frac{dT_1}{T_1} = C_2 \int_{T_2^\circ}^{T_f} \frac{dT_2}{T_2}. \tag{6}$$

Integrating this expression yields

$$-C_1 \ln\left[\frac{T_f}{T_1^o}\right] = C_2 \ln\left[\frac{T_f}{T_2^o}\right]. \quad (7)$$

Exponentiation of both sides of Eq. 7 gives

$$\left[\frac{T_f}{T_1^o}\right]^{-C_1} = \left[\frac{T_f}{T_2^o}\right]^{C_2}. \quad (8)$$

We may now solve Eq. 8 to obtain T_f. We first write the equation in the form

$$\left[\frac{T_1^o}{T_f}\right]^{C_1} = \left[\frac{T_f}{T_2^o}\right]^{C_2}. \quad (9)$$

Cross multiplication of the fractions gives

$$T_f^{C_1+C_2} = [T_1^o]^{C_1}[T_2^o]^{C_2}. \quad (10)$$

Raising both sides of Eq. 10 to the $(C_1 + C_2)^{-1}$ power, we obtain

$$\boxed{T_f = [T_1^o]^{\alpha}[T_2^o]^{\beta}}, \quad (11)$$

where $\alpha = C_1/(C_1 + C_2)$ and $\beta = C_2/(C_1 + C_2)$.

The total work can be obtained from the First law. The total energy extracted from the hot reservoir is

$$[q_1]_{\text{total}} = -C_1[T_f - T_1^o]. \quad (12)$$

The total heat discharged into the surroundings is

$$[q_3]_{\text{total}} = -C_2[T_f - T_2^o]. \quad (13)$$

The sum must be the total work done. Thus,

$$\boxed{\begin{aligned} w_{\text{total}} &= [q_1]_{\text{total}} + [q_3]_{\text{total}} = -C_1[T_f - T_1^o] - C_2[T_f - T_2^o] \\ &= -T_f[C_1 + C_2] + C_1T_1^o + C_2T_2^o \end{aligned}}, \quad (14)$$

where T_f is given by Eq. 11. Equations 11 and 14 are the desired results.

(B) If $C_1 = C_2$, then

$$\alpha = \frac{C_1}{C_1 + C_2} = 0.5 \quad (15)$$

and

$$\beta = \frac{C_2}{C_1 + C_2} = 0.5. \quad (16)$$

Substituting these results into Eq. 11 gives

$$T_f = [T_1^oT_2^o]^{1/2} = [(650)(298)]^{1/2} = \underline{440.1\text{ K}} \quad (17)$$

The final temperature is, therefore, the geometric mean of the two initial temperatures.

4.13 A three-stroke, reversible engine like that described in Problem 4.12 contains 10 moles of an ideal gas as a working material. The initial conditions are $T_1 = 500$ K and $V_1 = 20$ L. At the end of Stroke 1, the volume, V_2, is 50 L. If C_v^m for the gas is $1.5R$, using the results of Problem 4.12, compute the lower temperature T_2 and the engine's efficiency.

Solution

The two volumes, V_1 and V_2, must be connected by the adiabatic compression in Stroke 3. For an adiabatic, reversible process for an ideal gas, we have

$$dU = nC_v^m dT = \delta w = -P\,dV = -\frac{nRT}{V}dV. \tag{1}$$

Separating variables and integrating between corresponding limits gives

$$C_v^m \int_{T_1}^{T_2} \frac{dT}{T} = -R\int_{V_1}^{V_2} \frac{dV}{V}, \tag{2}$$

or

$$C_v^m \ln\left[\frac{T_2}{T_1}\right] - R\ln\left[\frac{V_2}{V_1}\right]. \tag{3}$$

Exponentiation of both sides produces

$$\left[\frac{T_2}{T_1}\right]^{C_v^m} = \left[\frac{V_2}{V_1}\right]^{-R} = \left[\frac{V_1}{V_2}\right]^{R}. \tag{4}$$

Raising each side to the $1/C_v^m$ power results in

$$\left[\frac{T_2}{T_1}\right] = \left[\frac{V_1}{V_2}\right]^{\frac{R}{C_v^m}} = \left[\frac{V_1}{V_2}\right]^{2/3}, \tag{5}$$

since $C_v^m = 1.5R$. Substituting values into Eq. 5 gives

$$T_2 = T_1\left[\frac{V_1}{V_2}\right]^{2/3} = 500\text{ K}\left[\frac{20}{50}\right]^{2/3} = \underline{271.44\text{ K}}. \tag{6}$$

The efficiency of the engine is, therefore,

$$\mathcal{E} = \left[1 - \frac{T_1 - T_2}{T_1\ln[T_1/T_2]}\right] = 1 - \frac{500 - 271.44}{500\ln\left[\frac{500}{271.44}\right]} = \underline{0.2517}. \tag{7}$$

4.15 One mole of $CO_2(g)$ is held on the left side of an insulated container similar to the one shown in Figure 4.15. The temperature of the gas is 300 K. The initial volume is 20 L. The dividing partition is removed, and the gas is allowed to expand into the vacuum on the right-hand side such that the final volume is 40 L. If the insulation makes the process adiabatic, compute the final temperature of the gas if

(A) $CO_2(g)$ is an ideal gas and

(B) $CO_2(g)$ obeys a van der Waals equation of state with the parameters being those given in Table 1.1. C_v^m for $CO_2(g)$ is 28.80 J mol^{-1} K^{-1}.

Solution

(A) If the gas is ideal, we have

$$dU = \delta q + \delta w = \delta q - P_{ext}dV = 0 = C_v^m dT, \tag{1}$$

since δq is zero for an adiabatic process and $P_{ext} = 0$ for expansion into a vacuum. Therefore, we must have $dT = 0$, and the final temperature is $\underline{300\text{ K}}$, the same as the initial temperature.

(B) If $CO_2(g)$ obeys a van der Waals equation of state, we have

$$dU = nC_v^m dT + \left(\frac{\partial U}{\partial V}\right)_T dV = 0, \tag{2}$$

since δq and δw are still zero. Rearranging Eq. 2 gives

$$nC_v^m dT = -\left(\frac{\partial U}{\partial V}\right)_T dV. \tag{3}$$

The rate of change of U with volume at $dT = 0$ is given by Eq. 4.92:

$$\left(\frac{\partial U}{\partial V}\right)_T = T\left(\frac{\partial P}{\partial T}\right)_V - P. \tag{4}$$

The pressure of a van der Waals gas is

$$P = \frac{nRT}{V - nb} - \frac{n^2 a}{V^2}. \tag{5}$$

Therefore,

$$T\left(\frac{\partial P}{\partial T}\right)_V = \frac{nRT}{V - nb}, \tag{6}$$

which gives

$$\left(\frac{\partial U}{\partial V}\right)_T = \frac{nRT}{V - nb} - \frac{nRT}{V - nb} + \frac{n^2 a}{V^2} = \frac{n^2 a}{V^2}. \tag{7}$$

Substituting Eq. 7 into Eq. 3 yields

$$nC_v^m dT = -\frac{n^2 a}{V^2} dV. \tag{8}$$

Integrating both sides of Eq. 9 between corresponding limits produces

$$nC_v^m \int_{300\text{ K}}^{T} dT = nC_v^m (T - 300) = -n^2 a \int_{20\text{ L}}^{40\text{ L}} \frac{dV}{V^2} = n^2 a\left[\frac{1}{40} - \frac{1}{20}\right] = -\frac{n^2 a}{40\text{ L}}. \tag{9}$$

Solving Eq. 9 for T yields

$$T = 300\text{ K} - \frac{n^2 a}{n(40\text{ L})C_v^m}. \tag{10}$$

To do the calculation, we must put C_v^m and a in the same units. Let us convert the heat capacity to L bar mol^{-1} K^{-1}:

$$C_v^m = 28.80\text{ J mol}^{-1}\text{ K}^{-1} \times \frac{0.08314\text{ L bar}}{8.314\text{ J}} = 0.2880\text{ L bar mol}^{-1}\text{ K}^{-1}. \tag{11}$$

The value of the parameter a in Table 1.1 is 3.658 L^2 bar mol^{-2}. Substituting into Eq. 10 gives

$$T = 300\text{ K} - \frac{(1\text{ mol})^2\, 3.658\text{ L}^2\text{ bar mol}^{-2}}{(40\text{ L})(0.2880\text{ L bar mol}^{-1}\text{ K}^{-1})(1\text{ mol})}$$
$$= [300 - 0.3159]\text{ K} = \underline{299.68\text{ K}}. \tag{12}$$

The van der Waals gas cools because some of the internal energy must be used to expand the gas against the restraining intermolecular forces. Consequently, the temperature decreases. Equation 10 shows that the magnitude of the effect is dependent upon the parameter a, which measures the magnitude of the internal forces.

4.17 Two moles of an ideal gas at 300 K and a volume of 35 L expand isothermally and irreversibly against a constantly varying external pressure until the final volume is 90 L. Compute, if possible, ΔU, ΔH, q, w, and ΔS for the process.

Solution

It is not possible to compute either q or w, since they have inexact differentials and the expansion path is not known (i.e., the external pressure at each point in the expansion is unknown). However, U, H, and S all have exact differentials, so their values can be obtained. We have

$$\boxed{dU = 2C_v^m dT = 0}, \qquad (1)$$

since $dT = 0$ for the isothermal process. Also,

$$\boxed{dH = 2C_p^m dT = 0}, \qquad (2)$$

since $dT = 0$.

Equation 4.77 shows that

$$\left(\frac{\partial S}{\partial V}\right)_T = \frac{P}{T} = \frac{nR}{V} \qquad (3)$$

for an ideal gas. Therefore,

$$\Delta S = \int dS = nR\int_{35\text{ L}}^{90\text{ L}} \frac{dV}{V} = (2\text{ mol})(8.314\text{ J mol}^{-1}\text{ K}^{-1})\ln\left(\frac{90}{35}\right) = \underline{15.70\text{ J K}^{-1}} \qquad (4)$$

4.19 n moles of a gas that obeys a van der Waals equation of state are contained in an insulated piston–cylinder arrangement. The initial state of the gas is $T = T_o$ and $V = V_o$. A reversible adiabatic expansion of the gas is carried out until the volume reaches $V = 2V_o$.

(A) Obtain the final temperature T_f of the gas as a function of n, C_v^m, T_o, V_o, and the van der Waals parameters a and b. Assume that C_v^m is independent of temperature.

(B) If $n = 2$ mol, $C_v^m = 28.80$ J mol^{-1} K^{-1}, $T_o = 350$ K, and $V_o = 40$ L, compute T_f if the gas is $CO_2(g)$.

Solution

(A) The basic equation for any adiabatic, reversible process is

$$dU = \delta q + \delta w = \delta w = -PdV. \qquad (1)$$

Rearranging terms gives

$$dU + PdV = nC_v^m dT + \left(\frac{\partial U}{\partial V}\right)_T dV + PdV = 0. \qquad (2)$$

If the gas were ideal, the second term would be zero. This is not the case for a van der Waals gas. Equation 4.92 shows that

$$\left(\frac{\partial U}{\partial V}\right)_T = T\left(\frac{\partial P}{\partial T}\right)_V - P. \qquad (3)$$

The equation of state gives

$$P = \frac{nRT}{V - nb} - \frac{n^2 a}{V^2}. \qquad (4)$$

Therefore, we have

$$\left(\frac{\partial P}{\partial T}\right)_V = \frac{nR}{V - nb} \qquad (5)$$

which gives

$$\left(\frac{\partial U}{\partial V}\right)_T = \frac{nRT}{V - nb} - \left[\frac{nRT}{V - nb} - \frac{n^2 a}{V^2}\right] = \frac{n^2 a}{V^2}. \qquad (6)$$

Combining Eqs. 2, 4, and 6 yields

$$nC_v^m dT + \left[\frac{n^2 a}{V^2} + \left\{\frac{nRT}{V - nb} - \frac{n^2 a}{V^2}\right\}\right] dV = nC_v^m dT + \frac{nRT}{V - nb} dV = 0. \qquad (7)$$

Canceling out n in the numerator of Eq. 7, dividing by T, and rearranging terms gives

$$C_v^m \frac{dT}{T} = -\frac{R\,dV}{V - nb}. \qquad (8)$$

Integrating both sides between corresponding limits produces

$$C_v^m \int_{T_o}^{T_f} \frac{dT}{T} = C_v^m \ln\left[\frac{T_f}{T_o}\right] = -R \int_{V_o}^{2V_o} \frac{dV}{V - nb} = -R \ln\left[\frac{2V_o - nb}{V_o - nb}\right]$$
$$= R \ln\left[\frac{V_o - nb}{2V_o - nb}\right]. \qquad (9)$$

Exponentiation of both sides gives

$$\left[\frac{T_f}{T_o}\right]^{C_v^m} = \left[\frac{V_o - nb}{2V_o - nb}\right]^R. \qquad (10)$$

Solving Eq. 10 for T_f, we obtain

$$\boxed{T_f = T_o\left[\frac{V_o - nb}{2V_o - nb}\right]^{R/C_v^m}} \qquad (11)$$

as the final result. Somewhat surprisingly, the final temperature is independent of the parameter a. Thus, the result for T_f is independent of the magnitude of the intermolecular forces.

(B) The value of b for $CO_2(g)$ is given in Table 1.2. Substituting values into Eq. 11 gives

$$T_f = (350\text{ K})\left[\frac{40 - (2)(0.04286)}{(2)(40) - (2)(0.04286)}\right]^{8.314/28.80}$$
$$= (350\text{ K})\,(0.8184) = \underline{286.4\text{ K}}. \qquad (12)$$

4.21 (A) Using the formula $dU = T\,dS - P\,dV$, show that, for a closed system along a reversible path, we must have

$$\left(\frac{\partial T}{\partial V}\right)_S = -\left(\frac{\partial P}{\partial S}\right)_V.$$

(B) Using the formula $dH = T\,dS + V\,dP$, show that, for a closed system along a reversible path, we must have

$$\left(\frac{\partial T}{\partial P}\right)_S = \left(\frac{\partial V}{\partial S}\right)_P.$$

The equations derived in (A) and (B), along with two others, constitute Maxwell's relationships.

Solution

Both equations can be obtained using the Euler condition for exactness. Since both dU and dH are exact differentials, the Euler condition must hold for each. Applying this condition to dU, we obtain

$$\left(\frac{\partial T}{\partial V}\right)_S = -\left(\frac{\partial P}{\partial S}\right)_V. \qquad \textbf{(1)}$$

Applying the condition to dH gives

$$\left(\frac{\partial T}{\partial P}\right)_S = \left(\frac{\partial V}{\partial S}\right)_P, \qquad \textbf{(2)}$$

as required.

4.23 A nonideal gas obeys the equation of state $PV = nRT - aP/T$, where a is a positive constant. Obtain an expression for the Joule–Thomson coefficient for this gas in terms of the constant a and the heat capacity of the gas. Does the temperature of the gas increase or decrease in a Joule–Thomson experiment?

Solution

In Chapter 2, it was shown that the Joule–Thomson coefficient is related to the rate of change of enthalpy with pressure at constant temperature by

$$\left(\frac{\partial H}{\partial P}\right)_T = -\mu C_p. \qquad \textbf{(1)}$$

Equation 4.88 shows that

$$\left(\frac{\partial H}{\partial P}\right)_T = V - T\left(\frac{\partial V}{\partial T}\right)_P. \qquad \textbf{(2)}$$

For the nonideal gas in this problem, we have

$$V = \frac{nRT}{P} - \frac{a}{T}. \qquad \textbf{(3)}$$

Taking the derivative with respect to temperature, we obtain

$$\left(\frac{\partial V}{\partial T}\right)_P = \frac{nR}{P} + \frac{a}{T^2}. \qquad \textbf{(4)}$$

Substituting Eqs. 3 and 4 into Eq. 2 produces

$$\left(\frac{\partial H}{\partial P}\right)_T = \frac{nRT}{P} - \frac{a}{T} - T\left[\frac{nR}{P} + \frac{a}{T^2}\right] = -\frac{2a}{T}. \qquad \textbf{(5)}$$

Combining Eqs. 1 and 5, we have

$$\boxed{\mu = \frac{2a}{TC_p}}, \qquad \textbf{(6)}$$

which is the desired result.

The Joule–Thomson coefficient is, by definition,

$$\mu = \left(\frac{\partial T}{\partial P}\right)_H. \tag{7}$$

Since a, T and C_p are all positive, the Joule–Thomson coefficient for this non-ideal gas must be positive. Therefore, as P decreases in a Joule–Thomson experiment, we must also observe T to decrease so that we will have $(\partial T/\partial P)_H > 0$.

4.25 Consider the change in entropy attending the heating of a substance at $dP = 0$ from T_1 to T_2 when $C_p^m = a + bT + cT^2$. What average constant value must the heat capacity have over the same temperature range to give the same entropy change? Express your answer in terms of a, b, c, T_1, and T_2.

Solution

The rate of change of entropy with temperature at constant pressure is given by Eq. 4.73:

$$\left(\frac{\partial S}{\partial T}\right)_P = \frac{nC_p^m}{T}. \tag{1}$$

Substituting the heat capacity from the problem and integrating both sides gives

$$\Delta S = \int dS = n\int_{T_1}^{T_2} \frac{(a + bT + cT^2)dT}{T}$$

$$= n\left[a\ln\left(\frac{T_2}{T_1}\right) + b(T_2 - T_1) + \left(\frac{c}{2}\right)(T_2^2 - T_1^2)\right]. \tag{2}$$

If the heat capacity were constant, we would have

$$\Delta S = n\langle C\rangle \int_{T_1}^{T_2} \frac{dT}{T} = n\langle C\rangle \ln\left(\frac{T_2}{T_1}\right). \tag{3}$$

If we equate the right-hand sides of Eqs. 2 and 3, we may solve for the average constant heat capacity needed to make ΔS the same in both cases:

$$\langle C\rangle = \frac{n\left[a\ln\left(\frac{T_2}{T_1}\right) + b(T_2 - T_1) + \left(\frac{c}{2}\right)(T_2^2 - T_1^2)\right]}{n\ln\left(\frac{T_2}{T_1}\right)} = a + \frac{b(T_2 - T_1)}{\ln\left(\frac{T_2}{T_1}\right)} + \frac{c(T_2^2 - T_1^2)}{2\ln\left(\frac{T_2}{T_1}\right)}. \tag{4}$$

Equation 4 is the desired expression.

4.27 One mole of an ideal gas is heated reversibly along a path such that $T = AV^2$, where A is a constant. If the initial temperature is 273 K, what must the final temperature be if the entropy change is equal to 20.785 J K^{-1}? $C_v^m = 1.5R$ J mol^{-1} K^{-1} for the gas.

Solution

Since the path is reversible, we may write

$$dS = \frac{\delta q_{\text{rev}}}{T} = \frac{CdT}{T}, \tag{1}$$

where C is the heat capacity for the system along the reversible heating path used in the problem. The heat capacity is given by

$$C = \frac{\delta q}{dT} = \frac{dU - \delta w}{dT} = C_v^m + \left[\left(\frac{\partial U}{\partial V}\right)_T + P\right]\frac{dV}{dT}. \quad (2)$$

However, for an ideal gas, $(\partial U/\partial V)_T = 0$, and Eq. 2 becomes

$$C = 1.5R + P\frac{dV}{dT}. \quad (3)$$

For the path given in the problem, $V = [T/A]^{1/2}$, so that

$$\frac{dV}{dT} = 0.5\left[\frac{T}{A}\right]^{-1/2} A^{-1} = \frac{1}{2AV}. \quad (4)$$

Therefore,

$$P\frac{dV}{dT} = \frac{P}{2AV}. \quad (5)$$

However, $P = RT/V$, so that

$$P\frac{dV}{dT} = \frac{RT}{2AV^2} = \frac{R}{2}, \quad (6)$$

since $AV^2 = T$.

The heat capacity is, therefore, given by

$$C = 1.5R + 0.5R = 2.0R. \quad (6)$$

From Eq. 1, we now compute

$$\Delta S = \int_{273\text{ K}}^{T_f} \frac{2R\,dT}{T} = 2R\ln\left(\frac{T_f}{273}\right) = 20.785\text{ J K}^{-1}, \quad (7)$$

where T_f is the final temperature. Exponentiation of both sides of Eq. 7 gives

$$\frac{T_f}{273\text{ K}} = \exp\left[\frac{20.785}{2(8.314)}\right] = \exp[1.25] = 3.490. \quad (8)$$

Solving for T_f, we obtain

$$T_f = 3.490(273) = \underline{952.9\text{ K}} \quad (9)$$

if the total entropy change is to be 20.785 J K^{-1}.

4.29* The heat capacities of Mg between 12 K and 298.15 K are given in the table on the next page.

(A) Plot C_p^m as a function of T. Then, plot C_p^m/T vs. T. Notice that C_p^m approaches $3R$ as T increases. We shall see in a later chapter that this is the limiting value of the heat capacity for an elemental solid.

(B) Assuming the T^3 law to be valid below 12 K, and neglecting the difference between C_p and C_v at these low temperatures, calculate the absolute entropy of Mg at 298.15 K. How is this result related to the plot of C_p^m/T versus T? Compare the result with the data given in Appendix A.

(C) Calculate the entropy of Mg at 550 K. State the approximations you have made. The normal melting point of Mg is about 923 K.

Solution

(A) The plots are given at the end of the problem. Obviously, C_p^m is approaching $3R = 5.961$ cal mol^{-1} K^{-1} as T increases.

T (K)	C_p^m (cal mol^{-1} K^{-1})	T (K)	C_p^m (cal mol^{-1} K^{-1})
12	0.016	130	4.527
14	0.026	140	4.718
16	0.042	150	4.876
18	0.065	160	5.013
20	0.086	170	5.133
25	0.188	180	5.236
30	0.341	190	5.331
35	0.550	200	5.418
40	0.803	210	5.487
45	1.076	220	5.550
50	1.367	230	5.611
60	1.953	240	5.667
70	2.498	250	5.719
80	2.981	260	5.766
90	3.404	270	5.811
100	3.753	280	5.853
110	4.052	290	5.896
120	4.307	298.15	5.929

(B) Since there are no phase transitions for Mg between 0 and 298.15 K, the entropy at 298.15 K is given by

$$S^o_{\mathrm{Mg}} = \int_{0\,\mathrm{K}}^{298.15\,\mathrm{K}} \frac{C_p^m\,dT}{T}. \qquad (1)$$

We may write the right-hand side as two integrals, one from 0 to 12 K and the other from 12 K to 298.15 K:

$$S^o_{\mathrm{Mg}} = \int_{0\,\mathrm{K}}^{12\,\mathrm{K}} \frac{C_p^m\,dT}{T} + \int_{12\,\mathrm{K}}^{298.15\,\mathrm{K}} \frac{C_p^m\,dT}{T}. \qquad (2)$$

In the first integral, the heat capacity is given by $C_p^m = aT^3$. We may obtain the value of a by equating this expression to the measured heat capacity at 12 K:

$$a = \frac{0.016\ \mathrm{cal\ mol^{-1}\ K^{-1}}}{12^3\ \mathrm{K}^3} = 9.3 \times 10^{-6}\ \mathrm{cal\ mol^{-1}\ K^{-4}}. \qquad (3)$$

Using this value, we obtain, for the first integral,

$$\int_{0\,\mathrm{K}}^{12\,\mathrm{K}} \frac{C_p^m\,dT}{T} = 9.3 \times 10^{-6}\ \mathrm{cal\ mol^{-1}\ K^{-4}} \int_{0\,\mathrm{K}}^{12\,\mathrm{K}} T^2\,dT$$
$$= 9.3 \times 10^{-6}(12^3)/3\ \mathrm{cal\ mol^{-1}\ K^{-1}} = 0.0054\ \mathrm{cal\ mol^{-1}\ K^{-1}}. \qquad (4)$$

The second integral must be found numerically. It is equal to the area under the plot of C_p^m/T vs. T from 12 K to 298.15 K. Numerical integration yields 7.768 cal mol^{-1} K^{-1}.

The entropy per mole of Mg at 298.15 K is, therefore,

$$S^o_{\mathrm{Mg}} = 0.0054 + 7.768\ \mathrm{cal\ K^{-1}} = 7.773\ \mathrm{cal\ mol^{-1}} = \underline{32.52\ \mathrm{J\ K^{-1}}}. \qquad (5)$$

The result given in Appendix A is $S^o_{Mg} = 32.68\ \mathrm{J\ K^{-1}}$. The percent error is

$$\%\ \text{error} = \frac{100(32.52 - 32.68)}{32.68} = \underline{-0.48\%}\,. \tag{6}$$

(C) The heat capacity is approaching the limiting value at 298.15. We will, therefore, assume that $C_p^m = 3R$ from 298.15 to 550 K. The entropy of Mg at 550 K can then be computed by calculating the entropy change from 298.15 K to 550 K:

$$\Delta S = \int_{298.15\ \mathrm{K}}^{550\ \mathrm{K}} dS^o = S^o(550\ \mathrm{K}) - S^o(298.15\ \mathrm{K}) = \int_{298.15\ \mathrm{K}}^{550\ \mathrm{K}} \frac{C_p^m\, dT}{T}$$

$$= 3R\ln\left(\frac{550}{298.15}\right) = 15.27\ \mathrm{J\ K^{-1}}. \tag{7}$$

The entropy at 550 K is thus

$$S^o(550\ \mathrm{K}) = S^o(298.15\ \mathrm{K}) + \Delta S = 32.52\ \mathrm{J\ K^{-1}} + 15.27\ \mathrm{J\ K^{-1}} = \underline{47.79\ \mathrm{J\ K^{-1}}}. \tag{8}$$

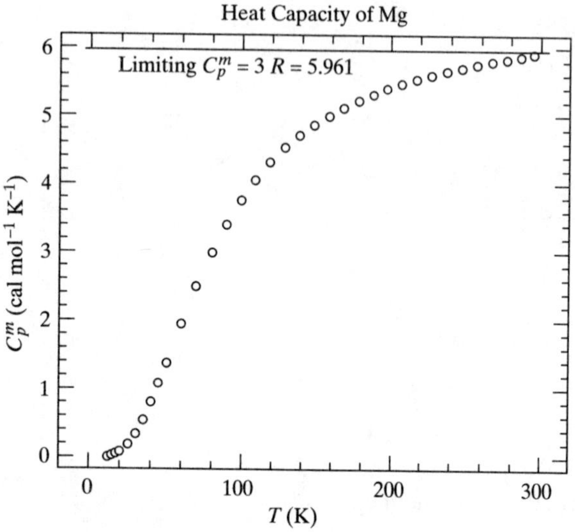

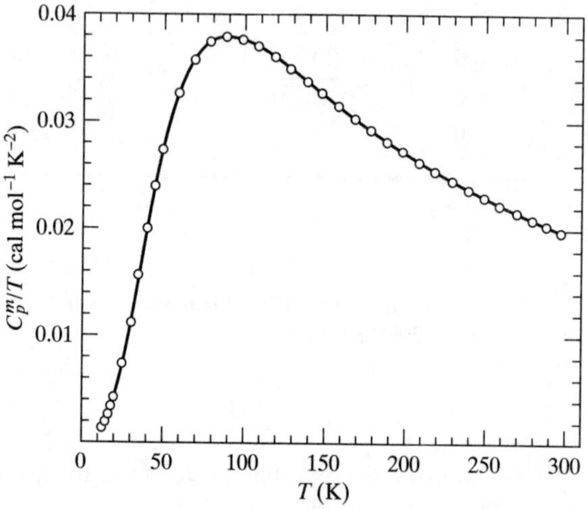

4.31 (A) Using the data in Appendix A, compute ΔS^o for the reaction

$$2\ \mathrm{NO(g)} \longrightarrow \mathrm{N_2O_4(g)}$$

at 298.15 K and 1 bar of pressure.

(B) Discuss the result in terms of order–disorder concepts.

(C) By computing the total entropy change, determine whether this reaction will be spontaneous at 298.15 K and 1 bar of pressure.

Solution

(A) The standard entropy change is

$$\Delta S = S^o_{N_2O_4(g)} - 2S^o_{NO_2(g)} = 304.29 - 2(240.06)\ \mathrm{J\ K^{-1}} = \underline{-175.83\ \mathrm{J\ K^{-1}}}. \tag{1}$$

(B) The 2 moles of $NO_2(g)$ were free to move at random. After they have combined into a single molecule, the two $NO_2(g)$ molecules are in the very ordered structure of the $N_2O_4(g)$ molecule. Consequently, order has increased and ΔS is negative, as expected. In most cases, if the number of moles of gas decreases in a reaction, we will observe $\Delta S < 0$.

(C) To determine the spontaneity of the reaction, we must compute $\Delta S_{\text{surroundings}}$. To do this, we must calculate ΔH at 298.15 K and 1 bar of pressure. This is given by

$$\Delta H = \Delta H^o = \overline{H}^o_{N_2O_4(g)} - 2\overline{H}^o_{NO_2(g)} = 9.16 \text{ kJ} - 2(33.18) \text{ kJ} = -57.20 \text{ kJ}$$
$$= -5.720 \times 10^4 \text{ J}. \quad (2)$$

This heat is added to the surroundings at a constant temperature of 298.15 K. Consequently, we have

$$\Delta S_{\text{surroundings}} = \int \frac{\delta q_{\text{rev}}}{T} = T^{-1}\delta q_{\text{rev}} = \frac{q_{\text{rev}}}{T} = \frac{|\Delta H|}{T} = \frac{5.720 \times 10^4 \text{ J}}{298.15 \text{ K}}$$
$$= +191.8 \text{ J K}^{-1}. \quad (3)$$

Therefore, we obtain

$$\Delta S_{\text{total}} = \Delta S_{\text{system}} + \Delta S_{\text{surroundings}} = -175.83 \text{ J K}^{-1} + 191.8 \text{ J K}^{-1} = \underline{16.0 \text{ J K}^{-1}}. \quad (4)$$

Consequently, $\Delta S_{\text{total}} > 0$, and the process is spontaneous under these conditions.

4.33 A gas obeys the Berthelot equation of state given in Table 1.2. A particular nonideal gas has the Berthelot parameters $a = 1{,}639 \text{ L}^2 \text{ atm mol}^{-2} \text{ K}$ and $b = 0.03049 \text{ L mol}^{-1}$. One mole of this gas is isothermally expanded from 20 L to 50 L at 300 K.

(A) Compute ΔU and ΔS for the process.

(B) If the gas is ideal, compute ΔU and ΔS.

Solution

(A) The Berthelot equation is

$$(V_m - b)\left(P + \frac{a}{TV_m^2}\right) = RT, \quad (1)$$

where V_m is the molar volume. Since we have 1 mole in this problem, we will drop the subscript. The rate of change of U with V when T is constant is given by Eq. 4.92:

$$\left(\frac{\partial U}{\partial V}\right)_T = T\left(\frac{\partial P}{\partial T}\right)_V - P. \quad (2)$$

Therefore, we need to solve Eq. 1 for P:

$$P = \frac{RT}{V - b} - \frac{a}{TV^2}. \quad (3)$$

Using Eq. 3, we obtain

$$\left(\frac{\partial P}{\partial T}\right)_V = \frac{R}{V - b} + \frac{a}{T^2V^2}. \quad (4)$$

Combining Eqs. 2, 3, and 4 gives

$$\left(\frac{\partial U}{\partial V}\right)_T = \frac{RT}{V - b} + \frac{a}{TV^2} - \frac{RT}{V - b} + \frac{a}{TV^2} = \frac{2a}{TV^2}. \quad (5)$$

Separating variables in Eq. 5 and integrating between 20 L and 50 L yields

$$\Delta U = \frac{2a}{T}\int_{20\text{ L}}^{50\text{ L}} \frac{dV}{V^2} = \frac{2a}{T}\left[\frac{1}{20} - \frac{1}{50}\right] = \frac{(2)(1{,}639\text{ L}^2\text{ atm mol}^{-2}\text{ K})}{300\text{ K}}$$

$$\times\left[\frac{1}{20} - \frac{1}{50}\right]\text{L}^{-1}\text{ mol} = 0.3278\text{ L atm mol}^{-1} = 33.21\text{ J mol}^{-1}. \quad \textbf{(6)}$$

Since we have 1 mole of gas, $\Delta U = 33.21\text{ J}$.

The rate of change of entropy with volume at constant temperature is given by Eq. 4.76:

$$\left(\frac{\partial S}{\partial V}\right)_T = T^{-1}\left[\left(\frac{\partial U}{\partial V}\right)_T + P\right]. \quad \textbf{(7)}$$

Combining Eqs. 2 and 7 yields

$$\left(\frac{\partial S}{\partial V}\right)_T = T^{-1}\left[T\left(\frac{\partial P}{\partial T}\right)_V - P + P\right] = \left(\frac{\partial P}{\partial T}\right)_V = \frac{R}{V-b} + \frac{a}{T^2V^2}. \quad \textbf{(8)}$$

We can now separate the variables and integrate between the initial and final volumes to obtain

$$\Delta S = \int_{20\text{ L}}^{50\text{ L}}\left[\frac{R}{V-b} + \frac{a}{T^2V^2}\right]dV = R\ln\left[\frac{50-b}{20-b}\right] + \frac{a}{T^2}\left[\frac{1}{20} - \frac{1}{50}\right]$$

$$= 0.08206\text{ L atm mol}^{-1}\text{ K}^{-1}\ln\left[\frac{50 - 0.03049}{20 - 0.03049}\right] + \frac{1{,}639\text{ L}^2\text{ atm mol}^{-2}\text{ K}}{300^2\text{ K}^2}$$

$$\times\left[\frac{1}{20} - \frac{1}{50}\right]\text{mol L}^{-1} = 0.07527\text{ L atm mol}^{-1}\text{ K}^{-1}$$

$$+\ 0.000546\text{ L atm mol}^{-1}\text{ K}^{-1} = 0.07582\text{ L atm mol}^{-1}\text{ K} \quad \textbf{(9)}$$

Since we have 1 mole of gas, $\Delta S = 0.07582\text{ L atm K}^{-1} = 7.681\text{ J K}^{-1}$.

(B) If the gas were ideal, we would have

$$dU = C_V dT = 0, \quad \textbf{(10)}$$

since the process is isothermal. This gives $\Delta U_{\text{ideal}} = 0$.

For an ideal gas, $(\partial U/\partial V)_T = 0$; therefore, Eq. 7 becomes

$$\left(\frac{\partial S}{\partial V}\right)_T = \frac{P}{T} = \frac{nR}{V}. \quad \textbf{(11)}$$

Integration shows the change in entropy to be

$$\Delta S = nR\int_{20\text{ L}}^{50\text{ L}}\frac{dV}{V} = (1\text{ mol})(8.314\text{ J mol K}^{-1})\ln\left(\frac{50}{20}\right) = 7.618\text{ J K}^{-1}. \quad \textbf{(12)}$$

4.35 Knowledge frequently permits you to perform tasks that you might think would be impossible. For example, Clausius stated the first and second laws of thermodynamics succinctly as follows:

Die Energie der Welt ist konstant.

Die Entropie der Welt strebt einem maximum zu.

The German word *Welt* means *world.* Regardless of whether you know any other German words or not, translate Clausius's statements. (Thanks to Fredrick L. Minn, M.D., Ph.D. for this contribution.)

Solution

The translation can be made with knowledge of what the first and second laws say. The first statement is "The energy of the world is constant." The second statement is "The entropy of the world strives toward a maximum."

4.37 In this problem, you will derive Eq. 4.88, starting with Eq. 4.87 and the knowledge that the second law requires that dS be an exact differential.

(A) Rewrite the left-hand side of Eq. 4.87, using the fact that $C_p = (\partial H/\partial T)_P$.

(B) Show that the left-hand side of Eq. 4.87 is given by

$$\frac{\partial}{\partial P}\left[\frac{C_p}{T}\right]_T = \frac{1}{T}\frac{\partial^2 H}{\partial P \partial T}.$$

(C) Show that the right-hand side of Eq. 4.87, when expanded using the standard rules for taking derivatives, is given by

$$\frac{\partial}{\partial T}\left[\left\{\left(\frac{1}{T}\right)\left[\left(\frac{\partial H}{\partial P}\right)_T - V\right]\right\}\right]_P = -\frac{1}{T^2}\left[\left(\frac{\partial H}{\partial P}\right)_T - V\right] + \frac{1}{T}\left[\frac{\partial^2 H}{\partial T \partial P} - \left(\frac{\partial V}{\partial T}\right)_P\right].$$

(D) By combining the results obtained in (B) and (C), show that we must have

$$-\frac{1}{T^2}\left[\left(\frac{\partial H}{\partial P}\right)_T - V\right] - \frac{1}{T}\left(\frac{\partial V}{\partial T}\right)_P = 0.$$

(*Hint:* Remember that the order of differentiation in a second derivative does not alter the value of the derivative.)

(E) Use the result obtained in (D) to show that Eq. 4.88 is valid. This completes the derivation.

Solution

(A) If we substitute $(\partial H/\partial T)_P$ for C_p in Eq. 4.87, the result is

$$\frac{\partial}{\partial P}\left[\frac{1}{T}\left(\frac{\partial H}{\partial T}\right)_P\right]_T = \frac{\partial}{\partial T}\left[\left\{\left(\frac{1}{T}\right)\left[\left(\frac{\partial H}{\partial P}\right)_T - V\right]\right\}\right]_P. \quad \textbf{(1)}$$

(B) Since the temperature is being held constant in the derivative, we obtain

$$\frac{\partial}{\partial P}\left[\frac{C_p}{T}\right]_T = \frac{\partial}{\partial P}\left[\frac{1}{T}\left(\frac{\partial H}{\partial T}\right)_P\right]_T = \frac{1}{T}\frac{\partial^2 H}{\partial P \partial T}, \quad \textbf{(2)}$$

which is the desired result.

(C) Differentiating the right-hand side of Eq. 1 gives

$$\frac{\partial}{\partial T}\left[\left\{\left(\frac{1}{T}\right)\left[\left(\frac{\partial H}{\partial P}\right)_T - V\right]\right\}\right]_P = \left[\left(\frac{\partial H}{\partial P}\right)_T - V\right]\frac{\partial}{\partial T}\left(\frac{1}{T}\right)$$

$$+ \frac{1}{T}\frac{\partial}{\partial T}\left[\left(\frac{\partial H}{\partial P}\right)_T - V\right]_P = -\frac{1}{T^2}\left[\left(\frac{\partial H}{\partial P}\right)_T - V\right] + \frac{1}{T}\frac{\partial^2 H}{\partial T \partial P} - \frac{1}{T}\left(\frac{\partial V}{\partial T}\right)_P$$

$$= -\frac{1}{T^2}\left[\left(\frac{\partial H}{\partial P}\right)_T - V\right] + \frac{1}{T}\left[\frac{\partial^2 H}{\partial T \partial P} - \left(\frac{\partial V}{\partial T}\right)_P\right], \quad \textbf{(3)}$$

which is the expression to be derived.

(D) Equating the left and right-hand sides of Eq. 4.87 and inserting the results from Eqs. 2 and 3, we obtain

$$\frac{1}{T}\frac{\partial^2 H}{\partial P \partial T} = -\frac{1}{T^2}\left[\left(\frac{\partial H}{\partial P}\right)_T - V\right] + \frac{1}{T}\left[\frac{\partial^2 H}{\partial T \partial P} - \left(\frac{\partial V}{\partial T}\right)_P\right]. \quad \textbf{(4)}$$

Since the order of differentiation does not alter the value of the second derivative, we have

$$\frac{\partial^2 H}{\partial P \partial T} = \frac{\partial^2 H}{\partial T \partial P}. \tag{5}$$

Combining Eqs. 4 and 5 produces

$$0 = -\frac{1}{T^2}\left[\left(\frac{\partial H}{\partial P}\right)_T - V\right] - \frac{1}{T}\left(\frac{\partial V}{\partial T}\right)_P. \tag{6}$$

(E) Rearranging terms in Eq. 6 gives

$$\left[\left(\frac{\partial H}{\partial P}\right)_T - V\right] = -T\left(\frac{\partial V}{\partial T}\right)_P. \tag{7}$$

Solving for $(\partial H/\partial P)_T$, we obtain

$$\boxed{\left(\frac{\partial H}{\partial P}\right)_T = V - T\left(\frac{\partial V}{\partial T}\right)_P}, \tag{8}$$

which is Eq. 4.88 in the text.

4.39 Sam is desperate. His grade in physical chemistry has fallen below the limits of detectability. However, he still feels that his creativity should compensate for his inability to take examinations. After thinking deeply about the philosophical implications of the second law of thermodynamics, he presents the following position paper to his professor:

The second law requires the total entropy change for any spontaneous process to be positive. Therefore, since the universe progresses spontaneously from one day to the next, we must have $\Delta S_{\text{universe}} = S_{\text{universe}}(\text{day } n + 1) - S_{\text{universe}}(\text{day } n) > 0$. That is, if we were to compute the total entropy of the universe on a given day, it would necessarily be greater than the total entropy of the universe on the previous day. Consequently, a plot of S_{universe} versus time must be a montonically increasing function. Such a plot might have the form given in Figure 4.23. Since the combination of the third law and the relationship between entropy and disorder requires that $S_{\text{universe}} \geq 0$, we know that the plot can never assume a negative value. Hence, if we extrapolate the curve shown in the figure to the point at which $S_{\text{universe}} = 0$, we must have the earliest point in time at which the universe could have been created. Of course, the creation point could have been later, because we do not know whether there was perfect order at the point of creation. The total entropy then might have been greater than zero. However, creation cannot have occurred at an earlier date. We have, therefore, determined an upper limit for the age of the universe. Of course, there are technical difficulties in computing the total entropy of the universe, but such difficulties have nothing to do with the fundamental theory.

Comment on Sam's position paper. Does he have a valid point? Are there any flaws in his analysis? Do you think Sam's professor should give his grade special consideration?

Solution

If we assume that the second and third laws of thermodynamics are indeed correct, then Sam has a valid point. If we could determine the point in time at which the universe had perfect order ($S = 0$), that point would be an upper limit to the age of the universe. There are no flaws in the concept, but there are two major difficulties. The first is mentioned by Sam in his position paper: We

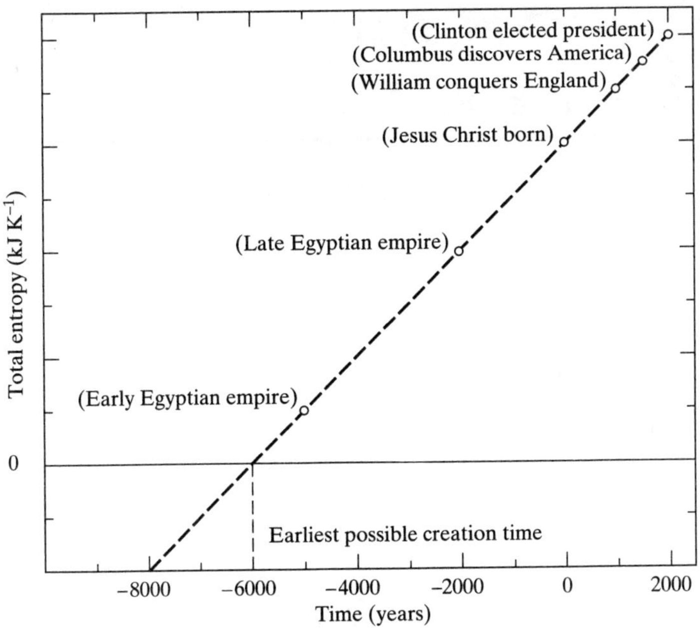

▲ FIGURE 4.23
Entropy of the universe vs. time.

do not have sufficient data to compute the total entropy of the universe. Under the present conditions, such a calculation is impossible. The second is the method used to extrapolate to the point of zero entropy. The plot of S versus time need not be linear, as it is in Sam's sample plot. If the plot were to approach $S = 0$ asymptotically, there would be no good way to determine the curve's intercept with the $S = 0$ line. However, as Sam points out, these are technical difficulties, not fundamental flaws in the concept. Perhaps the professor *should* think more carefully about Sam's grade.

CHAPTER 5

Chemical Equilibrium

5.1 Two moles of an ideal gas are expanded isothermally from 25 L to 125 L at 320 K. Calculate ΔS, ΔG, and ΔA for the process.

Solution

The differential for G is

$$dG = -S\,dT + V\,dP. \tag{1}$$

Since we have $dT = 0$, the change in G is given by

$$\Delta G = \int dG = \int_{P_1}^{P_2} V\,dP = nRT \int_{P_1}^{P_2} \frac{dP}{P} = nRT \ln\left[\frac{P_2}{P_1}\right]. \tag{2}$$

The gas is ideal, so that $P_2/P_1 = V_1/V_2$ when $dT = 0$ (Boyle's law). Thus,

$$\Delta G = nRT \ln\left[\frac{V_1}{V_2}\right] = (2\text{ mol})(8.314\text{ J mol}^{-1}\text{ K}^{-1})(320\text{ K}) \ln\left(\frac{25}{125}\right) = \underline{-8{,}564\text{ J}}. \tag{3}$$

The change in A is related to ΔG by the equation

$$\Delta G = \Delta H - T\,\Delta S = \Delta U + \Delta(PV) - T\,\Delta S = \Delta A - \Delta(PV). \tag{4}$$

Therefore,

$$\Delta A = \Delta G - \Delta(PV) = \Delta G = \underline{-8{,}564\text{ J}}, \tag{5}$$

since $\Delta(PV) = 0$ for an isothermal process for an ideal gas. We could also compute ΔA from

$$dA = -S\,dT - P\,dV. \tag{6}$$

At constant temperature, we have

$$\Delta A = -\int_{25\text{ L}}^{125\text{ L}} \frac{nRT\,dV}{V} = -nRT \ln\left(\frac{125}{25}\right) = \underline{-8{,}564\text{ J}}, \tag{7}$$

which is the same as in Eq. 5.

Since $dH = C_p dT = 0$ for an isothermal process for an ideal gas, we have $\Delta H = 0$. Using Eq. 4, we obtain

$$\Delta S = \frac{\Delta H - \Delta G}{T} = -\frac{\Delta G}{T} = -\frac{-8{,}564\text{ J}}{320\text{ K}} = \underline{26.76\text{ J K}^{-1}}. \tag{8}$$

We could also compute ΔS using

$$\left(\frac{\partial S}{\partial V}\right)_T = \left(\frac{\partial P}{\partial T}\right)_V = \frac{nR}{V}, \tag{9}$$

where the first equality is the Maxwell relationship from dA. Using Eq. 9, we obtain

$$\Delta S = nR \int_{25\text{ L}}^{125\text{ L}} \frac{dV}{V} = nR \ln\left(\frac{125}{25}\right)$$

$$= (2\text{ mol})(8.314\text{ J mol}^{-1}\text{ K}^{-1})\ln(5) = \underline{26.76\text{ J K}^{-1}}. \tag{10}$$

5.3 One mole of a nonideal gas is described by the virial expansion

$$P = RT\left[\frac{1}{V} + \frac{C_2(T)}{V^2} + \frac{C_3(T)}{V^3} + \cdots\right].$$

If the gas is isothermally expanded from volume V_1 to volume V_2, obtain an expression for ΔA for the process in terms of V_1, V_2, T, and the virial coefficients.

Solution

Since the expansion is isothermal, we need $(\partial A/\partial V)_T$. Using the fundamental equation $dA = -S\,dT - P\,dV$, we obtain

$$\left(\frac{\partial A}{\partial V}\right)_T = -P = -RT\left[\frac{1}{V} + \frac{C_2(T)}{V^2} + \frac{C_3(T)}{V^3} + \cdots\right]. \quad (1)$$

We may now separate variables and integrate from V_1 to V_2 to get the total change in A:

$$\Delta A = \int dA = -\int_{V_1}^{V_2} RT\left[\frac{1}{V} + \frac{C_2(T)}{V^2} + \frac{C_3(T)}{V^3} + \cdots\right] dV$$

$$= -RT\ln\left[\frac{V_2}{V_1}\right] + RTC_2(T)\left[\frac{1}{V_2} - \frac{1}{V_1}\right] + \frac{RTC_3(T)}{2}\left[\frac{1}{V_2^2} - \frac{1}{V_1^2}\right] + \cdots$$

$$= \boxed{RT\ln\left[\frac{V_1}{V_2}\right] + RT\sum_{n=1}^{\infty}\frac{C_{n+1}(T)}{n}\left[V_2^{-n} - V_1^{-n}\right]}, \quad (2)$$

which is the required expression for ΔA.

5.5 (A) Using the data in Appendix A, compute the chemical potential of CO(g) at 400 K and 10 bar of pressure. Assume that the heat capacity of CO(g) is a constant. (B) How big is the error if we assume that $\overline{H}^o_{CO}(g)$ is a constant equal to its value at 298.15 K?

Solution

(A) We need the standard chemical potential at 400 K. At 298.15 K, we have $\mu^o = -137.17$ kJ mol^{-1}. Equation 5.80 shows that

$$\frac{\partial}{\partial T}\left[\frac{\mu^o}{T}\right] = -\frac{\overline{H}}{T^2}. \quad (1)$$

Therefore, we need $\overline{H}$ as a function of T. The variation of $\overline{H}$ with temperature is given by

$$\left(\frac{\partial \overline{H}}{\partial T}\right)_P = C_p^m. \quad (2)$$

If the heat capacity is constant, we may integrate Eq. 2 to obtain

$$\int_{\overline{H}(298.15\text{ K})}^{\overline{H}(T)} d\overline{H} = \overline{H}(T) - \overline{H}(298.15) = \int_{298.15\text{ K}}^{T} C_p^m\,dT = C_p^m[T - 298.15]. \quad (3)$$

Rearranging Eq. 3 produces

$$\overline{H}(T) = \overline{H}(298.15) - 298.15C_p^m + C_p^m T. \quad (4)$$

Appendix A gives $\overline{H}(298.15) = -110.53$ kJ mol^{-1} and $C_p^m = 0.02914$ kJ mol^{-1} K^{-1}. Substituting these values into Eq. 4 results in

$$\overline{H}(T) = -110.53\text{ kJ mol}^{-1} - 298.15(0.02914)\text{ kJ mol}^{-1} + 0.02914T$$

$$= -119.22 + 0.02914T\text{ kJ mol}^{-1}. \quad (5)$$

Substituting Eq. 5 into Eq. 1 produces

$$\frac{\partial}{\partial T}\left[\frac{\mu^o}{T}\right] = \frac{119.22}{T^2} - \frac{0.02914}{T}. \tag{6}$$

Separating the variables and then integrating both sides between corresponding limits gives

$$\int_{\mu^o(298.15)/298.15}^{\mu^o(T)/T} d\left[\frac{\mu^o}{T}\right] = \frac{\mu^o(T)}{T} - \frac{\mu^o(298.15)}{298.15} = \int_{298.15}^{T} \frac{-\overline{H}}{T^2} dT$$

$$= \int_{298.15}^{T}\left[\frac{119.22}{T^2} - \frac{0.02914}{T}\right]dT$$

$$= -119.22\left[\frac{1}{T} - \frac{1}{298.15}\right] - 0.02914 \ln\left[\frac{T}{298.15}\right]. \tag{7}$$

Solving for $[\mu^o(T)]/T$ and setting $T = 400$ K, we obtain

$$\frac{\mu^o(400)}{400} = \frac{-137.17}{298.15} - 119.22\left[\frac{1}{400} - \frac{1}{298.15}\right] - 0.02914 \ln\left[\frac{400}{298.15}\right]$$

$$= -0.46007 + 0.10182 - 0.008563 = -0.36681 \text{ kJ mol K}^{-1}. \tag{8}$$

Multiplying both sides of Eq. 8 by 400 K yields

$$\mu^o(400) = 400 \text{ K}(-0.36681 \text{ kJ mol}^{-1} \text{ K}^{-1}) = -146.72 \text{ kJ mol}^{-1}. \tag{9}$$

The chemical potential at 400 K and 10 bar of pressure is, therefore,

$$\mu(400 \text{ K}) = \mu^o(400) + RT \ln(P)$$

$$= -146.72 \text{ kJ mol}^{-1} + (0.008314 \text{ kJ mol}^{-1} \text{ K}^{-1})(400 \text{ K}) \ln(10)$$

$$= \underline{-139.06 \text{ kJ mol}^{-1}}. \tag{10}$$

(B) If $\overline{H}$ is constant at the value $\overline{H}(298.15) = -110.53$ kJ mol^{-1}, integrating Eq. 7 produces

$$\int_{\mu^o(298.15)/298.15}^{\mu^o(T)/T} d\left[\frac{\mu^o}{T}\right] = \frac{\mu^o(T)}{T} - \frac{\mu^o(298.15)}{298.15} = \int_{298.15}^{T} \frac{-\overline{H}}{T^2} dT$$

$$= -110.53\left[\frac{1}{T} - \frac{1}{298.15}\right]. \tag{11}$$

At $T = 400$ K, Eq. 1 gives

$$\frac{\mu^o(400)}{400} = \frac{-137.17}{298.15} - 110.53\left[\frac{1}{400} - \frac{1}{298.15}\right] = -0.46007 + 0.094394$$

$$= -0.36568 \text{ kJ mol}^{-1} \text{ K}^{-1}, \tag{12}$$

or

$$\mu^o(400) = 400 \text{ K}(-0.36568 \text{ kJ mol}^{-1} \text{ K}^{-1}) = -146.27 \text{ kJ mol}^{-1}. \tag{13}$$

The chemical potential is, therefore,

$$\mu(400 \text{ K}) = \mu^o(400) + RT \ln(P)$$

$$= -146.27 \text{ kJ mol}^{-1} + (0.008314 \text{ kJ mol}^{-1} \text{ K}^{-1})(400 \text{ K}) \ln(10)$$

$$= \underline{-138.61 \text{ kJ mol}^{-1}}. \tag{14}$$

The percent error produced by assuming that $\overline{H}$ is constant is

$$\% \text{ error} = 100 \times \frac{(-138.61) - (-139.06)}{139.06} = \underline{0.324\%}. \tag{15}$$

5.7 The equilibrium constant K_p for the Haber reaction (see Problem 5.6) at 500 K is 0.1744. Suppose that 0.1 mole of $H_2(g)$ is mixed with 0.15 mole of $N_2(g)$ and 1.5 moles of $NH_3(g)$ in a 2-L fixed-volume reactor at 500 K.

(A) Compute ΔG for the reaction of 0.001 mole of $N_2(g)$ with $H_2(g)$ to form $NH_3(g)$ under these conditions. Ignore the change in the partial pressures produced by this amount of reaction.

(B) Is the process spontaneous under these conditions? (Assume that all gases are ideal.)

Solution

(A) At constant volume and temperature, Eq. 5.51 shows that we have

$$dA = \sum_{i=1}^{3} \mu_i dn_i. \tag{1}$$

For the Haber reaction, $N_2(g) + 3\,H_2(g) \longrightarrow 2\,NH_3(g)$. If $d\varepsilon$ moles of $N_2(g)$ react, the change in the Helmholtz free energy will be

$$\begin{aligned} dA &= [\mu_{N_2(g)}(-d\varepsilon) + \mu_{H_2(g)}(-3d\varepsilon) + \mu_{NH_3(g)}(2d\varepsilon)] \\ &= d\varepsilon[2\mu_{NH_3(g)} - \mu_{N_2(g)} - 2\mu_{H_2(g)}], \end{aligned} \tag{2}$$

since 3 moles of $H_2(g)$ react and 2 moles of $NH_3(g)$ are formed for each mole of $N_2(g)$ that reacts. Because the gases are ideal, the chemical potentials are given by

$$\mu_i = \mu_i^o + RT \ln P_i \tag{3}$$

where P_i is expressed in bars. Substituting Eq. 3 into Eq. 2 gives

$$\begin{aligned} dA &= d\varepsilon[2\mu^o_{NH_3(g)} - \mu^o_{N_2(g)} - 2\mu^o_{H_2(g)}] + d\varepsilon RT \ln\left[\frac{P^2_{NH_3(g)}}{P_{N_2(g)}P^3_{H_2(g)}}\right] \\ &= d\varepsilon\left\{\Delta\mu^o + RT \ln\left[\frac{P^2_{NH_3(g)}}{P_{N_2(g)}P^3_{H_2(g)}}\right]\right\} \\ &= d\varepsilon\left\{-RT \ln K_p + RT \ln\left[\frac{P^2_{NH_3(g)}}{P_{N_2(g)}P^3_{H_2(g)}}\right]\right\}. \end{aligned} \tag{4}$$

The partial pressures, in bars, are

$$P_{H_2(g)} = \frac{nRT}{V} = \frac{(0.1\text{ mol})(0.083144\text{ L bar mol}^{-1}\text{ K}^{-1})(500\text{ K})}{2\text{ L}} = 2.0786\text{ bar}, \tag{5}$$

$$P_{N_2(g)} = \frac{nRT}{V} = \frac{(0.15\text{ mol})(0.083144\text{ L bar mol}^{-1}\text{ K}^{-1})(500\text{ K})}{2\text{ L}} = 3.1179\text{ bar}, \tag{6}$$

and

$$P_{NH_3(g)} = \frac{nRT}{V} = \frac{(1.5\text{ mol})(0.083144\text{ L bar mol}^{-1}\text{ K}^{-1})(500\text{ K})}{2\text{ L}} = 31.179\text{ bar}. \tag{7}$$

Substituting into Eq. 4 with $d\varepsilon = 0.001$ mol, we obtain

$$\begin{aligned} dA &= 0.001(8.314\text{ J mol}^{-1}\text{ K}^{-1})(500\text{ K})\left[-\ln(0.1744) + \ln\left[\frac{31.179^2}{(3.1179)(2.0786)^3}\right]\right. \\ &= 4.157[1.7464 + 3.5472]\text{ J mol}^{-1} = \underline{22.01\text{ J mol}^{-1}}. \end{aligned} \tag{8}$$

The foregoing calculation assumes that the reaction of 0.001 mole of $N_2(g)$ does not appreciably change the partial pressures of reactants and products present in the two-liter container. dG can now be obtained from

$$dG = d(A + PV) = dA + d(PV) = dA + d(nRT) = dA + RT\,dn, \tag{9}$$

where dn is the change in the number of moles of gas in the process. In our case, 0.001 mole of $N_2(g)$ reacts with 0.003 mole of $H_2(g)$ to form 0.002 mole of $NH_3(g)$. Therefore, $dn = -0.002$ mol. Hence,

$$dG = dA - 0.002\,RT = 22.01\text{ J mol}^{-1} - 0.002\text{ mol}(8.314\text{ J mol}^{-1}\text{ K}^{-1})(500\text{ K})$$

$$= \underline{13.70\text{ J mol}^{-1}}. \tag{10}$$

(B) Since we have $dA > 0$ and $\Delta\mu > 0$ for a process carried out under conditions of constant volume and temperature, the process is nonspontaneous.

5.9* The equilibrium constant K_p for the Haber reaction (see Problem 5.6) at 500 K is 0.1744. Suppose that 3 moles of $H_2(g)$ and 1 mole of $N_2(g)$ are mixed at 500 K in a closed, fixed-volume 2-L container. When equilibrium is reached, what are the partial pressures of $H_2(g)$, $N_2(g)$, and $NH_3(g)$? What is the percent conversion to $NH_3(g)$? (Assume that all gases are ideal.)

Solution

Let 2x equal the number of moles of $NH_3(g)$ present at equilibrium. In this case, the stoichiometry of the reaction gives

$$n_{H_2(g)} \text{ at equilibrium} = 3 - 3x \tag{1}$$

and

$$n_{N_2(g)} \text{ at equilibrium} = 1 - x. \tag{2}$$

The partial pressures are

$$P_{H_2(g)} = \frac{nRT}{V} = \frac{(3 - 3x)(0.083144\text{ L bar mol}^{-1}\text{ K}^{-1})(500\text{ K})}{2\text{ L}}$$

$$= 62.358(1 - x)\text{ bar}, \tag{3}$$

$$P_{N_2(g)} = \frac{nRT}{V} = \frac{(1 - x)(0.083144\text{ L bar mol}^{-1}\text{ K}^{-1})(500\text{ K})}{2\text{ L}}$$

$$= 20.786(1 - x)\text{ bar}, \tag{4}$$

and

$$P_{NH_3(g)} = \frac{nRT}{V} = \frac{2x(0.083144\text{ L bar mol}^{-1}\text{ K}^{-1})(500\text{ K})}{2\text{ L}} = 41.572\,x\text{ bar}. \tag{5}$$

Substituting into the expression for K_p gives

$$K_p = \frac{P^2_{NH_3(g)}}{P_{N_2(g)}P^3_{H_2(g)}} = \frac{41.572^2x^2}{20.786(1 - x)\,62.358^3(1 - x)^3} = \frac{0.00034289x^2}{(1 - x)^4} = 0.1744. \tag{6}$$

Equation 6 may be written in the form

$$F(x) = \frac{x^2}{(1 - x)^4} - 508.62 = 0. \tag{7}$$

Since we have only 1 mole of $N_2(g)$, 2x must lie in the range from 0 to 1. All we need do, then, is conduct a grid search on the function $F(x)$ to find the value at which $F(x) = 0$. The initial search shows that $F(0.8098) = -7.417$ and $F(0.8108) = 4.541$. Consequently, the root lies between these two values. A second search gives $F(0.8104345) = 0.000$. Thus, the desired solution to Eq. 7 is $x = 0.8104$ mol, correct to four significant digits. The equilibrium partial pressures can now be computed from Eqs. 3 through 5, using this value of x:

$$P_{H_2(g)} = \frac{nRT}{V} = \frac{(3 - 3x)(0.083144 \text{ L bar mol}^{-1} \text{ K}^{-1})(500 \text{ K})}{2 \text{ L}} = \underline{11.82 \text{ bar}}; \quad \textbf{(8)}$$

$$P_{N_2(g)} = \frac{nRT}{V} = \frac{(1 - x)(0.083144 \text{ L bar mol}^{-1} \text{ K}^{-1})(500 \text{ K})}{2 \text{ L}} = \underline{3.941 \text{ bar}}; \quad \textbf{(9)}$$

$$P_{NH_3(g)} = \frac{nRT}{V} = \frac{2x(0.083144 \text{ L bar mol}^{-1} \text{ K}^{-1})(500 \text{ K})}{2 \text{ L}} = \underline{33.69 \text{ bar}}. \quad \textbf{(10)}$$

As a check, we can see if these values satisfy the equilibrium constant. We have

$$K_p = \frac{(33.69)^2}{3.941(11.82)^3} = 0.1744, \quad \textbf{(11)}$$

which is the correct result to four significant digits.

(B) The percent conversion to $NH_3(g)$ is given by

$$\% \text{ conversion} = \frac{\text{number moles } NH_3(g) \text{ formed}}{\text{number moles of } NH_3(g) \text{ if all } N_2(g) \text{ is converted}}(100)$$

$$= \frac{2x}{2}(100) = 100x = \underline{81.04\%}. \quad \textbf{(12)}$$

5.11* Consider the hypothetical reaction

$$A(g) + B(g) \longrightarrow C(g) + 10.00 \text{ kJ}.$$

Suppose the reaction is conducted at constant temperature and pressure between gases that may be accurately described by the ideal-gas law. The standard chemical potentials are $\mu^o = -12.00$ kJ mol^{-1}, -3.00 kJ mol^{-1}, and -4.00 kJ mol^{-1} for C(g), A(g), and B(g), respectively, at 298.15 K.

(A) Compute K_p for this reaction at 298.15 K.

(B) Express K_p in terms of the numbers of moles of A(g), B(g), and C(g) present, R, T, and the volume V of the container.

(C) 2 moles of A(g) and 2 moles of B(g) are placed in a 100-L container at 298.15 K. When equilibrium is reached, how many moles of C(g) will be present in the container?

(D) After equilibrium is reached, 1 mole of A(g) and 1 mole of B(g) are added to the container. The temperature of the system is then raised to a point such that when equilibrium is again established, the number of moles of C(g) in the container is exactly the same as it was before the addition of more A(g)and B(g) and the elevation of the temperature. Determine the new temperature of the gas in the container. Assume that ΔH^o is a constant.

Solution

(A) $\Delta\mu^o$ for the reaction is

$$\Delta\mu^o = \mu^o_{C(g)} - \mu^o_{A(g)} - \mu^o_{B(g)} = -12.00 - (-3.00) - (-4.00) \text{ kJ mol}^{-1}$$
$$= -5.00 \text{ kJ mol}^{-1}. \quad \textbf{(1)}$$

The thermodynamic equilibrium constant is given by

$$K_p = \exp\left[-\frac{\Delta\mu^o}{RT}\right] = \exp\left[\frac{5{,}000 \text{ J mol}^{-1}}{(8.314 \text{ J mol}^{-1}\text{ K}^{-1})(298.15 \text{ K})}\right] = \underline{7.516}. \quad \textbf{(2)}$$

(B) K_p is given by

$$K_p = \left[\frac{P_{C(g)}}{P_{A(g)}P_{B(g)}}\right]_{eq}. \quad \textbf{(3)}$$

But we have $P = (nRT)/V$, so that

$$\boxed{K_p = \left[\frac{n_{C(g)}}{n_{A(g)}n_{B(g)}}\right]_{eq}\left[\frac{V}{RT}\right]}, \quad \textbf{(4)}$$

which is the required expression.

(C) Let the number of moles of $C(g)$ at equilibrium be x. The stoichiometry then requires that the number of moles of $A(g)$ and $B(g)$ each equal $2 - x$ at equilibrium. Substituting into Eq. 4 gives

$$K_p = \frac{x}{(2-x)^2}\left[\frac{V}{RT}\right]. \quad \textbf{(5)}$$

Putting Eq. 5 into quadratic form, we obtain

$$\left(\frac{K_pRT}{V}\right)x^2 - x\left[4\left(\frac{K_pRT}{V}\right) + 1\right] + 4\left(\frac{K_pRT}{V}\right) = 0. \quad \textbf{(6)}$$

Substituting values gives

$$\left(\frac{K_pRT}{V}\right) = \frac{7.516(0.083144 \text{ L mol}^{-1}\text{ K}^{-1})(298.15 \text{ K})}{100 \text{ L}} = 1.863 \text{ mol}^{-1}, \quad \textbf{(7)}$$

where R is expressed in units of L bar mol^{-1} K^{-1}, but with the pressure unit dropped, since K_p is unitless. Combining Eqs. 6 and 7 gives

$$1.863x^2 - 8.452x + 7.452 = 0. \quad \textbf{(8)}$$

Using the quadratic formula, we obtain

$$x = \frac{8.452 \pm [8.452^2 - 4(1.863)(7.452)]^{1/2}}{2(1.863)} = 1.198 \text{ or } 3.339. \quad \textbf{(9)}$$

The second root is physically meaningless, since we must have $x < 2$ mol because we have only 2 moles of $A(g)$ and $B(g)$. Therefore, the number of moles of $C(g)$ present at equilibrium is $\underline{1.198 \text{ mol}}$. We may check this result to see if we obtain the correct value of K_p. We have

$$K_p = \frac{x}{(2-x)^2}\left[\frac{V}{RT}\right] = \frac{1.198}{(2-1.198)^2} \times \left(\frac{100 \text{ L}}{(0.083144)(298.15)}\right) = 7.514, \quad \textbf{(10)}$$

which is correct except for rounding errors in the fourth significant digit.

(D) Since the temperature is raised, we will observe a change in K_p. The new equilibrium constant is given by Eq. 5.87, because ΔH^o is constant. Using this equation, we have

$$K_p(T) = K_p(298.15)\exp\left[-\frac{\Delta H^o}{R}\left\{\frac{1}{T} - \frac{1}{298.15}\right\}\right]. \quad \textbf{(11)}$$

The heat of reaction is given as $q = -10.00\ \text{kJ mol}^{-1} = -10{,}000\ \text{J mol}^{-1}$. At $dP = 0$, we have $q = q_p = \Delta H^o = -10{,}000\ \text{J mol}^{-1}$. Substituting values into Eq. 11 yields

$$K_p(T) = 7.516 \exp\left[-\frac{-10{,}000\ \text{J mol}^{-1}}{8.314\ \text{J mol}^{-1}\ \text{K}^{-1}}\left\{\frac{1}{T} - \frac{1}{298.15}\right\}\right]$$

$$= 7.516 \exp[-4.034179] \exp\left[\frac{1{,}202.8}{T}\right] = 0.1330 \exp\left[\frac{1{,}202.8}{T}\right]. \quad \textbf{(12)}$$

The conditions of the problem tell us that the new equilibrium number of moles of $C(g) = 1.198$ mol, the same as in Part (B). Consequently, no additional $A(g)$ and $B(g)$ reacted to form more $C(g)$ nor did any $C(g)$ react to form more $A(g)$ and $B(g)$. Therefore, we must have at equilibrium

$$n_{C(g)} = 1.198\ \text{mol},\ n_{A(g)} = n_{B(g)} = (2 - 1.198 + 1) = 1.802\ \text{mol}. \quad \textbf{(13)}$$

Substituting these results into Eq. 4 gives

$$K_p = \frac{1.198}{(1.802)^2}\ \text{mol}^{-1} \times \frac{100\ \text{L}}{(0.083144\ \text{L mol}^{-1}\ \text{K}^{-1})T} = \frac{443.73}{\text{T}}. \quad \textbf{(14)}$$

Combining Eqs. 12 and 14 produces

$$0.1330 \exp\left[\frac{1{,}202.8}{T}\right] = \frac{443.73}{T}, \quad \textbf{(15)}$$

which must be solved for T to find the new temperature of the system. Since T appears both inside and outside the exponential function, there is no analytic solution to Eq. 15. We can, however, easily solve the equation numerically by conducting a search for the root of the equation

$$F(T) = 0.1330 \exp\left[\frac{1{,}202.8}{T}\right] - \frac{443.73}{T} = 0. \quad \textbf{(16)}$$

A simple search, repeated twice, gives $\underline{T = 990.2\ \text{K}}$ as the new temperature of the system.

5.13 The equation of state for 1 mole of a van der Waals gas is

$$\left(p + \frac{a}{V^2}\right)(V - b) = RT.$$

The gas is expanded isothermally from volume V_1 to volume V_2 at temperature T. Determine ΔA for the process, and then use the Gibbs–Helmholtz equation to obtain ΔU for the process.

Solution

The differential for the Helmholtz free energy is

$$dA = -S dT - P dV. \quad \textbf{(1)}$$

The van der Waals equation of state for 1 mole of gas is

$$\left(P + \frac{a}{V^2}\right)(V - b) = RT. \quad \textbf{(2)}$$

Solving for P, we obtain

$$P = \frac{RT}{V - b} - \frac{a}{V^2}. \quad \textbf{(3)}$$

Since we have $dT = 0$, Eq. 1 becomes

$$dA = -PdV = -\left[\frac{RT}{V-b} - \frac{a}{V^2}\right]dV. \quad (4)$$

Integration produces

$$\boxed{\Delta A = \int dA = -\int_{V_1}^{V_2}\left[\frac{RT}{V-b} - \frac{a}{V^2}\right]dV = -RT\ln\left[\frac{V_2 - b}{V_1 - b}\right] - a\left[\frac{1}{V_2} - \frac{1}{V_1}\right]}, \quad (5)$$

which is the required expression. The Gibbs–Helmholtz equation, Eq. 5.78 now gives us

$$\frac{\partial}{\partial T}\left[\frac{\Delta A}{T}\right]_V = -\frac{\Delta U}{T^2}. \quad (6)$$

Using Eq. 5, this becomes

$$\frac{\partial}{\partial T}\left[R\ln\left[\frac{V_2 - b}{V_1 - b}\right] + \frac{a}{T}\left[\frac{1}{V_2} - \frac{1}{V_1}\right]\right]_V = \frac{\Delta U}{T^2} = -\frac{a}{T^2}\left[\frac{1}{V_2} - \frac{1}{V_1}\right]. \quad (7)$$

Solving for ΔU, we obtain

$$\boxed{\Delta U = -a\left[\frac{1}{V_2} - \frac{1}{V_1}\right]}, \quad (8)$$

which is the desired result.

5.15 Show that $\sum_{i=1}^{K}\overline{H}_i\, dn_i = \sum_{i=1}^{K}\overline{G}_i\, dn_i$.

Solution

From the definitions of G and H, we have

$$G = H - TS. \quad (1)$$

Taking differentials of both sides, we obtain

$$dG = dH - TdS - SdT. \quad (2)$$

Substituting for dG and dH using Eqs. 5.30 and 5.28, respectively, gives

$$-SdT + VdP + \sum_{i=1}^{K}\overline{G}_i dn_i = TdS + VdP + \sum_{i=1}^{K}\overline{H}_i dn_i - TdS - SdT. \quad (3)$$

Adding out identical terms on the right-hand side produces

$$-SdT + VdP + \sum_{i=1}^{K}\overline{G}_i dn_i = -SdT + VdP + \sum_{i=1}^{K}\overline{H}_i dn_i. \quad (4)$$

Subtracting $[-SdT + VdP]$ from both sides gives the desired result:

$$\boxed{\sum_{i=1}^{K}\overline{G}_i dn_i = \sum_{i=1}^{K}\overline{H}_i dn_i}. \quad (5)$$

5.17 The fugacity of a gas is found to obey the equation

$$f = P\exp\left[\frac{AP + BP^2/2}{RT}\right],$$

where A and B are constants. Determine the equation of state for this gas.

Solution

The Gibbs free energy for 1 mole of gas is given by

$$G = \mu^o + RT\ln(f). \quad \textbf{(1)}$$

Substituting the fugacity yields

$$G = \mu^o + RT\ln(P) + AP + \frac{BP^2}{2}, \quad \textbf{(2)}$$

since $\ln(e^x) = x$. From the fundamental equation for dG, $dG = -S\,dT + V\,dP$, we have

$$\left(\frac{\partial G}{\partial P}\right)_T = V. \quad \textbf{(3)}$$

From Eq. 2, the pressure derivative of G is given by

$$\left(\frac{\partial G}{\partial P}\right)_T = \frac{RT}{P} + A + BP, \quad \textbf{(4)}$$

since μ^o is independent of pressure. Combining Eqs. 3 and 4 produces

$$V = \frac{RT}{P} + A + BP. \quad \textbf{(5)}$$

Multiplying of Eq. 5 by P gives

$$\boxed{PV = RT + AP + BP^2} \quad \textbf{(6)}$$

as the equation of state of the gas.

5.19* In this problem, we shall treat both $NO_2(g)$ and $N_2O_4(g)$ as nonideal. The equilibrium reaction of concern is the same as in Problem 5.18: $2\ NO_2(g) \longrightarrow N_2O_4(g)$. The data required in the various parts of the problem either are given directly or can be found in Appendix A. To simplify the calculations, we shall make the following approximations:

1. Both $NO_2(g)$ and $N_2O_4(g)$ obey a modified truncated virial equation of state as described in (C) to follow.
2. Dalton's law holds; that is,

$$P_i = X_i\,P_T,$$

where

P_i = partial pressure of gas i

P_T = total pressure = $P_{NO_2(g)} + P_{N_2O_4(g)}$

and

X_i = mole fraction of gas i.

3. The fugacity of gas i depends only upon the temperature and the partial pressure of gas i; it is independent of the partial pressures of any other gases present.

(A) If

$$C_p^m \text{ for } NO_2(g) = 6.37 + 0.0101T$$
$$- 34.05 \times 10^{-7}T^2 \text{ cal mol}^{-1}\text{ K}^{-1}$$

and

$$C_p^m \text{ for } N_2O_4(g) = 10.719 + 0.0286T$$
$$- 87.26 \times 10^{-7}T^2 \text{ cal mol}^{-1}\text{ K}^{-1},$$

obtain ΔH° for the reaction $2\ NO_2(g) \longrightarrow N_2O_4(g)$ as a function of T. Plot ΔH as a function of T over the range $298\text{ K} \leq T \leq 400\text{ K}$.

(B) Obtain K_p for the reaction $2\ NO_2(g) \longrightarrow N_2O_4(g)$ as a function of T. Plot K_p as a function of T over the range $298\text{ K} \leq T \leq 400\text{ K}$.

(C) We shall use a modified form of the virial equation for the equation of state for both gases. If we truncate the virial expansion after the third term, we have

$$PV = RT\left[1 + \frac{C_2(T)}{V} + \frac{C_3(T)}{V^2}\right]$$

for 1 mole of gas. Equation 1.66 shows that if we employed a van der Waals equation of state, $C_2(T) = b - a/RT$ and $C_3(T) = b^2$. To simplify the calculations, let us replace $1/V$ in the second and third terms with its ideal-gas form, P/RT. Then the truncated virial expression becomes $PV = RT + C_2(T)P + C_3(T)P^2/RT$. The latter expression will represent the equation of state for both gases. This type of expansion, in which the PV product is expressed as a power series in the pressure, is called the *Reichsanstalt equation.* $C_2(T)$ and $C_3(T)$ will be estimated from van der Waals parameters. For $NO_2(g)$, we have $a = 5.354\text{ L}^2\text{ bar mol}^{-2}$ and $b = 0.04424\text{ L mol}^{-1}$. For $N_2O_4(g)$, we shall use $a = 6.550\text{ L}^2\text{ bar mol}^{-2}$ and $b = 0.05636\text{ L mol}^{-1}$. Obtain an analytic function giving the fugacities of both gases in terms of $C_2(T)$, $C_3(T)$, and the pressure. Prepare plots of the fugacities of both gases at 298.15 K over the pressure range $1\text{ bar} \leq P \leq 400\text{ bar}$.

(D) Compute the fugacity coefficients for both gases at 298.15 K over the pressure range $1\text{ bar} \leq P \leq 400\text{ bar}$. Plot the results.

(E) Assuming both gases to be ideal, obtain the mole fraction of $N_2O_4(g)$ present in the equilibrium system at 298.15 K as a function of the total pressure P_T over the range $1\text{ bar} \leq P_T \leq 400\text{ bar}$. Plot the results, with one graph showing $X_{N_2O_4(g)}$ from 1 to 400 bar of pressure and a second showing $X_{N_2O_4(g)}$ from 1 to 50 bar of pressure.

(F) Assuming $NO_2(g)$ and $N_2O_4(g)$ to be nonideal gases described by the truncated virial equation developed in (C), compute the mole fraction of $N_2O_4(g)$ present at equilibrium as a function of total pressure over the range $1\text{ bar} \leq P_T \leq 400\text{ bar}$.

(G) Prepare a plot of the mole fraction of $N_2O_4(g)$ present at equilibrium versus the total pressure over the range $1\text{ bar} \leq P_T \leq 400\text{ bar}$ for the case of ideal and nonideal gases. Compute the percent error made by the ideal-gas assumption as a function of total pressure.

Solution

(A) ΔH^o for the reaction at 1 bar of pressure and 298.15 K is given by

$$\Delta H = \Delta H^o = \overline{H}^o_{N_2O_4(g)} - 2\overline{H}^o_{NO_2(g)} = 9.16\text{ kJ mol}^{-1} - 2(33.18)\text{ kJ mol}^{-1}$$
$$= -57.20\text{ kJ mol}^{-1}. \qquad \textbf{(1)}$$

The temperature dependence of ΔH^o is given by

$$\left(\frac{\partial \Delta H^o}{\partial T}\right)_P = \Delta C_p = [C_p^m]_{N_2O_4(g)} - 2[C_p^m]_{NO_2(g)}$$

$$= 10.719 + 0.0286T - 87.26 \times 10^{-7}T^2 \text{ cal mol}^{-1}\text{ K}^{-1}$$

$$- (2)[6.37 + 0.0101T - 34.05 \times 10^{-7}T^2] \text{ cal mol}^{-1}\text{ K}^{-1}$$

$$= -2.021 + 0.0084T - 19.16 \times 10^{-7}T^2 \text{ cal mol}^{-1}\text{ K}^{-1}$$

$$= -8.456 + 0.03515T - 80.17 \times 10^{-7}T^2 \text{ J mol}^{-1}\text{ K}^{-1}. \quad (2)$$

Separating the variables in Eq. 2 and then integrating between corresponding limits gives

$$\int_{\Delta H^o(298.15)}^{\Delta H^o(T)} d\Delta H^o = \Delta H^o(T) - \Delta H^o(298.15) = \int_{298.15}^{T} \Delta C_p \partial T$$

$$= -8.456(T - 298.15) + \frac{0.03515}{2}(T^2 - 298.15^2)$$

$$- \frac{80.17 \times 10^{-7}}{3}(T^3 - 298.15^3). \quad (3)$$

Rearranging Eq. 3 and collecting constants produces

$$\Delta H^o(T) = -57{,}200 + 8.456(298.15) - \frac{0.03515(298.15^2)}{2}$$

$$+ \frac{80.17 \times 10^{-7}(298.15^3)}{3}$$

$$- 8.456T + 0.01758T^2 - 26.72 \times 10^{-7}T^3. \quad (4)$$

$$\boxed{\Delta H^o(T) = -5.624 \times 10^4 - 8.456T + 0.01758T^2 - 26.72 \times 10^{-7}T^3 \text{ J mol}^{-1}}. \quad (5)$$

Equation 5 is the desired result. The plot follows.

(B) The change in μ^o is

$$\Delta\mu^o = \mu^o_{N_2O_4(g)} - 2\mu^o_{NO_2(g)} = 97.89 - 2(51.31) \text{ kJ mol}^{-1}$$

$$= -4.73 \text{ kJ mol}^{-1} = -4{,}730 \text{ J mol}^{-1}. \quad (6)$$

At 298.15 K,

$$K_p(298.15) = \exp\left[\frac{-\Delta\mu^o}{RT}\right] = \exp\left[\frac{4{,}730 \text{ J mol}^{-1}}{(8.314 \text{ J mol}^{-1}\text{ K}^{-1})(298.15 \text{ K})}\right] = 6.741. \quad (7)$$

The temperature dependence of K_p is given by

$$\left(\frac{\partial \ln K_p}{\partial T}\right)_p = \frac{\Delta H^o}{RT^2} = \frac{-5.624 \times 10^4 - 8.456T + 0.01758T^2 - 26.72 \times 10^{-7}T^3}{RT^2}. \quad (8)$$

Separating variables and integrating between corresponding limits yields

$$\int_{K_p(298.15)}^{K_p(T)} d \ln K_p = \ln K_p(T) - \ln K_p(298.15)$$

$$= \int_{298.15}^{T} \frac{-5.624 \times 10^4 - 8.456T + 0.01758T^2 - 26.72 \times 10^{-7}T^3}{RT^2} dT, \quad (9)$$

and

$$\ln K_p(T) = \ln(6.741) + \frac{5.624 \times 10^4}{8.314}\left[\frac{1}{T} - \frac{1}{298.15}\right] - \frac{8.456}{8.314}\ln\left[\frac{T}{298.15}\right]$$

$$+ \frac{0.01758}{8.314}[T - 298.15] - \frac{26.72 \times 10^{-7}}{2(8.314)}[T^2 - 298.15^2]. \quad \textbf{(10)}$$

Collecting constants yields

$$\boxed{\ln K_p(T) = -15.60 + \frac{6{,}764}{T} - 1.017\ln(T) + 0.002115T - 1.607 \times 10^{-7}T^2}, \quad \textbf{(11)}$$

which is the required function of temperature. (See plot below.)

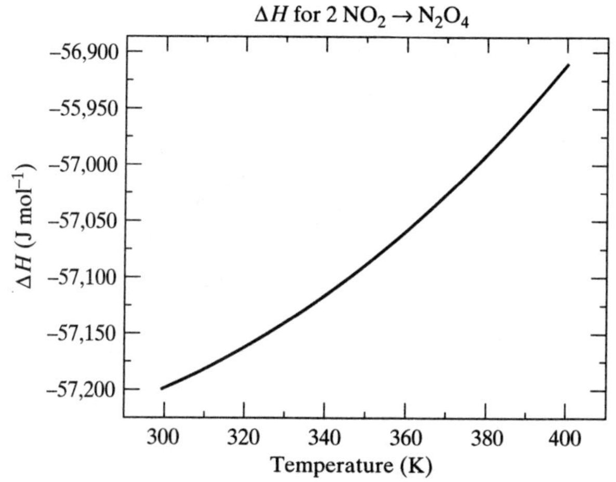

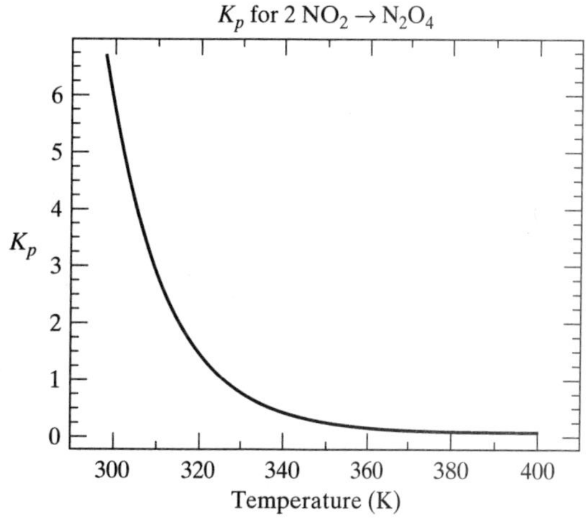

(C) The fugacity is given by Eq. 5.110:

$$\ln(f) = \ln(P) + \frac{1}{RT}\int_0^P \left[V_m - \frac{RT}{P}\right] dP. \quad \textbf{(12)}$$

Therefore, we need an expression for $V_m - (RT)/P$. For 1 mole of the gas, we have

$$V_m - \frac{RT}{P} = C_2(T) + \frac{C_3(T)P}{RT}. \quad \textbf{(13)}$$

Substituting into Eq. 1 produces

$$\ln(f) = \ln(P) + \frac{C_2(T)}{RT}\int_0^P dP + \frac{C_3(T)}{(RT)^2}\int_0^P P\,dP = \ln(P) + \frac{C_2P}{RT} + \frac{C_3(T)P^2}{2(RT)^2}. \quad \textbf{(14)}$$

Exponentiation of both sides gives

$$\boxed{f = P\exp\left[\frac{C_2(T)P}{RT} + \frac{C_3(T)P^2}{2(RT)^2}\right]}. \quad \textbf{(15)}$$

Equation 15 is the required relationship. The values of $C_2(T)$ for each gas may be computed from the data given in the statement of the problem.
For $NO_2(g)$,

$$C_2(T) = b - \frac{a}{RT} = 0.04424 \text{ L mol}^{-1} - \frac{5.354}{(0.083144)(298.15 \text{ K})} \text{ L mol}^{-1}$$

$$= -0.1717 \text{ L mol}^{-1}, \quad \textbf{(16)}$$

and

$$C_3(T) = 0.04424^2 \text{ L}^2 \text{ mol}^{-2} = 0.001957 \text{ L}^2 \text{ mol}^{-2}. \quad \textbf{(17)}$$

For $N_2O_4(g)$,

$$C_2(T) = b - \frac{a}{RT} = 0.05636 \text{ L mol}^{-1} - \frac{6.550}{(0.083144)(298.15 \text{ K})} \text{ L mol}^{-1}$$

$$= -0.2079 \text{ L mol}^{-1}, \quad \textbf{(18)}$$

and

$$C_3(T) = b^2 = 0.05636^2 \text{ L}^2 \text{ mol}^{-2} = 0.003176 \text{ L}^2 \text{ mol}^{-2}. \quad \textbf{(19)}$$

The fugacity plots follow.

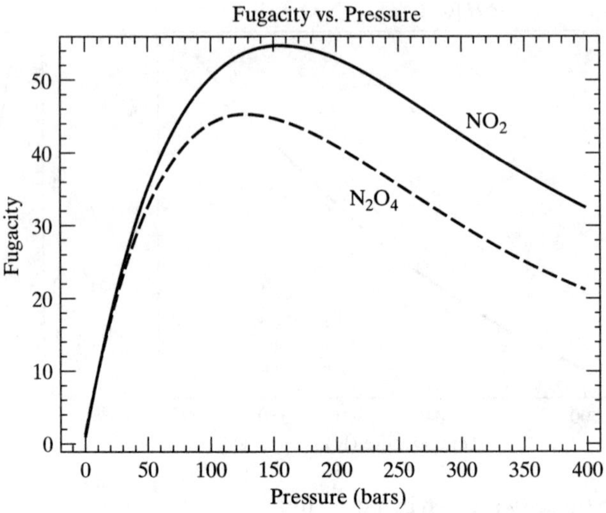

(D) The fugacity coefficients are given by

$$\boxed{\gamma = \frac{f}{P} = \exp\left[\frac{C_2(T)P}{RT} + \frac{C_3(T)P^2}{2(RT)^2}\right]}. \quad \textbf{(20)}$$

The plots follow.

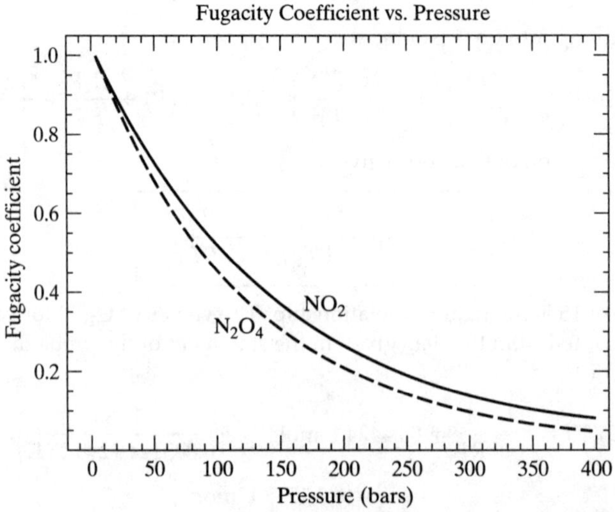

(E) For the ideal system, we have

$$K_p = \left[\frac{P_{N_2O_4(g)}}{P^2_{NO_2(g)}}\right]_{eq} = \left[\frac{XP_T}{(1-X)^2P_T^2}\right]_{eq} = \left[\frac{X}{(1-X)^2P_T}\right]_{eq}, \tag{21}$$

where X is the mole fraction of $N_2O_4(g)$ present at equilibrium and P_T is the total pressure. Putting Eq. 21 in quadratic form, we obtain

$$K_pP_TX^2 - X(2P_TK_p + 1) + K_pP_T = 0. \tag{22}$$

Solving for X, we have

$$X = \frac{(2K_pP_T + 1) - [(2K_pP_T + 1)^2 - 4K_p^2P_T^2]^{1/2}}{2K_pP_T}. \tag{23}$$

We know that we need a negative sign in front of the square root in Eq. 23, since we must have $X \longrightarrow 0$ when $K_p \longrightarrow 0$. (We obtain this result with the negative sign, but not with a positive sign.) At $T = 298.15$ K, $K_p = 6.741$. Substituting this value into Eq. 2 gives

$$\boxed{X = \frac{(13.482\,P_T + 1) - [(13.482\,P_T + 1)^2 - 181.76\,P_T^2]^{1/2}}{13.482\,P_T}}, \tag{24}$$

which is the required function. The plots follow.

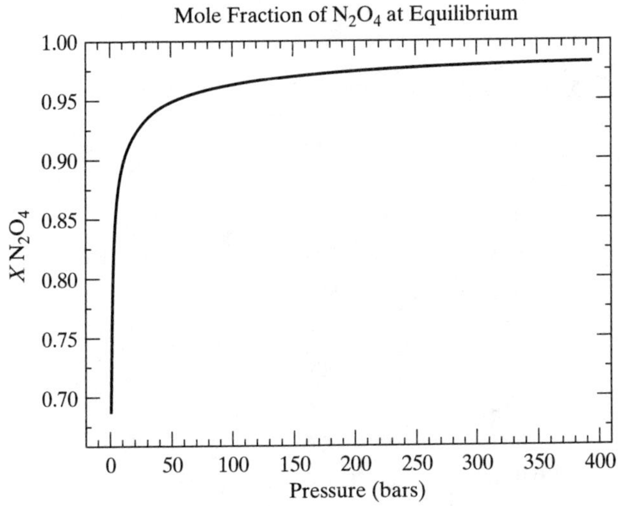

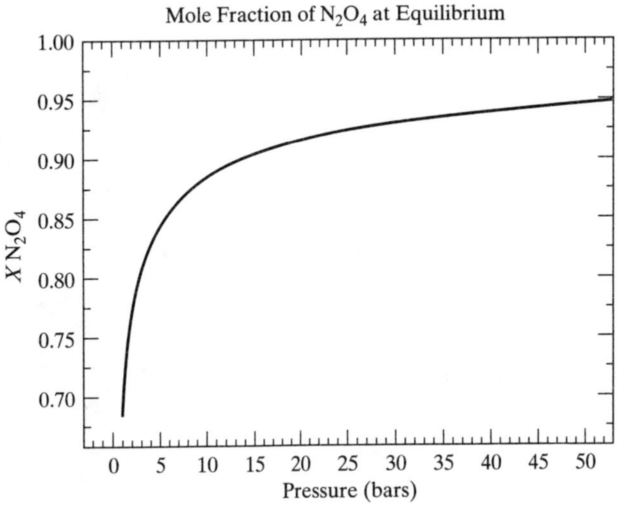

(F) For the nonideal system, we have

$$K_p = \left[\frac{f_{N_2O_4(g)}}{f^2_{NO_2(g)}}\right]_{eq}. \tag{25}$$

Substituting the fugacity from Eq. 15 gives

$$K_p = \frac{\left\{P\exp\left[\frac{C_2(T)P}{RT} + \frac{C_3(T)P^2}{2(RT)^2}\right]\right\}_{N_2O_4(g)}}{\left\{P\exp\left[\frac{C_2(T)P}{RT} + \frac{C_3(T)P^2}{2(RT)^2}\right]\right\}^2_{NO_2(g)}}. \tag{26}$$

Using Dalton's law, we have $P_i = X_i P_T$. Let X = mole fraction of $N_2O_4(g)$ at equilibrium. Substituting into Eq. 26 yields

$$K_p = \frac{\left\{XP \exp\left[\frac{C_2(T)XP}{RT} + \frac{C_3(T)X^2P^2}{2(RT)^2}\right]\right\}_{N_2O_4(g)}}{\left\{(1-X)P \exp\left[\frac{C_2(T)(1-X)P}{RT} + \frac{C_3(T)(1-X)^2P^2}{2(RT)^2}\right]\right\}^2_{NO_2(g)}}, \quad (27)$$

where P in Eq. 27 represents the total pressure in the system, P_T. Substituting values at 298.15 K gives

$$6.741 = \frac{X \exp[-0.008387XP + 0.000002584X^2P^2]}{(1-X)^2P \exp[-0.01385(1-X)P + 0.000003185(1-X)^2P^2]}. \quad (28)$$

We now seek the values of X that satisfy this equation for pressures between 1 and 400 bar. To find these solutions, we execute a grid search for the root of

$$F(X) = 6.741 - \frac{X \exp[-0.008387XP + 0.000002584X^2P^2]}{(1-X)^2P \exp[-0.01385(1-X)P + 0.000003185(1-X)^2P^2} = 0. \quad (29)$$

The results of the search are given in the following table:

P_T (bar) Z	*X (nonideal)*	*X (ideal)*	*% error*
1	0.682	0.682	0.0
2	0.762	0.762	0.0
3	0.802	0.800	0.2
4	0.826	0.824	0.2
5	0.844	0.842	0.2
10	0.888	0.884	0.4
15	0.908	0.904	0.4
20	0.922	0.916	0.6
25	0.930	0.926	0.4
35	0.944	0.936	0.8
40	0.948	0.940	0.8
45	0.952	0.944	0.8
50	0.954	0.946	0.8
100	0.975	0.962	1.3
200	0.987	0.972	1.5
300	0.993	0.978	1.5
400	0.995	0.981	1.4

The requested plot is at the top of the next page.

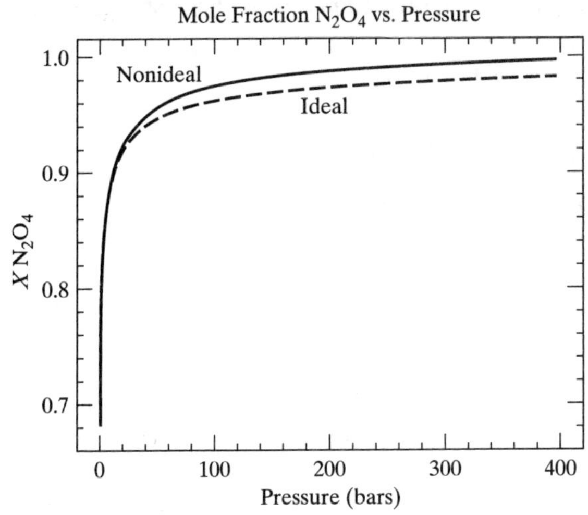

5.21 The reference point chosen for the fugacity is

$$\lim_{P\longrightarrow 0}\left[\frac{f}{P}\right] = 1.$$

Let us assume that there exists a nonideal gas for which $f = P/(1 + bP)$, with b equal to a constant for this choice of reference point.
(A) If we had instead chosen

$$\lim_{P\longrightarrow 0}\left[\frac{f}{P}\right] = 2$$

as our reference point, determine the form the fugacity would have for this gas.
(B) Show that ΔG for the isothermal expansion of the gas from pressure P_1 to pressure P_2 is the same for either choice of reference point.

Solution

(A) If we reproduce the derivation given in the text from Eq. 5.104 to 5.109, we obtain

$$\ln(f) - \ln(P) - \lim_{P^*\longrightarrow 0}\ln\left[\frac{f^*}{P^*}\right] = (RT)^{-1}\lim_{P^*\longrightarrow 0}\int_{P^*}^{P}\left[V_m - \frac{RT}{P}\right]dP. \qquad (1)$$

With our new choice of reference point, we have

$$\lim_{P\longrightarrow 0}\left[\frac{f^*}{P^*}\right] = 2. \qquad (2)$$

Therefore, Eq. 1 becomes

$$\ln(f) - \ln(P) - \ln(2) = (RT)^{-1}\int_{0}^{P}\left[V_m - \frac{RT}{P}\right]dP. \qquad (3)$$

Rearranging terms produces

$$\ln(f) = \ln(P) + \ln(2) + (RT)^{-1}\int_{0}^{P}\left[V_m - \frac{RT}{P}\right]dP. \qquad (4)$$

With the usual choice for the reference point, we obtained Eq. 5.110:

$$\ln(f) = \ln(P) + (RT)^{-1}\int_0^P \left[V_m - \frac{RT}{P}\right] dP = \ln\left[\frac{P}{1 + bP}\right]$$

$$= \ln(P) - \ln(1 + bP). \tag{5}$$

We may, therefore, use Eq. 5 to replace the first and third terms on the right side of Eq. 4 to obtain

$$\ln(f) = \ln(P) - \ln(1 + bP) + \ln(2) = \ln\left[\frac{2P}{1 + bP}\right], \tag{6}$$

so that we have

$$\boxed{f = \frac{2P}{1 + bP}} \tag{7}$$

with the new choice of reference point.

(B) The Gibbs free energy is given by Eq. 5.95 and is

$$G = n\mu^o(T) + nRT\ln(f), \tag{8}$$

where we have replaced the activity with the fugacity, as is customary. For the usual choice of reference point, where

$$\lim_{P \longrightarrow 0}\left[\frac{f}{P}\right] = 1,$$

we have $f = \dfrac{P}{1 + bP}$, so that

$$G(T, P_1) = n\mu^o + nRT\ln\left[\frac{P_1}{1 + bP_1}\right]. \tag{9}$$

At pressure P_2, the Gibbs free energy is

$$G(T, P_2) = n\mu^o + nRT\ln\left[\frac{P_2}{1 + bP_2}\right]. \tag{10}$$

Therefore,

$$\Delta G = G(T, P_2) - G(T, P_1)$$

$$= n\mu^o + nRT\ln\left[\frac{P_1}{1 + bP_1}\right] - n\mu^o - nRT\ln\left[\frac{P_2}{1 + bP_2}\right]$$

$$= nRT\ln\left[\frac{P_1(1 + P_2)}{P_2(1 + bP_1)}\right]. \tag{11}$$

If our reference point is chosen so that the limit is equal to 2, Eq. 7 shows that the Gibbs free energies will now be given by

$$G(T, P_1) = n\mu^o + nRT\ln\left[\frac{2P_1}{1 + bP_1}\right] \tag{12}$$

and

$$G(T, P_2) = n\mu^o + nRT\ln\left[\frac{2P_2}{1 + bP_2}\right]. \tag{13}$$

ΔG is the difference of these two quantities; that is,

$$\Delta G = G(T, P_2) - G(T, P_1)$$

$$= n\mu^o + nRT \ln\left[\frac{2P_1}{1 + bP_1}\right] - n\mu^o - nRT \ln\left[\frac{2P_2}{1 + bP_2}\right]$$

$$= nRT \ln\left[\frac{P_1(1 + P_2)}{P_2(1 + bP_1)}\right], \quad (14)$$

since the factors of 2 divide out.

Equation 14 is identical to Eq. 11. The change in G is, therefore, independent of the choice of reference point, as we knew it had to be.

5.23* Consider the equilibrium system

$$A(g) + 2B(g) = 2C(g) + D(g),$$

where all gases are ideal. If the equilibrium constant for this system is 10 at 300 K, and if 0.5 mol L^{-1} of $A(g)$, 5 mol L^{-1} of $B(g)$, 2 mol L^{-1} of $C(g)$, and 0.2 mol L^{-1} of $D(g)$ are simultaneously mixed together in a 1-L container at 300 K,

(A) will $A(g) + 2B(g)$ spontaneously react to form $2C(g) + D(g)$, or will $2C(g) + D(g)$ spontaneously react to form $A(g) + 2B(g)$? Show that your answer is correct.

(B) When equilibrium is reached, how many moles of each gas will be present?

Solution

(A) The change in A at constant volume and temperature is given by Eq. 5.51:

$$dA = \sum_{i=1}^{4} \mu_i dn_i. \quad (1)$$

For the given reaction, if $d\varepsilon$ moles of $A(g)$ react, the change in the Helmholtz free energy will be

$$dA = [\mu_{A(g)}(-d\varepsilon) + \mu_{B(g)}(-2d\varepsilon) + \mu_{C(g)}(2d\varepsilon) + \mu_{D(g)}(d\varepsilon)], \quad (2)$$

since $2d\varepsilon$ moles of $B(g)$ react and $2d\varepsilon$ moles of $C(g)$ and $d\varepsilon$ moles of $D(g)$ are formed each time $d\varepsilon$ moles of $A(g)$ react. Because the gases are all ideal, the chemical potentials are given by

$$\mu_i = \mu_i^o + RT \ln P_i \quad (3)$$

where P_i is expressed in bars. Substituting Eq. 3 into Eq. 2 gives

$$dA = d\varepsilon[2\mu^o_{C(g)} + \mu^o_{D(g)} - \mu^o_{A(g)} - 2\mu^o_{B(g)}] + d\varepsilon RT \ln\left[\frac{P_C^2 P_D}{P_A P_B^2}\right]$$

$$= d\varepsilon \Delta\mu^o + d\varepsilon RT \ln\left[\frac{P_C^2 P_D}{P_A P_B^2}\right]. \quad (4)$$

$\Delta\mu^o$ can be expressed in terms of the equilibrium constant. When this is done, Eq. 4 becomes

$$dA = d\varepsilon RT\left\{-\ln K_p + \ln\left[\frac{P_C^2 P_D}{P_A P_B^2}\right]\right\}. \quad (5)$$

Since the number of moles of gaseous compounds is the same on both sides of the reaction, we have $K_p = K_c$, and partial pressures can be replaced with concentrations in mol L^{-1}. Therefore, for the given reaction, Eq. 5 may be written as

$$dA = d\varepsilon RT\left\{-\ln(K_p) + \ln\left[\frac{[C]^2[D]}{[A][B]^2}\right]\right\}. \tag{6}$$

Inserting the given values from the problem produces

$$dA = d\varepsilon(8.314\ \text{J mol}^{-1}\ \text{K}^{-1})(300\ \text{K})\left\{-\ln(10) + \ln\left[\frac{[2.0]^2[0.2]}{[0.5][5.0]^2}\right]\right\}$$

$$= -1.260 \times 10^4 d\varepsilon\ \text{J mol}^{-1}. \tag{7}$$

Since $d\varepsilon > 0$, we have $dA < 0$, and the reaction to give more $C(g)$ and $D(g)$ will be spontaneous.

(B) Let x moles of $A(g)$ react to form more $C(g)$ and $D(g)$. When equilibrium is reached, the stoichiometry shows that the number of moles of each compound will be

$$n_A = 0.5 - x,$$

$$n_B = 5 - 2x,$$

$$n_C = 2 + 2x,$$

and

$$n_D = 0.2 + x. \tag{8}$$

Therefore, at equilibrium, we must have

$$K = 10 = \frac{(0.2 + x)(2 + 2x)^2}{(0.5 - x)(5.0 - 2x)^2}. \tag{9}$$

Consequently, we seek the root to the equation

$$F(x) = 10 - \frac{(0.2 + x)(2 + 2x)^2}{(0.5 - x)(5.0 - 2x)^2} = 0 \tag{10}$$

that lies between 0 and 0.5 mol, since the initial $A(g)$ concentration is 0.5 mol L^{-1}. A simple grid search gives the result $x = 0.4655$ mol. Hence, the equilibrium concentrations are

$$n_A = \underline{0.0345\ \text{mol L}^{-1}},\quad n_B = \underline{4.069\ \text{mol L}^{-1}},\quad n_C = \underline{2.931\ \text{mol L}^{-1}}, \tag{11}$$

and

$$n_D = \underline{0.6655\ \text{mol L}^{-1}}.$$

We may check the accuracy of this result by substituting into the equilibrium constant expression. We then obtain

$$K = \frac{(2.931)^2(0.6655)}{(0.0345)(4.069)^2} = 10.01, \tag{12}$$

as required.

5.25 State the direction of the equilibrium shift (if any) in each of the cases on the next page, and stipulate whether the shift is taking place because the equilibrium constant is increasing or decreasing or because of the necessity to maintain the equilibrium constant's value unchanged.

(A) In the reaction $2\,NO_2(g) \longrightarrow N_2O_4(g) + 57.2$ kJ, the total pressure is decreased.

(B) In the reaction $2\,NO_2(g) \longrightarrow N_2O_4(g) + 57.2$ kJ, the temperature is decreased.

(C) In the reaction $2\,NO_2(g) \longrightarrow N_2O_4(g) + 57.2$ kJ, some of the $N_2O_4(g)$ is removed.

Solution

(A) The equilibrium will shift toward the reactants to increase the number of moles of gas, thereby relieving the stress produced by the pressure reduction. Thermodynamically, this shift occurs in such manner to maintain a constant value for K_p, which would otherwise increase.

(B) The equilibrium shifts towards the products to produce more heat and thereby relieve the stress caused by the removal of heat. Thermodynamically, we have $\Delta H < 0$, so that K_p must increase as T decreases. Therefore, the value of K_p rises and the equilibrium shifts toward the product side.

(C) The equilibrium will shift toward the products to increase the number of moles of $N_2O_4(g)$, thereby relieving the stress produced by reducing $P_{N_2O_4(g)}$. Thermodynamically, this shift occurs in such manner to maintain a constant value for K_p, which would otherwise decrease.

5.27 Using the data in Appendix A, Sam computes $\Delta\mu^o$ for the reaction

$$2\,HI(g) \longrightarrow H_2(g) + I_2(g)$$

at 298.15 K. He obtains the result $\Delta\mu^o = -3.40$ kJ mol^{-1}. Since $\Delta\mu^o < 0$, Sam concludes that the reaction at 298.15 K will be spontaneous and irreversible if carried out at either constant volume or constant pressure.

(A) Has Sam done the calculation of $\Delta\mu^o$ correctly?

(B) Are his conclusions concerning the spontaneity of the reaction correct? Explain.

Solution

(A) The change in the standard chemical potential is given by

$$\Delta\mu^o = \mu^o_{H_2(g)} + \mu^o_{I_2(g)} - 2\mu^o_{HI(g)} = 0 + 0 - 2(1.70 \text{ kJ mol}^{-1}) = -3.40 \text{ kJ mol}^{-1}. \quad (1)$$

Therefore, Sam has correctly computed the value of $\Delta\mu^o$ at 298.15 K.

(B) The criteria for spontaneity, nonspontaneity, and equilibrium depend upon $\Delta\mu$, not upon $\Delta\mu^o$. Therefore, Sam's conclusions concerning the spontaneity of the reaction are not necessarily correct. The reaction will not be spontaneous under all conditions at 298.15 K with either constant volume or pressure.

5.29 If the reaction in Problem 5.27 is conducted at constant volume with $T = 298.15$ K and initial partial pressures of $H_2(g)$ and $I_2(g)$ of 0.100 bar, under what conditions of partial pressure for HI(g) will the reaction to form $H_2(g) + I_2(g)$ be spontaneous? Assume that $H_2(g)$ and $I_2(g)$ are ideal, but that HI(g) is described by the equation of state given in Example 5.15, where the pressure refers to the partial pressure of HI(g) and $\beta = -0.0300$ at 298.15 K.

Solution

For a reaction conducted at constant temperature and volume, we have

$$dA = d\Delta\mu = \sum_{i=1}^{3} \mu_i dn_i. \tag{1}$$

For the reaction in Problem 5.27, if $d\varepsilon$ moles of HI(g) react, the change in the chemical potential is

$$dA = d\Delta\mu = d\varepsilon[\mu_{H_2(g)} + \mu_{I_2(g)} - 2\mu_{HI(g)}]. \tag{2}$$

For an ideal gas, we have

$$\mu_i = \mu_i^o + RT \ln P_i. \tag{3}$$

Since $H_2(g)$ and $I_2(g)$ are ideal, we may replace $\mu_{H_2(g)}$ and $\mu_{I_2(g)}$ with Eq. 3. However, for HI(g), which is described by the equation of state in Example 5.15, the fugacity is given by

$$f = \frac{P}{1 - \beta P}. \tag{4}$$

Therefore, the chemical potential for HI(g) is

$$\mu_{HI(g)} = \mu_{HI(g)}^o + RT \ln f = \mu_{HI(g)}^o + RT \ln P - RT \ln(1 - \beta P). \tag{5}$$

Combining Eqs. 2, 3, and 5 gives

$$dA = d\Delta\mu = d\varepsilon\{[\mu_{H_2(g)}^o + \mu_{I_2(g)}^o - 2\mu_{HI(g)}^o] + RT \ln\left[\frac{P_{H_2(g)}P_{I_2(g)}}{P_{HI(g)}^2}\right] + 2RT \ln[1 - \beta P_{HI(g)}]\}. \tag{6}$$

The change in the standard chemical potentials can be obtained from Appendix A. The result is

$$\Delta\mu^o = \mu_{H_2(g)}^o + \mu_{I_2(g)}^o - 2\mu_{HI(g)}^o = 0 + 0 - 2(1.70 \text{ kJ mol}^{-1}) = -3.40 \text{ kJ mol}^{-1}. \tag{7}$$

Combining Eqs. 6 and 7 gives

$$dA = d\Delta\mu = d\varepsilon\left[-3.40 \times 10^3 \text{ J mol}^{-1} + RT \ln\left\{\frac{0.01}{P_{HI(g)}^2}\right\} + 2RT \ln[1 - \beta P_{HI(g)}]\right] \tag{8}$$

when the partial pressures of $H_2(g)$ and $I_2(g)$ are both 0.100 bar. In order for the reaction to be spontaneous, we must have dA and $d\Delta\mu < 0$. This requires that

$$-3.40 \times 10^3 + RT \ln(0.01) + 2RT \ln\left[\frac{1 - \beta P_{HI(g)}}{P_{HI(g)}}\right] < 0. \tag{9}$$

The inequality in Eq. 9 requires that we have

$$\ln\left[\frac{1 - \beta P_{HI(g)}}{P_{HI(g)}}\right] < \frac{3.40 \times 10^3 - (8.314)(298.15) \ln(0.01)}{2(8.314)(298.15)} = 2.988. \tag{10}$$

Exponentiation of both sides of Eq. 10 produces

$$\left[\frac{1 + 0.1P_{HI(g)}}{P_{HI(g)}}\right] < \exp(2.988) = 19.85, \quad \textbf{(11)}$$

since the problem gives the value of β as -0.100. Solving the inequality for $P_{HI(g)}$, we obtain

$$1 < 19.85P_{HI(g)} - 0.100P_{HI(g)} = 19.75P_{HI(g)}. \quad \textbf{(12)}$$

The reaction will be spontaneous whenever the partial pressure of HI(g) satisfies the inequality

$$P_{HI(g)} > \frac{1}{19.75}\text{ bar} = \underline{0.0506\text{ bar}}. \quad \textbf{(13)}$$

CHAPTER 6

Phase Equilibrium

6.1 Using the data given in Appendix A, along with the assumption that the partial molar entropies of solid and liquid aluminum are constants with a value equal to their values at 298.15 K and 1 bar of pressure, obtain the chemical potentials of Al(s) and Al(l) as a function of temperature, and compute the temperature at which these two quantities are equal. Compare your results with those shown in Figure 6.1.

Solution

The rate of change of the chemical potential with temperature at constant pressure is given by Eq. 5.44:

$$\left(\frac{\partial \mu}{\partial T}\right)_P = -\bar{S}. \quad (1)$$

If $\bar{S}$ is constant, we can separate variables and integrate to obtain

$$\int_{\mu^o(298.15)}^{\mu^o(T)} d\mu = \mu^o(T) - \mu^o(298.15) = -\bar{S}\int_{298.15}^{T} dT = -\bar{S}(T - 298.15). \quad (2)$$

In Appendix A, we find that $\mu^o(298.15)$ for Al(s) and Al(l) are zero and 7.20×10^3 J mol^{-1}, respectively. The absolute partial molar entropies at 298.15 K are $\bar{S}^o(\text{Al(s)}) = 28.33$ J K^{-1} mol^{-1} and $\bar{S}^o(\text{Al(l)}) = 39.55$ J K^{-1} mol^{-1}. Substituting these data into Eq. 2 yields

$$\mu^o(T)_{\text{Al(s)}} = 0 - 28.33(T - 298.15) \text{ J mol}^{-1} \quad (3)$$

and

$$\mu^o(T)_{\text{Al(l)}} = 7.20 \times 10^3 - 39.55(T - 298.15) \text{ J mol}^{-1}. \quad (4)$$

Equations 3 and 4 yield the straight lines shown in Figure 6.4. When the two chemical potentials are equal, we have

$$-28.33(T - 298.15) = 7.20 \times 10^3 - 39.55(T - 298.15). \quad (5)$$

Solving for T, we obtain

$$T(39.55 - 28.33) \text{ J K}^{-1} \text{ mol}^{-1} = 7.20 \times 10^3 + [39.55 - 28.33](298.15)$$

$$= 1.055 \times 10^4 \text{ J mol}^{-1}, \quad (6)$$

which gives

$$T = \frac{1.055 \times 10^4 \text{ J mol}^{-1}}{11.22 \text{ J K}^{-1} \text{ mol}^{-1}} = \underline{940 \text{ K}}. \quad (7)$$

This is the result shown in Figure 6.4.

6.3 Using the data in Appendix A, along with the assumption that the partial molar volumes of liquid and solid H_2O are independent of pressure, obtain the chemical potentials of H_2O(s) and H_2O(l) as a function of pressure in bars at 262.65 K. Determine the pressure required to melt ice at 262.65 K. Compare your result with the data given in Figure 6.6. Assume that the density of solid H_2O is 0.9168 g cm^{-3}, the density of liquid H_2O is 0.9998 g cm^{-3}, μ^o(262.65 K) for liquid H_2O is 195.6 J mol^{-1}, and μ^o(262.65 K) for solid H_2O is 0.000 J mol^{-1}.

Solution

The rate of change of the chemical potential with pressure at constant temperature is given by Eq. 5.44:

$$\left(\frac{\partial \mu}{\partial P}\right)_T = \overline{V}. \tag{1}$$

If $\overline{V}$ is constant, we can separate variables and integrate to obtain

$$\int_{\mu(1\text{ bar})}^{\mu(P)} d\mu = \mu(P) - \mu^o = \overline{V}\int_1^P dP = \overline{V}(P-1). \tag{2}$$

The partial molar volumes can be computed from the densities:

$$\overline{V}_{H_2O(s)} = \frac{18.016\text{ g mol}^{-1}}{0.9168\text{ g cm}^{-3}} \times \frac{1\text{ L}}{1{,}000\text{ cm}^3} = 1.965 \times 10^{-2}\text{ L mol}^{-1} \tag{3}$$

and

$$\overline{V}_{H_2O(l)} = \frac{18.016\text{ g mol}^{-1}}{0.9998\text{ g cm}^{-3}} \times \frac{1\text{ L}}{1{,}000\text{ cm}^3} = 1.802 \times 10^{-2}\text{ L mol}^{-1}. \tag{4}$$

Substituting these data into Eq. 2 yields

$$\mu(P)_{H_2O(s)} - \mu^o_{H_2O(s)}$$

$$= (1.965 \times 10^{-2}\text{ L mol}^{-1})(P-1)\text{ bar} \times \frac{8.314\text{ J mol}^{-1}\text{ K}^{-1}}{0.083144\text{ L bar mol}^{-1}\text{ K}^{-1}}$$

$$= 1.965(P-1)\text{ J mol}^{-1} \tag{5}$$

and

$$\mu(P)_{H_2O(l)} - \mu^o_{H_2O(l)}$$

$$= (1.802 \times 10^{-2}\text{ L mol}^{-1})(P-1)\text{ bar} \times \frac{8.314\text{ J mol}^{-1}\text{ K}^{-1}}{0.083144\text{ L bar mol}^{-1}\text{ K}^{-1}}$$

$$= 1.802(P-1)\text{ J mol}^{-1}. \tag{6}$$

If the pressure is that which makes $\mu(P)_{H_2O(s)} = \mu(P)_{H_2O(l)}$, then subtracting Eq. 6 from Eq. 5 produces

$$\mu^o_{H_2O(l)} - \mu^o_{H_2O(s)} = (1.965 - 1.802)(P-1), \tag{7}$$

where the units on both sides of Eq. 7 are J mol^{-1}. The data given in the problem tell us that we have $\mu^o_{H_2O(l)} - \mu^o_{H_2O(s)} = 195.6$ J mol^{-1}. Substituting this value into Eq. 7 yields

$$195.6 = 0.163(P-1). \tag{8}$$

Solving Eq. 8 for P, we obtain

$$P = \frac{195.6\text{ J mol}^{-1}}{(1.965 - 1.802)\text{ J mol}^{-1}\text{ bar}^{-1}} + 1\text{ bar} = \underline{1{,}201\text{ bar}}, \tag{9}$$

which is essentially the result shown in Figure 6.6.

6.5 The normal boiling point of ethanol is 352.6 K. Using the data in Appendix A, along with the assumption that $\Delta C_p = 0$ for the vaporization process, compute the equilibrium vapor pressure of ethanol at 298 K.

Solution

We first need the partial molar enthalpy of vaporization. This is the enthalpy change for the reaction $C_2H_5OH(l) \longrightarrow C_2H_5OH(g)$. Since we are assuming that $\Delta C_p = 0$, we may evaluate $\Delta \overline{H}_{vap}$ at any temperature, as its value will be a constant. At 298.15 K, we have

$$\Delta\overline{H} = \Delta\overline{H}^o = \overline{H}^o_{C_2H_5OH(g)} - \overline{H}^o_{C_2H_5OH(l)} = -235.10 - (-277.69)\ \text{kJ mol}^{-1}$$
$$= 42.59\ \text{kJ mol}^{-1}. \qquad (1)$$

Using the result of Eq. 1 in Eq. 6.13A gives

$$\ln P^{eq} = \ln(760) - \frac{\Delta\overline{H}_{vap}}{R}\left[\frac{1}{T^{eq}} - \frac{1}{T_b^o}\right]$$

$$= \ln(760) - \frac{4.269 \times 10^4\ \text{J mol}^{-1}}{8.314\ \text{J mol}^{-1}\ \text{K}^{-1}}\left[\frac{1}{298.15} - \frac{1}{352.6}\right] = 6.633 - 2.653 = 3.980. \qquad (2)$$

Exponentiation then yields

$$P^{eq} = \underline{53.52\ \text{torr}} = \underline{0.07042\ \text{atm}} = \underline{0.07135\ \text{bar}}. \qquad (3)$$

6.7 Carbon disulfide (CS_2) has vapor pressures of 40 and 100 torr at 250.65 K and 268.05 K, respectively.
(A) Compute the enthalpy vaporization and the normal boiling point of CS_2, assuming that $\Delta C_p = 0$.
(B) The standard partial molar enthalpy of $CS_2(l)$ is 89.70 kJ mol^{-1}. What does the calculation in (A) predict the standard partial molar enthalpy of $CS_2(g)$ to be?

Solution

(A) When $\Delta C_p = 0$, $\Delta\overline{H}_{vap}$ is a constant and Eq. 6.13A holds:

$$\ln P^{eq} = \ln(760) - \frac{\Delta\overline{H}_{vap}}{R}\left[\frac{1}{T^{eq}} - \frac{1}{T_b^o}\right]. \qquad (1)$$

If we have $P^{eq} = P_1$ at $T^{eq} = T_1$ and $P^{eq} = P_2$ at $T^{eq} = T_2$, Eq. 1 gives us two equations:

$$\ln P_1 = \ln(760) - \frac{\Delta\overline{H}_{vap}}{R}\left[\frac{1}{T_1} - \frac{1}{T_b^o}\right] \qquad (2)$$

and

$$\ln P_2 = \ln(760) - \frac{\Delta\overline{H}_{vap}}{R}\left[\frac{1}{T_2} - \frac{1}{T_b^o}\right]. \qquad (3)$$

Subtracting Eq. 2 from Eq. 3 yields

$$\ln P_2 - \ln P_1 = \ln\left[\frac{P_2}{P_1}\right] = -\frac{\Delta\overline{H}_{vap}}{R}\left[\frac{1}{T_2} - \frac{1}{T_1}\right]. \qquad (4)$$

Solving for $\Delta\overline{H}_{vap}$, we obtain

$$\Delta\overline{H}_{vap} = -\frac{R\ln\left[\frac{P_2}{P_1}\right]}{\left[\frac{1}{T_2}-\frac{1}{T_1}\right]}. \quad (5)$$

If $T_1 = 250.65$ K and $T_2 = 268.05$ K, Eq. 5 gives

$$\Delta\overline{H}_{vap} = -\frac{(8.314\ \text{J mol}^{-1}\ \text{K}^{-1})\ln\left[\frac{100}{40}\right]}{\left[\frac{1}{268.05}-\frac{1}{250.65}\right]}$$

$$= 2.942 \times 10^4\ \text{J mol}^{-1} = \underline{29.42\ \text{kJ mol}^{-1}}. \quad (6)$$

The boiling point may now be computed from Eq. 3:

$$\frac{1}{T_b^o} = \frac{1}{T_2} + \frac{R\ln\left[\frac{P_2}{760}\right]}{\Delta\overline{H}_{vap}} = \frac{1}{268.05\ \text{K}} + \frac{(8.314\ \text{J mol}^{-1}\ \text{K}^{-1})\ln\left(\frac{100}{760}\right)}{2.942 \times 10^4\ \text{J mol}^{-1}}$$

$$= 0.003157\ \text{K}^{-1}. \quad (7)$$

Therefore, inverting both sides, we obtain

$$T_b^o = \underline{316.8\ \text{K}}. \quad (8)$$

The measured boiling point of CS_2 is 319.65 K. The −2.8 K error is due primarily to our assumption that $\Delta C_p = 0$.

(B) The enthalpy of vaporization is given by

$$\Delta\overline{H}_{vap} = \overline{H}^o_{CS_2(g)} - \overline{H}^o_{CS_2(l)} = \overline{H}^o_{CS_2(g)} - 89.70\ \text{kJ mol}^{-1} = 29.42\ \text{kJ mol}^{-1}. \quad (9)$$

Therefore,

$$\overline{H}^o_{CS_2(g)} = 89.70 + 29.42\ \text{kJ mol}^{-1} = \underline{119.12\ \text{kJ mol}^{-1}}. \quad (10)$$

T (K)	Vapor pressure (torr)
253.55	10
277.45	40
296.15	100
330.95	400

6.9 Some measured vapor pressures of $CCl_4(l)$ are given in the table at the left.

(A) Determine $\Delta\overline{H}_{vap}$ for $CCl_4(l)$ from a plot of $\ln P^{eq}$ vs. $1/T$. Use a least-squares fitting procedure.

(B) The measured heat of fusion of $CCl_4(s)$ at 249.15 K is 2,677 J mol^{-1}. If we assume that the heat capacities of solid, liquid, and vapor CCl_4 are all equal, we will have $\Delta\overline{H}_{sub} = \Delta\overline{H}_{vap} + \Delta\overline{H}_{fus}$. The measured vapor pressure over $CCl_4(s)$ at 223.15 K is 1 torr. With these data, the results obtained in (A), and the assumption that the heat capacities of all phases are equal, compute the melting point of CCl_4. The measured result is 250.55 K. Compute the percent error introduced by the heat capacity assumption.

Solution

(A) The plot at the end of the problem shows $\ln P^{eq}$ versus T^{-1}. The least-squares fit gives

$$\ln P^{eq} = 18.094 - \frac{4{,}000.2(\text{K})}{T}. \quad (1)$$

Equation 6.13A shows that the slope is given by

$$\text{slope} = -4{,}000.2\ \text{K} = -\frac{\Delta\overline{H}_{vap}}{R}. \quad (2)$$

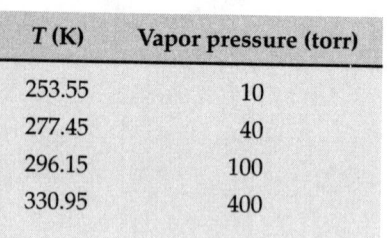

Therefore,

$$\Delta\overline{H}_{vap} = -R(-4{,}000.2\text{ K})$$
$$= (8.314\text{ J mol}^{-1}\text{ K}^{-1})(4{,}000.2\text{ K}) = \underline{3.326 \times 10^4\text{ J mol}^{-1}}. \quad (3)$$

(B) Assuming that all heat capacities are equal, we have

$$\Delta\overline{H}_{sub} = \Delta\overline{H}_{vap} + \Delta\overline{H}_{fus} = 3.326 \times 10^4 + 2{,}677 = 3.594 \times 10^4\text{ J mol}^{-1}. \quad (4)$$

Now let the melting point be T_m. Since the solid and liquid are in equilibrium at $T = T_m$, they must have equal vapor pressures, which we denote by P_m^{eq}. Therefore, using Eq. 6.12, we may write

$$\ln\left[\frac{P_2^{eq}}{P_1^{eq}}\right] = -\frac{\Delta\overline{H}_{vap}}{R}\left[\frac{1}{T_2^{eq}} - \frac{1}{T_1^{eq}}\right]. \quad (5)$$

We take $T_2^{eq} = T_m$. For CCl_4(l), we take $T_1^{eq} = T_1^{(l)} = 253.55$ K, so that $P_1^{(l)} = 10$ torr. For CCl_4(s), we take $T_1 = T_1^{(s)} = 223.15$ K, which gives $P_1^{(s)} = 1$ torr. Thus,

$$\ln P_2^{eq} = \ln P_m^{eq} = \ln[P_1^{(l)}] - \frac{\Delta\overline{H}_{vap}}{R}\left[\frac{1}{T_m} - \frac{1}{T_1^{(l)}}\right]$$
$$= \ln[P_1^{(s)}] - \frac{\Delta\overline{H}_{sub}}{R}\left[\frac{1}{T_m} - \frac{1}{T_1^{(s)}}\right]. \quad (6)$$

We may now solve Eq. 6 for the melting point T_m. The solution is the same as Eq. 6.26 for the triple point:

$$T_m = \frac{\Delta\overline{H}_{vap} - \Delta\overline{H}_{sub}}{R\ln\left[\frac{P_1^{(l)}}{P_1^{(s)}}\right] - \frac{\Delta\overline{H}_{sub}}{T_1^{(s)}} + \frac{\Delta\overline{H}_{vap}}{T_1^{(l)}}}$$

$$= \frac{(3.326 \times 10^4 - 3.594 \times 10^4)\text{ J mol}^{-1}}{(8.314\text{ J mol}^{-1}\text{ K}^{-1})\ln\left(\frac{10}{1}\right) - \frac{3.594 \times 10^4\text{ J mol}^{-1}}{223.15\text{ K}} + \frac{3.326 \times 10^4\text{ J mol}^{-1}}{253.55\text{ K}}}$$
$$= \underline{249.6\text{ K}}. \quad (7)$$

The percent error in the calculation is

$$\%\text{ error} = \frac{100 \times (249.6 - 250.55)}{250.55} = \underline{-0.38\%\text{ error}}. \quad (8)$$

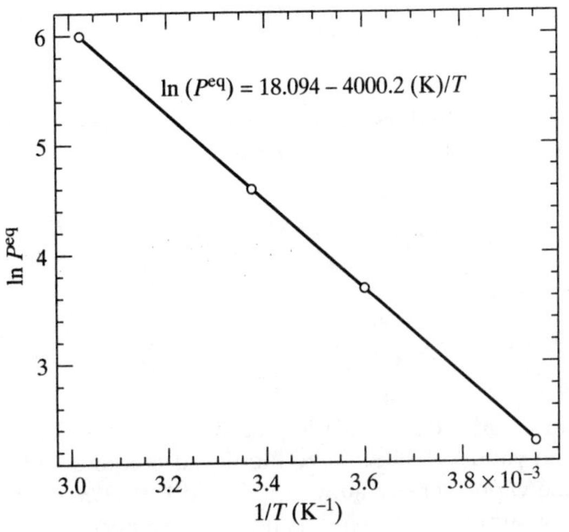

6.11 Consider a vapor whose equation of state for 1 mole is $PV(1 - bP) = RT$, where b is a constant. For this vapor, derive an expression for the equilibrium pressure as a function of temperature, assuming that the enthalpy of vaporization is independent of temperature and that the partial molar volume of the liquid is negligible relative to that of the vapor.

Solution

The Clapeyron equation for the system is given by Eq. 6.7:

$$dP^{\text{eq}} = \frac{\Delta\overline{H}_{\text{vap}}}{T^{\text{eq}}\Delta\overline{V}}dT^{\text{eq}}. \tag{1}$$

If we neglect the molar volume of the liquid relative to that of the vapor, Eq. 1 becomes

$$dP^{\text{eq}} = \frac{\Delta\overline{H}_{\text{vap}}}{T^{\text{eq}}\overline{V}(\text{g})}dT^{\text{eq}}. \tag{2}$$

Solving the equation of state for the molar volume, we obtain

$$\overline{V} = \frac{RT}{P(1 - bP)}. \tag{3}$$

Substituting Eq. 3 into Eq. 2 and dividing by $P^{\text{eq}}(1 - bP^{\text{eq}})$ gives

$$\frac{dP^{\text{eq}}}{P^{\text{eq}}(1 - bP^{\text{eq}})} = \frac{\Delta\overline{H}_{\text{vap}}}{R(T^{\text{eq}})^2}dT^{\text{eq}}. \tag{4}$$

The left-hand side of Eq. 4 can be written in the form

$$\frac{dP^{\text{eq}}}{P^{\text{eq}}(1 - bP^{\text{eq}})} = \left[\frac{1}{P^{\text{eq}}} + \frac{b}{1 - bP^{\text{eq}}}\right]dP^{\text{eq}}. \tag{5}$$

Substituting Eq. 5 into Eq. 4 and integrating between corresponding limits produces

$$\int_{P_1}^{P_2}\left[\frac{1}{P^{\text{eq}}} + \frac{b}{1 - bP^{\text{eq}}}\right]dP^{\text{eq}} = \frac{\Delta\overline{H}_{\text{vap}}}{R}\int_{T_1}^{T_2}\frac{dT^{\text{eq}}}{(T^{\text{eq}})^2}dT^{\text{eq}}$$

$$= -\frac{\Delta\overline{H}_{\text{vap}}}{R}\left[\frac{1}{T_2} - \frac{1}{T_1}\right]. \tag{6}$$

Integrating on the left gives

$$\boxed{\ln\left[\frac{P_2}{P_1}\right] - \ln\left[\frac{1 - bP_2}{1 - bP_1}\right] = \ln\left[\frac{P_2(1 - bP_1)}{P_1(1 - bP_2)}\right] = -\frac{\Delta\overline{H}_{\text{vap}}}{R}\left[\frac{1}{T_2} - \frac{1}{T_1}\right]}. \tag{7}$$

Equation 7 is the desired result. If we take T_1 to be the normal boiling temperature T_o, so that $P_1 = 1$ atm, Eq. 7 becomes

$$\ln\left[\frac{P(1 - b)}{(1 - bP)}\right] = -\frac{\Delta\overline{H}_{\text{vap}}}{R}\left[\frac{1}{T} - \frac{1}{T_o}\right], \tag{8}$$

where P is in atm.

6.13 Some liquid ethanol (C_2H_5OH) is placed inside a sealed 10-L container. At 292.15 K, the vapor density above the liquid is found to be 0.1011 g L^{-1}. At 308.05 K, the vapor density above the liquid is 0.2398 g L^{-1}. Compute $\Delta\overline{H}_{\text{vap}}$ for ethanol, assuming it to be independent of temperature.

Solution

Using the ideal gas equation of state for the vapor, we have

$$P = \frac{nRT}{V} = \frac{mRT}{MV}, \quad (1)$$

where M is the molar mass of ethanol. Since the density is $\rho = m/V$, Eq. 1 may be written in the form

$$P = \frac{RT\rho}{M}. \quad (2)$$

The equilibrium vapor pressures at 292.15 and 308.05 K are, therefore,

$$P^{eq}_{292.15} = \frac{(0.08206 \text{ L atm mol}^{-1} \text{ K}^{-1})(292.15 \text{ K})(0.1011 \text{ g L}^{-1})}{46.07 \text{ g mol}^{-1}} = 0.05261 \text{ atm} \quad (3)$$

and

$$P^{eq}_{308.05} = \frac{(0.08206 \text{ L atm mol}^{-1} \text{ K}^{-1})(308.05 \text{ K})(0.2398 \text{ g L}^{-1})}{46.07 \text{ g mol}^{-1}} = 0.1316 \text{ atm}. \quad (4)$$

If $\Delta\overline{H}_{vap}$ is a constant, the vapor pressure ratio is given by

$$\ln\left[\frac{P^{eq}_{308.95}}{P^{eq}_{292.15}}\right] = \ln\left[\frac{0.1316}{0.05261}\right] = -\frac{\Delta\overline{H}_{vap}}{R}\left[\frac{1}{308.05} - \frac{1}{292.15}\right]. \quad (5)$$

Solving for $\Delta\overline{H}_{vap}$, we obtain

$$\Delta\overline{H}_{vap} = \frac{-R\ln\left[\frac{0.1316}{0.05261}\right]}{\left[\frac{1}{308.05} - \frac{1}{292.15}\right]} = \underline{4.315 \times 10^4 \text{ J mol}^{-1}}. \quad (6)$$

6.15 If y is a function of x and we wish to measure y, it may be advantageous to measure x and then compute y using the relationship $y = f(x)$. This will be the case if it is much easier to measure x than y or if small changes in y produce large changes in x that are more easily measured. A chemist wishes to use this fact to determine the temperature of a substance by measuring its vapor pressure. To execute this measurement, she constructs a small closed sphere containing pure water and its equilibrium vapor. The device is equipped with a pressure-measuring device that can easily detect pressure changes of 10^{-3} torr. (Actually, it is possible to do much better than this.) She places the device in a gas-filled chamber whose temperature is to be monitored.

(A) Assuming that the enthalpy of vaporization of water is a constant equal to 44,010 J mol^{-1}, obtain the equation relating the measured vapor pressure inside the sphere to the chamber's temperature. The equilibrium vapor pressure of water at 298.15 K is 23.756 torr.

(B) Use the result obtained in (A) to compute the temperature inside the chamber if the pressure is 475 torr.

(C) Use the Clausius–Clapeyron equation to show that if the percent error in the measurement of pressure is X%, the error in the temperature will be only 0.057X% at 300 K and 0.076X% at 400 K. Consequently, the pressure measurement affords a very accurate means of measuring the temperature. (*Hint:* The percent error in x is $100\, dx/x$, where x is the measured value and dx is the uncertainty in the measurement.)

Solution

(A) The calibration curve is obtained directly from Eq. 6.12:

$$\ln\left[\frac{P_2^{eq}}{P_1^{eq}}\right] = -\frac{\Delta\overline{H}_{vap}}{R}\left[\frac{1}{T_2^{eq}} - \frac{1}{T_1^{eq}}\right]. \quad (1)$$

If we take $T_1^{eq} = 298.15$ K, Eq. 1 becomes

$$\ln\left[\frac{P^{eq}}{23.756}\right] = -\frac{44{,}010 \text{ J mol}^{-1}}{8.314 \text{ J mol}^{-1} \text{K}^{-1}}\left[\frac{1}{T} - \frac{1}{298.15}\right] = -5{,}293.5\left[\frac{1}{T} - \frac{1}{298.15}\right]. \quad (2)$$

Solving Eq. 2 for T, we obtain

$$\boxed{T = \frac{(298.15)(5{,}293.5)}{5{,}293.5 - 298.15 \ln\left[\frac{P^{eq}}{23.756}\right]} \text{K} = \frac{1.5783 \times 10^6}{5{,}293.5 - 298.15 \ln\left[\frac{P^{eq}}{23.756}\right]} \text{K}}. \quad (3)$$

Equation 3 is the desired equation. The measured equilibrium pressure is inserted in the right-hand side of Eq. 3, and the resulting temperature is that within the chamber.

(B) If the device reads a pressure of 475 torr, the temperature inside the chamber will be

$$T = \frac{1.5783 \times 10^6}{5{,}293.48 - 298.15 \ln\left[\frac{475}{23.756}\right]} \text{K} = \underline{358.67 \text{ K}}. \quad (4)$$

(C) The Clausius–Clapeyron equation is given by Eq. 6.9:

$$\frac{dP^{eq}}{P^{eq}} = \frac{\Delta\overline{H}_{vap}}{R(T^{eq})^2} dT^{eq}. \quad (5)$$

Rewriting this equation and multiplying both sides by 100, we obtain

$$\frac{100\, dP^{eq}}{P^{eq}} = \frac{\Delta\overline{H}_{vap}}{RT^{eq}}\left[\frac{100\, dT^{eq}}{T^{eq}}\right]. \quad (6)$$

If we identify dP^{eq} and dT^{eq} as the uncertainties in the pressure and temperature measurements, respectively, the term on the left side is the percent error in the pressure measurement, while the factor in brackets on the right side is the percent error in the temperature measurement:

$$\% \text{ error in pressure} = \frac{\Delta\overline{H}_{vap}}{RT^{eq}}[\% \text{ error in the temperature measurement}]. \quad (7)$$

Therefore,

$$\% \text{ error in the temperature} = \frac{RT^{eq}}{\Delta\overline{H}_{vap}}[\% \text{ error in the pressure}]. \quad (8)$$

If the percent error in the pressure measurement is X at 300 K, we have

$$\% \text{ error in temperature} = \frac{(8.314 \text{ J mol}^{-1} \text{K}^{-1})(300 \text{ K})}{44{,}010 \text{ J mol}^{-1}} X = \underline{0.057\, X}. \quad (9)$$

If the temperature is 400 K, we have

$$\% \text{ error in temperature} = \frac{(8.314 \text{ J mol}^{-1} \text{ K}^{-1})(400 \text{ K})}{44{,}010 \text{ J mol}^{-1}} X = \underline{0.076 X}. \quad \textbf{(10)}$$

Thus, the percent error in T is more than an order of magnitude less than the percent error in pressure.

6.17* A closed piston–cylinder arrangement containing pure H_2O is equipped with temperature and pressure controls. By careful manipulation of the temperature and pressure controls, the system is cooled from 300 K to 257 K along a line whose equation is

$$P = [0.46345 \text{ torr K}^{-1}]\, T - 119.035 \text{ torr}.$$

Describe all the phase changes that occur as the system is cooled along this path. Give the temperatures and pressures at which these changes take place.

Solution

At 310 K, the pressure is

$$P = [0.46345 \text{ torr K}^{-1}]T - 119.035 \text{ torr}$$

$$= 0.46345 \text{ torr K}^{-1}(310 \text{ K}) - 119.035 \text{ torr} = 24.63 \text{ torr}. \quad \textbf{(1)}$$

Inspection of the H_2O diagram in Figure 6.23A shows that the point $T = 310$ K and $P = 24.63$ torr lies in the vapor region. Consequently, at the start of the cooling process, the cylinder contains only water vapor.

The equilibrium vapor pressure of H_2O is given by Eq. 6.32:

$$P^{eq}_{(l)} = 8.609 \exp\left\{-\frac{5.651 \times 10^4}{R}\left[\frac{1}{T^{eq}} - \frac{1}{282.15}\right] - \frac{41.71}{R}\ln\left[\frac{T^{eq}}{282.15}\right]\right\} \text{ torr}. \quad \textbf{(2)}$$

Liquid water will appear at the point where the pressures given by Eqs. 1 and 2 are equal. That is, we must have

$$0.46345 T^{eq} - 119.035$$

$$= 8.609 \exp\left\{-\frac{5.651 \times 10^4}{R}\left[\frac{1}{T^{eq}} - \frac{1}{282.15}\right] - \frac{41.71}{R}\ln\left[\frac{T^{eq}}{282.15}\right]\right\}. \quad \textbf{(3)}$$

Consequently, we conduct a grid search for the root of the function $F(T^{eq})$, where

$$F(T) = 0.46345 T - 119.035$$

$$- 8.609 \exp\left\{-\frac{5.651 \times 10^4}{R}\left[\frac{1}{T} - \frac{1}{282.15}\right] - \frac{41.71}{R}\ln\left[\frac{T}{282.15}\right]\right\} = 0. \quad \textbf{(4)}$$

Two searches yield a root at $T^{eq} = \underline{291.92 \text{ K}}$. Accordingly, at this temperature, the vapor begins to condense to a liquid, and both phases are present in equilibrium. The pressure at that temperature is

$$P = [0.46345\ (291.92) - 119.035] \text{ torr} = \underline{16.26 \text{ torr}}. \quad \textbf{(5)}$$

The temperature and pressure remain at these levels until all of the vapor has condensed to a liquid. The system then begins to cool once more. When the temperature reaches $\underline{273.16 \text{ K}}$, the liquid begins to freeze to a solid, with which it is in equilibrium. The pressure at this point is

$$P = [0.46345\,(273.16) - 119.035] \text{ torr} = \underline{7.56 \text{ torr}}. \quad \textbf{(6)}$$

After all of the liquid has frozen, the system again begins to cool with only the solid phase present. When the pressure reaches the equilibrium pressure between solid and vapor, the vapor will reappear. The equilibrium vapor pressure over the solid is given by Eq. 6.33:

$$P^{eq}_{(s)} = 1.950 \exp\left\{-\frac{5.101 \times 10^4}{R}\left[\frac{1}{T^{eq}} - \frac{1}{263.15}\right]\right\} \text{ torr.} \quad \textbf{(7)}$$

Consequently, we need the temperature at which

$$[0.46345 \text{ torr K}^{-1}]T^{eq} - 119.035 \text{ torr}$$
$$= 1.950 \exp\left\{-\frac{5.101 \times 10^4}{R}\left[\frac{1}{T^{eq}} - \frac{1}{263.15}\right]\right\} \text{ torr.} \quad \textbf{(8)}$$

Therefore, we seek the root of the equation

$$F(T) = 0.46345T - 119.035 - 1.950 \exp\left\{-\frac{5.101 \times 10^4}{R}\left[\frac{1}{T} - \frac{1}{263.15}\right]\right\} = 0. \quad \textbf{(9)}$$

Two grid searches yield a root at $T^{eq} = 260.02$ K, at which point the pressure is

$$P^{eq} = [0.46345(260.02) - 119.035] \text{ torr} = 1.471 \text{ torr.} \quad \textbf{(10)}$$

The pressure and temperature remain at these levels until all of the solid has sublimed. At temperatures below 260.02 K, where the pressures decreases to near zero, we have only the vapor phase.

Summary

T (K)	Pressure (torr)	Phase(s) Present
310.00	$16.26 < P \leq 24.63$	Vapor Only
291.92	16.26	Liquid and Vapor in Equilibrium
$273.16 < T < 291.92$	$7.56 < P < 16.26$	Liquid Only
273.16	7.56	Solid and Liquid in Equilibrium
$260.02 < T < 273.16$	$1.471 < P < 7.56$	Solid Only
260.02	1.471	Solid and Vapor in Equilibrium
$T < 260.02$	$0 < P < 1.471$	Vapor Only

The figure that follows shows the cooling path superimposed on the H_2O phase diagram. The transition regions we have just computed are seen as the crossing

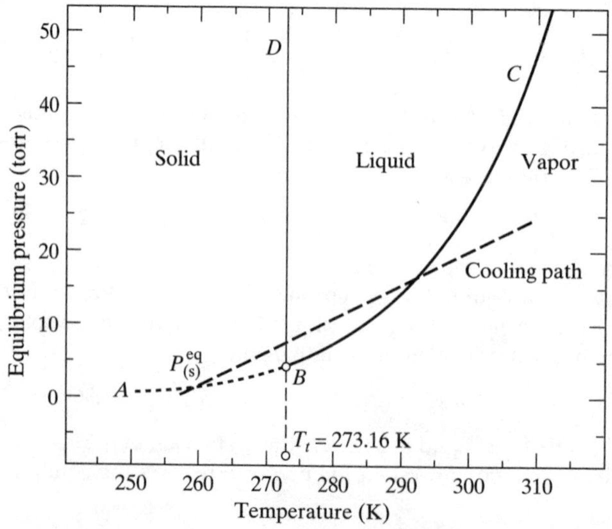

points of the cooling line with the equilibrium curves of the phase diagram.

6.19 In the dead of winter, Sam finds his driveway covered with ice. The outside temperature is 263 K. Being a student of physical chemistry, Sam knows that the melting point of ice decreases when pressure is exerted on the solid. Therefore, he decides to clear his driveway by exerting sufficient pressure, using a sledgehammer to cause the ice to melt. Later in the day, his physical chemistry classmates find Sam collapsed from exhaustion on his ice-covered, shattered driveway.

(A) Was Sam incorrect in his reasoning concerning the effect of pressure on the melting point of ice?

(B) Why has Sam collapsed from exhaustion, and why is his driveway still covered with ice?

Solution

(A) Sam is correct in his analysis that since $\Delta\overline{V} = \overline{V}(l) - \overline{V}(s) < 0$, the melting point of ice decreases with increasing pressure. The governing equation is Eq. 6.21. If this equation is solved for the equilibrium melting temperature, T^{eq}, we obtain

$$T^{eq} = T_m^o \exp\left[\frac{\Delta\overline{V}(P^{eq} - 1)}{\Delta\overline{H}_{fus}}\right]. \quad \textbf{(1)}$$

Since $\Delta\overline{V} < 0$ and $\Delta\overline{H}_{fus} > 0$, T^{eq} will decrease as P^{eq} increases.

(B) Sam has collapsed from exhaustion after crushing his driveway without success in melting the ice because he has overestimated the magnitude of the effect of pressure on the melting point of the ice. In order to reduce the melting point to the outside temperature of 263 K so that the ice might melt, the external pressure exerted would have to be

$$P^{eq} = 1 \text{ atm} + \frac{\Delta\overline{H}_{fus}}{\Delta\overline{V}} \ln\left[\frac{263}{273.15}\right]. \quad \textbf{(2)}$$

Using the data given in Example 6.4, we have $\Delta\overline{H}_{fus} = 59.32\ L$ atm and $\Delta\overline{V} = -0.001630\ L\ mol^{-1}$. Therefore,

$$P^{eq} = 1 \text{ atm} + \frac{59.32 \text{ L atm}}{-0.001630 \text{ L mol}^{-1}} \ln\left[\frac{263}{273.15}\right] = 1{,}379 \text{ atm}. \quad \textbf{(3)}$$

This pressure is equivalent to 2.026×10^4 lbs/in^2. Sam just can't manage to hit the ice hard enough with the sledgehammer. In addition, even if he did manage to melt some of the ice in this manner, as soon as the pressure was relieved, the ice would refreeze.

6.21* It is possible to use vapor pressure data to determine the temperature dependence of the difference in heat capacity between a vapor and a liquid or a vapor and a solid. If the temperature dependence of either phase is known, such a determination provides the temperature dependence of the heat capacity for the other phase. This problem serves as an example of one possible method for making such a determination.

Some measured equilibrium vapor pressures for $H_2O(l)$ are given in the table to the left.

T (K)	P^{eq} (torr)
303.15	31.824
313.15	55.324
323.15	92.510
333.15	149.38
343.15	233.70
353.15	355.10
363.15	525.76
373.15	760.00

(A) Assume that ΔC_p for the vaporization of water can be regarded as being constant over a small range of temperatures. Using standard partial molar enthalpies in Appendix A, obtain $\Delta\overline{H}_{vap}$ as a function of ΔC_p and the temperature.

(B) By substituting the result of (A) into Eq. 6.11 for $\Delta\overline{H}_{vap}$, obtain an expression for $\ln[P_2/P_1]_{eq}$ in terms of ΔC_p and the temperatures T_1 and T_2 at which the vapor pressures are P_1 and P_2, respectively.

(C) If we take T_1 and T_2 to be adjacent temperatures in the table on the previous page of temperatures and pressures, we can easily compute ΔC_p by using the result obtained in (B). We can now repeat this calculation for each pair of adjacent temperatures in the table. This will give ΔC_p over each temperature interval. Execute these calculations for all temperature intervals in the table, and plot the resulting values of ΔC_p versus the temperature at the center point of each interval considered. Obtain the best straight-line fit to the data, and thereby obtain ΔC_p as a linear function of temperature.

(D) Using the results obtained in (C) that give $\Delta C_p = A + BT$, calculate the equilibrium vapor pressure of water at 393.15 K. The experimental result at this temperature is 1489.14 torr. Calculate the percent error in your result. How much error would be present if you assumed $\Delta\overline{H}_{vap}$ to be constant at its value at 298.15 K?

Solution

(A) The temperature dependence of $\Delta\overline{H}$ is given by

$$\left(\frac{\partial \Delta\overline{H}}{\partial T}\right)_p = \Delta C_p^m. \tag{1}$$

Integrating both sides between limits gives

$$\int_{\Delta\overline{H}(298.15)}^{\Delta\overline{H}(T)} d\Delta\overline{H} = \Delta\overline{H}(T) - \Delta\overline{H}(298.15)$$

$$= \int_{298.15}^{T} \Delta C_p^m dT = \Delta C_p^m(T - 298.15). \tag{2}$$

The standard enthalpy change for the vaporization of $H_2O(l)$ at 298.15 K is

$$\Delta\overline{H}(298.15) = \overline{H}^o_{H_2O(g)} - \overline{H}^o_{H_2O(l)} = -241.82 - (285.83) \text{ kJ mol}^{-1}$$

$$= 44.01 \text{ kJ mol}^{-1} = \underline{44{,}010 \text{ J mol}^{-1}}. \tag{3}$$

Substituting into Eq. 2 produces

$$\Delta\overline{H}(T) = 44{,}010 + \Delta C_p^m(T - 298.15)$$

$$\boxed{[44{,}010 - 298.15\Delta C_p^m] + \Delta C_p^m T = D + \Delta C_p^m T}, \tag{4}$$

where $D = [44{,}010 - 298.15\Delta C_p^m]$. This is the required expression for $\Delta\overline{H}(T)$.

(B) Equation 6.10 is

$$\ln\left[\frac{P_2^{eq}}{P_1^{eq}}\right] = \frac{1}{R}\int_{T_1}^{T_2} \frac{\Delta\overline{H}_{vap} dT}{T^2} = \frac{1}{R}\int_{T_1}^{T_2} \frac{[D + \Delta C_p^m T]dT}{T^2}$$

$$= -\frac{D}{R}\left[\frac{1}{T_2} - \frac{1}{T_1}\right] + \frac{\Delta C_p^m}{R}\ln\left[\frac{T_2}{T_1}\right]. \tag{5}$$

Substituting the definition of D into Eq. 5 and then collecting terms yields

$$\boxed{\ln\left[\frac{P_2^{eq}}{P_1^{eq}}\right] = \Delta C_p^m\left\{\frac{298.15}{R}\left[\frac{1}{T_2} - \frac{1}{T_1}\right] + \frac{1}{R}\ln\left[\frac{T_2}{T_1}\right]\right\} - \frac{44{,}010}{R}\left[\frac{1}{T_2} - \frac{1}{T_1}\right].} \tag{6}$$

Equation 6 is the desired result.

(C) Equation 6 may be easily solved for ΔC_p^m:

$$\Delta C_p^m = \frac{\ln\left[\frac{P_2^{eq}}{P_1^{eq}}\right] + \frac{44{,}010}{R}\left[\frac{1}{T_2} - \frac{1}{T_1}\right]}{\left\{\frac{298.15}{R}\left[\frac{1}{T_2} - \frac{1}{T_1}\right] + \frac{1}{R}\ln\left[\frac{T_2}{T_1}\right]\right\}}. \tag{7}$$

For the first interval, with $T_1 = 303.15$ K, $P_1^{eq} = 31.824$ torr, $T_2 = 313.15$ K, and $P_2^{eq} = 55.324$ torr, the value of ΔC_p^m is given by

$$\Delta C_p^m = \frac{\ln\left[\frac{55.324}{31.824}\right]_{eq} + \frac{44{,}010}{8.314}\left[\frac{1}{313.15} - \frac{1}{303.15}\right]}{\left\{\frac{298.15}{8.314}\left[\frac{1}{313.15} - \frac{1}{303.15}\right] + \frac{1}{R}\ln\left[\frac{313.15}{303.15}\right]\right\}} = -36.701 \text{ J mol}^{-1}\text{ K}^{-1}. \quad (8)$$

Similar calculations in each of the remaining intervals yields the following results:

Temperature Interval (K)	Center Point (K)	ΔC_p^m (J mol^{-1} K^{-1})
303.15 – 313.15	308.15	−36.701
313.15 – 323.15	318.15	−37.918
323.15 – 333.15	328.15	−37.418
333.15 – 343.15	338.15	−36.862
343.15 – 353.15	348.15	−37.222
353.15 – 363.15	358.15	−36.126
363.15 – 373.15	368.15	−35.692

The plot of these data and the best straight-line fit are given at the end of the problem. The best fit yields

$$\boxed{\Delta C_p^m = [-45.069 + (0.0243 \text{ K}^{-1})T](\text{J mol}^{-1}\text{ K}^{-1})}.$$

(D) The predicted equilibrium pressure at 393.15 K is given by

$$\ln\left[\frac{P_2^{eq}}{P_1^{eq}}\right] = \frac{1}{R}\int_{T_1}^{T_2} \frac{\Delta \overline{H}_{vap}\, dT}{T^2}, \quad (9)$$

where ΔH_{vap} is given by

$$\int_{\Delta\overline{H}(298.15)}^{\Delta\overline{H}(T)} d\Delta\overline{H} = \Delta\overline{H}(T) - \Delta\overline{H}(298.15) = \int_{298.15}^{T} \Delta C_p^m dT$$
$$= A(T - 298.15) + (B/2)[T^2 - 298.15^2], \quad (10)$$

in which A and B are the least-squares parameters determined in Part (C). Therefore,

$$\Delta\overline{H}_{vap} = \Delta\overline{H}(298.15) - 298.15A - \frac{298.15^2 B}{2} + AT + \frac{BT^2}{2}. \quad (11)$$

Inserting $A = -45.069$ J mol^{-1} K^{-1}, $B = 0.0243$ J mol^{-1} K^{-2}, and $\Delta\overline{H}(298.15) = 44{,}010$ J mol^{-1}gives

$$\Delta\overline{H}_{vap} = [5.637 \times 10^4 - 45.069T + 0.01215T^2] \text{ J mol}^{-1}. \quad (12)$$

Combining of Eqs. 9 and 12 produces

$$\ln\left[\frac{P_2^{eq}}{P_1^{eq}}\right] = \frac{1}{R}\int_{T_1}^{T_2} \frac{[5.637 \times 10^4 - 45.069T + 0.01215T^2]dT}{T^2}$$
$$= -\frac{5.637 \times 10^4}{R}\left[\frac{1}{T_2} - \frac{1}{T_1}\right] - \frac{45.069}{R}\ln\left[\frac{T_2}{T_1}\right] + \frac{0.01215}{R}[T_2 - T_1]. \quad (13)$$

Let us now take $T_1 = 373.15$ K with $P_1^{eq} = 760.00$ torr and $T_2 = 393.15$ K. We may then use Eq. 13 to compute P_2^{eq}, the equilibrium vapor pressure over $H_2O(l)$ at 393.15 K:

$$P_2^{eq} = P_1^{eq}\exp\left\{-\frac{5.637\times10^4}{R}\left[\frac{1}{T_2}-\frac{1}{T_1}\right]-\frac{45.069}{R}\ln\left[\frac{T_2}{T_1}\right]+\frac{0.01215}{R}[T_2-T_1]\right\}$$

$$= 760\exp\left\{-\frac{5.637\times10^4}{R}\left[\frac{1}{393.15}-\frac{1}{373.15}\right]-\frac{45.069}{R}\ln\left[\frac{393.15}{373.15}\right]\right.$$

$$\left.+\frac{0.01215}{R}[393.15-373.15]\right\} = \underline{1{,}485\ \text{torr}}. \quad \textbf{(14)}$$

The percent error in the result is

$$\%\ \text{error} = \frac{100\times(1{,}485-1{,}489.14)}{1{,}489.14} = \underline{-0.28\%}. \quad \textbf{(15)}$$

If we had assumed $\Delta\overline{H}_{vap}$ to be constant at 298.15 K, the equilibrium vapor pressure would be given by Eq. 6.13A with $\Delta\overline{H}_{vap} = 44{,}010\ \text{J mol}^{-1}$. This would produce

$$P^{eq} = 760\exp\left[-\frac{44{,}010}{8.314}\left\{\frac{1}{393.15}-\frac{1}{373.15}\right\}\right] = 1{,}564\ \text{torr}. \quad \textbf{(16)}$$

The error in this result is 74.9 torr, which corresponds to a percent error of

$$\%\ \text{error for constant}\ \Delta\overline{H}_{vap} = \frac{100\times(1{,}564-1{,}489.14)}{1{,}489.14} = \underline{5.03\%}. \quad \textbf{(17)}$$

Thus, the magnitude of the percent error is now about 18 times larger.

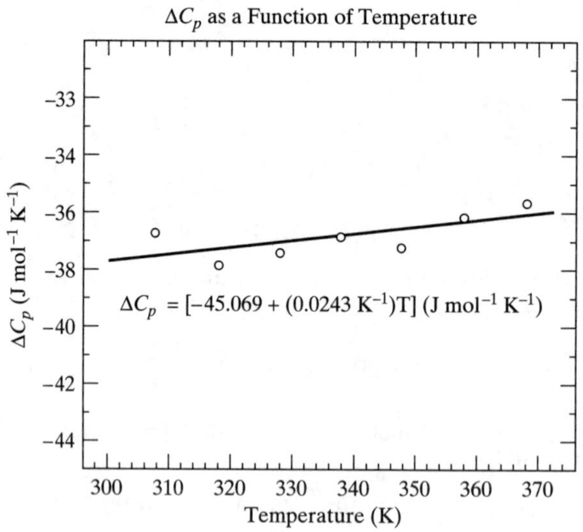

6.23 The partial molar enthalpy of fusion of $CO_2(s)$ is 7,950 J mol^{-1} at the melting point. At this same temperature, the partial molar enthalpy of sublimation is about 25,505 J mol^{-1}. At 220 K, the equilibrium vapor pressure over liquid CO_2 is 590.6 kPa. Use the Clausius–Clapeyron equation to obtain an analytic expression for the vapor pressure over liquid CO_2, assuming the enthalpy of vaporization to be independent of temperature.

Solution

The enthalpy of vaporization is given by

$$\Delta\overline{H}_{\text{vap}} = \Delta\overline{H}_{\text{sub}} - \Delta\overline{H}_{\text{fus}} = 2.550 \times 10^4 - 7{,}950 \text{ J mol}^{-1} = 1.756 \times 10^4 \text{ J mol}^{-1}. \quad \textbf{(1)}$$

If we assume this value to be constant over a narrow temperature range around the melting point, the equilibrium vapor pressure will be given by Eq. 6.12:

$$\ln\left[\frac{P_2^{\text{eq}}}{P_1^{\text{eq}}}\right] = -\frac{\Delta\overline{H}_{\text{vap}}}{R}\left[\frac{1}{T_2^{\text{eq}}} - \frac{1}{T_1^{\text{eq}}}\right]. \quad \textbf{(2)}$$

Taking $T_1^{\text{eq}} = 220$ K and $P_1^{\text{eq}} = 590.6$ kPa and dropping the subscripts on P_2^{eq} and T_2^{eq}, we obtain

$$\ln P^{\text{eq}} = \ln(590.6) - \frac{1.756 \times 10^4}{8.314}\left[\frac{1}{T^{\text{eq}}} - \frac{1}{220}\right] = 15.98 - \frac{2{,}111}{T^{\text{eq}}}. \quad \textbf{(3)}$$

This gives

$$\boxed{P = \exp\left[15.98 - \frac{2{,}111}{T^{\text{eq}}}\right] \text{ kPa}}, \quad \textbf{(4)}$$

which is the desired expression.

6.25 The densities of solid and liquid CO_2 are 1.56 kg L^{-1} and 1.101 kg L^{-1}, respectively. The triple point was predicted in Problem 6.24 to occur at $T = 216.8$ K, with a pressure of 5.19 bar. The partial molar enthalpy of fusion of $CO_2(s)$ is 7,950 J mol^{-1} = 79.504 L bar mol^{-1}. Use these data to obtain an analytical equation showing the dependence of the equilibrium temperature (melting point) of solid $CO_2(s)$ on pressure. At what pressure will the $CO_2(s)$ melting point be 222 K?

Solution

Equation 6.20 gives the dependence of the equilibrium temperature upon pressure:

$$[P_2^{\text{eq}} - P_1^{\text{eq}}] = \frac{\Delta\overline{H}_{\text{fus}}}{\Delta\overline{V}}\ln\left[\frac{T_2^{\text{eq}}}{T_1^{\text{eq}}}\right]. \quad \textbf{(1)}$$

Let us take $P_1^{\text{eq}} = 5.19$ bar and $T_1^{\text{eq}} = 217.0$ K, since we know that liquid and solid CO_2 are in equilibrium at this point. The molar volume of the solid is

$$\overline{V}(\text{s}) = \frac{\text{mass of 1 mole}}{\text{density}} = \frac{0.04401 \text{ kg mol}^{-1}}{1.56 \text{ kg L}^{-1}} = 0.02821 \text{ L mol}^{-1}. \quad \textbf{(2)}$$

For the liquid,

$$\overline{V}(\text{l}) = \frac{0.04401 \text{ kg mol}^{-1}}{1.101 \text{ kg L}^{-1}} = 0.03997 \text{ L mol}^{-1}. \quad \textbf{(3)}$$

Therefore,

$$\Delta\overline{V} = \overline{V}(\text{l}) - \overline{V}(\text{s}) = 0.03997 - 0.02821 \text{ L mol}^{-1} = 0.01176 \text{ L mol}^{-1}. \quad \textbf{(4)}$$

After dropping the subscript "2," substituting into Eq. 1 gives

$$\boxed{P^{\text{eq}} = 5.19 \text{ bar} + \frac{79.504 \text{ L bar mol}^{-1}}{0.01176 \text{ L mol}^{-1}}\ln\left[\frac{T^{\text{eq}}}{217.0}\right] = 5.19 + 6{,}760.5\ln\left[\frac{T^{\text{eq}}}{217.0}\right] \text{ bar}}, \quad \textbf{(5)}$$

which is the required expression.

The pressure required to make the equilibrium temperature (melting point) be 222 K is

$$P^{eq} = 5.19 + 6{,}760.5 \ln\left(\frac{222}{217.0}\right) \text{bar} = \underline{159.2 \text{ bar}} = \underline{1.592 \times 10^5 \text{ kPa}}. \qquad (6)$$

6.27 10 moles of $H_2O(l)$ at 330 K are inserted into an evacuated container that is insulated so as to prevent heat transfer to or from the surroundings. The volume of vacuum above the water is 20 L. As the $H_2O(l)$ vaporizes to establish equilibrium with $H_2O(g)$, the temperature drops because the heat of vaporization must be provided by the internal energy present in the $H_2O(l)$, since there is no heat transfer from the surroundings. If we ignore the change in the volume, mass, and heat capacity of the liquid, what is the final temperature of the system when equilibrium is established between $H_2O(l)$ and $H_2O(g)$? Assume that $\Delta\overline{H}_{vap}$ for $H_2O(l)$ is constant and equal to 44,010 J mol^{-1} and that the total heat capacity of the $H_2O(l)$ is 752.9 J K^{-1}.

Solution

Let dn moles of $H_2O(1)$ vaporize into the gas phase at temperature T. The total loss of internal energy by the water is

$$\text{Energy loss by water} = -\Delta\overline{H}_{vap}\, dn. \qquad (1)$$

This energy loss will cause a decrease in the water's temperature given by $C\,dT$, where C is the total heat capacity of the water and dT is the differential temperature change caused by the vaporization of dn moles of water. Therefore, we have

$$C\,dT = -\Delta\overline{H}_{vap}\, dn. \qquad (2)$$

The change in the gas-phase pressure produced by the vaporization is

$$dP = \frac{RT\,dn}{V}. \qquad (3)$$

We may now solve for dn in Eq. 3 and substitute the result into Eq. 2. This operation gives

$$C\,dT = -\frac{V\Delta\overline{H}_{vap}\, dP}{RT}. \qquad (4)$$

Rearranging terms produces

$$dP = -\frac{RTC}{V\Delta\overline{H}_{vap}}\, dT. \qquad (5)$$

Initially, we had $T = 330$ K and $P = 0$. After equilibrium is established, we will have $P = P^{eq}$ and $T = T_f$, where T_f is the final system temperature and P^{eq} is the equilibrium vapor pressure at $T = T_f$. Integrating Eq. 5 between these limits gives

$$\int_0^{P^{eq}} dP = P^{eq} = -\frac{RC}{V\Delta\overline{H}_{vap}} \int_{330}^{T_f} T\,dT = \frac{RC}{2V\Delta\overline{H}_{vap}}[330^2 - T_f^2]. \qquad (6)$$

Removing C and $\Delta\overline{H}_{vap}$ from the scope of the integral in Eq. 6 invokes the approximations that $\Delta\overline{H}_{vap}$ is independent of temperature and that the liquid's heat capacity is constant and independent of the vaporization process. This assumption requires that the mass of water be large relative to the final mass of vapor. The equilibrium vapor pressure, P^{eq}, is given by Eq. 6.13B. Use of this equation yields

$$\exp\left[-\frac{\Delta\overline{H}_{vap}}{R}\left\{\frac{1}{T_f} - \frac{1}{373.15}\right\}\right] \text{atm} = \frac{RC}{2V\Delta\overline{H}_{vap}}[330^2 - T_f^2], \qquad (7)$$

since the normal boiling point of water is 373.15 K. Substituting the data given in the problem gives

$$\exp\left[-\frac{44{,}010}{8.314}\left\{\frac{1}{T_f}-\frac{1}{373.15}\right\}\right]\text{ atm}$$
$$=\frac{(0.08206\text{ L atm mol}^{-1}\text{ K}^{-1})\,752.9\text{ J K}^{-1}}{2(20\text{ L})44{,}010\text{ J mol}^{-1}}[330^2-T_f^2]\text{ K}^2,$$
$$=3.5096\times10^{-5}[330^2-T_f^2]\text{ atm.} \qquad \textbf{(8)}$$

Equation 8 must now be solved for T_f. We seek a root of

$$F(T_f)=\exp\left[-\frac{44{,}010}{8.314}\left\{\frac{1}{T_f}-\frac{1}{373.15}\right\}\right]-3.5096\times10^{-5}[330^2-T_f^2]=0. \qquad \textbf{(9)}$$

Two passes through a grid search yields a root at $\underline{T_f = 324.75\text{ K}}$.

We may check the energy balance to be certain that our solution is correct and that our approximations are reasonable. The final pressure in the container will be

$$p^{eq}=\exp\left[-\frac{44{,}010}{8.314}\left\{\frac{1}{324.75}-\frac{1}{373.15}\right\}\right]=0.12073\text{ atm.} \qquad \textbf{(10)}$$

Therefore, the number of moles of gas will be

$$n_f=\frac{p^{eq}V}{RT}=\frac{(0.12073\text{ atm})(20\text{ L})}{(0.08206\text{ L atm mol}^{-1}\text{ K}^{-1})(324.75\text{ K})}=0.09061\text{ mole of vapor.} \qquad \textbf{(11)}$$

Thus, we see that our approximation of ignoring the mass of the vapor is very good. Only 0.906% of the water has vaporized. The vaporization of this many moles of water requires a total heat of 0.09061 mole (44,010 J mol^{-1}) = 3,987.7 J. The change in the temperature of the water produced by this enthalpy change is 3,987.7 J/752.9 J K^{-1} = 5.296 K. Consequently, the final temperature is 330 − 5.296 K = 324.7 K, which is correct. A total change in temperature of less than 6 K fully justifies the assumption that $\Delta\overline{H}_{vap}$ is constant.

6.29# A scientist has water in an enclosed container at 300 K. (See accompanying figure.) At this temperature, the equilibrium vapor pressure is 0.0328 atm. The scientist wishes to conduct an experiment in which the temperature of the system will be gradually lowered, but he desires to keep the equilibrium vapor pressure fixed at 0.0328 atm. He intends to accomplish this by introducing argon gas into the chamber in sufficient quantity to hold the equilibrium vapor pressure constant as the temperature is lowered.

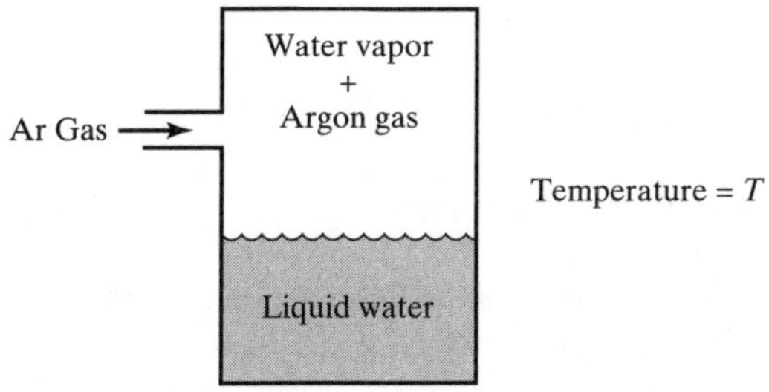

(A) Obtain an expression for the argon pressure that must be present in the container as a function of temperature. Assume that the partial molar enthalpy of vaporization of water is constant and that the liquid is incompressible.

(B) Plot the pressure of argon needed versus T over the range 290 K $\leq T \leq$ 300 K. Prepare a second plot over the temperature range 50 K $\leq T \leq$ 300 K.

Solution

(A) The general condition for phase equilibrium is given by Eq. 6.1:

$$d\mu_c = -\overline{S}_c dT_c^{\rm eq} + \overline{V}_c dP_c^{\rm eq} = d\mu_v = -\overline{S}_v dT_v^{\rm eq} + \overline{V}_v dP_v^{\rm eq}, \quad (1)$$

where the subscripts c and v denote the condensed and vapor phases, respectively. We wish to keep $P_v^{\rm eq}$ fixed. Therefore, we have $dP_v^{\rm eq} = 0$. The experimental setup guarantees that both phases will be at the same temperature. Consequently, $dT_c^{\rm eq} = dT_v^{\rm eq} = dT^{\rm eq}$. Rearranging Eq. 1 gives

$$\overline{V}_c dP_c^{\rm eq} = -[\overline{S}_v - \overline{S}_c]dT^{\rm eq} = -\frac{\Delta \overline{H}_{\rm vap}}{T^{\rm eq}} dT^{\rm eq}. \quad (2)$$

Therefore,

$$dP_c^{\rm eq} = -\frac{\Delta \overline{H}_{\rm vap}}{\overline{V}_c} \frac{dT^{\rm eq}}{T^{\rm eq}}. \quad (3)$$

Integrating both sides between corresponding limits yields

$$\int_{0.0328\ \rm atm}^{P^{\rm eq}_c} dP_c^{\rm eq} = P_c^{\rm eq} - 0.0328\ {\rm atm} = -\frac{\Delta \overline{H}_{\rm vap}}{\overline{V}_c} \int_{300\ \rm K}^{T^{\rm eq}} \frac{dT^{\rm eq}}{T^{\rm eq}} = \frac{\Delta \overline{H}_{\rm vap}}{\overline{V}_c} \ln\left[\frac{300}{T^{\rm eq}}\right]. \quad (4)$$

Hence, the required total pressure is

$$P_c^{\rm eq} = 0.0328\ {\rm atm} + \frac{\Delta \overline{H}_{\rm vap}}{\overline{V}_c} \ln\left[\frac{300}{T^{\rm eq}}\right]. \quad (5)$$

The argon pressure needed in the experiment to produce this total pressure is

$$\boxed{P_{\rm Ar}^{\rm eq} = \frac{\Delta \overline{H}_{\rm vap}}{\overline{V}_c} \ln\left[\frac{300}{T^{\rm eq}}\right]}. \quad (6)$$

(B) For the H_2O system, we have

$$\Delta \overline{H}_{\rm vap} = \overline{H}^o_{\rm H_2O(g)} - \overline{H}^o_{\rm H_2O(l)} = -241.82\ {\rm kJ\ mol^{-1}} - (-285.83\ {\rm kJ\ mol^{-1}})$$

$$= 44.10\ {\rm kJ\ mol^{-1}} = 44{,}010\ {\rm J\ mol^{-1}} \quad (7)$$

and

$$\overline{V}_c = 0.018\ {\rm L\ mol^{-1}}. \quad (8)$$

Substituting into Eq. 6 gives

$$P_{\rm Ar}^{\rm eq} = \frac{44{,}010\ {\rm J\ mol^{-1}}}{0.018\ {\rm L\ mol^{-1}}} \frac{0.08206\ {\rm L\ atm\ mol^{-1}\ K^{-1}}}{8.314\ {\rm J\ mol^{-1}\ K^{-1}}} \ln\left[\frac{300}{T^{\rm eq}}\right]$$

$$\boxed{2.41 \times 10^4 \ln\left[\frac{300}{T^{\rm eq}}\right]\ {\rm atm}}. \quad (8)$$

The requested plots are as shown below:

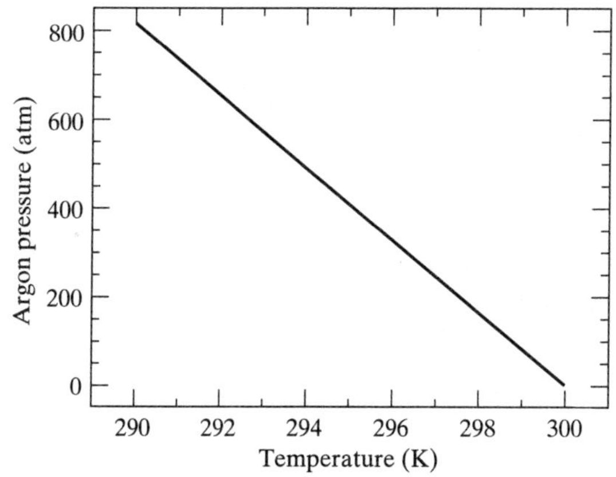

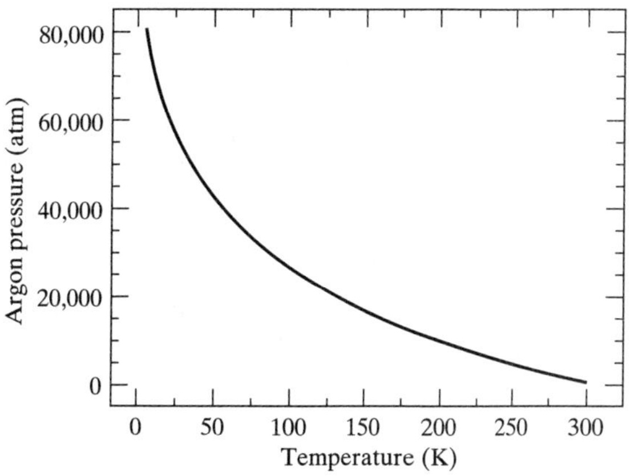

6.31 The following thermodynamic data for liquid and gaseous mercury are available at 298.15 K and 1 bar of pressure:

Substance	μ^o (J mol^{-1})	$\bar{S}$ (J mol^{-1} K^{-1})
Hg(l)	0	76.02
Hg(g)	31,820	174.96

Using only these data, obtain a reasonably accurate estimate of the standard boiling point of mercury. State any assumptions you are making in your calculations. The experimentally measured standard boiling temperature of mercury is 629.88 K. Compute the percent error in your result. What is the major source of the error?

Solution

At the standard boiling temperature T_b, the solid and liquid phases are in equilibrium. At this point, we must have $\Delta\mu^o(T_b) = 0$. For the process Hg(l) $\longrightarrow$ Hg(g), $\Delta\mu^o$ at 298.15 K can be obtained from the data in the problem:

$$\Delta\mu^o(298.15\text{ K}) = [\mu_o(298.15\text{ K})]_{\text{gas}} - [\mu_o(298.15\text{ K})]_{\text{liquid}}$$

$$= 31{,}820 - 0 = 31{,}820\text{ J mol}^{-1}. \quad \textbf{(1)}$$

Thus, we know that Hg(l) has the lower chemical potential at 298.15 K, so that mercury will be in the liquid state at this temperature. We seek the temperature T_b at which we have $\Delta\mu^o(T_b) = 0$. The total differential of the chemical potential is

$$d\Delta\mu^o = -\Delta\bar{S}^o dT + \Delta\bar{V}^o dp. \quad \textbf{(2)}$$

Vaporization occurs at constant pressure, so that we have $dp = 0$. Under this condition, Eq. 2 becomes

$$d\Delta\mu^o = -\Delta\bar{S}^o dT. \quad \textbf{(3)}$$

We may now take integrals of both sides of Eq. 3 between 298.15 K and T_b. This operation gives

$$\int_{\Delta\mu^o(298.15\text{ K})}^{\Delta\mu^o(T_b)} d\Delta\mu^o = \Delta\mu^o(T_b) - \Delta\mu^o(298.15\text{ K}) = \int_{298.15\text{ K}}^{T_b} -\Delta\overline{S}^o dT. \qquad (4)$$

Since we have no heat capacity data available, we are forced to assume that $\Delta\overline{S}^o$ is nearly constant over the temperature interval from 298.15 K to T_b. With this assumption, Eq. 4 becomes

$$\Delta\mu^o(T_b) - \Delta\mu^o(298.15\text{ K}) = -\Delta\overline{S}^o[T_b - 298.15]. \qquad (5)$$

Substituting the data given in the problem for the entropies and $\Delta\mu^o(298.15\text{ K})$ from Eq. 1 into Eq. 5 produces

$$\Delta\mu^o(T_b) - 31{,}820 = -[174.96 - 76.02][T_b - 298.15]. \qquad (6)$$

At $T = T_b$, $\Delta\mu^o(T_b) = 0$. Solving Eq. 6 for T_b yields

$$T_b = 298.15\text{ K} + \frac{-31{,}820\text{ J mol}^{-1}}{-[174.96 - 76.02]\text{ J mol}^{-1}\text{ K}^{-1}} = \underline{619.8\text{ K}}. \qquad (7)$$

The percent error in this estimate is

$$\%\text{ error} = 100 \times \frac{619.8 - 629.88}{629.88} = \underline{-1.60\%}. \qquad (8)$$

Virtually all of this error is due to our assumption that $\Delta\overline{S}^o$ is independent of temperature.

6.33 The following thermodynamic data for liquid and gaseous mercury are available at 298.15 K and 1 bar of pressure.

Substance	μ^o (J mol^{-1})	$\overline{H}^o$ (J mol^{-1})
Hg(l)	0	0
Hg(g)	31,820	61,320

Using only these data, obtain a reasonably accurate estimate for the equilibrium vapor pressure of mercury at 400 K. The experimentally measured equilibrium vapor pressure at 400 K is 0.00138 bar. Compute the percent error in your estimate. What is the major source of the error in your calculations?

Solution

We may obtain a reasonably accurate estimate of the equilibrium vapor pressure for liquid mercury at 400 K using the Clausius–Clapeyron equation,

$$P_2^{eq} = P_1^{eq}\exp\left[-\frac{\Delta\overline{H}_{vap}}{R}\left(\frac{1}{T_2} - \frac{1}{T_1}\right)\right], \qquad (1)$$

where P_2^{eq} and P_1^{eq} are the equilbrium vapor pressures over liquid mercury at temperatures T_2 and T_1, respectively. At 298.15 K and 1 bar pressure,

$$\Delta\overline{H}_{vap} = \Delta\overline{H}^o_{vap} = \overline{H}^o_{gas} - \overline{H}^o_{liquid} = 61{,}320 - 0 = 61{,}320\text{ J mol}^{-1}. \qquad (2)$$

Since we have no heat capacity data in this problem, we shall assume that $\Delta\overline{H}_{vap}$ is independent of temperature. We now need the equilibrium vapor pressure at some temperature T_1. The equilibrium constant for the process Hg(l) = Hg(g) is

$$K_p = P^{eq} = \exp\left[-\frac{\Delta\mu^o(T)}{RT}\right]. \quad \textbf{(3)}$$

The data provides $\Delta\mu^o$ at 298.15 K. At this temperature, we have

$$K_p = P^{eq}_{298} = \exp\left[-\frac{31{,}820}{(8.314)(298.15)}\right] = 2.661 \times 10^{-6} \text{ bar}, \quad \textbf{(4)}$$

since the pressure must be expressed in bars. Combining Eqs. 1, 2, and 4 gives

$$P^{eq}_{400} = P^{eq}_{298} \exp\left[-\frac{\Delta\overline{H}_{vap}}{R}\left(\frac{1}{T_2} - \frac{1}{T_1}\right)\right]$$

$$= (2.661 \times 10^{-6}) \exp\left[-\frac{61{,}320}{8.314}\left(\frac{1}{400} - \frac{1}{298.15}\right)\right] = \underline{0.00145 \text{ bar}}. \quad \textbf{(5)}$$

The percent error in this computed value is

$$\% \text{ error} = 100 \times \frac{0.00145 - 0.00138}{0.00138} = \underline{5.07\%}. \quad \textbf{(6)}$$

The assumption that $\Delta\overline{H}_{vap}$ is independent of temperature is the major source of error. Since this enthalpy appears inside an exponential function, small errors in $\overline{H}_{vap}$ create very large errors in the computed vapor pressures.

CHAPTER 7

Thermodynamics of Solids

7.1 A chemist obtains the following data:

Compound	Normal Melting Point (°C)
Na_2O	1275
K_2O	decomposes before melting
CaO	2614
BaO	1918

Using these data alone, answer the following questions and give reasons for your answers:

(A) Is the normal melting point of Li_2O higher or lower than 1,275°C?

(B) Can you provide upper and lower limits for the normal melting point of SrO? If not, why not? If so, provide these limits and state why they are appropriate.

(C) Can we determine whether Li_2O or MgO has the higher normal melting point? Explain.

(D) Can we determine whether Li_2O or BaO has the higher normal melting point? Explain.

Solution

(A) Li_2O will exhibit a higher melting point than Na_2O. The fact that the binding forces in K_2O are small enough to allow the decomposition of the compound before it melts indicates that the binding forces are decreasing as the atomic number of the alkali metal increases. This suggests that, for this homologous series, ionic size is the dominating factor.

(B) Yes. The data show that the normal melting points are decreasing as the atomic number of the alkaline earth metal increases. Therefore, the normal melting temperature for SrO must be below that for CaO and above that for BaO. That is, $\boxed{1{,}918°C < T_m^o < 2{,}614°C}$. In fact, the measured normal melting point is 2,430°C.

(C) Yes. MgO will melt at a significantly higher temperature than Li_2O, since (1) in MgO the q_1q_2 product in the coulombic potential is always 4, whereas in Li_2O this product can be 1, 2, or 4, depending upon the ions involved. This makes the coulombic interaction stronger for MgO than for Li_2O. The ionic size of Mg^{2+} (0.72 Å) listed in Table 7.2 is nearly the same as that for Li^{1+} (0.59 Å). It is unlikely that this difference is sufficient to compensate for the difference in charges. Measurements show that the melting temperatures for MgO and Li_2O are 2,852°C and 1,700°C, respectively.

(D) This is the same question as in Part (C). Here, however, the situation is uncertain. The difference in charges favors a higher melting temperature for BaO, and the difference in ionic radii (0.59 Å for Li^{1+} and 1.36 Å for Ba^{2+}) favors a lower melting temperature for BaO. As a result, the two melting points should be close, and qualitative determination is uncertain. In fact, BaO melts at 1,918°C, 218°C above Li_2O. Note how much closer these two values are than is the case for Li_2O and MgO, considered in Part (C).

7.3* The normal boiling points of various aliphatic 2-ketones are given in the table on the top of next page.

Ketone	Normal Boiling Point (K)
acetone	329.4
2-butanone	352.8
2-pentanone	375.2
2-hexanone	401.2
2-heptanone	424.6
2-nonanone	468.4
2-decanone	483.2

(A) By plotting the normal boiling points against the molar masses of the ketones, determine whether a correlation exists between molar mass and boiling point. Using a least-squares method, find the best straight-line fit to the data.
(B) Use the results obtained in (A) to predict the normal boiling point for 2–octanone. Compute the percent error in your prediction.

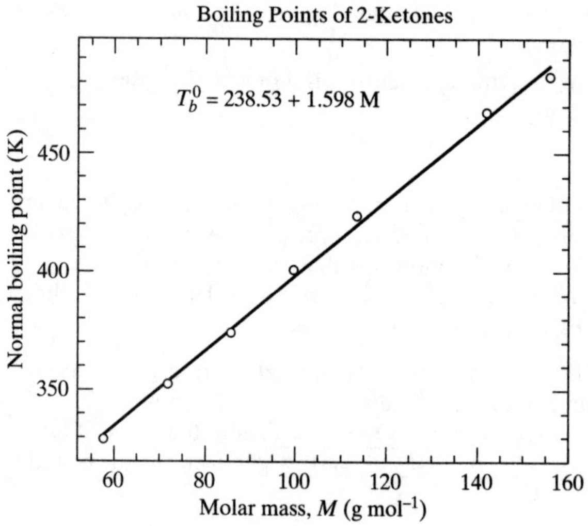

Solution

(A) A correlation obviously exists. The line shown in the above is the least-squares linear fit to the data points. The resulting equation is

$$T_b^\circ = 238.53 + 1.598\, M \text{ K}, \quad (1)$$

where M is the molar mass in units of g mol^{-1}.

(B) From the linear correlation given in Eq. 1, the predicted boiling point of 2-octanone is

$$T_b^\circ(\text{2-octanone}) = 238.53 + 1.598(128.208) = \underline{443.4 \text{ K}}. \quad (2)$$

The measured normal boiling point of 2-octanone is 446.2 K. The percent error in the prediction is, therefore,

$$\% \text{ error} = \frac{100 \times (443.4 - 446.2)}{446.2} = \underline{-0.63\%}. \quad (3)$$

7.5 The alkali metals all have one electron in a valence *s* orbital. All of them form high-melting ionic crystals with the halogens, whose melting points increase as the atomic number decreases. (See Table 7.1.) Hydrogen also has a single electron in its valence *s* orbital and is listed in Group IA in most periodic tables. However, hydrogen does not form an ionic crystal with any of the halogens. Why not?

Solution

As noted in the text, in Table 7.2, and in Fig. 7.1, there is only a small change in ionization potential of the alkali metals as we move from Li to Cs, and all of the ionization potentials are less than 5.4 eV. Consequently, ionic size variations are the dominant factor in controlling the magnitude of the coulombic forces. However, when we move upward to hydrogen, the ionization potential jumps to 13.60 eV. This is so large that the transfer of the electron to a halogen atom is an energetically unfavorable process. As a result, the extent of the charge transfer is insufficient to support ionic bonding, and HCl(c) does not form.

7.7 (A) Using the method described in the text with a repulsive potential of the form $V_{rep}(d) = K/d^n$ with $n = 11.67$, compute the crystal enthalpy of KCl(c) at 298.15 K.

(B) Use the Born–Haber cycle, along with the result obtained in (A), the data in Appendix A, and Table 7.2, to obtain an estimate of the electron affinity of Cl. Compute the percent error in your estimate. KCl forms a face-centered cubic crystal, with the shortest distance between ions being 3.14×10^{-10} m.

Solution

(A) From Eq. 7.14, the crystal enthalpy at 298.15 K is

$$\Delta H_c(\text{KCl}) = \frac{N\mathcal{M}q^2}{4\pi\varepsilon_o d_o}\left[1 - \frac{1}{n}\right] + 2RT$$

$$= \frac{(6.022 \times 10^{23})(1.74756)(1.602 \times 10^{-19})^2}{4(3.14159)(8.85419 \times 10^{-12})(3.140 \times 10^{-10})}\left[1 - \frac{1}{11.67}\right]$$

$$+ 2(8.314)(298.15)\text{ J mol}^{-1} = 7.118 \times 10^5\text{ J mol}^{-1}$$

$$= \underline{711.8\text{ kJ mol}^{-1}}. \qquad (1)$$

(B) The counterclockwise steps around the Born–Haber Cycle are as follows:

Step 1: Dissociation of KCl(c) to K(s) + 0.5 Cl_2(g): $\Delta H_1 = -\overline{H}^o_{KCl(c)}$.
Step 2: Sublimation of K(s) to K(g): $\Delta H_2 = \overline{H}^o_{K(g)}$.
Step 3: Dissociation of 0.5 Cl_2(g) to Cl(g): $\Delta H_3 = \overline{H}^o_{Cl(g)}$.
Step 4: Ionization of K(g) to give K^+(g): $\Delta H_4 = I_1^K + RT$.
Step 5: Addition of an electron to Cl(g) to give Cl^{1-}(g): $\Delta H_5 = -A_{Cl(g)} - RT$.
Step 6: Combination of K^+(g) and Cl^-(g) to give KCl(c): $\Delta H_6 = -\Delta Hc(\text{KCl})$.

The summation of all changes in enthalpy around the closed cycle must be zero. Therefore,

$$\Delta H_1 + \Delta H_2 + \Delta H_3 + \Delta H_4 + \Delta H_5 + \Delta H_6 = 0, \qquad (2)$$

and we have

$$A_{Cl(g)} = -\overline{H}^o_{KCl(c)} + \overline{H}^o_{K(g)} + \overline{H}^o_{Cl(g)} + I_1^K - \Delta H_c(\text{KCl})$$

$$= 436.75 + 89.24 + 121.68 + 4.318(96.48) - 711.8\text{ kJ mol}^{-1}$$

$$= \underline{352.5\text{ kJ mol}^{-1}}. \qquad (3)$$

Using the data given in Table 7.4, we compute the percent error:

$$\% \text{ error} = \frac{100 \times (352.5 - 348.7)}{348.7} = \underline{1.09\%}. \quad \textbf{(4)}$$

7.9* The crystal enthalpy of a 1:1 ionic crystal is

$$\Delta H_c = \frac{N\mathcal{M}q^2}{4\pi d_o \varepsilon_o}\left[1 - \frac{\rho}{d_o}\right] + 2RT$$

if an exponential repulsive potential form is assumed. Use a least-squares procedure with the following data at 298.15 K to obtain the best value of the parameter ρ in the foregoing expression:

Crystal	Crystal Enthalpy (kJ mol^{-1}K^{-1})	d_o(m)
NaCl	787	2.814×10^{-10}
KCl	717	3.138×10^{-10}
NaBr	752	2.981×10^{-10}

All of these crystals have face-centered cubic form. Compute ΔH_c for each of these crystals, using the value of ρ determined in the least-squares analysis.

Solution

The constants in the calculation are:

$$\frac{N\mathcal{M}q^2}{4\pi d_o \varepsilon_o} = \frac{(6.022 \times 10^{23})(1.74756)(1.602 \times 10^{-19})^2}{4(3.14159)(8.85419 \times 10^{-12})d_o} = \frac{2.427 \times 10^{-4}}{d_o} \text{ J mol}^{-1} \quad \textbf{(1)}$$

and

$$2RT = 2(8.314)(298.15) = 4{,}957.6 \text{ J mol}^{-1}. \quad \textbf{(2)}$$

Therefore, the crystal enthalpy may be written in the form

$$\Delta H_c = \frac{2.427 \times 10^6}{d_o}\left[1 - \frac{\rho}{d_o}\right] + 4{,}957.6 \text{ J mol}^{-1}. \quad \textbf{(3)}$$

The sum of the squared deviations of the computed crystal enthalpies from the experimental results obtained with the Born–Haber cycle is

$$S = \left\{787{,}000 - \frac{2.427 \times 10^6}{2.814}\left[1 - \frac{\rho}{2.814}\right] - 4{,}957.6\right\}^2$$
$$+ \left\{717{,}000 - \frac{2.427 \times 10^6}{3.138}\left[1 - \frac{\rho}{3.138}\right] - 4{,}957.6\right\}^2$$
$$+ \left\{752{,}000 - \frac{2.427 \times 10^6}{2.981}\left[1 - \frac{\rho}{2.981}\right] - 4{,}957.6\right\}^2, \quad \textbf{(4)}$$

provided that we express ρ and d_o in the ratio ρ/d_o in Å. We wish to obtain a minimum for S. At the minimum, we have $dS/d\rho = 0$. The derivative is a linear equation that is simple to solve. The first derivative of S gives

$$\frac{dS}{d\rho} = 2\left\{787{,}000 - \frac{2.427 \times 10^6}{2.814}\left[1 - \frac{\rho}{2.814}\right] - 4{,}957.6\right\}\frac{2.427 \times 10^6}{(2.814)^2}$$

$$+ 2\left\{717{,}000 - \frac{2.427 \times 10^6}{3.138}\left[1 - \frac{\rho}{3.138}\right] - 4{,}957.6\right\}\frac{2.427 \times 10^6}{(3.138)^2}$$

$$+ 2\left\{752{,}000 - \frac{2.427 \times 10^6}{2.981}\left[1 - \frac{\rho}{2.981}\right] - 4{,}957.6\right\}\frac{2.427 \times 10^6}{(2.981)^2} = 0. \quad \textbf{(5)}$$

Dividing out the constant terms and collecting terms produces

$$2.427 \times 10^6 \rho\left[\frac{1}{(2.814)^4} + \frac{1}{(3.138)^4} + \frac{1}{(2.918)^4}\right]$$

$$+ \left\{\frac{782{,}042}{(2.814)^2} + \frac{712{,}042}{(3.138)^2} + \frac{747{,}042}{(2.981)^2}\right.$$

$$\left. - 2{,}427{,}000\left[\frac{1}{(2.814)^3} + \frac{1}{(3.138)^3} + \frac{1}{(2.918)^3}\right]\right\} = 0. \quad \textbf{(6)}$$

The solution of this linear equation, after doing some arithmetic manipulations, is

$$\boxed{\rho_{\text{min}} = 0.246 \text{ Å} = 0.246 \times 10^{-10} \text{ m}}. \quad \textbf{(7)}$$

The values of ΔH_c for the three crystals used in the analysis are

$$\Delta H_c(\text{NaCl}) = \frac{2.427 \times 10^6}{2.814}\left[1 - \frac{0.246}{2.814}\right] + 4{,}957.6$$

$$= 7.92 \times 10^5 \text{ J mol}^{-1} = \underline{792 \text{ kJ mol}^{-1}}, \quad \textbf{(8)}$$

$$\Delta H_c(\text{KCl}) = \frac{2.427 \times 10^6}{3.138}\left[1 - \frac{0.246}{3.138}\right] + 4{,}957.6$$

$$= 7.18 \times 10^5 \text{ J mol}^{-1} = \underline{718 \text{ kJ mol}^{-1}}, \quad \textbf{(9)}$$

and

$$\Delta H_c(\text{NaBr}) = \frac{2.427 \times 10^6}{2.981}\left[1 - \frac{0.246}{2.981}\right] + 4{,}957.6$$

$$= 7.52 \times 10^5 \text{ J mol}^{-1} = \underline{752 \text{ kJ mol}^{-1}}. \quad \textbf{(10)}$$

7.11* In the text, we computed the value of the Madelung constant for a hypothetical linear crystal. Let us assume that we are in the universe of the Flatlanders, where everything is two dimensional. (See Edwin A. Abbot, *Flatland* (6th edition, Dover, New York, 1952; 1st edition, Seeley & Co., Great Britian, 1884).) In this universe, a common crystal system is the square lattice. One of the subclasses of such a system is the primitive square, which has an ion at each vertex. The figure illustrates this subclass on the top of the next page.

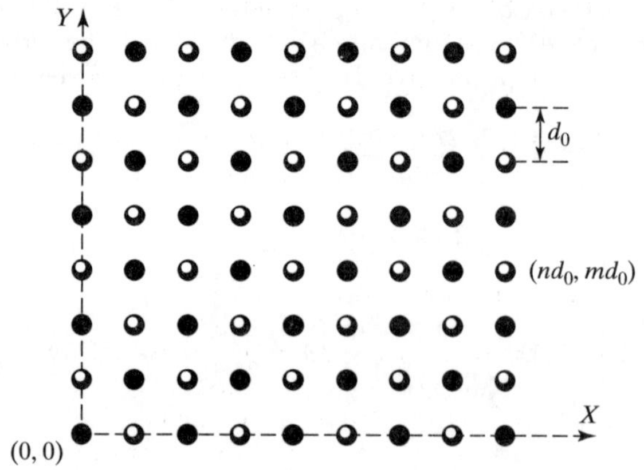

▲ A primitive square lattice in the Flatland universe. The black dots denote positive cations. The lighter dots indicate negatively charged anions in the crystal. The lattice spacing is denoted by d_o.

Calculate the Madelung constant for the square primitive lattice for the case where the anions and the cations have the same magnitude of charge. You will need to use a PC, a workstation, or some type of spreadsheet for this calculation. The various parts of the problem serve as a procedure guide. [*Note to instructor:* If you wish to omit the numerical evaluation of $\mathcal{M}$, just skip Part (E).]

(A) To obtain the Madelung constant, we need to sum all of the interactions with ion i which we shall take to be the ion at the origin of the coordinate system shown in the preceding figure. For interactions involving ions of the same charge, we sum with a negative sign; interactions between ions of different charge we sum with a positive sign. Obtain an equation for the term in the Madelung sum involving ion i and the ion in the figure whose (X, Y) coordinates are (nd_o, md_o). Show that this term is a function of n and m alone.

(B) Show that the sign on the term in the Madelung sum will be negative if $(n + m)$ is even and positive if $(n + m)$ is odd.

(C) Note that the terms in the Madelung sum for the interactions with the ions whose coordinates are (md_o, nd_o) or $(-nd_o, md_o)$ are identical to the term for the ion whose coordinates are (nd_o, md_o). If $n \neq m$ and neither n nor m is zero, how many terms have the same value as the term for the (nd_o, md_o) ion? If $n = m \neq 0$, how many terms have the same value as the term for the (nd_o, md_o) ion? If either n or $m = 0$, how many terms have the same value as the term for the (nd_o, md_o) ion? Let us call these quantities the *degeneracy* for the (nd_o, md_o) lattice point and denote it by the notation $g(n, m)$.

(D) If we multiply the interaction for the (nd_o, md_o) ion by its degeneracy factor $g(n, m)$, we need sum only over ions for which $n \geq 0$ and $m \geq 0$ and for which $m \geq n$. Using this fact and the results of (A), (B), and (C), express the Madelung constant as a double summation over n and m with $m \geq n$.

(E) Use a PC, a workstation, or a spreadsheet to execute the summation obtained in (D) with the upper limits on each sum equal to K. Execute the summation for $K = 500, 750, 1{,}000, 1{,}500, 3{,}000$, and $5{,}000$. From the results, determine the number of significant digits to which $\mathcal{M}$ has been evaluated for the square primitive lattice.

Solution

(A) The Madelung constant is given by

$$\mathcal{M} = \sum_{j=1}^{2N}{}' \pm d_o r_{ij}^{-1}. \tag{1}$$

From the Pythagorean theorem, the distance between the origin and the ion at coordinates (nd_o, md_o) is

$$r_{ij} = [m^2 d_o^2 + n^2 d_o^2]^{1/2} = d_o[m^2 + n^2]^{1/2}. \quad (2)$$

Consequently, this term in the sum has the form

$$\boxed{\pm d_o r_{ij}^{-1} = \pm[m^2 + n^2]^{-1/2}}, \quad (3)$$

and the term depends only upon n and m.

(B) Reference to the figure shows that each time we move one lattice site in either direction, the charge changes. The first step produces ions of different signs, for which we wish the sign on the term to be positive. The second step will, therefore, produce an interaction between ions of the same charge, for which we need a negative sign. The sign needed is negative if $(n + m)$ is even and positive if $(n + m)$ is odd. We can produce this sign by using the factor $(-1)^{n+m+1}$, which is negative if $(n + m)$ is even and positive if $(n + m)$ is odd. Consequently, the term in Eq. 3 can be written in the form

$$\boxed{\pm d_o r_{ij}^{-1} = (-1)^{n+m+1}[n^2 + m^2]^{-1/2}}. \quad (4)$$

(C) If $n \neq m$ and neither is zero, there are eight terms in the Madelung sum whose values are identical. These terms are for the ions at the coordinates (nd_o, md_o), $(-nd_o, -md_o)$, $(nd_o, -md_o)$, $(-nd_o, md_o)$, (md_o, nd_o), $(-md_o, -nd_o)$, $(-md_o, nd_o)$, and $(md_o, -nd_o)$. Therefore, we have $g(n, m) = 8$ for $n \neq m \neq 0$. If $n = m \neq 0$, there are four terms whose values are the same. They are the terms for ions at the coordinates (nd_o, nd_o), $(-nd_o, -nd_o)$, $(-nd_o, nd_o)$ and $(nd_o, -nd_o)$. Therefore, $g(n, n) = 4$. If either $n = 0$ or $m = 0$, there are four terms with same value. They are the terms for ions at $(nd_o, 0)$, $(-nd_o, 0)$, $(0, nd_o)$, and $(0, -nd_o)$. Thus, $g(n, 0) = g(0, m) = 4$.

(D) Using the results obtained in Parts A, B, and C, we can write the Madelung constant in the form

$$\boxed{\mathcal{M} = \sum_{n=0}^{n=\infty} \sum_{m=n}^{m=\infty} g(n, m)(-1)^{n+m+1}[n^2 + m^2]^{-1/2}}, \quad (5)$$

where we define

$$g(0, 0) = 0,$$

so that we prevent ion i at the origin from interacting with itself,

$$g(n, n) = 4,$$

$$g(0, n) = g(n, 0) = 4,$$

and $$g(n, m) = 8 \quad \text{for } n \neq m \neq 0.$$

(E) In this part, we replace the infinite upper limit in Eq. 5 with K and then execute the sum. The results are given in the table to the right. Therefore, it appears that we have about four significant digits, and $\mathcal{M} \approx 1.6155$.

K	$\mathcal{M}$
500	1.6141298
750	1.6146004
1,000	1.6148359
1,500	1.6150714
3,000	1.6153070
5,000	1.6154012

7.13 (A) Use the data in Appendix A and Tables 7.2, 7.4, and 7.6, along with the fact that the first two ionization potentials of Zn are $I_1 = 9.394$ eV and $I_2 = 17.964$ eV, to compute the crystal enthalpy of ZnO at 298.15 K.

(B) Using the results of (A) and Eq. 7.14, determine the shortest distance between the Zn^{2+} and O^{2-} ions in a crystal of ZnO.

Solution

(A) Calculate the steps in the Born–Haber cycle whose sum yields the crystal enthalpy of ZnO:

Step 1: Dissociation of ZnO(c) to Zn(s) + 0.5 O_2(g): $\Delta H_1 = -\overline{H}^o_{ZnO(c)}$.

Step 2: Sublimation of Zn(s) to Zn(g): $\Delta H_2 = \overline{H}^o_{Zn(g)}$.

Step 3: Dissociation of 0.5 O_2(g) to O(g): $\Delta H_3 = \overline{H}^o_{O(g)}$.

Step 4: Ionization of Zn(g) to give Zn^{2+}(g): $\Delta H_4 = I_1^{Zn} + I_2^{Zn} + 2RT$.

Step 5: Addition of two electrons to O(g) to give O^{2-}(g): $\Delta H_5 = -A_1^o - A_2^o - 2RT$.

Thus,

$$\Delta H_c(\text{ZnO}) = -\overline{H}^o_{\text{ZnO(c)}} + \overline{H}^o_{\text{Zn(g)}} + \overline{H}^o_{\text{O(g)}} + I_1^{\text{Zn}} + I_2^{\text{Zn}} + 2RT - A_1^o - A_2^o - 2RT$$

$$= 348.28 + 130.73 + 249.17 + 9.394(96.48) + 17.964(96.48)$$

$$- 141 + 844 \text{ kJ mol}^{-1} = \underline{4{,}071 \text{ kJ mol}^{-1}}. \quad (1)$$

(B) Equation 7.14 gives

$$\Delta H_c(\text{ZnO}) = \frac{N\mathcal{M}q^2}{4\pi\varepsilon_o d_o}\left[1 - \frac{1}{n}\right] + 2RT$$

$$= \frac{(6.022 \times 10^{23})(1.4985)(2 \times 1.602 \times 10^{-19})^2}{4(3.14159)(8.85419 \times 10^{-12})d_o}\left[1 - \frac{1}{11.67}\right]$$

$$+ 2(8.314)(298.15) \text{ J mol}^{-1}$$

$$= \frac{7.6123 \times 10^{-4}}{d_o} + 4{,}957.6 \text{ J mol}^{-1}, \quad (2)$$

since Table 7.6 shows ZnO(c) to be a hexogonal crystal with $\mathcal{M} = 1.4985$. Equating Eqs. 1 and 2, we obtain

$$\frac{7.6123 \times 10^{-4}}{d_o} + 4{,}957.6 = 4{,}071{,}000. \quad (3)$$

Solving Eq. 3 for d_o, we obtain

$$d_o = \frac{7.6123 \times 10^{-4}}{4{,}071{,}000 - 4{,}957.6} \text{ m} = \underline{1.872 \times 10^{-10} \text{ m}}. \quad (4)$$

7.15 (A) By fitting the low-temperature limiting form of the Debye heat capacity to the measured value at 30 K for Cu given in Problem 7.14, obtain an expression for C_v^m for Cu for temperatures in the range 0 K to 30 K.

(B) Use the expression for C_v^m that you obtained in (A) to compute the absolute entropy of Cu(s) at 30 K.

Solution

(A) The limiting form of the Debye heat capacity is given by Eq. 7.43:

$$\lim_{T \to 0} C_v^m = 1{,}943\left[\frac{T}{\theta}\right]^3 \text{ J mol}^{-1}\text{ K}^{-1}. \quad (1)$$

By equating this expression to the measured heat capacity at $T = 30$ K, we obtain

$$1{,}943\left[\frac{T}{\theta}\right]^3 = 1.693. \tag{2}$$

Solving for θ yields

$$\theta = \left[\frac{1{,}943}{1.693}\right]^{1/3}(30) = 314.1 \text{ K}. \tag{3}$$

Therefore, we have

$$\boxed{C_v^m = 1{,}943\left[\frac{T}{314.1}\right]^3} \quad \text{for } T \leq 30 \text{ K}. \tag{4}$$

(B) The absolute entropy is given by

$$\bar{S}(T) = \int_{T=0\text{ K}}^{T} \frac{C_v^m dT}{T} = \frac{1{,}943}{(314.1)^3}\int_0^T T^2 dT = \frac{1{,}943}{314.1^3}\left[\frac{T^3}{3}\right] \text{J mol}^{-1}\,\text{K}^{-1}. \tag{5}$$

At $T = 30$ K, we have

$$\bar{S}(30\text{ K}) = \frac{1{,}943(30)^3}{3(314.1)^3} = \underline{0.5643 \text{ J mol}^{-1}\,\text{K}^{-1}}. \tag{6}$$

7.17* The measured heat capacities of Ag at four temperatures are as follows:

Temperature (K)	$C_v^m \longrightarrow \text{J mol}^{-1}\text{K}^{-1}\vert$
20	1.647
35	6.612
60	14.27
130	22.07

(A) Obtain an expression for the sum of the squares of the differences between the data given in the table and the results predicted by the Einstein heat capacity equation.
(B) Plot the expression obtained in (A) as a function of the characteristic temperature.
(C) Find the value of the characteristic temperature that minimizes the value of the expression plotted in (B). (*Hint:* See Example 7.5.)

Solution

(A) This problem is essentially the same as Example 7.5. Let $C_v^{\text{expt}}(T)$ represent the experimental data given in the table and $C_v^E(T)$ the Einstein prediction of the heat capacity. The desired sum is

$$\boxed{S = \sum_{i=1}^{i=4}[C_v^{\text{expt}}(T_i) - C_v^E(T_i)]^2 = \sum_{i=1}^{i=4}\left[C_v^{\text{expt}}(T_i) - \frac{3R(\theta/T_i)^2\exp[\theta/T_i]}{\{\exp[\theta/T_i]-1\}^2}\right]^2}, \tag{1}$$

where the summations in Eq. 1 run over the four data points given in the table.

(B) and (C) The requested plot of Eq. 1 accompanies this solution. As can be seen, the best fit of the Einstein formula to the measured data occurs with θ = 153.0 K. At that point, the value of S is 1.525. This corresponds to the a root-mean-square deviation of ± 0.36 J mol^{-1} K^{-1}.

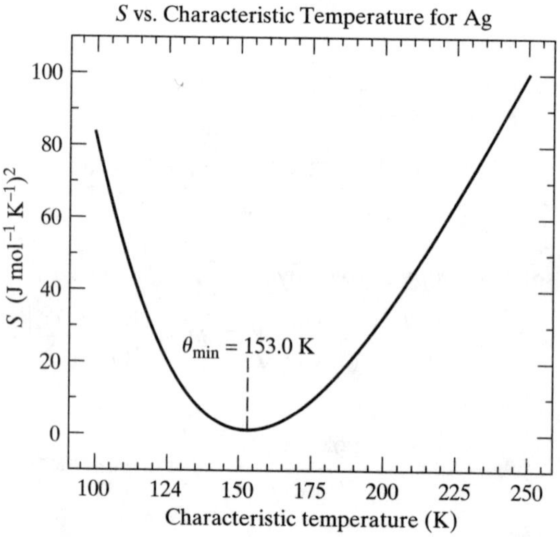

7.19 Equation 7.46 gives the first four terms of the high-temperature limiting form for the Debye heat capacity. By expanding the derivation given in the text, obtain the fifth term of this expansion.

Solution

The first step is to expand the long division of the fraction $x^3/(x + x^2/2! + x^3/3! + \cdots)$ to include terms in x^{10} and x^{11}. Proceeding as outlined in the text, the division produces

$$x^2 - x^3/2 + x^4/12 - x^6/720 + x^8/30{,}240 - x^{10}/1{,}209{,}600 + \cdots$$

$$\sum_{n=1}^{n=\infty} x^n/n! \;\Big)\; x^3$$

$$-(x^3 + x^4/2 + x^5/6 + x^6/24 + x^7/120 + x^8/720 + x^9/5{,}040 + x^{10}/40{,}320 + x^{11}/362{,}880$$

$$-x^4/2 - x^5/6 - x^6/24 - x^7/120 - x^8/720 - x^9/5{,}040 - x^{10}/40{,}320 - x^{11}/362{,}880$$

$$-(-x^4/2 - x^5/4 - x^6/12 - x^7/48 - x^8/240 - x^9/1{,}440 - x^{10}/10{,}080 - x^{11}/80{,}640$$

$$x^5/12 + x^6/24 + x^7/80 + x^8/360 + x^9/2{,}016 + x^{10}/13{,}440 + x^{11}/103{,}680$$

$$-(x^5/12 + x^6/24 + x^7/72 + x^8/288 + x^9/1{,}440 + x^{10}/8{,}640 + x^{11}/60{,}480$$

$$-x^7/720 - x^8/1{,}440 - x^9/5{,}040 - x^{10}/24{,}192 - x^{11}/145{,}152$$

$$-(-x^7/720 - x^8/1{,}440 - x^9/4{,}320 - x^{10}/17{,}280 - x^{11}/86{,}400$$

$$x^9/30{,}240 + x^{10}/60{,}480 + 17x^{11}/3{,}628{,}800$$

$$-(x^9/30{,}240 + x^{10}/60{,}480 + x^{11}/181{,}440$$

$$-x^{11}/1{,}209{,}600$$

Thus, the first six terms in the expansion of $F(x)$ are given by

$$F(x) = \frac{x^3}{e^x - 1} = x^2 - \frac{x^3}{2} + \frac{x^4}{12} - \frac{x^6}{720} + \frac{x^8}{30{,}240} - \frac{x^{10}}{1{,}209{,}600} + \cdots. \quad (1)$$

Substituting Eq. 1 into Eq. 7.37 yields

$$U_{\text{total}} = \frac{9R\theta}{8} + \frac{9RT^4}{\theta^3}\int_{x=0}^{x=\frac{\theta}{T}}\left[x^2 - \frac{x^3}{2} + \frac{x^4}{12} - \frac{x^6}{720} + \frac{x^8}{30{,}240} - \frac{x^{10}}{1{,}209{,}600} + \cdots\right]dx$$

$$= \frac{9R\theta}{8} + \frac{9RT^4}{\theta^3}\left[\frac{\theta^3}{3T^3} - \frac{\theta^4}{8T^4} + \frac{\theta^5}{60T^5} - \frac{\theta^7}{5{,}040T^7} + \frac{\theta^9}{272{,}160T^9} - \frac{\theta^{11}}{13{,}305{,}600T^{11}} + \cdots\right]$$

$$= \frac{9R\theta}{8} + 3R\left[T - \frac{3\theta}{8} + \frac{\theta^2}{20T} - \frac{\theta^4}{1{,}680T^3} + \frac{\theta^6}{92{,}720T^5} - \frac{\theta^8}{4{,}435{,}200T^7} + \cdots\right]. \quad (2)$$

Taking the partial derivative of U_{total} with respect to T yields C_v^m:

$$C_v^m = 3R\left[1 - \frac{\theta^2}{20T^2} + \frac{\theta^4}{560T^4} - \frac{\theta^6}{18{,}544T^6} + \frac{\theta^8}{633{,}600T^8} - \cdots\right]. \quad (3)$$

Thus, the fifth term in the high-temperature expansion of the Debye heat capacity is

$$\boxed{+\frac{\theta^8}{633{,}600T^8}}.$$

7.21 By treating translations and rotations by means of the classical equipartition theorem and vibrations with quantum/statistical theory, compute the gas-phase heat capacities of (A) $I_2(g)$ and (B) $Cl_2(g)$ at 298.15 K. The fundamental vibrational frequencies for $I_2(g)$ and $Cl_2(g)$ are $6.431 \times 10^{12}\,s^{-1}$ and $1.6921 \times 10^{13}\,s^{-1}$, respectively. Use the data given in Appendix A to compute the percent error in your results.

Solution

The diatomic gas-phase heat capacity is given by Eq. 7.50, viz.,

$$C_p^m = 3.5R + \frac{Rx^2e^x}{\{e^x - 1\}^2}, \quad (1)$$

where $x = h\nu/(k_bT)$.

Insertion of the data gives

$$x(I_2) = \frac{(6.62608 \times 10^{-34}\,\text{J s})(6.431 \times 10^{12}\,\text{s}^{-1})}{(1.38066 \times 10^{-23}\,\text{J K}^{-1})(298.15\,\text{K})} = 1.0352 \quad (2)$$

and

$$x(Cl_2) = \frac{(6.62608 \times 10^{-34}\,\text{J s})(1.6921 \times 10^{13}\,\text{s}^{-1})}{(1.38066 \times 10^{-23}\,\text{J K}^{-1})(298.15\,\text{K})} = 2.7238. \quad (3)$$

The corresponding heat capacities are, for I_2,

$$C_p^m(I_2) = 3.5(8.314) + \frac{(8.314)(1.0352)^2e^{1.0352}}{(e^{1.0352} - 1)^2} = 29.099 + 7.6097\,\text{J mol}^{-1}\,\text{K}^{-1}$$

$$= \underline{36.71\,\text{J mol}^{-1}\,\text{K}^{-1}} \quad (4)$$

and, for Cl_2,

$$C_p^m(Cl_2) = 3.5(8.314) + \frac{(8.314)(2.7238)^2 e^{2.7238}}{(e^{2.7238} - 1)^2} = 29.099 + 4.6365\ \mathrm{J\ mol^{-1}\ K^{-1}}$$

$$= \underline{33.74\ \mathrm{J\ mol^{-1}\ K^{-1}}}. \tag{5}$$

Hence,

$$\%\ \text{error for } C_p^m(I_2) = \frac{100 \times (36.71 - 36.90)}{36.90} = \underline{-0.51\%}. \tag{6}$$

and

$$\%\ \text{error for } C_p^m(Cl_2) = \frac{100 \times (33.74 - 33.91)}{33.91} = \underline{-0.50\%}. \tag{7}$$

7.23* Sam is beginning to get the hang of physical chemistry. His grades are improving rapidly, and his creative ideas are starting to move on target. His latest proposal is for a "heat capacity spectrometer." He correctly notes that Raman and IR spectrometers needed to obtain molecular spectra are very expensive. On the other hand, calorimeters needed to measure heat capacities are much less expensive. Sam suggests that we place the compound whose gas-phase spectrum is to be measured in a calorimeter and measure its heat capacity at some convenient temperature. The data are then fed into a dedicated computer, which computes the vibrational frequencies of the molecule.

(A) Obtain a calibration curve at 298.15 K for the vibrational frequency of gas-phase diatomic molecules, which could be used to convert measured heat capacity data into vibration frequencies. That is, plot ν versus C_p^m at $T = 298.15$ K.

(B) Compute the fundamental vibrational frequencies of HCl(g), S_2(g), and I_2(g), from their measured heat capacities given in Appendix A. Comment on Sam's proposed heat capacity spectrometer.

Solution

(A) C_p^m for diatomic molecules is given by Eq. 7.50:

$$C_p^m = 3.5R + \frac{Rx^2e^x}{\{e^x - 1\}^2} \tag{1}$$

with $x = h\nu/(k_bT)$. At $T = 298.15$ K, we have

$$\frac{h}{k_bT} = \frac{(6.62608 \times 10^{-34}\ \mathrm{J\ s})}{(1.38066 \times 10^{-23}\ \mathrm{J\ K^{-1}})(298.15\ \mathrm{K})} = 1.60966 \times 10^{-13}\ \mathrm{s}. \tag{2}$$

Let $C = 1.60966 \times 10^{-13}$ s. In these terms, Eq. 1 can be written in the form

$$\boxed{\frac{\nu^2 e^{C\nu}}{(e^{C\nu} - 1)^2} = \frac{C_p^m - 3.5R}{C^2R}}. \tag{3}$$

The accompanying figures show the right-hand side of Eq. 3, plotted as a function of ν over various frequency ranges. These curves are the requested calibration curves.

The measured heat capacities of HCl(g), S_2(g), and I_2(g) are 29.12 J mol^{-1} K^{-1}, 32.47 J mol^{-1} K^{-1}, and 36.90 J mol^{-1} K^{-1}, respectively. Consequently, the right-hand sides of Eq. 3 for these compounds are $0.00097486 \times 10^{26}\ \mathrm{s^{-2}}$, $0.15649 \times 10^{26}\ \mathrm{s^{-2}}$, and $0.36214 \times 10^{26}\ \mathrm{s^{-2}}$, respectively. We could obtain approximate values for the frequencies simply by reading from the conversion curves. For HCl(g), ν is about $6.6 \times 10^{13}\ \mathrm{s^{-1}}$. For S_2(g), the result is about $2.1 \times 10^{13}\ \mathrm{s^{-1}}$, and for I_2(g), it is about $0.6 \times 10^{13}\ \mathrm{s^{-1}}$.

By conducting a grid search on the left-hand side of Eq. 3, we can obtain much more accurate values of the frequencies. Such searches yield the following results:

$$\nu(\mathrm{HCl(g)}) = 6.664 \times 10^{13}\ \mathrm{s}^{-1}; \tag{4}$$

$$\nu(\mathrm{S_2(g)}) = 2.136 \times 10^{13}\ \mathrm{s}^{-1}; \tag{5}$$

$$\nu(\mathrm{I_2(g)}) = 0.5448 \times 10^{13}\ \mathrm{s}^{-1}. \tag{6}$$

The measured frequencies for these molecules are $\nu(\mathrm{HCl(g)}) = 8.961 \times 10^{13}\ \mathrm{s}^{-1}$, $\nu(\mathrm{S_2(g)}) = 2.175 \times 10^{13}\ \mathrm{s}^{-1}$, and $\nu(\mathrm{I_2(g)}) = 0.6431 \times 10^{13}\ \mathrm{s}^{-1}$. The percent errors are as follows:

$$\%\ \text{error for HCl(g)} = \frac{100(6.664 - 8.961)}{8.961} = -25.63\%; \tag{7}$$

$$\%\ \text{error for } \mathrm{S_2(g)} = \frac{100(2.136 - 2.175)}{2.175} = -1.79\%; \tag{8}$$

$$\%\ \text{error for } \mathrm{I_2(g)} = \frac{100(.5448 - .6431)}{.6431} = -15.28\%. \tag{9}$$

Obviously, the accuracy of the heat-capacity spectrometer leaves a great deal to be desired. The problem is the exponential dependence of the heat capac-

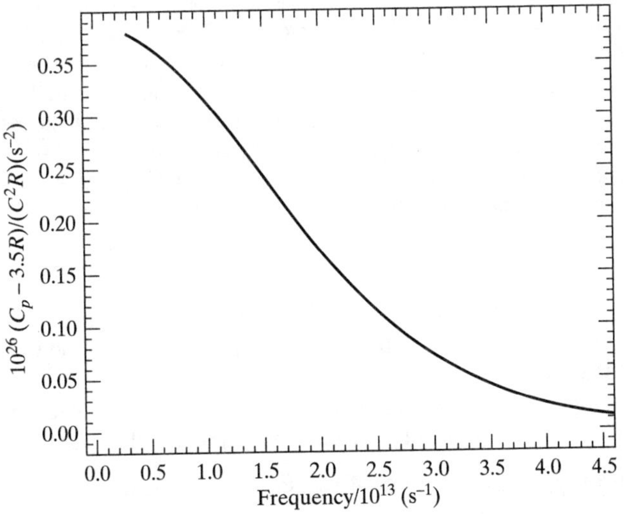

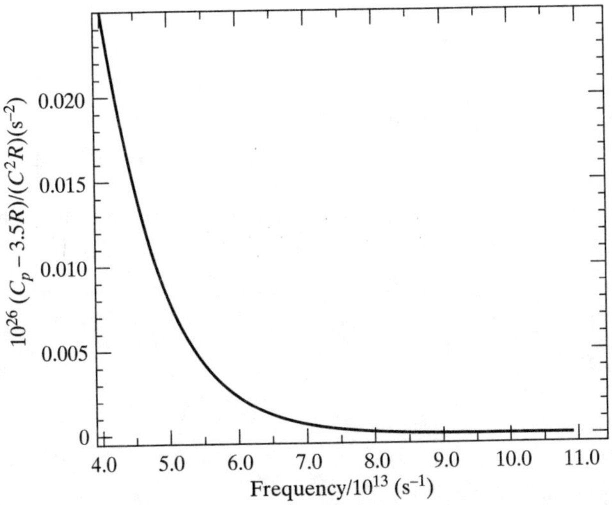

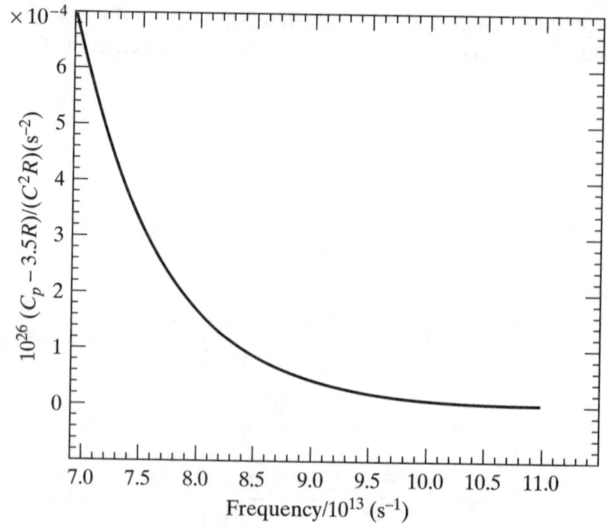

ity upon the frequency. A small error in measuring C_p^m or small deviations of the gas from ideality can produce serious errors in the calculated value of the frequency. For example, if the measured heat capacity of HCl(g) were 29.10 J mol^{-1} K^{-1} rather than 29.12 J mol^{-1} K^{-1}, the calculated HCl(g) frequency would be 8.92×10^{13} s^{-1}, a value that is within -0.46% of the measured frequency. Thus, an error of 0.02 J mol^{-1} K^{-1} in the measured heat capacity produces a 25.2 % error in the frequency in this case. That means we can use measured frequencies to compute accurate heat capacities, but we cannot generally employ measured heat capacities to obtain frequencies. Except for this unfortunate accuracy problem, Sam's idea would be excellent.

7.25 One mole of a metal is heated from a temperature $T = \alpha\theta$ to $T = \beta\theta$, where α and β are constants greater than or equal to unity and θ is the Debye temperature of the metal.

(A) Use the high-temperature limiting form of the Debye heat capacity expression to show that $\Delta\overline{S}$ for the process is the same for all metals.

(B) For $\alpha = 1$ and $\beta = 2$, compute $\Delta\overline{S}$ for the process.

Solution

(A) The high-temperature limiting form of the Debye heat capacity is given by Eq. 7.46:

$$C_v^m = 3R\left[1 - \frac{\theta^2}{20T^2} + \frac{\theta^4}{560T^4} - \frac{\theta^6}{18{,}544T^6} + \cdots\right]. \quad \textbf{(1)}$$

The rate of change of entropy with temperature for constant-volume heating was shown in Chapter 4 to be

$$\left(\frac{\partial S}{\partial T}\right)_V = \frac{C_v}{T} = \frac{nC_v^m}{T}. \quad \textbf{(2)}$$

ΔS for the heating process is therefore given by

$$\Delta\bar{S} = \int d\bar{S} = \int_{\alpha\theta}^{\beta\theta} \frac{nC_v^m dT}{T} = 3nR \int_{\alpha\theta}^{\beta\theta} \left[1 - \frac{\theta^2}{20T^2} + \frac{\theta^4}{560T^4} - \frac{\theta^6}{18{,}544T^6} + \cdots\right] \frac{dT}{T}$$

$$= 3nR\left[\ln\left(\frac{\beta}{\alpha}\right) + \frac{\theta^2}{40T^2}\Bigg|_{\alpha\theta}^{\beta\theta} - \frac{\theta^4}{2{,}240T^4}\Bigg|_{\alpha\theta}^{\beta\theta} + \frac{\theta^6}{111{,}264T^6}\Bigg|_{\alpha\theta}^{\beta\theta} \cdots\right] \quad (3)$$

Equation 4 is independent of θ; therefore, $\Delta\bar{S}$ is the same for all metals.

$$\boxed{\Delta\bar{S} = 3nR\left[\ln\left(\frac{\beta}{\alpha}\right) + \frac{1}{40}\left\{\frac{1}{\beta^2} - \frac{1}{\alpha^2}\right\} - \frac{1}{2{,}240}\left\{\frac{1}{\beta^4} - \frac{1}{\alpha^4}\right\} + \frac{1}{111{,}264}\left\{\frac{1}{\beta^6} - \frac{1}{\alpha^6}\right\} + \cdots\right]}. \quad (4)$$

(B) If $\alpha = 1$ and $\beta = 2$, Eq. 4 becomes

$$\Delta\bar{S} = 3R\left[\ln(2) - \frac{3}{160} + \frac{15}{35{,}840} - \frac{63}{7{,}120{,}896} + \cdots\right]$$

$$\approx 2.024\,R = \underline{16.83\ \mathrm{J\ mol^{-1}\ K^{-1}}}. \quad (5)$$

CHAPTER 8

Thermodynamics of Nonelectrolytic Solutions

8.1 An ideal binary solution is formed from liquids A and B. The pure component vapor pressures at 300 K are $P_A^o = 36.5$ torr and $P_B^o = 19.5$ torr. Determine the composition of the equilibrium vapor for an equimolar mixture of A and B.

Solution

For an equimolar mixture, we have $X_A = X_B = 0.5$. The composition of the vapor is given by Eq. 8.13:

$$Y_A = \frac{X_A P_A^o}{X_A P_A^o + X_B P_B^o}. \tag{1}$$

Substituting data gives

$$Y_A = \frac{0.5(36.5)}{0.5(36.5) + 0.5(19.5)} = \underline{0.652}. \tag{2}$$

The mole fraction of B in the vapor is easily obtained from the fact that $Y_A + Y_B$ must be unity:

$$Y_B = 1 - Y_A = \underline{0.348}. \tag{3}$$

8.3 Consider an ideal binary solution of A and B.

(A) Derive a general equation in terms of P_A^o and P_B^o that gives the mole fraction of A at which $P_A = P_B$. Note that this point corresponds to the crossing point of the two lines representing Raoult's law on a plot of ideal vapor pressure vs. mole fraction.

(B) Show that the total equilibrium vapor pressure over the solution at the point where $P_A = P_B$ is given by

$$P_T = \frac{2P_A^o P_B^o}{P_A^o + P_B^o}.$$

Solution

(A) At the point where $P_A = P_B$, we must have

$$X_A P_A^o = X_B P_B^o = (1 - X_A)P_B^o. \tag{1}$$

Rearranging terms in Eq. 1 produces

$$X_A(P_A^o + P_B^o) = P_B^o. \tag{2}$$

Thus, the mole fraction at which $P_A = P_B$ is

$$\boxed{X_A = \frac{P_B^o}{P_A^o + P_B^o}}. \tag{3}$$

(B) Dalton's law tells us that the total equilibrium vapor pressure will be

$$P_T = P_A + P_B = X_A P_A^o + X_B P_B^o = X_A P_A^o + (1 - X_A)P_B^o. \tag{4}$$

Collecting terms gives

$$P_T = X_A(P_A^o - P_B^o) + P_B^o. \tag{5}$$

At the point that $P_A = P_B$, Eq. 3 gives the value of X_A. Substituting this result into Eq. 5 yields

$$P_T = \frac{P_B^o}{P_A^o + P_B^o}(P_A^o - P_B^o) + P_B^o = \frac{P_A^o P_B^o - (P_B^o)^2}{P_A^o + P_B^o} + P_B^o. \tag{6}$$

Taking a common denominator in the right-hand side of Eq. 6 produces

$$\boxed{P_T = \frac{2P_A^o P_B^o}{P_A^o + P_B^o}}, \tag{7}$$

which is the desired expression.

8.5 Forty grams of $Ba(NO_3)_2$ are dissolved in 1 kilogram of water.

(A) If no ionization of the $Ba(NO_3)_2$ takes place, compute the vapor pressure of the solution at 313.15 K, where the equilibrium vapor pressure of pure water is 55.324 torr. Assume that the solution is ideal.

(B) If the vapor pressure of the solution is found to be 54.909 torr, compute the fraction of the $Ba(NO_3)_2$ that is ionized in the solution.

(C) What would the equilibrium H_2O vapor pressure be if we had 100% ionization of the $Ba(NO_3)_2$?

(D) What percent error is made in computating of the equilibrium vapor pressure using the limiting form given by Eq. 8.11?

Solution

(A) The vapor pressure lowering for an ideal solution is given by Eq. 8.6, viz.,

$$\Delta P_{H_2O} = -X_B P^o_{H_2O}, \tag{1}$$

where X_B is the mole fraction for all solute species combined. If there is no ionization of $Ba(NO_3)_2$, then

$$X_B = \frac{n_1}{n_1 + n_o}, \tag{2}$$

where n_1 is the number of moles of $Ba(NO_3)_2$ and n_o is the number of moles of H_2O. Now,

$$n_o = \frac{1{,}000 \text{ g}}{18.016 \text{ g mol}^{-1}} = 55.506 \text{ mol } H_2O, \tag{3}$$

and

$$n_1 = \frac{40 \text{ g}}{261.34 \text{ g mol}^{-1}} = 0.15306 \text{ mol of } Ba(NO_3)_2. \tag{4}$$

Therefore, we have

$$X_B = \frac{0.15306}{0.15306 + 55.506} = 0.0027500. \tag{5}$$

The vapor pressure lowering is

$$\Delta P_{H_2O} = -0.0027500(55.324) \text{ torr} = -0.15214 \text{ torr}. \tag{6}$$

The corresponding equilibrium vapor pressure is

$$P_{H_2O} = P^o_{H_2O} + \Delta P_{H_2O} = 55.324 - 0.15214 = \underline{55.172} \text{ torr}. \tag{7}$$

(B) The ionization reaction is $Ba(NO_3)_2 \longrightarrow Ba^{2+}_{(aq)} + 2\,NO^-_{3(aq)}$. If a fraction f ionizes, then the number of moles of each species is $Ba^{2+}_{(aq)} = fn_1$, $NO^-_{3\,(aq)} = 2fn_1$, and $Ba(NO_3)_2 = n_1 - fn_1$, where n_1 is given by Eq. 4. The total mole fraction of solute species is, therefore,

$$X_B = \frac{n_1 - fn_1 + fn_1 + 2fn_1}{n_1 - fn_1 + fn_1 + 2fn_1 + n_o} = \frac{n_1 + 2fn_1}{n_1 + 2fn_1 + n_o}. \tag{8}$$

The associated vapor pressure lowering is

$$\Delta P_{H_2O} = -X_B P^o_{H_2O} = -\frac{n_1 + 2fn_1}{n_1 + 2fn_1 + n_o} P^o_{H_2O}. \quad \textbf{(9)}$$

Solving Eq. 9 for f, we obtain

$$-2fn_1[P^o_{H_2O} + \Delta P_{H_2O}] = n_1[\Delta P_{H_2O} + P^o_{H_2O}] + n_o \Delta P_{H_2O}, \quad \textbf{(10)}$$

or

$$f = -\frac{n_o \Delta P_{H_2O}}{2n_1[P^o_{H_2O} + \Delta P_{H_2O}]} - \frac{1}{2}. \quad \textbf{(11)}$$

The data given in the problem show that $\Delta P_{H_2O} = 54.909 - 55.324$ torr $= -0.415$ torr. Substituting this result into Eq. 11 gives

$$f = -\frac{55.506(-0.415)}{2(0.15306)[55.324 - 0.415]} - \frac{1}{2} = 1.370 - 0.50 = \underline{0.870}. \quad \textbf{(12)}$$

(C) If we have 100% ionization, we would have $f = 1$ in Eq. 9. Setting $f = 1$, we obtain

$$\Delta P_{H_2O} = -\frac{3n_1}{3n_1 + n_o} P^o_{H_2O} = -\frac{3(.15306)}{3(0.15306) + 55.506}(55.324) = -0.4539 \text{ torr}. \quad \textbf{(13)}$$

In this situation, the vapor pressure would be

$$P_{H_2O} = P^o_{H_2O} + \Delta P_{H_2O} = 55.342 - 0.4539 = \underline{54.888} \text{ torr}. \quad \textbf{(14)}$$

(D) In the dilute solution limit, the vapor pressure lowering for complete ionization is

$$\Delta P_{H_2O} = -iX_B P^o_{H_2O}, \quad \textbf{(15)}$$

where X_B is given by Eq. 5. For the ionization of $Ba(NO_3)_2$, $i = 3$, so that

$$\Delta P_{H_2O} = -3(0.0027500)(55.324) \text{ torr} = -0.4564 \text{ torr}, \quad \textbf{(16)}$$

which gives

$$P_{H_2O} = P^o_{H_2O} + \Delta P_{H_2O} = 55.324 - 0.4564 = 54.868 \text{ torr}. \quad \textbf{(17)}$$

The percent error in the vapor pressure lowering is

$$\% \text{ error in } \Delta P_{H_2O} = \frac{100 \times (0.4564 - 0.4539)}{0.4539} = \underline{0.55\%}. \quad \textbf{(18)}$$

The percent error in the vapor pressure itself is

$$\% \text{ error in } P_{H_2O} = \frac{100 \times (54.868 - 54.888)}{54.888} = \underline{-0.036\%}. \quad \textbf{(19)}$$

8.7* Chloroform ($CHCl_3$) and carbon tetrachloride (CCl_4) have similar structures, and their boiling points are within 16 K of one another. We might, therefore, expect that a mixture of the two compounds will be nearly ideal. Chloroform boils at 334.41 K and has a partial molar enthalpy of vaporization of 29,469 J mol^{-1}. Carbon tetrachloride boils at 349.95 K and has a partial molar enthalpy of vaporization of 29,863 J mol^{-1}. Compute the boiling points of mixtures of these two compounds with mole fractions of CCl_4 from $X_{CCl_4} = 0$ to $X_{CCl_4} = 1$ in increments of 0.1. Assume that the applied pressure is 760 torr. Plot your results as a function of X_{CCl_4}, and connect the data points by fitting them to a quadratic function of X_{CCl_4} using a least-squares method. Assume that the solution is ideal, that the molar enthalpies of vaporization are constant, and that the volume of the liquid is negligible relative to that of its equilibrium vapor.

Solution

With the preceding assumptions, the Clausius–Clapeyron equation holds, and the pure-component equilibrium vapor pressures are given by

$$P^o_{CCl_4} = 760 \exp\left[-\frac{\Delta\overline{H}^{CCl_4}_{vap}}{R}\left\{\frac{1}{T} - \frac{1}{T_b}\right\}\right] \text{ torr,} \quad \textbf{(1)}$$

(where T_b is the normal boiling point) and

$$P^o_{CHCl_3} = 760 \exp\left[-\frac{\Delta\overline{H}^{CHCl_3}_{vap}}{R}\left\{\frac{1}{T} - \frac{1}{T_b}\right\}\right] \text{ torr.} \quad \textbf{(2)}$$

The solution boils when

$$P_{Total} = P_o = \text{applied pressure} = X_{CCl_4}P^o_{CCl_4} + X_{CHCl_3}P^o_{CHCl_3}$$
$$= X_{CCl_4}P^o_{CCl_4} + (1 - X_{CCl_4})P^o_{CHCl_3} = X_{CCl_4}(P^o_{CCl_4} - P^o_{CHCl_3}) + P^o_{CHCl_3}. \quad \textbf{(3)}$$

Substituting Eqs. 1 and 2 into 3 yields

$$\frac{P_o}{760} = X_{CCl_4}\left[\exp\left[-\frac{\Delta\overline{H}^{CCl_4}_{vap}}{R}\left\{\frac{1}{T} - \frac{1}{T_b}\right\}\right] - \exp\left[-\frac{\Delta\overline{H}^{CHCl_3}_{vap}}{R}\left\{\frac{1}{T} - \frac{1}{T_b}\right\}\right]\right]$$
$$+ \exp\left[-\frac{\Delta\overline{H}^{CHCl_3}_{vap}}{R}\left\{\frac{1}{T} - \frac{1}{T_b}\right\}\right]. \quad \textbf{(4)}$$

Substituting the relevant data gives

$$X_{CCl_4}\left[\exp\left[-\frac{29{,}863}{8.314}\left\{\frac{1}{T} - \frac{1}{349.95}\right\}\right] - \exp\left[-\frac{29{,}469}{8.314}\left\{\frac{1}{T} - \frac{1}{334.41}\right\}\right]\right]$$
$$+ \exp\left[-\frac{29{,}469}{8.314}\left\{\frac{1}{T} - \frac{1}{334.41}\right\}\right] - \frac{P_o}{760} = 0. \quad \textbf{(5)}$$

X_{CCl_4}	T_b (K)
0.00	334.41
0.10	335.63
0.20	336.91
0.30	338.24
0.40	339.67
0.50	341.14
0.60	342.70
0.70	344.36
0.80	346.10
0.90	347.97
1.00	349.95

We now seek solutions to Eq. 5 with $P_o = 760$ torr for various values of X_{CCl_4}. These solutions can easily be located using a one-dimensional grid search. The results are in the table at the left.

These data points are plotted on the graph below. The solid line is a quadratic least-squares fit of the data. Its equation is $T_b = 334.453 + 11.298\,X_{CCl_4} + 4.1562\,X^2_{CCl_4}$, where all the fitting coefficients are in kelvins.

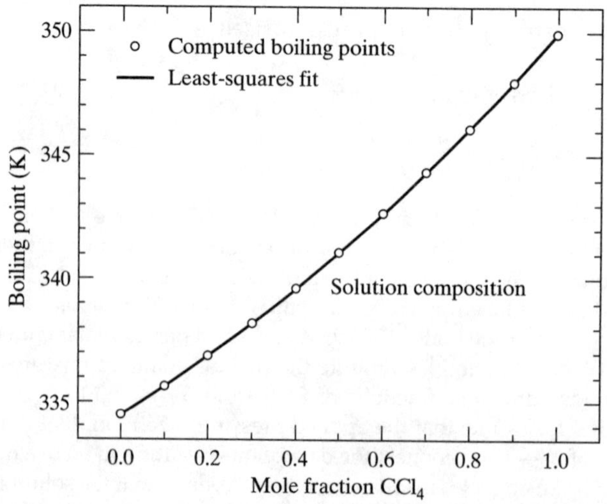

8.9 The calculations of the solution boiling point as a function of X_{CCl_4} in Problem 8.7 show that the data are accurately fitted by the equation

$$T_b = 334.453 + 11.298X_{CCl_4} + 4.1562X^2_{CCl_4}(K),$$

where X_{CCl_4} is the mole fraction of CCl_4 in the CCl_4–$CHCl_3$ solution. The computations carried out in Problem 8.8 show that the boiling points as a function of the mole fraction of CCl_4 in the vapor phase, Y_{CCl_4}, are accurately represented by the equation

$$T_b = 334.44 + 18.634Y_{CCl_4} - 3.1590Y^2_{CCl_4}\ (K).$$

Using these results, determine the composition of the distillate of a CCl_4–$CHCl_3$ solution that originally had $X_{CCl_4} = X_{CHCl_3} = 0.50$ if the fractionating column has two theoretical plates. At what temperature does this solution boil? (*Note:* Least-squares fitting is extremely sensitive to round-off errors. Consequently, the fits obtained will vary slightly, depending upon the computer software employed).

Solution

By equating the two expressions for the boiling point of the solution, we can obtain a simple expression relating the composition of the equilibrium vapor to that of the solution:

$$334.44 + 18.634Y_{CCl_4} - 3.1590Y^2_{CCl_4} = 334.453 + 11.298X_{CCl_4} + 4.1562X^2_{CCl_4}. \quad (1)$$

Rearranging Eq. 1 produces

$$3.159Y^2_{CCl_4} - 18.634Y_{CCl_4} + [0.013 + 11.298X_{CCl_4} + 4.1562X^2_{CCl_4}] = 0. \quad (2)$$

Solving this quadratic for Y_{CCl_4} gives

$$Y_{CCl_4} = \frac{18.634 - \{(18.634)^2 - 4(3.159)(11.298X_{CCl_4} + 4.1562X^2_{CCl_4} + 0.013)\}^{1/2}}{(2)(3.159)}. \quad (3)$$

Substituting $X_{CCl_4} = 0.50$ into Eq. 3 yields $Y_{CCl_4} = 0.3847$ after the first theoretical plate. This vapor is now condensed to a new liquid solution, and a second vaporization–condensation cycle is executed. In the second cycle, $X_{CCl_4} = 0.3847$, since the new solution is that resulting from condensing the vapor in the first step. Inserting this value of X_{CCl_4} into Eq. 3 gives $Y_{CCl_4} = 0.2803$. The composition of the distillate after two theoretical plates of fractionation is, therefore, $X_{CCl_4} = \underline{0.2803}$ and $X_{CHCl_3} = \underline{0.7197}$.

The boiling point of the final solution may be computed using

$$T_b = 334.453 + 11.298X_{CCl_4} + 4.1562X^2_{CCl_4}$$
$$= 334.453 + 11.298(0.2803) + 4.1562(0.2803)^2 = \underline{337.95\ K}. \quad (4)$$

8.11* (A) Compute the composition of the equilibrium vapor over solutions of CCl_4 and $CHCl_3$ at each of the boiling points determined in Problem 8.10. That is, determine the mole fraction of CCl_4 in the vapor, Y_{CCl_4}, at each boiling point. These boiling temperatures are given in the table at the right. Assume that the solution is ideal, that the partial molar enthalpies of vaporization given in Problem 8.10 are constant, and that the volume of the liquid is negligible relative to that of its equilibrium vapor. The normal boiling points are given in Problem 8.10.
(B) Plot the boiling point vs. X_{CCl_4} and Y_{CCl_4} on the same graph. Fit the new data to a quadratic function of Y_{CCl_4}.

X_{CCl_4}	T_b (K)
0.00	280.71
0.10	281.61
0.20	282.55
0.30	283.53
0.40	284.57
0.50	285.67
0.60	286.82
0.70	288.04
0.80	289.34
0.90	290.72
1.00	292.20

Solution

The mole fraction of CCl_4 in the equilibrium vapor is given by Eq. 8.13, viz.,

$$Y_{CCl_4} = \frac{X_{CCl_4}P^o_{CCl_4}}{X_{CCl_4}P^o_{CCl_4} + X_{CHCl_3}P^o_{CHCl_3}} = \frac{X_{CCl_4}P^o_{CCl_4}}{100}, \quad (1)$$

where the denominator equals 100 torr, since we are at the boiling point with $P_o = 100$ torr. The pure-component vapor pressure of CCl_4 is given by the Clausius–Clapeyron equation:

$$P^o_{CCl_4} = = 760 \exp\left[-\frac{\Delta \overline{H}^{CCl_4}_{vap}}{R}\left\{\frac{1}{T} - \frac{1}{T_b}\right\}\right] = 760 \exp\left[-\frac{29{,}863}{8.314}\left\{\frac{1}{T} - \frac{1}{349.95}\right\}\right]. \quad \textbf{(2)}$$

Substituting Eq. 2 into Eq. 1 and then evaluating at the points given in the table to the left.

X_{CCl_4}	Y_{CCl_4}	T_b (K)
0.00	0.00000	280.71
0.10	0.06044	281.61
0.20	0.12592	282.55
0.30	0.19706	283.53
0.40	0.28757	284.57
0.50	0.37736	285.67
0.60	0.47624	286.82
0.70	0.58589	288.04
0.80	0.70817	289.34
0.90	0.84505	290.72
1.00	1.00000	292.20

(B) The plots of T_b vs. X_{CCl_4} and T_b vs. Y_{CCl_4} are shown in the accompanying graph. The computed points are accurately fitted by quadratic functions whose equations are

$$T_b = 280.75 + 8.2521\,X_{CCl_4} + 3.1597\,X^2_{CCl_4}\ (\text{K})$$

and

$$T_b = 280.788 + 13.9031\,Y_{CCl_4} - 2.5248\,Y^2_{CCl_4}\ (\text{K}).$$

Note that least-squares fitting is very sensitive to rounding errors. Consequently, the fits obtained will vary slightly, depending upon the computer and the computer software employed to execute the fit.

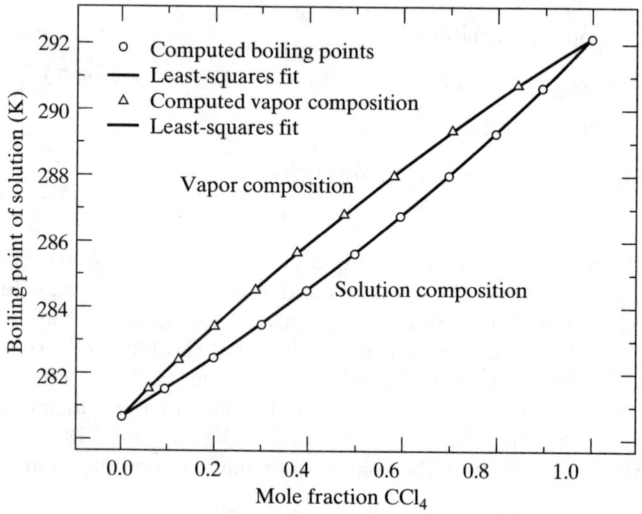

8.13 Seventy-five grams of CCl_4 are mixed with 10 grams of $CHCl_3$ at 298 K to form a solution. If the solution is ideal, compute ΔG_{mix}, ΔS_{mix}, ΔH_{mix}, ΔU_{mix}, ΔV_{mix}, and ΔA_{mix} for the process.

Solution

The discussion in the text shows that, for an ideal solution, we have

$$\Delta V_{mix} = \Delta H_{mix} = \Delta U_{mix} = 0. \quad \textbf{(1)}$$

The Gibbs free energy of mixing is given by Eq. 8.28:

$$\Delta G_{mix} = RT \sum_{i=1}^{i=K} n_i \ln X_i = RT[n_{CCl_4} \ln X_{CCl_4} + n_{CHCl_3} \ln X_{CHCl_3}]. \quad \textbf{(2)}$$

The number of moles of each compound is

$$n_{CCl_4} = \frac{75\text{ g}}{153.82\text{ g mol}^{-1}} = 0.4876\text{ mol } CCl_4 \quad \textbf{(3)}$$

and

$$n_{CHCl_3} = \frac{10\text{ g}}{119.38\text{ g mol}^{-1}} = 0.08377\text{ mol CHCl}_3. \qquad (4)$$

The mole fractions are

$$X_{CCl_4} = \frac{0.4876}{0.4876 + 0.08377} = 0.8534 \qquad (5)$$

and

$$X_{CHCl_3} = 1 - 0.8534 = 0.1466. \qquad (6)$$

The corresponding Gibbs free energy of mixing is

$$\begin{aligned}\Delta G_{mix} &= 8.314\text{ J mol}^{-1}\text{ K}^{-1}(298\text{ K})\,[0.4876\ln(.8534) + 0.08377\ln(0.1466)]\text{ mol} \\ &= \underline{-590.01\text{ J}}.\end{aligned} \qquad (7)$$

The entropy of mixing is given by Eq. 8.31:

$$\Delta S_{mix} = -R\sum_{i=1}^{i=K} n_i \ln X_i = -\frac{\Delta G_{mix}}{T} = \frac{590.01}{298}\text{ J K}^{-1} = \underline{1.980\text{ J K}^{-1}}. \qquad (8)$$

As Eq. 8.39 demonstrates, for an ideal solution, we have

$$\Delta A_{mix} = \Delta G_{mix} = \underline{-590.01\text{ J}}. \qquad (9)$$

8.15 An ideal solution of compounds A and B is to be formed by mixing a total of n_o moles of material, divided such that we have n_A and n_B moles of A and B, respectively. Prove that the entropy of mixing is a maximum if

$$f = \text{fraction of moles of type } A = \frac{n_A}{n_A + n_B} = 0.5.$$

That is, prove that, for a fixed number of total moles, n_o, the entropy of mixing is a maximum if A and B are mixed in equal molar amounts.

Solution

The entropy of mixing for an ideal solution is given by Eq. 8.31:

$$\Delta S_{mix} = -R\sum_{i=1}^{i=K} n_i \ln X_i = -R\,[n_A \ln X_A + n_B \ln X_B]. \qquad (1)$$

Let the total number of moles be n_o. That is, $n_o = n_A + n_B$. Let f be the fraction of n_o that are moles of A. We then have

$$n_A = f n_o \qquad (2)$$

and

$$n_B = (1 - f)n_o. \qquad (3)$$

With these substitutions, the mole fractions are

$$X_A = \frac{f n_o}{n_o} = f \text{ and } X_B = 1 - f. \qquad (4)$$

Substituting into Eq. 1 gives

$$\begin{aligned}\Delta S_{mix} &= -R[f n_o \ln(f) + (1 - f)n_o \ln(1 - f)] \\ &= -R n_o[f\ln(f) + \ln(1 - f) - f\ln(1 - f)].\end{aligned} \qquad (5)$$

The condition for an extremum in $\Delta S_{\rm mix}$ is $\partial \Delta S_{mix}/\partial f = 0$. The derivative of $\Delta S_{\rm mix}$ with respect to f is given by

$$\frac{\partial \Delta S_{\rm mix}}{\partial f} = -Rn_o\left[\ln(f) + \frac{f}{f} - \frac{1}{1-f} - \ln(1-f) + \frac{f}{1-f}\right]$$

$$= -Rn_o\left[\ln(f) - \ln(1-f) + 1 + \frac{f-1}{1-f}\right]$$

$$= -Rn_o[\ln(f) - \ln(1-f) + 1 - 1] = -Rn_o[\ln(f) - \ln(1-f)], \quad \textbf{(6)}$$

since $(f-1)/(1-f) = -1$. For the right-hand side of Eq. 6 to be equal to zero, we must have

$$\ln(f) - \ln(1-f) = 0. \quad \textbf{(7)}$$

This requires that $\ln(f) = \ln(1-f)$, which can be true only if we have

$$f = 1 - f \quad \textbf{(8)}$$

which implies that $f = 0.5$. Thus, an extremum in $\Delta S_{\rm mix}$ is obtained when we mix equal molar quantities of A and B if n_o is fixed. The extremum can easily be shown to be a maximum by examining the sign of the second derivative. Taking the derivative of Eq. 6, we get

$$\frac{\partial^2 \Delta S_{\rm mix}}{\partial f^2} = -Rn_o\left[\frac{1}{f} + \frac{1}{1-f}\right]. \quad \textbf{(9)}$$

At the point $f = 0.5$, we have $\partial^2 \Delta S_{\rm mix}/\partial f^2 = -4Rn_o$, which is negative. The negative value of the second derivative shows the extremum to be a maximum.

8.17 Gaseous A forms an ideal solution with liquid solvent B. At 350 K, a saturated solution is found to have a mole fraction of A of 0.3017 when the pressure of A above the solution is 760 torr. At a temperature of 200 K, gaseous A has condensed to a liquid that is found to have an equilibrium vapor pressure of 11.466 torr.

(A) Compute the partial molar enthalpy of vaporization of A. Assume that the vapor is ideal and that the heat capacity of the liquid and gaseous A are the same at all temperatures.

(B) Compute the normal boiling point of compound A. The same assumptions made in (A) may be made here.

Solution

(A) The gaseous solubility is given by Eq. 8.60, viz.,

$$X_A^{\rm sat} = \exp\left[\frac{\Delta \overline{H}_{\rm vap}}{R}\left\{\frac{1}{T} - \frac{1}{T_b}\right\}\right], \quad \textbf{(1)}$$

where T_b is the normal boiling point of A if the applied pressure is 760 torr. The equilibrium vapor pressure over liquid A is given by the Clausius–Clapeyron Equation:

$$P_A = 760 \exp\left[-\frac{\Delta \overline{H}_{\rm vap}}{R}\left\{\frac{1}{T} - \frac{1}{T_b}\right\}\right]. \quad \textbf{(2)}$$

Both Eqs. 1 and 2 require that $\Delta \overline{H}_{\rm vap}$ be constant. This will be the case if the heat capacities of liquid and gaseous A are identical at all temperatures, since we will then have $\Delta C_p = 0$ at all T. We have two unknowns, $\Delta \overline{H}_{\rm vap}$ and T_b, and two equations. Therefore, the problem should be straightforward. Let us first solve Eq. 2 for T_b:

$$\frac{1}{T_b} = \frac{1}{T} + \frac{R\ln(P_A/760)}{\Delta \overline{H}_{\rm vap}} = \frac{\Delta \overline{H}_{\rm vap} + RT\ln(P_A/760)}{T\Delta \overline{H}_{\rm vap}}. \quad \textbf{(3)}$$

Taking reciprocals gives T_b:

$$T_b = \frac{T\Delta\overline{H}_{vap}}{\Delta\overline{H}_{vap} + RT\ln(P_A/760)} = \frac{200\Delta\overline{H}_{vap}}{\Delta\overline{H}_{vap} + (8.314)(200)\ln(11.466/760)}$$

$$= \frac{200\Delta\overline{H}_{vap}}{\Delta\overline{H}_{vap} - 6{,}973.7}. \qquad (4)$$

Substituting the result obtained in Eq. 4 into Eq. 1 yields

$$X_A^{sat} = \exp\left[\frac{\Delta\overline{H}_{vap}}{R}\left\{\frac{1}{T} - \frac{\Delta\overline{H}_{vap} - 6{,}973.7}{200\Delta\overline{H}_{vap}}\right\}\right]. \qquad (5)$$

Taking logarithms of both sides and then rearranging terms in the resulting equation produces

$$\Delta\overline{H}_{vap}\left[\frac{1}{RT} - \frac{1}{200R}\right] = \ln X_A^{sat} - \frac{6{,}973.7}{200R}. \qquad (6)$$

Solving for $\Delta\overline{H}_{vap}$, we obtain

$$\Delta\overline{H}_{vap} = \frac{\left[\ln X_A^{sat} - \frac{6{,}973.7}{200R}\right]}{\left[\frac{1}{RT} - \frac{1}{200R}\right]}. \qquad (7)$$

Substituting the data given in the problem produces

$$\Delta\overline{H}_{vap} = (8.314\text{ J mol}^{-1}\text{ K}^{-1})\frac{\left[\ln(0.3017) - \frac{6{,}973.7}{200(8.314)}\right]}{\left[\frac{1}{350} - \frac{1}{200}\right]\text{K}^{-1}} = \underline{2.092 \times 10^4\text{ J mol}^{-1}}. \qquad (8)$$

(B) We may now use Eq. 4 to obtain

$$T_b = \frac{200(2.092 \times 10^4)}{(2.092 \times 10^4 - 6{,}973.7)} = \underline{300.0\text{ K}}. \qquad (9)$$

8.19 The following table gives solubility data for $CO_2(g)$ in H_2O as a function of temperature when the partial pressure of $CO_2(g)$ over the solution is 760 torr:

T (K)	Solubility (g of CO_2 per 100 g of H_2O)
273.15	0.3346
283.15	0.2318
288.15	0.1970
293.15	0.1688
298.15	0.1449
303.15	0.1257
313.15	0.0973
323.15	0.0761
333.15	0.0576

(A) Plot $\ln X_{CO_2}^{sat}$ vs. T^{-1}. Does the degree of linearity indicate that the solution is ideal?

(B) Assuming the solution to be ideal, what is the predicted partial molar enthalpy of vaporization of CO_2?

(C) *The Handbook of Chemistry and Physics* lists the partial molar enthalpy of fusion of CO_2 as 9,020 J mol^{-1}. It also lists the sublimation pressure of CO_2 (s) as 227.1 kPa and 518.0 kPa at 205 K and 216.58 K, respectively. Use these data to obtain $\Delta \overline{H}_{vap}$ for CO_2, assuming that $\Delta \overline{H}_{vap}$ and $\Delta \overline{H}_{fus}$ are independent of temperature. Using this result, compute the percent error for the result obtained in (B), where we assumed a solution of CO_2 and water to be ideal. Does this result suggest near ideality?

(D) Use the results of (A), (B), and (C) to obtain an estimate for the enthalpy of mixing of 1 mole of liquid CO_2 with sufficient $H_2O(l)$ to form a saturated solution.

Solution

(A) If the solution is ideal, the mole fraction of CO_2 at saturation is given by Eq. 8.60 which may be written in logarithmic form as

$$\ln X_{CO_2}^{sat} = \frac{\Delta \overline{H}_{vap}}{RT} - \frac{\Delta \overline{H}_{vap}}{RT_b}. \tag{1}$$

This equation tells us that a plot of $\ln X_{CO_2}^{sat}$ vs. T^{-1} should be linear. $X_{CO_2}^{sat}$ may be computed from the given solubility data using

$$X_{CO_2}^{sat} = \frac{G/44.01}{G/44.01 + 100/18.016}, \tag{2}$$

where G is the solubility of CO_2 in grams per 100 g of H_2O. The data needed for such a plot are given in the following table:

T (K)	1/T	G (g per 100 g of H_2O)	$\ln X_{CO_2}^{sat}$
273.15	0.003661	0.3346	−6.5945
283.15	0.003532	0.2318	−6.9611
288.15	0.003470	0.1970	−7.1237
293.15	0.003411	0.1688	−7.2781
298.15	0.003354	0.1449	−7.4306
303.15	0.003299	0.1257	−7.5727
313.15	0.003193	0.0973	−7.8287
323.15	0.003096	0.0761	−8.0743
333.15	0.003002	0.0576	−8.3528

The plot of $\ln X_{CO_2}^{sat}$ vs. T^{-1} appears at the end of the problem. The linearity of the plot is reasonably good. The solid line is a least-squares fit of a straight line. Its equation is

$$\ln X_{CO_2(g)}^{sat} = -16.211 + 2{,}621.2\frac{(\mathrm{K})}{T}. \tag{3}$$

(B) If the solution were ideal, the partial-molar enthalpy of vaporization would be

$$\Delta \overline{H}_{vap} = (\text{slope})R = 2{,}621.2\ \mathrm{K}(8.314\ \mathrm{J\ mol^{-1}\ K^{-1}}) = \underline{2.179 \times 10^4\ \mathrm{J\ mol^{-1}}}. \tag{4}$$

(C) The molar enthalpy of sublimation of $CO_2(s)$ may be computed using the Clausius–Clapeyron Equation:

$$\ln\left[\frac{P_1^{eq}}{P_2^{eq}}\right] = -\frac{\Delta \overline{H}_{sub}}{R}\left[\frac{1}{T_1} - \frac{1}{T_2}\right]. \quad \textbf{(5)}$$

Solving for $\Delta \overline{H}_{sub}$, we obtain

$$\Delta \overline{H}_{sub} = -\frac{R\ln\left[\frac{P_1^{eq}}{P_2^{eq}}\right]}{\left[\frac{1}{T_1} - \frac{1}{T_2}\right]} = -\frac{8.314\ln\left(\frac{518.0}{227.1}\right)}{\left[\frac{1}{216.58} - \frac{1}{205.0}\right]} = 2.629 \times 10^4 \text{ J mol}^{-1}. \quad \textbf{(6)}$$

If $\Delta \overline{H}_{sub}$ and $\Delta \overline{H}_{fus}$ are independent of temperature, we will have

$$\Delta \overline{H}_{vap} = \Delta \overline{H}_{sub} - \Delta \overline{H}_{fus} = 2.629 \times 10^4 - 9{,}020 \text{ J mol}^{-1} = \underline{1.727 \times 10^4 \text{ J mol}^{-1}}. \quad \textbf{(7)}$$

The percent error in $\Delta \overline{H}_{vap}$ resulting from the assumption of an ideal solution is, therefore,

$$\% \text{ error} = \frac{100 \times (2.179 - 1.727)10^4}{1.727 \times 10^4} = \underline{26.2\%}, \quad \textbf{(8)}$$

and the solution exhibits substantial deviation from ideality.

(D) The discussion in the text indicates that the difference between the result obtained in Eq. 4 and the partial-molar enthalpy of vaporization calculated in Eq. 7 is associated primarily with a nonzero enthalpy of mixing of $CO_2(l)$ and $H_2O(l)$. In this case, the difference is

$$(\Delta \overline{H}_{vap})_{slope} - (\Delta \overline{H}_{vap})_{expt} \approx -\Delta \overline{H}_{mix} = 2.179 \times 10^4 - 1.727 \times 10^4$$
$$= \underline{4.52 \times 10^3 \text{ J mol}^{-1}}. \quad \textbf{(9)}$$

Consequently, we estimate that mixing 1 mole of liquid CO_2 with sufficient water to form a saturated solution will result in an enthalpy change of about -4.5 kJ.

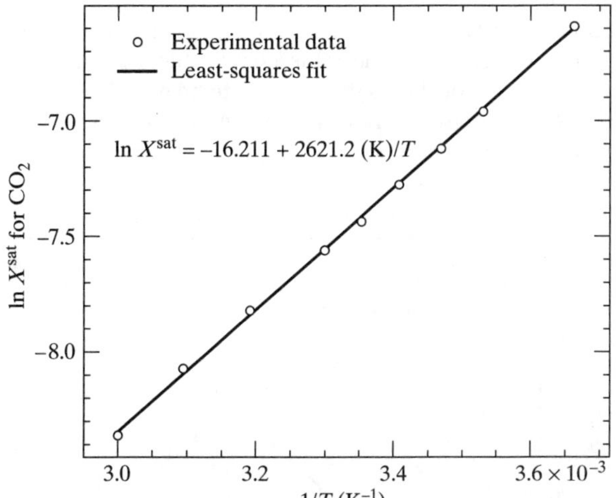

8.21 Compound A with a normal melting point of 400 K and a partial molar enthalpy of fusion of 3,500 cal mol^{-1}, forms ideal solutions with solvent B. It is found that, at $T = 300$ K, dissolving 150 g of A in 10 moles of B forms a saturated solution. Compute the molar mass of compound A.

Solution

If the solution is ideal, Eq. 8.52 gives the saturation mole fraction of the solute:

$$X_A^{\text{sat}} = \exp\left[-\frac{\Delta \overline{H}_{\text{fus}}}{R}\left\{\frac{1}{T} - \frac{1}{T_m}\right\}\right]$$

$$= \exp\left[-\frac{3{,}500 \text{ cal mol}^{-1}}{1.987 \text{ cal mol}^{-1}\text{ K}^{-1}}\left\{\frac{1}{300 \text{ K}} - \frac{1}{400 \text{ K}}\right\}\right] = 0.2304. \quad \textbf{(1)}$$

The mole fraction of A is given by

$$X_A^{\text{sat}} = \frac{150/M}{150/M + 10}, \quad \textbf{(2)}$$

where M is the molar mass of compound A in g mol^{-1}. Solving for M in Eq. 2, we obtain

$$\frac{150\,X_A^{\text{sat}}}{M} + 10\,X_A^{\text{sat}} = \frac{150}{M}. \quad \textbf{(3)}$$

Rearranging terms yields

$$\frac{150(1 - X_A^{\text{sat}})}{M} = 10\,X_A^{\text{sat}}, \quad \textbf{(4)}$$

so that

$$M = \frac{150(1 - X_A^{\text{sat}})}{10\,X_A^{\text{sat}}} = \frac{150(1 - 0.2304)}{10\,(0.2304)} = \underline{50.1 \text{ g mol}-1}. \quad \textbf{(5)}$$

X_B	$P_{B(\text{soln})}$ (torr)
0.000	0.000
0.050	1.000
0.100	2.200
0.150	3.500
0.200	5.000
0.250	7.000
0.300	8.500
0.350	10.000
0.400	13.000
0.450	15.200
0.500	18.600
0.550	21.800
0.600	25.500
0.650	28.900
0.700	33.300
0.750	36.900
0.800	40.600
0.850	43.300
0.900	46.100
0.950	49.240
1.000	51.858

8.23 Solute A is dissolved in solvent B at 320 K. The table at the left gives the measured equilibrium partial pressures of B over the solution as a function of its mole fraction in the solution, X_B. Is the solution ideal? Plot $P_{B(\text{soln})}$ vs. X_B at 320 K, and on the same graph plot the ideal-solution result. What can we conclude about the nature of the A–B intermolecular forces relative to the pure component intermolecular forces?

Solution

The accompanying figure shows a plot of $P_{B(\text{soln})}$ vs. X_B and the corresponding ideal-solution result where the table shows that $P_B^o = 51.858$ torr at 320 K. Examination of the plot shows the solution to be nonideal. The negative deviations from Raoult's law suggest that the A–B interactions are stronger than those between the pure components.

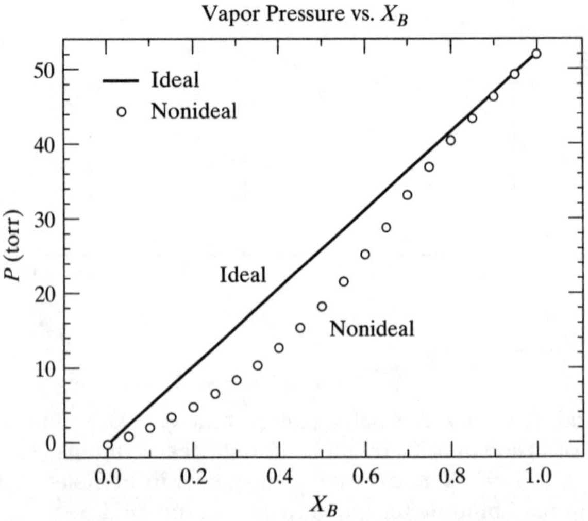

8.25# The equilibrium partial vapor pressures of compound B at 320 K above a binary solution of compound A in solvent B are measured and found to be accurately described by the analytical function

$$P_{B(\text{soln})} = P_B^o[0.442067X_B - 0.311495X_B^2 + 2.60907X_B^3 - 1.74685X_B^4] \text{ torr},$$

where P_B^o is the pure component equilibrium vapor pressure at 320 K, which has been measured and found to be 51.858 torr. If the mole fraction of B is taken as the reference function and $X_B \to 1$ is taken as the reference state, determine the activity and the activity coefficient of B in the A–B solution at 320 K as a function of X_B. Prepare a plot of γ_B vs. X_B for this system. Based on the appearance of the plot, comment on the accuracy of the analytical fit .

Solution

Equation 8.117 shows that, with the stated choices of reference function and reference state, the activity of component B is given by

$$a_B = \frac{P_{B(\text{soln})}}{P_B^o}. \quad \textbf{(1)}$$

Using the analytical fit to the data, we have

$$\boxed{a_B = [0.442067X_B - 0.311495X_B^2 + 2.60907X_B^3 - 1.74685X_B^4]}. \quad \textbf{(2)}$$

The activity coefficient is defined by Eq. 8.113:

$$\gamma_B = \frac{a_B}{X_B}. \quad \textbf{(3)}$$

Substituting Eq. 2 into Eq. 3 yields

$$\boxed{\gamma_B = [0.442067 - 0.311495X_B + 2.60907X_B^2 - 1.74685X_B^3]}. \quad \textbf{(4)}$$

γ_B is shown in the accompanying plot as a function of X_B at 320 K. The analytical fit is very good for most values of X_B. As $X_B \longrightarrow 1$, however, we see that γ_B tends to fall off from its expected value of unity. This behavior reflects inaccuracy of the analytical fit to the data.

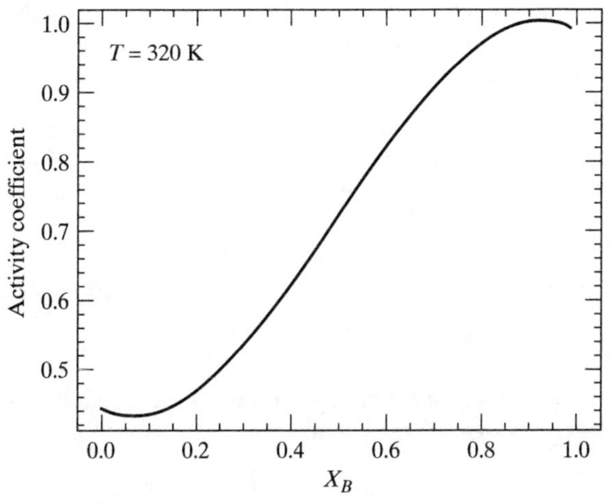

8.27 The equilibrium partial vapor pressures of compound B at 320 K above a binary solution of compound A in solvent B are measured and found to be accurately described by the analytical function

$$P_{B(\text{soln})} = P_B^o[0.442067X_B - 0.311495X_B^2 + 2.60907X_B^3 - 1.74685X_B^4] \text{ torr},$$

where P_B^o is the pure component equilibrium vapor pressure at 320 K, which has been measured and found to be 51.858 torr. Compute the osmotic pressure for this solution at 320 K when $X_B = 0.85$. The partial molar volume of B is 0.089 L mol^{-1}. What would the osmotic pressure be if the solution were ideal?

Solution

If we take X_B as the reference function and pure liquid B as the reference state, we have

$$a_B = \frac{P_{B(\text{soln})}}{P_B^o} = [0.442067\,X_B - 0.311495\,X_B^2 + 2.60907\,X_B^3 - 1.74685\,X_B^4]. \quad \textbf{(1)}$$

With $X_B = 0.85$, Eq. 1 gives

$$a_B = [0.442067(0.85) - 0.311495(0.85)^2 + 2.60907(0.85)^3 - 1.74685\,(0.85)^4]$$
$$= 0.84113. \quad \textbf{(2)}$$

The osmotic pressure for a nonideal solution is given by Eq. 8.134:

$$\pi = \frac{-RT\ln a_B}{\overline{V}_B(l)} = -\frac{(0.08206 \text{ L atm mol}^{-1}\text{ K}^{-1})(320 \text{ K})\ln(0.84113)}{0.089 \text{ L mol}^{-1}}$$
$$= \underline{51.05 \text{ atm}}. \quad \textbf{(3)}$$

If the solution were ideal, we would have

$$\pi = \frac{-RT\ln X_B}{\overline{V}_B(l)} = -\frac{(0.08206 \text{ L atm mol}^{-1}\text{ K}^{-1})(320 \text{ K})\ln(0.85)}{0.089 \text{ L mol}^{-1}} = \underline{47.95 \text{ atm}}. \quad \textbf{(4)}$$

8.29 The activities of the solvent B at 320 K have been determined by vapor pressure measurements. The results for $\ln[a_B/X_B]$ have been fitted by linear least-square methods to a polynomial function. The result is

$$\ln[a_B/X_B] = -0.359427X_A^2 - 3.428316X_A^3 + 3.156032X_A^4 - 0.1863966X_A^5.$$

From these data, together with the Gibbs–Duhem equation, the activity coefficients of solute A are computed as a function of the mole fraction of A in Problem 8.28. The result is

$$\ln\gamma_A = 0.718854\,X_A + 4.783047X_A^2 - 7.636359X_A^3 + 3.389028X_A^4 - 0.186397X_A^5.$$

Compute ΔG_{mix} when 3 moles of solute A are mixed with 20 moles of solvent B at 320 K. Compare your result with that expected for an ideal solution.

Solution

When n_A moles of solute A are in the pure state, the Gibbs free energy is

$$G_{A(\text{pure})} = n_A\mu_A = n_A[\mu_A^o + RT\ln a_A^o], \quad \textbf{(1)}$$

where a_A^o is the activity of A in the pure state. After the solute is mixed with the solvent, the Gibbs free energy becomes

$$G_{A(\text{soln})} = n_A\mu_{A(\text{soln})} = n_A[\mu_A^o + RT\ln a_A]. \quad \textbf{(2)}$$

The same equations hold for component B. Therefore, the total Gibbs free energy change upon mixing is

$$\Delta G_{\text{mix}} = [G_{A(\text{soln})} - G_{A(\text{pure})}] + [G_{B(\text{soln})} - G_{B(\text{pure})}]. \quad \textbf{(3)}$$

Substituting Eqs. 1 and 2 into Eq. 3 and using equivalent equations for component B gives

$$\begin{aligned}\Delta G_{\text{mix}} &= n_A[\mu_A^o + RT\ln a_A] - n_A[\mu_A^o + RT\ln a_A^o]\\ &\quad + n_B[\mu_B^o + RT\ln a_B] - n_B[\mu_B^o + RT\ln a_B^o]\\ &= n_A RT\ln[a_A/a_A^o] + n_B RT\ln[a_B/a_B^o]. \qquad \textbf{(4)}\end{aligned}$$

We now need the mole fractions for the actual mixture. With the given data, we have

$$X_A = \frac{3}{3+20} = 0.1304 \text{ and } X_B = 0.8696. \qquad \textbf{(5)}$$

For component B, the choice of reference state yields $a_B^o = 1$. When $X_B = 0.8696$, a_B may be computed from the analytical fit given in the problem, as follows:

$$\begin{aligned}\ln[a_B/X_B]&-0.359427X_A^2 - 3.428316X_A^3 + 3.156032X_A^4 - 0.1863966X_A^5\\ &= -0.359427(0.1304)^2 - 3.428316(0.1304)^3 + 3.156032(0.1304)^4\\ &\quad -0.1863966(0.1304)^5\\ &= -0.01281. \qquad \textbf{(6)}\end{aligned}$$

Therefore,

$$a_B = 0.8696\,e^{-0.01281} = 0.8585 \qquad \textbf{(7)}$$

at $X_B = 0.8696$. For solute A, the values obtained using the Gibbs–Duhem equation are

$$\begin{aligned}\ln\gamma_A &= 0.718854X_A + 4.783047X_A^2 - 7.636359X_A^3\\ &\quad + 3.389028X_A^4 - 0.186397X_A^5\\ &= 0.718854(0.1304) + 4.783047(0.1304)^2 - 7.636359(0.1304)^3\\ &\quad + 3.389028(0.1304)^4 - 0.186397(0.1304)^5 = 0.1591. \qquad \textbf{(8)}\end{aligned}$$

The corresponding activity coefficient is

$$\gamma_A = e^{0.1591} = 1.1725. \qquad \textbf{(9)}$$

The activity of A is

$$a_A = \gamma_A X_A = 1.1725(0.1304) = 0.1529. \qquad \textbf{(10)}$$

When $X_A = 1$, the results are

$$\begin{aligned}\ln\gamma_A &= 0.718854(1) + 4.783047(1)^2 - 7.636359(1)^3 + 3.389028\,(1)^4\\ &\quad - 0.186397(1)^5 = 1.06817, \qquad \textbf{(11)}\end{aligned}$$

so that

$$\gamma_A = e^{1.06817} = 2.910 \qquad \textbf{(12)}$$

and

$$a_A = \gamma_A X_A = 2.910. \qquad \textbf{(13)}$$

Substituting these results into Eq. 4 yields

$$\begin{aligned}\Delta G_{\text{mix}} &= 3\text{ mol }(8.314\text{ J mol}^{-1}\text{ K}^{-1})(320\text{ K})\ln\left(\frac{0.1529}{2.910}\right)\\ &\quad + 20\text{ mol }(8.314\text{ J mol}^{-1}\text{ K}^{-1})(320\text{ K})\ln\left(\frac{.8585}{1.000}\right) = -2.351\times10^4\text{ J} - 8{,}118\text{ J}\\ &= \underline{-3.163\times10^4\text{ J}}. \qquad \textbf{(14)}\end{aligned}$$

If the solution is ideal, the Gibbs free energy of mixing is given by Eq. 8.28:

$$\Delta G_{\text{mix}}^{\text{ideal}} = RT \sum_{i=1}^{i=K} n_i \ln X_i$$
$$= (8.314\text{ J mol}^{-1}\text{ K}^{-1})(320\text{ K})[3\ln(0.1304) + 20\ln(0.8696)]$$
$$= \underline{-2.369 \times 10^4\text{ J}}. \quad \textbf{(15)}$$

8.31 For the solution considered in Problem 8.29, the activity coefficient of solute A at a mole fraction $X_A = 0.360$ is 1.783. If we change the reference function for solute A from mole fraction to molality, with the reference state taken to be $m \to 0$, compute the activity for solute A. The molar mass of the solvent is 78.11 g mol^{-1}.

Solution

The conversion equation between activity coefficients with the mole fraction as the reference function, γ_x, and those with the molality as the reference function, γ_m, is given by Eq. 8.151, viz.,

$$\frac{\gamma_x}{\gamma_m} = 1 + 0.001\,M_o m, \quad \textbf{(1)}$$

where m is the solution molality and M_o is the molar mass of the solvent in g mol^{-1}. A solution whose molality is m contains m moles of solute in each 1,000 g of solvent. The connecting equation between mole fraction and molality is, therefore,

$$X_A = \frac{m}{m + 1{,}000/M_o}, \quad \textbf{(2)}$$

which is Eq. 8.148 in the text. Inverting Eq. 2, we obtain

$$m = \frac{1{,}000\,X_A}{M_o(1 - X_A)}. \quad \textbf{(3)}$$

Therefore, for the foregoing solution with $X_A = 0.360$, we have

$$m = \frac{1{,}000\text{ g }(0.360)}{78.11\text{ g mol}^{-1}(1 - 0.360)} = 7.201\text{ molal}. \quad \textbf{(4)}$$

Substituting into Eq. 1 yields

$$\gamma_m = \frac{\gamma_x}{1 + 0.001\,M_o m} = \frac{1.783}{1 + 0.001\text{ g}^{-1}(78.11\text{ g mol}^{-1})(7.201\text{ mol})} = 1.141. \quad \textbf{(5)}$$

The activity when molality is chosen as the reference function is

$$a_A = \gamma_m\,\text{m} = 1.141(7.201) = \underline{8.216}. \quad \textbf{(6)}$$

8.33 Osmotic-pressure measurements on solutions containing a high-molecular-weight polymer are taken as a function of the mass of polymer dissolved per liter of solution, w, at 298 K. The data are in the table at the left.

w (g L^{-1})	π (torr)
1.000	0.02746
1.897	0.06162
2.170	0.07381
3.106	0.1219
4.513	0.2127

Determine the molar mass of the polymer.

Solution

The dilute-solution limiting form of the osmotic pressure equation, Eq. 8.97, is

$$\frac{\pi}{w} = \frac{RT}{M_A V_B}, \tag{1}$$

where M_A is the molar mass of the polymer in g mol^{-1} and V_B is the total volume of the solvent, which is equal to the solution volume in the limit of dilute solutions. This equation shows that if the solution is ideal π/w will be a constant for a given solution volume at a fixed temperature. For nonideal solutions, Eq. 1 will hold only in the limit as $w \longrightarrow 0$. For large polymers, we expect this to be the case. Therefore, we must plot π/w and extrapolate the result to the limit as $w \longrightarrow 0$, where we expect Eq. 1 to be valid. The data required for the plot are in the table at the right.

w (g L^{-1})	π / w (torr L g^{-1})
1.000	0.02746
1.897	0.03248
2.170	0.03401
3.106	0.03925
4.513	0.04713

The plot appears at the end of the problem. The linear least-squares fit is shown as the solid line. Its equation is $\pi/w = [0.02186 + 0.00560(\text{L g}^{-1})w]$ (torr L g^{-1}), so that the limit of π/w at $w \longrightarrow 0$ is 0.02186 torr L g^{-1}. Using this value in Eq. 1, we obtain

$$M_A = \frac{RT}{V_B(\pi/w)} = \frac{(0.08206 \text{ L atm mol}^{-1}\text{ K}^{-1})(298 \text{ K})(760 \text{ torr atm}^{-1})}{1 \text{ L}(0.02186 \text{ torr L g}^{-1})}$$

$$= \underline{8.502 \times 10^5 \text{ g mol}^{-1}}. \tag{2}$$

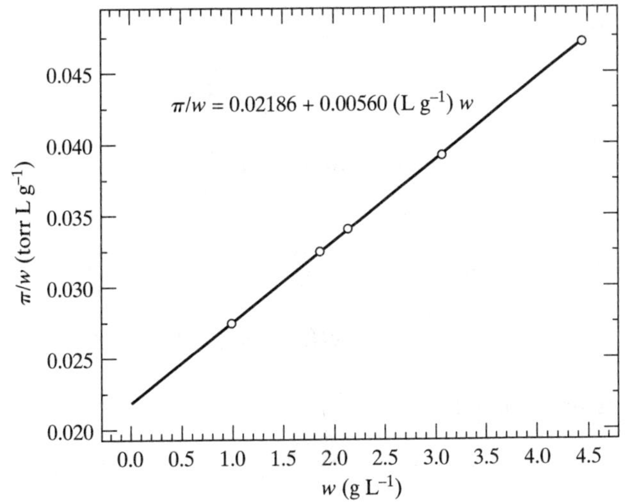

8.35 A chemist mixes 20 g of n-butanol with 80 g of H_2O at a temperature of 410 K.

(A) How many phases are present? Are any of the solutions saturated? How do you know?

(B) The chemist lowers the temperature of the mixture to 373.15 K. How many phases are now present? If there are two phases, what masses of H_2O and n-butanol are present in each phase? If there is only one phase, at what temperature will two phases appear. (See the phase diagram shown in Figure 8.21. The calculations need be done only as accurately as permitted by this diagram.)

Solution

(A) Figure 8.21 shows that a 20%-by-mass mixture of n-butanol in 80% H_2O at 410 K corresponds to a point that lies above the solubility curve of n-butanol in water. Therefore, only a single phase is present at 410 K. The solution is not saturated. If it were, the point would have to lie on the solubility curve shown in the figure, and it does not.

(B) When the temperature of the mixture is lowered to 373.15 K, the point now lies below the solubility curve in the figure. Therefore, we are in the two-phase region, where we have a water solution saturated with n-butanol and an n-butanol solution saturated with water. The tie line at 373.15 K in the phase diagram intersects the solubility curve at approximately 9% n-butanol by mass in the water solution and 33% water in the n-butanol solution. Let w_1 be the amount of water present in the water-saturated n-butanol solution, and let w_2 be the mass of n-butanol in this solution. The phase diagram tells us that we must have

$$\frac{w_1}{w_2} = \frac{33}{67}. \tag{1}$$

By mass balance, the masses of water and n-butanol in the n-butanol-saturated water solution must be $m_{\text{water}} = 80 - w_1$ and $m_{\text{n-butanol}} = 20 - w_2$. The phase diagram shows that, in the water solution, we have

$$\frac{80 - w_1}{20 - w_2} = \frac{91}{9}. \tag{2}$$

Using Eq. 1, we get

$$w_1 = \frac{33w_2}{67}. \tag{3}$$

From Eq 2, by cross multiplication, we obtain

$$9(80 - w_1) = 91(20 - w_2). \tag{4}$$

Substituting Eq. 3 into Eq. 4 produces

$$9\left[80 - \frac{33w_2}{67}\right] = 91(20 - w_2) = 1{,}820 - 91w_2. \tag{5}$$

Rearranging terms in Eq. 5 yields

$$-4.433w_2 + 91w_2 = 86.57w_2 = 1{,}820 - 720 = 1{,}100. \tag{6}$$

Solving for w_2, we obtain

$$w_2 = \frac{1{,}100}{86.57}\text{ g} = \underline{12.71\text{ g}}\; of\; n - \text{butanol}. \tag{7}$$

Using this result in Eq. 3 yields

$$w_1 = \frac{(33)(12.71)}{67}\text{ g} = \underline{6.26\text{ g}}\; of\; H_2O. \tag{8}$$

Thus, the water-saturated n-butanol solution contains 12.71 g of n-butanol and 6.26 g of water, so that its total mass is 18.97 g, which gives 67% by mass of butanol and 33% by mass of water, as required. Mass balance now tells us that in the n-butanol-saturated water solution, we have 73.74 g of water and 7.29 g of n-butanol, for a total mass of 81.03 g. This solution is, therefore, $73.74/81.03 \times 100 = 91\%$ water, as indicated by the phase diagram.

8.37 A chemist mixes Mg and Zn. Starting at a temperature of 950 K, she obtains the cooling curve shown in Figure 8.29 for her mixture.

(A) If we know that the number of moles of Mg in the mixture is greater than the number of moles of Zn, what is the mole percent of magnesium in the mixture, and at what temperature does the flat portion of the cooling curve in the figure occur?

(B) If we know that the number of moles of Mg in the mixture is less than the number of moles of Zn, can we determine the mole percent of Mg present? If so, obtain this percentage. If not, state what information we can obtain about the mole percent of Mg present and the temperature at which the flat portion of the cooling occurs. (See the phase diagram in Figure 8.25.)

Solution

(A) Since there is only one break in the cooling curve, we must be cooling a Mg–Zn mixture whose composition corresponds to one of the two eutectics shown in Fig. 8.25, or to the compound $MgZn_2$ at the maximum of the freezing curve. If we know that the mole percent of Mg is greater than 50%, so that it exceeds the mole percent of Zn, the only possibility is that we are cooling the eutectic mixture corresponding to about 74 mole percent of Mg. The phase diagram indicates that this mixture freezes at 620 K, so that must be the temperature corresponding to the flat portion of the cooling curve shown in Figure 8.29.

(B) If we know that the mole percent of Mg is less than 50%, so that we have a greater mole percent of Zn present, there are two possibilities. First, we could have the mixture corresponding to the compound $MgZn_2$, for which the mole percent of Mg is 33.333%. If this is the case, the flat portion of the cooling curve occurs at 863 K, the freezing point for $MgZn_2$, as shown on the phase diagram. The second possibility is that we have a composition corresponding to the

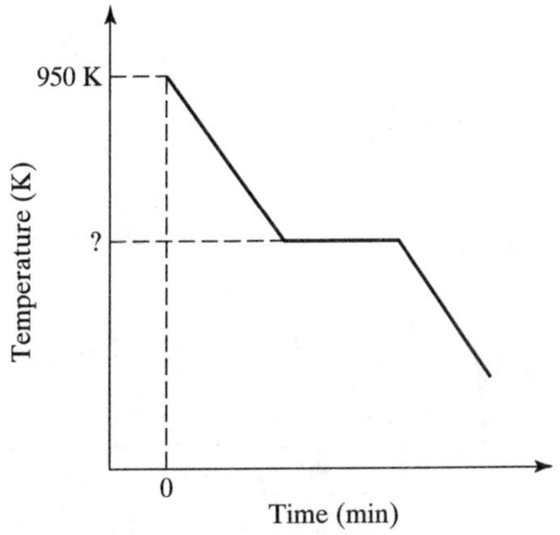

▲ FIGURE 8.29
Cooling curve obtained by a chemist using a Mg–Zn solution with an initial temperature of 950 K.

eutectic, with about 10 mole percent of Mg. This mixture freezes at 653 K. In that case, the flat portion of the cooling curve will be at 653 K.

8.39 The H_2O–H_2SO_4 phase diagram is shown in Figure 8.30 on the next page. How many compounds of H_2O and H_2SO_4 can be formed? What are their chemical formulae? Label the phases present in each region of the diagram.

Solution

The phase diagram shown in Figure 8.30 exhibits three distinct maxima. Therefore, three compounds of water and sulfuric acid are formed. The phase diagram shows that the first maximum has 20 mole percent sulfuric acid. Consequently, its molecular formula must be $H_2SO_4 \cdot 4H_2O$. The second maximum occurs at 33.33% mole percent sulfuric acid. This means one-third of the moles in the compound are sulfuric acid. Therefore, the molecular formula of this compound is $H_2SO_4 \cdot 2H_2O$. The final maximum has an equal molar mixture, so the formula is $H_2SO_4 \cdot H_2O$.

The contents of various regions of the phase diagram in Figure 8.30 are given as follows:

Region A: One phase, a homogenous solution of water and sulfuric acid.

Region B: Two phases. One is solid H_2O. The other is an H_2O–$H_2SO_4 \cdot 4H_2O$ solution that is saturated in H_2O.

Region C: Two phases. One is solid crystals of sulfuric acid tetrahydrate ($H_2SO_4 \cdot 4H_2O$). The other is an H_2O–$H_2SO_4 \cdot 4H_2O$ solution that is saturated in $H_2SO_4 \cdot 4H_2O$.

Region D: Two phases. Solid crystals of sulfuric acid tetrahydrate ($H_2SO_4 \cdot 4H_2O$) and an $H_2SO_4 \cdot 4H_2O$–$H_2SO_4 \cdot 2H_2O$ solution that is saturated with $H_2SO_4 \cdot 4H_2O$.

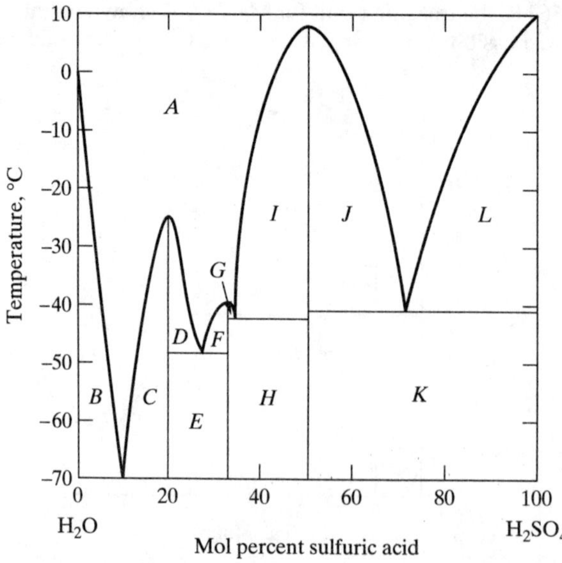

▲ FIGURE 8.30
The H_2O–H_2SO_4 phase diagram. The letters mark the various regions of the diagram. Note that all temperatures are given in degrees Celsius.

Region E: One solid phase containing a mixture of solid $H_2SO_4 \cdot 4H_2O$ and solid $H_2SO_4 \cdot 2H_2O$.

Region F: Two phases. Solid crystals of sulfuric acid dihydrate ($H_2SO_4 \cdot 2H_2O$) and an $H_2SO_4 \cdot 4H_2O$–$H_2SO_4 \cdot 2H_2O$ solution that is saturated with $H_2SO_4 \cdot 2H_2O$.

Region G: Two phases. Solid crystals of $H_2SO_4 \cdot 2H_2O$ and a solution of $H_2SO_4 \cdot 2H_2O$ and $H_2SO_4 \cdot H_2O$ that is saturated in $H_2SO_4 \cdot 2H_2O$.

Region H: One solid phase containing a mixture of solid $H_2SO_4 \cdot 2H_2O$ and solid $H_2SO_4 \cdot 2H_2O$.

Region I: Two phases. Solid $H_2SO_4 \cdot H_2O$ and a solution of $H_2SO_4 \cdot H_2O$ and $H_2SO_4 \cdot 2H_2O$ that is saturated with $H_2SO_4 \cdot H_2O$.

Region J: Two phases. Solid $H_2SO_4 \cdot H_2O$ and a solution of $H_2SO_4 \cdot H_2O$ and H_2SO_4 that is saturated with $H_2SO_4 \cdot H_2O$.

Region K: One solid phase containing a mixture of solid $H_2SO_4 \cdot H_2O$ and solid H_2SO_4.

Region L: Two phases. Solid H_2SO_4 and a solution of $H_2SO_4 \cdot H_2O$ and H_2SO_4 that is saturated with H_2SO_4.

8.41 Sam's grades in physical chemistry have been increasing rapidly. He is now one of the better students in the class. He has noticed that a popular laboratory manual for freshman chemistry contains the following simple boiling-point elevation experiment: Students place 100 grams of an unknown compound, which is actually sucrose ($C_{12}H_{22}O_{11}$), into 1000 grams of H_2O and then measure the elevation of the boiling point. Using the limiting form of the boiling-point elevation equation, $\Delta T_b = K_b m$ with $K_b = 0.51$ K mol^{-1}, the students are required to determine the molar mass of the unknown solute. Sam points out to the class that when students in New Orleans and Miami perform this experiment, they generally obtain molar masses in the range 336.7 to 343.5, an error of about $\pm 1\%$. However, when students in Lhasa, Tibet, perform this same experiment using the same equipment, they usually obtain results in the range 364.6 to 372.0, an error of about $\pm 7.6\%$. Before beginning the study of the thermodynamics of electrolytic solutions, Sam offers to buy lunch for anyone in the class who can explain why the students in Lhasa are getting such poor results, provided that those who cannot explain this unexpected result buy him lunch. Are there any gamblers in the class? (The partial molar enthalpy of vaporization of H_2O is 40,657 J mol^{-1}.)

Solution

New Orleans, Louisiana and Miami, Florida are at sea level, where the applied pressure is 760 torr. Therefore, the boiling point of pure water in those locales is 373.15 K. The molal boiling-point elevation constant given by Eq. 8.83 is

$$K_b = \frac{RT_{bo}^2 M_B}{1{,}000\Delta \overline{H}_{\text{vap}}^B} = \frac{8.314(373.15)^2 18.016}{1{,}000(40{,}657)} = 0.513 \text{ K mol}^{-1}. \quad \textbf{(1)}$$

The molality of the solution being used is

$$m = \frac{100/342.3}{1.000} = 0.2921, \quad \textbf{(2)}$$

where 342.3 is the molar mass in g mol^{-1} of sucrose.

Consequently, the students in New Orleans and Miami will measure boiling-point elevations of about

$$\Delta T_b = 0.513(0.2921) = 0.150 \text{ K}, \quad (3)$$

plus or minus a small amount due to experimental error. They then compute the molar mass of their unknown as follows:

$$m = \frac{\Delta T_b}{0.51} = \frac{0.150}{0.51} = 0.294. \quad (4)$$

If M is the unknown molar mass, then

$$m = \frac{100/M}{1.000} = 0.294. \quad (5)$$

This yields $M = 340.1$. Some students have small errors in their measurement of the boiling-point elevation and, therefore, obtain somewhat different answers. Most, however, are in the range given in the problem.

Lhasa is the capital city of Tibet. Its elevation is 12,004 ft above sea level. At that elevation, the barometric pressure is about 449 torr. Using the Clausius–Clapeyron equation, we may easily determine the boiling point at this applied pressure. Water will boil when its equilibrium vapor pressure is equal to the applied pressure. Therefore, in Lhasa, we have

$$\ln\left(\frac{449}{760}\right) = -\frac{40{,}657}{8{,}314}\left(\frac{1}{T_b} - \frac{1}{T_b^o}\right), \quad (6)$$

where T_b^o is the normal H_2O boiling point and T_b is its boiling point in Lhasa. This gives

$$\frac{1}{T_b} = \frac{1}{373.15} - \frac{8.314 \ln(449/760)}{40{,}657} = 0.002788 \quad (7)$$

so that

$$T_b = 358.7 \text{ K}. \quad (8)$$

Under these conditions, the correct molal boiling-point elevation constant in Lhasa is

$$K_b = \frac{RT_{bo}^2 M_B}{1{,}000\Delta H_{vap}^B} = \frac{8.314(358.7)^2 18.016}{1{,}000(40{,}657)} = 0.474 \text{ K mol}^{-1}. \quad (9)$$

The students in Lhasa will, therefore, observe a boiling-point elevation equal to

$$\Delta T_b(\text{Lhasa}) = 0.474 \text{ m} = 0.474(0.2921) = 0.138 \text{ K}. \quad (10)$$

When they employ the erroneous value of 0.51 K mol^{-1} for K_b, their calculation of the molar mass gives

$$m = \frac{\Delta T_b}{0.51} = \frac{0.138}{0.51} = 0.271. \quad (11)$$

Therefore, they obtain a molar mass

$$M = \frac{100}{(1.000)(0.271)} = 369, \quad (12)$$

plus or minus a small amount due to their experimental error. The ±7.6% error obtained by Lhasa students is not their fault; rather, it is the responsibility of the person who devised the experiment. That person did not understand the dependence of K_b upon applied pressure. Did you? Do you owe Sam lunch, or vice versa?

CHAPTER 9

Thermodynamics of Electrolytic Solutions

9.1 Show that, for a one-to-one electrolyte such as NaCl with $\nu_+ = \nu_- = 1$, the mean ionic activity, the mean ionic activity coefficient, and the mean stoichiometric coefficient correspond to the geometric mean value of the individual ionic activities, activity coefficients, and ionic stoichiometric coefficients, respectively.

Solution

The mean ionic activity, activity coefficient, and stiochiometric coefficient are defined by Eq. 9.10; thus,

$$a_\pm^\nu = a_+^{\nu_+} a_-^{\nu_-}, \tag{1}$$

$$\gamma_\pm^\nu = \gamma_+^{\nu_+} \gamma_-^{\nu_-}, \tag{2}$$

and

$$\nu_\pm^\nu = \nu_+^{\nu_+} \nu_-^{\nu_-}, \tag{3}$$

with $\nu = \nu_+ + \nu_-$. For a 1 : 1 electrolyte, we have $\nu = 2$, which gives

$$a_\pm^2 = a_+ a_- \tag{4}$$

so that

$$\boxed{a_\pm = [a_+ a_-]^{1/2}} \tag{5}$$

which is the geometric mean of a_+ and a_-. The same results are obtained for $\gamma_\pm$ and $\nu_\pm$. We have

$$\gamma_\pm^2 = \gamma_+ \gamma_- \tag{6}$$

so that

$$\boxed{\gamma_\pm = [\gamma_+ \gamma_-]^{1/2}}, \tag{7}$$

which is the geometric mean of γ_+ and γ_-; and

$$\nu_\pm^2 = \nu_+ \nu_- \tag{8}$$

so that

$$\boxed{\nu_\pm = [\nu_+ \nu_-]^{1/2}} \tag{9}$$

which is the geometric mean of ν_+ and ν_-.

9.3 Suppose it is found that an aqueous solution of $BaCl_2$ completely dissociates into ions and that 10% of the Ba^{2+} ions form ion pairs with Cl^- ions. Show that, under these conditions, we will have $\gamma_i = 1.2696\gamma_\pm$.

Solution

The ionization reaction is $BaCl_2 + xH_2O \longrightarrow Ba^{2+}_{(aq)} + 2\,Cl^-_{(aq)}$; hence, $\nu_+ = 1$, $\nu_- = 2$, and $\nu = 1 + 2 = 3$. Therefore,

$$\nu_\pm^\nu = \nu_\pm^3 = \nu_+^{\nu_+} \nu_-^{\nu_-} = (1)^1(2)^2 = 4, \tag{1}$$

which gives

$$\nu_\pm = 4^{1/3} = 1.587401. \tag{2}$$

If 10% of the Ba^{2+} ions form ion pairs, then $f = 0.90$, since f is defined to be the fraction of the cations that do not form ion pairs. Equation 9.21 relates γ_i to $\gamma_\pm$:

$$\gamma_i^\nu = \nu_\pm^{-1}[f\nu_+]^{\nu_+}[\nu_- - (1-f)\nu_+]^{\nu_-}\gamma_\pm^\nu. \quad (3)$$

Substituting the foregoing values into Eq. 3 yields

$$\gamma_i^3 = (1.587401)^{-1}[0.9(1)]^1[2-(1-0.9)(1)]^2\gamma_\pm^3 = 2.04674\gamma_\pm^3, \quad (4)$$

so that we have

$$\boxed{\gamma_i = (2.04674)^{1/3}\gamma_\pm = 1.26966\gamma_\pm}, \quad (5)$$

as required.

9.5 (A) Show that the solution of $\partial^2F/\partial r^2 = -\kappa^2 F$ has the form $F = C_1 \sin[\kappa r] + C_2 \cos[\kappa r]$, where C_1 and C_2 are constants. (B) Show that we may also express this solution in the complex form $F = c_1 \exp[i\kappa r] + c_2 \exp[-i\kappa r]$, where i is the imaginary quantity $(-1)^{1/2}$. Obtain the relationship between the C_i and the c_i.

Solution

(A) If we have $F = C_1 \sin[\kappa r] + C_2 \cos[\kappa r]$, where C_1 and C_2 are constants, the first derivative of F with respect to r is given by

$$\frac{\partial F}{\partial r} = \kappa C_1 \cos[\kappa r] - \kappa C_2 \sin[\kappa r]. \quad (1)$$

The second derivative of F with respect to r is given by the derivative of the first derivative:

$$\frac{\partial^2 F}{\partial r^2} = \frac{\partial}{dr}\frac{\partial F}{\partial r} = -\kappa^2 C_1 \sin[\kappa r] - \kappa^2 C_2 \cos[\kappa r] = -\kappa^2\{C_1 \sin[\kappa r] + C_2 \cos[\kappa r]\}$$

$$= -\kappa^2 F, \quad (2)$$

as required. Therefore, the function $F = C_1 \sin[\kappa r] + C_2 \cos[\kappa r]$ is a solution of the differential equation.

(B) The complex exponential e^{ix} can be written in the following form:

$$e^{ix} = \cos(x) + i\sin(x). \quad (3)$$

Therefore,

$$e^{-ix} = \cos(-x) + i\sin(-x) = \cos(x) - i\sin(x). \quad (4)$$

By using Eqs. 3 and 4, we can expand F as follows:

$$\begin{aligned} F &= c_1 \exp[i\kappa r] + c_2 \exp[-i\kappa r] \\ &= c_1[\cos(\kappa r) + i\sin(\kappa r)] + c_2[\cos(\kappa r) - i\sin(\kappa r)] \\ &= [c_1 + c_2]\cos(\kappa r) + i[c_1 - c_2]\sin(\kappa r). \end{aligned} \quad (5)$$

We now identify the constants by means of the equations

$$c_1 + c_2 = C_1 \quad \text{and} \quad i[c_1 - c_2] = C_2, \quad (6)$$

so that

$$F = C_1 \cos(\kappa r) + C_2 \sin(\kappa r). \quad (7)$$

Therefore, the functions $F = C_1\cos(\kappa r) + C_2\sin(\kappa r)$ and $F = c_1\exp[i\kappa r] + c_2\exp[-i\kappa r]$ are equivalent.

9.7 Consider a spherical cavity of radius r_o. It is possible to set up spherically symmetric standing waves inside this cavity, which we shall see are very closely related to the quantum mechanical description of translational motion. The differential equation these standing waves must satisfy is

$$\nabla^2\Psi = -\beta^2\Psi.$$

The boundary conditions on the standing waves require that $\Psi = \beta$ when $r = 0$ and that $\Psi = 0$ at $r = r_o$. Using the methods employed to solve the Poisson equation, find the equation for the spherical symmetric standing waves inside the cavity.

Solution

Inserting the Laplacian makes the differential equation become

$$\left[\frac{1}{r^2}\frac{\partial}{\partial r}\left[r^2\frac{\partial}{\partial r}\right] + \frac{1}{r^2\sin\theta}\frac{\partial}{\partial\theta}\left[\sin\theta\frac{\partial}{\partial\theta}\right] + \frac{1}{r^2\sin^2\theta}\frac{\partial^2}{\partial\phi^2}\right]\Psi = -\beta^2\Psi. \quad \textbf{(1)}$$

Ψ is spherically symmetric, so it does not depend upon the spherical polar angles θ and ϕ. Consequently, the derivatives with respect to these variables vanish, and we have

$$\left[\frac{1}{r^2}\frac{\partial}{\partial r}\left[r^2\frac{\partial}{\partial r}\right]\right]\Psi = -\beta^2\Psi. \quad \textbf{(2)}$$

We now make the substitution $\Psi = F/r$. The left side of the equation then becomes

$$\left[\frac{1}{r^2}\frac{\partial}{\partial r}\left[r^2\frac{\partial}{\partial r}\right]\right]\frac{F}{r} = \frac{1}{r^2}\frac{\partial}{\partial r}\left[r^2\left\{-\frac{F}{r^2} + \frac{1}{r}\frac{\partial F}{\partial r}\right\}\right] = \frac{1}{r^2}\frac{\partial}{\partial r}\left[-F + r\frac{\partial F}{\partial r}\right]$$

$$= \frac{1}{r^2}\left[-\frac{\partial F}{\partial r} + \frac{\partial F}{\partial r} + r\frac{\partial^2 F}{\partial r^2}\right] = \frac{1}{r}\frac{\partial^2 F}{\partial r^2}. \quad \textbf{(3)}$$

Combining Eqs. 2 and 3 produces

$$\frac{1}{r}\frac{\partial^2 F}{\partial r^2} = -\beta^2\frac{F}{r}. \quad \textbf{(4)}$$

Multiplying both sides by r yields

$$\frac{\partial^2 F}{\partial r^2} = -\beta^2 F. \quad \textbf{(5)}$$

The solution of this equation is

$$F = c_1\sin(\beta r) + c_2\cos(\beta r), \quad \textbf{(6)}$$

where c_1 and c_2 are constants.

We can easily verify the correctness of this solution by extracting the second derivative of the function given in Eq. 6. The first derivative is

$$\frac{\partial F}{\partial r} = \beta c_1\cos(\beta r) - \beta c_2\sin(\beta r). \quad \textbf{(7)}$$

The second derivative is, therefore,

$$\frac{\partial^2 F}{\partial r^2} = -\beta^2 c_1\sin(\beta r) - \beta^2 c_2\cos(\beta r) = -\beta^2 F, \quad \textbf{(8)}$$

as required. With F given by Eq. 6, we have

$$\Psi = \frac{c_1 \sin(\beta r) + c_2 \cos(\beta r)}{r}. \tag{9}$$

When $r = 0$, Eq. 9 gives

$$\Psi = \frac{c_2 \cos(\beta r)}{0}, \tag{10}$$

which will go to infinity unless we have $c_2 = 0$. With this choice, the limiting value of Ψ as $r \longrightarrow 0$ is

$$\lim_{r \longrightarrow 0} \Psi = \lim_{r \longrightarrow 0} \frac{c_1 \sin(\beta r)}{r}. \tag{11}$$

This expression is indeterminate, so we employ L'Hôpital's rule:

$$\lim_{r \longrightarrow 0} \Psi = c_1 \lim_{r \longrightarrow 0} \frac{\beta \cos(\beta r)}{1} = c_1 \beta. \tag{12}$$

Since the boundary condition requires that $\Psi = \beta$ at $r = 0$, we must take $c_1 = 1$. The second boundary condition requires that $\Psi = 0$ at $r = r_o$, and this condition in turn requires that

$$\Psi = \frac{\sin(\beta r_o)}{r_o} = 0. \tag{13}$$

To satisfy this condition, we must have

$$\beta = \frac{n\pi}{r_o}, \tag{14}$$

where n is a positive integer: 1, 2, 3, 4, ...

The solutions for the spherical waves are, therefore,

$$\boxed{\Psi = \frac{\sin\left[\dfrac{n\pi r}{r_o}\right]}{r}}. \tag{15}$$

Plots of the spherical waves for $n = 1$, $n = 2$, and $n = 3$ are shown in the graph below for the specific case of $r_o = 10$ Å.

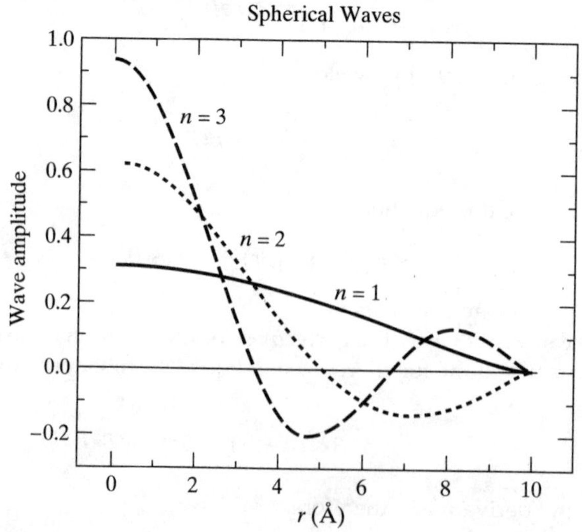

9.9 The Debye–Hückel theory predicts that the mean ionic activity coefficient is a monotonically decreasing function of the ionic strength (i.e., as S increases, $\gamma_\pm$ steadily decreases). In contrast, the empirical relationship given in Eq. 9.68 exhibits a minimum when $\gamma_\pm$ is plotted against the molality of the solution. In many cases, the experimentally measured activity coefficients also exhibit such a minimum.

(A) Show that the minimum in $\gamma_\pm$ predicted by the equation occurs at the point when the molality of the solute is

$$c_e = \frac{0.789959\ldots}{\sum_{i=1}^{i=K} \nu_i z_i^2}$$

(B) At what concentration would we obtain a minimum value of $\gamma_\pm$ for $CaCl_2$?

Solution

(A) Equation 9.68 predicts the mean ionic activity coefficient to be given by

$$\log(\gamma_i) = -0.50926\, z_+ z_- \left[\frac{S^{1/2}}{1 + S^{1/2}} - 0.30S\right], \qquad \textbf{(1)}$$

where

$$S = 0.5 c_e \sum_{i=1}^{i=K} \nu_i z_i^2. \qquad \textbf{(2)}$$

First, we note that whenever γ_i is a minimum, $\log(\gamma_i)$ is also a minimum. We can, therefore, minimize $\log(\gamma_i)$ and get the same result as we would obtain if we were to minimize γ_i itself. The criterion for a minimum is that we must have

$$\frac{\partial \log(\gamma_i)}{\partial c_e} = -0.50926 z_+ z_- \left[\frac{(1 + S^{1/2}) 0.5 S^{-1/2} \dfrac{\partial S}{\partial c_e} - S^{1/2}(0.5) S^{-1/2} \dfrac{\partial S}{\partial c_e}}{(1 + S^{1/2})^2} - 0.3 \frac{\partial S}{\partial c_e}\right] = 0. \qquad \textbf{(3)}$$

Simplifying Eq. 3 produces

$$-0.50926\, z_+ z_- \frac{\partial S}{\partial c_e}\left[\frac{0.5(1 + S^{-1/2}) - 0.5}{(1 + S^{1/2})^2} - 0.3\right] = 0. \qquad \textbf{(4)}$$

Dividing by $-0.50926\, z_+ z_- (\partial S/\partial c_e)$, we obtain

$$\left[\frac{0.5 S^{-1/2}}{(1 + S^{1/2})^2} - 0.3\right] = 0 = \left[\frac{0.5}{S^{1/2}(1 + 2S^{1/2} + S)} - 0.3\right]. \qquad \textbf{(5)}$$

We now let $x = S^{1/2}$, so that Eq. 5 becomes

$$\frac{0.5}{x(1 + 2x + x^2)} - 0.3 = 0. \qquad \textbf{(6)}$$

Rearranging terms in this equation gives

$$0.3x^3 + 0.6x^2 + 0.3x - 0.5 = 0. \qquad \textbf{(7)}$$

We now need to solve Eq. 7. A one-dimensional grid search, repeated twice, shows that the root of Eq. 7 is $x = 0.628474\ldots$. The minimum in γ_i will, therefore, occur whenever

$$S^{1/2} = 0.628474, \tag{8}$$

that is, whenever we have

$$S = 0.3949796\ldots = 0.5c_e \sum_{i=1}^{i=K} \nu_i z_i^2. \tag{9}$$

Equation 9 shows that the concentration at which we will have a minimum is

$$\boxed{[c_e]_{\min} = 0.789959\ldots\left[\sum_{i=1}^{i=K} \nu_i z_i^2\right]^{-1}}, \tag{10}$$

as was required in the problem.

(B) For $CaCl_2$, we have

$$\left[\sum_{i=1}^{i=K} \nu_i z_i^2\right]^{-1} = \frac{1}{(1)(2)^2 + (2)(1)^2} = \frac{1}{6}. \tag{11}$$

Therefore, the minimum occurs at

$$[c_e]_{\min} = \frac{0.789959\ldots}{6} = \underline{0.131660\ldots \text{ molal.}} \tag{12}$$

The data given in Problem 9.8 show that, experimentally, the minimum occurs between $m = 0.5$ mol kg^{-1} and $m = 1.00$ mol kg^{-1}, a concentration significantly higher than predicted by the Davies empirical equation. This is usually the case. The minimum predicted by Eq. 9.68 occurs at a molality that usually lies below the experimentally observed minimum.

m (kg mol^{-1})	γ_i
0.00001	0.917
0.00010	0.761
0.00100	0.422

9.11 An investigator determines the activity coefficients for a salt and finds the data in the table to the left. If the salt is represented by the molecular formula $C_{\nu_+}A_{\nu_-}$, what are the possible values of ν_+ and ν_-? Show that your answer is correct.

Solution

The Debye–Hückel limiting law requires that, for low concentrations we have

$$\log(\gamma_i) = -0.50926 z_+ z_- S^{1/2} = -0.50926 z_+ z_- [0.5\,\text{m}\{\nu_+ z_+^2 + \nu_- z_-^2\}]^{1/2}. \tag{1}$$

However, charge neutrality requires that we have $z_+ = \nu_-$ and $z_- = \nu_+$. Inserting these equalities into Eq. 1 gives

$$\log(\gamma_i) = -0.36010 v_- \nu_+ [\nu_+ \nu_-^2 + \nu_- \nu_+^2]^{1/2}\,\text{m}^{1/2}. \tag{2}$$

Rearranging terms in Eq. 2 produces

$$\nu_-^{3/2} \nu_+^{3/2} [\nu_- + \nu_+]^{1/2} = \frac{\log(\gamma_i)}{-0.36010\,\text{m}^{1/2}}. \tag{3}$$

Substituting the data yields

$$\nu_-^{3/2}\nu_+^{3/2}[\nu_- + \nu_+]^{1/2} = \frac{\log(0.917)}{-0.36010(0.00001)^{1/2}} = 33.05, \tag{4}$$

$$\nu_-^{3/2}\nu_+^{3/2}[\nu_- + \nu_+]^{1/2} = \frac{\log(0.761)}{-0.36010(0.0001)^{1/2}} = 32.94, \tag{5}$$

and

$$\nu_-^{3/2}\nu_+^{3/2}[\nu_- + \nu_+]^{1/2} = \frac{\log(0.422)}{-0.36010(0.001)^{1/2}} = 32.90. \tag{6}$$

The left-hand side of Eq. 3 is symmetric in ν_+ and ν_-; therefore, the possibilities are as follows:

$$\nu_+ = \nu_- = 1,$$

for which we have $\nu_-^{3/2}\nu_+^{3/2}[\nu_- + \nu_+]^{1/2} = 1.4142\ldots$;

$$\nu_+ = 2 \quad \text{and} \quad \nu_- = 1 \quad \text{or} \quad \nu_+ = 1 \quad \text{and} \quad \nu_- = 2,$$

for which we have $\nu_-^{3/2}\nu_+^{3/2}[\nu_- + \nu_+]^{1/2} = 4.899$;

$\nu_+ = \nu_- = 2$, for which we have $\nu_-^{3/2}\nu_+^{3/2}[\nu_- + \nu_+]^{1/2} = 16.0000$;

and

$$\nu_+ = 2 \quad \text{and} \quad \nu_- = 3 \quad \text{or} \quad \nu_+ = 3 \quad \text{and} \quad \nu_- = 2,$$

for which we have $\nu_-^{3/2}\nu_+^{3/2}[\nu_- + \nu_+]^{1/2} = 32.86$.

It is obvious that only the last case fits the data. We conclude that the salt is either $\boxed{C_2A_3 \text{ or } C_3A_2}$.

9.13[#] An investigator chooses to employ a least-squares method to fit the freezing-point depression data given in Example 9.5 between $m = 0$ and $m = 0.02$ molal to an analytic function whose form is that given in Problem 9.12. The result using $\nu\, K_f = 5.58$ K mol^{-1} is

$$\Delta T_f = \nu K_f\,[\,m - 23.84066m^2 + 1{,}980.933m^3 - 53{,}995.094m^4].$$

Use this equation to compute the activity coefficient for the solute at molalities between 0 and 0.02. Plot the results, and on the same plot, show those obtained from the numerical integration carried out in Example 9.5.

Solution

The activity coefficient is given by Eq. 9.78:

$$\ln(\gamma_i) = -\left[1 - \frac{\Delta T_f}{K_f\nu m}\right] - \int_0^m \left[m^o - \frac{\Delta T_f}{K_f\nu m}\right]\frac{dm}{m}. \tag{1}$$

Let us first consider the integral. Substituting the polynomial expression for ΔT_f and setting $m^o = 1$, we obtain

$$\int_0^m \left[1 - \frac{\Delta T_f}{K_f\nu m}\right]\frac{dm}{m} = \int_0^m [-(a_2 m + a_3 m^2 + a_4 m^3)]\frac{dm}{m}$$

$$= -\left[a_2 m + \frac{a_3 m^2}{2} + \frac{a_4 m^3}{3}\right]. \tag{2}$$

The first term is given by

$$\left[1 - \frac{\Delta T_f}{K_f \nu m}\right] = -[a_2 m + a_3 m^2 + a_4 m^3], \tag{3}$$

where $a_2 = -23.84066\ \text{mol}^{-1}$, $a_3 = 1{,}980.933\ \text{mol}^{-2}$, and $a_4 = -53{,}995.094\ \text{mol}^{-3}$, from the data given in the problem. Accordingly, by combining Eqs. 1, 2, and 3, we see that the activity coefficient is given by

$$\ln(\gamma_i) = 2a_2 m + \frac{3a_3 m^2}{2} + \frac{4a_4 m^3}{3}. \tag{4}$$

Substituting the values of a_2, a_3, and a_4 gives the results shown in the accompanying figure. The points are results obtained from numerical integration methods in Example 9.5. The deviations between the results are due both to numerical errors in the integration in Example 9.5 and to fitting errors in this procedure. One problem is that the activity coefficient is very sensitive to small errors in ΔT_f. Another problem is that the form chosen to fit the ΔT_f data causes $\ln(\gamma_i)$ to depend upon integral powers of the molality, whereas the Debye–Hückel theory implies that $\ln(\gamma_i)$ should depend upon $m^{1/2}$ at low values of m.

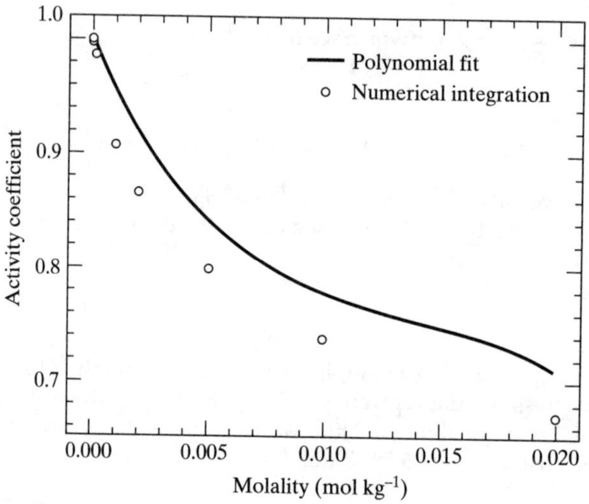

m (mol kg^{-1})	γ_i
0.000	1.000
0.001	0.891
0.002	0.855
0.005	0.797
0.010	0.745
0.020	0.690
0.050	0.621
0.100	0.582
0.200	0.567
0.500	0.639
1.000	0.946

9.15# The *Handbook of Chemistry and Physics* provides activity coefficients for $Ni(ClO_4)_2$. This problem demonstrates how we may extract freezing-point depression data from measured activity coefficients. In effect, the problem shows how to work the usual activity coefficient problem in reverse. The measured values of γ_i for $Ni(ClO_4)_2$ are listed in the table to the left. In Problem 9.14, it was found that if freezing-point depression data are fitted to the function $\Delta T_f = \nu K_f [m + b_1 m^{3/2} + b_2 m^2 + b_3 m^{5/2} + b_4 m^3]$, the activity coefficient is given by

$$\ln(\gamma_i) = \left[3b_1 m^{1/2} + 2b_2 m + \frac{5b_3 m^{3/2}}{3} + \frac{3b_4 m^2}{2}\right],$$

where

$$3b_1 = -1.1726 z_+ z_- [0.5\{\nu_+ z_+^2 + \nu_- z_-^2\}]^{1/2}$$

if the result is to be compatible with the Debye–Hückel theory.

(A) For the $Ni(ClO_4)_2$ system, determine the value of $3b_1$ if the fitted results are to agree with the Debye–Hückel theory.

(B) Using the value of $3b_1$ determined in (A), we now conduct a least-squares fitting of the expression for $\ln(\gamma_i)$ to the data given in the foregoing table. The result of this fitting is $b_2 = 5.4184\ \mathrm{mol}^{-1}\ \mathrm{kg}$, $b_3 = -7.50582\ \mathrm{mol}^{-3/2}\ \mathrm{kg}^{3/2}$, and $b_4 = 3.78647\ \mathrm{mol}^{-2}\ \mathrm{kg}^2$. Plot the fitted results for γ_i as a function of m. On the same graph, show the data obtained from the *Handbook of Chemistry and Physics*.

(C) Using the fitted results, obtain the expected freezing-point depression for solutions of $Ni(ClO_4)_2$ with molalities between zero and 1 molal. Plot ΔT_f versus m.

(D) At what temperature would an aqueous solution of $Ni(ClO_4)_2$ containing 69.55 grams of $Ni(ClO_4)_2$ per 1,000 grams of H_2O freeze under an applied pressure of 1 bar?

Solution

(A) Since $\nu_+ = 1, \nu_- = 2, z_+ = 2$, and $z_- = 1$ for $Ni(ClO_4)_2$, we have

$$1.1726 z_+ z_- [0.5\{\nu_+ z_+^2 + \nu_- z_-^2\}]^{1/2} = 1.1726(2)(1)[0.5\{(1)(2)^2 + (2)(1)^2\}]^{1/2}$$
$$= 1.1726(2)[3]^{1/2} = 4.0620, \quad (1)$$

so that

$$3b_1 = \underline{-4.0620}. \quad (2)$$

(B) The equation we wish to fit to the measured activity coefficients is

$$\gamma_i = \exp\left[3b_1 m^{1/2} + 2b_2 m + \frac{5b_3 m^{3/2}}{3} + \frac{3b_4 m^2}{2}\right]$$
$$= \exp\left[-4.06204 m^{1/2} + 2b_2 m + \frac{5b_3 m^{3/2}}{3} + \frac{3b_4 m^2}{2}\right]. \quad (3)$$

Using the results given in the problem for this fitting, we obtain

$$\gamma_i = \exp[-4.0620 m^{1/2} + 10.8368 m - 12.5091 m^{3/2} + 5.67941 m^2]. \quad (4)$$

The graph at the top of the next page shows the results of this fit compared to the actual data. As can be seen, the fit is very good except for a small deviation in the region of $m \approx 0.2$ molal.

(C) Substituting the b_i coefficients into the expression for ΔT_f, we obtain

$$\Delta T_f = \nu K_f [m + b_1 m^{3/2} + b_2 m^2 + b_3 m^{5/2} + b_4 m^3]$$
$$= (3)(-1.86)[m - 1.354 m^{3/2} + 5.4184 m^2 - 7.5055 m^{5/2} + 3.7863 m^3]$$
$$= -5.58[m - 1.354 m^{3/2} + 5.4184 m^2 - 7.5055 m^{5/2} + 3.7863 m^3]. \quad (5)$$

The requested plot of Eq. 5 accompanies this solution.

(D) The molality of the solution is

$$m = \frac{\#\ \text{moles of Ni(ClO}_4)_2}{\#\ \text{kg H}_2\text{O}} = \frac{69.55\ \text{g} \times (1\ \text{mol}/257.59\ \text{g})}{1\ \text{kg}} = 0.2700\ \text{molal}. \quad (6)$$

Substituting $m = 0.2700$ molal into Eq. 5 gives

$$\Delta T_f = -5.58[0.2700 - 1.354(0.2700)^{3/2} + 5.4184(0.2700)^2 - 7.5055(0.2700)^{5/2}$$
$$+ 3.7863(0.2700)^3] = -1.480\ \text{K}. \quad (7)$$

The corresponding freezing point of the aqueous solution is

$$T_f = T_f^\circ + \Delta T_f = T_f^\circ - 1.480\text{ K} = 273.15 - 1.480\text{ K} = \underline{271.67\text{ K}}. \qquad (8)$$

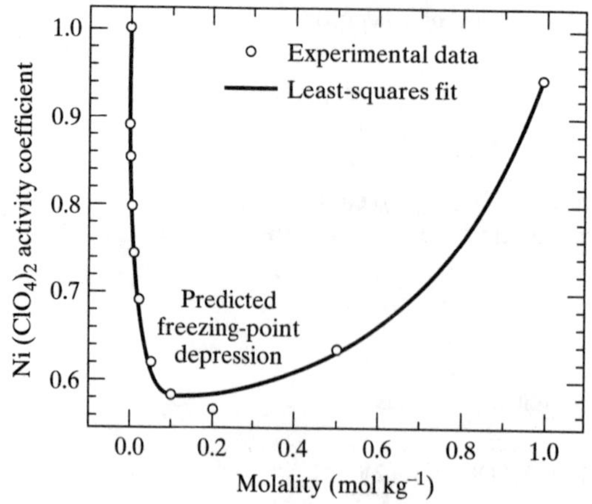

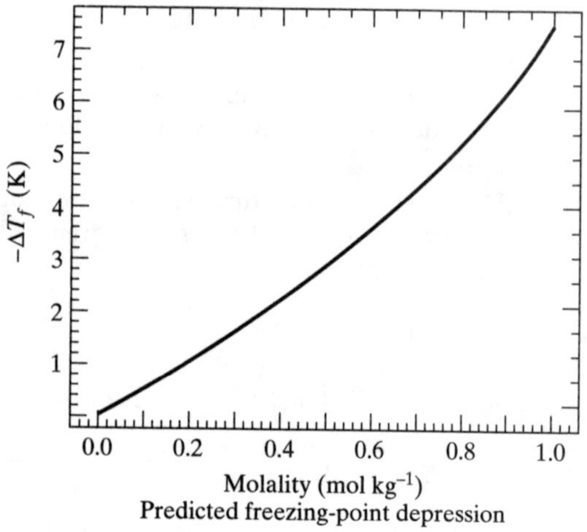

9.17 0.1 mole of formic acid ($HCHO_2$) and 0.15 mole of lactic acid ($HC_3H_5O_3$) are added to 1 kg of water at 298.15 K in the same container. The ionization constants are $K_a(HCHO_2) = 1.8 \times 10^{-4}$ and $K_a(HC_3H_5O_3) = 1.4 \times 10^{-4}$. Compute the H_3O^+ concentration in the solution. You may use the Debye–Hückel theory to compute mean ionic activity coefficients if needed.

Solution

The first step is to determine whether we need to consider the self-ionization of water. In the present example, we have $C_o K_a(HCHO_2) = (0.1)(1.8 \times 10^{-4}) = 1.8 \times 10^{-5}$ and $C_o K_a(HC_3H_5O_3) = (0.15)(1.4 \times 10^{-4}) = 2.1 \times 10^{-5}$. Both of these results are very large compared to K_w at 298.15 K, so we do not need to consider the self-ionization of water. There are, therefore, two equilibria of importance:

$$HCHO_2 + H_2O = H_3O^+ + CHO_2^- \text{ and}$$

$$HC_3H_5O_3 + H_2O = H_3O^+ + C_3H_5O_3^-.$$

The equilibrium-constant expressions for these two processes are:

$$K_a(HCHO_2) = \frac{a_{H_3O^+} a_{CHO_2^-}}{a_{HCHO_2} a_{H_2O}} = \frac{\gamma_\pm^2 m(H_3O^+) m(CHO_2^-)}{\gamma_{HCHO_2} m(HCHO_2) a_{H_2O}}$$

$$= \frac{\gamma_\pm^2 m(H_3O^+) m(CHO_2^-)}{m(HCHO_2)} \qquad (1)$$

and

$$K_a(HC_3H_5O_3) = \frac{a_{H_3O^+} a_{C_3H_5O_3^-}}{a_{HC_3H_5O_3} a_{H_2O}} = \frac{\gamma_\pm^2 m(H_3O^+) m(C_3H_5O_3^-)}{\gamma HC_3H_5O_3 m(HC_3H_5O_3) a_{H_2O}}$$

$$= \frac{\gamma_\pm^2 m(H_3O^+) m(C_3H_5O_3^-)}{m(HC_3H_5O_3)}, \qquad (2)$$

provided that we take both the activity of the solvent and the activity coefficient for the uncharged species to be unity. We now define two variables that represent the contribution of each reaction to the concentration of H_3O^+. Let

x = number of moles of H_3O^+ and CHO_2^- obtained from the first ionization and y = number of moles of H_3O^+ and $C_3H_5O_3^-$ obtained from the second ionization reaction. In these terms, we have

$$m(H_3O^+) = x + y,\ m(CHO_2^-) = x,\ m(HCHO_2) = C_o(HCHO_2) - x,$$
$$m(C_3H_5O_3^-) = y,$$

and

$$m(HC_3H_5O_3) = C_o(HC_3H_5O_3) - y.$$

Substituting these values into Eqs. 1 and 2 produces

$$K_a(HCHO_2) = \frac{\gamma_\pm^2(HCHO_2)(x+y)x}{C_o(HCHO_2) - x} \tag{3}$$

and

$$K_a(HC_3H_5O_3) = \frac{\gamma_\pm^2(HC_3H_5O_3)(x+y)y}{C_o(HC_3H_5O_3) - y}. \tag{4}$$

These equations must be solved iteratively. In the first iteration, we take both mean ionic activity coefficients to be unity, $C_o(HCHO_2) - x \approx C_o(HCHO_2) = C_{o1}$, and $C_o(HC_3H_5O_3) - y \approx C_o(HC_3H_5O_3) = C_{o2}$. With these approximations, Eqs. 3 and 4 become

$$K_a(HCHO_2) = K_{a1} = \frac{(x+y)x}{C_{o1}} \tag{5}$$

and

$$K_a(HC_3H_5O_3) = K_{a2} = \frac{(x+y)y}{C_{o2}}. \tag{6}$$

The solution for x and y in Eqs. 5 and 6 is easy to obtain. We first divide Eq. 5 by Eq. 6. This gives

$$\frac{K_{a1}}{K_{a2}} = \frac{C_{o2}x}{C_{o1}y}, \tag{7}$$

so that

$$x = \left[\frac{C_{o1}K_{a1}}{C_{o2}K_{a2}}\right]y. \tag{8}$$

Substituting Eq. 8 into Eq. 6 then yields

$$C_{o2}K_{a2} = y^2\left[\left(\frac{C_{o1}K_{a1}}{C_{o2}K_{a2}}\right) + 1\right]. \tag{9}$$

Solving Eq. 9 for y^2 produces

$$y^2 = \frac{C_{o2}^2K_{a2}^2}{C_{o1}K_{a1} + C_{o2}K_{a2}}. \tag{10}$$

Therefore,

$$y = \frac{C_{o2}K_{a2}}{[C_{o1}K_{a1} + C_{o2}K_{a2}]^{1/2}}. \tag{11}$$

Substituting Eq. 11 into Eq. 8 gives

$$x = \frac{C_{o1}K_{a1}}{[C_{o1}K_{a1} + C_{o2}K_{a2}]^{1/2}}. \tag{12}$$

The H_3O^+ is the sum $x + y$. This is

$$m(H_3O^+) = x + y = \frac{C_{o2}K_{a2} + C_{o1}K_{a1}}{[C_{o1}K_{a1} + C_{o2}K_{a2}]^{1/2}} = [C_{o1}K_{a1} + C_{o2}K_{a2}]^{1/2}. \quad \textbf{(13)}$$

Equations 11–13 are the formal solutions of the first iteration. Substituting the data gives

$$x = \frac{(0.1)(1.8 \times 10^{-4})}{[(0.1)(1.8 \times 10^{-4}) + (0.15)(1.4 \times 10^{-4})]^{1/2}} = 0.2882 \times 10^{-2} \text{ mol kg}^{-1}, \quad \textbf{(14)}$$

$$y = \frac{(0.15)(1.4 \times 10^{-4})}{[(0.1)(1.8 \times 10^{-4}) + (0.15)(1.4 \times 10^{-4})]^{1/2}} = 0.3363 \times 10^{-2} \text{ mol kg}^{-1}, \quad \textbf{(15)}$$

and

$$m(H_3O^+) = x + y = 0.6245 \times 10^{-2} \text{ mol kg}^{-1}. \quad \textbf{(16)}$$

In the second iteration, we use the Debye–Hückel theory to calculate the mean ionic activity coefficients. Since all charges are ± 1, both activity coefficients are equal to

$$\log(\gamma_\pm) = -0.50926(1)(1)S^{1/2} = -0.50926S^{1/2}, \quad \textbf{(17)}$$

where the ionic strength is given by

$$S = 0.5 \sum_{i=1}^{i=K} c_i z_i^2$$
$$= 0.5[0.2882 \times 10^{-2}(1)^2 + 0.3363 \times 10^{-2}(1)^2 + 0.6245 \times 10^{-2}(1)^2]$$
$$= 0.6245 \times 10^{-2}. \quad \textbf{(18)}$$

Substituting into Eq. 17 gives

$$\log(\gamma_\pm) = -0.04024. \quad \textbf{(19)}$$

Consequently, $\gamma_\pm = 0.9115$ and $\gamma_\pm^2 = 0.8308$.

In the second iteration, we replace K_{a1} with $K_{a1}/0.8308 = 2.167 \times 10^{-4}$ and K_{a2} with $K_{a2}/0.8308 = 1.685 \times 10^{-4}$. We also replace C_{o1} with $C_{o1} - x = 0.1 - 0.002882 = 0.09712$ and C_{o2} with $C_{o2} - y = 0.15 - 0.003363 = 0.1466$. With these replacements, the solutions given by Eqs. 11, 12, and 13 are

$$x = 0.003111 \text{ mol kg}^{-1}, \quad \textbf{(20)}$$

$$y = 0.003652 \text{ mol kg}^{-1}, \quad \textbf{(21)}$$

and

$$m(H_3O^+) = x + y = 0.006763 \text{ mol kg}^{-1}. \quad \textbf{(22)}$$

This last value differs by 8.3% from that obtained in the first iteration. We could regard this result as being nearly converged, with a final answer of $m(H_3O^+) \approx \underline{0.00676 \text{ mol kg}^{-1}}$, or we could execute a third iteration.

Note that this answer is very different from that which would be obtained if the two equilibria were regarded as being independent, as they would be regarded if the formic and lactic acid were in different containers. Under these conditions, the H_3O^+ concentration in each container is given approximately by

$$m(H_3O^+) = [C_o K_a]^{1/2} \quad \textbf{(23)}$$

provided that we set the activity coefficients to unity and that $C_o - x \approx C_o$. The results would then be

$$m(H_3O^+)_{\text{formic acid}} = [(0.1)1.8 \times 10^{-4}]^{1/2} = 0.004243 \text{ mol kg}^{-1} \quad \textbf{(24)}$$

and

$$m(H_3O^+)_{\text{lactic acid}} = [(0.15)1.4 \times 10^{-4}]^{1/2} = 0.004583 \text{ mol kg}^{-1}. \quad \textbf{(25)}$$

The total H_3O^+ would then be

$$m(H_3O^+)_{\text{formic acid}} + m(H_3O^+)_{\text{lactic acid}} = 0.008826 \text{ mol kg}^{-1}. \quad \textbf{(26)}$$

This result is higher than the correct answer by 30.5%.

9.19 Compute the H_3O^+ concentration in mol kg^{-1} if 10^{-7} mol kg^{-1} of HCl is added to water at 298.15 K. Assume that the HCl is completely ionized and that the mean ionic activity coefficient is unity at these low concentrations of ionic species.

Solution

The only equilibrium involved in this problem is that for the self-ionization of water, viz.,

$$H_2O + H_2O = H_3O^+_{(aq)} + OH^{1-}_{(aq)},$$

for which we have

$$K_w = 1.0 \times 10^{-14} = \gamma_\pm^2 m(H_3O^+)m(OH^-) = m(H_3O^+)m(OH^-) \quad \textbf{(1)}$$

if we may assume that $\gamma_\pm^2 = 1$.

Let x = number of mol kg^{-1} of H_3O^+ and OH^- produced by the ionization of water. The total H_3O^+ concentration is, therefore,

$$m(H_3O^+) = 10^{-7} + x \quad \textbf{(2)}$$

and

$$m(OH^-) = x. \quad \textbf{(3)}$$

Substituting Eqs. 2 and 3 into Eq. 1 gives

$$[10^{-7} + x]x = 1.0 \times 10^{-14} = x^2 + 10^{-7}x. \quad \textbf{(4)}$$

Solving Eq. 4 produces

$$x = \frac{-10^{-7} + [10^{-14} + 4 \times 10^{-14}]^{1/2}}{2} = \frac{(5^{1/2} - 1) \times 10^{-7}}{2}$$

$$= 0.61803 \times 10^{-7} \text{ mol kg}^{-1} \quad \textbf{(5)}$$

The total H_3O^+ concentration given by Eq. 2 is thus

$$m(H_3O^+) = 10^{-7} + x = \underline{1.61803 \times 10^{-7} \text{ mol kg}^{-1}}. \quad \textbf{(6)}$$

9.21 A quantity of 0.2 mol kg^{-1} of trimethylammonium chloride [$(CH_3)_3NHCl$] is added to water at 298.15 K. Compute the H_3O^+ concentration in the solution. Assume complete ionization of the $(CH_3)_3NHCl$ salt. K_b for $(CH_3)_3N$ is 6.25×10^{-10} at 298.15 K.

Solution

(A) The important reactions are

$$(CH_3)_3NHCl + x\,H_2O \longrightarrow (CH_3)_3NH^+_{(aq)} + Cl^-_{(aq)},$$

$$(CH_3)_3NH^+_{(aq)} + H_2O = H_3O^+_{(aq)} + (CH_3)_3N,$$

and

$$H_2O + H_2O = H_3O^+_{(aq)} + OH^-_{(aq)}.$$

The problem states that we may assume that the first reaction goes to completion. Since we have added 0.2 mol kg^{-1} of $(CH_3)_3NHCl$, we get 0.2 mol kg^{-1} of $(CH_3)_3NH^+_{(aq)}$ and $Cl^-_{(aq)}$. The hydrolysis constant for the second reaction is given by Eq. 9.92 for strong conjugate acids:

$$K_h = \frac{K_w}{K_b} = \frac{1.0 \times 10^{-14}}{6.25 \times 10^{-5}} = 1.6 \times 10^{-10}. \tag{1}$$

Consequently, $C_o K_h = 3.2 \times 10^{-11} \gg K_w = 1.0 \times 10^{-14}$, so we may ignore the effect of the self-ionization of water on the H_3O^+ concentration. The development leading up to Eq. 9.94 shows that we expect to have

$$K_h = \frac{m(H_3O^+)m((CH_3)_3N)}{m((CH_3)_3NH^+)}. \tag{2}$$

If we now let x = number of mol kg^{-1} of H_3O^+ and $(CH_3)_3N$, the stoichiometry of the second reaction shows that $m[(CH_3)_3NH^+_{(aq)}] = 0.2 - x$. Substituting these values into Eq. 2 gives

$$K_h = \frac{x^2}{0.2 - x}. \tag{3}$$

Writing Eq. 3 in quadratic form produces

$$x^2 + K_h x - 0.2K_h = 0. \tag{4}$$

The solution of Eq. 4 is

$$x = m(H_3O^+) = \frac{-K_h + [K_h^2 + 0.8K_h]^{1/2}}{2}. \tag{5}$$

Substituting K_h from Eq. 1 gives

$$m(H_3O^+) = \underline{5.66 \times 10^{-6}\ \text{mol kg}^{-1}}. \tag{6}$$

(B) For the self-ionization of water,

$$K_w = \gamma_\pm^2 m(H_3O^+)m(OH^-). \tag{7}$$

Therefore, the $OH^-_{(aq)}$ concentration is

$$m(OH^-) = \frac{K_w}{\gamma_\pm^2 m(H_3O^+)}. \tag{8}$$

The ionic strength of the solution is determined almost solely by the $(CH_3)_3NH^+_{(aq)}$ and $Cl^-_{(aq)}$ ions, whose concentrations are each 0.20 mol kg^{-1}. The amount of $(CH_3)_3NH^+_{(aq)}$ that has hydrolyzed (5.66×10^{-6} mol kg^{-1}) makes no difference, since the hydrolysis of one $(CH_3)_3NH^+_{(aq)}$ ion produces one $H_3O^+_{(aq)}$ ion in its place. The ionic strength of the solution is, therefore, unaffected. Thus,

$$S = 0.5 \sum_{i=1}^{i=K} c_i z_i^2 = 0.5[(0.2)(1)^2 + (0.2)(1)^2] = 0.20. \tag{9}$$

At this concentration, we need to use the Davies empirical formula to estimate $\gamma_\pm^2$. Equation 9.68 is

$$\log(\gamma_i) = -0.50926 z_+ z_- \left[\frac{S^{1/2}}{1 + S^{1/2}} - 0.30S\right]$$

$$= -0.50926(1)(1)\left[\frac{(0.2)^{1/2}}{1 + (0.2)^{1/2}} - 0.30(0.2)\right] = -0.12681. \tag{10}$$

Hence, $\gamma_\pm = 10^{-0.12681} = 0.74677$ and $\gamma_\pm^2 = 0.55766$. Substituting into Eq. 8 produces

$$m(\text{OH}^-) = \frac{1.0 \times 10^{-14}}{(0.55766)(5.66 \times 10^{-6})} = \underline{3.17 \times 10^{-9}\ \text{mol kg}^{-1}}. \tag{11}$$

9.23 A quantity of 5×10^{-4} mol kg^{-1} of trimethylammonium chloride $[(CH_3)_3NHCl]$ is added to a water solution at 298.15 K containing 0.2 mol kg^{-1} of NaCl. Compute the H_3O^+ concentration in the solution. Assume complete ionization of NaCl and $(CH_3)_3NHCl$. K_b for $(CH_3)_3N$ is 6.25×10^{-10} at 298.15 K.

Solution

The important reactions are

$$(\text{CH}_3)_3\text{NHCl} + x\text{H}_2\text{O} \longrightarrow (\text{CH}_3)_3\text{NH}^+_{(aq)} + \text{Cl}^-_{(aq)},$$

$$\text{NaCl} + x\text{H}_2\text{O} \longrightarrow \text{Na}^+_{(aq)} + \text{Cl}^-_{(aq)}$$

$$(\text{CH}_3)_3\text{NH}^+_{(aq)} + \text{H}_2\text{O} \longrightarrow \text{H}_3\text{O}^+_{(aq)} + (\text{CH}_3)_3\text{N},$$

and

$$\text{H}_2\text{O} + \text{H}_2\text{O} = \text{H}_3\text{O}^+_{(aq)} + \text{OH}^-_{(aq)}.$$

The problem states that we may assume that the first two reactions go to completion. Since we have added 5×10^{-4} mol kg^{-1} of $(CH_3)_3NHCl$, we get 5×10^{-4} mol kg^{-1} of $(CH_3)_3NH^+_{(aq)}$ and $Cl^-_{(aq)}$. We also get 0.2 mol kg^{-1} of $Na^+_{(aq)}$ and $Cl^-_{(aq)}$ from ionization of the NaCl. The hydrolysis constant for the third reaction is given by Eq. 9.92 for strong conjugate acids:

$$K_h = \frac{K_w}{K_b} = \frac{1.0 \times 10^{-14}}{6.25 \times 10^{-5}} = 1.6 \times 10^{-10}. \tag{1}$$

We see that $C_oK_h = 8.0 \times 10^{-14}$. Therefore, we do not have $C_oK_a \gg K_w$. Consequently, we must consider the effect of the self-ionization of water on the H_3O^+ concentration. The presence of the NaCl means we cannot set the mean ionic activity coefficient for the ionization of water equal to unity. The ionic strength of the solution is given by

$$\begin{aligned} S &= 0.5 \sum_{i=1}^{i=K} c_i z_i^2 \\ &= 0.5[(0.2)(1)^2 + (0.2)(1)^2 + 5 \times 10^{-4}(1)^2 + 5 \times 10^{-4}(1)^2] = 0.2005, \end{aligned} \tag{2}$$

since the concentrations of $Na^+_{(aq)}$ and $Cl^-_{(aq)}$ are each 0.20 mol kg^{-1} and there are 5×10^{-4} mol kg^{-1} of $(CH_3)_3NH^+_{(aq)}$ and $Cl^-_{(aq)}$ present from the ionization of the $(CH_3)_3NHCl$ salt. At this concentration, we need to use the Davies empirical formula to estimate $\gamma_\pm^2$. Equation 9.68 is

$$\begin{aligned} \log(\gamma_i) &= -0.50926 z_+ z_- \left[\frac{S^{1/2}}{1 + S^{1/2}} - 0.30S\right] \\ &= -0.50926(1)(1)\left[\frac{(0.2005)^{1/2}}{1 + (0.2005)^{1/2}} - 0.30(0.2005)\right] = -0.12687. \end{aligned} \tag{3}$$

Thus, $\gamma_\pm = 10^{-0.12687} = 0.7467$ and $\gamma_\pm^2 = 0.5576$.

For the water ionization equilibrium, we have

$$K_w = \gamma_\pm^2 m(\text{H}_3\text{O}^+) m(\text{OH}^-), \tag{4}$$

which we can write in the form

$$\frac{K_w}{\gamma_\pm^2} = K'_w = m(H_3O^+)m(OH^-) = \frac{1.0 \times 10^{-14}}{0.5576} = 1.793 \times 10^{-14}. \quad \textbf{(5)}$$

The development leading up to Eq. 9.97 shows that, for a strong conjugate acid, we expect to have

$$m(H_3O^+) = \frac{K_w}{[K_w + C_oK_h]^{1/2}} + \frac{C_oK_h}{[K_w + C_oK_h]^{1/2}} = [K_w + C_oK_h]^{1/2} \quad \textbf{(6)}$$

when the contribution from the self-ionization of H_2O is considered. The same result will hold for this problem, wherein we must include the effect of $\gamma_\pm^2$, provided that we replace K_w in Eq. 6 with K'_w. This gives

$$m(H_3O^+) = [1.793 \times 10^{-14} + (5 \times 10^{-4})(1.6 \times 10^{-10})]^{1/2} = \underline{3.13 \times 10^{-7}\ \text{mol kg}^{-1}}. \quad \textbf{(7)}$$

9.25 Using the data in Appendix A, compute the K_{sp} values for the following compounds at 298.15 K:

(A) calcite ($CaCO_3(s)$);

(B) NaCl(s);

(C) $BaCl_2(s)$.

Solution

(A) The reaction in question is $CaCO_3(s) + xH_2O = Ca^{2+}_{(aq)} + CO^{2-}_{3(aq)}$. The change in the standard chemical potential for this reaction is

$$\Delta\mu^o = \mu^o_{CO_3^{2-}(aq)} + \mu^o_{Ca^{2+}}(aq) - \mu^o_{CaCO_3(s)} = -527.81 - 553.58 - (-1{,}128.8)\ \text{kJ mol}^{-1}$$

$$= 47.41\ \text{kJ mol}^{-1}. \quad \textbf{(1)}$$

The solubility product is related to $\Delta\mu^o$ by

$$\Delta\mu^o = -RT\ln(K_{sp}), \quad \textbf{(2)}$$

with $T = 298.15$ K. Solving for K_{sp}, we obtain

$$K_{sp} = \exp\left[-\frac{47{,}410\ \text{J mol}^{-1}}{(8.314\ \text{J mol}^{-1}\ \text{K}^{-1})(298.15\ \text{K})}\right] = e^{-19.126} = \underline{4.94 \times 10^{-9}}. \quad \textbf{(3)}$$

(B) We now have $NaCl(s) + xH_2O = Na^+_{(aq)} + Cl^-_{(aq)}$, for which

$$\Delta\mu^o = \mu^o_{Na^+(aq)} + \mu^o_{Cl^-(aq)} - \mu^o_{NaCl(s)} = -261.91 - 131.23 - (-384.14)$$

$$= -9.000\ \text{kJ mol}^{-1}. \quad \textbf{(4)}$$

This yields

$$K_{sp} = \exp\left[-\frac{-9{,}000\ \text{J mol}^{-1}}{(8.314\ \text{J mol}^{-1}\ \text{K}^{-1})(298.15\ \text{K})}\right] = e^{3.63076} = \underline{37.74}. \quad \textbf{(5)}$$

(C) The solution reaction is $BaCl_2(s) + xH_2O = Ba^{2+}_{(aq)} + 2\,Cl^-_{(aq)}$. The corresponding change in the chemical potential is

$$\Delta\mu^o = \mu^o_{Ba^{2+}(aq)} + 2\,\mu^o_{Cl^-(aq)} - \mu^o_{BaCl_2(s)} = -560.77 + 2(-131.23) - (-810.4)$$

$$= -12.83\ \text{kJ mol}^{-1}, \quad \textbf{(4)}$$

so that

$$K_{sp} = \exp\left[-\frac{-12{,}830 \text{ J mol}^{-1}}{(8.314 \text{ J mol}^{-1} \text{ K}^{-1})(298.15 \text{ K})}\right] = e^{5.17585} = \underline{176.94}. \quad (5)$$

9.27[#] The measured mean ionic activity coefficients for NaCl in aqueous solution at 298.15 K are reported in the *Handbook of Chemistry and Physics*. The table to the left shows the data at different concentrations. In Problem 9.14, it was shown that we may obtain an accurate fit of measured activity coefficients by using the function

m (mol kg^{-1})	$\gamma_\pm$
0.000	1.000
0.001	0.965
0.002	0.952
0.005	0.928
0.010	0.903
0.020	0.872
0.050	0.822
0.100	0.779
0.200	0.734
0.500	0.681
1.000	0.657
2.000	0.668
5.000	0.874

$$\gamma_i = \exp\left[3b_1 m^{1/2} + 2b_2 m + \frac{5b_3 m^{3/2}}{3} + \frac{3b_4 m^2}{2}\right],$$

where, for compatibility with the Debye-Hückel theory, we must have

$$3b_1 = -1.17261 z_+ z_- [0.5\{\nu_+ z_+^2 + \nu_- z^2\}]^{1/2}$$

When the data for NaCl are fitted to the function

$$\gamma_\pm = \exp[-1.17261 m^{1/2} + c_2 m + c_3 m^{3/2} + c_4 m^2]$$

using least-squares methods, the result is

$$\gamma_\pm = \exp[-1.17261 m^{1/2} + 1.31903 m - 0.704482 m^{3/2} + 0.150754 m^2].$$

The plot at the end of the problem shows the quality of this fit. The solubility product constant for NaCl(s) is 37.74 at 298.15 K. Compute the solubility of NaCl(s) in H_2O at 298.15 K in terms of grams per kg of water. Use the fitted experimental data for NaCl to obtain activity coefficients if needed. The datum reported in the *Handbook of Chemistry and Physics* is 359.62 g NaCl per kg of water at 298.15 K. Compute the percent error in your answer.

Solution

The solution reaction is NaCl(s) + $x H_2O$ = $Na^+_{(aq)} + Cl^-_{(aq)}$.
The solubility product constant is given by Eq. 9.102

$$K_{sp} = \gamma_\pm^{(\nu_+ + \nu_-)} [\nu_+ (m/m^o)]^{\nu_+} [\nu_- (m/m^o)]^{\nu_-}. \quad (1)$$

For NaCl(s), this equation has the form

$$K_{sp} = \gamma_\pm^2 (m/m^o)^2. \quad (2)$$

Canceling the units and setting $m^o = 1$, we have

$$K_{sp} = \gamma_\pm^2 m^2. \quad (3)$$

Substituting the fitted expression for the activity coefficient into Eq. 3, we obtain

$$m^2\{\exp[-1.17261 m^{1/2} + 1.31903 m - 0.704482 m^{3/2} + 0.150754\, m^2]\}^2 - K_{sp} = 0. \quad (4)$$

By using $K_{sp} = 37.74$, we can execute a one-dimensional grid search for the solution of this equation. The result is $m = 5.800$ mol kg^{-1}, correct to four significant digits. Since the molar mass of NaCl is 58.443 g mol^{-1}, the solubility is <u>339.0 g</u> of NaCl per 1,000 g of water. The percent error in this result is

$$\% \text{ error} = \frac{100 \times (339.0 - 359.62)}{359.62} = \underline{-5.73\%}. \quad (5)$$

Most of the error is probably due to our use of the fitted expression for $\gamma_{\pm}$ at $m = 5.800$ mol kg^{-1}, which is outside the range of the experimental data from which the analytic fit was obtained.

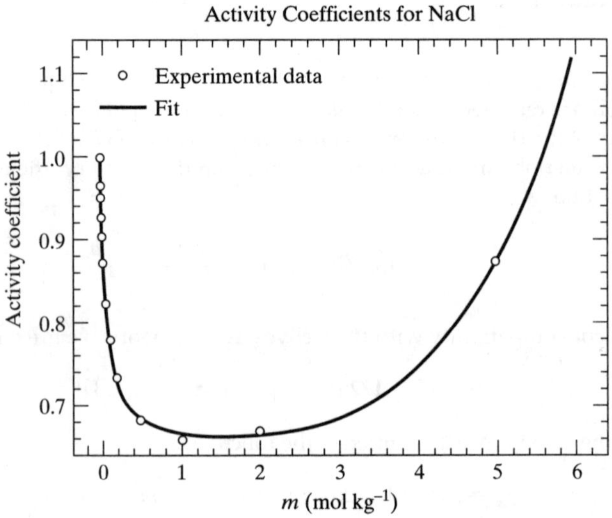

9.29 The solubility product constant for $BaCl_2(s)$ is 176.94 at 298.15 K. The measured solubility of $BaCl_2(s)$ in water at the temperature is 370.43 g kg^{-1} water. Determine the mean ionic activity coefficient for $BaCl_2$ at saturation.

Solution

The solution reaction is $BaCl_2(s) + xH_2O = Ba^{2+}_{(aq)} + 2\,Cl^{-}_{(aq)}$.
The solubility product constant is given by Eq. 9.102:

$$K_{sp} = \gamma_{\pm}^{(\nu_+ + \nu_-)}[\nu_+(m/m^o)]^{\nu_+}[\nu_-(m/m^o)]^{\nu_-}. \qquad (1)$$

For $BaCl_2(s)$, this equation has the form

$$K_{sp} = 4\gamma_{\pm}^3(m/m^o)^3. \qquad (2)$$

Canceling the units and setting $m^o = 1$, we have

$$K_{sp} = 4\gamma_{\pm}^3 m^3. \qquad (3)$$

A solubility of 370.43 g kg^{-1} corresponds to

$$m = [370.43\text{ g kg}^{-1}]\left[\frac{1\text{ mol}}{208.21\text{ g}}\right] = 1.7791\text{ mol kg}^{-1}. \qquad (4)$$

Solving Eq. 3 for $\gamma_{\pm}^3$, we obtain

$$\gamma_{\pm}^3 = \frac{K_{sp}}{4m^3} = \frac{176.94}{4(1.7791)^3} = 7.8553. \qquad (5)$$

This gives

$$\gamma_{\pm} = \underline{1.988} \qquad (6)$$

at saturation.

9.31 The measured molar conductivities of $AgNO_3$ at 298.15 K are given in the table below.

Data for Problem 9.31

c (mol L^{-1})	Λ_m (S cm^2 mol^{-1})
0.0005	131.29
0.0010	130.45
0.0050	127.14
0.0100	124.70
0.0200	121.35

Using an appropriate method, obtain Λ_m^∞ for $AgNO_3$ in aqueous solution at 298.15 K.

Solution

The discussion in the text suggests that a plot of Λ_m versus $c^{1/2}$ should be nearly linear at low concentrations, since the Debye–Hückel theory shows that the electrostatic interaction between ions varies as $c^{1/2}$ at low concentration. The plot given at the end of the problem shows that this expectation holds reasonably well. The equation of the linear least-squares fit to the given data is

$$\Lambda_m = 133.10 - 83.469c^{1/2}\ S\ \text{cm}^2\ \text{mol}^{-1}. \qquad \textbf{(1)}$$

Therefore, the infinite-dilution extrapolated value of Λ_m^∞ is $\underline{133.10\ S\ \text{cm}^2\ \text{mol}^{-1}}$. The value reported in the *Handbook of Chemistry and Physics* is $\underline{133.29\ S\ \text{cm}^2\ \text{mol}^{-1}}$.

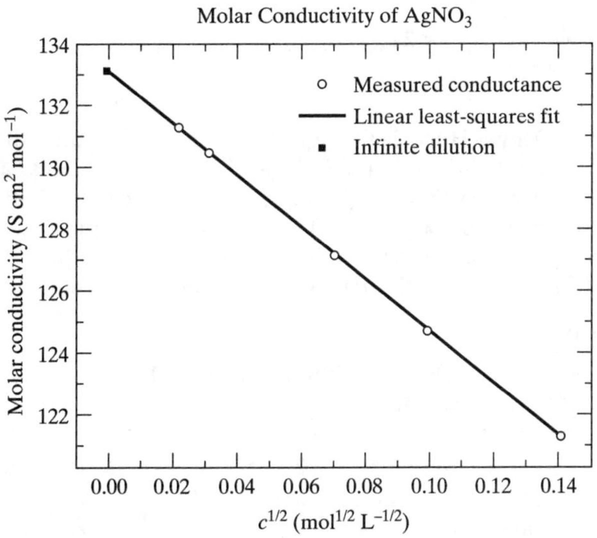

9.33 The electric mobility of Mg^{2+}(aq) and Cl^-(aq) at infinite dilution and 298.15 K are 55.0×10^{-5} cm^2 V^{-1} s^{-1} and 79.1×10^{-5} cm^2 V^{-1} s^{-1}, respectively. Compute the transference numbers and molar ionic conductances of Mg^{2+}(aq) and Cl^-(aq) at infinite dilution. What is the value of $\Lambda_{MgCl_2}^\infty$ at 298.15 K?

Solution

The transference numbers are given by Eq. 9.124:

$$t_{Mg^{2+}} = \frac{u_{Mg^{2+}}}{u_{Mg^{2+}} + u_{Cl^-}} = \frac{55.0 \times 10^{-5}}{55.0 \times 10^{-5} + 79.1 \times 10^{-5}} = \underline{0.410}. \quad (1)$$

Since transference numbers must add to unity, we have

$$t_{Cl^-} = 1 - t_{Mg^{2+}} = 1 - 0.410 = \underline{0.590}. \quad (2)$$

The ionic conductances are given by

$$\kappa_i = \mathcal{F} c_i z_i u_i. \quad (3)$$

The molar ionic conductances are given by Eq. 9.125:

$$\lambda_i = \kappa_i / c_i = \mathcal{F} z_i u_i. \quad (4)$$

Therefore, at infinite dilution, we have

$$\begin{aligned} \lambda^{\infty}_{Mg^{2+}} &= (96{,}485 \text{ C mol}^{-1})(2)(55.0 \times 10^{-5} \text{ cm}^2 \text{ V}^{-1} \text{ s}^{-1}) \\ &= (2)(96{,}485 \text{ amp s mol}^{-1})(55.0 \times 10^{-5} \text{ cm}^2 \text{ V}^{-1} \text{ s}^{-1}) \\ &= 106.1 \text{ amp cm}^2\, V^{-1} \text{ mol}^{-1}. \end{aligned} \quad (5)$$

But an ohm (Ω) is a volt amp^{-1}, so 1 amp V^{-1} = 1 Ω^{-1} = 1 S. Therefore,

$$\lambda^{\infty}_{Mg^{2+}} = \underline{106.1 \text{ S cm}^2 \text{ mol}^{-1}}. \quad (6)$$

For the Cl^- ion, we obtain

$$\lambda^{\infty}_{Cl^-} = (96{,}485 \text{ C mol}^{-1})(1)(79.1 \times 10^{-5} \text{ cm}^2 \text{ V}^{-1} \text{ s}^{-1}) = \underline{76.32 \text{ S cm}^2 \text{ mol}^{-1}}. \quad (7)$$

The molar conductivity is given by Eq. 9.126:

$$\begin{aligned} \Lambda^{\infty}_{MgCl_2} &= c^{-1}[c_{Mg^{2+}}\lambda^{\infty}_{Mg^{2+}} + c_{Cl^-}\lambda^{\infty}_{Cl^-}] = \lambda^{\infty}_{Mg^{2+}} + 2\,\lambda^{\infty}_{Cl^-} \\ &= 106.1 + (2)\,76.32 \text{ S cm}^2 \text{ mol}^{-1} = \underline{258.7 \text{ S cm}^2 \text{ mol}^{-1}}. \end{aligned} \quad (8)$$

In Eq. 8, we have $c_{Cl^-} = 2\,c_{Mg^{2+}} = 2c$.

9.35 Using the Debye–Hückel–Onsager limiting law, compute the ionic molar conductivity for Zn^{2+}(aq) and SO_4^{2-}(aq) ions in a 0.01-molar aqueous solution of $ZnSO_4$ at 298.15 K. The limiting ionic molar conductivities are $\lambda^{\infty}_{Zn^{2+}} =$ 105.6 S cm^2 mol^{-1} and $\lambda^{\infty}_{SO_4^-} = 160.0$ S cm^2 mol^{-1}. Use your results to compute $\Lambda_m(ZnSO_4)$ at the given concentration. The value listed in the *Handbook of Chemistry and Physics* is 169.74 S cm^2 mol^{-1}. Calculate the percent error in your result.

Solution

The Debye–Hückel–Onsager limiting law for a 1:1 electrolyte in aqueous solution at 298.15 K is given by Eq. 9.129:

$$\lambda_m = \lambda^{\infty}_m - [30.32 z^3 + 0.2290 z^3 \lambda^{\infty}_m]\left[\frac{c}{c^o}\right]^{1/2} \Omega^{-1} \text{ cm}^2 \text{ mol}^{-1}. \quad (1)$$

For the Zn^{2+} ion,

$$\begin{aligned} \lambda_{Zn^{2+}} &= 105.6 - [30.32(2)^3 + 0.2290(2)^3(105.6)](0.01)^{1/2} \\ &= 62.00\ \Omega^{-1} \text{ cm}^2 \text{ mol}^{-1}. \end{aligned} \quad (2)$$

For the SO_4^{2-} ion

$$\begin{aligned} \lambda_{SO^{2-}{}_4} &= 160.0 - [30.32(2)^3 + 0.2290(2)^3(160.0)](0.01)^{1/2} \\ &= 106.43\ \Omega^{-1} \text{ cm}^2 \text{ mol}^{-1}. \end{aligned} \quad (3)$$

The molar conductivity is related to the individual ionic molar conductivities by Eq. 9.126:

$$\Lambda_m = c^{-1}\sum_{i=1}^{K} c_i\lambda_i = \lambda_{Zn^{2+}} + \lambda_{SO^{2-}_4} = 62.00 + 106.43 = \underline{168.43\ \Omega^{-1}\ \text{cm}^2\ \text{mol}^{-1}}. \quad (4)$$

The percent error in this result is

$$\%\ \text{error} = \frac{100(168.43 - 169.74)}{169.74} = \underline{-0.77\%}. \quad (5)$$

9.37 When a 0.005-molar aqueous solution of $BaCl_2$ at 298.15 K is placed in a conductivity cell whose cell constant is 0.280 cm^{-1}, its resistance is found to be 218.82 ohms. The solution is now placed above LiCl in a moving-boundary apparatus whose cross-sectional area is 25 cm^2. An electric potential is applied across the apparatus such that a current of 2.5 amperes flows. After 135 seconds have elapsed, how far has the $BaCl_2$–LiCl boundary moved if the electric mobility of Ba^{2+} in the solution is 0.0006029 V^{-1} cm^2 s^{-1}?

Solution

The specific conductivity of the $BaCl_2$ solution at 298.15 K is

$$\kappa = \frac{C}{R} = \frac{0.280\ \text{cm}^{-1}}{218.82\ \Omega} = 0.0012796\ \Omega^{-1}\ \text{cm}^{-1} = 0.0012796\ \text{amp V}^{-1}\ \text{cm}^{-1}, \quad (1)$$

since 1 Ω^{-1} = 1 amp V^{-1}. The relationship between the distance the boundary has moved and the Ba^{2+} electric mobility is given by Eq. 9.137:

$$u(Ba^{2+}) = \frac{\kappa A d}{Q}. \quad (2)$$

Solving Eq. 2 for d, we obtain

$$d = \frac{u(Ba^{2+})Q}{\kappa A} = \frac{u(Ba^{2+})It}{\kappa A}. \quad (3)$$

Substituting the data yields

$$d = \frac{(0.0006029\ \text{V}^{-1}\ \text{cm}^2\ \text{s}^{-1})(2.5\ \text{amp})(135\ \text{s})}{(0.0012796\ \text{amp V}^{-1}\ \text{cm}^{-1})(25\ \text{cm}^2)} = \underline{6.36\ \text{cm}}. \quad (4)$$

9.39 A 0.01-molar aqueous solution of $CuSO_4$ at 298.15 K is placed in the Hittorf apparatus shown in Figure 9.21. A current of 0.1 amp is passed through the cell for 30 minutes, at which point the concentration of Cu^{2+} in the anode compartment is found to be 0.0286 molar. If the volume of the anode compartment is 30 ml, determine the transference numbers of Cu^{2+} and SO_4^{2-}.

Solution

Using Eq. 9.141, we have

$$c = c_o + \frac{t_- Q}{2\mathcal{F}V}. \quad (1)$$

Solving this equation for t_- for the SO_4^{2-} ion, we obtain

$$t_- = \frac{(c - c_o)2\mathcal{F}V}{Q} = \frac{(c - c_o)2\mathcal{F}V}{It}. \quad (2)$$

Substituting the given data results in

$$t_- = \frac{(0.0286 \text{ mol L}^{-1} - 0.010 \text{ mol L}^{-1})(2)(96{,}485 \text{ amp s mol}^{-1})(0.030 \text{ L})}{(0.1 \text{ amp})(1{,}800 \text{ s})} = \underline{0.598} \tag{3}$$

for the SO_4^{2-} ion. For the Cu^{2+} ion,

$$t_{Ba^{2+}} = 1 - t_{SO_4^{2-}} = 1 - 0.598 = \underline{0.402}. \tag{4}$$

9.41# One liter of 0.01-molar HCl is being titrated with 0.01-molar NaOH.

(A) Obtain expressions giving the composition of the solution in terms of the molarity of HCl, NaCl, and NaOH as a function of the volume V of NaOH added. Assume that the volumes are additive, and ignore the contribution of the self-ionization of H_2O.

(B) The molar conductivities of HCl, NaCl, and NaOH are given in the following table in units of Ω^{-1} cm^2 mol^{-1}. The Debye–Hückel–Onsager theory indicates that the molar conductivities should vary linearly at low concentrations with $c^{1/2}$. Plot the data sets given in the table against $c^{1/2}$, and obtain the best linear fit for each compound.

c (mol L^{-1})	Λ_{HCl}	Λ_{NaCl}	Λ_{NaOH}
0.000	425.95	126.39	247.7
0.0005	422.53	124.44	245.5
0.001	421.15	123.68	244.6
0.005	415.59	120.59	240.7
0.010	411.80	118.45	237.9

Data from the *Handbook of Chemistry and Physics*, 78th ed., CRC Press, Boca Raton, FL, 1997–1998.

(C) Using the results of (A) and (B), compute the conductivity of the solution as a function of V, the volume in liters of NaOH added, from $V = 0$ to $V = 2$ L. Assume that the conductivities of HCl, NaCl, and NaOH are additive at all concentrations. Plot the calculated conductivity versus V over the range $0 \le V \le 2$ L.

Solution

(A) The stoichiometry of the titration reaction is such that 1 mole of OH^- neutralizes 1 mole of H_3O^+ formed by the ionization of the HCl in aqueous solution. Therefore,

$$\text{Number moles of HCl} = N_{HCl}$$

$$= \text{initial number of moles of HCl} - \text{moles of HCl titrated.} \tag{1}$$

The initial number of moles of HCl present is given by the initial volume of the HCl solution times the molarity, (1 L)(0.01 mol L^{-1}) = 0.01 mol HCl. The number of moles of HCl titrated is equal to the number of moles of NaOH added, which is the volume in liters of NaOH added times the molarity of the NaOH solution. Thus, we have

$$N_{\text{HCL}} = 0.01 - 0.01V = 0.01(1 - V), \tag{2}$$

for $V \leq 1.00$ L.

The corresponding concentration is

$$C_{\text{HCl}} = \frac{N_{\text{HCl}}}{V_T}, \tag{3}$$

where V_T is the total volume of the solution. Since we have $V_T = 1 + V$,

$$\boxed{C_{\text{HCl}} = \frac{0.01(1 - V)}{(1 + V)} \text{ mol L}^{-1}}, \tag{4}$$

for $V \leq 1.00$ L.

The number of moles of NaCl present is equal to the number of moles of HCl neutralized. Thus, the concentration of NaCl in the container will be

$$\boxed{C_{\text{NaCl}} = \frac{N_{\text{NaCl}}}{V_T} = \frac{0.01V}{1 + V}} \tag{5}$$

for $V \leq 1.00$ L.

Equations (4) and (5) are valid only when the volume of NaOH added is less than or equal to 1.00 L. Beyond this, all of the HCl has been titrated, and no more NaCl can form. For the same reason, we will get NaOH in the container only after all the HCl has been titrated. Therefore, the number of moles of NaOH present is the NaOH molarity times the volume of NaOH added beyond 1.00 L. This gives

$$\boxed{C_{\text{NaOH}} = \frac{N_{\text{NaOH}}}{V_T} = \frac{0.01(V - 1)}{1 + V}} \tag{6}$$

for $V > 1.00$ L.

Equations 4, 5, and 6 are the desired expressions.

(B) The requested plots are shown in the graphs on the next page. The equations of the linear least-squares fits are as follows:

$$\Lambda_{\text{HCl}} = 525.76 - 141.346\, c^{1/2} \text{ S cm}^2 \text{ mol}^{-1}; \tag{7}$$

$$\Lambda_{\text{NaCl}} = 126.26 - 78.988\, c^{1/2} \text{ S cm}^2 \text{ mol}^{-1}; \tag{8}$$

$$\Lambda_{\text{NaOH}} = 247.7 - 98.085\, c^{1/2} \text{ S cm}^2 \text{ mol}^{-1}. \tag{9}$$

(C) The specific conductivity is given by

$$\kappa = c\Lambda_m. \tag{10}$$

If the conductivities are additive, we will have

$$\kappa_{\text{total}} = \kappa_{\text{HCl}} + \kappa_{\text{NaCl}} + \kappa_{\text{NaOH}} = C_{\text{HCl}}\Lambda_{\text{HCl}} + C_{\text{NaCl}}\Lambda_{\text{NaCl}} + C_{\text{NaOH}}\Lambda_{\text{NaOH}}. \tag{11}$$

Substituting the results from Parts (A) and (B) yields

$$\kappa_{\text{total}} = \frac{0.01(1 - V)}{(1 + V)}\left[525.76 - 141.346\left\{\frac{0.01(1 - V)}{(1 + V)}\right\}^{1/2}\right]$$

$$+ \frac{0.01V}{1 + V}\left[126.26 - 78.988\left\{\frac{0.01V}{1 + V}\right\}^{1/2}\right] m\text{S cm}^{-1} \quad \text{for } V \leq 1.00 \text{ L}, \quad \textbf{(12)}$$

since the only species present with $V \leq 1$ L are HCl and NaCl. When $V > 1$ L, we have 0.01 mole of NaCl in a volume V_T, plus the concentration of NaOH given by Eq. 6. Therefore,

$$\kappa_{\text{total}} = \frac{0.01}{1 + V}\left[126.26 - 78.988\left\{\frac{0.01}{1 + V}\right\}^{1/2}\right]$$

$$+ \frac{0.01(V - 1)}{1 + V}\left[247.7 - 98.085\left\{\frac{0.01(V - 1)}{1 + V}\right\}^{1/2}\right] m\text{S cm}^{-1} \quad \text{for } V \geq 1.00 \text{ L}. \quad \textbf{(13)}$$

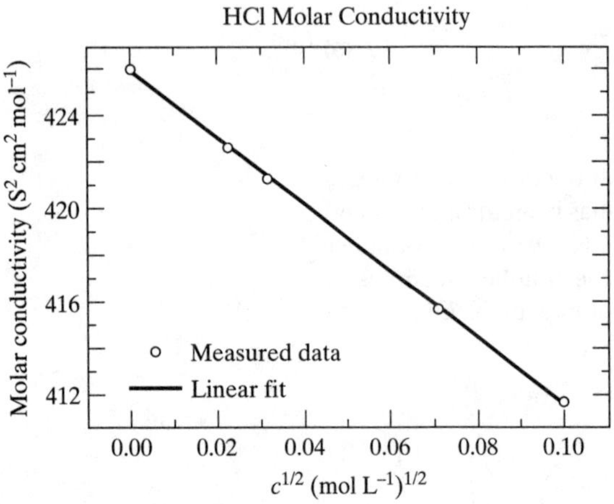

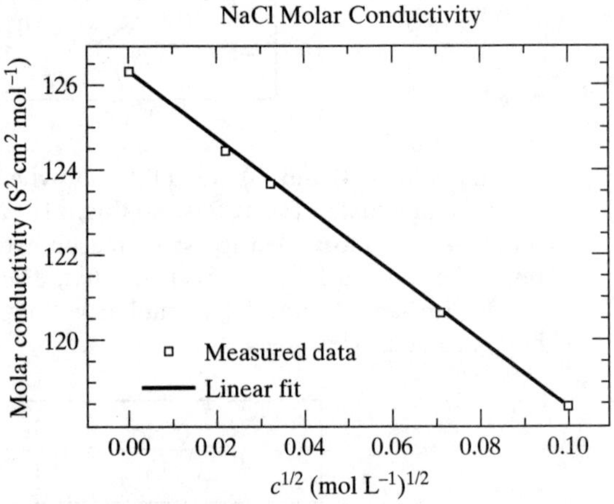

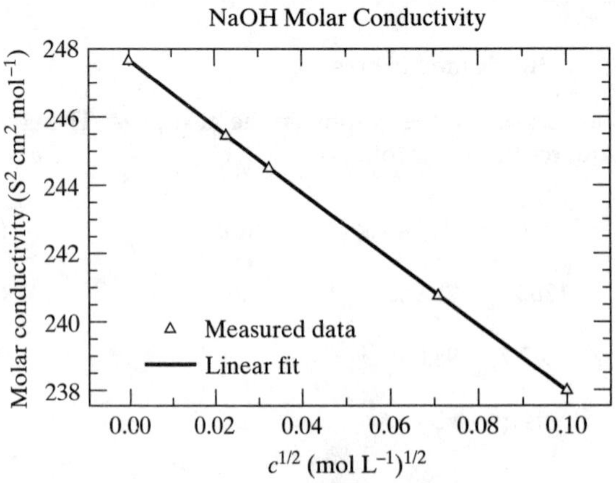

In Eqs. 12 and 13, the units are millisiemens cm^{-1}. The requested plot appears at the top of the next page. The abrupt change of slope at the endpoint, $V = 1.00$ L, is obvious. This is the principle upon which conductivity titrations rest.

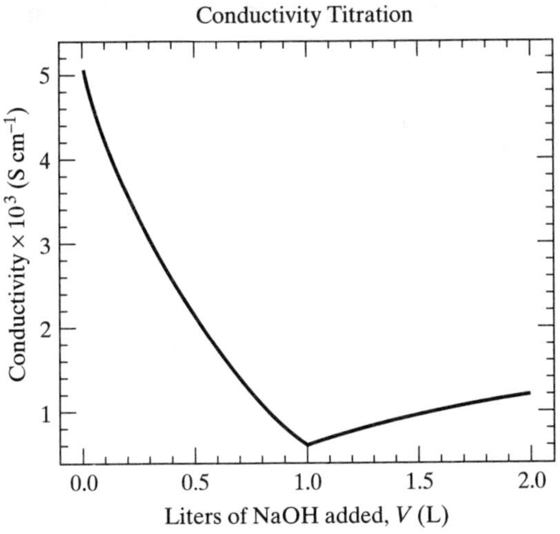

9.43 The ionization constant for HF in aqueous solution at 298.15 K is reported to be 3.5×10^{-4}. Use the Debye–Hückel theory and the Debye–Hückel–Onsager equation to estimate the specific conductivity of a 0.02-molar HF solution at 298.15 K. The molar ionic conductance of H_3O^+ and F^- at infinite dilution are listed in the *Handbook of Chemistry and Physics* as 349.65 S cm^2 mol^{-1} and 55.4 S cm^2 mol^{-1}, respectively.

Solution

The ionization process is $HF + x\,H_2O = H_3O^+_{(aq)} + F^-_{(aq)}$. The ionization constant for this reaction is

$$K_i = \frac{\gamma_\pm^2 m(H_3O^+)m(F^-)}{\gamma_{HF}m(HF)a_{H_2O}}. \tag{1}$$

At the low concentrations of HF in this problem, we can take a_{H_2O} to be unity, since X_{H_2O} is nearly unity, the reference state for the solvent, at which point the activity is unity. We can also take the activity coefficient for the uncharged HF to be unity. This gives

$$K_i = \frac{\gamma_\pm^2 m(H_3O^+)m(F^-)}{m(HF)} = 3.5 \times 10^{-4}. \tag{2}$$

To obtain the equilibrium concentrations of the ions, we must solve Eq. 2 iteratively. In the first iteration, we take $\gamma_\pm^2$ to be unity. Let $m(H_3O^+) = m(F^-) = x$, so that we will have $m(HF) = 0.02 - x$. Substituting into Eq. 2 gives

$$\frac{x^2}{0.02 - x} = K_i. \tag{3}$$

The solution of Eq. 3 is the root of the quadratic

$$x^2 + K_i x - 0.02K_i = 0. \tag{4}$$

This is

$$\begin{aligned} x &= \frac{-K_i + [K_i^2 + 0.08K_i]^{1/2}}{2} \\ &= \frac{-3.5 \times 10^{-4} + [(3.5 \times 10^{-4})^2 + 0.08(3.5 \times 10^{-4})]^{1/2}}{2} \\ &= 2.476 \times 10^{-3} \text{ mol kg}^{-1}. \end{aligned} \tag{5}$$

Using the result obtained in Eq. 5 in the Debye–Hückel equation, we obtain

$$\log \gamma_{\pm} = -0.50926(1)(1)(2.476 \times 10^{-3})^{1/2} = -0.02534. \quad (6)$$

This yields $\gamma_{\pm} = 0.9433$ and $\gamma_{\pm}^2 = 0.8898$.

Substituting this result into Eq. 2 produces

$$\frac{x^2}{0.02 - x} = \frac{K_i}{\gamma_{\pm}^2} = \frac{3.5 \times 10^{-4}}{0.8898} = 3.933 \times 10^{-4}. \quad (7)$$

Solving the quadratic in Eq. 4 gives

$$x = 2.615 \times 10^{-3} \text{ mol kg}^{-1}. \quad (8)$$

A third iteration produces

$$\log \gamma_{\pm} = -0.50926(1)(1)(2.615 \times 10^{-3})^{1/2} = -0.026042, \quad (9)$$

which yields $\gamma_{\pm} = 0.9418$ and $\gamma_{\pm}^2 = 0.8870$. Substituting of this result into Eq. 2 produces

$$\frac{x^2}{0.02 - x} = \frac{K_i}{\gamma_{\pm}^2} = \frac{3.5 \times 10^{-4}}{0.8870} = 3.946 \times 10^{-4}. \quad (10)$$

Solving the quadratic in Eq. 4 gives

$$x = m(H_3O^+) = m(F^-) = 2.619 \times 10^{-3} \text{ mol kg}^{-1}. \quad (11)$$

The result is now converged to nearly four significant digits. The specific conductance of the HF solution is given by

$$\kappa_{total} = \kappa_{H^+} + \kappa_{F^-}. \quad (12)$$

Using the Debye–Hückel–Onsager equation, we obtain

$$\kappa_{total} = 10^{-3}c\{\lambda_m^{\infty} - [30.32z^3 + 0.2290z^3\lambda_m^{\infty}]c^{1/2}\}_{H^+}$$
$$+ 10^{-3}c\{\lambda_m^{\infty} - [30.32z^3 + 0.2290z^3\lambda_m^{\infty}]c^{1/2}\}_{F^-}. \quad (13)$$

At the low ionic concentrations present in the problem, the molality and molarity are nearly identical in water. Therefore, substituting data into Eq. 13 yields

$$\kappa_{total} =$$
$$10^{-3}(2.619 \times 10^{-3})\{349.65 - [30.32(1)^3 + 0.2290(1)^3(349.65))](2.619 \times 10^{-3})^{1/2}\}_{H^+}$$
$$+ 10^{-3}(2.619 \times 10^{-3})\{55.4 - [30.32(1)^3 + 0.2290(1)^3(55.4)](2.619 \times 10^{-3})^{1/2}\}_{F^-}$$
$$= \underline{0.00104 \text{ S cm}^{-1}}. \quad (14)$$

9.45 Using the data given in Table 9.3, determine the voltage that would be obtained from the galvanic cell

$$Pt|Zn|ZnSO_4\ (m = 0.005)\|PbSO_4(s)|Pb(s)$$

at 298.15 K if junction potentials are ignored. You may use the Debye–Hückel theory to estimate any activity coefficients needed.

Solution

The half-reaction occurring at the anode is $Zn \longrightarrow Zn^{2+} + 2e^-$. At the cathode, the half-reaction is $PbSO_4(s) + 2e^- \longrightarrow Pb(s) + SO_4^{2-}$. The total cell reaction is the sum of these:

$$Zn(s) + PbSO_4(s) \longrightarrow ZnSO_4(m = 0.005) + Pb(s).$$

The cell potential is given by Eq. 9.159:

$$\Phi_{cell} = \Phi^o_{cell} - \frac{0.02569}{n}\ln\left[\frac{a_{ZnSO_4}a_{Pb(s)}}{a_{Zn(s)}a_{PbSO_4(s)}}\right]. \quad (1)$$

In this case, we have $n = 2$. If the activities of the solids are set to unity and that for $ZnSO_4$ is written in terms of the mean ionic activity coefficient and molality, Eq. 1 becomes

$$\Phi_{cell} = \Phi^o_{cell} - 0.012845\ln[\gamma^2_\pm m^2]. \quad (2)$$

The difference in standard half-reaction potentials can be obtained from Eq. 9.161 and the data in Table 9.3:

$$\Phi^o_{cell} = |\Phi^o_a - \Phi^o_c| = \Phi^o_c - \Phi^o_a = -0.3588 - (-0.7618)\text{ volts} = 0.403\text{ volt}. \quad (3)$$

The mean ionic activity coefficient is given by

$$\log\gamma_\pm = -0.50926z_+z_-S^{1/2}, \quad (4)$$

where the ionic strength is obtained from Eq. 9.64:

$$S = 0.5[0.005(2)^2 + 0.005(2)^2] = 0.02. \quad (5)$$

Combining Eqs. 4 and 5, we obtain

$$\log\gamma_\pm = -0.50926(2)(2)(0.02)^{1/2} = -0.28808. \quad (6)$$

This gives $\gamma_\pm = 0.51513$. Substituting into Eq. 2 then produces

$$\Phi_{cell} = 0.403 - 0.012845\ln[(0.51513)^2(0.005)^2] = 0.403 + 0.153 = \underline{0.556\text{ volt}}. \quad (7)$$

The accuracy of this result is questionable, because of the use of Debye–Hückel theory to compute the activity coefficient. A more accurate result could be obtained if a measured value were employed. If the Davies empirical equation, Eq. 9.68, is employed to compute the activity coefficient, the result is $\Phi_{cell} = \underline{0.553\text{ volt}}$.

9.47 Consider a Daniel cell such as that shown in Figure 9.23 with KCl replacing the NaCl in the salt bridge to eliminate the liquid junction potential more completely. The cell potential at 298.15 K will, therefore, be that given in Eq. 9.159, plus the contact potential between the Cu and Zn electrodes, which we denote as Φ_J. The cell potential at 298.15 K, Φ_1, is now measured with a potentiometer under conditions of zero current flow when the concentrations of $ZnCl_2$ and $CuCl_2$ are m_o and m_1, respectively. A second cell potential at 298.15 K, Φ_2, is obtained in the same manner, with m_1 changed to m_2 for the same concentration of $ZnCl_2$.

(A) Obtain an expression giving the difference $\Phi_2 - \Phi_1$ in terms of m_1, m_2, and the corresponding Cu^{2+} activity coefficients.

(B) If m_1 is adjusted to 0.0001 mol kg^{-1} so that the Debye–Hückel theory may be accurately used to compute the single-ion activity coefficient, obtain an expression giving that coefficient when the molality of the $CuCl_2$ electrolyte is m_2.

(C) When $m_2 = 1.0$ mol kg^{-1}, the measured cell voltage is 0.0961 volt. Compute γ_+ for the Cu^{2+} ion.

Solution

(A) If we ignore the contact potential between the Cu and Zn electrodes, the cell potential is given by Eq. 9.157. When the contact potential is added, we obtain

$$\Phi_1 = \Phi^o_{cell} + \Phi_J - \frac{RT}{2\mathcal{F}}\ln\left[\frac{a_{Zn^{2+}}(m_o)}{a_{Cu^{2+}}(m_1)}\right]_{eq} \quad (1)$$

if the $CuCl_2$ concentration is m_1. The notation $a_{Zn^{2+}}(m_o)$ denotes the Zn^{2+} activity when the Zn^{2+} concentration is m_o. A similar notation is employed for the activity of the Cu^{2+} ion. When the $CuCl_2$ concentration is changed to m_2, the result is

$$\Phi_2 = \Phi^o_{cell} + \Phi_J - \frac{RT}{2\mathcal{F}} \ln\left[\frac{a_{Zn^{2+}}(m_o)}{a_{Cu^{2+}}(m_2)}\right]_{eq}. \tag{2}$$

Therefore, the difference is

$$\Phi_2 - \Phi_1 = -\frac{RT}{2\mathcal{F}} \ln\left[\frac{a_{Cu^{2+}}(m_1)}{a_{Cu^{2+}}(m_2)}\right]_{eq}. \tag{3}$$

Substituting $\gamma_{+}m$ for the activities gives

$$\boxed{\Phi_2 - \Phi_1 = -\frac{RT}{2\mathcal{F}} \ln\left[\frac{\gamma_{+(m_1)}m_1}{\gamma_{+(m_2)}m_2}\right]_{eq} = -0.012845 \ln\left[\frac{\gamma_{+(m_1)}m_1}{\gamma_{+(m_2)}m_2}\right]_{eq}}, \tag{4}$$

which is the desired expression.

(B) If $m_1 = 0.0001$, we can use Eq. 9.67B to compute γ_{+}:

$$\log[\gamma_{+(m_1)}] = -0.50926 z_+^2 S^{1/2} = -0.50926(2)^2 S^{1/2} = -2.03704 S^{1/2}. \tag{5}$$

The ionic strength of the $CuCl_2$ solution is given by applying Eq. 9.64:

$$S = 0.5[(0.0001)(2)^2 + (0.0002)(1)^2] = 0.0003. \tag{6}$$

Therefore,

$$\log[\gamma_{+(m_1)}] = -2.03704(0.0003)^{1/2} = -0.035283. \tag{7}$$

This gives $\gamma_{+(m_1)} = 0.9220$ and $m_1\gamma_{+(m_1)} = (0.9220)(0.0001) = 0.00009220$. Substituting these results into Eq. 4 yields

$$\Phi_2 - \Phi_1 = -0.012845 \ln\left[\frac{0.00009220}{\gamma_{+(m_2)}m_2}\right]. \tag{8}$$

Dividing by -0.012845 and taking exponents of both sides produces

$$\frac{0.000092197}{\gamma_{+(m_2)}m_2} = \exp\left[-\frac{\Phi_2 - \Phi_1}{0.012845}\right]. \tag{9}$$

Solving for $\gamma_{+(m_2)}$, we obtain

$$\boxed{\gamma_{+(m_2)} = \frac{0.000092197}{m_2} \exp\left[\frac{\Phi_2 - \Phi_1}{0.012845}\right]}, \tag{10}$$

which is the desired expression.

(C) Substituting the measured voltage into Eq. 12 produces

$$\gamma_{+(m_2=1\ \text{mol kg}^{-1})} = \frac{0.000092197}{1} \exp\left[\frac{0.0961}{0.012845}\right] = \underline{0.1636}. \tag{11}$$

9.49 If we know the activity coefficient at one concentration, then the effect of junction potentials may be eliminated by working with the difference in voltage obtained at two concentrations. From previous measurements, the activity coefficient for HCl at $m = 0.500 \text{ mol kg}^{-1}$, is known to be 0.759. The voltage of the galvanic cell

$$\text{Pt}|\text{H}_2(\text{g})(P = 1 \text{ bar}), \text{HCl } (m), \text{AgCl(s)}|\text{Ag(s)}$$

is measured at 298.15 K with $m = 0.500 \text{ mol kg}^{-1}$ and again with $m = 2.000 \text{ mol kg}^{-1}$, using a bridge circuit so that no current flows during the measurement. The cell voltage with $m = 2.000 \text{ mol kg}^{-1}$ is found to be higher by 0.08586 volt than the voltage for the cell with $m = 0.5000 \text{ mol kg}^{-1}$.

(A) Show that the voltage difference is independent of junction potentials.

(B) Determine the activity coefficient for HCl at a concentration of $2.000 \text{ mol kg}^{-1}$ at 298.15 K.

Solution

(A) The reaction at the anode is $0.5\, \text{H}_2(\text{g})(P = 1 \text{ bar}) \longrightarrow \text{H}^+(m) + 1e^-$. At the cathode, the reaction is $\text{AgCl(s)} + 1e^- \longrightarrow \text{Ag(s)} + \text{Cl}^-(m)$. The overall cell reaction is, therefore,

$$0.5\, \text{H}_2(\text{g})(P = 1 \text{ bar}) + \text{AgCl(s)} \longrightarrow \text{HCl}(m) + \text{Ag(s)}.$$

If junction potentials are ignored, the cell voltage is given by Eq. 9.159:

$$\Phi_{\text{cell}} = \Phi^o_{\text{cell}} - \frac{0.02569}{n} \ln\left[\frac{a_{\text{HCl}} a_{\text{Ag}}}{f_{\text{H}_2}^{1/2} a_{\text{AgCl}}}\right]. \quad \textbf{(1)}$$

If Φ_J is the total junction potential for the cell, then the measured voltage will be

$$\Phi_{\text{cell}} = \Phi^o_{\text{cell}} + \Phi_J - \frac{0.02569}{n} \ln\left[\frac{a_{\text{HCl}} a_{\text{Ag}}}{f_{\text{H}_2}^{1/2} a_{\text{AgCl}}}\right]. \quad \textbf{(2)}$$

For this cell reaction, $n = 1$, and the fugacity of the H_2 gas can be equated to its pressure at this low value of the pressure. If we take the pure-component activities of Ag(s) and AgCl(s) to be unity, Eq. 2 becomes

$$\begin{aligned} \Phi_{\text{cell}} &= \Phi^o_{\text{cell}} + \Phi_J - 0.02569 \ln[a_{\text{HCl}}] \\ &= |\Phi^o_{\text{AgCl|Ag}} - \Phi^o_{\text{H}_2}| + \Phi_J - 0.02569 \ln[\gamma_\pm^2 m^2] \text{ volts.} \end{aligned} \quad \textbf{(3)}$$

Let Φ_1 and Φ_2 be the measured voltages when the HCl concentrations are 0.5000 and $2.000 \text{ mol kg}^{-1}$, respectively. With this definition and the data given in the problem, we have

$$\begin{aligned} \Phi_1 &= \Phi^o_{\text{cell}} + \Phi_J - 0.02569 \ln[\gamma_\pm^2 m^2] = \Phi^o_{\text{cell}} + \Phi_J - 0.05138 \ln[\gamma_\pm m] \\ &= \Phi^o_{\text{cell}} + \Phi_J - 0.05138 \ln[(0.759)(0.500)] = \Phi^o_{\text{cell}} + \Phi_J + 0.049782 \text{ volts} \end{aligned} \quad \textbf{(4)}$$

and

$$\Phi_2 = \Phi^o_{\text{cell}} + \Phi_J - 0.02569 \ln[\gamma_\pm^2 m^2] = \Phi^o_{\text{cell}} + \Phi_J - 0.05138 \ln[2\gamma_\pm] \text{ volts.} \quad \textbf{(5)}$$

The measured difference in voltage is $\Phi_1 - \Phi_2$. This can be obtained from Eqs. 4 and 5:

$$\boxed{\Phi_1 - \Phi_2 = 0.049782 + 0.05138 \ln[2\gamma_\pm]|\text{volts}}. \quad \textbf{(6)}$$

Since Φ_J no longer appears in Eq. 6, the difference in the measured potentials is clearly independent of the junction potentials present in the cell.

(B) Using the measured voltage difference, we may use Eq. 6 to calculate the value of $\gamma\pm$:

$$0.08586 \text{ volt} = 0.049782 + 0.05138 \ln[2\gamma_\pm]|\text{volt}. \quad \textbf{(7)}$$

Solving Eq. 7 for $\gamma_\pm$, we obtain

$$\gamma_\pm = 0.5\exp\left[\frac{(0.08586 - 0.049782)}{0.05138}\right] = \underline{1.009}. \quad \textbf{(8)}$$

9.51 Φ^o_{cell} for the cell

$$Ag(s)|H_2(g)\ (P = 1\ \text{bar}),\ HCl\ (m = 1.00\ \text{molal}),\ AgCl(s)|Ag(s)$$

is measured at different temperatures. The data are fitted to a Taylor series expansion. The result of the fitting is

$$\Phi^o_{cell} = 0.22233 - 0.0006477\,(T - 298.15)$$
$$- 3.241 \times 10^{-6}(T - 298.15)^2\ \text{volts.}$$

Determine the equilibrium constant, $\Delta\bar{S}^o$, $\Delta\bar{H}^o$, and ΔC^o_p for the cell reaction at 280 K.

Solution

The cell reaction is 0.5 $H_2(g)(P = 1\ \text{bar}) + AgCl(s) = HCl(m = 1\ \text{molal}) + Ag(s)$. For this reaction,

$$\Delta\mu^o = -n\mathcal{F}\Phi^o_{cell}$$
$$= -n\mathcal{F}[0.22213 - 0.0006477(T - 298.15) - 3.241 \times 10^{-6}(T - 298.15)^2]. \quad \textbf{(1)}$$

At 280 K, we obtain

$$\Delta\mu^o(T = 280\ \text{K}) = -(1)(96{,}485\ \text{C mol}^{-1})[0.22213 - 0.0006477(-18.15)$$
$$- 3.241 \times 10^{-6}(-18.15)^2]\ \text{volts} = -22{,}463\ \text{J mol}^{-1}. \quad \textbf{(2)}$$

The corresponding equilibrium constant is

$$K = \exp\left[-\frac{\Delta\mu^o}{RT}\right] = \exp\left[\frac{22{,}463\ \text{J mol}^{-1}}{(8.314\ \text{J mol}^{-1}\ \text{K}^{-1})(280\ \text{K})}\right] = \underline{1.55 \times 10^4}. \quad \textbf{(3)}$$

The change in the standard partial molar entropy for the reaction is given by Eq. 9.170:

$$\Delta\bar{S}^o = n\mathcal{F}\left(\frac{\partial\Phi^o_{cell}}{\partial T}\right)_P = n\mathcal{F}[-0.0006477 - 6.482 \times 10^{-6}(T - 298.15)]\ \text{volts K}^{-1}. \quad \textbf{(4)}$$

At $T = 280$ K, we obtain

$$\Delta\bar{S}^o = (1)(96{,}485\ \text{C mol}^{-1})[-0.0006477 - 6.482 \times 10^{-6}(-18.15)]\text{volts K}^{-1}$$
$$= \underline{-51.14\ \text{J mol}^{-1}\ \text{K}^{-1}}. \quad \textbf{(5)}$$

The change in the standard partial molar enthalpy for the reaction is given by Eq. 9.171:

$$\Delta\bar{H}^o = \Delta\mu^o + T\Delta\bar{S}^o = -22{,}463\ \text{J mol}^{-1} + (280\ \text{K})(-51.14\ \text{J mol}^{-1}\ \text{K}^{-1})$$
$$= \underline{-3.678 \times 10^4\ \text{J mol}^{-1}}. \quad \textbf{(6)}$$

ΔC^o_p is given by Eq. 9.172:

$$\Delta C^o_p = n\mathcal{F}T\left(\frac{\partial^2\Phi^o_{cell}}{\partial T^2}\right)_P = n\mathcal{F}T[-6.482 \times 10^{-6}\ \text{volt K}^{-2}]$$
$$= (1)(96{,}485\ \text{C mol}^{-1})(280\ \text{K})(-6.482 \times 10^{-6}\ \text{volt K}^{-2}) = \underline{-175.1\ \text{J mol}^{-1}\ \text{K}^{-1}}. \quad \textbf{(7)}$$

9.53 In this problem, you will prove the theorem embodied in Eqs. 9.173 and 9.174. The various steps in the problem serve as a road map for the proof.

(A) Let the function $y = f(x)$ be represented by a polynomial in x of order 6. That is, let $y = \sum_{i=0}^{6} a_i x^i$, where we will take the origin of the x-coordinate system to be at the point at which we wish to evaluate the derivative dy/dx. Show that dy/dx at the point $x = 0$ is equal to a_1.

(B) Let us assume that we know the value of y at the points $x = -3h, -2h, -h, h, 2h$, and $3h$. Let these values be denoted by $y_{-3}, y_{-2}, y_{-1}, y_1, y_2$, and y_3, respectively. Obtain expressions for each of the y_i in terms of h and the a_i.

(C) Using the result of (B), obtain expressions for S_1, S_2, and S_3 as defined in Eq. 9.174. Be certain to collect terms in the various powers of h.

(D) We now seek to find a linear combination of S_1, S_2, and S_3 that is exactly equal to the desired derivative, which we have shown in (A) to be equal to a_1. That is, we write

$$AS_1 + BS_2 + CS_3 = a_1 = \frac{dy}{dx} \text{ at the point } x = 0$$

and ask what must be the values of A, B, and C to make this equation correct. Using the results obtained in (C), find the values of A, B, and C, and show that Eq. 9.173 is correct as written.

Solution

(A) The polynomial representation of y gives us

$$y = a_0 + a_1x + a_2x^2 + a_3x^3 + a_4x^4 + a_5x^5 + a_6x^6. \quad (1)$$

The derivative is given by

$$\frac{dy}{dx} = a_1 + 2a_2x + 3a_3x^2 + 4a_4x^3 + 5a_5x^4 + 6a_6x^5. \quad (2)$$

Clearly, at the point $x = 0$, we have

$$\boxed{\left.\frac{dy}{dx}\right|_{x=0} = a_1}, \quad (3)$$

as required.

(B) Direct substitution of the values of x in terms of h into Eq. 1 gives the following results:

$$\begin{aligned} y_{-3} &= a_0 + a_1(-3h) + a_2(-3h)^2 + a_3(-3h)^3 \\ &\quad + a_4(-3h)^4 + a_5(-3h)^5 + a_6(-3h)^6 \\ &= a_0 - 3a_1h + 9a_2h^2 - 27a_3h^3 + 81a_4h^4 - 243a_5h^5 + 729a_6h^6; \quad (4) \end{aligned}$$

$$\begin{aligned} y_{-2} &= a_0 + a_1(-2h) + a_2(-2h)^2 + a_3(-2h)^3 \\ &\quad + a_4(-2h)^4 + a_5(-2h)^5 + a_6(-2h)^6 \\ &= a_0 - 2a_1h + 4a_2h^2 - 8a_3h^3 + 16a_4h^4 - 32a_5h^5 + 64a_6h^6; \quad (5) \end{aligned}$$

$$\begin{aligned} y_{-1} &= a_0 + a_1(-h) + a_2(-h)^2 + a_3(-h)^3 + a_4(-h)^4 + a_5(-h)^5 + a_6(-h)^6 \\ &= a_0 - a_1h + a_2h^2 - a_3h^3 + a_4h^4 - a_5h^5 + a_6h^6; \quad (6) \end{aligned}$$

$$\begin{aligned} y_3 &= a_0 + a_1(3h) + a_2(3h)^2 + a_3(3h)^3 + a_4(3h)^4 + a_5(3h)^5 + a_6(3h)^6 \\ &= a_0 + 3a_1h + 9a_2h^2 + 27a_3h^3 + 81a_4h^4 + 243a_5h^5 + 729a_6h^6; \quad (7) \end{aligned}$$

$$y_2 = a_o + a_1(2h) + a_2(2h)^2 + a_3(2h)^3 + a_4(2h)^4 + a_5(2h)^5 + a_6(2h)^6$$
$$= a_o + 2a_1h + 4a_2h^2 + 8a_3h^3 + 16a_4h^4 + 32a_5h^5 + 64a_6h^6; \quad (8)$$
$$y_1 = a_o + a_1(h) + a_2(h)^2 + a_3(h)^3 + a_4(h)^4 + a_5(h)^5 + a_6(h)^6$$
$$= a_o + a_1h + a_2h^2 + a_3h^3 + a_4h^4 + a_5h^5 + a_6h^6. \quad (9)$$

Equations 4–9 are the required expressions.

(C) Using the definitions of S_1, S_2, and S_3 given in Eq. 9.174, we obtain

$$S_1 = \frac{[y_1 - y_{-1}]}{h} = \{[a_o + a_1h + a_2h^2 + a_3h^3 + a_4h^4 + a_5h^5 + a_6h^6]$$
$$- [a_o - a_1h + a_2h^2 - a_3h^3 + a_4h^4 - a_5h^5 + a_6h^6]\}/h = \frac{2}{h}[a_1h + a_3h^3 + a_5h^5]$$
$$= \boxed{2[a_1 + a_3h^2 + a_5h^4]}, \quad (10)$$

$$S_2 = \frac{[y_2 - y_{-2}]}{h} = \{[a_o + 2a_1h + 4a_2h^2 + 8a_3h^3 + 16a_4h^4 + 32a_5h^5 + 64a_6h^6]$$
$$- [a_o - 2a_1h + 4a_2h^2 - 8a_3h^3 + 16a_4h^4 - 32a_5h^5 + 64a_6h^6]\}/h$$
$$= \frac{2}{h}[2a_1h + 8a_3h^3 + 32a_5h^5] = \boxed{4a_1 + 16a_3h^2 + 64a_5h^4}, \quad (11)$$

and

$$S_3 = \frac{[y_3 - y_{-3}]}{h}$$
$$= \{[a_o + 3a_1h + 9a_2h^2 + 27a_3h^3 + 81a_4h^4 + 243a_5h^5 + 729a_6h^6]$$
$$- [a_o - 3a_1h + 9a_2h^2 - 27a_3h^3 + 81a_4h^4 - 243a_5h^5 + 729a_6h^6]\}/h$$
$$= \frac{2}{h}[3a_1h + 27a_3h^{3+}243a_5h^5] = \boxed{6a_1 + 54a_3h^2 + 486a_5h^4}. \quad (12)$$

Equations 10, 11, and 12 are the required relationships.

(D) We now wish to find A, B, and C such that we have

$$a_1 = AS_1 + BS_2 + CS_3. \quad (13)$$

Substituting Eqs. 10–12 into Eq. 13 gives

$$a_1 = 2A[a_1 + a_3h^2 + a_5h^4] + B[4a_1 + 16a_3h^2$$
$$+ 64a_5h^4] + C[6a_1 + 54a_3h^2 + 486a_5h^4]. \quad (14)$$

Collecting terms on the right for the various coefficients, we obtain

$$a_1 = a_1[2A + 4B + 6C] + a_3h^2[2A + 16B + 54C] + a_5h^4[2A + 64B + 486C]. \quad (15)$$

Equation 15 will hold only if we have

$$2A + 4B + 6C = 1, \quad (16)$$
$$2A + 16B + 54C = 0, \quad (17)$$

and

$$2A + 64B + 486C = 0. \quad (18)$$

We now have three linear equations in three unknowns, A, B, and C. All that remains is to solve Eqs. 16–18 for the solutions for A, B, and C. From Eq. 17, we have

$$2A = -16B - 54C. \tag{19}$$

Substituting this result into Eq. 18 gives

$$-16B - 54C + 64B + 486C = 48B + 432C = 0. \tag{20}$$

Solving for B, we obtain

$$B = -\frac{432C}{48} = -9C. \tag{21}$$

Substituting this result into Eq. 19 produces

$$2A = -16(-9C) - 54C = 144C - 54C = 90C. \tag{22}$$

Thus, we have $A = 45C$. Substituting the results in Eqs. 21 and 22 into Eq. 16 yields

$$90C + 4(-9C) + 6C = 90C - 36C + 6C = 60C = 1. \tag{23}$$

Therefore,

$$\boxed{C = \frac{1}{60}}.$$

Using this solution in Eq. 21 shows that we must have

$$\boxed{B = -9C = -\frac{9}{60} = -\frac{3}{20} = 0.150000\ldots}. \tag{24}$$

Finally, substituting into Eq. 22 gives

$$\boxed{A = 45C = \frac{45}{60} = \frac{3}{4} = 0.7500000\ldots}. \tag{25}$$

Therefore, the final result is

$$\boxed{\frac{dy}{dx} \text{ at } x = 0 = a_1 = 0.75S_1 - 0.15S_2 + \left(\frac{1}{60}\right)S_3}, \tag{26}$$

which is exactly Eq. 9.173.

CHAPTER 10

The Mathematics of Chance

10.1 There are 10 students in a class. On a 100-point hour examination, their grades are as follows:

58, 26, 88, 95, 71, 40, 15, 52, 85, 64.

(A) Compute the average grade.

(B) Compute the square uncertainty in this distribution of grades.

Solution

(A) There are 101 possible outcomes on the examinations (grades from 0 to 100). Because we have executed a measurement by grading the papers, it is now known that the probability factors for these outcomes are $f_{58} = f_{26} = f_{88} = f_{95} = f_{71} = f_{40} = f_{15} = f_{52} = f_{85} = f_{64} = 1$, and all other $f_i = 0$. The normalization constant is given by Eq. 10.5:

$$C = \left[\sum_{i=0}^{100} f_i \right]^{-1} = (10)^{-1} = \frac{1}{10}. \quad (1)$$

The average grade may now be computed using Eq. 10.13:

$$\langle g \rangle = C \sum_{i=0}^{100} g_i f_i = 10^{-1}[58 + 26 + 88 + 95 + 71 + 40 + 15 + 52 + 85 + 64]$$

$$= \frac{594}{10} = \underline{59.4}. \quad (2)$$

(B) The square uncertainty in the grade distribution is given by

$$\langle \Delta g^2 \rangle = \langle g^2 \rangle - \langle g \rangle^2, \quad (3)$$

where

$$\langle g^2 \rangle = C \sum_{i=0}^{100} g_i^2 f_i$$

$$= 10^{-1}[(58)^2 + (26)^2 + (88)^2 + (95)^2 + (71)^2$$

$$+ (40)^2 + (15)^2 + (52)^2 + (85)^2 + (64)^2]$$

$$= 10^{-1}(41{,}700) = \frac{41{,}700}{10} = 4{,}170.0 \quad (4)$$

Therefore, we obtain

$$\langle \Delta g^2 \rangle = 4{,}170.0 - (59.4)^2 = 4{,}170.0 - 3{,}528.36 = \underline{641.64}. \quad (5)$$

10.3 Jim, John, and Bob are reporters for a newspaper. Jim gets the facts correct 88% of the time, John gets them right 80% of the time, and Bob gets them right 55% of the time (Bob has now gone on to a political career). Jim and John each cover a story independently and submit stories that agree as to the facts.

(A) What is the probability that Jim and John have the facts of the story correct?

(B) The editor, wishing confirmation, sends Bob to cover the same story, and Bob's facts also agree with those of Jim and John. What is the probability that the facts, as reported, are correct?

(C) If Bob's facts had disagreed with those of Jim and John, what would be the probability that the facts, as reported by Jim and John, are correct?

Solution

(A) This problem is simple if you are careful to take into account the effect of measurement. If Jim and John each cover a story and we have no further information, the possible outcomes are given in the table below.

Note that f_i is a result of the probabilities of two successive events being independent. (John reports the facts, and then Jim reports the facts.) The normalization constant when Jim and John reporting the facts is viewed as a single event is

$$C = \left[\sum_{i=1}^{4} f_i\right]^{-1} = [(0.704) + (0.096) + (0.176) + (0.024)]^{-1} = (1)^{-1} = 1. \quad \textbf{(1)}$$

Outcome No.	f_i
1. John right, Jim right	$(0.80)(0.88) = 0.704$
2. John right, Jim wrong	$(0.80)(0.12) = 0.096$
3. John wrong, Jim right	$(0.20)(0.88) = 0.176$
4. John wrong, Jim wrong	$(0.20)(0.12) = 0.024$

The normalized probability distribution for the four outcomes listed in the table are, therefore,

$$P(X_1) = Cf_1 = (1)(0.704) = 0.704, P(X_2) = Cf_2 = 0.096$$
$$P(X_3) = Cf_3 = 0.176, \text{ and } P(X_4) = Cf_4 = 0.024. \quad \textbf{(2)}$$

Thus, if we were to ask what the probability is that both Jim and John will get the facts correct, the answer is $P(X_1) = 0.704$. However, that is not the answer to this problem, since a measurement has been made that changes the probabilities. The measurement is the editor reading the reports and finding that they agree. Therefore, we now know that $P(X_2) = P(X_4) = 0$, since, if the reports agree, it is impossible for one to be right and the other wrong. Both reporters must be either right or wrong. The new probability factors are then:

Outcome No.	f_i
1. John right, Jim right	$(0.80)(0.88) = 0.704$
2. John right, Jim wrong	0
3. John wrong, Jim right	0
4. John wrong, Jim wrong	$(0.20)(0.12) = 0.024$

The new normalization constant is

$$C\sum_{i=1}^{4} f(i) = 1 = C[0.704 + 0 + 0 + 0.024] = 0.728C, \quad \textbf{(3)}$$

which gives

$$C = \frac{1}{0.728}. \quad \textbf{(4)}$$

With this normalization constant,

$$P(X_1) = Cf_1 = \frac{0.704}{0.728} = \underline{0.9670}, \quad \textbf{(5)}$$

which is the probability that the reports are correct.

(B) With the second measurement in place, there are only two outcomes:

Outcome 1	f_i
1. John, Jim, and Bob all right	(0.80)(0.88)(0.55) = 0.3872
2. John, Jim, and Bob all wrong	(0.20)(0.12)(0.45) = 0.0108

Normalization of this distribution requires that

$$C \sum_{i=1}^{2} f_i = 1 = C[0.3872 + 0.0108] = 0.3980C. \tag{6}$$

The normalization constant is

$$C = \frac{1}{0.3980}. \tag{7}$$

The probability that the reports are all correct is

$$P(X_1) = Cf(1) = \frac{0.3872}{0.3980} = \underline{0.9729}. \tag{8}$$

(C) In this case, the second observation leaves the following two outcomes:

Outcome 1	$f(i)$
1. John, Jim right; Bob wrong	(0.80)(0.88)(0.45) = 0.3168
2. John, Jim wrong; Bob right	(0.20)(0.12)(0.55) = 0.0132

Normalization of this distribution requires that

$$C \sum_{i=1}^{2} f_i = 1 = C[0.3168 + 0.0132] = 0.3300C. \tag{9}$$

The new normalization constant is

$$C = \frac{1}{0.3300}. \tag{10}$$

The probability that John and Jim are correct while Bob is wrong is

$$P(X_1) = Cf_i = \frac{0.3168}{0.3300} = \underline{0.9600}. \tag{11}$$

10.5 Consider three playing cards. One card is white on both sides. A second card is white on one side and black on the other. The third card is black on both sides. The cards are shuffled and one card is chosen randomly. With eyes closed, you place this card on the table. Upon opening your eyes, you find that the card on the table has a black side face up. What is the probability that the down side of the card on the table is also black?

Solution

If no measurements are made, there are six outcomes, all equally probable:

Z Card on Table	Side Up	f_i
1. White–white card,	side 1 white	1
2. White–white card,	side 2 white	1
3. Black–white card,	side 1 black	1
4. Black–white card,	side 2 white	1
5. Black–black card,	side 1 black	1
6. Black–black card,	side 2 black	1

The normalized probability distribution in the absence of any measurement is

$$C\sum_{i=1}^{6} f_i = C[1+1+1+1+1+1] = 6C = 1, \tag{1}$$

so that $C = \frac{1}{6}$ and

$$P(X_i) = \frac{f_i}{6} = \frac{1}{6}. \tag{2}$$

A measurement is now made. We look at the card on the table and find a black side up. This measurement changes the probability of events 1, 2, and 4, all of which have a white side up. We now know that $f_1 = f_2 = f_4 = 0$. The new probability factors are, therefore, the following:

Card on Table	Side Up	f_i
1. White–white card,	side 1 white	0
2. White–white card,	side 2 white	0
3. Black–white card,	side 1 black	1
4. Black–white card,	side 2 white	0
5. Black–black card,	side 1 black	1
6. Black–black card,	side 2 black	1

The new normalization constant is

$$C\sum_{i=1}^{6} f_i = C[0+0+1+0+1+1] = 3C = 1. \tag{3}$$

Therefore, $C = \frac{1}{3}$ and $P(X_3) = P(X_5) = P(X_6) = \frac{1}{3}$.
The hidden side of the card will be black if either event 5 or event 6 occurs. It will be white if event 3 occurs. Since this is a single event involving independent probabilities, the probability of having a black face hidden is

$$\boxed{P(\text{black face hidden}) = P(X_5) + P(X_6) = \frac{2}{3}}. \tag{4}$$

The number of individuals who cannot work this problem correctly is astounding. Virtually all those who obtain the incorrect answer will claim that the probability that the hidden face is black is $\frac{1}{2}$. This type of error occurs because of a failure to consider all possibilities and a failure to renormalize the distribution properly.

10.7 Mr. Jones is known to have two children. You are told in advance that Mr. Jones' oldest child is a boy. What is the probability that both children are boys?

Solution

This problem is worked in the same manner as Problem 10.6, but because the measured data are different, the effect on the probability distribution is different. In the absence of any measurement, there are four possibilities, all equally probable. See the following table:

Event No.	Older Child	Younger Child	f_i
1	Girl	Girl	1
2	Girl	Boy	1
3	Boy	Girl	1
4	Boy	Boy	1

The normalization condition requires that

$$C\sum_{i=1}^{4} f_i = C[1+1+1+1] = 4C = 1, \tag{1}$$

so that $C = \frac{1}{4}$. Someone now makes a measurement and finds that the older of Mr. Jones's children is a boy. This information changes the probabilities of Events 1 and 2, for which the older child is a girl. We now know that $f_1 = f_2 = 0$. Consequently, the new probability factors are as follows:

Event No.	Older Child	Younger Child	f_i
1	Girl	Girl	0
2	Girl	Boy	0
3	Boy	Girl	1
4	Boy	Boy	1

The new normalization constant is

$$C\sum_{i=1}^{4} f_i = C[0+0+1+1] = 2C = 1. \tag{2}$$

Therefore, $C = \frac{1}{2}$. The probability that both children are boys is given by

$$\boxed{P(X_4) = Cf_4 = \frac{1}{2}}. \tag{3}$$

A professor wishes to evaluate the probability that a student who has scored X_1 points in first-semester physical chemistry out of a possible score of 667 will score X_2 points during the second semester using the same grading system. The system has 300 points possible on three 100-point hour exams, 167 points on the final examination, and 200 points on the homework.

The professor makes the assumptions on the next page to estimate the desired probability.

(A) The variation in the student's score will come entirely from the examinations. The homework score will be identical to that made during the first semester.

(B) The probability distribution for the exam scores will be a normal Gaussian distribution centered at the student's total exam score during the first semester, X_{T1}. The functional form of this distribution is

$$P(X_e) = C \exp\left[-\frac{(X_e - X_{T1})^2}{2\sigma^2}\right],$$

where X_e is the number of points the student will score on exams during the second semester, C is the normalization constant, and σ^2 is the square of the uncertainty in the distribution of the student's exam scores during the first semester. That is, if F_i is the fraction of total points scored by the student on exam i during the first semester, then

$$\sigma^2 = [\langle F^2\rangle - \langle F\rangle^2](467)^2,$$

where $\langle\ \rangle$ stands for the average value. As an example, if the student has scores of 75, 84, and 71 on the three hour exams and 101 on the final, then $F_1 = 75/100 = 0.75$, $F_2 = 84/100 = 0.84$, $F_3 = 71/100 = 0.71$, and $F_4 = 101/167 = 0.6048$. We then have $\langle F\rangle = (0.75 + 0.84 + 0.71 + 0.6048)/4 = 0.7262$ and $\langle F^2\rangle = [(0.75)^2 + (0.84)^2 + (0.71)^2 + (0.6048)^2]/4 = 0.5345$. Therefore, $\sigma^2 = [0.5345 - (.7262)^2](467)^2 = 1{,}556$.

(C) The distribution $P(X_e)$ can be treated as if X_e is a continuous variable over the range $0 \le X_e \le 467$.

10.9* What is the probability that the student described in the previous problem will make a grade of A in the second-semester course if the cutoff lines are as follows:

Range	Grade
$500 \le X_2 \le 667$	A
$400 \le X_2 \le 499$	B
$320 \le X_2 \le 399$	C
$265 \le X_2 \le 319$	D
$0 \le X_2 \le 264$	F

Solution

Since the professor is assuming that the homework score of the student will remain unchanged, the student will need to score at least $500 - 180 = 320$ points on the examinations during the second semester to make a grade of A. The cumulative probability that his grade on the examinations will be at least 320 is

$$P(X_e \ge 320) = \int_{X_e=320}^{467} P(X_e)dX_e = 0.010117 \int_{320}^{467} \exp\left[-\frac{(X_e - 331)^2}{3{,}111.6}\right] dX_e. \quad (1)$$

Numerical evaluation of this integral gives

$$P(X_e \ge 320) = \underline{0.6097}. \quad (2)$$

The student has about a 61% chance to make an A in the second-semester course.

10.11* Let us now consider a second student in the class, whom we shall call Student 2. The student in the previous three problems will be Student 1. Student 2 made a B during the first semester, having a total homework score of 165 and exam scores of 67, 73, and 69, on the hour exams and 112 on the final.

(A) Obtain the probability distribution for Student 2's exam grades during the second semester.

(B) Which student, 1 or 2, is the more consistent performer on examinations? How do you know?

(C) What is the probability that Student 2 will make a higher grade than Student 1 during the second semester? (Grade cutoff lines are as given in Problem 10.9.)

Solution

(A) The grade distribution for Student 2 depends upon the uncertainty of his exam grades during the first semester. This is computed as follows:

$$F_1 = \frac{67}{100} = 0.67; \tag{1}$$

$$F_2 = \frac{73}{100} = 0.73; \tag{2}$$

$$F_3 = \frac{69}{100} = 0.69; \tag{3}$$

$$F_4 = \frac{112}{167} = 0.6707. \tag{4}$$

Therefore,

$$\langle F \rangle = \frac{1}{4}\sum_{i=1}^{4} F_i = \frac{2.7607}{4} = 0.6902 \tag{5}$$

and

$$\langle F^2 \rangle = \frac{1}{4}\sum_{i=1}^{4} F_i^2 = [(0.67)^2 + (0.73)^2 + (0.69)^2 + (0.6707)^2]/4$$

$$= \frac{1.9077}{4} = 0.4769. \tag{6}$$

Hence, we obtain

$$\sigma^2 = [(0.4769) - (.6902)^2](467)^2 = 114.3. \tag{7}$$

The associated probability distribution is

$$P(X_e) = C\exp\left[-\frac{(X_e - 321)^2}{(2)(114.3)}\right] = C\exp\left[-\frac{(X_e - 321)^2}{228.6}\right]. \tag{8}$$

Since we are assuming that X_e is a continuous variable, the normalization constant is given by

$$\int_{X_e=0}^{467} P(X_e)dX_e = C\int_0^{467} \exp\left[-\frac{(X_e - 321)^2}{228.6}\right]dX_e = 1. \tag{9}$$

Solving for C, we obtain

$$C = \left[\int_0^{467} \exp\left[-\frac{(X_e - 321)^2}{228.6}\right]dX_e\right]^{-1}. \tag{10}$$

Equation 10 can be integrated numerically. The result is

$$C = [26.799]^{-1} = 0.037315. \tag{11}$$

Thus, Student 2's normalized exam grade distribution is

$$P_2(X_e) = 0.037315 \exp\left[-\frac{(X_e - 321)^2}{228.6}\right]. \quad \textbf{(12)}$$

(B) Student 2 is the more consistent performer. The uncertainty in his or her grades is

$$\sigma_2 = (\sigma_2^2)^{1/2} = (114.3)^{1/2} = \pm 10.69 \text{ points.} \quad \textbf{(13)}$$

In contrast, Student 1 has an uncertainty of

$$\sigma_1 = (\sigma_1^2)^{1/2} = (1{,}555.8)^{1/2} = \pm 39.44 \text{ points.} \quad \textbf{(14)}$$

This greater uncertainty is reflected in the following plot:

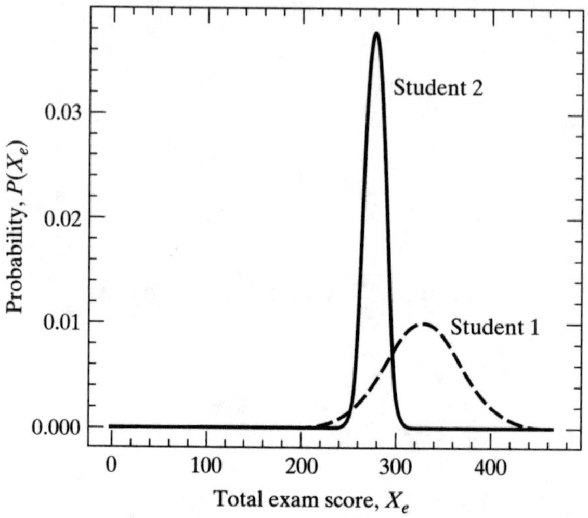

The distribution of exam grades for Student 1 is much broader than that for Student 2, reflecting the more consistent performance of Student 2.

(C) There are four ways that Student 2 can make a grade higher than Student 1:

Case No.	Student 2's Grade	Student 1's Grade
1	A	B, C, D, or F
2	B	C, D, or F
3	C	D or F
4	D	F

The total probability is the sum of the probabilities for each of these four cases, since we have a single event and independent probabilities. To execute the calculations, we need the value of eight different probabilities. First, we need the probability that Student 2 makes grades of A, B, C, and D. Second, we need the probability that Student 1 makes a B, C, D, or F, the probability that Student 1 makes C, D, or F, the probability that Student 1 makes D or F, and, finally, the probability that Student 1 makes an F. We begin with Student 2.

Since Student 2 will, by assumption, score 165 points on the homework, he or she needs 335 or more points to make an A. To make a B, he or she must score between 235 and 334 points. For a grade of C, Student 2's total exam scores must lie in the range from 155 to 234. A grade of D requires that student 2 have between 100 and 154 exam points. The required probabilities are, therefore,

$$P_2(\geq 335) = 0.037315 \int_{335}^{467} \exp\left[-\frac{(X_e - 321)^2}{228.6}\right] dX_e = 0.0952, \quad \textbf{(15)}$$

$$P_2(235 \leq X_e \leq 334) = 0.037315 \int_{235}^{334} \exp\left[-\frac{(X_e - 321)^2}{228.6}\right] dX_e = 0.8880, \quad \textbf{(16)}$$

$$P_2(155 \leq X_e \leq 234) = 0.037315 \int_{155}^{234} \exp\left[-\frac{(X_e - 321)^2}{228.6}\right] dX_e = 2.02 \times 10^{-16}, \quad \textbf{(17)}$$

and

$$P_2(100 \leq X_e \leq 154) = 0.037315 \int_{100}^{154} \exp\left[-\frac{(X_e - 321)^2}{228.6}\right] dX_e = 2.64 \times 10^{-55}. \quad \textbf{(18)}$$

The effect of Student 2's consistency on exams is seen in these probabilities. Student 2 is virtually certain to make a grade of A or B during the second semester. Therefore, the answer to the question depends almost entirely upon Cases 1 and 2. Consequently, we need only the probability that Student 1 will make a grade of B, C, D, or F and the probability that Student 1 will make a grade of C, D, or F. The first of these is the probability that Student 1 will score less than 320 points on the exams, since, by assumption, he or she will make 180 points on the homework. This total probability is (see problem 10.8 for the normalized probability distribution)

$$P_1(\text{B, C, D, or F}) = 0.010117 \int_{0}^{319} \exp\left[-\frac{(X_e - 331)^2}{3{,}111.6}\right] dX_e = 0.03806. \quad \textbf{(19)}$$

For Student 1 to make a grade of C, D, or F, he or she must score between 0 to 219 points. The probability that this will occur is

$$P_1(\text{C, D, or F}) = 0.010117 \int_{0}^{219} \exp\left[-\frac{(X_e - 331)^2}{3{,}111.6}\right] dX_e = 0.00226. \quad \textbf{(20)}$$

The total probability for Cases 1 and 2 is thus

$$P(\text{Grade of \#2} > \text{Grade \#1}) = P_2(A)P_1(\text{B, C, D, or F}) + P_2(B)P_1(\text{C, D, or F})$$
$$= (0.0952)(0.03806) + (0.8880)(0.00226) = 0.00563. \quad \textbf{(21)}$$

There is, then, about a 0.56% chance that Student 2 will make a grade higher than that of Student 1.

10.13* In Problems 10.8 through 10.12, the professor treated the probability distribution as if the total exam score X_e were a continuous variable when, in reality, it is discrete. (That is, exam scores are usually integers.) However, if the difference $|P(X_e + 1) - P(X_e)|$ is very small relative to the total integrated probability (i.e., if the probabilities are closely spaced), the results obtained by treating the distribution as if it involved a continuous variable will be nearly identical to those obtained by treating the variable as discrete. In this problem, we illustrate this point by repeating Problems 10.8 and 10.9 with X_e being a discrete variable. When X_e is continuous over the range $0 \le X_e \le 467$, the normalized probability distribution is

$$P(X_e) = 0.010117 \exp\left[-\frac{(X_e - 331)^2}{3111.6}\right].$$

(A) Compute the nomalization constant for the probability distribution

$$P(X_e) = C \exp\left[-\frac{(X_e - 331)^2}{3111.6}\right],$$

assuming that X_e is a discrete variable which takes the values 0, 1, 2, 3, 4, ..., 467.
(B) Compute the probability that the student will make a grade of A during the second semester. The result using the assumption that X_e may be treated as a continuous variable is 0.6097. How much error was made by the continuous-variable assumption? The grade cutoffs are given in Problem 10.9. In Problem 10.8, it was stated that the student will make 180 points on the homework.

Solution

(A) The normalization condition now requires that we have

$$C \sum_{X_i=0}^{467} \exp\left[-\frac{(X_i - 331)^2}{3{,}111.6}\right] = 1. \tag{1}$$

The normalization constant is, therefore, given by

$$C = \left[\sum_{X_i=0}^{467} \exp\left[-\frac{(X_i - 331)^2}{3{,}111.6}\right]\right]^{-1}. \tag{2}$$

This sum may be evaluated using simple FORTRAN or C code or an Excel spreadsheet. The result is

$$C = (98.8439)^{-1} = \underline{0.010117}. \tag{3}$$

The corresponding discrete probability distribution is

$$P(X_e) = 0.010117 \exp\left[-\frac{(X_e - 331)^2}{3{,}111.6}\right] \quad \text{for } X_e = 0, 1, 2, \ldots, 467, \tag{4}$$

which is essentially the same as that obtained if one assumes that X_e is a continuous variable.

(B) In order to make a grade of A, the student must score 320 or more points on exams, since his or her homework grade is, by assumption, 180. The probability of doing this is now given by

$$P(X_e \ge 320) = 0.010117 \sum_{i=320}^{467} \exp\left[-\frac{(X_e - 331)^2}{3{,}111.6}\right]. \tag{5}$$

Again, we may compute the sum using an Excel spreadsheet to obtain

$$P(X_e \ge 320) = (0.010117)(60.748) = \underline{0.6146}. \tag{6}$$

The percent error due to the assumption that X_e is a continuous variable is

$$\%\ \text{error} = \frac{100(0.6097 - 0.6146)}{0.6146} = \underline{-0.80\%}. \tag{7}$$

The result would have been even better if the range of X_e had been larger. This type of approximation is frequently made in statistical mechanics. The problem illustrates the fact that the accuracy is usually very good.

10.15 In Chapter 17, which deals with statistical mechanics, we shall find that the probability that a system will occupy an energy state with energy E is proportional to $\exp[-E/RT]$ if E is given in units of energy per mole. Consider a simple system having only four possible energy states that we denote as States 1, 2, 3, and 4. The energies of these states are $E_1 = 0\ \text{J mol}^{-1}$, $E_2 = 1{,}000\ \text{J mol}^{-1}$, $E_3 = 2{,}000\ \text{J mol}^{-1}$, and $E_4 = 3{,}000\ \text{J mol}^{-1}$.

(A) Obtain the normalized probability distribution for the system in state i. Evaluate the normalization constant if $T = 298$ K.

(B) Calculate the average energy per system that a large number of such systems would have at 298 K.

(C) Which energy state makes the greatest contribution to the average energy?

(D) Which energy state is the most heavily populated?

(E) What is the probability that the combined energy of two separate systems described by the given probability distribution will add to 3000 joules?

(F) Compute the average total energy the two systems in (E) will have at 298 K. What relationship does this result have to that obtained in (B)?

Solution

(A) The probability distribution is discrete. So we have

$$P(E_i) = C \exp\left[-\frac{E_i}{RT}\right]. \tag{1}$$

The normalization constant is obtained from

$$\sum_{i=1}^{4} P(E_i) = C \sum_{i=1}^{4} \exp\left[-\frac{E_i}{RT}\right] = 1. \tag{2}$$

Solving for C, we obtain

$$C = \left[\sum_{i=1}^{4} \exp\left[-\frac{E_i}{RT}\right]\right]^{-1}. \tag{3}$$

The expression within brackets in Eq. 3 is an extremely important quantity in statistical mechanics. It is called the *molecular partition function* and is given the symbol z. Therefore, we might write $C = z^{-1}$. The partition function is, in essence, the reciprocal of a normalization constant. If $T = 298$ K, we have

$$C = \left[\exp\left[-\frac{0}{298R}\right] + \exp\left[-\frac{1{,}000}{(8.314)(298)}\right] + \exp\left[-\frac{2{,}000}{(8{,}314)(298)}\right] + \exp\left[-\frac{3{,}000}{(8{,}314)(298)}\right]\right]^{-1}$$

$$= [1 + 0.6679 + 0.4461 + 0.2979]^{-1} = (2.412)^{-1} = \underline{0.4146}. \tag{4}$$

The corresponding normalized probability distribution function is

$$\boxed{P(E_i) = 0.4146 \exp\left[-\frac{E_i}{RT}\right]}. \tag{5}$$

(B) The average value of the energy is given by

$$\langle E \rangle = \sum_{i=1}^{4} E_i P(E_i) = 0.4146 \sum_{i=1}^{4} E_i \exp\left[-\frac{E_i}{RT}\right]. \tag{6}$$

Substituting directly into Eq. 6 yields

$$\langle E \rangle = 0.4146\left[0 + 1{,}000 \exp\left[-\frac{1{,}000}{(8.314)(298)}\right] + 2{,}000 \exp\left[-\frac{2{,}000}{(8{,}314)(298)}\right]\right.$$

$$+ 3{,}000 \exp\left[-\frac{3{,}000}{(8{,}314)(298)}\right]\Bigg] \text{ joules}$$

$$= 0.4146[0 + 1{,}000(0.6679) + 2{,}000(0.4461) + 3{,}000(.2979)] \text{ joules}$$

$$= 0.4146(2{,}453.8) = \underline{1{,}017.3 \text{ J}}. \quad \textbf{(7)}$$

(C) The contribution of the four states to the average energy is given by the four terms in Eq. 7. State 1, with $E_1 = 0$, makes a zero contribution to the average energy of the system. The other contributions are 276.9 joules, 369.9 joules, and 370.5 joules for States 2, 3, and 4, respectively. Therefore, State 4 makes the largest contribution by a very slight margin over State 3.

(D) As can be seen from the size of the terms in Eq. 4, State 1 has the highest probability. It will, therefore, be the most heavily populated. The lowest-energy state is called the *ground state*, and it is often the most heavily populated state.

(E) There are four ways for the energy of two systems to add to 3,000 J. These are described in the following table, in which the probabilities are computed using Eq. 5 with $T = 298$ K:

System 1	*E* (J)	Probability	System 2	*E* (J)	Probability	Combined Probability
State 1	0	0.4146	State 4	3,000	0.1235	(0.4146)(0.1235) = 0.05120
State 4	3,000	0.1235	State 1	0	0.4146	(0.4146)(0.1235) = 0.05120
State 2	1,000	0.2769	State 3	2,000	0.1850	(0.2769)(0.1850) = 0.05121
State 3	2,000	0.1850	State 2	1,000	0.2769	(0.1850)(0.2769) = 0.05121

The sum of these four cases is 4(0.0512) = 0.205, correct to three significant digits. There is, therefore, a 20.5% chance that the total energy of the two systems will add to 3,000 J.

(F) The combined probability of any combination of State *i* for System 1 and State *j* for System 2 is

$$P(E_i, E_j) = P(E_i)P(E_j) = (0.4146)^2 \exp\left[-\frac{(E_i + E_j)}{RT}\right]. \quad \textbf{(8)}$$

There are seven possible sums for $E_i + E_j$: 0, 1,000, 2,000, 3,000, 4,000, 5,000, and 6,000 J. When the total energy adds to one of these sums, the combined probability will be as follows:

$$P(0) = (0.4146)^2 = 0.1719; \quad \textbf{(9)}$$

$$P(1{,}000 \text{ J}) = (0.4146)^2 \exp\left[-\frac{1{,}000}{(8.314)(298)}\right] = 0.1148; \quad \textbf{(10)}$$

$$P(2{,}000 \text{ J}) = (0.4146)^2 \exp\left[-\frac{2{,}000}{(8.314)(298)}\right] = 0.07668; \quad \textbf{(11)}$$

$$P(3{,}000 \text{ J}) = (0.4146)^2 \exp\left[-\frac{3{,}000}{(8.314)(298)}\right] = 0.05121; \quad \textbf{(12)}$$

$$P(4{,}000 \text{ J}) = (0.4146)^2 \exp\left[-\frac{4{,}000}{(8.314)(298)}\right] = 0.03420; \quad \textbf{(13)}$$

$$P(5{,}000 \text{ J}) = (0.4146)^2 \exp\left[-\frac{5{,}000}{(8.314)(298)}\right] = 0.02284; \quad \textbf{(14)}$$

$$P(6{,}000 \text{ J}) = (0.4146)^2 \exp\left[-\frac{6{,}000}{(8.314)(298)}\right] = 0.01526. \quad \textbf{(15)}$$

There are 16 possible combinations for System 1 and System 2, as shown in the following table:

State of System 1	State of System 2	Total Energy (J), $E(i, j)$	$P(E_i, E_j)$
1	1	0	0.1719
1	2	1,000	0.1148
1	3	2,000	0.07668
1	4	3,000	0.05121
2	1	1,000	0.1148
2	2	2,000	0.07668
2	3	3,000	0.05121
2	4	4,000	0.03420
3	1	2,000	0.07668
3	2	3,000	0.05121
3	3	4,000	0.03420
3	4	5,000	0.02284
4	1	3,000	0.05121
4	2	4,000	0.03420
4	3	5,000	0.02284
4	4	6,000	0.01526

If E_T is the total energy of the two systems, then the average total energy is

$$\langle E_T \rangle = \sum_{i=1}^{16} (E_i + E_j) P(E_i, E_j), \tag{16}$$

where $(E_i + E_j)$ and $P(E_i, E_j)$ are given in the last two columns of the table. Summing, we obtain

$$\langle E_T \rangle = \underline{2{,}034.6 \text{ J}}. \tag{17}$$

This result is exactly twice the result in Part (B). Thus, we see that the average energy of two systems is just $2\langle E \rangle$, where $\langle E \rangle$ is the average energy of one system. If we have N systems, the average energy would be $N\langle E \rangle$. Consequently, if the system is a molecule, we expect that the average energy of 1 mole of molecules will be the average energy per molecule times Avogadro's number.

10.17 Example 10.5 examined the probabilities involved in a game of chance in which there were an equal number of each of five different-sized holes in a game board. (see Figure 10.4.) Without informing the players, an enterprising gambler changes the ratios of different sizes of holes. The new game board contains 11 25-cm holes and only 10 of each of the other sizes. Nothing else is changed. Compute the average return to the players on this new game board.

Solution

The addition of the extra 25-cm hole changes all of the f_i. The probability of falling into a hole of radius r_i depends upon the total area occupied on the game board by such holes. This quantity is proportional to $n_i r_i^2$, where n_i is the number of holes of radius r_i on the board. Therefore, we take $f_1 = 10(5)^2 = 250$ for the 5-cm hole, $f_2 = 10(10)^2 = 1{,}000$ for the 10-cm hole, $f_3 = 10(15)^2 = 2{,}250$ for the 15-cm hole, $f_4 = 10(20)^2 = 4{,}000$ for the 20-cm hole, and $f_5 = 11(25)^2 = 6{,}875$ for the 25-cm hole. The normalization constant given by Eq. 10.5 is now

$$C = \left[\sum_{i=1}^{5} f_i\right]^{-1} = (250 + 1{,}000 + 2{,}250 + 4{,}000 + 6{,}875)^{-1} = (14{,}375)^{-1}. \quad \textbf{(1)}$$

The average return is given by Eq. 10.13 where $h(x) = X$:

$$\langle X \rangle = \frac{\sum_{i=1}^{5} X_i f_i}{14{,}375}$$

$$= \frac{(0)(6{,}875) + (\$1.00)(4{,}000) + (\$2.00)(2{,}250) + (\$3.00)(1{,}000) + (\$5.00)(250)}{14{,}375}$$

$$= \underline{\$0.8869} \ldots . \quad \textbf{(2)}$$

The player, therefore, loses an average of 11.31 cents each time the game is played. The gambler has increased his profit per play by 55.6 % by adding the extra large hole in the board.

10.19 The sole bet available at the craps table which is better for the player than betting that a "pass" will be made is for the player to bet that a pass will not be made. If the player bets in this manner and rolls a 12 on the first roll, there is no action. That is, it is as if the dice had not been thrown at all. The player does not lose or win; instead, he or she rolls the dice again, and this new roll is counted as the first roll. Compute the probability that a player who bets on "Don't Pass" will win the bet at craps. (See Problem 10.18 for the rules of craps.)

Solution

Care must be taken in solving this problem to take into account the fact that the combination of sixes on both dice "no longer exists." That is, when a 12 is thrown, it is as if the roll never occurred. The probability distribution must, therefore, be renormalized. The table on the next page gives the weights f_i of the various combinations. The distribution is now not normalized, since we have

$$\sum_T f_T = \frac{35}{36}, \quad \textbf{(1)}$$

which does not add to unity. Renormalization gives

$$C \sum_T f_T = \frac{35C}{36} = 1. \quad \textbf{(2)}$$

Thus, $C = 36/35$, and the table of probabilities is as shown on the following page. The player will win when betting on "Don't Pass" if

1. a 2 is thrown on the first roll, *or*
2. a 3 is thrown on the first roll, *or*
3. a 4 is thrown on the first roll and a 7 is then thrown before a 4, *or*
4. a 5 is thrown on the first roll and a 7 is then thrown before a 5, *or*
5. a 6 is thrown on the first roll and a 7 is then thrown before a 6, *or*
6. an 8 is thrown on the first roll and a 7 is then thrown before an 8, *or*

Die 1	Die 2	Total of dice	Combined Probability	f_T
1	1	2	$P(1)P(1) = 1/36$	1/36
1	2	3	$P(1)P(2) = 1/36$	
2	1	3	$P(2)P(1) = 1/36$	2/36
2	2	4	$P(2)P(2) = 1/36$	
1	3	4	$P(1)P(3) = 1/36$	
3	1	4	$P(3)P(1) = 1/36$	3/36
2	3	5	$P(2)P(3) = 1/36$	
3	2	5	$P(3)P(2) = 1/36$	
1	4	5	$P(1)P(4) = 1/36$	
4	1	5	$P(4)P(1) = 1/36$	4/36
1	5	6	$P(1)P(5) = 1/36$	
5	1	6	$P(5)P(1) = 1/36$	
2	4	6	$P(2)P(4) = 1/36$	
4	2	6	$P(4)P(2) = 1/36$	
3	3	6	$P(3)P(3) = 1/36$	5/36
1	6	7	$P(1)P(6) = 1/36$	
6	1	7	$P(6)P(1) = 1/36$	
2	5	7	$P(2)P(5) = 1/36$	
5	2	7	$P(5)P(2) = 1/36$	
3	4	7	$P(3)P(4) = 1/36$	
4	3	7	$P(4)P(3) = 1/36$	6/36
2	6	8	$P(2)P(6) = 1/36$	
6	2	8	$P(6)P(2) = 1/36$	
3	5	8	$P(3)P(5) = 1/36$	
5	3	8	$P(5)P(3) = 1/36$	
4	4	8	$P(4)P(4) = 1/36$	5/36
3	6	9	$P(3)P(6) = 1/36$	
6	3	9	$P(6)P(3) = 1/36$	
4	5	9	$P(4)P(5) = 1/36$	
5	4	9	$P(5)P(4) = 1/36$	4/36
4	6	10	$P(4)P(6) = 1/36$	
6	4	10	$P(6)P(4) = 1/36$	
5	5	10	$P(5)P(5) = 1/36$	3/36
5	6	11	$P(5)P(6) = 1/36$	
6	5	11	$P(6)P(5) = 1/36$	2/36
6	6	12	0	0

7. a 9 is thrown on the first roll and a 7 is then thrown before a 9, *or*
8. a 10 is thrown on the first roll and a 7 is then thrown before a 10.

The sum of the probabilities for these eight outcomes is the probability of winning a "Don't Pass" bet. Let the probability for these outcomes be denoted by $p(i)$. The values of $p(1)$ and $p(2)$ are given in the table on the next page.

$$p(1) = P_2 = \frac{1}{35}; \tag{1}$$

$$p(2)p = P_3 = \frac{2}{35}; \tag{2}$$

$$p(3) = P_4[\text{Probability of getting a 7 before we get a 4}]. \tag{3}$$

Once a 4 is thrown on the first roll, the only two numbers of significance on subsequent rolls are the 4 and the 7. As the table shows, there are nine rolls that give either a 4 or a 7. Three of these give a 4, and six give a 7. Therefore, we have six chances out of nine to get a 7 before we get a 4. Hence,

$$p(3) = \left(\frac{3}{35}\right)\left(\frac{6}{9}\right) = \frac{2}{35} \tag{4}$$

and

$$p(4) = P_5[\text{Probability of getting a 7 before we get a 5}] = \left(\frac{4}{35}\right)\left(\frac{6}{10}\right) = \frac{12}{175}, \tag{5}$$

since there are 10 ways to get either a 5 or a 7. Four of these give a 5, and the other six give a 7. Therefore, we have 6 chances out of 10 to obtain a 7 before we obtain a 5:

$$p(5) = P_6[\text{Probability of getting a 7 before we get a 6}] = \left(\frac{5}{35}\right)\left(\frac{6}{11}\right) = \frac{30}{385}, \tag{6}$$

since there are 11 ways to get either a 6 or a 7. Five of these give a 6, and the other six give a 7. Therefore, we have 6 chances out of 11 to obtain a 7 before we obtain a 6. The logic for the remaining cases is identical to that given for the preceding ones. The results are

$$p(6) = P_8[\text{Probability of getting a 7 before we get a 8}] = \left(\frac{5}{35}\right)\left(\frac{6}{11}\right) = \frac{30}{385}; \tag{7}$$

$$p(7) = P_9[\text{Probability of getting a 7 before we get a 9}] = \left(\frac{4}{35}\right)\left(\frac{6}{10}\right) = \frac{12}{175}; \tag{8}$$

$$p(8) = P_{10}[\text{Probability of getting a 7 before we get a 10}] = \left(\frac{3}{35}\right)\left(\frac{6}{9}\right) = \frac{2}{35}. \tag{9}$$

The probability of winning by betting on "Don't Pass" is, therefore,

$$\begin{aligned} P(\text{Don't Pass}) &= \sum_{i=1}^{8} p(i) = \frac{1}{35} + \frac{2}{35} + \frac{2}{35} + \frac{12}{175} + \frac{30}{385} + \frac{30}{385} + \frac{12}{175} + \frac{2}{35} \\ &= \underline{0.492987} \ldots . \end{aligned} \tag{10}$$

Die 1	Die 2	Total of dice	Combined Probability	P_T
1	1	2	$CP(1)P(1) = 1/35$	1/35
1	2	3	$CP(1)P(2) = 1/35$	
2	1	3	$CP(2)P(1) = 1/35$	2/35
2	2	4	$CP(2)P(2) = 1/35$	
1	3	4	$CP(1)P(3) = 1/35$	
3	1	4	$CP(3)P(1) = 1/35$	3/35
2	3	5	$CP(2)P(3) = 1/35$	
3	2	5	$CP(3)P(2) = 1/35$	
1	4	5	$CP(1)P(4) = 1/35$	
4	1	5	$CP(4)P(1) = 1/35$	4/35
1	5	6	$CP(1)P(5) = 1/35$	
5	1	6	$CP(5)P(1) = 1/35$	
2	4	6	$CP(2)P(4) = 1/35$	
4	2	6	$CP(4)P(2) = 1/35$	
3	3	6	$CP(3)P(3) = 1/35$	5/35
1	6	7	$CP(1)P(6) = 1/35$	
6	1	7	$CP(6)P(1) = 1/35$	
2	5	7	$CP(2)P(5) = 1/35$	
5	2	7	$CP(5)P(2) = 1/35$	
3	4	7	$CP(3)P(4) = 1/35$	
4	3	7	$CP(4)P(3) = 1/35$	6/35
2	6	8	$CP(2)P(6) = 1/35$	
6	2	8	$CP(6)P(2) = 1/35$	
3	5	8	$CP(3)P(5) = 1/35$	
5	3	8	$CP(5)P(3) = 1/35$	
4	4	8	$CP(4)P(4) = 1/35$	5/35
3	6	9	$CP(3)P(6) = 1/35$	
6	3	9	$CP(6)P(3) = 1/35$	
4	5	9	$CP(4)P(5) = 1/35$	
5	4	9	$CP(5)P(4) = 1/35$	4/35
4	6	10	$CP(4)P(6) = 1/35$	
6	4	10	$CP(6)P(4) = 1/35$	
5	5	10	$CP(5)P(5) = 1/35$	3/35
5	6	11	$CP(5)P(6) = 1/35$	
6	5	11	$CP(6)P(5) = 1/35$	2/35

Thus, the bet is almost even money: The player will win about 4,930 times for every 10,000 times he plays. Comparing the winning chance for bets on "Don't Pass" with the chance for "Pass" bets shows that the chance of winning on the former is greater by 0.00005772 .... A bet on "Don't Pass" is the best bet available in any major casino if we exclude blackjack and poker, in which the presence of additional information in the form of cards revealed and actions of the other players permits the player to actually swing the odds to his or her favor if the player is sufficiently skilled.

10.21* (A) Show that the probability that x will lie in the range $\langle x\rangle - \sigma \leq x \leq \langle x\rangle + \sigma$ for a normal Gaussian probability distribution, as given in Eq. 10.24, is approximately 0.683.

(B) Compute the probability that x will lie in the range $\langle x\rangle - 2\sigma \leq x \leq \langle x\rangle + 2\sigma$.

(C) Compute the probability that x will lie in the range $\langle x\rangle - 3\sigma \leq x \leq \langle x\rangle + 3\sigma$ if the probability distribution is that given in Eq. 10.24. (See hint in Problem 10.20; as an additional hint, use a table of error function integrals, or do the integrations numerically.)

Solution

(A) The normalized probability distribution is

$$P(x)dx = [(2\pi)^{1/2}\sigma]^{-1}\exp\left[-\frac{(x-\langle x\rangle)^2}{2\sigma^2}\right]dx \quad \text{for } -\infty \leq x \leq \infty. \tag{1}$$

Following the hint given in Problem 10.20, we let

$$z = \frac{(x-\langle x\rangle)}{\sigma}, \tag{2}$$

so that

$$dz = \frac{dx}{\sigma}, \tag{3}$$

which gives $dx = \sigma\, dz$. Substituting Eqs. 2 and 3 into Eq. 1 produces

$$P(z)dz = (2\pi)^{-1/2}\exp\left[-\frac{z^2}{2}\right]dz. \tag{4}$$

When $x - \langle x\rangle = -\sigma, z = -1$ and when $x - \langle x\rangle = \sigma$, we have $z = 1$. The probability that z lies in the range from -1 to 1 is given by

$$P(-1 \leq z \leq 1) = \int_{-1}^{1} P(z)dz = (2\pi)^{-1/2}\int_{-1}^{1}\exp\left[-\frac{z^2}{2}\right]dz. \tag{5}$$

The error function of z is defined as

$$\text{erf}(z) = \frac{2}{\pi^{1/2}}\int_0^z \exp(-t^2)dt. \tag{6}$$

We may put Eq. 5 into this form using the transformation

$$t = \frac{z}{2^{1/2}}, \tag{7}$$

so that we have

$$t^2 = \frac{z^2}{2} \quad \text{and} \quad dz = 2^{1/2}dt. \tag{8}$$

Substituting Eqs. 7 and 8 into Eq. 5 then produces

$$P(-1 \le z \le 1) = (\pi)^{-1/2} \int_{-2^{-1/2}}^{2^{-1/2}} \exp[-t^2]\,dt$$

$$= \frac{2}{\pi^{1/2}} \int_0^{2^{-1/2}} \exp[-t^2]\,dt = \text{erf}(2^{-1/2}). \qquad (9)$$

Using a table of error function values, we find that

$$P(-1 \le z \le 1) = \text{erf}(0.7071\ldots) \approx \underline{0.6827}\ldots. \qquad (10)$$

The problem can also be solved by using any convenient numerical method to integrate Eq. 5.

(B) Using the analysis given in Part (A), we now need the integrated probability between $z = -2$ and $z = 2$. That is, we have

$$P(-2 \le z \le 2) = \int_{-2}^{2} P(z)\,dz = (2\pi)^{-1/2} \int_{-2}^{2} \exp\left[-\frac{z^2}{2}\right] dz. \qquad (11)$$

Equation 1 can be integrated numerically, or we may transform it to the error function using the methods shown in Part (A). The result is

$$P(-2 \le z \le 2) = (\pi)^{-1/2} \int_{-2^{1/2}}^{2^{1/2}} \exp[-t^2]\,dt = \frac{2}{\pi^{1/2}} \int_0^{2^{1/2}} \exp[-t^2]\,dt = \text{erf}(2^{1/2}).$$

$$(12)$$

Thus,

$$P(-2 \le z \le 2) = \text{erf}(1.414\ldots) \approx \underline{0.9545}. \qquad (13)$$

(C) The probability that z will lie in the range $-3 \le z \le 3$ is given by

$$P(-3 \le z \le 3) = (\pi)^{-1/2} \int_{-3/\sqrt{2}}^{3/\sqrt{2}} \exp[-t^2]\,dt$$

$$= \frac{2}{\pi^{1/2}} \int_0^{3/\sqrt{2}} \exp[-t^2]\,dt = \text{erf}\left(\frac{3}{2^{1/2}}\right). \qquad (14)$$

The result is

$$P(-3 \le z \le 3) = \text{erf}\left(\frac{3}{2^{1/2}}\right) = \text{erf}(2.1213\ldots) \approx \underline{0.9973}. \qquad (15)$$

10.23* There are N randomly chosen people in a room.

(A) Ignoring the effect of leap years and assuming that all years have 365 days, obtain an expression as a function of N giving the probabiliity that no two birthdays will fall on the same day of the year. Plot the result as a function of N over the range $2 \le N \le 60$.

(B) What value of N makes the probability as close to 0.5 as possible? If $N = 365$, what is the probability that no two persons in the room have their birthdays on the same day of the year. (*Hint*: This problem is similar to Example 10.3, but it is much easier to solve the problem if it is viewed as comprising N separate experiments rather than one experiment in which all N persons state their birthdays. The solution to the problem is rather surprising—so much so, that many persons refuse to believe it is correct!)

Solution

Following the hint in the problem, we view the process of collecting birthdays as being N separate events. From this viewpoint, the problem is relatively simple. First, suppose that $N = 2$. Then, the probability that the second person will not have his or her birthday fall on the same day as the first person's is clearly 364/365. That is, if P_N is the probability that no two birthdays will fall on the same day when there are N persons in the room, then

$$P_2 = \frac{364}{365}. \tag{1}$$

When $N = 3$, the probability that no birthdays will match is the combined probability that there will be no match among the first two persons, P_2, *and* that the third person's birthday will not match either of the first two. Since there are 363 days that permit this to happen out of 365, the probability that the third person will not match either of the other two is 363/365. The combined probability that the first two will not match *and* that the third person will not match either of the other two gives

$$P_3 = P_2\frac{364}{365} = \frac{(364)(363)}{(365)^2}. \tag{2}$$

When $N = 4$, the probability that no birthdays will match is the combined probability that there will be no match among the first three persons, P_3, *and* that the fourth person's birthday will not match any of the first three. Since there are 362 days that permit this to happen out of 365, the probability that the fourth person will not match any of the other three is 362/365. The combined probability that the first three will not match *and* that the fourth person will not match any of the other three gives

$$P_4 = P_3\frac{362}{365} = \frac{(364)(363)(362)}{(365)^3}. \tag{3}$$

It is now easy to generalize the result to N persons. Inspecting Eqs. 1, 2, and 3 yields

$$P_N = \frac{(364)(363)(362)(360)\cdots(366-N)}{(365)^{N-1}}. \tag{4}$$

Equation 4 is the desired result. We may write it in more compact form using factorials:

$$P_N = \frac{364!}{(365)^{N-1}(365-N)!}. \tag{5}$$

The plot on the top of the next page shows P_N as a function of N over the range $2 \leq N \leq 60$.

It is worth noting that this problem becomes extremely difficult to solve if the process is viewed as consisting of a single experiment in which one attempts to enumerate all of the possibilities and then adds the probabilities of those events which satisfy the conditions of the problem. The difficulty is the huge number of possible combinations for N persons, each of whom can have any of 365 different birthdays.

(B) The plot of P_N versus N shows that the probability is closer to 0.5 when $N = 23$ than for any other number. Substituting $N = 23$ into Eq. 5 gives

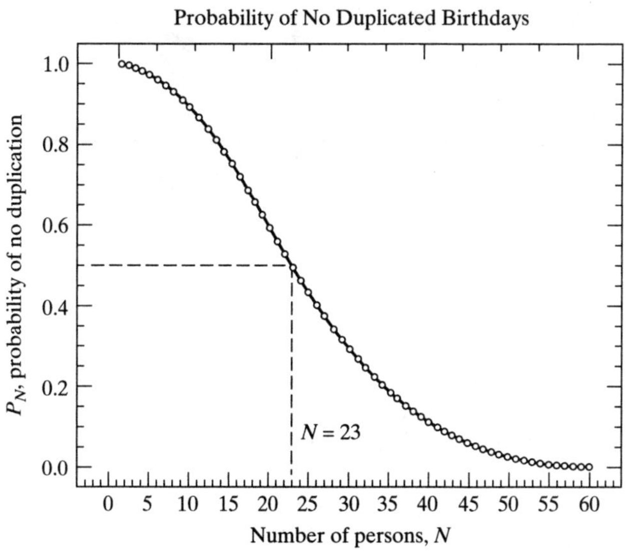

$$P_{23} = \frac{364!}{(364)^{22}(342!)} = \underline{0.4927}\ldots. \tag{6}$$

Thus, if 23 persons are in a room, there is about a 50% chance that at least two of them have their birthday on the same day of the year. This is a result that many people find difficult to believe.

If $N = 365$, the probability that there will be no matching birthdays is

$$P_{365} = \frac{364!}{(365)^{364}} \approx \underline{1.455 \times 10^{-157}}. \tag{7}$$

Shall we just call it zero?

CHAPTER 11

Introduction to Quantum Mechanics

11.1 This problem is based on the results obtained in Example 11.1. A quarterback is poised to throw a football in the x–y plane with an initial velocity v_o that we will assume is the result of the maximum force he can generate. Let us assume that a receiver is located at the point $x = x_o$, $y = 0$, and $z = -z_o$ at time $t = 0$. The receiver is running at a constant velocity equal to S_o in the $+z$ direction. The quarterback wishes to complete a pass to the receiver when he reaches the point $x = x_o$ and $y = z = 0$. (See the diagram below.)

(A) Show that if $x_o > v_o^2/g$, the receiver is out of range of the quarterback's ability to throw the football.
(B) Show that if $z_o/S_o < (2v_o/g)\sin[0.5\sin^{-1}\{gx_o/v_o^2\}]$, the receiver is moving too fast for the quarterback to complete the pass at the designated point.
(C) If $x_o < v_o^2/g$ and $z_o/S_o > (2v_o/g)\sin[0.5\sin^{-1}\{gx_o/v_o^2\}]$, show that the pass may be completed if the quarterback throws the ball when

$$t = (z_o/S_o) - (2v_o/g)\sin[0.5\sin^{-1}\{gx_o/v_o^2\}]$$

at an angle given by $\theta_o = 0.5\sin^{-1}\{gx_o/v_o^2\}$. Figure 11.20 illustrates the situation. You may use the results obtained in Example 11.1. Ignore the effects of air resistance.

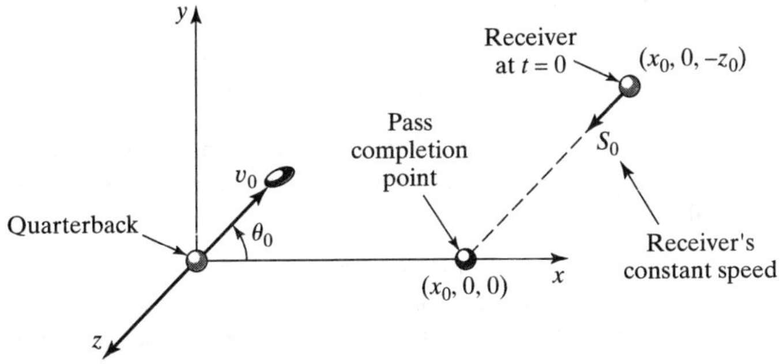

▲ FIGURE 11.20
Diagram of quarterback and receiver in Problem 11.1 who are attempting to complete a pass at the point $x = x_o$.

Solution

(A) In Example 11.1, it was shown that the time of impact with the ground, t_I, is given by

$$t_I = \frac{2v_o\sin(\theta_o)}{g}. \tag{1}$$

The impact point along the x axis x_I is, therefore,

$$x_I = v_o\cos(\theta_o)t_I = v_o\cos(\theta_o)\frac{2v_o\sin(\theta_o)}{g} = \frac{v_o^2\sin(2\theta_o)}{g}. \tag{2}$$

The maximum value of x_I is v_o^2/g, since $\sin(2\theta_o)$ has a maximum value of unity at $\theta_o = \pi/4$. We conclude that if $x_o > v_o^2/g$, the receiver will be out of range of the quarterback's throwing ability.

(B) The flight time of the football to move from the quarterback's hand to the impact point is given by Eq. 1. The time the receiver takes to reach the point $x = x_o$, $y = z = 0$, is

$$t_{\text{receiver}} = \frac{z_o}{S_o}, \tag{3}$$

since z_o is the distance that must be traveled at speed S_o. If $t_I > t_{receiver}$, the receiver will have crossed the impact point and left before the football reaches the target, and the pass will be incomplete as it falls behind the receiver. Hence, if we are to be successful, we must have

$$t_{receiver} = \frac{z_o}{S_o} = t_I = \frac{2v_o \sin(\theta_o)}{g}. \tag{4}$$

To hit the receiver, the football must be thrown at an angle θ_o so that the impact point is equal to x_o. That is, we must have

$$x_o = \frac{v_o^2 \sin(2\theta_o)}{g}. \tag{5}$$

Therefore, the required elevation occurs when we have

$$\sin(2\theta_o) = \frac{gx_o}{v_o^2}. \tag{6}$$

The required angle is

$$\theta_o = 0.5 \sin^{-1}\left[\frac{gx_o}{v_o^2}\right]. \tag{7}$$

The condition on the time necessary to complete the pass is given by combining Eqs. 7 and 4 to produce

$$\frac{z_o}{S_o} = \frac{2v_o}{g} \sin\left\{0.5 \sin^{-1}\left[\frac{gx_o}{v_o^2}\right]\right\}. \tag{8}$$

If Z_o/S_o is less than the expression on the right-hand side of Eq. 8, the receiver will be moving too fast for the quarterback to get off a pass that will hit him at the point $x = x_o$.

(C) If the range and the receiver's speed are such that a completed pass at point x_o is possible, Eq. 7 gives the elevation the quarterback needs to use to hit his receiver. After elevating to this angle, he wishes to throw the football at a time t^* such that $t^* + t_I$ will be exactly equal to the time required for his receiver to reach the point $x = x_o, y = z = 0$. Thus, the condition to be satisfied is

$$t^* + t_I = t^* + \frac{2v_o \sin(\theta_o)}{g} = t_{receiver} = \frac{z_o}{S_o}. \tag{9}$$

Therefore,

$$t^* = t_{receiver} - t_I = \frac{z_o}{S_o} - \frac{2v_o \sin(\theta_o)}{g}. \tag{10}$$

Substituting Eq. 7 for θ_o gives

$$\boxed{t^* = \frac{z_o}{S_o} - \frac{2v_o}{g} \sin\left\{0.5 \sin^{-1}\left[\frac{gx_o}{v_o^2}\right]\right\}}, \tag{11}$$

which is the required expression. Note that unless t^* is positive, it will be too late, and the quarterback will not be able to hit the receiver.

11.3 Example 11.1 requested that we solve the classical equations of motion for a projectile moving in the x–y plane subject to the gravitational potential $V = mgy$ with $x = y = z = 0$ at $t = 0$ and with

$$\frac{dx}{dt} = v_o \cos\theta_o \quad \text{and} \quad \frac{dy}{dt} = v_o \sin\theta_o \quad \text{at} \quad t = 0.$$

The effect of air friction was ignored.

(A) Obtain the solution to this problem if the potential field is $V(x, y) = mgy + mkx$ instead of the gravitational potential. (Take m, g, and k to be constants.)

(B) For the potential in (A), determine x at the point of impact with the ground in terms of v_o, θ, k, and g.

(C) For the potential in (A), determine the maximum height attained by the projectile in terms of v_o, θ, k, and g.

Solution

(A) The y and z results are identical to those in Example 11.1, since we still have

$$\frac{\partial V}{\partial y} = \text{mg and } \frac{\partial V}{\partial z} = 0. \tag{1}$$

Therefore,

$$\boxed{y = v_o \sin(\theta_o)t - \frac{gt^2}{2}}, \quad \boxed{\frac{\partial y}{dt} = v_y = v_o \sin(\theta_o) - gt}, \tag{2}$$

and

$$\boxed{z = 0 \text{ and } \frac{\partial z}{dt} = v_z = 0}, \tag{3}$$

for all t. The x equation is different, however, since we now have $F_x = -\partial V/\partial x = -mk$. Equation 11.6A in the text gives

$$m\frac{d^2x}{dt^2} + \frac{\partial V}{\partial x} = m\frac{d^2x}{dt^2} + mk = 0. \tag{4}$$

Dividing by m and integrating once, we obtain

$$\int d\left(\frac{dx}{dt}\right) = -k \int dt, \tag{5}$$

which gives

$$\left(\frac{dx}{dt}\right) = -kt + C = v_x. \tag{6}$$

At $t = 0$, we have $v_x = v_o \cos(\theta_o)$. Therefore, $C = v_o \cos(\theta_o)$. Substituting into Eq. 6 produces

$$\boxed{\left(\frac{dx}{dt}\right) = -kt + v_o \cos(\theta_o)}. \tag{7}$$

The second integration gives

$$\int dx = x = \int [-kt + v_o \cos(\theta_o)]\,dt = -\frac{kt^2}{2} + v_o \cos(\theta_o)t + C'. \tag{8}$$

Since $x = 0$ at $t = 0$, we must have $C' = 0$ and

$$\boxed{x = -\frac{kt^2}{2} + v_o \cos(\theta_o)t}. \tag{9}$$

(B) The time of impact with the ground is the time at which $y = 0$. This is the same as in Example 11.1, since the y equation is unchanged. The result is

$$t_I = \text{impact time} = \frac{2v_o \sin(\theta_o)}{g}. \quad \textbf{(10)}$$

Substituting this time into Eq. 9 gives the x-coordinate at impact:

$$x_I = v_o \cos(\theta_o)\frac{2v_o \sin(\theta_o)}{g} - \frac{k}{2}\left[\frac{2v_o \sin(\theta_o)}{g}\right]^2$$

$$= \boxed{\frac{v_o^2 \sin(2\theta_o)}{g} - \frac{2v_o^2 k \sin^2(\theta_o)}{g^2}}. \quad \textbf{(11)}$$

Since the second term is negative if $k > 0$, the range of the projectile is now less.

(C) This result is the same as that obtained in Example 11.1. The added force in the x direction does nothing to affect the height in the y direction. Thus, $y_{\max}$ occurs at $t_{\max}$ where

$$\frac{dy}{dt} = -gt + v_o \sin(\theta_o)\bigg|_{t=t_{max}} = 0, \quad \textbf{(12)}$$

so that

$$t_{\max} = \frac{v_o \sin(\theta_o)}{g}. \quad \textbf{(13)}$$

Substituting into the expression for y gives

$$y_{\max} = -\frac{gt_{\max}^2}{2} + v_o \sin(\theta_o)t_{\max} = -\frac{g}{2}\left[\frac{v_o \sin(\theta_o)}{g}\right]^2 + v_o \sin(\theta_o)\frac{v_o \sin(\theta_o)}{g}$$

$$= \boxed{\frac{v_o^2 \sin^2(\theta_o)}{2g}}, \quad \textbf{(14)}$$

as in Example 11.1.

11.5 Assume the particle illustrated in Figure 11.5 is moving in three-dimensional space. Then, in rectangular Cartesian coordinates, the kinetic energy of the particle is

$$T = \frac{m}{2}[v_x^2 + v_y^2 + v_z^2].$$

Suppose we wish to transform to a spherical polar coordinate system with coordinates (R, θ, ϕ) instead of (x, y, z). This coordinate system is illustrated in the figure given in Problem 11.2 if we replace F with R. The transformation equations between the spherical polar coordinate system and a rectangular Cartesian system, given in Chapter 9, Figure 9.3, are

$$x = R \sin\theta \cos\phi,$$

$$y = R \sin\theta \sin\phi,$$

and

$$z = R \cos\theta.$$

(A) Obtain the kinetic energy of the particle in terms of the spherical polar coordinates and their rates of change with time.

(B) Obtain expressions for the momenta conjugate to the spherical polar coordinates.

(C) Show that the momentum conjugate to ϕ, P_ϕ, is the same as the Cartesian z-component of angular momentum.

(D) Express the kinetic energy in terms of the spherical polar conjugate momenta and coordinates.

Solution

(A) We need expressions for the Cartesian velocities in terms of the rates of change of the spherical polar coordinates. If we take derivatives of the transformation equations with respect to time, we obtain

$$\begin{aligned}\frac{dx}{dt} &= \frac{dR}{dt}\sin\theta\cos\phi + R\frac{d\theta}{dt}\cos\theta\cos\phi - R\frac{d\phi}{dt}\sin\theta\sin\phi \\ &= v_R\sin\theta\cos\phi + Rv_\theta\cos\theta\cos\phi - Rv_\phi\sin\theta\sin\phi = v_x, \end{aligned} \tag{1}$$

$$\begin{aligned}\frac{dy}{dt} &= \frac{dR}{dt}\sin\theta\sin\phi + R\frac{d\theta}{dt}\cos\theta\sin\phi + R\frac{d\phi}{dt}\sin\theta\cos\phi \\ &= v_R\sin\theta\sin\phi + Rv_\theta\cos\theta\sin\phi + Rv_\phi\sin\theta\cos\phi = v_y, \end{aligned} \tag{2}$$

and

$$\frac{dz}{dt} = \frac{dR}{dt}\cos\theta - R\frac{d\theta}{dt}\sin\theta = v_R\cos\theta - Rv_\theta\sin\theta = v_z. \tag{3}$$

Squaring each expression, yields

$$\begin{aligned} v_x^2 &= v_R^2\sin^2\theta\cos^2\phi + R^2v_\theta^2\cos^2\theta\cos^2\phi + R^2v_\phi^2\sin^2\theta\sin^2\phi \\ &\quad + 2v_RRv_\theta\sin\theta\cos\theta\cos^2\phi - 2v_RRv_\phi\sin^2\theta\cos\phi\sin\phi \\ &\quad - 2R^2v_\theta v_\phi\cos\theta\cos\phi\sin\theta\sin\phi, \end{aligned} \tag{4}$$

$$\begin{aligned} v_y^2 &= v_R^2\sin^2\theta\sin^2\phi + R^2v_\theta^2\cos^2\theta\sin^2\phi + R^2v_\phi^2\sin^2\theta\cos^2\phi \\ &\quad + 2v_RRv_\theta\sin\theta\cos\theta\sin^2\phi + 2v_RRv_\phi\sin^2\theta\cos\phi\sin\phi \\ &\quad + 2R^2v_\theta v_\phi\cos\theta\cos\phi\sin\theta\sin\phi, \end{aligned} \tag{5}$$

and

$$v_z^2 = v_R^2\cos^2\theta + R^2v_\theta^2\sin^2\theta - 2v_RRv_\theta\sin\theta\cos\theta. \tag{6}$$

We must now sum Eqs. 4–6 to obtain $v_x^2 + v_y^2 + v_z^2$:

$$\begin{aligned} v_x^2 + v_y^2 + v_z^2 &= v_R^2\sin^2\theta[\cos^2\phi + \sin^2\phi] + R^2v_\theta^2\cos^2\theta[\cos^2\phi + \sin^2\phi] \\ &\quad + R^2v_\theta^2\sin^2\theta + v_R^2\cos^2\theta + R^2v_\phi^2\sin^2\theta[\cos^2\phi + \sin^2\phi] \\ &\quad + 2v_RRv_\theta\sin\theta\cos\theta[\cos^2\phi + \sin^2\phi] - 2v_RRv_\theta\sin\theta\cos\theta. \end{aligned} \tag{7}$$

The remaining cross terms have different signs and cancel. Using the fact that $\sin^2 z + \cos^2 z = 1$, we can write Eq. 7 in the form

$$v_x^2 + v_y^2 + v_z^2 = v_R^2[\sin^2\theta + \cos^2\theta] + R^2v_\theta^2[\sin^2\theta + \cos^2\theta] + R^2v_\phi^2\sin^2\theta. \tag{8}$$

Once again making use of the fact that $\sin^2 z + \cos^2 z = 1$, we obtain

$$v_x^2 + v_y^2 + v_z^2 = v_R^2 + R^2v_\theta^2 + R^2v_\phi^2\sin^2\theta. \tag{9}$$

Consequently, the kinetic energy, expressed in terms of spherical polar coordinates and velocities, is

$$\boxed{T = \frac{m}{2}[v_R^2 + R^2 v_\theta^2 + R^2 v_\phi^2 \sin^2\theta]}. \tag{10}$$

This is the desired expression.

(B) The momentum conjugate to coordinate q is given by Eq. 11.13:

$$P_q = \frac{\partial T}{\partial v_q}. \tag{11}$$

Therefore,

$$\boxed{P_R = \frac{\partial T}{\partial v_R} = m v_R}. \tag{12}$$

The radial momentum has the same form as the Cartesian momenta. It is just the particle mass times the radial velocity v_R. The angular momentum conjugate to θ is

$$\boxed{P_\theta = \frac{\partial T}{\partial v_\theta} = mR^2 v_\theta}. \tag{13}$$

We see that this angular momentum has the same form as that obtained for the two-dimensional system shown in Figure 11.5 and treated in the text. The momentum conjugate to ϕ is

$$\boxed{P_\phi = \frac{\partial T}{\partial v_\phi} = mR^2 v_\phi \sin^2\theta}. \tag{14}$$

Since the angle ϕ measures the rotation about the z-axis, this conjugate momentum is equivalent to M_z for the three-dimensional system. In Part (C), we will prove this equivalence.

(C) The z component of angular momentum about the origin is given by Eq. 11.26C:

$$M_z = xP_y - yP_x. \tag{15}$$

Substituting the transformation equations for x and y and using Eqs. 1 and 2 to obtain P_x and P_y, we have

$$M_z = R\sin\theta\cos\phi\, m[v_R \sin\theta\sin\phi + Rv_\theta\cos\theta\sin\phi + Rv_\phi\sin\theta\cos\phi]$$
$$- R\sin\theta\sin\phi\, m[v_R\sin\theta\cos\phi + Rv_\theta\cos\theta\cos\phi - Rv_\phi\sin\theta\sin\phi]. \tag{16}$$

Collecting terms, we obtain

$$M_z = mR[v_R\sin^2\theta\cos\phi\sin\phi + Rv_\theta\sin\theta\cos\phi\cos\theta\sin\phi + Rv_\phi\sin^2\theta\cos^2\phi$$
$$- v_R\sin^2\theta\sin\phi\cos\phi - Rv_\theta\sin\theta\sin\phi\cos\theta\cos\phi + Rv_\phi\sin^2\theta\sin^2\phi]. \tag{17}$$

The first and second terms inside the brackets cancel with the fourth and fifth terms. The result is

$$\boxed{M_z = mR^2 v_\phi \sin^2\theta[\sin^2\phi + \cos^2\phi] = mR^2 v_\phi\sin^2\theta}, \tag{18}$$

which is identical to the expression for P_ϕ obtained in Eq. 14.

(D) Using the expressions for the conjugate momenta, we obtain

$$\frac{mv_R^2}{2} = \frac{P_R^2}{2m}, \tag{19}$$

$$\frac{mR^2v_\theta^2}{2} = \frac{P_\theta^2}{2mR^2}, \tag{20}$$

and

$$\frac{mR^2v_\phi^2 \sin^2\theta}{2} = \frac{P_\phi^2}{2mR^2\sin^2\theta}. \tag{21}$$

Substituting the results from Eqs. 19–21 into Eq. 10 yields

$$T = \frac{m}{2}[v_R^2 + R^2v_\theta^2 + R^2v_\phi^2\sin^2\theta] = \boxed{\frac{P_R^2}{2m} + \frac{P_\theta^2}{2mR^2} + \frac{P_\phi^2}{2mR^2\sin^2\theta}}, \tag{22}$$

which is the desired expression.

11.7 Show that Planck's equation for the energy density of blackbody radiation reduces to the equation suggested by Wien for small values of λT.

Solution

When λT is small, the exponential $\exp[hc/\lambda k_b T]$ becomes very large relative to unity. Therefore,

$$\exp\left[\frac{hc}{\lambda k_b T}\right] - 1 \approx \exp\left[\frac{hc}{\lambda k_b T}\right]. \tag{1}$$

Under this condition, the Planck equation becomes

$$u(T, \lambda) = \frac{8\pi hc}{\lambda^5}\exp\left[-\frac{hc}{\lambda k_b T}\right]. \tag{2}$$

If we take c_1 of the Wien equation to be $8\pi hc$ and c_2 to be hc/k_b, Eq. 2 becomes identical to the one proposed by Wien.

11.9 Show that Planck's equation predicts a maximum in the energy density of blackbody radiation at the point $\lambda \approx 0.290$ cm K/T when $u(T, \lambda)$ is plotted against λ. [*Hint*: You may solve the problem either by a one-dimensional grid search or by iterative methods.]

Solution

The Planck equation is

$$u(T, \lambda) = \frac{8\pi hc}{\lambda^5} \frac{1}{\exp\left[\frac{hc}{\lambda k_b T}\right] - 1}. \tag{1}$$

The condition for a maximum is that $\partial u(T,\lambda)/\partial\lambda = 0$. The derivative of Eq. 1 is

$$\frac{\partial u(T,\lambda)}{\partial\lambda} = -\frac{40\pi hc}{\lambda^6}\frac{1}{\exp\left[\frac{hc}{\lambda k_b T}\right]-1} - \frac{8\pi hc}{\lambda^5}\frac{-hc}{\lambda^2 k_b T}\frac{\exp\left[\frac{hc}{\lambda k_b T}\right]}{\left\{\exp\left[\frac{hc}{\lambda k_b T}\right]-1\right\}^2}$$

$$= -\frac{40\pi hc}{\lambda^6}\frac{1}{\exp\left[\frac{hc}{\lambda k_b T}\right]-1} + \frac{8\pi h^2c^2}{\lambda^7 k_b T}\frac{\exp\left[\frac{hc}{\lambda k_b T}\right]}{\left\{\exp\left[\frac{hc}{\lambda k_b T}\right]-1\right\}^2}. \quad \textbf{(2)}$$

Setting the right-hand side of Eq. 2 to zero and multiplying both sides by $\lambda^7 k_b T\{\exp[hc/(\lambda k_b T)] - 1\}$, we obtain

$$-40\pi hck_bT\lambda + 8\pi h^2c^2\frac{\exp\left[\frac{hc}{\lambda k_b T}\right]}{\left\{\exp\left[\frac{hc}{\lambda k_b T}\right]-1\right\}} = 0. \quad \textbf{(3)}$$

Now let $z = \lambda T$. Substituting into Eq. 3 and rearranging terms yields

$$z = \frac{hc}{5k_b}\frac{\exp\left[\frac{hc}{k_b z}\right]}{\left\{\exp\left[\frac{hc}{k_b z}\right]-1\right\}}. \quad \textbf{(4)}$$

The value of z that produces the maximum seen in Figure 11.5 is the solution of Eq. 4. There is no analytic solution, but we can obtain a numerical solution either by iteration or by a one-dimensional grid search for z. Let us illustrate the iterative method here. If $hc/(k_b z)$ is large, the factor containing the ratio of exponential functions will approach unity, so, in the first iteration, we take the ratio to be unity. This gives

$$z \approx \frac{hc}{5k_b} = \frac{(6.62608\times 10^{-27}\text{ ergs s})(2.997\times 10^{10}\text{ cm s}^{-1})}{5(1.3807\times 10^{-16}\text{ ergs K}^{-1})} = 0.2877\text{ cm K}. \quad \textbf{(5)}$$

In the second iteration, we set z in the exponential factor to 0.2877 cm K. This yields

$$\exp\left[\frac{hc}{k_b z}\right] = \exp\left[\frac{(6.62608\times 10^{-27}\text{ ergs s})(2.997\times 10^{10}\text{ cm s}^{-1})}{(0.2877\text{ cm K})(1.3807\times 10^{-16}\text{ ergs K}^{-1})}\right]$$

$$= \exp(4.999) = 148.3. \quad \textbf{(6)}$$

Therefore, the exponential factor is

$$\frac{\exp\left[\frac{hc}{k_b z}\right]}{\left\{\exp\left[\frac{hc}{k_b z}\right]-1\right\}} = \frac{148.3}{147.3} = 1.0068. \quad \textbf{(7)}$$

Substituting this result into Eq. 4 gives

$$z = \frac{hc}{5k_b}(1.0068). \quad \textbf{(8)}$$

Solving for z, we obtain

$$z = \frac{(6.62608 \times 10^{-27}\ \text{ergs s})(2.997 \times 10^{10}\ \text{cm s}^{-1})}{5(1.3807 \times 10^{-16}\ \text{ergs K}^{-1})}(1.0068) = 0.290\ \text{cm K}. \quad \textbf{(9)}$$

At this point, the solution has converged to three significant digits. Therefore, we obtain a maximum at the point

$$z = \lambda T = 0.290\ \text{cm K}. \quad \textbf{(10)}$$

Thus,

$$\boxed{\lambda_{max} = \frac{0.290}{T}\ \text{cm}}. \quad \textbf{(11)}$$

The less rigorous equation suggested by Wien gave the result $\lambda_{max} = 0.294/T$ cm. A one-dimensional grid search yields the result given in Eq. 11.

11.11 Compute the possible range of emisson energies seen for the Lyman, Balmer, and Paschen series for the hydrogen atom.

Solution

The emission wavelengths are given by

$$\lambda^{-1} = 109{,}677\left[\frac{1}{n^2} - \frac{1}{m^2}\right]\text{cm}^{-1}. \quad \textbf{(1)}$$

The Lyman series of lines has $n = 1$. Therefore, the shortest wavelength (most energetic) is obtained when $m = \infty$. The longest wavelength (least energetic) is obtained for $m = 2$. The shortest wavelength is

$$(\lambda)^{-1}_{\text{shortest}} = \frac{109{,}677}{1^2}\ \text{cm}^{-1} = 109{,}677\ \text{cm}^{-1}. \quad \textbf{(2)}$$

The energy corresponding to this wavelength is

$$E_{\text{largest}} = hc(\lambda)^{-1}_{\text{shortest}} = (6.62608 \times 10^{-34}\ \text{J s})(2.99792 \times 10^{10}\ \text{cm s}^{-1})(109{,}677\ \text{cm}^{-1})$$
$$= 2.1769 \times 10^{-18}\ \text{J}. \quad \textbf{(3)}$$

The longest wavelength is

$$(\lambda)^{-1}_{\text{longest}} = 109{,}677\left[\frac{1}{1^2} - \frac{1}{2^2}\right]\text{cm}^{-1} = 82{,}257.8\ \text{cm}^{-1}. \quad \textbf{(4)}$$

The corresponding energy is

$$E_{\text{smallest}} = hc(\lambda)^{-1}_{\text{longest}} = (6.62608 \times 10^{-34}\ \text{J s})(2.99792 \times 10^{10}\ \text{cm s}^{-1})(82{,}257.8\ \text{cm}^{-1})$$
$$= 1.6327 \times 10^{-18}\ \text{J}. \quad \textbf{(5)}$$

The energy range for the Lyman series is, therefore,

$$\boxed{1.6327 \times 10^{-18}\ \text{J} \le E_{\text{emission}} \le 2.1769 \times 10^{-18}\ \text{J}}.$$

The Balmer series of lines has $n = 2$. Therefore, the shortest wavelength (most energetic) is obtained when $m = \infty$. The longest wavelength (least energetic) is obtained for $m = 3$. The shortest wavelength is

$$(\lambda)^{-1}_{\text{shortest}} = \frac{109{,}677}{2^2}\ \text{cm}^{-1} = 27{,}419.2\ \text{cm}^{-1}. \tag{6}$$

The energy corresponding to this wavelength is

$$E_{\text{largest}} = hc(\lambda)^{-1}_{\text{shortest}} = (6.62608 \times 10^{-34}\ \text{J s})(2.99792 \times 10^{10}\ \text{cm s}^{-1})(27{,}419.2\ \text{cm}^{-1}) = 5.4423 \times 10^{-19}\ \text{J}. \tag{7}$$

The longest wavelength is

$$(\lambda)^{-1}_{\text{longest}} = 109{,}677\left[\frac{1}{2^2} - \frac{1}{3^2}\right]\ \text{cm}^{-1} = 15{,}232.9\ \text{cm}^{-1}. \tag{8}$$

The corresponding energy is

$$E_{\text{smallest}} = hc(\lambda)^{-1}_{\text{longest}} = (6.62608 \times 10^{-34}\ \text{J s})(2.99792 \times 10^{10}\ \text{cm s}^{-1})(15{,}232.9\ \text{cm}^{-1}) = 3.0235 \times 10^{-19}\ \text{J}. \tag{9}$$

The energy range for the Balmer series is, therefore,

$$\boxed{3.0235 \times 10^{-19}\ \text{J} \le E_{\text{emission}} \le 5.4423 \times 10^{-19}\ \text{J}}.$$

The Paschen series of lines has $n = 3$. The shortest wavelength (most energetic) is obtained when $m = \infty$. The longest wavelength (least energetic) is obtained for $m = 4$. The shortest wavelength is

$$(\lambda)^{-1}_{\text{shortest}} = \frac{109{,}677}{3^2}\ \text{cm}^{-1} = 12{,}186.3\ \text{cm}^{-1}. \tag{10}$$

The energy corresponding to this wavelength is

$$E_{\text{largest}} = hc(\lambda)^{-1}_{\text{shortest}} = (6.62608 \times 10^{-34}\ \text{J s})(2.99792 \times 10^{10}\ \text{cm s}^{-1})(12{,}186.3\ \text{cm}^{-1}) = 2.4188 \times 10^{-19}\ \text{J}. \tag{11}$$

The longest wavelength is

$$(\lambda)^{-1}_{\text{longest}} = 109{,}677\left[\frac{1}{3^2} - \frac{1}{4^2}\right]\ \text{cm}^{-1} = 5{,}331.52\ \text{cm}^{-1}. \tag{12}$$

The corresponding energy is

$$E_{\text{smallest}} = hc(\lambda)^{-1}_{\text{longest}} = (6.62608 \times 10^{-34}\ \text{J s})(2.99792 \times 10^{10}\ \text{cm s}^{-1})(5{,}331.52\ \text{cm}^{-1}) = 1.0582 \times 10^{-19}\ \text{J}. \tag{13}$$

The energy range for the Paschen series is thus

$$\boxed{1.0582 \times 10^{-19}\ \text{J} \le E_{\text{emission}} \le 2.4188 \times 10^{-19}\ \text{J}}.$$

11.13 (A) What is the magnitude of the centrifugal force on an electron in the second Bohr orbit ($n = 2$) for a hydrogen atom?

(B) What is this centrifugal force for the He^+ ion?

Solution

(A) The centrifugal force is given by the first term in Eq. 11.23C:

$$F_{\text{cent}} = \frac{P_\theta^2}{mR^3} = \frac{n^2\hbar^2}{mR^3}. \tag{1}$$

The radius of the orbit is

$$R = \frac{n^2 a_o}{Z}. \tag{2}$$

Combining Eqs. 1 and 2, we obtain

$$F_{\text{cent}} = \frac{\hbar^2 Z^3}{mn^4 a_o^3}. \tag{3}$$

Inserting the constants for hydrogen gives

$$F_{\text{cent}}(\text{H atom}) = \frac{(1.054573 \times 10^{-34}\ \text{J s})^2(1)^3}{(9.10939 \times 10^{-31}\ \text{kg})(2)^4(5.29178 \times 10^{-11}\ \text{m})^3}$$

$$= 5.1492 \times 10^{-9}\ \text{kg m s}^{-2} = \underline{5.1492 \times 10^{-9}\ \text{newtons}}. \tag{4}$$

(B) The only change for He^+ is the fact that the atomic number of He is 2. Therefore, the result, which depends upon Z^3, is a factor of $2^3 = 8$ larger for He^+. Thus,

$$F_{\text{cent}}(\text{He}^+) = 8F_{\text{cent}}(\text{H atom}) = \underline{4.1194 \times 10^{-8}\ \text{newtons}}. \tag{5}$$

11.15 (A) Compute the energy of an electron in the $n = 1$ Bohr orbit. The linear velocity v of a particle moving in a circular orbit of radius R is related to its angular velocity v_θ by $v = v_\theta R$.
(B) What is the de Broglie wavelength of this $n = 1$ electron?

Solution

(A) The energy of the nth orbit is given by Eq. 11.41:

$$E_n(n = 1) = -\frac{e^2}{8\pi n^2 \varepsilon_o a_o} = -\frac{e^2}{8\pi\varepsilon_o a_o}$$

$$= -\frac{(1.602177 \times 10^{-19}\ \text{C})^2}{8(3.1415927)(8.85419 \times 10^{-12}\ \text{J}^{-1}\ \text{C}^2\ \text{m}^{-1})(5.29178 \times 10^{-11}\ \text{m})}$$

$$= -2.17987 \times 10^{-18}\ \text{J}. \tag{1}$$

(B) The angular momentum of the electron is given by Eq. 11.21:

$$P_\theta = mv_\theta R^2. \tag{2}$$

The Bohr theory requires that $P_\theta = n\hbar = nh/(2\pi)$. For $n = 1$, we have

$$mv_\theta R^2 = mvR = \frac{h}{2\pi}. \tag{3}$$

Therefore,

$$v = \frac{h}{2\pi mR}. \tag{4}$$

The linear momentum of the electron is $P = mv$. Consequently,

$$P = mv = \frac{h}{2\pi R}. \tag{5}$$

The wavelength of the matter wave is given by Eq. 11.51, which is

$$\lambda = \frac{h}{P} = 2\pi R = 2(3.1415927)(5.29178 \times 10^{-11}\ \text{m})$$

$$= \underline{3.3249 \times 10^{-10}\ \text{m}} = \underline{3.3249\ \text{Å}}, \tag{6}$$

since $R = a_o = 5.29178 \times 10^{-11}$ m in the first Bohr orbit.

11.17 An incident X ray with wavelength 1.540 Å undergoes Compton scattering at an angle θ relative to the direction of motion of the ray. The wavelength of the scattered electron is 3.000 Å. Compute the wavelength of the scattered X ray and the scattering angle of the X ray.

Solution

Using the de Broglie hypothesis, we can obtain the momentum of the scattered electron:

$$P = \frac{h}{\lambda} = \frac{6.62608 \times 10^{-34}\ \text{J s}}{3.000 \times 10^{-10}\ \text{m}} = 2.209 \times 10^{-24}\ \text{kg m s}^{-1}. \tag{1}$$

By energy conservation, we must have

$$h\nu = \frac{hc}{\lambda} = h\nu' + 0.5mv^2 = \frac{hc}{\lambda'} + \frac{P^2}{2m}. \tag{2}$$

Solving for $1/\lambda'$, we obtain

$$\frac{1}{\lambda'} = \frac{1}{\lambda} - \frac{P^2}{2mhc}. \tag{3}$$

The factor $1/2mhc$ is given by

$$(2mhc)^{-1} = [2(9.10939 \times 10^{-31}\ \text{kg})(6.62608 \times 10^{-34}\ \text{J s})(2.997924 \times 10^{8}\ \text{m s}^{-1})]^{-1}$$

$$= 2.76314 \times 10^{54}\ \text{kg}^{-2}\ \text{m}^{-3}\ \text{s}^2, \tag{4}$$

so that

$$\frac{1}{\lambda'} = \frac{1}{1.540 \times 10^{-10}\ \text{m}} - (2.209 \times 10^{-24}\ \text{kg m s}^{-1})^2(2.76314 \times 10^{54}\ \text{kg}^{-2}\ \text{m}^{-3}\ \text{s}^2)$$

$$= (6.494 \times 10^{9} - 1.349 \times 10^{7})\ \text{m}^{-1} = 6.480 \times 10^{9}\ \text{m}^{-1}, \tag{5}$$

which gives

$$\lambda' = \underline{1.543 \times 10^{-10}\ \text{m}}. \tag{6}$$

The scattering angle is given by the Compton equation:

$$\lambda' = \lambda + 2.4263 \times 10^{-12}\ \text{m}\ [1 - \cos\theta]. \tag{7}$$

Solving Eq. 7 for $\cos\theta$, we obtain

$$\cos\theta = 1 - \frac{\lambda' - \lambda}{2.4263 \times 10^{-12}} = 1 - \frac{1.543 \times 10^{-10} - 1.540 \times 10^{-10}}{2.4263 \times 10^{-12}} \approx 0.9. \tag{8}$$

Therefore,

$$\theta \approx \cos^{-1}(0.9) \approx 25°. \qquad (9)$$

The scattering angle for the X ray is somewhat uncertain, because the difference $\lambda' - \lambda$ contains only one significant digit.

11.19 (A) Show that the function $\phi = A[e^{ax} + e^{-ax}]$, where A and a are constants, is an eigenfunction of the operator $\mathcal{G} = \partial^2/\partial x^2$, but not of the operator $\mathcal{F} = \partial/\partial x$.
(B) What is the eigenvalue of $\partial^2/\partial x^2$ associated with the eigenfunction ϕ?

Solution

(A) First, let $\mathcal{G} = \partial^2/\partial x^2$. We let $\mathcal{G}$ operate on ϕ to obtain

$$\mathcal{G}\phi = \frac{\partial^2}{\partial x^2}\{A[e^{ax} + e^{-ax}]\} = A\frac{\partial}{\partial x}\{ae^{ax} - ae^{-ax}\} = A[a^2e^{ax} + a^2e^{-ax}]$$

$$= a^2A[e^{ax} + e^{-ax}] = a^2\phi. \qquad (1)$$

Equation 1 shows that, when $\mathcal{G}$ operates on ϕ, it produces a constant (a^2) times ϕ. This is the condition that must hold if ϕ is to be an eigenfunction $\mathcal{G}$. Thus, ϕ is an eigenfunction of $\mathcal{G}$ if $\mathcal{G} = \partial^2/\partial x^2$.

If, however, $\mathcal{G} = \partial/\partial x$, the result is

$$\mathcal{G}\phi = \frac{\partial}{\partial x}A[e^{ax} + e^{-ax}] = A[ae^{ax} - ae^{-ax}] = aA[e^{ax} - e^{-ax}], \qquad (2)$$

which is not a constant times ϕ, since $A[e^{ax} + e^{-ax}] \neq A[e^{ax} - e^{-ax}]$. We conclude that ϕ is not an eigenfunction of $\mathcal{G}$ when $\mathcal{G} = \partial/\partial x$.

(B) The eigenvalue of $\mathcal{G}$ when $\mathcal{G} = \partial^2/\partial x^2$ is given by Eq. 1. The constant on the right-hand side is a^2. Therefore,

$$\boxed{\text{the eigenvalue is } a^2}.$$

11.21 Energy and time are conjugate variables in quantum mechanics. Using the operators given in Table 11.1, show that the commutator of t and $i\hbar\,(\partial/\partial t)$ obeys the requirements of the second postulate in that it is equal to $-i\hbar$.

Solution

The commutator of t and $i\hbar\,\partial/\partial t$ is given by

$$\left[t, i\hbar\frac{\partial}{\partial t}\right]\Psi = \left(t\left\{i\hbar\frac{\partial}{\partial t}\right\} - \left\{i\hbar\frac{\partial}{\partial t}\right\}t\right)\Psi. \qquad (1)$$

Letting each term on the right-hand side of Eq. 1 operate on Ψ produces

$$\left[t, i\hbar\frac{\partial}{\partial t}\right]\Psi = i\hbar t\frac{\partial\Psi}{\partial t} - i\hbar\frac{\partial}{\partial t}(t\Psi). \qquad (2)$$

Expanding the second term gives

$$\left[t, i\hbar\frac{\partial}{\partial t}\right]\Psi = i\hbar t\frac{\partial\Psi}{\partial t} - i\hbar t\frac{\partial\Psi}{\partial t} - i\hbar\Psi = -i\hbar\Psi. \qquad (3)$$

Equation 3 shows that the commutator of t and $i\hbar\,\partial/\partial t$ is

$$\boxed{\left[t, i\hbar\frac{\partial}{\partial t}\right] = -i\hbar}, \tag{4}$$

as required.

11.23 Let ψ and ϕ be degenerate eigenfunctions of the operator $\mathcal{G}$. That is, the eigenvalues of $\mathcal{G}$ associated with ψ and ϕ are both the same. Show that any linear combination of ψ and ϕ is also an eigenfunction of $\mathcal{G}$ with the same eigenvalue.

Solution

Since ψ and ϕ are both eigenfunctions of $\mathcal{G}$ with the same eigenvalue, we have

$$\mathcal{G}\psi = a\psi \tag{1}$$

and

$$\mathcal{G}\phi = a\phi, \tag{2}$$

where a is the eigenvalue. An arbitrary linear combination of ψ and ϕ can be written in the form $\omega = c_1\psi + c_2\phi$, where c_1 and c_2 represent arbitrary constants. When $\mathcal{G}$ operates on this linear combination, we obtain

$$\mathcal{G}\omega = \mathcal{G}[c_1\psi + c_2\phi] = c_1\mathcal{G}\psi + c_2\mathcal{G}\phi. \tag{3}$$

Using Eqs. 1 and 2, we see that the right-hand side of Eq. 3 becomes

$$\boxed{\mathcal{G}\omega = \mathcal{G}[c_1\psi + c_2\phi] = ac_1\psi + ac_2\phi = a[c_1\psi + c_2\phi] = a\omega}, \tag{4}$$

and ω is, therefore, an eigenfunction of $\mathcal{G}$ with eigenvalue a.

11.25 Classical electrodynamics requires that the frequency of radiation emitted from a charged particle accelerating through an electric field be equal to the frequency of the particle's rotation. In this problem, you will show that the Bohr correspondence principle does indeed make the orbital frequency of the electron in the Bohr model equal to the radiation frequency as n becomes large. Consequently, we expect the correspondence principle to hold as the energy becomes large.

(A) Compute the angular momentum of the hydrogen-atom electron in the first Bohr orbit ($n = 1$). Use Eq. 11.21 to obtain the angular velocity v_θ of the electron in this orbit. What is the rotational frequency of the electron about the nucleus in this orbit?

Next, assume that emission occurs from a hydrogen-atom electron excited to the $n = 2$ orbit undergoing a transition to the adjacent $n = 1$ orbit. Compute the frequency of the radiation emitted. How does this frequency compare with the orbital frequency of the electron?

(B) Repeat (A) with the electron in the $n = 100$ orbit and the emission occurring from a transition from the $n = 101$ orbit to the $n = 100$ orbit. How does the emission frequency compare with the electron's orbital frequency? What does this tell us about the Bohr correspondence principle?

Solution

(A) The angular momentum is quantized in units of $\hbar$. Therefore, for the $n = 1$ state,

$$P_\theta = n\hbar = \hbar = \underline{1.054573 \times 10^{-34}\text{ J s}}. \quad (1)$$

From Eq. 11.21, we have

$$P_\theta = mv_\theta R^2 = mv_\theta a_o^2, \quad (2)$$

since, for $n = 1$, the radius of the orbit is $R = a_o$. The angular velocity is, therefore,

$$v_\theta = \frac{P_\theta}{ma_o^2} = \frac{1.054573 \times 10^{-34}\text{ J s}}{(9.10939 \times 10^{-31}\text{ kg})(5.29178 \times 10^{-11}\text{ m})^2} = \underline{4.13413 \times 10^{16}\text{ s}^{-1}}. \quad (3)$$

Since one rotation is 2π radians, the rotational frequency of the electron about the nucleus is

$$\nu = \frac{v_\theta}{2\pi} = \underline{6.57967 \times 10^{15}\text{ s}^{-1}}. \quad (4)$$

If emission occurs from a transition from the $n = 2$ orbit to the $n = 1$ orbit, the wavelength of the emitted radiation is given by Eq. 11.45, or

$$\lambda^{-1} = 109{,}677\left[\frac{1}{n^2} - \frac{1}{m^2}\right]\text{ cm}^{-1}, \quad (5)$$

where we have used the correct value of the Rydberg constant. With $n = 1$ and $m = 2$, the result is

$$\lambda^{-1} = 109{,}677\left[\frac{1}{1^2} - \frac{1}{2^2}\right]\text{ cm}^{-1} = 82{,}257.8\text{ cm}^{-1}. \quad (6)$$

The emission frequency is

$$\nu_e = \frac{c}{\lambda} = c\lambda^{-1} = 2.997 \times 10^{10}\text{ cm s}^{-1}(82{,}257.8\text{ cm}^{-1}) = \underline{2.465 \times 10^{15}\text{ s}^{-1}}. \quad (7)$$

Comparing Eqs. 4 and 7 shows that the orbital frequency is the same order of magnitude as the emission frequency, but they differ by 167%. This large difference shows that the Bohr correspondence principle does not hold for the low-energy $n = 1$ orbit.

(B) When the electron is in the $n = 100$ orbit, the angular momentum is

$$P_\theta = n\hbar = 100\hbar = \underline{1.054573 \times 10^{-32}\text{ J s}}. \quad (8)$$

Using Eqs. 2 and 11.21, we have

$$P_\theta = mv_\theta R^2 = mv_\theta(n^2a_o)^2 = mv_\theta n^4 a_o^2, \quad (9)$$

since the radius of the orbit is $R = n^2a_o$. Thus, the angular velocity is

$$v_\theta = \frac{P_\theta}{mn^4a_o^2} = \frac{(100)1.054573 \times 10^{-34}\text{ J s}}{(9.10939 \times 10^{-31}\text{ kg})(100)^4(5.29178 \times 10^{-11}\text{ m})^2}$$

$$= \underline{4.13413 \times 10^{10}\text{ s}^{-1}}. \quad (10)$$

The corresponding rotational frequency is

$$\nu = \frac{v_\theta}{2\pi} = \underline{6.57967 \times 10^9 \text{ s}^{-1}}. \tag{11}$$

The emission now occurs from the $n = 101$ orbit to the $n = 100$ orbit. The wavelength corresponding to this emission is

$$\lambda^{-1} = 109{,}677\left[\frac{1}{n^2} - \frac{1}{m^2}\right] \text{cm}^{-1} = 109{,}677\left[\frac{1}{100^2} - \frac{1}{101^2}\right] \text{cm}^{-1}$$

$$= 0.216107 \text{ cm}^{-1}. \tag{12}$$

The emission frequency is

$$\nu_e = \frac{c}{\lambda} = c\lambda^{-1} = 2.997 \times 10^{10} \text{ cm s}^{-1}(0.216107 \text{ cm}^{-1}) = \underline{6.477 \times 10^9 \text{ s}^{-1}}. \tag{13}$$

The percent difference between ν and ν_e given in Eqs. 11 and 13, respectively, is

$$\% \text{ difference} = \frac{100(6.57967 \times 10^9 \text{ s}^{-1} - 6.477 \times 10^9 \text{ s}^{-1})}{6.477 \times 10^9 \text{ s}^{-1}} = \underline{1.59\%}. \tag{14}$$

We see that the correspondence principle is now more accurate. It becomes increasingly so at even higher energies.

11.27 The discussion of the Rutherford atom in the text points out that such a system cannot exist in a classical world. In this problem, we will quantitatively determine what would happen to the electron in a hydrogen-like atom if classical mechanics and electrodynamics were valid. The various parts of the problem serve as a procedure guide.

(A) For a circular orbit, the radial momentum P_R and its derivative dP_R/dt are both zero. Use this fact together with Eqs. 11.22 and 11.23C to show that the classical Hamiltonian, and hence the energy E, of a hydrogen-like electron is given by

$$E = -\frac{Ze^2}{8\pi\varepsilon_o R}.$$

(B) Using Eq. 11.21 and the equations obtained in (A), show that the angular velocity is given by $v_\theta = [Ze^2/(4\pi\varepsilon_o mR^3)]^{1/2}$.

(C) Use the fact that classical electrodynamics requires that the rate of energy loss due to radiation as the electron accelerates through the electric field be

$$\frac{dE}{dt} = -\frac{e^2 v_\theta^4 R^2}{6\pi\varepsilon_o c_3}$$

to show that we must have

$$\frac{dE}{dt} = -\frac{Z^2e^6}{96\pi^3\varepsilon_o^3 c^3 m^2 R^4}.$$

(D) Use the expression for E obtained in (A) to derive an expression for dE/dt in terms of R and dR/dt.

(E) Combine the results of (C) and (D) to show that we must have

$$R^2\,dR = -\frac{Ze^4}{12\pi^2\varepsilon_o^2 c^3 m^2}dt.$$

(F) Assuming that the radial position of the electron is at a distance equal to the Bohr radius a_o at time $t = 0$, integrate the result of (E) to obtain R as a function of time.

(G) Compute how long it will take for the electron in a hydrogen atom to collapse into the nucleus.

Solution

(A) Since $P_R = 0$, the classical Hamiltonian and the energy are given by

$$H = E = \frac{P_\theta^2}{2mR^2} - \frac{Ze^2}{4\pi\varepsilon_o R}. \tag{1}$$

Because the radial force $-dP_R/dt$ must be zero, Eq. 11.23C shows that

$$\frac{\partial V}{\partial R} = \frac{P_\theta^2}{mR^3} = \frac{Ze^2}{4\pi\varepsilon_o R^2}. \tag{2}$$

Therefore, we must have

$$\frac{P_\theta^2}{2mR^2} = \frac{Ze^2}{8\pi\varepsilon_o R}. \tag{3}$$

Combining Eqs. 1 and 3, we obtain

$$\boxed{E = \frac{Ze^2}{8\pi\varepsilon_o R} - \frac{Ze^2}{4\pi\varepsilon_o R} = -\frac{Ze^2}{8\pi\varepsilon_o R}}, \tag{4}$$

as required.

(B) Equation 11.21 is

$$P_\theta = mv_\theta R^2. \tag{5}$$

Combining Eqs. 3 and 5 produces

$$\frac{m^2 v_\theta^2 R^4}{2mR^2} = \frac{mR^2 v_\theta^2}{2} = \frac{Ze^2}{8\pi\varepsilon_o R}. \tag{6}$$

Solving Eq. 6 for v_θ, we obtain

$$\boxed{v_\theta = \left[\frac{Ze^2}{4\pi\varepsilon_o m R^3}\right]^{1/2}}, \tag{7}$$

as required by the problem.

(C) Electrodynamics requires that we have

$$\frac{dE}{dt} = -\frac{e^2 v_\theta^4 R^2}{6\pi\varepsilon_o c^3}. \tag{8}$$

Combining Eqs. 7 and 8 gives

$$\boxed{\frac{dE}{dt} = -\frac{e^2R^2}{6\pi\varepsilon_o c^3}\left[\frac{Ze^2}{4\pi\varepsilon_o mR^3}\right]^2 = -\frac{Z^2e^6}{96\pi^3\varepsilon_o^3c^3m^2R^4}}, \tag{9}$$

as required.

(D) Differentiating both sides of Eq. 4 yields

$$\boxed{\frac{dE}{dt} = \frac{Ze^2}{8\pi\varepsilon_o R^2}\frac{dR}{dt}}. \tag{10}$$

(E) The rate of change of energy with time is given by Eqs. 9 and 10. Therefore, the right-hand sides of these equations must be equal. Equating the right-hand sides, we obtain

$$\frac{Ze^2}{8\pi\varepsilon_o R^2}\frac{dR}{dt} = -\frac{Z^2e^6}{96\pi^3\varepsilon_o^3c^3m^2R^4}. \tag{11}$$

Multiplying both sides of Eq. 11 by $8\pi\varepsilon_o R^4 dt/(Ze^2)$ gives

$$\boxed{R^2dR = -\frac{Ze^4}{12\pi^2\varepsilon_o^2c^3m^2}dt}, \tag{12}$$

as required by the problem.

(F) Integrating both sides of Eq. 12 between corresponding limits yields

$$\int_{R=a_0}^{R} R^2dR = -\frac{Ze^4}{12\pi^2\varepsilon_o^2c^3m^2}\int_{t=0}^{t} dt, \tag{13}$$

where the constant on the right-hand side may be factored out of the integral. To simplify the notation, let

$$B = \frac{Ze^4}{12\pi^2\varepsilon_o^2c^3m^2}. \tag{14}$$

Integrating Eq. 13 then produces

$$\frac{R^3}{3} - \frac{a_o^3}{3} = -Bt, \tag{15}$$

so that

$$\boxed{R = [(a_o)^3 - 3Bt]^{1/3}}. \tag{16}$$

(G) For the hydrogen atom, $Z = 1$. We wish to determine the time required for R to become zero. Setting $R = 0$ in Eq. 16 gives

$$3Bt = a_o^3, \tag{17}$$

so that

$$t = \frac{a_o^3}{3B}. \tag{18}$$

Hence,

$$B = \frac{(1.602177 \times 10^{-19}\,\text{C})^4}{12(3.1415927)^2(8.85419 \times 10^{-12}\,\text{J}^{-1}\,\text{C}^2\,\text{m}^{-1})^2(2.99792 \times 10^8\,\text{m s}^{-1})^3(9.10939 \times 10^{-31}\,\text{kg})^2}$$

$$= 3.1741 \times 10^{-21}\,\text{m}^3\,\text{s}^{-1}. \tag{19}$$

Combining Eqs. 18 and 19, we obtain

$$t = \frac{(5.29178 \times 10^{-11}\,\text{m})^3}{3(3.1741 \times 10^{-21}\,\text{m}^3\,\text{s}^{-1})} = \underline{1.556 \times 10^{-11}\,\text{s}}. \tag{20}$$

In other words, if the world were classical, the lifetime of the hydrogen atom would be about 15.56 ps. All atoms would rapidly collapse, and the universe as we know it would not exist.

11.29 An eigenvalue equation that we will encounter in Chapter 12 has the form

$$\frac{d^2}{dx^2}\,\psi(x) = -C\,\psi(x),$$

where C is a positive constant. Find the general form for the eigenfunction $\psi(x)$. To the extent possible, evaluate the constants that appear in the equation in terms of C. (*Hint*: The only two common functions whose second derivative is a constant times the function itself are exponentials and trigonometric sine or cosine functions.)

Solution

The eigenvalue equation tells us that if we differentiate $\psi(x)$ twice, we obtain $\psi(x)$ times a constant. Making use of the hint given in the problem, we might reasonably investigate two possibilities:

Possibility 1: $\psi_1(x) = A\sin(bx) + B\cos(bx)$;

Possibility 2: $\psi_2(x) = A\exp(bx) + B\exp(-bx)$.

In both possibilities, b is real.

We first examine Possibility 1. The first derivative of $\psi_1(x)$ is

$$\frac{d}{dx}\psi_1(x) = \frac{d}{dx}[A\sin(bx) + B\cos(bx)] = Ab\cos(bx) - Bb\sin(bx). \tag{1}$$

The second derivative is the derivative of the right-hand side of Eq. 1, or

$$\frac{d^2}{dx^2}\psi_1(x) = \frac{d}{dx}[Ab\cos(bx) - Bb\sin(bx)] = -Ab^2\sin(bx) - Bb^2\cos(bx)$$

$$= -b^2[A\sin(bx) + B\cos(bx)] = -b^2\psi_1(x). \tag{2}$$

Equation 2 demonstrates that $\psi_1(x)$ is an eigenfunction of the operator d^2/dx^2 and that it has a negative eigenvalue, $-b^2$. Thus, $\psi_1(x)$ satisfies the conditions of the problem, provided that we take $b^2 = C$ so that $b = \sqrt{C}$.

We now examine Possibility 2. The first derivative of $\psi_2(x)$ is

$$\frac{d}{dx}\psi_2(x) = \frac{d}{dx}[A\exp(bx) + B\exp(-bx)] = Ab\exp(bx) - Bb\exp(-bx). \quad \textbf{(3)}$$

The second derivative is the derivative of the right-hand side of Eq. 3, or

$$\frac{d^2}{dx^2}\psi_2(x) = \frac{d}{dx}[Ab\exp(bx) - Bb\exp(-bx)] = Ab^2\exp(bx) + Bb^2\exp(-bx)$$

$$= b^2[A\exp(bx) + B\exp(-bx)] = b^2\psi_2(x). \quad \textbf{(4)}$$

Equation 4 shows that $\psi_2(x)$ is an eigenfunction of the operator d^2/dx^2 with a positive eigenvalue b^2. Since the conditions of the problem require that the eigenvalue be negative, $\psi_2(x)$ is not a satisfactory eigenfunction.

The foregoing analysis demonstrates that the general form we seek is

$$\boxed{\psi(x) = A\sin[\sqrt{C}x] + B\cos[\sqrt{C}x]}. \quad \textbf{(5)}$$

11.31 Table 11.1 shows that the quantum mechanical operator for P_x is $(\hbar/i)(\partial/\partial x)$.

(A) Write down an eigenvalue equation with eigenvalue a whose solution will give the eigenfunctions of the operator for P_x.

(B) Solve the eigenvalue equation developed in (A) for the eigenfunctions of the momentum operator P_x.

(C) Separate the eigenfunction obtained in (B) into real and imaginary parts, and plot each on a separate graph as a function of the variable $z = ax/(\pi\hbar)$. Show the plots for the range $0 \le z \le 8$. Take the amplitudes of the waves to be unity.

(D) Determine the wavelength for both the real and imaginary parts of the wave in terms of z. What is the wavelength in terms of the distance variable x? Use the de Broglie hypothesis to interpret the physical significance of the eigenvalue of P_x.

Solution

(A) The appropriate form for the eigenvalue equation is given in Eq. 11.53, viz.,

$$\boxed{\frac{\hbar}{i}\frac{\partial}{\partial x}\psi(x) = a\psi(x)}, \quad \textbf{(1)}$$

where a is the eigenvalue.

(B) This part of the problem is very similar to Problem 11.30. We first multiply both sides of Eq. 1 by $i/\hbar$ and then divide both sides by $\psi(x)$. These operations produce

$$\frac{d\psi(x)}{\psi(x)} = \frac{ia\,dx}{\hbar}. \quad \textbf{(2)}$$

Taking indefinite integrals of both sides, we obtain

$$\int\frac{d\psi(x)}{\psi(x)} = \ln\psi(x) = \int\frac{ia\,dx}{\hbar} = \frac{iax}{\hbar} + k, \quad \textbf{(3)}$$

where k is a constant. Exponentiation of both sides of Eq. 3 gives

$$\boxed{\psi(x) = K \exp\left[\frac{iax}{\hbar}\right],} \tag{4}$$

where K is a constant that is equal to $\exp(k)$. Equation 3 is the general form for the eigenfunction of p_x whose eigenvalue is a.

(C) By writing the complex exponential in Eq. 3 in terms of sine and cosine functions, we can separate the real and imaginary portions of the eigenfunction. The results are

$$\text{Real } \psi(x) = K \cos\left[\frac{ax}{\hbar}\right] \tag{5}$$

and

$$\text{Im } \psi(x) = K \sin\left[\frac{ax}{\hbar}\right]. \tag{6}$$

If we take the wave amplitude to be unity ($K = 1$) and substitute $z = ax/(\pi\hbar)$ into Eqs. 4 and 5, we obtain

$$\text{Real } \psi(x) = \cos(\pi z) \tag{7}$$

and

$$\text{Im } \psi(x) = \sin(\pi z). \tag{8}$$

Plots of these two functions are shown below.

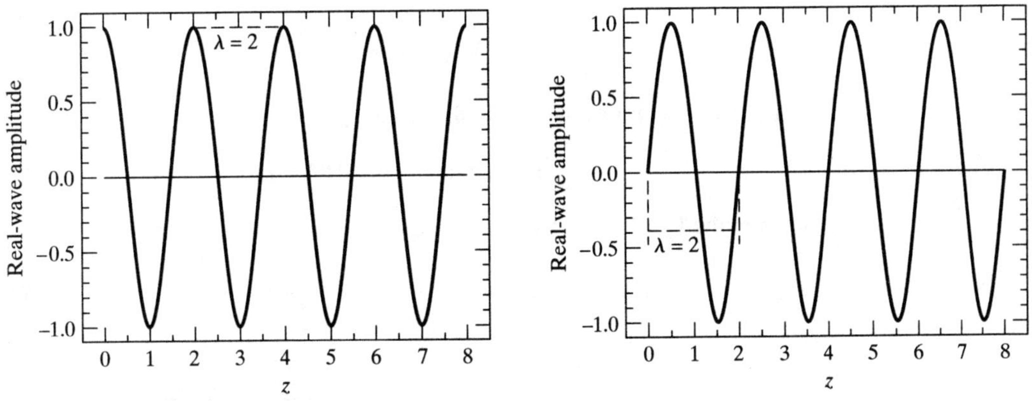

(D) Equations 6 and 7 and the two plots show that the wavelength in terms of z for both real and imaginary parts of the eigenfunction is 2. Using the transformation equation between z and x, we obtain

$$\boxed{x(\text{for } z = 2) = \lambda = \frac{2\pi\hbar}{a} = \frac{h}{a}.} \tag{9}$$

The de Broglie hypothesis tells us that the wavelength of a matter wave is

$$\lambda = \frac{h}{p}. \tag{10}$$

Comparing Eqs. 8 and 9 reveals the fact that we must have $p_x = a$ for consistency. This tells us that the eigenvalue is the momentum in the x direction.

11.33 Sam has just noticed an unusual expression in an article he was reading. The expression involved the sine of an imaginary quantity. Specifically, the expression in question was

$$y = A \sin[iax].$$

(A) Is it necessary that this expression be a misprint? Does the sine of an imaginary quantity have any meaning? If so, what is the meaning? Is y real? Is y imaginary? Is y complex? What is the complex conjugate of y? Is $|y|^2$ real? Prove that all your responses are correct. (*Hint:* See Section 11.3.1.)

Solution

The series expansion for sin(z) is given in Section 11.3.1 of Chapter 11. This expansion is

$$\sin(z) = z - \frac{z^3}{3!} + \frac{z^5}{5!} - \frac{z^7}{7!} + \ldots. \tag{1}$$

Therefore, we can write sin(iax) in the form

$$\sin(iax) = iax - \frac{i^3a^3x^3}{3!} + \frac{i^5a^5x^5}{5!} - \frac{i^7a^7x^7}{7!} + \frac{i^9a^9x^9}{9!} - \ldots. \tag{2}$$

Since $i = \sqrt{-1}$, we have

$$i^3 = i^7 = i^{11} = \cdots = i^{3+4k} = \cdots = -i, \tag{3}$$

where k is a positive integer. We also have

$$i = i^5 = i^9 = \cdots = i^{1+4k} = i. \tag{4}$$

Combining Eqs. 2, 3, and 4 yields

$$\sin(iax) = i\left[ax + \frac{a^3x^3}{3!} + \frac{a^5x^5}{5!} + \frac{a^7x^7}{7!} + \frac{a^9x^9}{9!} - \cdots\right]. \tag{5}$$

Thus,

$$\boxed{A\sin(iax) = iA\left[ax + \frac{a^3x^3}{3!} + \frac{a^5x^5}{5!} + \frac{a^7x^7}{7!} + \frac{a^9x^9}{9!} - \cdots\right]}, \tag{6}$$

which is imaginary. We see that the expression $y = A\sin(iax)$ is meaningful; y is a purely imaginary number. The complex conjugate of y is

$$\boxed{y^* = A\sin(-iax) = -iA\left[ax + \frac{a^3x^3}{3!} + \frac{a^5x^5}{5!} + \frac{a^7x^7}{7!} + \frac{a^9x^9}{9!} - \cdots\right]}. \tag{7}$$

Also, $|y|^2$ is real, since we have

$$|y|^2 = y^*y = -i^2A^2\left[ax + \frac{a^3x^3}{3!} + \frac{a^5x^5}{5!} + \frac{a^7x^7}{7!} + \frac{a^9x^9}{9!} - \cdots\right]^2$$

$$= \boxed{A^2\left[ax + \frac{a^3x^3}{3!} + \frac{a^5x^5}{5!} + \frac{a^7x^7}{7!} + \frac{a^9x^9}{9!} - \cdots\right]^2}, \qquad \textbf{(8)}$$

which is real and positive.

CHAPTER 12

Translational, Rotational, and Vibrational Energies of Molecular Systems

12.1 Verify that Eq. 12.9 is a solution of the stationary-state Schrödinger equation for the one-dimensional free particle.

Solution

The Schrödinger equation for the one-dimensional free "particle" is given by Eq. 12.6:

$$\frac{\partial^2}{\partial x^2}\psi(x) + k^2\psi(x) = 0. \tag{1}$$

The solution suggested by Eq. 12.9 is

$$\psi(x) = Ae^{ikx} + Be^{-ikx}. \tag{2}$$

Substituting this wave function for $\psi(x)$ in Eq. 1 gives

$$\frac{\partial^2}{\partial x^2}\psi(x) = \frac{\partial^2}{\partial x^2}[Ae^{ikx} + Be^{-ikx}] = i^2Ak^2e^{ikx} + i^2Bk^2e^{-ikx}$$

$$= -k^2[Ae^{ikx} + Be^{-ikx}] = -k^2\psi(x). \tag{3}$$

Combining the result obtained in Eq. 3 with Eq. 1 produces

$$\frac{\partial^2}{\partial x^2}\psi(x) + k^2\psi(x) = -k^2\psi(x) + k^2\psi(x) = 0, \tag{4}$$

as required.

12.3 Show that a quantum free particle whose wave function is $\psi(x) = Be^{-ikx}$ has an average momentum $-k\hbar$ and an average square momentum $k^2\hbar^2$, so that the square uncertainty in the momentum is exactly zero.

Solution

The average momentum is given by Eq. 12.13 in the text:

$$\langle p_x\rangle = \frac{\int_{\text{all } x} \psi^*(x)\frac{\hbar\partial}{i\partial x}\psi(x)dx}{\int_{\text{all } x} |\psi(x)|^2 dx}. \tag{1}$$

The numerator integrand is given by

$$\psi^*(x)\frac{\hbar\partial}{i\partial x}\psi(x) = B^*e^{ikx}\frac{\hbar}{i}(-ik)Be^{-ikx} = -k\hbar|B|^2e^0 = -k\hbar|B|^2. \tag{2}$$

Substituting Eq. 2 into Eq. 1 produces

$$\langle p_x\rangle = \frac{-k\hbar|B|^2\int_{\text{all } x} dx}{\int_{\text{all } x} B^*e^{ikx}Be^{-ikx}dx} = \frac{-k\hbar|B|^2\int_{\text{all } x} dx}{|B|^2\int_{\text{all } x} dx} = -k\hbar. \tag{3}$$

The quantum mechanical operator for p_x^2 is

$$p_x^2 = p_xp_x = \left[\frac{\hbar}{i}\frac{\partial}{\partial x}\right]\left[\frac{\hbar}{i}\frac{\partial}{\partial x}\right] = -\hbar^2\frac{\partial^2}{\partial x^2}. \tag{4}$$

Therefore,

$$\langle p_x^2 \rangle = -\hbar^2 \frac{\int_{\text{all } x} \psi^*(x) \frac{\partial^2}{\partial x^2} \psi(x)\,dx}{\int_{\text{all } x} |\psi(x)|^2 dx}. \tag{5}$$

The numerator integrand is given by

$$\psi^*(x) \frac{\partial^2}{\partial x^2} \psi(x) = B^* e^{ikx}(-ik)^2 B e^{-ikx} = -k^2|B|^2 e^0 = -k^2|B|^2. \tag{6}$$

Substituting Eq. 6 into Eq. 5 produces

$$\langle p_x^2 \rangle = \frac{k^2\hbar^2|B|^2 \int_{\text{all } x} dx}{\int_{\text{all } x} B^* e^{ikx} B e^{-ikx} dx} = \frac{k^2\hbar^2|B|^2 \int_{\text{all } x} dx}{|B|^2 \int_{\text{all } x} dx} = k^2\hbar^2. \tag{7}$$

The average square uncertainty in p_x is

$$\langle \Delta p_x^2 \rangle = \langle p_x^2 \rangle - \langle p_x \rangle^2 = k^2\hbar^2 - (-k\hbar)^2 = 0. \tag{8}$$

Consequently, the probability distribution for p_x is a delta function whose only nonzero value occurs at $p_x = -k\hbar$. This means that every measurement of p_x will produce $-k\hbar$. In quantum theory, such quantities are called "sharp" quantities or "constants of the motion."

12.5 The functions $\psi_2(x) = [2/a]^{1/2} \sin[2\pi x/a]$ and $\psi_3(x) = [2/a]^{1/2} \sin[3\pi x/a]$ are eigenfunctions for a particle in an infinite one-dimensional well. Show that these eigenfunctions are orthogonal. Could their orthogonality have been deduced without integrating? How? [*Hint:* $\sin(ax)$ can be written in the form $(e^{iax} - e^{-iax})/(2i)$.]

Solution

If the two eigenfunctions are orthogonal, we must have

$$\langle \psi_2(x) | \psi_3(x) \rangle = 0. \tag{1}$$

Since the wave functions are all zero for $x < 0$ or $x > a$, we need consider only the range $0 \leq x \leq a$. Therefore, we must show that

$$\frac{2}{a} \int_0^a \sin\left[\frac{2\pi x}{a}\right] \sin\left[\frac{3\pi x}{a}\right] dx = 0. \tag{2}$$

The $\sin(ax)$ factor may be written as

$$\sin(ax) = \frac{1}{2i}[e^{iax} - e^{-iax}] = \frac{1}{2i}\{[\cos(ax) + i\sin(ax)] - [\cos(ax) - i\sin(ax)]\}. \tag{3}$$

This identity allows us to write

$$\sin\left[\frac{2\pi x}{a}\right] = \frac{1}{2i}\left[\exp\left\{\frac{i2\pi x}{a}\right\} - \exp\left\{-\frac{i2\pi x}{a}\right\}\right]. \tag{4}$$

and

$$\sin\left[\frac{3\pi x}{a}\right] = \frac{1}{2i}\left[\exp\left\{\frac{i3\pi x}{a}\right\} - \exp\left\{-\frac{i3\pi x}{a}\right\}\right]. \tag{5}$$

Consequently, the integrand in Eq. 2 is

$$\sin\left[\frac{2\pi x}{a}\right]\sin\left[\frac{3\pi x}{a}\right]$$
$$= -\frac{1}{4}\left[\exp\left\{\frac{i5\pi x}{a}\right\} + \exp\left\{-\frac{i5\pi x}{a}\right\} - \exp\left\{\frac{i\pi x}{a}\right\} - \exp\left\{-\frac{i\pi x}{a}\right\}\right]. \quad (6)$$

But the sum of the first two terms on the right-hand side of Eq. 6 is

$$-\frac{1}{4}\left[\exp\left\{\frac{i5\pi x}{a}\right\} + \exp\left\{-\frac{i5\pi x}{a}\right\}\right] = -\frac{1}{2}\cos\left[\frac{5\pi x}{a}\right], \quad (7)$$

and the sum of the last two terms on the right-hand side of Eq. 6 may be written in the form

$$-\frac{1}{4}\left[-\exp\left\{\frac{i\pi x}{a}\right\} - \exp\left\{-\frac{i\pi x}{a}\right\}\right] = \frac{1}{2}\cos\left[\frac{\pi x}{a}\right]. \quad (8)$$

Substituting Eqs. 6, 7, and 8 into Eq. 2 produces

$$\frac{1}{a}\int_0^a \left\{\cos\left[\frac{\pi x}{a}\right] - \cos\left[\frac{5\pi x}{a}\right]\right\} dx = \frac{1}{\pi}\sin\left[\frac{\pi x}{a}\right]_0^a$$
$$-\frac{1}{5\pi}\sin\left[\frac{5\pi x}{a}\right]_0^a = \frac{1}{\pi}[0-0] - \frac{1}{5\pi}[0-0] = 0. \quad (9)$$

Yes, the energy eigenvalues of $\psi_2(x)$ and $\psi_3(x)$ are not equal. Since $\mathcal{H}$ is Hermitian, we must have $\psi_2(x)$ orthogonal to $\psi_3(x)$.

12.7 The classical kinetic energy of a one-dimensional free particle moving in the x direction is $p_x^2/(2m)$, where p_x is a continuous variable. Is there any similarity between the classical system and the quantum free particle? Discuss the similarities and differences.

Solution

There is a great deal of similarity between the classical and quantum systems. The quantum mechanical energies of the free particle are given by Eq. 12.7:

$$E_{\text{QM}} = \frac{k^2\hbar^2}{2m}. \quad (1)$$

It is shown in the text that k can assume any value. Therefore, the energy for both the quantum and classical systems is a continuous variable. The general form of the eigenfunctions for the quantum free particle is given by Eq. 12.9:

$$\psi(x) = Ae^{ikx} + Be^{-ikx}. \quad (2)$$

The quantum mechanical operator for p_x^2 is

$$p_x^2 = p_x p_x = \left[\frac{\hbar}{i}\frac{\partial}{\partial x}\right]\left[\frac{\hbar}{i}\frac{\partial}{\partial x}\right] = -\hbar^2\frac{\partial^2}{\partial x^2}. \quad (3)$$

Therefore,

$$\langle p_x^2\rangle = -\hbar^2\frac{\int_{\text{all } x}\psi^*(x)\frac{\partial^2}{\partial x^2}\psi(x)\,dx}{\int_{\text{all } x}|\psi(x)|^2\,dx}. \quad (4)$$

The integral in the numerator of Eq. 4 is

$$\int_{\text{all } x} [A^*e^{-ikx} + B^*e^{ikx}]\frac{\partial^2}{\partial x^2}[Ae^{ikx} + Be^{-ikx}]\,dx$$

$$= -k^2 \int_{\text{all } x} [A^*e^{-ikx} + B^*e^{ikx}][Ae^{ikx} + Be^{-ikx}]\,dx$$

$$= -k^2 \int_{\text{all } x} [|A|^2 + |B|^2 + B^*Ae^{2ikx} + A^*Be^{-2ikx}]\,dx. \quad (5)$$

The integral in the denominator of Eq. 4 is

$$\int_{\text{all } x} |\psi(x)|^2 dx = \int_{\text{all } x} [A^*e^{-ikx} + B^*e^{ikx}][Ae^{ikx} + Be^{-ikx}]\,dx$$

$$= \int_{\text{all } x} [|A|^2 + |B|^2 + B^*Ae^{2ikx} + A^*Be^{-2ikx}]\,dx. \quad (6)$$

Combining Eqs. 5 and 6 with Eq. 4 gives

$$\langle p_x^2 \rangle = -\hbar^2 \frac{-k^2 \int_{\text{all } x} [|A|^2 + |B|^2 + B^*Ae^{2ikx} + A^*Be^{-2ikx}]\,dx}{\int_{\text{all } x} [|A|^2 + |B|^2 + B^*Ae^{2ikx} + A^*Be^{-2ikx}]\,dx} = k^2\hbar^2. \quad (7)$$

Combining Eqs. 1 and 9, we obtain

$$E_{\text{QM}} = \frac{\langle p_x^2 \rangle}{2m}, \quad (8)$$

so that both the classical and quantum mechanical systems have similar functional forms for the kinetic energy of the system.

The differences between the classical and quantum systems reside primarily in the wave description of the quantum system, as opposed to the particulate nature of the classical system, which has a well-defined momentum and position for the particle. The quantum system has a well-defined square momentum and sometimes a well-defined momentum, but the position is not totally specified. Instead, there is a probability distribution for the position of the quantum free particle. The quantum system is associated with a wavelength, whereas there is no such concept for the classical system.

12.9 A particle whose mass is 1.67×10^{-27} kg is in a one-dimensional infinite potential well of width 3 Å. The eigenfunction for the quantum state of this particle has a wavelength of 0.4 Å.

(A) Compute the particle's translational energy.

(B) Calculate the magnitude of the momentum of the particle.

Solution

(A) Equation 12.40 relates the wavelength to the translational quantum number for this system. We have

$$\lambda = \frac{2a}{n_x}, \quad (1)$$

where a is the width of the well. Solving for n_x, we obtain

$$n_x = \frac{2a}{\lambda} = \frac{2(3 \times 10^{-10}\text{ m})}{0.4 \times 10^{-10}\text{ m}} = 15. \quad (2)$$

Using Eq. 12.39, we have

$$E_x = \frac{n_x^2 h^2}{8ma^2} = \frac{(15)^2(6.62608 \times 10^{-34}\text{ J s})^2}{8(1.67 \times 10^{-27}\text{ kg})(3 \times 10^{-10}\text{ m})^2} = \underline{8.22 \times 10^{-20}\text{ J}}. \quad (3)$$

(B) The magnitude of the momentum for this particle can be computed using the de Broglie hypothesis:

$$p = \frac{h}{\lambda} = \frac{6.62608 \times 10^{-34}\text{ J s}}{0.400 \times 10^{-10}\text{ m}} = \underline{1.66 \times 10^{-23}\text{ kg m s}^{-1}}. \quad (4)$$

We may also obtain the magnitude by using the fact that

$$E_x = \frac{p^2}{2m}, \quad (5)$$

so that

$$p = (2mE_x)^{1/2} = [(2)(1.67 \times 10^{-27}\text{ kg})(8.22 \times 10^{-20}\text{ J})]^{1/2}$$
$$= \underline{1.66 \times 10^{-23}\text{ kg m s}^{-1}}. \quad (6)$$

12.11[#] A Las Vegas casino sets up a gambling game using a "particle" in a one-dimensional infinite potential well as the device. Players place bets that, upon measurement, the "particle" will be found within a distance b of the center of the box, located at the point $x = a/2$.

(A) Plot the probability that the "particle" will be found within a distance b of the center as a function of the ratio b/a for the $n_x = 1$ ground state of the system. On the same graph, plot the result if the system is in the first excited state, $n_x = 2$.

(B) The casino desires to fix the value of b such that the player has a 48% chance of winning an even-money bet. If the casino announces that the "particle" will be in its ground state during the game, what value of b should the casino use as the cutoff. That is, if the "particle" is found within distance b of the center, the player wins. If the "particle" is beyond this distance, the casino wins.

(C) The casino, wishing to increase its advantage, has a hidden device that permits the "particle" to be excited into the $n_x = 2$, first excited state without the player's knowledge. If the particle is so excited, what are the player's chances of winning the bet?

Solution

(A) The ground-state eigenfunction is

$$\psi(x) = \left(\frac{2}{a}\right)^{1/2} \sin\left[\frac{\pi x}{a}\right]. \quad (1)$$

Since the probability distribution function is $\psi^*(x)\psi(x)\,dx$, the probability that the "particle" will be within a distance b of the center of the box is

$$P_1\left(\frac{b}{a}\right) = \int_{a/2-b}^{a/2+b} \psi^*(x)\psi(x)\,dx = \frac{2}{a}\int_{a/2-b}^{a/2+b} \sin^2\left[\frac{\pi x}{a}\right] dx. \quad (2)$$

Since $\cos(2x) = \cos^2 x - \sin^2 x = 1 - 2\sin^2 x$, we have

$$\sin^2\left[\frac{\pi x}{a}\right] = \frac{1 - \cos\left[\frac{2\pi x}{a}\right]}{2}. \quad (3)$$

Substituting Eq. 3 into Eq. 2 and integrating, we obtain

$$P_1\left(\frac{b}{a}\right) = \frac{2}{a}\left[\frac{x}{2} - \frac{a}{4\pi}\sin\left[\frac{2\pi x}{a}\right]\right]_{a/2-b}^{a/2+b} = \left[\frac{x}{a} - \frac{1}{2\pi}\sin\left[\frac{2\pi x}{a}\right]\right]_{a/2-b}^{a/2+b}$$

$$= \frac{(a/2+b) - (a/2-b)}{a} - \frac{1}{2\pi}\left[\sin\left(\pi + \frac{2\pi b}{a}\right) - \sin\left(\pi - \frac{2\pi b}{a}\right)\right]$$

$$= \frac{2b}{a} + \frac{\sin\left(\frac{2\pi b}{a}\right)}{\pi}, \tag{4}$$

since sin (A ± B) = sin A cos B ± cos A sin B. Now let $z = b/a$. In terms of z, Eq. 4 becomes

$$\boxed{P(z) = 2z + \frac{\sin(2\pi z)}{\pi}}. \tag{5}$$

Since the range of b is $0 \le b \le a/2$, z varies over the range $0 \le z \le 0.5$. The requested plot appears at the end of the problem.

(B) We wish to determine the value of z for which $P(z) = 0.48$. Thus, we need

$$P_1(z) = 2z + \frac{\sin(2\pi z)}{\pi} = 0.48. \tag{6}$$

This is a transcendental equation that can only be solved numerically. A one-dimensional grid search shows the solution, to be $\underline{Z_{\text{cutoff}} = 0.12644}$, correct to five digits. This cutoff point is shown in the graph.

(C) The first excited-state eigenfunction is

$$\psi(x) = \left(\frac{2}{a}\right)^{1/2}\sin\left[\frac{2\pi x}{a}\right]. \tag{7}$$

With this wave function, we have

$$P_2\left(\frac{b}{a}\right) = \int_{a/2-b}^{a/2+b}\psi^*(x)\psi(x)\,dx = \frac{2}{a}\int_{a/2-b}^{a/2+b}\sin^2\left[\frac{2\pi x}{a}\right]dx. \tag{8}$$

Using the same trigonometric expression we employed in Part (A), we obtain

$$\sin^2\left[\frac{2\pi x}{a}\right] = \frac{1 - \cos\left[\frac{4\pi x}{a}\right]}{2}. \tag{9}$$

Therefore,

$$P_2\left(\frac{b}{a}\right) = \frac{2}{a}\left[\frac{x}{2} - \frac{a}{8\pi}\sin\left[\frac{4\pi x}{a}\right]\right]_{a/2-b}^{a/2+b} = \left[\frac{x}{a} - \frac{1}{4\pi}\sin\left[\frac{4\pi x}{a}\right]\right]_{a/2-b}^{a/2+b}$$

$$= \frac{(a/2+b) - (a/2-b)}{a} - \frac{1}{4\pi}\left[\sin\left(2\pi + \frac{4\pi b}{a}\right) - \sin\left(2\pi - \frac{4\pi b}{a}\right)\right]$$

$$= \frac{2b}{a} - \frac{\sin\left(\frac{4\pi b}{a}\right)}{2\pi}. \tag{10}$$

In terms of z, we have

$$\boxed{P_2(z) = 2z - \frac{\sin(4\pi z)}{2\pi}}. \tag{11}$$

A plot of $P_2(z)$ is also shown in the aforementioned graph. At the cutoff point $z = 0.12644$, the probability that the player will win is

$$P_2(0.12644) = 2(0.12644) - \frac{\sin[4\pi(0.12644)]}{2\pi} = 0.25288 - 0.15913 = \underline{0.09375}. \tag{12}$$

The player's presumed 48% chance is now down to 9.38 %. Caveat emptor!!

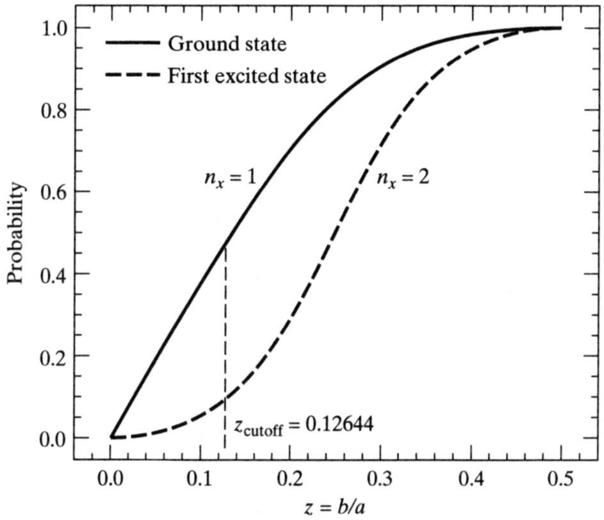

12.13 Solve Eq. 12.47 to obtain the eigenfunctions for a three-dimensional particle in an infinite rectangular parallelepiped well. Show that the total translational energy is given by

$$E = \frac{h^2}{8m}\left[\frac{n_x^2}{a^2} + \frac{n_y^2}{b^2} + \frac{n_z^2}{c^2}\right].$$

Solution

The Schrödinger equation for the system is given by 12.47:

$$-\frac{\hbar^2}{2m}\left[\frac{\partial^2}{\partial x^2} + \frac{\partial^2}{\partial y^2} + \frac{\partial^2}{\partial z^2}\right]\psi(x, y, z) - E\psi(x, y, z) = 0. \tag{1}$$

Following the procedure employed for the free particle, we assume that a separable solution exists:

$$\psi(x, y, z) = F(x)G(y)H(z). \tag{2}$$

Substituting Eq. 2 into Eq. 1, factoring those functions not affected by the differential operators to the left, and then dividing both sides by $F(x)G(y)H(z)$ produces

$$\left\{\frac{\left[-\frac{\hbar^2}{2m}\right]\frac{\partial^2}{\partial x^2}F(x)}{F(x)}\right\} + \left\{\frac{\left[-\frac{\hbar^2}{2m}\right]\frac{\partial^2}{\partial y^2}G(y)}{G(y)}\right\} + \left\{\frac{\left[-\frac{\hbar^2}{2m}\right]\frac{\partial^2}{\partial z^2}H(z)}{H(z)}\right\} - E = 0. \tag{3}$$

Inspection of Eq. 3 shows that the variables are now separated. The first term in Eq. 3 depends only upon x, the second only upon y, and the third only upon z. Thus, Eq. 3 must hold for all values of x, y, and z. The only way this can be true is for each of the first three terms to be equal to a constant such that the sum of the three constants is exactly E. That is, we must have

$$\left\{\frac{\left[-\frac{\hbar^2}{2m}\right]\frac{\partial^2}{\partial x^2}F(x)}{F(x)}\right\} = E_x, \tag{4}$$

$$\left\{\frac{\left[-\frac{\hbar^2}{2m}\right]\frac{\partial^2}{\partial y^2}G(y)}{G(y)}\right\} = E_y, \tag{5}$$

and

$$\left\{\frac{\left[-\frac{\hbar^2}{2m}\right]\frac{\partial^2}{\partial z^2}H(z)}{H(z)}\right\} = E_z, \tag{6}$$

with

$$E_x + E_y + E_z = E. \tag{7}$$

Equations 4–6 may be written in the forms

$$\left[-\frac{\hbar^2}{2m}\right]\frac{\partial^2}{\partial x^2}F(x) = E_x F(x), \tag{8}$$

$$\left[-\frac{\hbar^2}{2m}\right]\frac{\partial^2}{\partial y^2}G(y) = E_y G(y), \tag{9}$$

and

$$\left[-\frac{\hbar^2}{2m}\right]\frac{\partial^2}{\partial z^2}H(z) = E_z H(z), \tag{10}$$

respectively. Equations 8–10 are identical to Eq. 12.24 for the particle in the one-dimensional infinite potential well. Except for the substitution of b and c for a in the $G(y)$ and $H(z)$ solutions, the results are identical to the one-dimensional case. Therefore, we have

$$F(x) = \left[\frac{2}{a}\right]^{1/2} \sin\left[\frac{n_x \pi x}{a}\right] \tag{11}$$

inside the well and $F(x) = 0$ for $x < 0$ or $x > a$,

$$G(y) = \left[\frac{2}{b}\right]^{1/2} \sin\left[\frac{n_y \pi y}{b}\right] \tag{12}$$

inside the well and $G(y) = 0$ for $y < 0$ or $y > b$,

$$H(z) = \left[\frac{2}{c}\right]^{1/2} \sin\left[\frac{n_z \pi z}{c}\right] \tag{13}$$

inside the well and $H(z) = 0$ for $z < 0$ or $z > c$, with eigenvalues

$$E_x = \frac{n_x^2 h^2}{8ma^2}, \tag{14}$$

$$E_y = \frac{n_y^2 h^2}{8mb^2}, \tag{15}$$

and

$$E_z = \frac{n_z^2 h^2}{8mc^2}, \tag{16}$$

so that the total energy is given by

$$\boxed{E = E_x + E_y + E_z = E = \frac{h^2}{8m}\left[\frac{n_x^2}{a^2} + \frac{n_y^2}{b^2} + \frac{n_z^2}{c^2}\right].} \tag{17}$$

The associated eigenfunction is, therefore,

$$\psi(x, y, z) = \left[\frac{8}{abc}\right]^{1/2} \sin\left[\frac{n_x \pi x}{a}\right] \sin\left[\frac{n_y \pi y}{b}\right] \sin\left[\frac{n_z \pi z}{c}\right]$$

$$\text{for } 0 \le x \le a,\ 0 \le y \le b, \text{ and } 0 \le z \le c \tag{18}$$

and $\psi(x, y, z) = 0$ otherwise, with $(n_x, n_y, n_z = 1, 2, 3, \ldots)$.

12.15 It is stated in the text that degenerate eigenfunctions of an operator need not be orthogonal, but in the event that they are not, linear combinations of the nonorthogonal eigenfunctions can always be developed that are orthogonal. In this problem, you will explore how this can be done. The first step is to repeat Problem 11.23.

(A) Let ϕ_1 and ϕ_2 be degenerate, normalized eigenfunctions of the operator $\mathcal{G}$. That is, the eigenvalues of $\mathcal{G}$ associated with ϕ_1 and ϕ_2 are both the same. Show that any linear combination of ϕ_1 and ϕ_2 is also an eigenfunction of $\mathcal{G}$ with the same eigenvalue.

(B) Assume that ϕ_1 is not orthogonal to ϕ_2. This means that the integral $\langle\phi_1|\phi_2\rangle$ will be equal to S, where S is not zero. In (A), we found that any arbitrary linear combination of ϕ_1 and ϕ_2 is also an eigenfunction of $\mathcal{G}$. Therefore, instead of using the eigenfunction ϕ_2, we are free to substitute an eigenfunction $\phi_3 = c_1\phi_1 + c_2\phi_2$, with c_1 and c_2 arbitrary constants. Find the values of c_1 and c_2 in terms of S that make ϕ_3 orthogonal to ϕ_1 and, at the same time, normalize ϕ_3.

Solution

(A) Since ϕ_1 and ϕ_2 are both eigenfunctions of $\mathcal{G}$ with the same eigenvalue, we have

$$\mathcal{G}\phi_1 = a\phi_1 \tag{1}$$

and

$$\mathcal{G}\phi_2 = a\phi_2, \tag{2}$$

where a is the eigenvalue. An arbitrary linear combination of ϕ_1 and ϕ_2 can be written in the form $\phi_3 = c_1\phi_1 + c_2\phi_2$, where c_1 and c_2 represent any two constants you desire. When $\mathcal{G}$ operates on this linear combination, we obtain

$$\mathcal{G}\phi_3 = \mathcal{G}[c_1\phi_1 + c_2\phi_2] = c_1\mathcal{G}\phi_1 + c_2\mathcal{G}\phi_2. \tag{3}$$

Using Eqs. 1 and 2, we find that the right-hand side of Eq. 3 becomes

$$\mathcal{G}\phi_3 = \mathcal{G}[c_1\phi_1 + c_2\phi_2] = ac_1\phi_1 + ac_2\phi_2 = a[c_1\phi_1 + c_2\phi_2] = a\phi_3, \tag{4}$$

and ϕ_3 is, therefore, an eigenfunction of $\mathcal{G}$ with eigenvalue a.

(B) We desire to have ϕ_3 orthogonal to ϕ_1. Therefore, we wish to have

$$\langle \phi_1 | \phi_3 \rangle = \langle \phi_1 | c_1\phi_1 + c_2\phi_2 \rangle = 0. \quad (5)$$

We also wish to have ϕ_3 normalized. This requires that

$$\langle \phi_3 | \phi_3 \rangle = 1. \quad (6)$$

Expanding Eq. 5 produces

$$\langle \phi_1 | \phi_3 \rangle = c_1\langle \phi_1 | \phi_1 \rangle + c_2\langle \phi_1 | \phi_2 \rangle = c_1 + c_2 S = 0, \quad (7)$$

since $\langle \phi_1 | \phi_1 \rangle = 1$ because ϕ_1 is normalized and we are told that $\langle \phi_1 | \phi_2 \rangle = S$. To ensure orthogonality, we must have

$$c_2 = -\frac{c_1}{S}. \quad (8)$$

This means that ϕ_3 can be written in the form

$$\phi_3 = c_1\phi_1 - \frac{c_1}{S}\phi_2 = c_1\left[\phi_1 - \frac{\phi_2}{S}\right]. \quad (9)$$

The normalization condition given by Eq. 6 requires that

$$\langle \phi_3 | \phi_3 \rangle = 1 = c_1^2\left\langle \left[\phi_1 - \frac{\phi_2}{S}\right] \middle| \left[\phi_1 - \frac{\phi_2}{S}\right] \right\rangle$$

$$= c_1^2\left[\langle \phi_1 | \phi_1 \rangle - \frac{2}{S}\langle \phi_1 | \phi_2 \rangle + \frac{1}{S^2}\langle \phi_2 | \phi_2 \rangle\right]. \quad (10)$$

We know that $\langle \phi_1 | \phi_1 \rangle = \langle \phi_2 | \phi_2 \rangle = 1$ because the eigenfunctions are normalized, and we also know that $\langle \phi_1 | \phi_2 \rangle = S$. Substituting these values into Eq. 10 produces

$$\langle \phi_3 | \phi_3 \rangle = c_1^2\left[1 - 2 + \frac{1}{S^2}\right] = c_1^2\left[\frac{1}{S^2} - 1\right] = 1, \quad (11)$$

which gives

$$c_1 = \frac{1}{\left[\frac{1}{S^2} - 1\right]^{1/2}} = \left[\frac{S^2}{1 - S^2}\right]^{1/2}. \quad (12)$$

Therefore, the two eigenfunctions ϕ_1 and ϕ_3 will be normalized and orthogonal, each with eigenvalue a, if we take

$$\boxed{\phi_3 = \left[\frac{S^2}{1 - S^2}\right]^{1/2}\left[\phi_1 - \frac{\phi_2}{S}\right]}. \quad (13)$$

12.17 A particle in an infinite potential well is known to be in either the $n_x = 2$ or $n_x = 3$ eigenstates. The eigenfunctions of these states are $\phi_2(x) = [2/a]^{1/2} \sin[2\pi x/a]$ and $\phi_3(x) = [2/a]^{1/2} \sin[3\pi x/a]$, respectively.

(A) Write an appropriate wave function for the system that reflects our knowledge of the state of the system.

(B) What energies might be obtained if the energy of the particle is measured? What is the probability of obtaining each of these values?

Solution

(A) Since, according to the conditions of the problem both states are equally probable, we need to have the wave function be a superposition of $\phi_2(x)$ and $\phi_3(x)$ with coefficients whose absolute squares are equal. Therefore, we could take $C_2 = 1$ and $C_3 = \pm 1$. An appropriate wave function is, therefore,

$$\psi(x) = N[\phi_2(x) \pm \phi_3(x)] = N\left\{\left[\frac{2}{a}\right]^{1/2} \sin\left[\frac{2\pi x}{a}\right] \pm \left[\frac{2}{a}\right]^{1/2} \sin\left[\frac{3\pi x}{a}\right]\right\}. \quad \textbf{(1)}$$

The condition that the wave function be normalized determines the value of N. First,

$$\langle\psi(x)|\psi(x)\rangle = N^2\langle[\phi_2(x) \pm \phi_3(x)]|[\phi_2(x) \pm \phi_3(x)]\rangle = 1. \quad \textbf{(2)}$$

Expanding Eq. 2 then produces

$$N^2[\langle\phi_2(x)|\phi_2(x)\rangle + \langle\phi_3(x)|\phi_3(x)\rangle \pm 2\langle\phi_2(x)|\phi_3(x)\rangle] = 1. \quad \textbf{(3)}$$

We have $\langle\phi_2(x)|\phi_2(x)\rangle = \langle\phi_3(x)|\phi_3(x)\rangle = 1$ and $\langle\phi_2(x)|\phi_3(x)\rangle = 0$, because $\phi_2(x)$ must be orthogonal to $\phi_3(x)$. Therefore,

$$N^2[1 + 1 + 0] = 1, \quad \textbf{(4)}$$

so that $N = (2)^{-1/2}$. This makes the wave function

$$\boxed{\psi(x) = (2)^{-1/2}[\phi_2(x) \pm \phi_3(x)] = (2)^{-1/2}\left\{\left[\frac{2}{a}\right]^{1/2} \sin\left[\frac{2\pi x}{a}\right] \pm \left[\frac{2}{a}\right]^{1/2} \sin\left[\frac{3\pi x}{a}\right]\right\}}. \quad \textbf{(5)}$$

Plots of the two possible superpositions and their absolute squares are shown in the accompanying graphs for the case in which $a = 1$ Å. Since the potential well is symmetric about $x = 0.5$ Å, it is seen that the probability distributions are the same.

(B) The only two energies that can be obtained in a measurement of E_x are the eigenvalues for the states $n_x = 2$ and $n_x = 3$. These values are, respectively,

$$\boxed{E_2 = \frac{h^2}{2ma^2} \quad \text{and} \quad E_3 = \frac{9h^2}{8ma^2}}. \quad \textbf{(6)}$$

Since $|C_2|^2 = |C_3|^2$, each of these energies is equally probable.

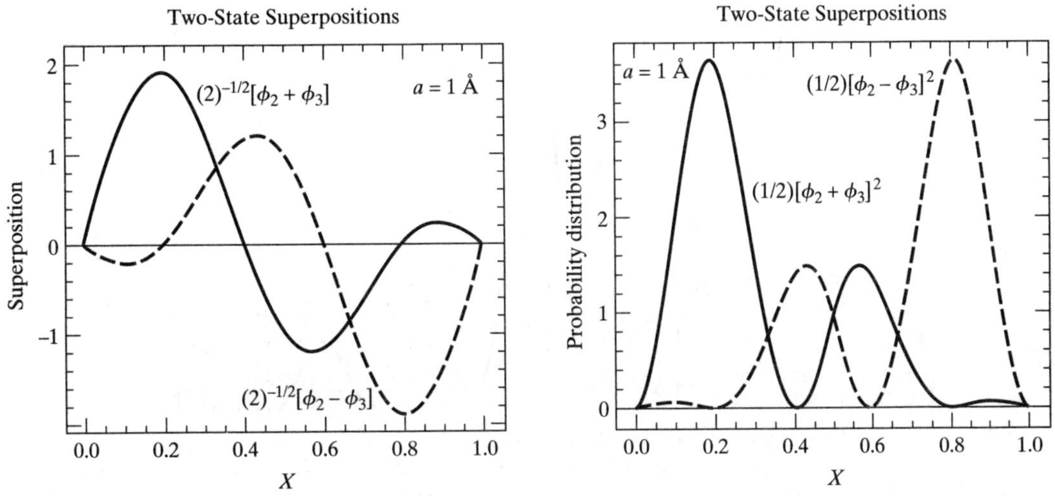

12.19* A particle in an infinite, one-dimensional potential well is equally likely to be in any of the six quantum states of lowest energy.

(A) Write down an appropriate form for the normalized wave function of this system.

(B) Plot the wave function over the range $0 \leq x \leq a$ if $a = 1$ Å.

(C) Are there other possibilities for the wave function that reflect our state of knowledge of the system? If not, why not? If so, illustrate two of them with appropriate equations and plots.

Solution

(A) Since each of the first six lowest-energy eigenfunctions is equally probable, they should all have coefficients of equal magnitude in the wave function superposition of states. That is, we might take the wave function to be

$$\psi(x) = N[\phi_1 + \phi_2 + \phi_3 + \phi_4 + \phi_5 + \phi_6], \qquad (1)$$

where

$$\phi_n = \left[\frac{2}{a}\right]^{1/2} \sin\left[\frac{n_x \pi x}{a}\right]. \qquad (2)$$

Normalization requires that we have

$$\langle\psi(x)|\psi(x)\rangle = 1 = N^2[\langle\phi_1|\phi_1\rangle + \langle\phi_2|\phi_2\rangle + \langle\phi_3|\phi_3\rangle + \langle\phi_4|\phi_4\rangle + \langle\phi_5|\phi_5\rangle + \langle\phi_6|\phi_6\rangle], \qquad (3)$$

since all integrals of the form $\langle\phi_i|\phi_j\rangle$ with $i \neq j$ are zero because the eigenfunctions must be orthogonal. The eigenfunctions are individually normalized. Therefore,

$$\langle\psi(x)|\psi(x)\rangle = 1 = N^2[1 + 1 + 1 + 1 + 1 + 1] = 6N^2, \qquad (4)$$

and we have

$$N = \left(\frac{1}{6}\right)^{1/2}. \qquad (5)$$

Consequently, one possible normalized wave function is

$$\psi(x) = \left(\frac{1}{6}\right)^{1/2} [\phi_1 + \phi_2 + \phi_3 + \phi_4 + \phi_5 + \phi_6]. \qquad (6)$$

(B) The requested plot of $y(x)$ over the range $0 \leq x \leq a$ for $a = 1$ Å is shown below. Note that the particle is now much more localized, so that the uncertainty in its position is greatly reduced. This type of result is directly connected to the uncertainty principle.

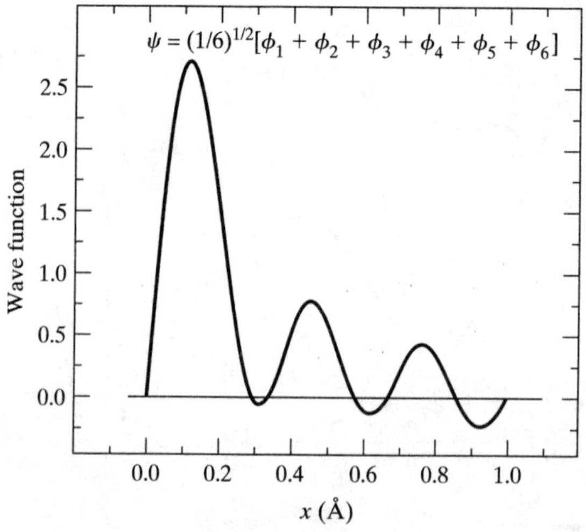

(C) We can take any linear combination of the first six eigenstates, provided that the coefficients are equal. Two alternative possibilities are

$$\psi_2(x) = \left(\frac{1}{6}\right)^{1/2} [\phi_1 - \phi_2 + \phi_3 - \phi_4 + \phi_5 - \phi_6] \tag{7}$$

and

$$\psi_3(x) = \left(\frac{1}{6}\right)^{1/2} [\phi_1 + \phi_2 + \phi_3 - \phi_4 - \phi_5 - \phi_6]. \tag{8}$$

Plots for each of these choices are given below; notice that in each case we have localized the particle to a much greater extent.

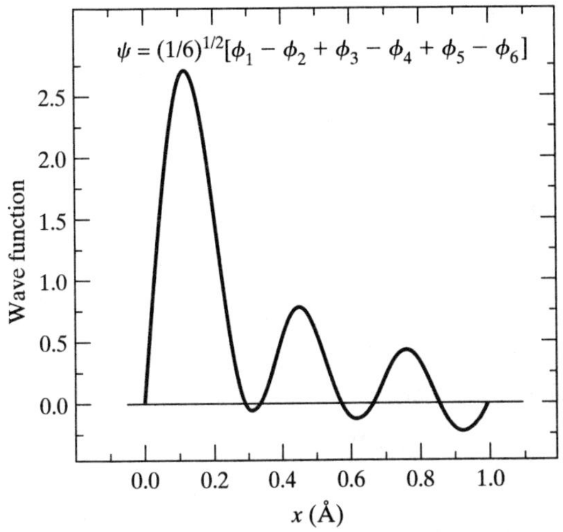

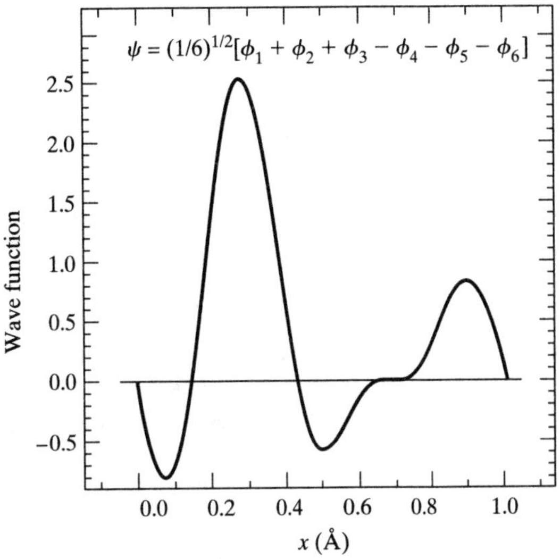

12.21 An electron is contained in a three-dimensional infinite potential well whose shape is that of a rectangular parallelepiped with dimensions $a = b = c = 1.0 \times 10^{-10}$ m. Use the uncertainty principle to estimate the minimum kinetic energy this electron can have.

Solution

In Problem 12.10, it is shown that the uncertainty in position for a particle in an infinite one-dimensional well is

$$\langle \Delta x^2 \rangle^{1/2} = \mathrm{a}\left[\frac{1}{12} - \frac{1}{2n^2\pi^2}\right]^{1/2}. \tag{1}$$

Therefore, the maximum value $\langle \Delta x^2 \rangle^{1/2}$ can have is $\mathrm{a}(12)^{-1/2}$. When $\mathrm{a} = 1.0 \times 10^{-10}$ m, we have

$$[\langle \Delta x^2 \rangle^{1/2}]_{\max} = 1.0 \times 10^{-10} \frac{\mathrm{m}}{(12)}^{1/2} = 2.89 \times 10^{-11}\ \mathrm{m}. \tag{2}$$

The same analysis for the y- and z-coordinates gives

$$[\langle \Delta y^2 \rangle^{1/2}]_{\max} = [\langle \Delta z^2 \rangle^{1/2}]_{\max} = 2.89 \times 10^{-11}\ \mathrm{m}. \tag{3}$$

The minimum uncertainty in momentum is, therefore,

$$[\langle \Delta p_x^2 \rangle^{1/2}]_{\min} = \frac{\hbar}{2[\langle \Delta x^2 \rangle^{1/2}]_{\max}} = \frac{6.626 \times 10^{-34}\ \mathrm{J\ s}}{4(3.14159)(2.89 \times 10^{-11}\ \mathrm{m})}$$

$$= 1.82 \times 10^{-24}\ \mathrm{kg\ m\ s^{-1}}. \tag{4}$$

The same result is obtained for $[\langle \Delta p_y^2 \rangle^{1/2}]_{min}$ and $[\langle \Delta p_z^2 \rangle^{1/2}]_{min}$. The momenta cannot be significantly less than this level of uncertainty, or the uncertainty will decrease below its minimum allowed value. Therefore, we have

$$(\text{Kinetic Energy})_{min} = T_{min} \approx \frac{p_x^2 + p_y^2 + p_z^2}{2m} = \frac{3(1.82 \times 10^{-24}\ \text{kg m s}^{-1})^2}{2(9.11 \times 10^{-31}\ \text{kg})}$$

$$= 5.45 \times 10^{-18}\ \text{J electron}^{-1}. \quad \textbf{(5)}$$

Multiplying by Avogadro's constant gives

$$T_{min} = 5.48 \times 10^{-18}\ \text{J electron}^{-1} \times (6.022 \times 10^{23}\ \text{electrons mol}^{-1})$$

$$= 3.28 \times 10^{6}\ \text{J mol}^{-1}$$

$$= \underline{3.28 \times 10^{3}\ \text{kJ mol}^{-1}}. \quad \textbf{(6)}$$

12.23 The boundary conditions at $x = 0$ and $x = a$ in Figure 12.12 lead to four equations involving the coefficients in the wave function describing tunneling through a square barrier. These equations, which are derived in the text, are

(a) $A + B = C + D$,

(b) $Ce^{Ra} + De^{-Ra} = Ee^{ika}$,

(c) $ikA - ikB = RC - RD$,

and

(d) $RCe^{Ra} - RDe^{-Ra} = ikEe^{ika}$.

(A) Combine Eqs. (a) and (c) to show that

$$A = \frac{C(R + ik) - D(R - ik)}{2ik}.$$

(B) Combine Eqs. (b) and (d) to show that

$$D = \frac{E(R - ik)e^{ika}}{2Re^{-Ra}}.$$

(C) Use Eq. (b) with the result from (B) to show that

$$C = \frac{Ee^{ika}}{e^{Ra}}\left[\frac{R + ik}{2R}\right].$$

(D) Use the results obtained from (A), (B), and (C) to show that

$$2ikA = \frac{Ee^{ika}}{2R}\left[\frac{(R + ik)^2}{e^{Ra}} + \frac{(R - ik)^2}{e^{-Ra}}\right].$$

(E) Let

$$F = \left[\frac{(R + ik)^2}{e^{Ra}} + \frac{(R - ik)^2}{e^{-Ra}}\right].$$

Show that

$$|F|^2 = F^*F = \frac{(R^2 + k^2)^2}{e^{2Ra}} + \frac{(R^2 + k^2)^2}{e^{-2Ra}}$$

$$+ 2(R^2 - k^2)^2 - 8k^2R^2.$$

(F) By taking the absolute squares of both sides of the expression derived in (D), show that

$$|A|^2 = \frac{|E|^2}{16k^2R^2}\left[\frac{(R^2 + k^2)^2}{e^{2Ra}} + \frac{(R^2 + k^2)^2}{e^{-2Ra}} + 2(R^2 - k^2)^2 - 8k^2R^2\right].$$

(G) Show that the fraction of particles transmitted through the square barrier shown in Figure 12.12 is

$$\tau = \frac{16k^2R^2}{\left[\dfrac{(R^2+k^2)^2}{e^{2Ra}} + \dfrac{(R^2+k^2)^2}{e^{-2Ra}} + 2(R^2-k^2)^2 - 8k^2R^2\right]}.$$

Solution

(A) Solving Eq. A for A, we obtain

$$A = C + D - B. \tag{1}$$

Solving Eq. C for B produces

$$B = \frac{RD - RC + ikA}{ik}. \tag{2}$$

Substituting Eq. 2 into Eq. 1 gives

$$A = C + D - \frac{RD - RC + ikA}{ik}. \tag{3}$$

Multiplying by ik and then rearranging terms gives

$$ikA = ikC + ikD - RD + RC - ikA = C(R + ik) - D(R - ik) - ikA. \tag{4}$$

Solving Eq. 4 for A, we obtain

$$\boxed{A = \frac{C(R + ik) - D(R - ik)}{2ik}}, \tag{5}$$

as required.

(B) From Eq. B, we can write

$$Ce^{Ra} = Ee^{ika} - De^{-Ra}. \tag{6}$$

Substituting this result into Eq. D produces

$$R[Ee^{ika} - De^{-Ra}] - RDe^{-Ra} = ikEe^{ika}. \tag{7}$$

Rearranging and collecting terms, we have

$$Ee^{ika}(R - ik) = 2RDe^{-Ra}. \tag{8}$$

Solving Eq. 8 for D, we obtain

$$\boxed{D = \frac{E(R - ik)e^{ika}}{2Re^{-Ra}}}, \tag{9}$$

which is the desired expression.

(C) From Eq. B, we may write

$$Ce^{Ra} = Ee^{ika} - De^{-Ra}, \tag{10}$$

so that

$$C = \frac{Ee^{ika} - De^{-Ra}}{e^{Ra}}. \tag{11}$$

Substituting D from Eq. 9 into Eq. 11 produces

$$C = \frac{Ee^{ika}}{e^{Ra}} - \frac{E(R - ik)e^{ika}}{2Re^{-Ra}}\frac{e^{-Ra}}{e^{Ra}} = \frac{Ee^{ika}}{e^{Ra}} - \frac{E(R - ik)e^{ika}}{2Re^{Ra}}. \quad \textbf{(12)}$$

Factoring out Ee^{ika}/e^{Ra} from the right-hand side of Eq. 12 produces

$$\boxed{C = \frac{Ee^{ika}}{e^{Ra}}\left[1 - \frac{(R - ik)}{2R}\right] = \frac{Ee^{ika}}{e^{Ra}}\left[\frac{(R + ik)}{2R}\right] = \frac{Ee^{ika}(R + ik)}{2Re^{Ra}}}, \quad \textbf{(13)}$$

as requested.

(D) From Eq. 5, the result obtained in Part (A), we have

$$2ikA = C(R + ik) - D(R - ik). \quad \textbf{(14)}$$

Substituting the result for C shown in Eq. 13 and for D given in Eq. 9 into Eq. 14 yields

$$\boxed{2ikA = \frac{Ee^{ika}(R + ik)^2}{2Re^{Ra}} - \frac{E(R - ik)^2e^{ika}}{2Re^{-Ra}} = \frac{Ee^{ika}}{2R}\left[\frac{(R + ik)^2}{e^{Ra}} + \frac{(R - ik)^2}{e^{-Ra}}\right]}, \quad \textbf{(15)}$$

which is the required relationship.

(E) With F defined as $[(R + ik)^2/e^{Ra} + (R - ik)^2/e^{-Ra}]$, we have

$$F = \left[\frac{R^2 - k^2 + 2ikR}{e^{Ra}} + \frac{R^2 - k^2 - 2ikR}{e^{-Ra}}\right] \quad \textbf{(16)}$$

by expanding the square terms.

The absolute square of F is now given by

$$F^*F = \left[\frac{R^2 - k^2 - 2ikR}{e^{Ra}} + \frac{R^2 - k^2 + 2ikR}{e^{-Ra}}\right]\left[\frac{R^2 - k^2 + 2ikR}{e^{Ra}} + \frac{R^2 - k^2 - 2ikR}{e^{-Ra}}\right]$$

$$= \frac{(R^2 - k^2)^2 + 4k^2R^2}{e^{2Ra}} + \frac{(R^2 - k^2)^2 + 4k^2R^2}{e^{-2Ra}} + \frac{R^2 - k^2 - 2ikR}{e^{Ra}}$$

$$\times \frac{R^2 - k^2 - 2ikR}{e^{-Ra}} + \frac{R^2 - k^2 + 2ikR}{e^{-Ra}} \times \frac{R^2 - k^2 + 2ikR}{e^{Ra}}$$

$$= \frac{R^4 + k^4 + 2k^2R^2}{e^{2Ra}} + \frac{R^4 + k^4 + 2k^2R^2}{e^{-2Ra}} + (R^2 - k^2)^2 - 4k^2R^2 - 4ikR(R^2 - k^2)$$

$$+ (R^2 - k^2)^2 - 4k^2R^2 + 4ikR(R^2 - k^2). \quad \textbf{(17)}$$

Collecting terms on the right and noting that $R^4 + k^4 + 2k^2R^2 = (R^2 + k^2)^2$, the result becomes

$$\boxed{|F|^2 = F^*F = \frac{(R^2 + k^2)^2}{e^{2Ra}} + \frac{(R^2 + k^2)^2}{e^{-2Ra}} + 2(R^2 - k^2)^2 - 8k^2R^2}, \quad \textbf{(18)}$$

which is the expression requested in Part (E).

(F) Equation 15 obtained in Part (D) can be written in the form

$$2ikA = \frac{Ee^{ika}}{2R}F, \quad \textbf{(19)}$$

where F is defined in Part (E). Taking the absolute squares of both sides gives

$$(-2ikA^*)(2ikA) = 4k^2|A|^2 = \frac{E^* e^{-ika}}{2R} F^* \times \frac{E e^{ika}}{2R} F = \frac{|E|^2|F|^2}{4R^2}. \quad \textbf{(20)}$$

Substituting the result given in Eq. 18 for $|F|^2$ into Eq. 20 produces

$$4k^2|A|^2 = \frac{|E|^2}{4R^2}\left[\frac{(R^2+k^2)^2}{e^{2Ra}} + \frac{(R^2+k^2)^2}{e^{-2Ra}} + 2(R^2-k^2)^2 - 8k^2R^2\right]. \quad \textbf{(21)}$$

Solving Eq. 21 for $|A|^2$ gives

$$\boxed{|A|^2 = \frac{|E|^2}{16k^2R^2}\left[\frac{(R^2+k^2)^2}{e^{2Ra}} + \frac{(R^2+k^2)^2}{e^{-2Ra}} + 2(R^2-k^2)^2 - 8k^2R^2\right]}, \quad \textbf{(22)}$$

as requested.

(G) The fraction of particles transmitted through the square barrier is shown in the text to be

$$\tau = \frac{|E|^2}{|A|^2}. \quad \textbf{(23)}$$

The ratio can be obtained directly from Eq. 22. The result is

$$\boxed{\tau = \frac{16k^2R^2}{\left[\dfrac{(R^2+k^2)^2}{e^{2Ra}} + \dfrac{(R^2+k^2)^2}{e^{-2Ra}} + 2(R^2-k^2)^2 - 8k^2R^2\right]}}. \quad \textbf{(24)}$$

12.25* It is known that the summation of all free-particle eigenfunctions of the form $\psi(x) = C_2\cos(kx)$ over the range $k_o - \Delta k \le k \le k_o + \Delta k$ with equal weight has the form

$$\Phi(x) = \frac{2C_2\cos(k_o x)\sin(\Delta k x)}{x}.$$

(See Problem 12.23.)

(A) For the case $k_o = 5.09$ Å^{-1} and $\Delta k = 1.00$ Å^{-1}, plot the unnormalized wave packet $\psi(x)$ over the range -4π Å $\le x \le 4\pi$ Å. [If Problem 12.24 has been done, this plot may be omitted, as it is the same as that requested in (C) of that problem.]

(B) For the specific case given in (A), obtain the probability distribution function for observing momenta in the range from p_x to $p_x + dp_x$. Normalize this distribution. [*Hint:* Remember, all magnitudes of momentum in the range $(k_o - \Delta k)\hbar \le p_x \le (k_o + \Delta k)\hbar$ are equally probable.]

(C) Determine the average momentum $\langle p_x\rangle$ and the average square momentum, $\langle p_x^2\rangle$, in terms of $\hbar$. Calculate the uncertainty in the momentum distribution in terms of $\hbar$.

Solution

(A) This is identical to Part (C) of Problem 12.24. The plot is given in the solution to that problem.

(B) As shown in the text and in Problem 12.24, an eigenfunction of the form $\psi(x) = C_2\cos(kx)$ describes a system for which two momenta are possible, $-k\hbar$ and $k\hbar$, each with equal probability of being observed in a measurement.

The wave function $\Phi(x)$ contains a sum of such free-particle eigenfunctions for all wave numbers in the range $k_o - \Delta k \le k \le k_o + \Delta k$. Therefore, the possible momenta lie in the ranges $-(k_o + \Delta k)\hbar \le p_x \le -(k_o - \Delta k)\hbar$ and $(k_o - \Delta k)\hbar \le p_x \le (k_o + \Delta k)\hbar$, with each momentum in the ranges being equally probable. For the specific case given in the problem, these ranges are

$$-6.09 \times 10^{10}\,\text{m}^{-1}\,\hbar \le p_x \le -4.09 \times 10^{10}\,\text{m}^{-1}\,\hbar$$

and

$$4.09 \times 10^{-10}\,\text{m}^{-1}\,\hbar \le p_x \le 6.09 \times 10^{10}\,\text{m}^{-1}\,\hbar,$$

where Å^{-1} have been converted to m^{-1} so that $\hbar$ will be in units of J s. The probability distribution for the momentum is, therefore,

$$\boxed{P(p_x)dp_x = C\,dp_x},$$

where C is a constant for p_x in either of the preceding two ranges, and

$$\boxed{P(p_x)dp_x = 0}$$

for any p_x that does not lie in one of those ranges. A plot of $P(p_x)$ is given at the end of the problem. The normalization requirement is that

$$\int_{\text{all momenta}} P(p_x)dp_x = \int_{-6.09\times 10^{10}\hbar}^{-4.09\times 10^{10}\hbar} P(p_x)dp_x + \int_{4.09\times 10^{10}\hbar}^{6.09\times 10^{10}\hbar} P(p_x)dp_x = 1. \quad \textbf{(1)}$$

Substituting $P(p_x)dp_x = C\,dp_x$ into Eq. 1 gives

$$C[-4.09 \times 10^{10}\hbar - (-6.09 \times 10^{10}\hbar)] + C[6.09 \times 10^{10}\hbar - 4.09 \times 10^{10}\hbar]$$
$$= 4 \times 10^{10}\hbar C = 1. \quad \textbf{(2)}$$

Therefore, the normalization constant is

$$C = \frac{1}{(4 \times 10^{10}\text{m}^{-1})\hbar}. \quad \textbf{(3)}$$

(C) The average momentum is

$$\langle p_x \rangle = \frac{\int_{-\infty}^{\infty} p_x P(p_x)dp_x}{\int_{-\infty}^{\infty} P(p_x)dp_x}. \quad \textbf{(4)}$$

Since the distribution is normalized, the denominator integral in Eq. 4 is unity, and we have

$$\langle p_x \rangle = \int_{-\infty}^{\infty} C p_x dp_x = C \int_{-6.09\times 10^{10}\hbar}^{-4.09\times 10^{10}\hbar} p_x dp_x + C \int_{4.09\times 10^{10}\hbar}^{6.09\times 10^{10}\hbar} p_x dp_x$$
$$= C\left[\frac{p_x^2}{2}\right]_{-6.09\times 10^{10}\hbar}^{-4.09\times 10^{10}\hbar} + C\left[\frac{p_x^2}{2}\right]_{4.09\times 10^{10}\hbar}^{6.09\times 10^{10}\hbar}$$
$$= \frac{C\hbar^2}{2}[(-4.09 \times 10^{10})^2 - (-6.09 \times 10^{10})^2]$$
$$+ \frac{C\hbar^2}{2}[(6.09 \times 10^{10})^2 - (4.09 \times 10^{10})^2] = 0, \quad \textbf{(5)}$$

as expected, since every positive contribution $k\hbar$ is canceled by an equal negative contribution $-k\hbar$. The average square momentum is

$$\langle p_x^2 \rangle = \frac{\int_{-\infty}^{\infty} p_x^2 P(p_x)\,dp_x}{\int_{-\infty}^{\infty} P(p_x)\,dp_x} = \int_{-\infty}^{\infty} C p_x^2\,dp_x, \qquad (6)$$

since the integral in the denominator is unity. Therefore,

$$\langle p_x^2 \rangle = C\int_{-6.09\times10^{10}\hbar}^{-4.09\times10^{10}\hbar} p_x^2\,dp_x + C\int_{4.09\times10^{10}\hbar}^{6.09\times10^{10}\hbar} p_x^2\,dp_x$$

$$= \frac{C}{3}[p_x^3]_{-6.09\times10^{10}\hbar}^{-4.09\times10^{10}\hbar} + \frac{C}{3}[p_x^3]_{4.09\times10^{10}\hbar}^{6.09\times10^{10}\hbar}$$

$$= \frac{C\hbar^3}{3}[(-4.09\times10^{10})^3-(-6.09\times10^{10})^3] + \frac{C\hbar^3}{3}[(6.09\times10^{10})^3-(4.09\times10^{10})^3]$$

$$= \frac{2C\hbar^3}{3}[(6.09\times10^{10})^3-(4.09\times10^{10})^3] = (1.574\times10^{32}\,\text{m}^{-3})\frac{2C\hbar^3}{3}. \qquad (7)$$

Substituting the value of C from Eq. 3 yields

$$\langle p_x^2 \rangle = \frac{(1.574\times10^{32}\,\text{m}^{-3})(2)\hbar^3}{3(4\times10^{10}\,\text{m}^{-1})\hbar} = (2.623\times10^{21}\,\text{m}^{-2})\hbar^2. \qquad (8)$$

The uncertainty in momentum is, therefore,

$$\langle \Delta p_x^2\rangle^{1/2} = [\langle p_x^2\rangle - \langle p_x\rangle^2]^{1/2} = [2.623\times10^{21}\hbar^2 - 0^2]^{1/2} = \underline{(5.122\times10^{10}\,\text{m}^{-1})\hbar}. \qquad (9)$$

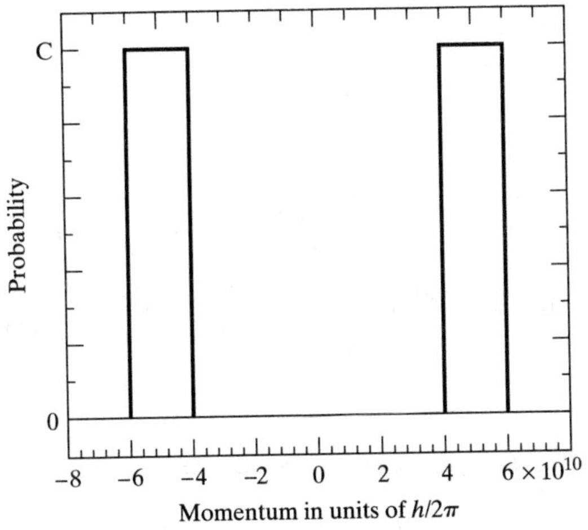

12.27 Let us assume that the potential barrier to electron tunneling in an STM is a one-dimensional square barrier whose height is 2.00×10^{-19} J. If the tunneling electron has an energy of 1.00×10^{-20} J, compute the value of the tunneling probability if the tunneling distance a is 0.500×10^{-10} m. What is the tunneling probability if the tunneling distance is 1.500×10^{-10} m? [*Hint:* Can you use the limiting form for the tunneling probability given by Eq. 12.81? Be certain to investigate this point.]

Solution

We first need to compute the value of R. Using Eq. 12.69, we have

$$R^2 = \frac{2m}{\hbar^2}[V_o - E] = \frac{(2)(9.109 \times 10^{-31}\,\text{kg})(2.00 \times 10^{-19} - 1.00 \times 10^{-20})\,\text{J}}{(1.055 \times 10^{-34}\,\text{J s})^2}$$

$$= 3.110 \times 10^{19}\,\text{m}^{-2}. \quad \textbf{(1)}$$

The value of R is, therefore, $5.577 \times 10^9\,\text{m}^{-1}$. If the tunneling distance is $0.500 \times 10^{-10}\,\text{m}$, we have

$$2Ra = 2(5.577 \times 10^9\,\text{m}^{-1})(0.500 \times 10^{-10}\,\text{m}) = 0.5577. \quad \textbf{(2)}$$

The exponential factor in Eq. 12.81 is

$$e^{-2Ra} = \exp\left[-2\left\{\frac{2m}{\hbar^2}[V_o - E]\right\}^{1/2} a\right] = \exp(-0.5577) = 0.5725, \quad \textbf{(3)}$$

and the positive exponential is

$$e^{2Ra} = \exp(0.5577) = 1.747, \quad \textbf{(4)}$$

so that it is no longer true that $e^{-2Ra} \ll e^{2Ra}$. Rather, both terms are approximately equal. For this reason, the limiting form represented by Eq. 12.81 is not very accurate. We need the full expression given by Eq. 12.79:

$$\tau = \frac{16k^2R^2}{\left[\dfrac{(R^2 + k^2)^2}{e^{2ra}} + \dfrac{(R^2 + k^2)^2}{e^{-2Ra}} + 2(R^2 - k^2)^2 - 8k^2R^2\right]}. \quad \textbf{(5)}$$

The value of k^2 is given by

$$k^2 = \frac{2m}{\hbar^2}E = \frac{(2)(9.109 \times 10^{-31}\,\text{kg})(1.00 \times 10^{-20})\,\text{J}}{(1.055 \times 10^{-34}\,\text{J s})^2} = 1.637 \times 10^{18}\,\text{m}^{-2}. \quad \textbf{(6)}$$

Therefore,

$$\frac{(R^2 + k^2)^2}{e^{2Ra}} = \frac{(3.110 \times 10^{19}\,\text{m}^{-2} + 1.637 \times 10^{18}\,\text{m}^{-2})^2}{1.747} = 6.135 \times 10^{38}\,\text{m}^{-4}, \quad \textbf{(7)}$$

$$\frac{(R^2 + k^2)^2}{e^{-2Ra}} = \frac{(3.110 \times 10^{19}\,\text{m}^{-2} + 1.637 \times 10^{18}\,\text{m}^{-2})^2}{0.5725} = 1.872 \times 10^{39}\,\text{m}^{-4}, \quad \textbf{(8)}$$

and

$$2(R^2 - k^2)^2 - 8k^2R^2 = 2[3.110 \times 10^{19} - 1.637 \times 10^{18}]^2$$
$$- 8(3.110 \times 10^{19})(1.637 \times 10^{18})$$
$$= 1.736 \times 10^{39} - 4.073 \times 10^{38}\,\text{m}^{-4} = 1.329 \times 10^{39}\,\text{m}^{-4}. \quad \textbf{(9)}$$

Combining Eqs. 5, 8, 9, and 10 yields

$$\tau = \frac{8.146 \times 10^{38}\,\text{m}^{-4}}{[6.135 \times 10^{38} + 1.872 \times 10^{39} + 1.329 \times 10^{39}]\,\text{m}^{-4}} = \underline{0.214}. \quad \textbf{(10)}$$

Note how much larger the electron tunneling probability is than that for the hydrogen atom in Example 12.12, even though the potential barrier is larger in this problem than in the example. This is due to the much smaller electron mass. If the tunneling distance is 1.500×10^{-10} m, the same calculation with $a = 1.500 \times 10^{-10}$ m gives a tunneling probability of 0.113. In this case, the limiting form of the tunneling probability, Eq. 12.81, is reasonably accurate. The use of this equation with $a = 1.500 \times 10^{-10}$ m gives a tunneling probability of 0.149.

12.29 An investigator wishes to represent a four-atom system by using an independent center-of-mass, relative coordinate system. She labels the Cartesian coordinates of the center of mass as (X_C, Y_C, Z_C). The Cartesian coordinates of the vector from atom B to atom A are denoted as (X_{R1}, Y_{R1}, Z_{R1}). The Cartesian coordinates of the vector **CD** are represented by (X_{R2}, Y_{R2}, Z_{R2}). Finally, the investigator denotes the Cartesian coordinates of the vector from the center of mass of AB to the center of mass of CD as (X_{R3}, Y_{R3}, Z_{R3}).

(A) Are these coordinates independent? If not, state why not.

(B) What is the form of the kinetic energy of the four-atom system, expressed in terms of the derivatives of these coordinates with respect to time (i.e., velocities)?

Solution

(A) The coordinate system is shown in the following diagram:

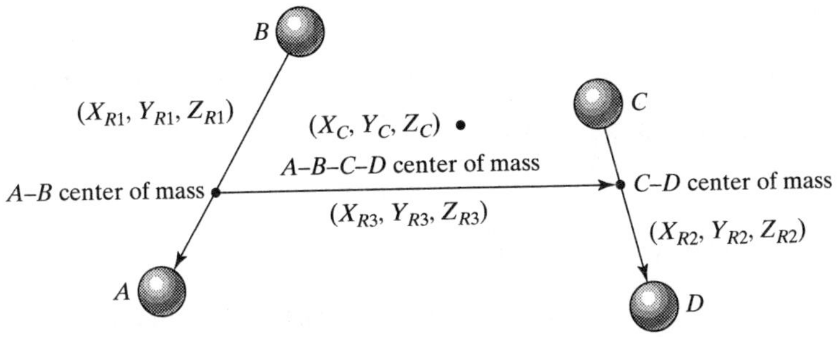

The coordinates in the diagram are all independent. We can change X_{R1}, Y_{R1}, and Z_{R1} by moving atoms A and B along the AB vector so as not to alter the position of either the center of mass of AB or the center of mass of $ABCD$. Thus, the other two relative vectors will not be changed. The same argument can be employed to show that all the coordinates are independent.

(B) Since the coordinates are independent, the kinetic energy has the form given in Eq. 12.91; viz.,

$$T = \frac{M}{2}[v_{XC}^2 + v_{YC}^2 + v_{ZC}^2] + \frac{\mu_{AB}}{2}[v_{X1}^2 + v_{Y1}^2 + v_{Z1}^2] + \frac{\mu_{CD}}{2}[v_{X2}^2 + v_{Y2}^2 + v_{Z2}^2]$$

$$+ \frac{\mu_{AB\text{-}CD}}{2}[v_{X3}^2 + v_{Y3}^2 + v_{Z3}^2], \quad (1)$$

where

$$\mu_{AB} = \frac{m_A m_B}{m_A + m_B}, \quad (2)$$

$$\mu_{CD} = \frac{m_C m_D}{m_C + m_D}, \quad (3)$$

and

$$\mu_{AB\text{-}CD} = \frac{(m_A + m_B)(m_C + m_D)}{M}, \quad (4)$$

with $M = m_A + m_B + m_C + m_D$.

12.31 Consider a linear triatomic molecule ABC with masses m_1, m_2, and m_3 and bond lengths R_1 and R_2, as shown in the following diagram:

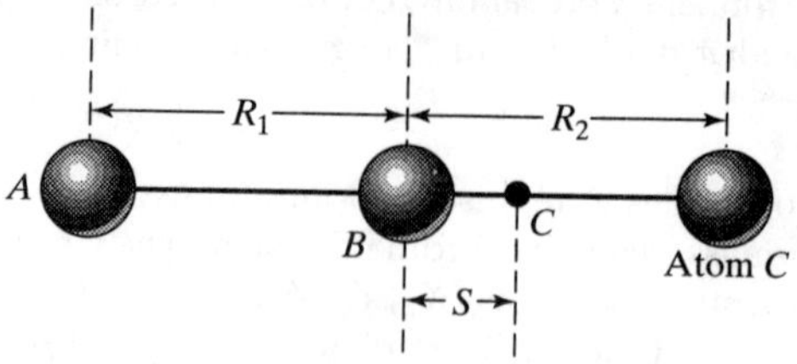

Let point C be the system center of mass, which lies at a distance S from atom B.

(A) Show that

$$S = \frac{(m_3R_2 - m_1R_1)}{M},$$

where $M = m_1 + m_2 + m_3$.

(B) The moment of inertia is given by

$$I = \sum_{i=1}^{3} m_i r_i^2$$

where r_i is the perpendicular distance from atom i to the principal axis, which will be through the center of mass C and perpendicular to the bond axis. From the figure, it may be seen that

$$I = m_1(R_1 + S)^2 + m_2S^2 + m_3(R_2 - S)^2.$$

Show that I may also be expressed as

$$I = m_1R_1^2 + m_3R_2^2 - M^{-1}(m_1 R_1 - m_3 R_2)^2.$$

Solution

(A) Since point C is the system's center of mass, we must have

$$m_1(R_1 + S) + m_2S = m_3(R_2 - S). \quad \textbf{(1)}$$

Rearranging terms gives

$$S(m_1 + m_2 + m_3) = m_3R_2 - m_1R_1. \quad \textbf{(2)}$$

Thus, we obtain

$$\boxed{S = \frac{m_3R_2 - m_1R_1}{M}}, \quad \textbf{(3)}$$

as required.

(B) Expanding the expression for I given in the problem produces

$$I = m_1R_1^2 + m_1S^2 + 2R_1Sm_1 + m_2S^2 + m_3R_2^2 + m_3S^2 - 2R_2Sm_3$$
$$= m_1R_1^2 + m_3R_2^2 + S^2[m_1 + m_2 + m_3] + 2S[m_1R_1 - m_3R_2]. \quad \textbf{(4)}$$

Substituting Eq. 3 into Eq. 4 and using the fact that $M = m_1 + m_2 + m_3$ gives

$$I = m_1R_1^2 + m_3R_2^2 + M\left[\frac{m_3R_2 - m_1R_1}{M}\right]^2 + 2\left[\frac{m_3R_2 - m_1R_1}{M}\right][m_1R_1 - m_3R_2], \quad \textbf{(5)}$$

or

$$I = m_1R_1^2 + m_3R_2^2 + \frac{1}{M}[m_3R_2 - m_1R_1]^2 - \frac{2}{M}[m_3R_2 - m_1R_1]^2$$

$$= m_1R_1^2 + m_3R_2^2 - \frac{1}{M}[m_3R_2 - m_1R_1]^2 =$$

$$\boxed{m_1R_1^2 + m_3R_2^2 - \frac{1}{M}[m_1R_1 - m_3R_2]^2}, \quad (6)$$

as required.

12.33 A member of the class suggests that the function $\Theta(\theta) = C\theta$, where C is a constant, might be a solution of the θ-equation for the diatomic rigid rotor. On the surface, this is a very reasonable suggestion. Show, however, that that function cannot be made to satisfy the θ-equation, regardless of the values used for E_R and m.

Solution

The θ-equation is

$$\sin\theta\frac{\partial}{\partial\theta}\left(\sin\theta\frac{\partial}{\partial\theta}\right)\Theta(\theta) + \frac{2IE}{\hbar^2}\sin^2\theta\,\Theta(\theta) - \mathrm{m}^2\Theta(\theta) = 0, \quad (1)$$

where we have written E for E_R. Letting $\Theta(\theta) = C\theta$, we obtain, from the first term,

$$\sin\theta\frac{\partial}{\partial\theta}\left(\sin\theta\frac{\partial}{\partial\theta}\right)\Theta(\theta) = \sin\theta\frac{\partial}{\partial\theta}\left(\sin\theta\frac{\partial}{\partial\theta}\right)C\theta = C\sin\theta\frac{\partial}{\partial\theta}[\sin\theta]$$

$$= C[\sin\theta\cos\theta]. \quad (2)$$

From the second term, we get

$$\frac{2IE}{\hbar^2}\sin^2\theta\,\Theta(\theta) = C\frac{2IE}{\hbar^2}(\sin^2\theta)\theta. \quad (3)$$

The third term yields

$$-m^2\Theta(\theta) = -Cm^2\theta. \quad (4)$$

Adding the three terms gives

$$C[\sin\theta\cos\theta] + C\frac{2IE}{\hbar^2}\theta\sin^2\theta - Cm^2\theta = 0. \quad (5)$$

The left side of Eq. 5 must be zero no matter what value is assigned to θ. Dividing by C produces

$$[\sin\theta\cos\theta] + \frac{2IE}{\hbar^2}\theta\sin^2\theta - m^2\theta = 0. \quad (6)$$

We could make the second term vanish by taking $E = 0$. The third term can be forced to be zero by requiring that $m = 0$, but there is no way to make the first term vanish for all values of θ. Nor can the three terms be combined into two, each of which is multiplied by a factor that depends only upon E or m. Therefore, $C\,\theta$ is not a solution of the θ-equation.

12.35 Consider the H_2 molecule. Assume that the molecule's rotational motion may be treated as that of a quantum mechanical rigid rotor. (The H_2 bond length is 0.740 Å.)

(A) Derive a formula giving the energy spacing between the Jth and the $(J + 1)$st rotational energy levels in terms of J and the moment of inertia of H_2.

(B) Evaluate the spacing found in (A) for $J = 12$. Is this energy sufficiently large to permit its measurement? Is the system essentially classical or not?

Solution

(A) The rotational energy levels of the rigid rotor are given by

$$E(\ell) = J(J + 1)\frac{\hbar^2}{2\mu R^2}, \quad (1)$$

where μ is the reduced mass of the rotor and R is the rotor's length. The moment of inertia is

$$I = \mu R^2, \quad (2)$$

so that

$$E(J) = J(J + 1)\frac{\hbar^2}{2I}. \quad (3)$$

The energy spacing between adjacent levels is, therefore,

$$\Delta E = E(J + 1) - E(J)$$

$$= [(J + 1)(J + 2) - J(J + 1)]\frac{\hbar^2}{2I} = 2(J + 1)\frac{\hbar^2}{2I} = \boxed{(J + 1)\frac{\hbar^2}{I}}. \quad (4)$$

(B) The moment of inertia for the H_2 molecule is

$$I = \mu R^2 = \frac{m_H R^2}{2}$$

$$= \frac{(0.001008 \text{ kg mol}^{-1})(0.740 \times 10^{-10} \text{ m})^2}{2(6.02214 \times 10^{23} \text{ mol}^{-1})} = 4.58 \times 10^{-48} \text{ kg m}^2. \quad (5)$$

When $J = 12$, we have

$$\Delta E = \frac{(12 + 1)(6.626 \times 10^{-34} \text{ J s})^2}{4(3.1415927)^2(4.58 \times 10^{-48}) \text{ kg m}^2} = 3.15 \times 10^{-20} \text{ J molecule}^{-1}. \quad (6)$$

In terms of energy per mole of H_2, the spacing is

$$\Delta E = 3.15 \times 10^{-20} \text{ J molecule}^{-1}\frac{6.022 \times 10^{23} \text{ molecules}}{1 \text{ mole}} = 1.90 \times 10^4 \text{ J mol}^{-1}$$

$$= \underline{19.0 \text{ kJ mol}^{-1}}. \quad (7)$$

This quantity of energy can be easily measured. Therefore, the system will not behave classically.

12.37 Determine the location of the nodal angular surfaces or axes for the $\Theta_3^0(\theta)$ function.

Solution

The $\Theta_3^0(\theta)$ function is

$$\Theta_3^0(\theta) = \left[\frac{7}{8}\right]^{1/2}[5 \cos^3 \theta - 3 \cos \theta]. \quad (1)$$

The nodal surfaces or axes occur at the values of θ that make the preceding function zero:

$$\left[\frac{7}{8}\right]^{1/2}[5\cos^3\theta - 3\cos\theta] = 0 = \left[\frac{7}{8}\right]^{1/2}\cos\theta[5\cos^2\theta - 3]. \quad \textbf{(2)}$$

Equation 2 is satisfied if we have either

$$\cos\theta = 0 \quad \textbf{(3)}$$

or

$$[5\cos^2\theta - 3] = 0. \quad \textbf{(4)}$$

Equation 3 has roots $\theta = \pi/2$ or $3\pi/2$, so the X–Y plane is a nodal plane. The roots of Eq. 4 are

$$\theta = \cos^{-1}\left[\left(\frac{3}{5}\right)^{1/2}\right] \quad \textbf{(5)}$$

and

$$\theta = \cos^{-1}\left[-\left(\frac{3}{5}\right)^{1/2}\right]. \quad \textbf{(6)}$$

Therefore, when $\theta = 39.23°$ or $219.23°$, we have one nodal surface. When $\theta = 140.77°$ or $320.77°$, we have a second nodal surface.

12.39 In this problem, we explore the concept of angular momentum in greater detail. We seek to show that the quantum mechanical operator corresponding to the Z component of angular momentum, when expressed in spherical polar coordinates, is $L_z = (\hbar/i)(\partial/\partial\phi)$. We begin with L_z in rectangular Cartesian coordinates. Using the second postulate, the text shows that

$$L_z = \frac{\hbar}{i}\left[x\frac{\partial}{\partial y} - y\frac{\partial}{\partial x}\right].$$

We now need to transform L_z into the corresponding equation in spherical polar coordinates. The transformation equations are

$$x = R\sin\theta\cos\phi,$$

$$y = R\sin\theta\sin\phi,$$

$$z = R\cos\theta,$$

and

$$R^2 = x^2 + y^2 + z^2.$$

The chain rule for the conversion of derivatives gives the result

$$\frac{\partial}{\partial y} = \frac{\partial}{\partial R}\frac{\partial R}{\partial y} + \frac{\partial}{\partial\theta}\frac{\partial\theta}{\partial y} + \frac{\partial}{\partial\phi}\frac{\partial\phi}{\partial y}$$

and

$$\frac{\partial}{\partial x} = \frac{\partial}{\partial R}\frac{\partial R}{\partial x} + \frac{\partial}{\partial\theta}\frac{\partial\theta}{\partial x} + \frac{\partial}{\partial\phi}\frac{\partial\phi}{\partial x}.$$

Use the chain rule in conjunction with the transformation equations to show that, in spherical polar coordinates,

$$L_z = \frac{\hbar}{i}\frac{\partial}{\partial\phi}.$$

Solution

Using the chain rule, we have

$$\left(\frac{\partial}{\partial y}\right) = \left(\frac{\partial}{\partial R}\right)\left(\frac{\partial R}{\partial y}\right) + \left(\frac{\partial}{\partial \theta}\right)\left(\frac{\partial \theta}{\partial y}\right) + \left(\frac{\partial}{\partial \phi}\right)\left(\frac{\partial \phi}{\partial y}\right). \tag{1}$$

The derivatives in Eq. 1 can all be obtained from the transformation equations. The radial coordinate R is given by

$$R^2 = x^2 + y^2 + z^2. \tag{2}$$

Therefore,

$$2R\frac{\partial R}{\partial y} = 2y \tag{3}$$

and

$$\frac{\partial R}{\partial y} = \frac{y}{R}. \tag{4}$$

Using the third transformation equation, we obtain

$$\cos\theta = \frac{z}{R}. \tag{5}$$

Differentiating both sides with respect to y gives

$$-\sin\theta\frac{\partial \theta}{\partial y} = -\frac{z}{R^2}\frac{\partial R}{\partial y} = -\frac{zy}{R^3}, \tag{6}$$

where we have used the result in Eq. 4 to replace $\partial R/\partial y$. Solving for the derivative, we obtain

$$\frac{\partial \theta}{\partial y} = \frac{zy}{R^3 \sin\theta}. \tag{7}$$

Dividing the y transformation equation by the x equation gives

$$\frac{y}{x} = \frac{R\sin\theta\sin\phi}{R\sin\theta\cos\phi} = \frac{\sin\phi}{\cos\phi} = \tan\phi. \tag{8}$$

Taking derivatives of both sides with respect to y yields

$$\sec^2\phi\frac{\partial \phi}{\partial y} = \frac{1}{\cos^2\phi}\frac{\partial \phi}{\partial y} = \frac{1}{x}. \tag{9}$$

The derivative is

$$\frac{\partial \phi}{\partial y} = \frac{\cos^2\phi}{x}. \tag{10}$$

Combining Eqs. 4, 7, and 10 with Eq. 1, we obtain

$$x\frac{\partial}{\partial y} = \frac{xy}{R}\frac{\partial}{\partial R} + \frac{xyz}{R^3\sin\theta}\frac{\partial}{\partial \theta} + \cos^2\phi\frac{\partial}{\partial \phi}. \tag{11}$$

The second term in the expression for L_z requires the derivative $\partial/\partial x$. The chain rule for this is

$$\left(\frac{\partial}{\partial x}\right) = \left(\frac{\partial}{\partial R}\right)\left(\frac{\partial R}{\partial x}\right) + \left(\frac{\partial}{\partial \theta}\right)\left(\frac{\partial \theta}{\partial x}\right) + \left(\frac{\partial}{\partial \phi}\right)\left(\frac{\partial \phi}{\partial x}\right). \tag{12}$$

We may obtain all the derivatives in the same manner as before. Using Eq. 5, we get

$$2R\frac{\partial R}{\partial x} = 2x \tag{13}$$

and

$$\frac{\partial R}{\partial x} = \frac{x}{R}. \tag{14}$$

Again from Eq. 5, we have

$$-\sin\theta\frac{\partial\theta}{\partial x} = -\frac{z}{R^2}\frac{\partial R}{\partial x} = -\frac{zx}{R^3} \tag{15}$$

by using the result in Eq. 14. Solving for the derivative, we obtain

$$\frac{\partial\theta}{\partial x} = \frac{zx}{R^3\sin\theta}. \tag{16}$$

Finally, from Eq. 8, we get

$$\sec^2\phi\frac{\partial\phi}{\partial x} = \frac{1}{\cos^2\phi}\frac{\partial\phi}{\partial x} = -\frac{y}{x^2}. \tag{17}$$

The derivative is

$$\frac{\partial\phi}{\partial x} = -\frac{y\cos^2\phi}{x^2}. \tag{18}$$

Combining Eqs. 14, 16, and 18 with Eq. 12, we obtain

$$y\frac{\partial}{\partial x} = \frac{xy}{R}\frac{\partial}{\partial R} + \frac{xyz}{R^3\sin\theta}\frac{\partial}{\partial\theta} + \frac{y^2\cos^2\phi}{x^2}\frac{\partial}{\partial\phi}. \tag{19}$$

Using the fact that $L_z = (\hbar/i)[x\partial/\partial y - y\partial/\partial x]$ with the results in Eqs. 11 and 19, we get

$$L_z = \frac{\hbar}{i}\left[\left(\frac{xy}{R} - \frac{xy}{R}\right)\frac{\partial}{\partial R} + \left(\frac{xyz}{R^3\sin\theta} - \frac{xyz}{R^3\sin\theta}\right)\frac{\partial}{\partial\theta} + \left(\cos^2\phi - \frac{y^2\cos^2\phi}{x^2}\right)\frac{\partial}{\partial\phi}\right]$$

$$= \frac{\hbar}{i}\left(\cos^2\phi - \frac{y^2\cos^2\phi}{x^2}\right)\frac{\partial}{\partial\phi}. \tag{20}$$

The last term in parentheses in Eq. 20 can be written in the form

$$\cos^2\phi - \frac{y^2\cos^2\phi}{x^2} = \cos^2\phi\left[1 - \frac{y^2}{x^2}\right] = \cos^2\phi\left[\frac{x^2+y^2}{x^2}\right]. \tag{21}$$

Using the transformation equations, we find that Eq. 21 becomes

$$\cos^2\phi\left[\frac{x^2+y^2}{x^2}\right] = \cos^2\phi\left[\frac{R^2\sin^2\theta\cos^2\phi + R^2\sin^2\theta\sin^2\phi}{R^2\sin^2\theta\cos^2\phi}\right]$$

$$= \cos^2\phi\left[\frac{\cos^2\phi + \sin^2\phi}{\cos^2\phi}\right] = 1. \tag{22}$$

Therefore, we obtain

$$\boxed{L_z = \frac{\hbar}{i}\frac{\partial}{\partial\phi}}, \tag{23}$$

as required.

12.41 A rigid rotor is known to be in a state whose eigenfunction is $Y_4^3(\theta, \phi)$.

(A) What is the rotational energy of the rotor in terms of I and $\hbar$?

(B) What is the magnitude of the angular momentum of the rotor?

(C) Determine the z component of the angular momentum.

(D) What angle does the angular momentum vector make with the Z-axis?

(E) Can we compute the angle ϕ associated with the angular momentum vector in this state? If so, determine the value of ϕ. If not, state why not.

Solution

(A) In this eigenstate, we have $J = 4$. Using Eq. 12.107, we find the rotational energy to be

$$\boxed{E_4 = \frac{4(4+1)\hbar^2}{2I} = \frac{20\hbar^2}{2I} = \frac{10\hbar^2}{I}}. \quad (1)$$

(B) The total angular momentum is given by Eq. 12.112:

$$L^2 = J(J+1)\hbar^2 = (4)(5)(6.626 \times 10^{-34}\text{ J s})^2/4\pi^2 = \underline{2.224 \times 10^{-67}\text{ kg}^2\text{ m}^4\text{ s}^{-2}}. \quad (2)$$

Taking square roots of both sides, we obtain

$$|L| = \underline{4.716 \times 10^{-34}\text{ kg m}^2\text{ s}^{-1}}. \quad (3)$$

(C) L_z is given by

$$L_z = M\hbar = \frac{3(6.626 \times 10^{-34}\text{ J s})}{2\pi} = \underline{3.164 \times 10^{-34}\text{ kg m}^2\text{ s}^{-1}}. \quad (4)$$

(D) The angle made by L with the Z-axis is

$$\cos\theta = \frac{L_z}{L} = \frac{M\hbar}{[J(J+1)]^{1/2}\hbar} = \frac{M}{[J(J+1)]^{1/2}} = \frac{3}{[(4)(5)]^{1/2}} = 0.6708. \quad (5)$$

Thus,

$$\theta = \cos^{-1}(0.6708) = \underline{47.87°}. \quad (6)$$

(E) No, we cannot compute the angle ϕ. If we could, it would mean that we simultaneously know L^2, L_z, L_y, and L_x with no uncertainty. However, the uncertainty principle requires that we have

$$\langle \Delta L_x^2 \rangle \langle \Delta L_y^2 \rangle \geq \frac{1}{4}|[L_x, L_y]|^2 = \frac{1}{4}|i\hbar L_z|^2 = \frac{M^2\hbar^4}{4}. \quad (7)$$

Since $M = 3$, we must have a nonzero uncertainty in L_x and L_y. Therefore, the angle ϕ cannot be determined.

12.43 A rigid rotor is in a $J = 6$ rotational state. What are the possible angles the angular momentum vector might make with the Z-axis? Can we compute the angle ϕ?

Solution

When $J = 6$, M can have the values $-6, -5, -4, -3, -2, -1, 0, 1, 2, 3, 4, 5$, and 6. There are, therefore, 13 possible angles **L** can make with the Z-axis. The magnitude of **L** is given by

$$L = [J(J+1)\hbar^2]^{1/2} = (42)^{1/2}\hbar. \quad (1)$$

The Z-component of L is

$$L_z = M\hbar. \quad \textbf{(2)}$$

Therefore, the cosine of the angle between **L** and the z-axis is

$$\cos\theta = \frac{L_z}{L}, \quad \textbf{(3)}$$

and it follows that

$$\theta = \cos^{-1}\left(\frac{L_z}{L}\right) = \cos^{-1}\left(\frac{M\hbar}{(42)^{1/2}\hbar}\right) = \cos^{-1}\left[\frac{M}{(42)^{1/2}}\right]. \quad \textbf{(4)}$$

Using the possible values of M in Eq. 4 gives the following angles:

$$
\begin{aligned}
M &= -6: & \theta &= 157.792°;\\
M &= -5: & \theta &= 140.490°;\\
M &= -4: & \theta &= 128.112°;\\
M &= -3: & \theta &= 117.575°;\\
M &= -2: & \theta &= 107.975°;\\
M &= -1: & \theta &= 98.876°;\\
M &= -0: & \theta &= 90.000°;\\
M &= 1: & \theta &= 81.124°;\\
M &= 2: & \theta &= 72.025°;\\
M &= 3: & \theta &= 62.425°;\\
M &= 4: & \theta &= 51.888°;\\
M &= 5: & \theta &= 39.510°;\\
M &= 6: & \theta &= 22.208°. \quad \textbf{(5)}
\end{aligned}
$$

A schematic diagram of the possible directions of **L** is as follows:

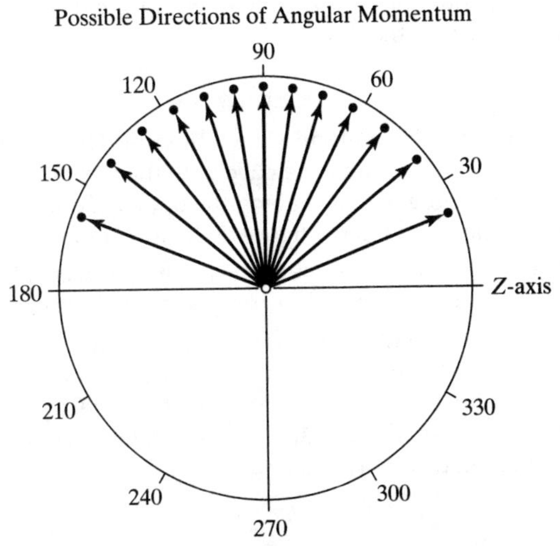

The angle ϕ cannot be determined for any of these states. If M is not required to be zero, the uncertainty principle forbids us to simultaneously know both L_x and L_y and, hence, ϕ. If M must be zero, we can determine that $L_z = L_x = L_y = 0$, but in this case ϕ is undefined, and therefore, it cannot be determined.

12.45 Use Eqs. 12.65A and 12.113 to show that the uncertainty product $\langle \Delta\phi^2 \rangle \langle \Delta L_z^2 \rangle$ is not zero, so that it is impossible to simultaneously know both the Z component of the angular momentum and the rotation angle about the Z-axis.

Solution

We know from Eq. 12.65A that the uncertainty product between two variables u and w has the form

$$\langle \Delta u^2 \rangle \langle \Delta w^2 \rangle \geq \frac{1}{4} |[u, w]|^2, \tag{1}$$

where $[u, w]$ is the commutator of the variables u and w. Therefore, in the present case, we have

$$\langle \Delta\phi^2 \rangle \langle \Delta L_z^2 \rangle \geq \frac{1}{4} |[\phi, L_z]|^2. \tag{2}$$

From Eq. 12.113, we know that the operator for L_z is $(\hbar/i)(\partial/\partial\phi)$. The commutator between ϕ and L_z is

$$\begin{aligned}[\phi, L_z]\psi(\phi) &= \left\{\phi \frac{\hbar}{i}\frac{\partial}{\partial\phi} - \frac{\hbar}{i}\frac{\partial}{\partial\phi}\phi\right\}\psi(\phi) = \phi\frac{\hbar}{i}\frac{\partial}{\partial\phi}\psi(\phi) - \frac{\hbar}{i}\frac{\partial}{\partial\phi}\{\phi\psi(\phi)\} \\ &= \frac{\hbar}{i}\left[\phi\frac{\partial\psi(\phi)}{\partial\phi} - \psi(\phi) - \phi\frac{\partial\psi(\phi)}{\partial\phi}\right] = -\frac{\hbar}{i}\psi(\phi). \end{aligned} \tag{3}$$

Comparing the left-hand and right-hand sides of Eq. 3 shows that we must have

$$[\phi, L_z] = -\frac{\hbar}{i} = i\hbar. \tag{4}$$

Substituting Eq. 4 into Eq. 2 produces

$$\boxed{\langle \Delta\phi^2 \rangle \langle \Delta L_z^2 \rangle \geq \frac{1}{4}|i\hbar|^2 = \frac{\hbar^2}{4}}. \tag{5}$$

Since the uncertainty product between ϕ and L_z is not zero, it is impossible to simultaneously know both the Z component of the angular momentum and the rotation angle about the Z-axis.

12.47 The geometry of the CH_4 molecule is tetrahedral. This configuration may be conveniently pictured by drawing a cube with the carbon atom in the center of the cube and the hydrogen atoms along the opposite diagonals of the top and bottom faces, as illustrated in the figure at the top of the next page.

(A) Show that X-, Y-, and Z-axes with an origin at the center of the cube and drawn perpendicular to the faces of the cube correspond to the principal axes of CH_4.

(B) Show that the moment of inertia about any principal axis is $I = 8m_H R^2/3$, where R is the C–H bond length and m_H is the mass of the hydrogen atom. Obtain an expression for the spacing between the $J = 0$ and the $J = 1$ rotational states of CH_4 in terms of m_H and R if the rotational states are accurately described by the rotation of a rigid body.

Solution

(A) To demonstrate that the X-, Y-, and Z-axes shown in the figure can serve as principal axes for CH_4, we must prove that the products of inertia, Eq. 12.129, are zero. Since the three Cartesian axes are all equivalent, due to symmetry, if we can demonstrate that either I_{xy}, I_{xz}, or I_{yz} is zero, the same

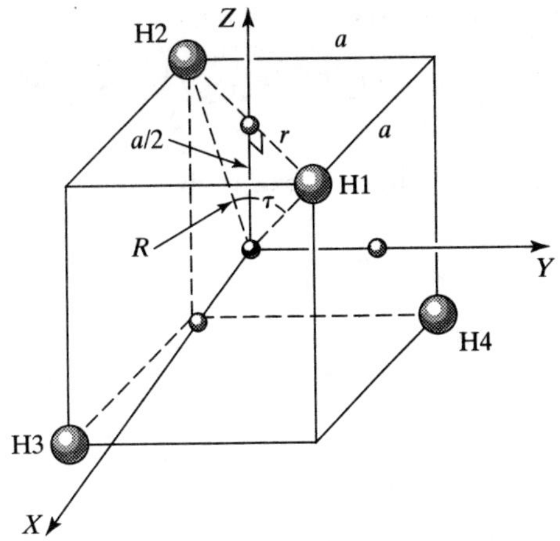

proof will show that they are all zero. Let us consider I_{xy}. This product of inertia is given by

$$I_{xy} = -\sum_{i-1}^{5} m_i x_i y_i. \tag{1}$$

The carbon atom is at the origin with $x_c = y_c = z_c = 0$, so it makes no contribution to the sum in Eq. 1. The remaining atoms are all hydrogen atoms with equal mass. Thus,

$$I_{xy} = -m_H[x_1y_1 + x_2y_2 + x_3y_3 + x_4y_4], \tag{2}$$

where x_i and y_i are the x- and y-coordinates of hydrogen atom i. Inspection of the figure given in the problem shows that we have

$$x_1 = x_3 = -x_2 = -x_4 \tag{3}$$

and

$$y_1 = y_4 = -y_2 = -y_3. \tag{4}$$

We can use Eqs. 3 and 4 to express the four terms in Eq. 2 in terms involving only x_1 and y_1. The result is

$$I_{xy} = -m_H[x_1y_1 + (-x_1)(-y_1) + (x_1)(-y_1) + (-x_1)(y_1)]. \tag{5}$$

Inspection of the right-hand side of Eq. 5 shows that it identically equals zero. Thus, $I_{xy} = 0$. A similar analysis shows that $I_{xz} = I_{yz} = 0$. Therefore, we may conclude that the X-, Y-, and Z-axes shown in the figure can serve as principal axes of rotation for CH_4.

(B) All the axes are equivalent, so we will derive the principal moment about the Z-axis. We now define r to be one-half the length of the diagonal across the face of the cube. In terms of the notation defined in the figure, we have

$$(2r)^2 = a^2 + a^2 = 2a^2, \tag{6}$$

which gives

$$r = \frac{a}{(2)^{1/2}}. \tag{7}$$

From the figure, we can see that

$$R^2 = \left[\frac{a}{2}\right]^2 + r^2. \tag{8}$$

Combining Eqs. 7 and 8 yields

$$R^2 = \left[\frac{(2)^{1/2} r}{2}\right]^2 + r^2 = \frac{r^2}{2} + r^2 = \frac{3r^2}{2}. \tag{9}$$

The moment of inertia about the Z-axis is given by

$$I_z = \sum_{i=1}^{5} m_i r_z^2 = m_C(0)^2 + 4m_H r^2 = 4m_H r^2, \tag{10}$$

since the carbon atom lies on the rotation axis and the distance of all four hydrogen atoms from the rotation axis is r. Combining Eqs. 9 and 10, we obtain

$$\boxed{I_z = 4m_H\left[\frac{2R^2}{3}\right] = \frac{8m_H R^2}{3}}, \tag{11}$$

as required. Since all moments of inertia are equal, we have

$$I = I_a = I_b = I_c = \frac{8m_H R^2}{3}, \tag{12}$$

and CH_4 is a spherical-top rotor. From Eq. 12.132, we have

$$E_J = \frac{J(J+1)\hbar^2}{2I}. \tag{13}$$

Therefore, the requested spacing is

$$\boxed{E_1 - E_o = \frac{2\hbar^2}{2I} - 0 = \frac{\hbar^2}{I} = \frac{3\hbar^2}{8m_H R^2}}. \tag{14}$$

12.49 Show that the total vibrational energy of a classical one-dimensional harmonic oscillator is $E_{vib} = kA^2/2$.

Solution

Equation 12.142 gives the oscillator displacement as a function of time:

$$x = A \sin\left[\left(\frac{k}{\mu}\right)^{1/2} t\right]. \tag{1}$$

The total vibrational energy of the classical oscillator is the sum of the kinetic and potential energies:

$$E_{vib} = \frac{\mu}{2}\left(\frac{dx}{dt}\right)^2 + \frac{kx^2}{2}. \tag{2}$$

Using Eq. 1, we obtain

$$\frac{dx}{dt} = A\left(\frac{k}{\mu}\right)^{1/2} \cos\left[\left(\frac{k}{\mu}\right)^{1/2} t\right] \tag{3}$$

and

$$\frac{kx^2}{2} = \frac{A^2 k}{2} \sin^2\left[\left(\frac{k}{\mu}\right)^{1/2} t\right]. \tag{4}$$

Combining Eqs. 2, 3, and 4 produces

$$E_{\text{vib}} = \frac{\mu}{2}A^2\left(\frac{k}{\mu}\right)\cos^2\left[\left(\frac{k}{\mu}\right)^{1/2}t\right] + \frac{A^2k}{2}\sin^2\left[\left(\frac{k}{\mu}\right)^{1/2}t\right]$$

$$= \frac{A^2k}{2}\left[\cos^2\left[\left(\frac{k}{\mu}\right)^{1/2}t\right] + \sin^2\left[\left(\frac{k}{\mu}\right)^{1/2}t\right]\right] = \boxed{\frac{A^2k}{2}}, \quad (5)$$

since $\sin^2 z + \cos^2 z = 1$.

12.51 (A) Compute the probability that a quantum harmonic oscillator in its ground state with a fundamental vibration frequency equal to $1.350 \times 10^{14}\ \text{s}^{-1}$ and a reduced mass equal to that of H_2 will be found in the classically forbidden region of space.

(B) Compute the probability that a quantum harmonic oscillator in its first excited state with a fundamental vibration frequency equal to $1.350 \times 10^{14}\ \text{s}^{-1}$ and a reduced mass equal to that of H_2 will be found in the classically forbidden region of space.

Solution

(A) At the classical turning points, $x = x_t$, all of the energy is in the form of potential energy. This fact permits us to obtain an expression for the turning points by equating the total energy to the potential at the turning points. Thus,

$$\frac{kx_t^2}{2} = E = \frac{h\nu_o}{2} \quad (1)$$

if the oscillator is in the ground state. Solving for x_t, we obtain

$$x_t = \pm\left[\frac{h\nu_o}{k}\right]^{1/2} = \pm\left[\frac{h}{2\pi k}(k/\mu)^{1/2}\right]^{1/2} = \pm\left[\frac{\hbar}{(k\mu)^{1/2}}\right]^{1/2} = \pm\alpha^{-1/2}. \quad (2)$$

If μ is that for H_2, we have

$$\mu = \frac{m_H m_H}{m_H + m_H} = \frac{m_H}{2} = \frac{1.0079\ \text{g mol}^{-1}}{2} \times \frac{1}{6.02214 \times 10^{23}\ \text{mol}^{-1}} = 8.3683 \times 10^{-25}\ \text{g}$$

$$= 8.3683 \times 10^{-28}\ \text{kg}. \quad (3)$$

The force constant is given by

$$k = 4\pi^2\mu\nu_o^2 = 4(3.14159)^2(8.3683 \times 10^{-28}\ \text{kg})(1.350 \times 10^{14}\ \text{s}^{-1})^2$$

$$= 602.1\ \text{kg s}^{-2}. \quad (4)$$

Therefore,

$$\alpha = \frac{[\mu k]^{1/2}}{\hbar} = \frac{[(8.3683 \times 10^{-28}\ \text{kg})(602.1\ \text{kg s}^{-2})]^{1/2}}{1.055 \times 10^{-34}\ \text{J s}} = 6.728 \times 10^{21}\ \text{m}^{-2}$$

$$= 6.728 \times 10^{17}\ \text{cm}^{-2} = 6.728 \times 10\ \text{Å}^{-2} = 67.28\ \text{Å}^{-2}. \quad (5)$$

The turning points are

$$x_t = \pm\alpha^{-1/2} = \pm 1.219 \times 10^{-11}\ \text{m} = \pm 1.219 \times 10^{-9}\ \text{cm} = \pm 0.1219\ \text{Å}. \quad (6)$$

The probability that the quantum oscillation will lie in the range $x \geq +\alpha^{-1/2}$ or $x \leq -\alpha^{-1/2}$ is

$$P(\text{forbidden}) = \int_{\alpha^{-1/2}}^{\infty} |\psi_o(x)|^2 dx + \int_{-\infty}^{-\alpha^{-1/2}} |\psi_o(x)|^2 dx = 2\int_{\alpha^{-1/2}}^{\infty} |\psi_o(x)|^2 dx, \quad (7)$$

because of the symmetry of the problem. Inserting the ground-state wave function gives

$$P(\text{forbidden}) = 2\left(\frac{\alpha}{\pi}\right)^{1/2} \int_{\alpha^{-1/2}}^{\infty} \exp[-\alpha x^2]\, dx. \quad (8)$$

Let $y = \alpha^{1/2}x$ so that $dy = \alpha^{1/2}dx$. Substituting this change of variable into Eq. 8 produces

$$P(\text{forbidden}) = 2\left(\frac{\alpha}{\pi}\right)^{1/2} \alpha^{-1/2} \int_1^{\infty} \exp[-y^2]\, dy = \frac{2}{\pi^{1/2}} \int_1^{\infty} \exp(-y^2)\, dy. \quad (9)$$

The integral in Eq. 9 can be expressed in terms of error functions. It can also be done numerically. Using the latter procedure, we obtain

$$P(\text{forbidden}) = \frac{2}{\pi^{1/2}}(0.13941) = \underline{0.1573}. \quad (10)$$

(B) In the first excited state, the total energy is $3h\nu_o/2$. Therefore, the classical turning points are given by

$$\frac{kx_t^2}{2} = E = \frac{3h\nu_o}{2}. \quad (11)$$

This yields

$$x_t = \pm\left[\frac{3h\nu_o}{k}\right]^{1/2} = \pm\left[\frac{3h}{2\pi k}\left(\frac{k}{\mu}\right)^{1/2}\right]^{1/2} = \pm\left[\frac{3\hbar}{(k\mu)^{1/2}}\right]^{1/2} = \pm 3^{1/2}\alpha^{-1/2}. \quad (12)$$

The probability of being in the classically forbidden region is

$$P(\text{forbidden}) = \int_{(3/\alpha)^{1/2}}^{\infty} |\psi_1(x)|^2 \mathrm{dx} + \int_{-\infty}^{-(3/\alpha)^{1/2}} |\psi_1(x)|^2 \mathrm{dx} = 2\int_{(3/\alpha)^{1/2}}^{\infty} |\psi_1(x)|^2 \mathrm{dx}. \quad (13)$$

The first excited-state wave function, is given by

$$\psi_1(x) = Nx \exp\left[-\frac{\alpha x^2}{2}\right]. \quad (14)$$

Normalization requires that we have

$$\int_{-\infty}^{\infty} |\psi_1(x)|^2 dx = 1 = N^2 \int_{-\infty}^{\infty} x^2 \exp[-\alpha x^2]\, dx = \frac{N^2}{2\alpha}\left(\frac{\pi}{\alpha}\right)^{1/2}. \quad (15)$$

Solving for the normalization constant, we obtain

$$N^2 = 2\alpha\left(\frac{\alpha}{\pi}\right)^{1/2}. \quad (16)$$

Again, we let $y = \alpha^{1/2}x$. Substituting this transformation and Eqs. 14 and 16 into Eq. 13 gives

$$P(\text{forbidden}) = \alpha^{-3/2}4\alpha\left(\frac{\alpha}{\pi}\right)^{1/2} \int_{3^{1/2}}^{\infty} y^2 \exp[-y^2]\, dy = \frac{4}{\pi^{1/2}} \int_{3^{1/2}}^{\infty} y^2 \exp[-y^2]\, dy. \quad (17)$$

Integrating numerically, we obtain

$$P(\text{forbidden}) = \frac{4}{\pi^{1/2}}(0.04946) = \underline{0.1116}. \quad \textbf{(18)}$$

Note that the tunneling probability is less for the $v = 1$ state than for the $v = 0$ ground state in spite of the fact that the total vibrational energy is larger in the former.

12.53 (A) Assume that the total vibrational and rotational energy of H_2 may be obtained simply by adding together the rotational energy of a rigid rotor with R equal to the equilibrium bond distance and the vibrational energy of a harmonic oscillator. Under these conditions, compute the total vibrational and rotational energy for H_2 in the following quantum states (the equilibrium bond distance for H_2 is 0.74000×10^{-10} m, and the fundamental vibration frequency is $1.3500 \times 10^{14}\ \text{s}^{-1}$):

$$v = 0,\ J = 0;\quad v = 1,\ J = 0.$$
$$v = 0,\ J = 1;\quad v = 1,\ J = 1.$$
$$v = 0,\ J = 2;\quad v = 1,\ J = 2.$$
$$v = 0,\ J = 3;\quad v = 1,\ J = 3.$$
$$v = 0,\ J = 4;\quad v = 1,\ J = 4.$$
$$v = 0,\ J = 5;\quad v = 1,\ J = 5.$$

(B) Compute the energy spacing between the following pairs of states:

$$(v = 0, J = 0)\ \text{and}\ (v = 1, J = 1).$$
$$(v = 0, J = 1)\ \text{and}\ (v = 1, J = 2).$$
$$(v = 0, J = 2)\ \text{and}\ (v = 1, J = 3).$$
$$(v = 0, J = 3)\ \text{and}\ (v = 1, J = 4).$$
$$(v = 0, J = 4)\ \text{and}\ (v = 1, J = 5).$$

What type of regular progression do you notice in the values obtained?

Solution

(A) The total energy of H_2 under the conditions of the problem is given by

$$E(v, J) = (v + 0.5)h\nu_o + J(J + 1)\frac{\hbar^2}{2\mu R^2}. \quad \textbf{(1)}$$

Therefore,

$$h\nu_o = (6.62608 \times 10^{-34}\ \text{J s})(1.3500 \times 10^{14}\ \text{s}^{-1}) = 8.945 \times 10^{-20}\ \text{J}. \quad \textbf{(2)}$$

The reduced mass of H_2 is

$$\mu = \frac{m_H m_H}{m_H + m_H} = \frac{m_H}{2} = \frac{1.00794\ \text{g mol}^{-1}}{2(6.02214 \times 10^{23}\ \text{mol}^{-1})} = 8.36862 \times 10^{-25}\ \text{g}$$
$$= 8.36862 \times 10^{-28}\ \text{kg}. \quad \textbf{(3)}$$

Hence,

$$\frac{\hbar^2}{2\mu R^2} = \frac{(6.62608 \times 10^{-34}\ \text{J s})^2}{8(3.1415927)^2(8.36862 \times 10^{-28}\ \text{kg})(0.74000 \times 10^{-10}\ \text{m})^2} = 1.213 \times 10^{-21}\ \text{J}. \quad \textbf{(4)}$$

The corresponding energies of the states are as follows:

v	J	Energy (J) $\times 10^{20}$
0	0	4.4725
0	1	4.7151
0	2	5.2003
0	3	5.9281
0	4	6.8985
0	5	8.1115
1	0	13.4175
1	1	13.6601
1	2	14.1453
1	3	14.8731
1	4	15.8435
1	5	17.0565

(B) The energy spacings are obtained by taking differences between the foregoing values. We obtain the following table:

State 1		State 2		$\Delta E(E$ State 2 $- E$ State 1$) \times 10^{20}$ (J)
v	J	v	J	
0	0	1	1	13.6601 − 4.4725 = 9.1876
				} 0.2426
0	1	1	2	14.1453 − 4.7151 = 9.4302
				} 0.2426
0	2	1	3	14.8731 − 5.2003 = 9.6728
				} 0.2426
0	3	1	4	15.8435 − 5.9281 = 9.9154
				} 0.2426
0	4	1	5	17.0565 − 6.8985 = 10.158

The energy gap between each successive difference is always equal to 0.2426 J. This point will be important in the analysis of vibrational–rotational spectra.

12.55. In this problem, we will compare the difference between the classical probability distribution for the position of a harmonic oscillator and that for a quantum mechanical harmonic oscillator in the ground state. It is shown in the text that the dependence of the displacement on time for a classical oscillator is $x(t) = A\sin[(k/\mu)^{1/2}t]$ for an oscillator with $x(t = 0) = 0$. The probability of finding the classical oscillator in the range from x to $x + dx$ is proportional to the time the oscillator spends in going from one end of the range to the other. This residence time is $t_r = dx(dx/dt)^{-1}$; that is, the time required to traverse the distance dx is dx divided by the velocity, which is (dx/dt). Use this relationship to obtain the normalized probability distribution function for the position of the classical harmonic oscillator. {*Hint:* Write cos(z) as $(1 - \sin^2 z)^{1/2}$, and then use the fact that $x = A\sin[(k/u)^{1/2}t]$.}

Solution

The classical velocity of the oscillator is

$$v_x = \frac{dx}{dt} = \left(\frac{k}{\mu}\right)^{1/2} A\cos\left[\left(\frac{k}{\mu}\right)^{1/2} t\right]. \quad \textbf{(1)}$$

Therefore, the time required to traverse a distance dx is

$$t_r = \frac{dx}{v_x} = \frac{dx}{(k/\mu)^{1/2} A\cos[(k/\mu)^{1/2}t]}. \quad \textbf{(2)}$$

Since the probability of finding the classical oscillator in the range from x to $x + dx$ is proportional to t_r, we have

$$P(x)dx = \frac{C\,dx}{(k/\mu)^{1/2} A\cos[(k/\mu)^{1/2}t]}, \quad \textbf{(3)}$$

where C is the constant of proportionality. Since $\cos z = (1 - \sin^2 z)^{1/2}$, Eq. 3 may be written in the form

$$P(x)dx = \frac{C\,dx}{(k/\mu)^{1/2}[A^2 - A^2\sin^2\{(k/\mu)^{1/2}t\}]^{1/2}}. \quad \textbf{(4)}$$

Using the fact that $x = A\sin[(k/\mu)^{1/2}t]$, we find that Eq. 4 becomes

$$P(x)dx = \frac{C\,dx}{(k/\mu)^{1/2}[A^2 - x^2]^{1/2}}. \quad \textbf{(5)}$$

The largest magnitude sin(z) can have is unity. Therefore, x must lie in the range $-A \le x \le A$. Outside this range, we have $P(x)dx = 0$. Normalization requires that

$$\int_{-A}^{A} P(x)dx = \frac{C}{(k/\mu)^{1/2}} \int_{-A}^{A} \frac{dx}{[A^2 - x^2]^{1/2}} = 1. \quad \textbf{(6)}$$

The integral in Eq. 6 is a standard form that can be found in any table of integrals. The result is

$$\frac{C}{(k/\mu)^{1/2}} \int_{-A}^{A} \frac{dx}{[A^2 - x^2]^{1/2}} = \frac{C}{(k/\mu)^{1/2}} \sin^{-1}\left(\frac{x}{A}\right)\Bigg|_{-A}^{A} = \frac{C}{(k/\mu)^{1/2}}[\sin^{-1}(1) - \sin^{-1}(-1)]$$

$$= \frac{C}{(k/\mu)^{1/2}}\left[\frac{\pi}{2} - \left(-\frac{\pi}{2}\right)\right] = \frac{C\pi}{(k/\mu)^{1/2}} = 1. \quad \textbf{(7)}$$

Solving for C, we obtain

$$C = \frac{(k/\mu)^{1/2}}{\pi}. \quad \textbf{(8)}$$

With C given by Eq. 8, Eq. 5 shows the normalized probability distribution to be

$$\boxed{P(x)dx = \frac{dx}{\pi[A^2 - x^2]^{1/2}} \quad \text{for } -A \leq x \leq A} \tag{9}$$

and

$$\boxed{P(x)dx = 0 \quad \text{for } x < -A \text{ or } x > A}.$$

12.57 The experimental well depth for the H_2 molecule is 7.607×10^{-19} J, and the vibrational frequency is $1.317 \times 10^{14}\ \text{s}^{-1}$. Use Eq. 12.156 to show that the $v = 17$ vibrational state is just about at the dissociation point.

Solution

The coefficient of the anharmonicity term in Eq. 12.156 is

$$\frac{h^2\nu_o^2}{4D} = \frac{(6.626 \times 10^{-34}\ \text{J s})^2(1.317 \times 10^{14}\ \text{s}^{-1})^2}{4(7.607 \times 10^{-19}\ \text{J})} = 2.503 \times 10^{-21}\ \text{J}. \tag{1}$$

The Morse vibrational energy levels are given by

$$E_v = (v + 0.5)A - B(v + 0.5)^2, \tag{2}$$

where

$$A = h\nu_o = (6.626 \times 10^{-34}\ \text{J s})(1.317 \times 10^{14}\ \text{s}^{-1}) = 8.726 \times 10^{-20}\ \text{J} \tag{3}$$

and

$$B = 2.503 \times 10^{-21}\ \text{J}. \tag{4}$$

Substituting $v = 17$ into Eq. 2 gives

$$E_{17} = 17.5(8.726 \times 10^{20}\ \text{J}) - (17.5)^2(2.503 \times 10^{-21}\ \text{J}) = 7.605 \times 10^{-19}\ \text{J}. \tag{5}$$

Since the well depth is 7.607×10^{-19} J, we are just about at the dissociation point.

12.59 Let us represent the interatomic potential for $H^{35}Cl$ by a Morse potential. The measured $H^{35}Cl$ vibration frequency and well depth are $8.960 \times 10^{13}\ \text{s}^{-1}$ and 7.394×10^{-19} J, respectively. Compute the ratio of the anharmonicity term to the harmonic term for the $v = 0, 1, 2,$ and 3 vibrational states of the oscillator.

Solution

Using Eq. 12.156, we see that the ratio of the anharmonicity to the harmonic term is

$$\text{Ratio} = \frac{\frac{h^2\nu_o^2}{4D}(v + 0.5)^2}{(v + 0.5)h\nu_o} = \frac{h\nu_o}{4D}(v + 0.5). \tag{1}$$

Thus,

$$\frac{h\nu_o}{4D} = \frac{(6.626 \times 10^{34}\text{ J s})(8.960 \times 10^{13}\text{ s}^{-1})}{4(7.394 \times 10^{-19}\text{ J})} = 0.02007. \quad \textbf{(2)}$$

Therefore, the ratios for the $v = 0, 1, 2,$ and 3 vibrational states are as follows:

$$\text{Ratio} = 0.02007(0.5) = \underline{0.01004} \quad \text{for the } v = 0 \text{ state;} \quad \textbf{(3)}$$

$$\text{Ratio} = 0.02007(1.5) = \underline{0.03011} \quad \text{for the } v = 1 \text{ state;} \quad \textbf{(4)}$$

$$\text{Ratio} = 0.02007(2.5) = \underline{0.05018} \quad \text{for the } v = 2 \text{ state;} \quad \textbf{(5)}$$

$$\text{Ratio} = 0.02007(3.5) = \underline{0.07026} \quad \text{for the } v = 3 \text{ state.} \quad \textbf{(6)}$$

The results show that the vibrational anharmonicity becomes more and more important relative to the harmonic term as the vibrational quantum number increases.

12.61 Show that $\psi(x) = Ae^{ikx}$ is an eigenfunction of p_x and also of p_x^2. What are the associated eigenvalues?

Solution

The operator for p_x is $(\hbar/i)(\partial/\partial x)$. Therefore, if $\psi(x) = Ae^{ikx}$ is eigenfunction of p_x, then

$$p_x\psi(x) = \frac{\hbar}{i}\frac{\partial}{\partial x}Ae^{ikx} = (\text{constant})\psi(x). \quad \textbf{(1)}$$

Performing the indicated operations, we obtain

$$p_x\psi(x) = \frac{\hbar}{i}\frac{\partial}{\partial x}Ae^{ikx} = \frac{A\hbar}{i}ike^{ikx} = k\hbar[Ae^{ikx}] = k\hbar\,\psi(x). \quad \textbf{(2)}$$

Therefore, $\boxed{\psi(x) = Ae^{ikx} \text{ is an eigenfunciton of } p_x \text{ with eigenvalue } k\hbar}$.

The operator for p_x^2 is

$$p_x^2 = \frac{\hbar}{i}\frac{\partial}{\partial x}\frac{\hbar}{i}\frac{\partial}{\partial x} = -\hbar^2\frac{\partial^2}{\partial x^2}. \quad \textbf{(3)}$$

If $\psi(x) = Ae^{ikx}$ is an eigenfunction of p_x^2, then

$$p_x^2\psi(x) = -\hbar^2\frac{\partial^2}{\partial x^2}Ae^{ikx} = (\text{constant})\psi(x). \quad \textbf{(4)}$$

Performing the indicated operations, we obtain

$$p_x^2\psi(x) = -\hbar^2\frac{\partial^2}{\partial x^2}Ae^{ikx} = -\hbar^2 A(ik)^2 e^{ikx} = k^2\hbar^2 Ae^{ikx} = k^2\hbar^2\psi(x). \quad \textbf{(5)}$$

Therefore, $\boxed{\psi(x) = Ae^{ikx} \text{ is an eigenfunction of } p_x^2 \text{ with eigenvalue } k^2\hbar^2}$.

When this situation holds, the distribution of p_x values will be a delta function (only one value of p_x will be possible), and we will have

$$\langle p_x\rangle = k\hbar \quad \textbf{(6)}$$

and

$$\langle p_x^2\rangle = k^2\hbar^2, \quad \textbf{(7)}$$

so that the uncertainty in p_x is

$$\langle \Delta p_x^2 \rangle^{1/2} = [\langle p_x^2 \rangle - \langle p_x \rangle^2]^{1/2} = [k^2\hbar^2 - k^2\hbar^2] = 0. \quad \textbf{(8)}$$

12.63 I was ready to quit and go to the next chapter on the electronic structure of atoms, but Sam wants to present one more problem to the class. He notes that the quantum solution for the ground state of a particle in a one-dimensional, infinite potential well of width 1 Å has the form shown in the following plot: Sam notes that at the points $x = 0.0$ and $x = 1.0$ Å, $\psi(x)$ has discontinuities (cusps), such as the one shown in Figure 11.16. But the first postulate forbids wave functions to have such discontinuities! Sam wishes to know if a mistake has been made in solving the Schrödinger equation. If so, he would like you to correct this mistake and present the right solution. If not, he would like you to explain why the first postulate is apparently violated. Has quantum theory fallen apart? Should we prepare a letter to the Nobel Committee in Sweden suggesting that the Nobel prizes awarded to Schrödinger and Heisenberg be posthumously rescinded?

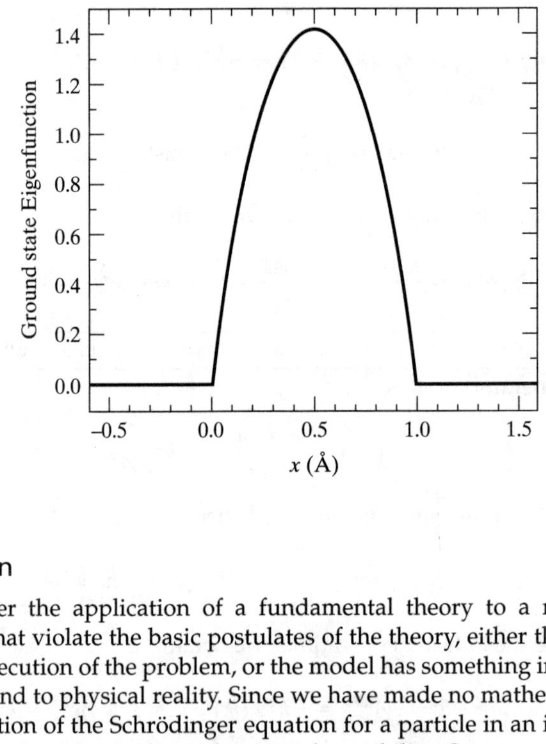

Solution

Whenever the application of a fundamental theory to a model produces results that violate the basic postulates of the theory, either there is a mistake in the execution of the problem, or the model has something in it that does not correspond to physical reality. Since we have made no mathematical errors in our solution of the Schrödinger equation for a particle in an infinite potential well, the answer must lie in the second possibility. Quantum theory is telling us that something is wrong with the model for a particle in an infinite potential well. Something in the model cannot represent the real physical world. We might think that it is the assumption of a one-dimensional system; however, this cannot be the answer, because the same discontinuity persists for the three-dimensional system. The problem is the assumption of infinite potential barriers at $x = 0$ and $x = a$. Such infinite potential barriers cannot exist. If we make the barriers extremely large, but finite, we can impose boundary conditions at $x = 0$ and $x = a$ on both the eigenfunction and its derivative, so that no discontinuities will exist. An example of this procedure may be seen by examining the solution to the tunneling problem for a one-dimensional square potential barrier. In this case, the barrier was finite and no discontinuities were present. The problem of a particle in a finite potential well is more difficult to solve, but the solutions will obey the first postulate, and in the limit of an extremely large barrier, the results will approach those we have obtained by assuming an infinite barrier.

Consequently, a letter to the Sweden Academy of Science suggesting that the Nobel prizes awarded to Schrödinger and Heisenberg be rescinded is probably not a good idea.

CHAPTER 13

Electronic Structure of Atoms

13.1 The Bohr model of the hydrogen atom [see Eq. 11.41] predicts the hydrogen atom ground-state energy to be $E = -e^2/(8\pi\varepsilon_o a_o)$, where e is in coulombs and a_o is the Bohr radius.

(A) What system of units is being used in this equation?

(B) What is the equivalent expression in electrostatic units?

(C) What is the value of the ground-state energy of the Bohr hydrogen atom in hartrees? in eV?

Solution

(A) SI units are being used. We know this because the vacuum permittivity appears in the equation.

(B) In electrostatic units, e will be in statC, and a factor of $4\pi\varepsilon_o$ will be missing from the denominator. Therefore, in electrostatic units, the result is

$$E = -\frac{e^2}{2a_o}. \tag{1}$$

(C) In atomic units, we have $e = 1$ and $a_o = 1$. Thus,

$$E = -\frac{(1)^2}{2(1)} = -\frac{1}{2}\text{ hartree}. \tag{2}$$

In eV, this result is

$$E = -(0.5\text{ hartrees}) \times \frac{27.211\text{ eV}}{\text{hartree}} = \underline{-13.606\text{ eV}}. \tag{3}$$

13.3 Molecular dynamics is the study of the motion of atoms and molecules in chemical reactions. It turns out that the atomic unit of energy is too large and the atomic mass unit is too small to be convenient for molecular dynamics calculations. A more commonly used set of units in such calculations defines the mass of the hydrogen atom to be equal to its atomic mass expressed in g mol^{-1}, 1.0079. With this choice, the masses of the other atoms are their respective atomic masses. The unit of distance is taken to be the angstrom (10^{-10} m), and the energy unit is the electron volt. These units are sometimes called molecular units.

(A) Determine the value of one molecular time unit in seconds.

(B) The H_2 vibration frequency is 1.350×10^{14} s^{-1}. What is its vibrational frequency in molecular time units?

(C) What is the H_2 vibrational period in molecular time units?

Solution

(A) We must satisfy the relationship $E = 0.5mv^2 = 0.5m[d/t]^2$. Therefore,

$$0.5\text{ energy units} = 0.5(1\text{ mass unit})\left[\frac{(1\text{ distance unit})}{(1\text{ time unit})}\right]^2. \tag{1}$$

Solving for the time unit, we obtain

$$1\text{ time unit} = \frac{(1\text{ mass unit})^{1/2}(1\text{ distance unit})}{(1\text{ energy unit})^{1/2}}. \tag{2}$$

Also,

$$1 \text{ mass unit} = 1.0079 \text{ g mol}^{-1} \times \frac{1 \text{ kg}}{1{,}000 \text{ g}} \times \frac{1}{6.022 \times 10^{23} \text{ mol}^{-1}}$$

$$= 1.674 \times 10^{-27} \text{ kg}, \qquad (3)$$

$$1 \text{ distance unit} = 10^{-10} \text{ m}, \qquad (4)$$

and

$$1 \text{ energy unit} = 1 \text{ eV} \times \frac{1 \text{ hartree}}{27.211 \text{ eV}} \times \frac{4.35975 \times 10^{-18} \text{ J}}{\text{hartree}} = 1.6022 \times 10^{-19} \text{ J}. \qquad (5)$$

Substituting Eqs. 3, 4, and 5 into Eq. 2 gives

$$1 \text{ time unit} = \frac{(1.674 \times 10^{-27} \text{ kg})^{1/2}(10^{-10} \text{ m})}{(1.6022 \times 10^{-19} \text{ kg m}^2 \text{ s}^{-2})^{1/2}} = \underline{1.022 \times 10^{-14} \text{ s}}. \qquad (6)$$

(B) The vibration frequency of H_2 is

$$\nu = 1.350 \times 10^{14} \text{ s}^{-1} \times \frac{1.022 \times 10^{-14} \text{ s}}{\text{time unit}} = \underline{1.380 \text{ (time units)}^{-1}}. \qquad (7)$$

Notice how close to unity the result is.

(C) The vibrational period is the reciprocal of the frequency. Therefore,

$$\tau = \frac{1}{1.380 \text{ time units}} = \underline{0.725 \text{ time unit}}. \qquad (8)$$

13.5 Show that the radial function $R(r) = Nr \exp[-br/2]$ is a solution of the radial equation for hydrogen-like atoms, provided that we have $b = Z/a_o$, $\ell = 1$, and $E = -Z^2e^2/(8a_o)$, where $a_o = \hbar^2/(\mu e^2)$, which is the Bohr radius to four significant digits.

Solution

In electrostatic units, the radial equation is

$$\frac{\partial}{\partial r}\left[r^2 \frac{\partial}{\partial r}\right] R(r) + \frac{2\mu r^2}{\hbar^2}\left[\frac{Ze^2}{r} + E\right] R(r) - \ell(\ell+1)R(r) = 0 \qquad (1)$$

for all values of r. Using the suggested radial function, we have

$$\frac{\partial}{\partial r}\left[r^2 \frac{\partial}{\partial r}\right] R(r) = \frac{\partial}{\partial r}\left[r^2 \frac{\partial}{\partial r}\right] Nr \exp\left[-\frac{br}{2}\right] = N \frac{\partial}{\partial r} r^2$$

$$\times \left[\exp\left[-\frac{br}{2}\right] - \frac{br}{2}\exp\left[-\frac{br}{2}\right]\right] = 2rN \exp\left[-\frac{br}{2}\right]\left[1 - \frac{br}{2}\right] + Nr^2$$

$$\times \left[-b \exp\left[-\frac{br}{2}\right] + \frac{b^2 r}{4}\exp\left[-\frac{br}{2}\right]\right] = N \exp\left[-\frac{br}{2}\right]$$

$$\times \left[2r - br^2 - br^2 + \frac{b^2 r^3}{4}\right] = N \exp\left[-\frac{br}{2}\right]\left[2r - 2br^2 + \frac{b^2 r^3}{4}\right]. \qquad (2)$$

Combining Eqs. 1 and 2 gives

$$N \exp\left[-\frac{br}{2}\right]\left[2r - 2br^2 + \frac{b^2r^3}{4}\right] + N \exp\left[-\frac{br}{2}\right]\frac{2Ze^2\mu r^2}{\hbar^2}$$

$$+ N \exp\left[-\frac{br}{2}\right]\frac{2\mu E r^3}{\hbar^2} - N \exp\left[-\frac{br}{2}\right] r\, \ell(\ell + 1) = 0. \qquad (3)$$

We may now factor out and divide by $N \exp[-(br)/2]$ and then collect terms in the various powers of r. The result is

$$r[2 - \ell(\ell + 1)] + r^2\left[-2b + \frac{2Ze^2\mu}{\hbar^2}\right] + r^3\left[\frac{b^2}{4} + \frac{2\mu E}{\hbar^2}\right] = 0. \qquad (4)$$

The only way Eq. 4 can hold for all values of r is for the quantities within brackets to all be zero. Therefore, we have, for the coefficient of the r^1 term,

$$2 - \ell(\ell + 1) = 0. \qquad (5)$$

For this equation to hold, we must have $\boxed{\ell = 1}$, as required. From the coefficient of the r^2 term, we obtain

$$-2b + \frac{2Ze^2\mu}{\hbar^2} = -2b + \frac{2Z}{a_o} = 0, \qquad (6)$$

since $a_o = \hbar^2/(\mu e^2)$. For Eq. 6 to be valid, we must have

$$\boxed{b = \frac{Z}{a_o}}, \qquad (7)$$

as required. From the coefficient of the r^3 term, we obtain

$$\frac{b^2}{4} + \frac{2\mu E}{\hbar^2} = 0, \qquad (8)$$

so that

$$E = -\frac{b^2\hbar^2}{8\mu} = -\frac{\hbar^2}{8\mu}\frac{\mu^2e^4Z^2}{\hbar^4} = -\frac{\mu e^4Z^2}{8\hbar^2} = \boxed{-\frac{Z^2e^2}{8a_o}}, \qquad (9)$$

as required.

13.7 With the approximation $a_o = \hbar^2/(\mu e^2)$, the ground-state energies of H, D, and T are all equal to $-\frac{1}{2}$ hartree. Compute the actual energies of these atoms and the percent error involved in the approximation.

Solution

The correct ground-state energies are given by Eq. 13.19, viz.,

$$E_1 = -\frac{\mu e^4}{2\hbar^2} \qquad (1)$$

since $Z = n = 1$ for the hydrogen, deuterium, and tritium ground states and we are assuming that $\mu = m_e = 1$. The reduced masses for hydrogen, deuterium, and tritium are, respectively,

$$\mu_H = \frac{m_H m_e}{m_H + m_e} = \frac{(1{,}836.1)(1)}{1{,}836.1 + 1} = \frac{1{,}836.1}{1{,}837.1} = 0.99946, \qquad (2)$$

$$\mu_D = \frac{m_D m_e}{m_D + m_e} = \frac{3{,}671.5(1)}{3{,}671.5 + 1} = 0.99973 \qquad (3)$$

and

$$\mu_T = \frac{m_T m_e}{m_T + m_e} = \frac{5{,}497.8(1)}{5{,}497.8 + 1} = 0.99982. \tag{4}$$

The corresponding energies of these atoms are

$$E_H = -\frac{0.99946(1)^4}{2(1)^2} = \underline{-0.49973 \text{ hartree}}, \tag{5}$$

$$E_D = -\frac{0.99973(1)^4}{2(1)^2} = \underline{-0.49986 \text{ hartree}}, \tag{6}$$

and

$$E_T = -\frac{0.99982(1)^4}{2(1)^2} = \underline{-0.49991 \text{ hartree}}. \tag{7}$$

The percent errors in the ground-state energies are, therefore,

$$\% \text{ error for H} = \frac{100 \times (-0.49973 - (-0.5000))}{0.50000} = \underline{0.054\%}, \tag{8}$$

$$\% \text{ error for D} = \frac{100 \times (-0.49986 - (-0.5000))}{0.50000} = \underline{0.028\%}, \tag{9}$$

and

$$\% \text{ error for T} = \frac{100 \times (-0.49991 - (-0.5000))}{0.50000} = \underline{0.018\%}. \tag{10}$$

13.9 Compute the average radial distance of the electron from the nucleus of a hydrogen-like atom or ion excited to the $2p$ eigenstate as a function of Z.

Solution

The average value of r in the $2p$ eigenstate is given by Eq. 13.28:

$$\langle r \rangle = \langle R_2^1(r) | r | R_2^1(r) \rangle. \tag{1}$$

The normalized radial eigenfunction for the $2p$ orbital is given in Table 13.3:

$$R_2^1(r) = \frac{1}{3^{1/2}} \left[\frac{Z}{2a_o}\right]^{3/2} \frac{Zr}{a_o} \exp\left[\frac{-Zr}{2a_o}\right]. \tag{2}$$

In atomic units, this is

$$R_2^1(r) = \frac{1}{3^{1/2}} \left[\frac{Z}{2}\right]^{3/2} Zr \exp\left[\frac{-Zr}{2}\right]. \tag{3}$$

Substituting Eq. 3 into Eq. 1 produces

$$\langle r \rangle = \frac{Z^5}{24} \int_0^\infty (r^2 e^{-Zr/2}) r (r^2 e^{-Zr/2}) r^2 \, dr = \frac{Z^5}{24} \int_0^\infty r^5 e^{-Zr} \, dr. \tag{4}$$

From the fact that

$$\int_0^\infty x^n e^{-ax} \, dx = \frac{n!}{a^{n+1}} \tag{5}$$

if $a \geq 0$ and n is an integer, Eq. 4 becomes

$$\boxed{\langle r \rangle = \frac{Z^5}{24} \frac{5!}{Z^6} = \frac{5}{Z} \text{ bohr} = \frac{2.646 \times 10^{-10} \text{ m}}{Z}}. \tag{6}$$

As expected, $\langle r \rangle$ decreases as the positive charge on the nucleus increases due to the increased attraction between the negatively charged electron and the positively charged nucleus.

13.11 Compute the most probable distance of the electron from the nucleus for the ground state of a hydrogen-like atom or ion as a function of Z, the atomic number of the atom or ion.

Solution

The radial probability distribution is given by Eq. 13.27:

$$P(r)dr = [R_n^\ell(r)]^* R_n^\ell(r) r^2 dr. \quad (1)$$

For the ground state, we have, from Table 13.3,

$$R_1^o(r) = 2\left[\frac{Z}{a_o}\right]^{3/2} \exp\left[\frac{-Zr}{a_o}\right]. \quad (2)$$

In atomic units, Eq. 2 becomes

$$R_1^o(r) = 2Z^{3/2}e^{-Zr}. \quad (3)$$

Substituting into Eq. 1 produces

$$P(r)dr = 4Z^3r^2e^{-2Zr}dr. \quad (4)$$

The most probable position occurs at the point where $P(r)$ attains a maximum value. The condition for a maximum in $P(r)$ is

$$\frac{dP(r)}{dr} = 4Z^3[2re^{-2Zr} - 2Zr^2e^{-2Zr}] = 0. \quad (5)$$

Equation 5 requires that we have

$$8Z^3e^{-2Zr}[r - Zr^2] = 0. \quad (6)$$

Hence,

$$r_{mp} = Zr_{mp}^2, \quad (7)$$

which yields

$$\boxed{r_{mp} = \text{most probably distance} = \frac{1}{Z}\text{ bohrs}}. \quad (8)$$

For hydrogen, $r_{mp} = 1$ bohr $= a_o = 0.5292 \times 10^{-10}$ m. Consequently, the radius of the first Bohr orbit is equal to the most probable radial position of the electron when the problem is treated quantum mechanically.

13.13 Without performing any integrations, compute the average kinetic energy of a hydrogen-like system when it is in the $n = 5$, $\ell = 2$, and $m = -1$ eigenstate.

Solution

Equation 13.20C shows the energy to be

$$E_5 = -\frac{Z^2}{2(5)^2} = -0.02000\, Z^2 \text{ hartree}, \quad (1)$$

in atomic units. This energy is related to the average potential and kinetic energies by Eq. 13.31:

$$E_5 = \langle V \rangle + \langle T \rangle \tag{2}$$

for the $n = 5$ state. The virial theorem tells us that

$$\langle V \rangle = -2\langle T \rangle. \tag{3}$$

Substituting Eq. 3 into Eq. 2 gives

$$E_5 = -2\langle T \rangle + \langle T \rangle = -\langle T \rangle. \tag{4}$$

Therefore, using the result in Eq. 1, we obtain

$$\boxed{\langle T \rangle = -E_5 = 0.02000\, Z^2 = \frac{Z^2}{50} \text{ hartree}}. \tag{5}$$

Note that it was not necessary to execute any integrals or even know the eigenfunctions for the eigenstate in question. Such is the power of the virial theorem.

13.15# In this problem, we shall investigate a problem similar to that of hydrogen-like atoms and ions. Consider a single particle of mass m in a spherical potential well whose radius is R_o. Let the potential within the well be zero and infinite outside the well. That is,

$$V(r) = 0 \quad \text{for } r \le R_o \quad \text{and } V(r) = \infty \quad \text{for } r > R_o.$$

Since the potential for $r > R_o$ is infinite, we know that we must have $\psi(r, \theta, \phi) = 0$ for $r > R_o$.

(A) Set up the Schrödinger equation for this system when $r \le R_o$ (inside the potential well), and show that the variables can be separated.

(B) Obtain the angular eigenfunctions for this system and the resulting form of the radial differential equation $R(r)$ that must be solved to obtain the radial eigenfunctions.

(C) Using the substitution $R(r) = F(r)/r$, make use of the first postulate to obtain the eigenfunctions and energy eigenvalues for the spherically symmetric case ($\ell = 0$). [*Hint:* At $r = 0$, we must have $F(r) = 0$ to avoid having an infinity in the wave function.]

(D) Plot $R(r)/A$ (A is the normalization constant) versus r if $R_o = 10$ bohr for the four lowest energy eigenstates.

(E) How is the number of radial nodes related to the principal quantum number for this system? (For a very similar example, see Problem 9.7.)

Solution

(A) The Schrödinger equation is identical to Eq. 13.8, except for the fact that the potential energy is now zero. Therefore, the Hamiltonian is given by

$$\mathcal{H} = -\frac{\hbar^2}{2m}\nabla^2. \tag{1}$$

The Schrödinger equation to be solved is

$$-\frac{\hbar^2}{2m}\left[\frac{1}{r^2}\frac{\partial}{\partial r}\left[r^2\frac{\partial}{\partial r}\right] + \frac{1}{r^2 \sin\theta}\frac{\partial}{\partial\theta}\left[\sin\theta\frac{\partial}{\partial\theta}\right] + \frac{1}{r^2\sin^2\theta}\frac{\partial^2}{\partial\phi^2}\right]\psi(r,\theta,\phi)$$
$$= E\psi(r,\theta,\phi). \tag{2}$$

To separate the variables, we assume that a solution of the form

$$\psi(r,\theta,\phi) = R(r)G(\theta,\phi) \tag{3}$$

exists. Following the procedure used for the hydrogen atom, we multiply by r^2 and rearrange terms to obtain

$$-\frac{\hbar^2}{2m}\left[\frac{\partial}{\partial r}\left[r^2\frac{\partial}{\partial r}\right]+\frac{1}{\sin\theta}\frac{\partial}{\partial\theta}\left[\sin\theta\frac{\partial}{\partial\theta}\right]+\frac{1}{\sin^2\theta}\frac{\partial^2}{\partial\phi^2}\right]\psi(r,\theta,\phi)$$
$$-r^2E\psi(r,\theta,\phi)=0. \quad (4)$$

Substituting Eq. 3 into Eq. 4 and then factoring those functions not affected by the differential operators to the left yields

$$G(\theta,\phi)\left\{-\frac{\hbar^2}{2m}\left[\frac{\partial}{\partial r}\left[r^2\frac{\partial}{\partial r}\right]\right]-r^2E\right\}R(r)$$
$$+R(r)\left\{-\frac{\hbar^2}{2m}\left[\frac{1}{\sin\theta}\frac{\partial}{\partial\theta}\left[\sin\theta\frac{\partial}{\partial\theta}\right]+\frac{1}{\sin^2\theta}\frac{\partial^2}{\partial\phi^2}\right]\right\}G(\theta,\phi)=0. \quad (5)$$

We now divide both sides of Eq. 5 by $R(r)\,G(\theta,\phi)$, giving

$$\frac{\left\{-\frac{\hbar^2}{2m}\left[\frac{\partial}{\partial r}\left[r^2\frac{\partial}{\partial r}\right]\right]-r^2E\right\}R(r)}{R(r)}$$
$$+\frac{\left\{-\frac{\hbar^2}{2m}\left[\frac{1}{\sin\theta}\frac{\partial}{\partial\theta}\left[\sin\theta\frac{\partial}{\partial\theta}\right]+\frac{1}{\sin^2\theta}\frac{\partial^2}{\partial\phi^2}\right]\right\}G(\theta,\phi)}{G(\theta,\phi)}=0. \quad (6)$$

The first term in Eq. 6 depends only upon the radial coordinate r, while the second term is a function only of the angular coordinates θ and ϕ. The variables have been separated, and a solution of the form of Eq. 3 exists.

(B) The only way Eq. 6 can hold for all values of r, θ, and ϕ is for the first term to be equal to a constant, say, $-K$, while the second term is equal to $+K$, so that the sum is always zero regardless of the values of the variables. This requirement produces two differential equations. The first depends only upon r and is, therefore, called the radial equation:

$$\left\{-\frac{\hbar^2}{2m}\left[\frac{\partial}{\partial r}\left[r^2\frac{\partial}{\partial r}\right]\right]-r^2E\right\}R(r)=-KR(r). \quad (7)$$

The second equation involves only the angular variables:

$$\left\{-\frac{\hbar^2}{2m}\left[\frac{1}{\sin\theta}\frac{\partial}{\partial\theta}\left[\sin\theta\frac{\partial}{\partial\theta}\right]+\frac{1}{\sin^2\theta}\frac{\partial^2}{\partial\phi^2}\right]\right\}G(\theta,\phi)=KG(\theta,\phi). \quad (8)$$

As we found in our solution of the hydrogen-atom problem, Eq. 8 is effectively the same as the differential equation for the rigid rotor. The eigenfunctions are, therefore,

$$\boxed{G(\theta,\phi)=Y_\ell^m(\theta,\phi)}, \quad (9)$$

and the value of K is

$$\boxed{K=\frac{\ell(\ell+1)\hbar^2}{2m}}, \quad (10)$$

where we have replaced the reduced mass for the two-particle system with the mass of our single particle. Substituting Eq. 10 into Eq. 7 yields the radial equation:

$$\left\{-\frac{\hbar^2}{2m}\left[\frac{\partial}{\partial r}\left[r^2\frac{\partial}{\partial r}\right]\right]-r^2E\right\}R(r)=-\frac{\ell(\ell+1)\hbar^2}{2m}R(r). \quad (11)$$

Multiplying both sides by $-2m/\hbar^2$ gives

$$\boxed{\left[\frac{\partial}{\partial r}\left[r^2\frac{\partial}{\partial r}\right]R(r) + \frac{2mEr^2}{\hbar^2}R(r) - \ell(\ell+1)R(r) = 0\right]}. \quad \textbf{(12)}$$

Equation 12 is the radial equation for a particle in a spherical infinite potential well.

(C) For the spherically symmetric case, $\ell = 0$, and Eq.12 becomes

$$\left[\frac{\partial}{\partial r}\left[r^2\frac{\partial}{\partial r}\right]R(r) + \frac{2mEr^2}{\hbar^2}R(r) = 0. \quad \textbf{(13)}$$

We now let

$$R(r) = \frac{F(r)}{r} \quad \textbf{(14)}$$

and seek a solution for $F(r)$. The first term in Eq. 13 becomes

$$\left[\frac{\partial}{\partial r}\left[r^2\frac{\partial}{\partial r}\right]\frac{F(r)}{r} = \frac{\partial}{\partial r}r^2\left[-\frac{F(r)}{r^2} + \frac{1}{r}\frac{\partial F(r)}{\partial r}\right] = \frac{\partial}{\partial r}\left[-F(r) + r\frac{\partial F(r)}{\partial r}\right]$$

$$= -\frac{\partial F(r)}{\partial r} + \frac{\partial F(r)}{\partial r} + r\frac{\partial^2 F(r)}{\partial r^2} = r\frac{\partial^2 F(r)}{\partial r^2}. \quad \textbf{(15)}$$

Substituting the result in Eq. 15 into Eq. 13 and replacing $R(r)$ in the second term with $F(r)/r$, we get

$$r\frac{\partial^2 F(r)}{\partial r^2} + \frac{2mEr^2}{\hbar^2}\frac{F(r)}{r} = r\frac{\partial^2 F(r)}{\partial r^2} + \frac{2mEr}{\hbar^2}F(r) = 0. \quad \textbf{(16)}$$

Dividing both sides of Eq. 16 by r gives

$$\frac{\partial^2 F(r)}{\partial r^2} + \frac{2mE}{\hbar^2}F(r) = 0. \quad \textbf{(17)}$$

Now let the constant $2mE/\hbar^2 = k^2$. This produces

$$\frac{\partial^2 F(r)}{\partial r^2} + k^2 F(r) = 0. \quad \textbf{(18)}$$

In this form, Eq. 18 is exactly the same equation we solved for the particle in a one-dimensional infinite potential well and for the ϕ part of the rigid rotor. The solution is

$$F(r) = A\sin(kr) + B\cos(kr). \quad \textbf{(19)}$$

Since $R(r) = F(r)/r$, we must have $F(r) = 0$ at $r = 0$; otherwise $R(r)$ will have an infinite value at $r = 0$, which is not permitted by the first postulate. Therefore,

$$F(r = 0) = A\sin(0) + B\cos(0) = B = 0. \quad \textbf{(20)}$$

With $B = 0$, $F(r)$ must have the form

$$F(r) = A\sin(kr). \quad \textbf{(21)}$$

Finally, at the point $r = R_o$, we must have $R(r = R_o) = 0$, because the wave function outside the well is zero and the wave function must be continuous at the boundary. Therefore,

$$F(r = R_o) = A\sin(kR_o) = 0. \quad \textbf{(22)}$$

We do not want the trivial solution, so we must have

$$kR_o = n\pi, \tag{23}$$

where $n = 1, 2, 3, \ldots$.

As a result, $k = n\pi/R_o$ and

$$F(r) = A \sin\left[\frac{n\pi r}{R_o}\right]. \tag{24}$$

The radial eigenfunctions are

$$\boxed{R(r) = \frac{F(r)}{r} = \frac{A \sin\left[\dfrac{n\pi r}{R_o}\right]}{r}}. \tag{25}$$

The eigenvalues are given by squaring both sides of Eq. 23:

$$k^2R_o^2 = n^2\pi^2 = \frac{2mE}{\hbar^2}R_o^2. \tag{26}$$

Solving for E, we obtain

$$\boxed{E_n = \frac{n^2\pi^2\hbar^2}{2mR_o^2} = \frac{n^2h^2}{8mR_o^2}} \tag{27}$$

for $n = 1, 2, 3 \ldots$.

The eigenfunctions are given by Eq. 25 and the corresponding eigenvalues by Eq. 27. It is interesting to note that the translational-energy eigenvalues have exactly the same form for the spherical well, with $\ell = 0$, as for the one-dimensional infinite potential well.

(D) Graphs of $R(r)/A = \sin[n\pi r/R_o]/r$ with $R_o = 10$ bohr for $n = 1, 2, 3$, and 4 are shown in the accompanying plot.

(E) Since the $\ell = 0$ eigenfunctions have no angular nodes, all the nodes must lie along the radial coordinate. Therefore, we expect $n - 1$ radial nodes.

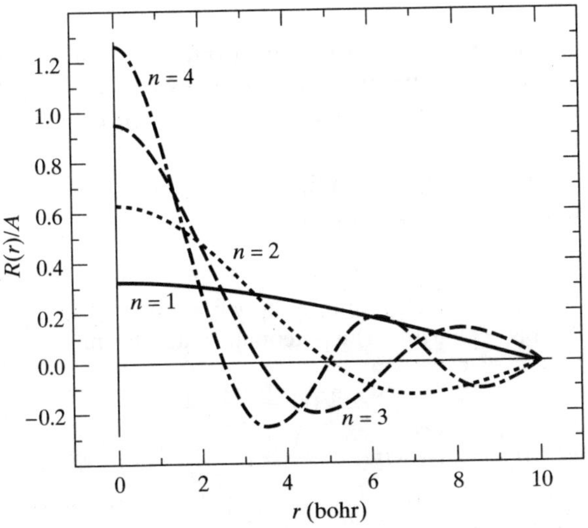

13.17 The normalized eigenfunctions for a particle in an infinite spherical well whose radius is R_o are

$$\psi(r,\theta,\phi) = R_n^0(r)\, Y_0^0(\theta,\phi) \text{ for } r \le R_o$$
$$\text{and } \psi(r,\theta,\phi) = 0 \text{ for } r > R_o.$$

See Problems 13.15 and 13.16. $Y_0^0(\theta,\phi)$ is a spherical harmonic equal to $(4\pi)^{-1/2}$, and $R_n^0(r) = [2/R_o]^{1/2}(\sin[n\pi r/R_o])/r$. Compute the average radial position of the particle, $\langle r\rangle$, as a function of the quantum number n.

Solution

The average radial position is given by

$$\langle r\rangle = \langle R_n^0(r)|r|R_n^0(r)\rangle = \left[\frac{2}{R_o}\right]\int_{r=0}^{r=R_o} \frac{\sin\left[\frac{n\pi r}{R_o}\right]}{r}(r)\frac{\sin\left[\frac{n\pi r}{R_o}\right]}{r} r^2\,dr, \tag{1}$$

or

$$\langle r\rangle = \left[\frac{2}{R_o}\right]\int_{r=0}^{r=R_o} r\sin^2\left[\frac{n\pi r}{R_o}\right]dr. \tag{2}$$

The integrand appearing in Eq. 2 is a standard form whose integral is

$$\int x\sin^2(ax)\,dx = \frac{x^2}{4} - \frac{x\sin(2ax)}{4a} - \frac{\cos(2ax)}{4a^2}. \tag{3}$$

Equation 2 has the same form as Eq. 3 if we set $a = n\pi/R_o$. Therefore,

$$\begin{aligned}\langle r\rangle &= \left[\frac{2}{R_o}\right]\left[\frac{r^2}{4} - \frac{r\sin(2n\pi r/R_o)}{4n\pi/R_o} - \frac{\cos(2n\pi r/R_o)}{4n^2\pi^2/R_o^2}\right]_0^{R_o}\\ &= \left[\frac{2}{R_o}\right]\left[\frac{R_o^2}{4} - 0 - \frac{R_o^2}{4n^2\pi^2} - \left\{0 - 0 - \frac{R_o^2}{4n^2\pi^2}\right\}\right],\end{aligned} \tag{4}$$

or

$$\boxed{\langle r\rangle = \frac{R_o}{2}}, \tag{5}$$

and the average radial position is a constant that is independent of the quantum state of the system.

13.19 Just as a harmonic oscillator can tunnel into classically forbidden regions where $V > E$, the electron can do the same in hydrogen-like systems.

(A) Determine the classical turning point for the hydrogen-atom electron in its ground state.

(B) Compute the probability that the electron will be at a value of r equal to or greater than its classical turning point.

Solution

(A) The ground-state energy of a hydrogen atom is given by Eq. 13.20C. In atomic units, it is $-\frac{1}{2}$ hartree. The potential, in atomic units, is

$$V(r) = -\frac{1}{r}. \tag{1}$$

We will have $V(r) > E$, when

$$-\frac{1}{r} > -\frac{1}{2}. \tag{2}$$

Therefore, the classical turning point is at $\underline{r = 2 \text{ bohr}}$.

(B) Table 13.3 indicates that the radial eigenfunction for the hydrogen atom in its ground state is

$$R_1^0(r) = 2e^{-r}. \tag{3}$$

The radial probability distribution function is

$$P(r)dr = [R_1^0(r)]^* R_1^0(r) r^2 dr = 4r^2 e^{-2r} dr. \tag{4}$$

The probability that the electron will be in the classically forbidden region is, therefore, given by

$$P(r \geq 2) = \int_2^\infty 4r^2 e^{-2r} dr = 4 \int_2^\infty r^2 e^{-2r} dr. \tag{5}$$

Equation 5 can be integrated by parts, or its solution can be found in a standard table. The result is

$$P(r \geq 2) = -4e^{-2r}\left[\frac{r^2}{2} + \frac{r}{2} + \frac{1}{4}\right]_2^\infty = 4e^{-4}\left[2 + 1 + \frac{1}{4}\right] = 13e^{-4} = \underline{0.2381}. \tag{6}$$

13.21 In Table 13.3, the $R_2^1(r)$ radial eigenfunction is shown to have the form

$$R_2^1(r) = N\frac{Zr}{a_o} e^{-Zr/a_o}$$

Show that the normalization constant for this eigenfunction is that given in the table.

Solution

Normalization requires that we have

$$\int_0^\infty [R_2^1(r)]^* R_2^1(r) r^2 dr = 1. \tag{1}$$

Substituting $R_2^1(r)$ into Eq. 1 gives

$$\frac{N^2 Z^2}{a_o^2} \int_0^\infty r^4 e^{-Zr/a_o} dr = 1. \tag{2}$$

Using the fact that $\int_0^\infty x^n e^{-ax} dx = n!/a^{n+1}$ for integer n and $a \geq 0$, we have

$$\frac{N^2 Z^2}{a_o^2}\left[\frac{4!}{(Z/a_o)^5}\right] = \frac{24 N^2 Z^2 a_o^5}{Z^5 a_o^2} = N^2 24\left[\frac{a_o^3}{Z^3}\right] = 1. \tag{3}$$

Solving for N, we obtain

$$\boxed{N = \left[\frac{Z^3}{24a_o^3}\right]^{1/2} = \frac{1}{3^{1/2}}\left[\frac{Z^3}{2^3 a_o^3}\right]^{1/2} = \frac{1}{3^{1/2}}\left[\frac{Z}{2a_o}\right]^{3/2}}. \tag{4}$$

The result in Eq. 4 is identical to that given in Table 13.3.

13.23 A hydrogen atom is in the ground state with energy $-\frac{1}{2}$ hartree. If this hydrogen atom were subjected to an applied external magnetic field of 10^5 gauss, by how much would the 1s energy state of the atom change?

Solution

The energy of the system would change because of the anomalous Zeeman effect—the interaction of the applied magnetic field with the magnetic moment produced by the spin angular momentum of the electron. The energy change is given by Eq. 13.40, viz.,

$$\Delta E_{\text{mag}} = -\gamma_s S_z \mathcal{B}, \tag{1}$$

where

$$\gamma_s = -\frac{e}{m_e c} \tag{2}$$

and $S_z = m_s \hbar$. There are two possible values of m_s: $-\frac{1}{2}$ and $+\frac{1}{2}$. Therefore,

$$\Delta E_{\text{mag}} = -\left[-\frac{e}{m_e c}\right]\hbar m_s \mathcal{B}, \tag{3}$$

or

$$\Delta E_{\text{mag}} = \left[\frac{e\hbar}{m_e c}\right] m_s \mathcal{B}. \tag{4}$$

The quantity in brackets in Eq. 4 is twice the Bohr magneton. Hence,

$$\Delta E_{\text{mag}} = 2\mu_B m_s \mathcal{B}. \tag{5}$$

The nondegenerate ground state splits into two energy states, one with $m_s = +\frac{1}{2}$ and the other with $m_s = -\frac{1}{2}$. The magnitude of the energy change is

$$2\mu_B \mathcal{B}\left(\frac{1}{2}\right) = \mu_B \mathcal{B} = 2.1270 \times 10^{-10} \text{ hartree gauss}^{-1}\,(10^5 \text{ gauss})$$

$$= \underline{2.1270 \times 10^{-5} \text{ hartree}}. \tag{6}$$

Therefore, we have one energy state with energy $-0.5000 + 2.127 \times 10^{-5} = \underline{-0.4999787 \text{ hartree}}$ and a second with energy $-0.500000 - 2.127 \times 10^{-5} = \underline{-0.50002127 \text{ hartree}}$.

13.25 Repeat Problem 13.24 for a hydrogen atom in an excited $4f$ eigenstate.

Solution

(A) In a $4f$ eigenstate, we have $\boxed{\ell = 3}$, so that m can have values $\boxed{3, 2, 1, 0, -1, -2, \text{and } -3}$.

(B) The possible values of m_s are $\pm\frac{1}{2}$. Consequently, the possible combinations of m and m_s that produce m_j by scalar addition ($m_j = m + m_s$) are given in the table to the left.

m	m_s	$m_j = m + m_s$
3	$\frac{1}{2}$	$\frac{7}{2}$
3	$-\frac{1}{2}$	$\frac{5}{2}$
2	$\frac{1}{2}$	$\frac{5}{2}$
2	$-\frac{1}{2}$	$\frac{3}{2}$
1	$\frac{1}{2}$	$\frac{3}{2}$
1	$-\frac{1}{2}$	$\frac{1}{2}$
0	$\frac{1}{2}$	$\frac{1}{2}$
0	$-\frac{1}{2}$	$-\frac{1}{2}$
−1	$\frac{1}{2}$	$-\frac{1}{2}$
−1	$-\frac{1}{2}$	$-\frac{3}{2}$
−2	$\frac{1}{2}$	$-\frac{3}{2}$
−2	$-\frac{1}{2}$	$-\frac{5}{2}$
−3	$\frac{1}{2}$	$-\frac{5}{2}$
−3	$-\frac{1}{2}$	$-\frac{7}{2}$

Since we can have $m_j = \pm\frac{7}{2}$, we must have a state with $\boxed{j = \frac{7}{2}}$. In this state, m_j can have values of $\frac{7}{2}, \frac{5}{2}, \frac{3}{2}, \frac{1}{2}, -\frac{1}{2}, -\frac{3}{2}, -\frac{5}{2}$, and $-\frac{7}{2}$. If these eight m_j states are eliminated from the table, we have 6 m_j states remaining, whose values are $\frac{5}{2}, \frac{3}{2}, \frac{1}{2}, -\frac{1}{2}, -\frac{3}{2}$, and $-\frac{5}{2}$. Therefore, we must have a second total-angular-momentum state with $\boxed{j = \frac{5}{2}}$.

(C) With $j = \frac{7}{2}$ or $\frac{5}{2}$, the two possible magnitudes of **J** are

$$\boxed{J = \left[\frac{7}{2}\left(\frac{7}{2} + 1\right)\hbar^2\right]^{1/2} = \frac{\sqrt{63}}{2}\hbar} \tag{1}$$

and

$$\boxed{J = \left[\frac{5}{2}\left(\frac{5}{2}+1\right)\hbar^2\right]^{1/2} = \frac{\sqrt{35}}{2}\hbar}. \quad \textbf{(2)}$$

(D) All angular momentum states have $L = \ell = 3$, so the L symbol will be F. All states have a total spin of $\frac{1}{2}$, so $2S + 1 = 2$ in all cases. The principal quantum number is 4; therefore, the term symbols for the two possible angular-momentum states are $\boxed{4\,^2F_{7/2}}$ and $\boxed{4\,^2F_{5/2}}$.

(E) The angle **J** makes with the Z-axis can be computed from the fact that

$$J_z = J\cos\theta. \quad \textbf{(3)}$$

Therefore,

$$\theta = \cos^{-1}\left[\frac{J_z}{J}\right] = \cos^{-1}\left[\frac{m_j\hbar}{\{j(j+1)\}^{1/2}\hbar}\right] = \cos^{-1}\left[\frac{m_j\hbar}{\{j(j+1)\}^{1/2}}\right]. \quad \textbf{(4)}$$

For the $j = \frac{5}{2}$, $m_j = \frac{5}{2}$ angular-momentum state, we have

$$\theta = \cos^{-1}\left[\frac{5/2}{\left\{\frac{5}{2}\left(\frac{5}{2}+1\right)\right\}^{1/2}}\right] = \cos^{-1}\left[\frac{5}{\{35\}^{1/2}}\right] = \cos^{-1}(0.84515) = \underline{32.31^\circ}. \quad \textbf{(5)}$$

Similar calculations for each of the 14 possible total-angular-momentum states yield the results given in the table to the right.

j	m_j	θ (deg)
$\frac{7}{2}$	$\frac{7}{2}$	28.13
$\frac{7}{2}$	$\frac{5}{2}$	50.95
$\frac{7}{2}$	$\frac{3}{2}$	67.79
$\frac{7}{2}$	$\frac{1}{2}$	82.76
$\frac{7}{2}$	$-\frac{1}{2}$	97.24
$\frac{7}{2}$	$-\frac{3}{2}$	112.21
$\frac{7}{2}$	$-\frac{5}{2}$	129.05
$\frac{7}{2}$	$-\frac{7}{2}$	151.87
$\frac{5}{2}$	$\frac{5}{2}$	32.31
$\frac{5}{2}$	$\frac{3}{2}$	59.53
$\frac{5}{2}$	$\frac{1}{2}$	80.27
$\frac{5}{2}$	$-\frac{1}{2}$	99.73
$\frac{5}{2}$	$-\frac{3}{2}$	120.47
$\frac{5}{2}$	$-\frac{5}{2}$	147.69

13.27 Write down the Hamiltonian for the lithium atom. Identify which term or terms are omitted when we make the Born–Oppenheimer approximation. Which terms are one-electron terms? How many two-electron terms are present? Which terms are omitted by the one-electron approximation?

Solution

There is one kinetic-energy term for the nucleus: $-[\hbar^2/(2m_{\text{Li}})]\,\nabla^2_{\text{Li}}$. Each of the three electrons has a kinetic-energy term, $-[\hbar^2/(2m_e)]\,\nabla^2_i$, and an electron–nuclear attraction, whose form is $-3e^2/r_i$, since the nuclear charge on Li is $+3e$. Finally, there are three electron–electron repulsion terms. Therefore, the complete Hamiltonian, in electrostatic units, is

$$\boxed{\mathcal{H} = -\frac{\hbar^2}{2m_{\text{Li}}}\nabla^2_{\text{Li}} - \frac{\hbar^2}{2m_e}\nabla^2_1 - \frac{\hbar^2}{2m_e}\nabla^2_2 - \frac{\hbar^2}{2m_e}\nabla^2_3 - \frac{3e^2}{r_1} - \frac{3e^2}{r_2} - \frac{3e^2}{r_e} + \frac{e^2}{r_{12}} + \frac{e^2}{r_{13}} + \frac{e^2}{r_{23}}}. \quad \textbf{(1)}$$

When we make the Born–Oppenheimer approximation, the kinetic-energy term for the nucleus, $-[\hbar^2/(2m_{\text{Li}})]\nabla^2_{\text{Li}}$, is omitted. The electron-kinetic-energy and electron–nuclear-attraction terms make up the one-electron terms, namely, $-[\hbar^2/(2m_e)]\nabla^2_1$, $-[\hbar^2/(2m_e)]\nabla^2_2$, $-[\hbar^2/(2m_e)]\nabla^2_3$, $-3e^2/r_1$, $-3e/r_2$, and $-3e^2/r_3$. In the one-electron approximation, we omit the three two-electron terms, e^2/r_{12}, e^2/r_{13}, and e^2/r_{23}.

13.29 Suppose we are in a universe in which the spin quantum number for the electron is $s = \frac{5}{2}$ instead of $s = \frac{1}{2}$, as it is in our universe.

(A) Describe in qualitative terms what the results of a Stern–Gerlach type of experiment on the hydrogen atom would be in the hypothetical universe.

(B) What are the possible values of S_z for the electron in that universe?

(C) What is the value of S^2 for the electron in that universe?

(D) What are the possible angles the spin angular momentum vector can make with the z-axis in the assumed universe?

(E) With $s = \frac{1}{2}$, there are 10 elements contained in each transition metal series in the periodic table. How many would there be if s for the electron were $\frac{5}{2}$?

Solution

(A) Since we now have $s = \frac{5}{2}$ and $-s \le m_s \le +s$, m_s must move by integer steps between $-\frac{5}{2} \le m_s \le +\frac{5}{2}$. Therefore, the possible values of m_s are $-\frac{5}{2}, -\frac{3}{2}, -\frac{1}{2}, \frac{1}{2}, \frac{3}{2}$ and $\frac{5}{2}$. Since there are six possible values of m_s, there will be six possible values of S_z and, therefore, six different directions for the spin-angular-momentum vector. Each of these will interact with the inhomogeneous magnetic field in a different manner. Consequently, the beam in a Stern–Gerlach type of experiment will split into six different components, each with equal intensity. Three of these components will be deflected toward higher magnetic field strength, while the remaining three are deflected toward the region of lower magnetic field strength.

(B) The values of the Z component of the spin angular momentum are given by

$$S_z = m_s \hbar. \tag{1}$$

Therefore, the possible values of S_z are

$$\boxed{S_z = \frac{5\hbar}{2}, \frac{3\hbar}{2}, \frac{\hbar}{2}, -\frac{\hbar}{2}, -\frac{3\hbar}{2}, \text{ and } -\frac{5\hbar}{2}}. \tag{2}$$

(C) The total spin angular momentum is given by

$$\boxed{S = [s(s+1)]^{1/2}\hbar = \left[\frac{5}{2}\left(\frac{5}{2}+1\right)\right]^{1/2}\hbar = \frac{\sqrt{35}}{2}\hbar}. \tag{3}$$

(D) The relationship between **S** and S_z is shown in the following diagram:

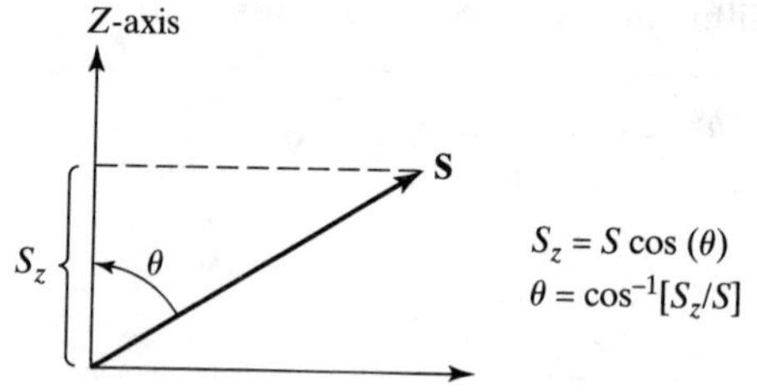

m_s	θ (deg)
$\frac{5}{2}$	32.31
$\frac{3}{2}$	59.53
$\frac{1}{2}$	80.27
$-\frac{1}{2}$	99.73
$-\frac{3}{2}$	120.47
$-\frac{5}{2}$	147.69

Therefore,

$$\theta = \cos^{-1}\left[\frac{S_z}{S}\right] = \cos^{-1}\left[\frac{m_s\hbar}{[s(s+1)]^{1/2}\hbar}\right]$$
$$= \cos^{-1}\left[\frac{m_s}{[s(s+1)]^{1/2}}\right] = \cos^{-1}\left[\frac{2m_s}{\sqrt{35}}\right]. \tag{4}$$

Inserting the permitted values of m_s yields the results shown in the table to the left.

(E) The transition series are filling the d ($\ell = 2$) orbitals. There are five of these with m equal to -2, -1, 0, 1, and 2. If m_s can take on six different values, the Pauli principle would permit each spatial orbital to hold six electrons. The five d orbitals could, therefore, accommodate $5 \times 6 = 30$ total electrons, and hence, there would be 30 elements in each transition metal series in our hypothetical universe.

13.31 Write down the Slater orbital representing a $3p$ orbital with $L_z = \hbar$. Normalize the orbital.

Solution

The form of the Slater orbital is

$$\psi_{\text{slater}}(r, \theta, \phi) = Nr^{n-1} \exp\left[-\frac{Z_e r}{na_o}\right] Y_\ell^m(\theta, \phi). \tag{1}$$

For the $3p$ orbital with $L_z = \hbar$, we have $n = 3$, $\ell = 1$, and $m = 1$. The Slater orbital is, therefore,

$$\boxed{\psi_{\text{slater}}(r, \theta, \phi) = Nr^2 \exp\left[-\frac{Z_e r}{3a_o}\right] \sin(\theta) e^{i\phi}}. \tag{2}$$

Normalization requires that we have

$$\langle \psi_{\text{slater}}(r, \theta, \phi) | \psi_{\text{slater}}(r, \theta, \phi) \rangle = 1. \tag{3}$$

Inserting Eq. 2 into Eq. 3 produces

$$\int_0^\infty \int_{\theta=0}^{\pi} \int_{\phi=0}^{2\pi} Nr^2 \exp\left[-\frac{Z_e r}{3a_o}\right] \sin(\theta) e^{-i\phi} Nr^2 \exp\left[-\frac{Z_e r}{3a_o}\right] \sin(\theta) e^{i\phi} r^2 dr \sin\theta \, d\theta \, d\phi$$
$$= 2\pi N^2 \int_{r=0}^\infty r^6 \exp\left[-\frac{2Z_e r}{3a_o}\right] dr \int_{\theta=0}^{\pi} \sin^3\theta \, d\theta = 1. \tag{4}$$

This gives

$$2\pi N^2 \left[\frac{6!}{(2Z_e/3a_o)^7}\right] \int_{\theta=0}^{\pi} \sin\theta [1 - \cos^2\theta] \, d\theta = 1. \tag{5}$$

Integrating, we obtain

$$2\pi N^2 \left[\frac{6! 3^7 a_o^7}{2^7 Z_e^7}\right] \left[-\cos\theta + \frac{\cos^3\theta}{3}\right]_0^\pi = 2\pi N^2 \left[\frac{6! 3^7 a_o^7}{2^7 Z_e^7}\right] \left[1 - \frac{1}{3} + 1 - \frac{1}{3}\right] = 1, \tag{6}$$

or

$$2\pi N^2 \left[\frac{6! 3^6 a_o^7}{2^5 Z_e^7}\right] = 1. \tag{7}$$

Therefore,

$$\boxed{N = \left[\frac{2^4 Z_e^7}{6! 3^6 \pi a_o^7}\right]^{1/2}}. \tag{8}$$

If we had done the problem using atomic units, a_o would have been omitted, and the result would have been

$$\boxed{N = \left[\frac{2^4 Z_e^7}{6!\,3^6 \pi}\right]^{1/2}}. \tag{9}$$

13.33 The one-electron electronic configuration of the Be atom (Z = 4) is usually written in the form $1s^2 2s^2$. Use a Slater determinant to write a one-electron type of wave function for the Be atom that has the proper symmetry with respect to electron exchange. What is the value of the normalization constant for the Slater determinant if the individual 1*s* and 2*s* Slater-type orbitals are normalized?

Solution

Let us represent the four atomic spin orbitals as $1s\alpha$, $1s\beta$, $2s\alpha$, and $2s\beta$, where α and β are the spin-up and spin-down functions and 1*s* and 2*s* denote a 1*s* or 2*s* orbital wave function. We may express a properly antisymmetric wave function by a Slater determinant; that is,

$$\psi(1,2,3,4) = N\begin{vmatrix} 1s\alpha(1) & 1s\alpha(2) & 1s\alpha(3) & 1s\alpha(4) \\ 1s\beta(1) & 1s\beta(2) & 1s\beta(3) & 1s\beta(4) \\ 2s\alpha(1) & 2s\alpha(2) & 2s\alpha(3) & 2s\alpha(4) \\ 2s\beta(1) & 2s\beta(2) & 2s\beta(3) & 2s\beta(4) \end{vmatrix}.$$

Interchanging any two electrons exchanges two columns of the determinant, which will change the sign, but not the magnitude, of $\psi(1,2,3,4)$. For example, if we interchange electrons 2 and 3, we interchange columns 2 and 3 and thereby change the sign on $\psi(1, 2, 3, 4)$. The wave function is, therefore, antisymmetric to electron exchange.

If the individual spatial 1*s* and 2*s* orbitals are normalized, the normalization constant for the determinant is $[1/N!]^{1/2}$, where *N* is the number of spin orbitals present. Thus, for Be, we have $\boxed{N = \text{normalization constant} = [\tfrac{1}{24}]^{1/2}}$.

13.35 Consider a hypothetical situation in which the electron is a boson with a spin quantum number of 1 instead of $\frac{1}{2}$. Would Mendeleev have noticed a periodic behavior for the elements if this were the case? What would the "periodic table" look like if the electron were a boson?

Solution

Under these conditions, Mendeleev would not have observed any periodic behavior in the chemical and physical properties of the elements. If the electron were a boson, there would be no Pauli principle. Therefore, the quantum number assignments for the electrons could all be the same. As a result, all electrons would be in the 1*s* orbital, since this is the most stable orbital. The new "periodic table" in this situation would be one long 1*s* series in which the electronic configuration of an element with atomic number Z and Z electrons would be $1s^z$.

13.37 Determine the term symbol in the ground state for the elements from Li (Z = 3) to Ar (Z = 18).

Solution

It is important to realize here that we do not need to find *all* the term states—just the one lowest in energy. This will be the one having maximum multiplicity and spin with a maximum L value and either a maximum or minimum J value, depending upon whether the shell is more or less than half filled.

Li: $1s^22s^1$. For Li, there is no choice: We must have $M = L = 0$ and $S = \frac{1}{2}$. So the ground-state, and only, term is $\boxed{^2S_{1/2}}$.

Be: $1s^22s^2$. All orbitals are filled. Therefore, $L = S = 0$, and the ground-state, and only, term is $\boxed{^1S_0}$.

B: $1s^22s^22p^1$. With only one electron in unfilled shells, we must have $S = \frac{1}{2}$. The maximum value of M is 1. So we have the terms $^2P_{3/2}$ and $^2P_{1/2}$. The subshell is less than half filled, so the ground-state term is $\boxed{^2P_{1/2}}$.

C: $1s^22s^22p^2$. To get maximum spin, the m_s values of the $2p^2$ electrons need to be the same to obtain $S = 1$. Since the m_s values are the same, the m values must be different. Therefore, the largest value of M that can be obtained is $M = 1$, when we have one electron in a p orbital with $m = 1$ and the other in the p orbital with $m = 0$. The ground-state term is, therefore, either $^3P_2$, $^3P_1$, or $^3P_0$. Since the shell is less than half filled, the lowest-energy-state term is $\boxed{^3P_0}$.

N: $1s^22s^22p^3$. Maximum spin is obtained when all three $2p$ electrons have $m_s = \frac{1}{2}$. This gives $S = \frac{3}{2}$. If all the m_s values are the same, all the m values must be different. Therefore, one electron must be placed in the $m = 1$ orbital, a second in the $m = 0$ orbital, and the third in the $m = -1$ spatial orbital. Consequently, $m_1 + m_2 + m_3 = 1 + 0 + (-1) = 0$, and we must have $L = 0$. Thus, the ground-state term is $\boxed{^4S_{3/2}}$.

O: $1s^22s^22p^4$. With only three $2p$ orbitals available, two of the $2p$ electrons must be in the same spatial orbital with paired spins. The maximum spin we can have occurs when the other two electrons both have $m_s = 1/2$, so that $S = 1$. With the same values of m_s, the spatial orbitals must be different. Therefore, the largest value of M we can obtain occurs when two electrons are in the $m = 1$ orbital, one is in the $m = 0$ orbital, and the third is in the $m = -1$ orbital. This gives $M = 2 + 0 - 1 = 1$, so that $L = 1$ and we have the terms $^3P_2$, $^3P_1$, and $^3P_0$. Since the shell is now more than half filled, the lowest-energy term has maximum J. Thus, the ground-state term is $\boxed{^3P_2}$.

F: $1s^22s^22p^5$. Four of the five $2p$ electrons must be in the same spatial orbital with paired spins. Therefore, the maximum value of M_s is $\frac{1}{2}$, and we have $S = \frac{1}{2}$. The largest value of M occurs when one pair of electrons is in the $m = 1$ orbital, a second pair in the $m = 0$ orbital, and the unpaired electron in the $m = -1$ orbital. This gives $M = 2 + 0 - 1 = 1$. The possible terms are, therefore, $^2P_{3/2}$ and $^2P_{1/2}$. Since the shell is more than half filled, the ground-state term is $\boxed{^2P_{3/2}}$.

Ne: $1s^22s^22p^6$. All orbitals are filled, so we have $L = S = 0$, and the ground-state, and only, term is $\boxed{^1S_0}$.

The second row is exactly like the first row. Na has $1s^22s^22p^63s^1$, so its ground-state term is identical to that for Li. Mg, with $1s^?2s^22p^63s^2$, is identical to Be. Al is the same as B, and so on. Note that all the elements in a given group of the periodic table have the same ground-state term. Thus, we can summarize our results in the following table:

Li	Be	B	C	N	O	F	Ne
$^2S_{1/2}$	$^1S_0$	$^2P_{1/2}$	$^3P_0$	$^4S_{3/2}$	$^3P_2$	$^2P_{3/2}$	$^1S_0$
Na	Mg	Al	Si	P	S	Cl	Ar

The fact that the ground-state term symbols for all elements in a given group are usually the same is the underlying reason for the chemical similarity of elements in a given group.

13.39 When carbon forms four bonds, its electronic configuration becomes $1s^2 2s^1 2p^3$. This configuration is called the *valence state* of carbon. Determine all the possible term states for the configuration.

Solution

This problem is similar to Example 13.11. The electrons in unfilled subshells are the $2s^1$ and $2p^3$ electrons. We must carefully consider the Pauli principle for the $2p^3$ electrons, but not for the $2s^1$ electron, since, for this electron, $\ell = 0$, which is different than the $\ell = 1$ values for the three $2p$ electrons. Therefore, the Pauli principle is already satisfied for the $2s^1$ electron. We shall, therefore, first find the term states for the $2p^3$ configuration and then couple each of these states to the $2s^1$ electron's angular momenta.

Using a diagram similar to Figure 13.22 for the $2p^3$ case, we can infer the Pauli-allowed assignments:

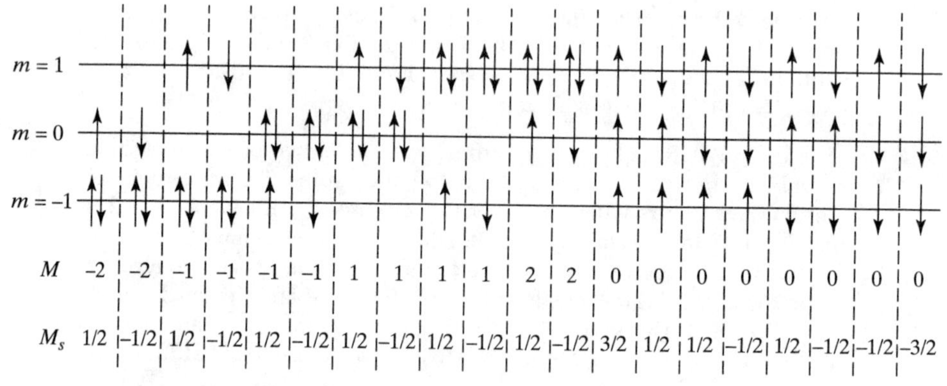

The largest value of M is 2, for which the corresponding M_s value is $\frac{1}{2}$. Therefore, we have one term state with $L = 2$ and $S = \frac{1}{2}$. This accounts for 10 of the 20 combinations shown in the diagram. When the $L = 2$, $S = \frac{1}{2}$ combinations are removed, the largest remaining value of M is 1, with $M_s = \frac{1}{2}$. Therefore, we also have an $L = 1$, $S = \frac{1}{2}$ term state. This accounts for six more combinations, leaving four. We also have a state with $M_s = \frac{3}{2}$ and an associated value of $M = 0$. This corresponds to an $L = 0$, $S = \frac{3}{2}$ term state, which accounts for the last four combinations. The results are, therefore, ^2D, ^2P, and ^4S for the $2p^3$ electrons.

We now wish to couple the angular momenta from the $2s^1$ electron to that of the terms for the $2p^3$ electrons. The $2s^1$ electron has $\ell = 0$, so the L states do not change. However, the $2s^1$ electron has $s = 1/2$. Therefore, the coupling is as follows:

$^2D(L = 2$ and $S = \frac{1}{2})$ with $\ell = 0$ and $s = \frac{1}{2}$; this gives $L = 2$ and $S = 1$ and 0.

$^2P(L = 1$ and $S = \frac{1}{2})$ with $\ell = 0$ and $s = \frac{1}{2}$; this gives $L = 1$ and $S = 1$ and 0.

$^4S(L = 0$ with $S = \frac{3}{2})$ with $\ell = 0$ and $s = \frac{1}{2}$; this gives $L = 0$ and $S = 2$ and 1.

The term states for the $2s^1 2p^3$ valence state of carbon are, therefore,

$$\boxed{^3D_3,\ ^3D_2,\ ^3D_1,\ ^1D_2,\ ^3P_2,\ ^3P_1,\ ^3P_0,\ ^1P_1,\ ^5S_2, \text{ and } ^3S_1}.$$

Each of these term states occurs exactly once. The number of combinations of m and m_s represented by each is 15 for ^3D, 5 for ^1D, 9 for ^3P, 3 for ^1P, 5 for ^5S, and 3 for ^3S. The total number of combinations is, therefore, 40. There are 20 combinations for the $2p^3$ configuration listed in the previous diagram. The $2s^1$ has two combinations with $m = 0$, $m_s = \frac{1}{2}$ and $m = 0$, $m_s = -\frac{1}{2}$. Therefore, the total should be $2 \times 20 = 40$, and we appear to have done the problem correctly.

13.41 Example 13.12 gives the spin–orbit splitting between the $^3P_0$, $^3P_1$, and $^3P_2$ states.
(A) Compute the magnitude of these splittings, in hartrees.
(B) Use Eq. 13.85 and the measured splitting between the $^3P_0$ and $^3P_1$ states to determine the spin–orbit splitting constant for carbon.
(C) Use the results obtained in (B) to compute the splitting between the $^3P_1$ and $^3P_2$ states in carbon. Calculate the percent error in your result.

Solution

(A) The splittings given in Example 13.12 are 16.4 and $(43.5 - 16.4) = 27.1\ \text{cm}^{-1}$. Using the conversion factors in Table 13.2, we obtain

$$^3P_o - {}^3P_1 \text{ splitting} = 16.4\ \text{cm}^{-1} \times \frac{1\ \text{eV}}{8{,}065.6\ \text{cm}^{-1}} \times \frac{1\ \text{hartree}}{27.211\ \text{eV}}$$

$$= \underline{7.47 \times 10^{-5}\ \text{hartree}}. \quad (1)$$

$$^3P_1 - {}^3P_2 \text{ splitting} = 27.1\ \text{cm}^{-1} \times \frac{1\ \text{eV}}{8{,}065.6\ \text{cm}^{-1}} \times \frac{1\ \text{hartree}}{27.211\ \text{eV}}$$

$$= \underline{1.23 \times 10^{-4}\ \text{hartree}}. \quad (2)$$

(B) The change in energy because of spin–orbit interactions is given approximately by Eq. 13.85. In atomic units, this equation is

$$\Delta E_{\text{spin-orbit}} = \frac{C'}{2}[J(J + 1) - L(L + 1) - S(S + 1)]. \quad (3)$$

Therefore, for the ^3P_o state, with $L = 1$, $J = 0$, and $S = 1$, we obtain

$$\Delta E(^3P_o) = \frac{C'}{2}[0 - 1(1 + 1) - 1(1 + 1)] = -2C'. \quad (4)$$

For the $^3P_1$ state, with $L = 1$, $J = 1$, and $S = 1$, the result is

$$\Delta E(^3P_1) = \frac{C'}{2}[1(1 + 1) - 1(1 + 1) - 1(1 + 1)] = -C'. \quad (5)$$

Therefore, the splitting is

$$\Delta E(^3P_1) - \Delta E(^3P_o) = -C' - (-2C') = \underline{C' = 7.47 \times 10^{-5}\ \text{hartree}}. \quad (6)$$

(C) The energy change for the $^3P_2$ state, with $L = 1$, $J = 2$, and $S = 1$ is

$$\Delta E(^3P_2) = \frac{C'}{2}[2(2 + 1) - 1(1 + 1) - 1(1 + 1)] = C'. \quad (7)$$

The spin–orbit splitting between the $^3P_2$ and $^3P_1$ states is

$$\Delta E(^3P_2) - \Delta E(^3P_1) = C' - (-C') = 2C' = 2(7.47 \times 10^{-5}\ \text{hartree})$$

$$= \underline{1.49 \times 10^{-4}\ \text{hartree}}. \quad (8)$$

The percent error in this result is

$$\% \text{ error} = \frac{100 \times (1.49 - 1.23)}{1.23} = \underline{21.1\%}. \quad \textbf{(9)}$$

13.43 Muonium is a transient atom with a proton nucleus and a negative muon. The muon is an elementary particle with a charge of $-e$ and a mass 206.77 times greater than that of the electron. Compute the ground-state energy of muonium in atomic units and in eV. (The author is indebted to Fredrick L. Minn, M.D., Ph.D, for providing this problem and the associated solution.)

Solution

The energy of all hydrogenlike atoms and ions is given by Eq. 13.19A:

$$E_n = -\frac{Z^2\mu e^4}{2n^2\hbar^2} \quad \text{for } n = 1, 2, 3, 4, \ldots. \quad \textbf{(1)}$$

For muonium, we have $Z = 1$, since the positive charge on the nucleus is $+e$. In the ground state, $n = 1$. In atomic units, $e = \hbar = 1$. The reduced mass of the system is

$$\mu = \frac{m_p m_\mu}{m_p + m_\mu}, \quad \textbf{(2)}$$

where m_p and m_μ are, respectively, the proton and muon masses, in atomic units. Using the data in Table 13.2, we have $m_p = 1{,}836.1$. Since the muon mass is $206.77m_e$ and $m_e = 1$ in atomic units, $m_\mu = 206.77$. Therefore,

$$\mu = \frac{(1{,}836.1)(206.77)}{1{,}836.1 + 206.77} = 185.84. \quad \textbf{(3)}$$

Substituting these data into Eq. 1 yields

$$E_1 = -\frac{(1)^2(185.84)(1)^4}{2(1)^2(1)^2} = \underline{-92.92 \text{ hartrees}} = -92.92 \text{ hartrees}$$

$$\times \frac{27.211 \text{ eV}}{\text{hartree}} = \underline{-2{,}528 \text{ eV}}. \quad \textbf{(4)}$$

Note that it would now be a disaster to write that $\mu \approx m_e$.

13.45 L–S coupling is most appropriate for lighter atoms with $Z \leq 36$. One member of the class has determined that one of the elements with $Z \leq 36$ has a ground-state term state $^7S_3$. Which element is it? Is there more than one possibility? If so, what are they? If there is only one, how do we know that?

Solution

The multiplicity is $2S + 1$. Therefore, if the multiplicity is 7, S must be 3. In order to have $S = 3$, it must be possible to have six unpaired electrons in the ground state. Table 13.7 shows that, for the elements with atomic numbers 36 or less, there is only one element whose ground state contains six unpaired electrons. This element is Cr, whose electronic configuration is $[\text{Ar}]4s^1 3d^5$. There are five $3d$ orbitals, with magnetic quantum numbers $-2, -1, 0, 1,$ and 2. If we assign one of the $3d$ electrons to each of these spatial orbitals, the ground state will contain six unpaired electrons, each with the same value of m_s. Consequently, the multiplicity will be $2(3) + 1 = 7$. Since the sum of the magnetic quantum numbers in this state is zero, the term state will be $^7S_3$.

Table 13.7 shows that no other element with $Z \leq 36$ can have six unpaired electrons and a multiplicity of 7.

13.47 I was about to leave this chapter and move to molecular electronic structure and bonding, but I see that Sam has his hand up again. "What is it, Sam?"

"Well, Sir, when we discussed thermodynamics, you never said that there was a limit to the amount of energy we could put into a system, but just look at the expression for the hydrogen-atom energy level. When the hydrogen atom is in its ground state, the energy is -0.5000 hartree. As we put in more and more energy, the atom becomes more and more excited, and the principal quantum number gets bigger and bigger. But n can never become infinite, and when we have $n = \infty$, the energy is zero. So the amount of energy we can add to the hydrogen atom seems to have an upper limit equal to the difference between zero and -0.5000 hartree, which is $+0.5000$ hartree. Does this mean that we can never add more than half a hartree to a hydrogen atom, no matter what kind of engineering design we might use? If this isn't right, what am I doing wrong? If it is right, why can't we add more energy than that?"

Can you respond to Sam's question?

Solution

Sam is correct. It is impossible to add more than $\frac{1}{2}$ hartree to a ground-state hydrogen atom. The reason for this can be seen by examining Figure 13.6. As the energy in the hydrogen atom increases, the classical turning point becomes larger and larger. When $n = 5$, the turning point is about 50 bohr. As the energy added approaches $\frac{1}{2}$ hartree, $n \longrightarrow \infty$ and the classical turning point approaches infinity. This is just another way of saying that the hydrogen atom loses its electron and ionizes to become a bare proton plus a free electron. It is possible to add more energy than $\frac{1}{2}$ hartree, but if we do, we will not have a hydrogen atom any longer. Instead, we will have a free proton and a free electron whose energy states form a continuum, since $E > V$ when the energy added exceeds $\frac{1}{2}$ hartree.

Thermodynamically, we are not limited in the amount of energy that can be added, but we are always limited in the amount of energy that can be added while still keeping the chemical composition of the system the same.

CHAPTER 14

Molecular Structure and Bonding

14.1 Consider a particle of mass m in a one-dimensional infinite potential well of width a for which the potential is $V(x) = 0$ for $0 \le x \le a$ and $V(x) = \infty$ for $x < 0$ or $x > a$. Suppose we alter the potential such that we now have

$$V(x) = \infty \quad \text{for } x < 0 \quad \text{or} \quad x > a,$$
$$V(x) = 0 \quad \text{for } 0 \le x < 0.8a,$$
$$V(x) = -V_o \quad \text{for } 0.8a \le x \le 0.9a$$

with

$$V_o \text{ a positive constant,}$$

and

$$V(x) = 0 \quad \text{for } 0.9a < x \le a.$$

That is, the potential now has the form shown in the following figure:

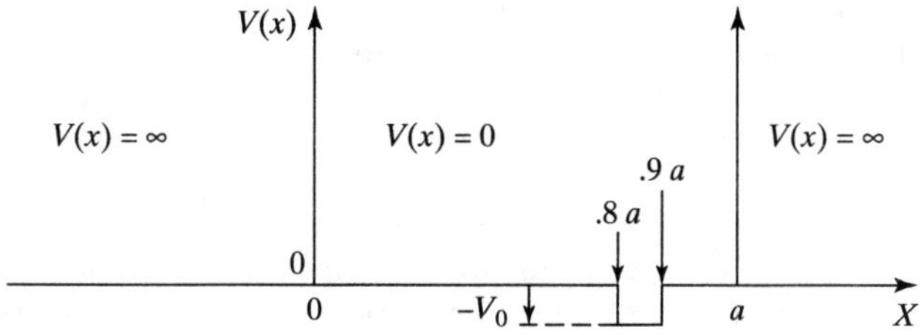

Use first-order perturbation theory to obtain an expression for the first-order ground-state energy for this system in terms of m, a, V_o, and h.

Solution

(A) The obvious perturbation procedure is to take $\mathcal{H}_o$ as the Hamiltonian for a particle in an infinite one-dimensional potential well and then treat the perturbation as being a constant equal to $-V_o$ between $x = 0.8a$ and $x = 0.9a$. If V_o is not too large, this will provide a good answer. The zeroth-order solutions, obtained in Chapter 12, are

$$\psi_o(x) = \left[\frac{2}{a}\right]^{1/2} \sin\left[\frac{\pi x}{a}\right] \quad \text{and} \quad E_o = \frac{h^2}{8ma^2}. \tag{1}$$

We now calculate the average of the perturbation over the zeroth-order eigenfunction:

$$\langle \mathcal{H}' \rangle = \int_0^a \psi_o^*(x)\mathcal{H}'\psi_o(x)\,dx. \tag{2}$$

However $\mathcal{H}'$ is zero except between $x = 0.8a$ and $x = 0.9a$, where it is the constant $-V_o$. As a result, the integral in Eq. 2 becomes

$$\langle \mathcal{H}' \rangle = -\frac{2V_o}{a}\int_{0.8a}^{0.9a} \sin^2\left[\frac{\pi x}{a}\right] dx. \tag{3}$$

This is a standard integral that can be found in any table. Integration produces

$$\langle \mathcal{H}' \rangle = -\frac{2V_o}{a}\left[\frac{x}{2} - \frac{1}{4\pi/a}\sin\left[\frac{2\pi x}{a}\right]\right]_{0.8a}^{0.9a}$$
$$= -\frac{2V_o}{a}\left\{\left[\frac{0.9a}{2} - \frac{1}{4\pi/a}\sin[1.8\pi]\right] - \left[\frac{0.8a}{2} - \frac{1}{4\pi/a}\sin[1.6\pi]\right]\right\}$$

$$= -\frac{2V_o}{a}\left[0.05a + \frac{a}{4\pi}\{\sin(1.6\pi) - \sin(1.8\pi)\}\right]$$

$$= -\frac{2V_o}{a}[0.05a - 0.0289a] = -0.04218\ V_o. \quad \textbf{(4)}$$

Therefore, the first-order energy is

$$\boxed{E = E_o + \langle \mathcal{H}' \rangle = \frac{h^2}{8ma^2} - 0.04218 V_o}. \quad \textbf{(5)}$$

As expected, the first-order correction is negative, since the perturbation is negative.

14.3 A phenomenon seen in many organizations, particularly among the faculty of a chemistry department, is the constant battle between the younger members of the organization (the Young Turks) and the more senior members (the Old Guard). Whenever a new problem arises, the Young Turks usually wish to dismantle the present system, start fresh, and obtain new, innovative solutions to the problem. The Old Guard is much more conservative. Its members usually wish to employ the old system and simply use it to generate acceptable solutions to the new problem. The Young Turks accuse the Old Guard of being mired in the past—of having lost the capacity for creative thought. The Old Guard looks upon the Young Turks as one might view a three-year-old child throwing a temper tantrum. Relate this situation to first-order perturbation theory. How are they similar? [*Author's advice:* Since "young" refers to your mental state, not to your chronological age, be a Young Turk all your life. First-order perturbation theory is not particularly accurate.]

Solution

The Old Guard is essentially using a perturbation approach to handle the "new problem," which is the perturbation $\mathcal{H}'$. The Old Guard intends to employ the old methods (the old zeroth-order eigenfunctions that existed before the perturbation came along) to obtain an "acceptable solution" to the new problem. This solution is the analogue of the first-order correction to the energy in which the perturbation is averaged over the zeroth-order functions and simply added on to the zeroth-order energy.

The Young Turks, on the other hand, regard the "new problem" as one that cannot be handled using the old methods. They feel that the perturbation is too large to permit accurate use of the old methods (the zeroth-order eigenfunctions). Their approach is to discard the old methods, start fresh (write down the new Hamiltonian, including the perturbation), and obtain new, innovative solutions (new eigenfunctions and eigenvalues for the new correct Hamiltonian). If the new solutions can be found, they will certainly be much better than a perturbation calculation.

14.5 Use the variational wave function $\lambda = N \exp[-ar^2]$, where N and a are constants, in a variational calculation of the energy of the hydrogen atom, and determine the value of a that minimizes the energy. Compare this minimum value with the true hydrogen-atom ground-state energy. Use atomic units. Is the variational energy too high or too low?

Solution

In atomic units, the hydrogen-atom Hamiltonian is

$$\mathcal{H} = -\frac{1}{2}\nabla^2 - \frac{1}{r}. \quad \textbf{(1)}$$

The variational energy is given by

$$E_v = \frac{\langle\lambda|\mathcal{H}|\lambda\rangle}{\langle\lambda|\lambda\rangle}. \tag{2}$$

We first integrate the denominator:

$$\langle\lambda|\lambda\rangle = N^2 \int_{r=0}^{\infty}\int_{\theta=0}^{\pi}\int_{\phi=0}^{2\pi} \exp[-2ar^2]r^2 dr \sin\theta\, d\theta\, d\phi$$

$$= 4\pi N^2 \int_0^{\infty} \exp[-2ar^2]r^2 dr. \tag{3}$$

Using the general form

$$\int_0^{\infty} x^{2n}\exp[-bx^2]dx = \frac{(1)(3)(5)\cdots(2n-1)}{2^{n+1}b^n}\left[\frac{\pi}{b}\right]^{1/2}, \tag{4}$$

we obtain

$$\langle\lambda|\lambda\rangle = 4\pi N^2 \frac{1}{4(2a)}\left[\frac{\pi}{2a}\right]^{1/2} = \frac{\pi N^2}{2a}\left[\frac{\pi}{2a}\right]^{1/2}. \tag{5}$$

The angular derivatives in the ∇^2 term vanish, because λ is a function of r only. Therefore, we have

$$-\frac{1}{2}\nabla^2\lambda = -\frac{N}{2}\frac{1}{r^2}\left[\frac{\partial}{\partial r}\left(r^2\frac{\partial}{\partial r}\right)\right]\exp[-ar^2] = -\frac{N}{2r^2}\frac{\partial}{\partial r}\{-2ar^3\exp[-ar^2]\}$$

$$= \frac{aN}{r^2}\{3r^2\exp[-ar^2] - 2ar^4\exp[-ar^2]\}$$

$$= 3aN\exp[-ar^2] - 2a^2Nr^2\exp[-ar^2]. \tag{6}$$

We may now use Eq. 6 to obtain the result when the Hamiltonian operates on λ:

$$\mathcal{H}\lambda = \left[-\frac{1}{2}\nabla^2 - \frac{1}{r}\right]\lambda = 3aN\exp[-ar^2] - 2a^2Nr^2\exp[-ar^2] - \frac{N}{r}\exp[-ar^2]$$

$$= N\exp[-ar^2]\left[3a - 2a^2r^2 - \frac{1}{r}\right]. \tag{7}$$

The numerator integral is, therefore,

$$\langle\lambda|\mathcal{H}|\lambda\rangle = 4\pi N^2 \int_0^{\infty}[3ar^2 - 2a^2r^4 - r]\exp[-2ar^2]dr \tag{8}$$

after integrating over the angles. From Eq. 4, the first two integrals in Eq. 8 become

$$4\pi N^2\int_0^{\infty}(3ar^2 - 2a^2r^4)\exp[-2ar^2]dr = 4\pi N^2\left[\frac{3a}{4(2a)}\left[\frac{\pi}{2a}\right]^{1/2} - \frac{6a^2}{8(2a)^2}\left[\frac{\pi}{2a}\right]^{1/2}\right]$$

$$= 4\pi N^2\left[\frac{\pi}{2a}\right]^{1/2}\left[\frac{3}{8} - \frac{3}{16}\right] = \frac{3\pi N^2}{4}\left[\frac{\pi}{2a}\right]^{1/2}. \tag{9}$$

The third integral in Eq. 8 is

$$-4\pi N^2\int_0^{\infty} r\exp[-2ar^2]dr = \frac{\pi N^2}{a}\exp[-2ar^2]_0^{\infty} = -\frac{\pi N^2}{a}. \tag{10}$$

Combining Eqs. 8, 9, and 10 yields

$$\langle\lambda|\mathcal{H}|\lambda\rangle = \frac{3\pi N^2}{4}\left[\frac{\pi}{2a}\right]^{1/2} - \frac{\pi N^2}{a}. \tag{11}$$

Combining Eqs. 2, 5, and 11, we obtain

$$E_v = \frac{\dfrac{3\pi N^2}{4}\left[\dfrac{\pi}{2a}\right]^{1/2} - \dfrac{\pi N^2}{a}}{\dfrac{\pi N^2}{2a}\left[\dfrac{\pi}{2a}\right]^{1/2}} = \frac{3a}{2} - 2\left[\frac{2a}{\pi}\right]^{1/2}. \quad (12)$$

The dependence of E_v upon a is shown in Figure 14.6 in the text. We now wish to minimize the variational energy with respect to a. The condition for a minimum in E_v is

$$\frac{dE_v}{da} = \frac{3}{2} - 2\left(\frac{1}{2}\right)\left(\frac{2}{\pi}\right)^{1/2} a^{-1/2} = 0. \quad (13)$$

Multiplying by $a^{1/2}$ gives

$$\frac{3}{2}a^{1/2} - \left(\frac{2}{\pi}\right)^{1/2} = 0, \quad (14)$$

so that

$$a_{\min}^{1/2} = \frac{2(2/\pi)^{1/2}}{3}. \quad (15)$$

Squaring both sides, we obtain

$$\boxed{a_{\min} = \frac{8}{9\pi}}. \quad (16)$$

The corresponding value of E_v is obtained by substituting Eq. 16 into Eq. 12:

$$(E_v)_{\min} = \frac{24}{18\pi} - 2\left(\frac{16}{9\pi^2}\right)^{1/2} = \frac{4}{3\pi} - 2\left(\frac{16}{9\pi^2}\right)^{1/2} = \underline{-0.424413\ldots \text{ hartree}}. \quad (17)$$

The exact answer for the hydrogen-atom ground state is $E_{\text{true}} = -0.5000$ hartree. Thus, the variational result is too high by 15.12%.

14.7 The ground-state eigenfunction for a harmonic oscillator is $\psi(x) = N \exp[-\alpha x^2/2]$, where $\alpha = (\mu k)^{1/2}/h$, in which μ and k are the reduced mass and vibrational force constant, respectively. The ground-state energy eigenvalue is $E = h\nu_o/2$, with the vibrational frequency given by $\nu_o = [1/(2\pi)][k/\mu]^{1/2}$. An investigator wishes to conduct a variational calculation on a harmonic oscillator using the trial eigenfunction

$$\lambda = N(A^2 - x^2) \quad \text{for } -A \le x \le A$$

and

$$\lambda = 0 \text{ for } x > A \text{ or } x < -A.$$

Using A as a variational parameter, obtain the best possible variational energy for this trial eigenfunction. Compare your result with the exact ground-state energy. Is the variational principle obeyed?

Solution

We first normalize the trial eigenfunction. This requires that

$$\langle\lambda|\lambda\rangle = 1 = N^2 \int_{-A}^{A} (A^2 - x^2)^2 dx = N^2 \int_{-A}^{A} [A^4 - 2A^2x^2 + x^4] dx$$

$$= N^2\left[A^4 x - \frac{2A^2x^3}{3} + \frac{x^5}{5}\right]_{-A}^{A} = \frac{16A^5N^2}{15}. \quad (1)$$

Therefore, the square of the normalization constant is $N^2 = 15/(16A^5)$. We must now compute the numerator of the variational energy expression, $\langle\lambda|\mathcal{H}|\lambda\rangle$. The Hamiltonian for a harmonic oscillator is $\mathcal{H} = -(\hbar^2/2\mu)(d^2/dx^2) + kx^2/2$. When this operates on λ, we obtain

$$\mathcal{H}\lambda = \left[-\frac{\hbar^2}{2\mu}\frac{d^2}{dx^2} + \frac{kx^2}{2}\right]N(A^2 - x^2) = \frac{N\hbar^2}{\mu} + \frac{Nkx^2(A^2 - x^2)}{2}. \quad (2)$$

Substituting Eq. 2 into $\langle\lambda|\mathcal{H}|\lambda\rangle$ gives

$$\begin{aligned} E_v = \langle\lambda|\mathcal{H}|\lambda\rangle &= N^2\int_{-A}^{A}(A^2 - x^2)\left[\frac{\hbar^2}{\mu} + \frac{kx^2(A^2 - x^2)}{2}\right]dx \\ &= N^2\left[\frac{A^2\hbar^2x}{\mu} - \frac{\hbar^2x^3}{3\mu} + \frac{A^4kx^3}{6} - \frac{A^2kx^5}{10} - \frac{A^2kx^5}{10} + \frac{kx^7}{14}\right]_{-A}^{A} \\ &= N^2\left[\frac{2A^3\hbar^2}{\mu} - \frac{2A^3\hbar^2}{3\mu} + \frac{A^7k}{3} - \frac{A^7k}{5} - \frac{A^7k}{5} + \frac{kA^7}{7}\right]. \end{aligned} \quad (3)$$

Collecting terms in various powers of A yields

$$E_v = N^2\left\{\frac{4A^3\hbar^2}{3\mu} + \frac{8A^7k}{105}\right\}. \quad (4)$$

Inserting the value of N^2 produces

$$\boxed{E_v = \frac{15}{16A^5}\left\{\frac{4A^3\hbar^2}{3\mu} + \frac{8A^7k}{105}\right\} = \frac{5\hbar^2}{4A^2\mu} + \frac{kA^2}{14}}. \quad (5)$$

We now wish to minimize E_v with respect to A to obtain the lowest possible result for the variational energy. Minimization requires that

$$\frac{dE_v}{dA} = 0 = -\frac{10\hbar^2}{4A^3\mu} + \frac{kA}{7}. \quad (6)$$

Solving for A, we obtain

$$\frac{kA}{7} = \frac{5\hbar^2}{2A^3\mu}, \quad (7)$$

which gives

$$3A^4 = \frac{35\hbar^2}{2k\mu}. \quad (8)$$

Taking square roots of both sides produces

$$\boxed{A^2 = \left[\frac{35}{2}\right]^{1/2}\frac{\hbar}{(k\mu)^{1/2}}}. \quad (9)$$

Substituting the result of Eq. 9 into Eq. 5 gives the lowest possible variational energy:

$$\begin{aligned} E_v &= \frac{5\hbar^2}{4\mu}\left[\frac{2}{35}\right]^{1/2}\frac{(k\mu)^{1/2}}{\hbar} + \frac{k}{14}\left[\frac{35}{2}\right]^{1/2}\frac{\hbar}{(k\mu)^{1/2}} \\ &= \frac{\hbar}{2}\left[\frac{k}{\mu}\right]^{1/2}\left[\frac{5}{(70)^{1/2}} + \frac{1}{7}\left[\frac{35}{2}\right]^{1/2}\right] = \frac{h}{2}\frac{1}{2\pi}\left[\frac{k}{\mu}\right]^{1/2}\left[2\left(\frac{5}{14}\right)^{1/2}\right] \\ &= \boxed{\left(\frac{10}{7}\right)^{1/2}\left[\frac{h\nu_o}{2}\right]}. \end{aligned} \quad (10)$$

We see that $(10/7)^{1/2}[h\nu_o/2] > [h\nu_o/2]$, so we have $E_v > E$, and the variational principle is obeyed, as expected.

14.9 Show that the higher energy root for E_v in Example 14.5 is an upper limit to the energy of the first excited state for a particle of mass m in a one-dimensional infinite potential well of width a. Obtain the approximate normalized eigenfunction for this excited state of the system. [See hint in Problem 14.8.]

Solution

The higher energy root for E_v is obtained by taking the plus sign on the radical in the solution of the quadratic resulting from the expansion of the secular equation in Example 14.5. This gives

$$E_v = \frac{0.006278589a^{20}\hbar^2/m + [(0.006278589)^2a^{20}\hbar^4/m^2 - 0.000016661358a^{20}\hbar^4/m^2]^{1/2}}{2(0.000150896)a^{22}}$$

$$= \frac{0.006278589 + 0.004770667}{2(0.000150896)}\frac{\hbar^2}{ma^2} = 36.6122\frac{\hbar^2}{ma^2} = \boxed{0.92740\frac{h^2}{ma^2}}. \quad \textbf{(1)}$$

The energy levels for a particle in an infinite, one-dimensional potential well are given by

$$E_n = \frac{n^2h^2}{8ma^2} \quad \text{for } n = 1, 2, 3, \ldots. \quad \textbf{(2)}$$

The first excited state has $n = 2$, so that its energy is

$$E_2 = 0.50000\frac{h^2}{ma^2}. \quad \textbf{(3)}$$

Comparing Eqs. 1 and 3 shows that we have E_v (higher root) $> E_2$. Although we have not proven that this must be the case, it can be proven with somewhat more quantum theory than we have available. In general, if we order the roots for E_v such that the lowest root is E_{v0}, the next lowest E_{v1}, and so on, we will always observe that E_{vn} is an upper limit to the exact energy of the excited state of the system.

The first excited-state eigenfunction is given by

$$\lambda_+ = c_1\left[\chi_1 + \frac{c_2}{c_1}\chi_2\right]. \quad \textbf{(4)}$$

Equation 14.32 shows that

$$\frac{c_2}{c_1} = \frac{H_{11} - E_vS_{11}}{E_vS_{12} - H_{12}}. \quad \textbf{(5)}$$

Substituting the matrix elements and overlap integrals from Example 14.5 and the result of Eq. 1 produces

$$\frac{c_2}{c_1} = \frac{\dfrac{2a^5\hbar^2}{5m} - 36.6122\dfrac{\hbar^2}{ma^2}\dfrac{8a^7}{105}}{36.6122\dfrac{\hbar^2}{ma^2}\dfrac{8a^{11}}{693} - \dfrac{6a^9\hbar^2}{63m}} = \frac{-7.298112219}{a^4}. \quad \textbf{(6)}$$

Normalization requires that we have

$$\langle\lambda_+|\lambda_+\rangle = 1. \quad \textbf{(7)}$$

Combining Eqs. 4, 6, and 7 gives

$$\langle\lambda_+|\lambda_+\rangle = c_1^2\left\langle\chi_1 + \frac{c_2}{c_1}\chi_2\middle|\chi_1 + \frac{c_2}{c_1}\chi_2\right\rangle = c_1^2[\langle\chi_1|\chi_1\rangle + \left(\frac{c_2}{c_1}\right)^2\langle\chi_2|\chi_2\rangle$$

$$+ 2\left(\frac{c_2}{c_1}\right)\langle\chi_1|\chi_2\rangle] = c_1^2\left[S_{11} + \left(\frac{c_2}{c_1}\right)^2S_{22} + 2\left(\frac{c_2}{c_1}\right)S_{12}\right]. \quad \textbf{(8)}$$

Substituting the values of the overlap integrals from Example 14.5 and the ratio c_2/c_1 for the first excited state produces

$$c_1^2\left[\frac{8a^7}{105} + \left\{\frac{-7.2981122}{a^4}\right\}^2 \frac{8a^{15}}{2{,}145} + 2\left\{\frac{-7.2981122}{a^4}\right\}\frac{8a^{11}}{693}\right]$$

$$= 0.10633929 c_1^2 a^7 = 1. \quad (9)$$

Solving for c_1, we obtain

$$c_1 = \left[\frac{1}{0.10633929a^7}\right]^{1/2} = 3.0665716a^{-7/2}. \quad (10)$$

The first excited-state variational eigenfunction is, therefore,

$$\boxed{\lambda_- = 3.06657a^{-7/2}\left[\chi_1 - \frac{7.2981122}{a^4}\chi_2\right]}, \quad (11)$$

where $\chi_1 = a^2x - x^3$ and $\chi_2 = a^2x^5 - x^7$.

14.11 Consider a one-dimensional particle in an infinite potential well of width a. The correct ground-state wave function for such a system is

$$\psi = \left[\frac{2}{a}\right]^{1/2} \sin\left[\frac{\pi x}{a}\right] \quad \text{for } 0 \le x \le a$$

and

$$\psi = 0 \text{ for } x < 0 \text{ or } x > a.$$

Show that the trial eigenfunction

$$\lambda = N[x^3 - a^2x] \quad \text{for } 0 \le x \le a$$

and

$$\lambda = 0 \text{ for } x < 0 \text{ or } x > a$$

gives a variational energy greater than that for the exact ground state.

Solution

The variational energy for the approximate wave function is

$$E_v = \frac{\langle\lambda|\mathcal{H}|\lambda\rangle}{\langle\lambda|\lambda\rangle}. \quad (1)$$

The denominator of Eq. 1 is given by

$$\langle\lambda|\lambda\rangle = \int_0^a N^2[x^3 - a^2x]^2 dx = N^2\int_0^a [x^6 - 2a^2x^4 + a^4x^2]dx$$

$$= N^2\left[\frac{x^7}{7} - \frac{2a^2x^5}{5} + \frac{a^4x^3}{3}\right]_0^a = N^2a^7\left[\frac{1}{7} - \frac{2}{5} + \frac{1}{3}\right] = \frac{8N^2a^7}{105}. \quad (2)$$

The Hamiltonian for the one-dimensional particle in an infinite well is

$$\mathcal{H} = -\frac{\hbar^2}{2m}\frac{d^2}{dx^2}. \quad (3)$$

The numerator of Eq. 1, therefore, is

$$\langle\lambda|\mathcal{H}|\lambda\rangle = N^2\int_0^a [x^3 - a^2x]\left\{-\frac{\hbar^2}{2m}\frac{d^2}{dx^2}\right\}[x^3 - a^2x]\,dx$$

$$= -\frac{6\hbar^2N^2}{2m}\int_0^a [x^3 - a^2x][x]\,dx = -\frac{3\hbar^2N^2}{m}\left[\frac{x^5}{5} - \frac{a^2x^3}{3}\right]_0^a$$

$$= -\frac{3\hbar^2N^2a^5}{m}\left(\frac{1}{5} - \frac{1}{3}\right) = \frac{2\hbar^2N^2a^5}{5m}. \qquad (4)$$

Substituting Eqs. 2 and 4 into Eq. 1 gives

$$\boxed{E_v \supseteq = \frac{\dfrac{2\hbar^2N^2a^5}{5m}}{\dfrac{8N^2a^7}{105}} = \frac{21\hbar^2}{4ma^2} = \frac{21h^2}{16\pi^2ma^2} = \frac{h^2}{7.5197ma^2}}. \qquad (5)$$

The true ground-state energy is $h^2/(8ma^2)$. Comparing E_v with this value shows that we have

$$\boxed{\frac{h^2}{7.5197ma^2} > \frac{h^2}{8ma^2}}, \qquad (6)$$

so that $E_v > E_{\text{exact}}$, and the variational principle is satisfied.

14.13 A linear variational calculation is carried out on a molecule using the expansion $\lambda = \sum_{i=1}^{15} a_i\chi_i$.

(A) How many variational energies will be obtained in the calculation? How many of these will be either imaginary or complex?

(B) How many molecular orbitals will result from the calculation?

(C) How many nodes will the MO corresponding to the next-lowest variational energy have?

(D) What is the value of the integral $\langle\lambda_i|\lambda_j\rangle$ if λ_i and λ_j are molecular orbitals corresponding to different variational energies? Explain.

Solution

(A) The secular equation will be a 15×15 determinant that will produce a 15th-order polynomial in E_v that has 15 roots. Thus, we will obtain 15 possible values of E_v. All of these values will be real, since $\mathcal{H}$ is a Hermitian operator and all the eigenvalues are real.

(B) Substituting each of the 15 possible values of E_v back into the linear equations that led to the secular equation will produce a molecular orbital. Therefore, we will obtain 15 different HMOs.

(C) The MO corresponding to the lowest value of E_v is the ground state and will, therefore, be nodeless. The next lowest value of E_v will produce an approximation to the first excited-state eigenfunction, which we expect to exhibit one node.

(D) The value of this integral is zero. The eigenfunctions of a Hermitian operator $\mathcal{H}$ must be orthogonal if the eigenvalues are not the same.

14.15 Show that the two molecular orbitals for the hydrogen-molecule ion derived in the text, λ_+ and λ_-, are orthogonal. Could this have been predicted without integrating? How?

Solution

The two MOs are

$$\lambda_+ = [2 + 2S_{12}]^{-1/2}[\chi_1 + \chi_2] \quad (1)$$

and

$$\lambda_- = [2 - 2S_{12}]^{-1/2}[\chi_1 - \chi_2]. \quad (2)$$

If these two orbitals are orthogonal, we must have

$$\langle\lambda_+|\lambda_-\rangle = 0. \quad (3)$$

Substituting Eqs. 1 and 2 into Eq. 3 produces

$$\begin{aligned}\langle\lambda_+|\lambda_-\rangle &= [2 - 2S_{12}]^{-1/2}[2 + 2S_{12}]^{-1/2}[\langle\chi_1 + \chi_2|\chi_1 - \chi_2\rangle] \\ &= [2 - 2S_{12}]^{-1/2}[2 + 2S_{12}]^{-1/2}[\langle\chi_1|\chi_1\rangle - \langle\chi_2|\chi_2\rangle - \langle\chi_1|\chi_2\rangle + \langle\chi_2|\chi_1\rangle] \\ &= [2 - 2S_{12}]^{-1/2}[2 + 2S_{12}]^{-1/2}[S_{11} - S_{22} - S_{12} + S_{12}] \\ &= [2 - 2S_{12}]^{-1/2}[2 + 2S_{12}]^{-1/2}[S_{11} - S_{22}]. \quad (4)\end{aligned}$$

Since both χ_1 and χ_2 are normalized, we have $S_{11} = S_{22} = 1$. Therefore, we obtain

$$\langle\lambda_+|\lambda_-\rangle = 0, \quad (5)$$

and λ_+ is orthogonal to λ_-, as we set out to show.

The preceding result could easily have been predicted without integrating even once. Since $\mathcal{H}$ is a Hermitian operator, its eigenvalues must be real and its eigenfunctions for nondegenerate eigenvalues orthogonal. The eigenvalues E_v^+ and E_v^- are not degenerate. Therefore, λ_+ must be orthogonal to λ_-.

14.17* In the text, it is shown that if the variational wave function $\lambda = a_1\chi_{1sA} + a_2\chi_{1sB}$, where

$$\chi_{1sA} = \left[\frac{Z_e^3}{\pi}\right]^{1/2} \exp[-Z_e r_a]$$

and

$$\chi_{1sB} = \left[\frac{Z_e^3}{\pi}\right]^{1/2} \exp[-Z_e r_b],$$

is used for the H_2^+ molecule, the two resulting MO energies are

$$E_v^+ = \frac{[H_{11} + H_{12}]}{[1 + S_{12}]}$$

and

$$E_v^- = \frac{[H_{11} - H_{12}]}{[1 - S_{12}]}.$$

Using Z_e as a variational parameter, determine the value of Z_e (to three significant digits) that yields the best result for the well depth of the hydrogen-molecule ion. (Numerical methods will need to be used for this part of the problem.) Compare your result with the experimental result given in Problem 14.16. What is the percent error in your result?

Solution

The three integrals, H_{11}, H_{12}, and S_{12}, required to compute E_v^+ and E_v^- are given in Eqs. 14.48, 14.49 and 14.50. In terms of R and Z_e, these integrals are

$$H_{11} = \frac{Z_e^2}{2} - Z_e + \frac{(Z_eR + 1)\exp[-2Z_eR]}{R}, \tag{1}$$

$$H_{12} = -\frac{S_{12}Z_e^2}{2} + (Z_e - 2)(Z_e^2R + Z_e)\exp[-Z_eR] + \frac{S_{12}}{R}, \tag{2}$$

and

$$S_{12} = \left[Z_eR + \frac{Z_e^2R^2}{3} + 1\right]\exp[-Z_eR]. \tag{3}$$

Direct substitution into the expression for E_v^+ with different values of Z_e, followed by numerical calculation, permits the value of R at which E_v^+ is a minimum for a given Z_e to be located. The results using different values of Z_e are as follows:

Z_e	R_{eq}	Well Depth = −0.5000 − E_{min} (hartrees)
1.000	2.4910	0.06483
1.100	2.2605	0.07922
1.200	2.0687	0.08595
1.300	2.0335	0.08379

The lower E_{min}, the closer to the exact answer the variational energy must be. Therefore, we know that the best value of Z_e lies between $Z_e = 1.2$ and $Z_e = 1.3$, because E_{min} is lowest somewhere in this range. Additional calculations yield the following results:

Z_e	R_{eq}	Well Depth = −0.5000 − E_{min} (hartrees)
1.230	2.0180	0.08646
1.240	2.0004	0.08650
1.250	1.9850	0.08645

The best well depth is therefore obtained for $\boxed{Z_e = 1.24}$ (correct to three significant digits). This well depth is $\boxed{0.08650 \text{ hartree}}$ at an equilibrium distance of $\boxed{R_{eq} = 2.0004 \text{ bohr}}$. The experimentally measured well depth is 0.1025 au. So the variational result is low by 15.6%. The computed equilibrium distance is nearly exact.

14.19* It is shown in the text that if the variational wave function $\lambda = a_1\chi_{1sA} + a_2\chi_{1sB}$, where

$$\chi_{1sA} = \left[\frac{Z_e^3}{\pi}\right]^{1/2}\exp[-Z_er_a]$$

and

$$\chi_{1sB} = \left[\frac{Z_e^3}{\pi}\right]^{1/2}\exp[-Z_er_b],$$

is used for the H_2^+ molecule, the two resulting MO energies are

$$E_v^+ = \frac{[H_{11} + H_{12}]}{[1 + S_{12}]}$$

and

$$E_v^- = \frac{[H_{11} - H_{12}]}{[1 - S_{12}]}.$$

The best result for the well depth of H_2^+ ion is obtained with $Z_e = 1.24$. With this effective charge, the computed equilibrium distance for H_2^+ is 2.001 bohr.

(A) Calculate the H_2^+ ground-state energy for R values of $R_{eq} + dx$, $R_{eq} + 2dx$, $R_{eq} - dx$, and $R_{eq} - 2dx$, where $dx = 0.01$ bohr. Assuming that the ground-state energy near the bottom of the well may be fitted accurately by a parabolic (harmonic) function, use least–squares methods to fit the function $E_v^+ = k(R - R_{eq})^2/2 + D$ to the results at $R = R_{eq}$ and the four values calculated in (A). That is, determine by least–squares methods the values of k and D that yield the best fit to your data.

(B) Assuming that the vibrational motion of H_2^+ in the ground state may be represented by a harmonic oscillator, use the results obtained in (A) to compute the fundamental vibration frequency for the hydrogen-molecule ion. Express your result in cm^{-1}. The measured vibrational frequency is 2,297 cm^{-1}. Calculate the percent error in your answer.

Solution

(A) The three integrals, H_{11}, H_{12}, and S_{12}, required to compute E_v^+ and E_v^- are given in Eqs. 14.48, 14.49 and 14.50. In terms of R and Z_e, these integrals are

$$H_{11} = \frac{Z_e^2}{2} - Z_e + \frac{(Z_eR + 1)\exp[-2Z_eR]}{R}, \tag{1}$$

$$H_{12} = -\frac{S_{12}Z_e^2}{2} + (Z_e - 2)(Z_e^2R + Z_e)\exp[-Z_eR] + \frac{S_{12}}{R}, \tag{2}$$

and

$$S_{12} = \left[Z_eR + \frac{Z_e^2R^2}{3} + 1\right]\exp[-Z_eR]. \tag{3}$$

Direct substitution into the expression for E_v^+ with $Z_e = 1.24$, followed by numerical calculation, gives E_v^+ at the distances $R = 2.021$, 2.011, 2.001, 1.991, and 1.981 bohr, as requested in the problem. The energies at these distances are as follows:

R (bohr)	$R - R_{eq}$ (bohr)	E_v^+ (hartrees)
1.98100000	−0.02000	−0.58647892
1.99100000	−0.01000	−0.58649918
2.00100000	−0.00000	−0.58650495
2.01100000	0.01000	−0.58649657
2.02100000	0.02000	−0.58647439

We now wish to fit these data to the function

$$E_v^+ = \frac{k(R - R_{eq})^2}{2} + D = \frac{kx^2}{2} + D, \tag{4}$$

where $x = R - R_{eq}$. A least-squares fit to the data yields $E_v^+ = -0.586505 + 0.0707387x^2$. A plot showing the quality of the fit appears at the end of the problem.

(B) The harmonic frequency is given by

$$\nu = \frac{1}{2\pi}\left[\frac{k}{\mu}\right]^{1/2}. \tag{5}$$

The reduced mass of H_2^+, in atomic units, is

$$\mu = \frac{m_p m_p}{m_p + m_p} = \frac{m_p}{2}, \tag{6}$$

where m_p is the mass of the proton. Using Table 13.2, we obtain

$$\mu = \frac{1{,}836.1}{2} = 918.05 \text{ atomic mass units (amu)}. \tag{7}$$

The force constant is twice the least-squares coefficient of x^2 in the harmonic fit. Thus,

$$k = 2(0.0707387) = 0.14148 \text{ amu tu}^{-2}. \tag{8}$$

Combining Eqs. 5, 7, and 8 produces

$$\nu = \frac{1}{2(3.14159)}\left[\frac{0.14148}{918.05}\right]^{1/2} = 0.0019758 \text{ tu}^{-1}. \tag{9}$$

The atomic time unit can be converted to s^{-1} using the conversion factor derived in the text and given in Table 13.1:

$$\nu = 0.0019758 \text{ tu}^{-1} \times \frac{1 \text{ time unit}}{2.41888 \times 10^{-17} \text{ s}} = 8.168 \times 10^{13} \text{ s}^{-1}. \tag{10}$$

The frequency, expressed in terms of cm^{-1}, is

$$\bar{\nu} = \frac{\nu}{c} = \frac{8.168 \times 10^{13} \text{ s}^{-1}}{2.998 \times 10^{10} \text{ cm s}^{-1}} = \underline{2{,}725 \text{ cm}^{-1}}. \tag{11}$$

The percent error in the MO result is

$$\% \text{ error} = 100 \times \frac{2{,}725 - 2{,}297}{2{,}297} = \underline{18.6\%}. \tag{12}$$

This is a general characteristic of MO calculations; the frequencies are usually too high by about 10% or so.

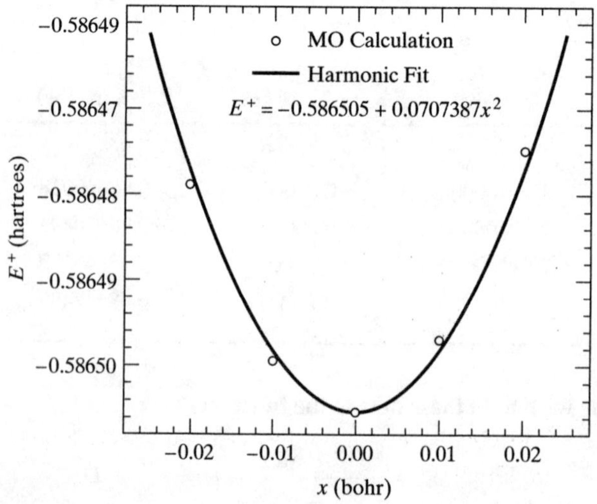

14.21 Write down an appropriate form for the ground-state wave function of the He_2^+ molecule ion. Is He_2 a bound molecule? Is He_2^+ a bound molecule? If we were to pass an electrical discharge through a container of He(g) so that we ionized a significant fraction of the atoms, discuss qualitatively what you might expect to find as the principal components inside the container a few moments after the discharge. At a much later time, what would you expect to find?

Solution

The simple MO electronic configuration for He_2^+ with three electrons is $(1s\sigma)^2(1s\sigma^*)^1$, where we have omitted the inversion symmetry designation for brevity of notation. Therefore, a properly antisymmetrized Slater determinant for the wave function might be

$$\psi(1,2,3) = \left[\frac{1}{3!}\right]^{1/2} \begin{vmatrix} 1s\sigma\alpha(1) & 1s\sigma\alpha(2) & 1s\sigma\alpha(3) \\ 1s\sigma\beta(1) & 1s\sigma\beta(2) & 1s\sigma\beta(3) \\ 1s\sigma^*\alpha(1) & 1s\sigma^*\alpha(2) & 1s\sigma^*\alpha(3) \end{vmatrix}. \tag{1}$$

Of course, the spin function for the electron in the $1s\sigma^*$ MO could equally well be β instead of α. In that case, we would have

$$\psi(1,2,3) = \left[\frac{1}{3!}\right]^{1/2} \begin{vmatrix} 1s\sigma\alpha(1) & 1s\sigma\alpha(2) & 1s\sigma\alpha(3) \\ 1s\sigma\beta(1) & 1s\sigma\beta(2) & 1s\sigma\beta(3) \\ 1s\sigma^*\beta(1) & 1s\sigma^*\beta(2) & 1s\sigma^*\beta(3) \end{vmatrix}. \tag{2}$$

The electronic configuration for He_2 is $(1s\sigma)^2(1s\sigma^*)^2$, so that its bond order is

$$\rho_{He_2} = \frac{2-2}{2} = 0. \tag{3}$$

Therefore, we do not expect He_2 to be bound. Indeed, it is observed to be a monatomic rare gas. The bond order of He_2^+, on the other hand, is

$$\rho_{He_2^{1+}} = \frac{2-1}{2} = 0.5. \tag{4}$$

Consequently, we expect He_2^+ to be a bound molecule.

Since He_2^+ forms a chemical bond, we would expect the helium ions to combine with neutral helium atoms to form He_2^+ via the reaction $He^+ + He \longrightarrow He_2^+$. Therefore, the principal components shortly after the discharge would be He_2^+ and some He atoms and He^+ ions due to the incompleteness of the reaction. At a much later time, the He_2^+ ions would have picked up an electron to become neutral He_2, which would dissociate into He atoms. Therefore, the only thing in the container after a long time interval would be He atoms.

14.23 (A) Describe a simple LCAO–MO expansion that might be employed to carry out a molecular orbital calculation for the NH molecule.

(B) Discuss qualitatively the expected nature of the molecular orbitals that would be obtained in the calculation. What would you expect the energy ordering of these orbitals to be? Write down an appropriate representation of the MO electronic configuration. What is the molecular term symbol for the ground state? What are the term symbols for the excited angular momentum states? Arrange all angular momentum states in order of increasing energy.

Solution

(A) The most obvious expansion includes the atomic orbitals in the ground-state configurations of the two atoms:

$$\lambda = a_1\chi_{1sN} + a_2\chi_{2sN} + a_3\chi_{2pzN} + a_4\chi_{2pxN} + a_5\chi_{2pyN} + a_6\chi_{1sH}. \tag{1}$$

(B) The nature of the MOs will be very similar to that shown in Figure 14.17:

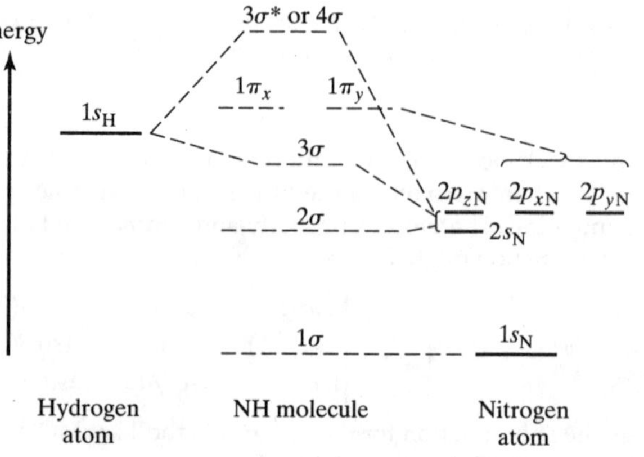

The energy of the nitrogen 1s orbital lies far below that of the remaining orbitals. Hence, the 1σ MO will be primarily a $1s_N$ orbital. $|a_1|$ will be large, $|a_2|$, $|a_3|$, and $|a_6|$ small, and $a_4 = a_5 = 0$. The $2s_N$ and $2p_{zN}$ orbitals will combine to form an *sp* hybrid. The 2σ MO will be primarily just this hybrid, so that $|a_2|$ and $|a_3|$ will be large, $|a_1|$ and $|a_6|$ small, and $a_4 = a_5 = 0$. 3σ will be a bonding combination of the *sp* nitrogen hybrid with the hydrogen 1s orbital, with $|a_2|$, $|a_3|$, and $|a_6|$ large, $|a_1|$ small, and a_4 and a_5 both zero. The next two MOs will be degenerate π orbitals, either $2p_{xN}$ or $2p_{yN}$, with either $|a_4|$ or $|a_5|$ large and the other one zero and all other coefficients zero. The highest-energy 4σ MO will be the antibonding combination of the hydrogen 1s and the nitrogen *sp* hybrid. This orbital will have $|a_2|$, $|a_3|$, and $|a_6|$ large, $|a_1|$ small, and a_4 and a_5 both zero. The electronic configuration for the eight electrons in the NH molecule is, therefore, $(1\sigma)^2(2\sigma)^2(3\sigma)^2(1\pi)^2$. The ground state requires that the multiplicity be maximum. Consequently, we want the spin functions of the two electrons in the unfilled π MOs both to be α or β. This means that they must have different values of m, one $+1$, the other -1, so that $|M| = 0$ while $S = 1$. The spin state is triplet, which makes the total spin function symmetric, so that the spatial orbitals must be antisymmetric to electron exchange. As a result, reflection through a plane containing the nuclei will exchange the π_{+1} and π_{-1} MOs and thereby produce a sign change on the total wave function. Hence, the reflection symmetry is negative, as shown in Table 14.5 in the text. These considerations yield a molecular term symbol $\boxed{^3\Sigma^-}$. There is no inversion symmetry for the heteronuclear diatomic molecule.

There are two excited angular-momentum states. Both of these have the spin functions of the two electrons in the unfilled π MOs paired, so that $S = 0$. In one case, we can have one electron with $m = -1$ and the other with $m = +1$, so that $|M| = 0$. In the second case, both electrons have either $m = -1$ or $+1$, which makes $|M| = 2$. The two excited term states are $^1\Sigma$ and $^1\Delta$. The reflection symmetry for the π^2 configuration with an antisymmetric singlet spin function is positive when in the Σ state. There is no reflection symmetry in a Δ state. Neither state has inversion symmetry. $\boxed{\text{The terms with symmetry included are } ^1\Sigma^+ \text{ and } ^1\Delta}$.

Using Hund's rules, we obtain the energy ordering

$$\boxed{E(^3\Sigma^-) < E(^1\Delta) < E(^1\Sigma^+)}.$$

14.25 The simple energy ordering given in the text for homonuclear diatomics predicts that the electronic configuration of B_2 should be $(1s\sigma_g)^2(1s\sigma_u^*)^2(2s\sigma_g)^2(2s\sigma_u)^2(2p_z\sigma_g)^2$. If this is the case, what would the full ground-state term symbol for B_2 be? Experimentally, it is found that the ground-state term symbol is actually $^3\Sigma_g^-$. What has occurred? Suggest a reasonable possibility for the actual electronic configuration for the ground state of B_2.

Solution

With the predicted configuration, all electrons are in filled sigma orbitals. This gives $|M| = S = 0$. Therefore, the angular-momentum state is $^1\Sigma$. The inversion symmetry for sigma orbitals is positive, and there is an even number of electrons in ungerade molecular orbitals, so that the overall inversion symmetry is g. Thus, we obtain $^1\Sigma_g^+$.

Since the observed ground-state term is $^3\Sigma_g^-$, it is clear that the electronic configuration predicted by the simple energy ordering given in the text has failed in this case. With g symmetry, we still have an even number of electrons in ungerade MOs. The negative reflection symmetry indicates that we must have a $(\pi)^2$ configuration, which will lead to negative symmetry when the spin state is triplet, as it is. Thus, it must be true that for B_2, the $2p_x\pi_u$ and $2p_y\pi_u$ orbitals have a lower energy than the $2p_z\sigma_g$ MO. In that case, we will have the ground-state configuration $(1s\sigma_g)^2(1s\sigma_u^*)^2(2s\sigma_g)^2(2s\sigma_u^*)^2(2p_x\pi_u)^1(2p_y\pi_u)^1$, which would give a ground-state term $^3\Sigma_g^-$, since the spins would both be $+\frac{1}{2}$ or $-\frac{1}{2}$ to give the maximum multiplicity. This would require that they have different values of m. One would have $m = +1$, while the other would have $m = -1$, so that $|M| = 0$ to give the $^3\Sigma$ state. There is an even number of electrons in ungerade MOs, yielding g overall symmetry, and the $(\pi)^2$ configuration with $S = 1$ gives negative reflection symmetry.

14.27# Consider a perturbed hydrogen atom whose Hamiltonian, in atomic units, is

$$\mathcal{H} = -\frac{1}{2}\nabla^2 - \frac{1}{r} + \frac{b}{r^2},$$

where b is a positive constant. The Schrödinger equation for this Hamiltonian can be solved exactly for the energy eigenvalues. The result for the ground state is

$$E = -\frac{1}{2\beta^2},$$

where

$$\beta = \frac{1 + \{1 + 8b\}^{1/2}}{2}.$$

Use first-order perturbation theory in which the perturbation is $\mathcal{H}' = b/r^2$ to compute the ground-state energy of the perturbed system for $b = 0.01, 0.05, 0.10, 0.50, 1.0$, and 10.0. In each case, compute the percent error in the perturbation calculation, and plot the percent error as a function of b.

Solution

The zeroth-order solution is the 1s hydrogen-atom eigenfunction:

$$\psi_0 = \left(\frac{1}{\pi}\right)^{1/2} e^{-r}. \qquad (1)$$

The average of the perturbation over the zeroth-order eigenfunction is, therefore,

$$\langle \mathcal{H}' \rangle = \left\langle \psi_o \left| \frac{b}{r^2} \right| \psi_o \right\rangle = \frac{b}{\pi} \int_{\phi=0}^{2\pi} \int_{\theta=0}^{\pi} \int_{r=0}^{\infty} \frac{e^{-2r}}{r^2} r^2 dr \sin\theta \, d\theta \, d\phi$$

$$= 4b \int_{r=0}^{\infty} e^{-2r} dr = -2b\, e^{-2r} \Big|_0^{\infty} = 2b. \qquad \textbf{(2)}$$

The zeroth-order, unperturbed ground-state energy of the hydrogen atom is $-\frac{1}{2}$ hartree. Thus, the first-order energy is

$$\boxed{E = -\frac{1}{2} + 2b \text{ hartrees}}. \qquad \textbf{(3)}$$

Using the exact result given in the problem, we obtain the following comparison:

b	β	E_{exact} (hartree)	$E_{\text{perturbation}}$ (hartree)	% error
0.01	1.0196	−0.48095	−0.48000	0.197
0.05	1.0916	−0.41961	−0.40000	4.673
0.10	1.1708	−0.36476	−0.30000	17.75
0.50	1.6180	−0.19099	0.50000	361.62
1.00	2.0000	−0.12500	1.50000	1,300.00

The accompanying plot illustrates the dependence of the perturbation result upon the size of the perturbation. When b is small, the result is very good. As b increases, the perturbation result becomes progressively worse. When the perturbation is even more important than the coulombic $-1/r$ term, the result is terrible.

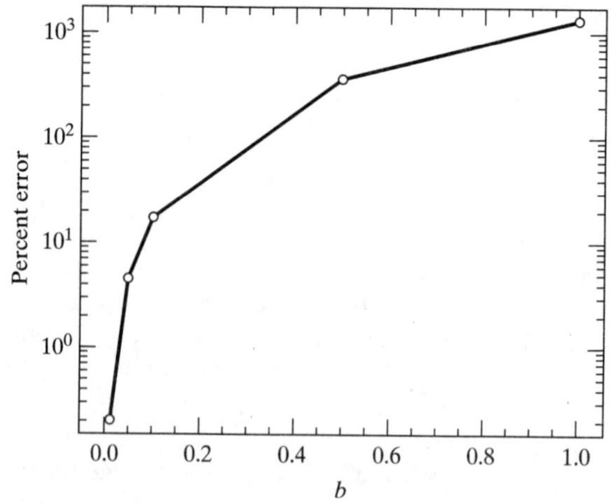

14.29 Prepare a diagram similar to that shown in Figure 14.23 for the λ_7 MO obtained in the minimal-basis-set SCF calculation for H_2O described in the text. The resulting variational coefficients are given in Table 14.6. Discuss why this orbital has antibonding characteristics that make it a high-energy orbital.

Solution

Like MO λ_3, λ_7 is a combination of the $2p_{yo}$ oxygen orbital and the 1*s* hydrogen orbitals. However, unlike λ_3, in this case the signs on the variational coefficients produce an antibonding situation. The coefficient multiplying the χ_{1sH2} orbital is negative, whereas that multiplying χ_{1sH3} is positive. The χ_{2pyo} coefficient is positive, so that we have the following situation:

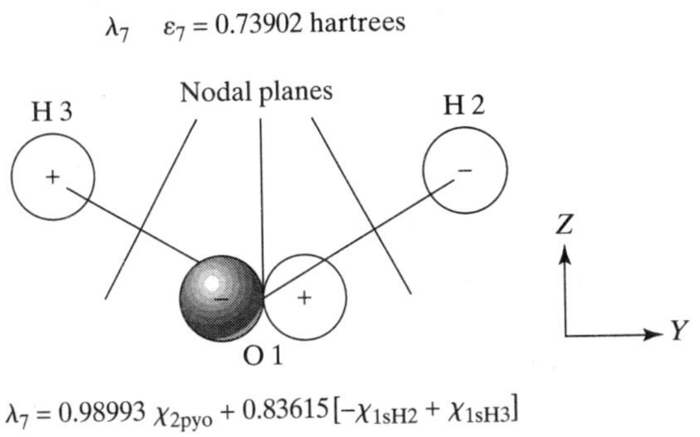

$$\lambda_7 = 0.98993\chi_{2pyo} + 0.83615[-\chi_{1sH2} + \chi_{1sH3}].$$

The orientation of the orbitals is clearly unfavorable to bonding. There are three nodal planes present. Two of these bisect the O–H bonding axes. As a result, the orbital has antibonding character and high energy.

14.31 How many total molecular orbitals were obtained in the LCAO–MO calculation discussed in the text in which 13 expansion functions were used for the H_2O molecule? How many of these orbitals are virtual orbitals?

Solution

The use of 13 expansion functions will produce a 13×13 secular equation, which will yield 13 orbital energies and, therefore, 13 MOs. H_2O contains 10 electrons, which will occupy 5 of these MO's with paired spins. There will, consequently, be eight virtual orbitals present.

14.33 The minimal-basis-set SCF calculation for H_2O described in the text produced seven molecular orbitals. Suppose an investigator carries out a CI calculation for H_2O employing this minimal-basis-set SCF calculation as a starting point. He then uses all possible single excitations with a multiplicity of unity in his CI expansion.

(A) How many total configurations are present in his CI expansion?

(B) What are these configurations?

(C) How many singlet-state energies would be obtained in the CI calculations?

(D) Which variational coefficient do you expect to have the largest magnitude for the eigenfunction corresponding to the lowest variational energy? Explain.

Solution

(A) and (B) Using the minimal basis set, the text shows the HF configuration of H_2O to be

$$D_o = (\lambda_1)^2(\lambda_2)^2(\lambda_3)^2(\lambda_4)^2(\lambda_5)^2. \quad \textbf{(1)}$$

We can generate 10 additional single excitations by taking one electron from one of the five occupied MOs and exciting it to either λ_6 or λ_7 with the same spin function. Since it doesn't matter whether the spin function is α or β, there are 10 possible singles that can be generated:

$$D_1 = (\lambda_1)^2(\lambda_2)^2(\lambda_3)^2(\lambda_4)^2(\lambda_5)^1(\lambda_6)^1, \qquad D_2 = (\lambda_1)^2(\lambda_2)^2(\lambda_3)^2(\lambda_4)^2(\lambda_5)^1(\lambda_7)^1,$$

$$D_3 = (\lambda_1)^2(\lambda_2)^2(\lambda_3)^2(\lambda_4)^1(\lambda_5)^2(\lambda_6)^1, \qquad D_4 = (\lambda_1)^2(\lambda_2)^2(\lambda_3)^2(\lambda_4)^1(\lambda_5)^2(\lambda_7)^1,$$

$$D_5 = (\lambda_1)^2(\lambda_2)^2(\lambda_3)^1(\lambda_4)^2(\lambda_5)^2(\lambda_6)^1, \qquad D_6 = (\lambda_1)^2(\lambda_2)^2(\lambda_3)^1(\lambda_4)^2(\lambda_5)^2(\lambda_7)^1,$$

$$D_7 = (\lambda_1)^2(\lambda_2)^1(\lambda_3)^2(\lambda_4)^2(\lambda_5)^2(\lambda_6)^1, \qquad D_8 = (\lambda_1)^2(\lambda_2)^1(\lambda_3)^2(\lambda_4)^2(\lambda_5)^2(\lambda_7)^1,$$

$$D_9 = (\lambda_1)^1(\lambda_2)^2(\lambda_3)^2(\lambda_4)^2(\lambda_5)^2(\lambda_6)^1, \text{ and } D_{10} = (\lambda_1)^1(\lambda_2)^2(\lambda_3)^2(\lambda_4)^2(\lambda_5)^2(\lambda_7)^1. \quad \textbf{(2)}$$

The CI expansion contains 11 terms:

$$\psi(1, 2, 3, \ldots, 10) = \sum_{i=0}^{10} C_i D_i. \quad \textbf{(3)}$$

(C) Since the CI contains 11 variational coefficients, the secular equation will be an 11×11 determinant that will yield 11 variational energies for various singlets states of H_2O.

(D) The C_o that multiplies the HF configuration will have the largest magnitude of any of the variational parameters, because the HF configuration will make the most important contribution to the ground-state energy and eigenfunction, since it uses the lowest-energy MOs.

14.35 Let us consider only the six π electrons in benzene. These six electrons are on six atoms arranged in a regular hexagon as shown in the following diagram:

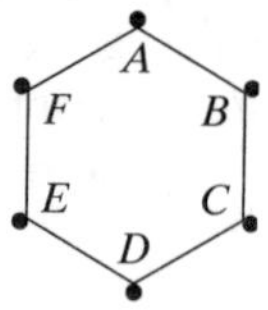

The black dots represent the six π electrons. In the perfect-pairing approximation used in VB theory, three of these electrons must have α spin functions and three β spin functions.

(A) How many Slater determinants can be constructed that satisfy the perfect-pairing approximation?

(B) Using the notation $(\alpha\alpha\alpha\beta\beta\beta)$ to represent the Slater determinant $|a\alpha\, b\alpha\, c\alpha\, d\beta\, e\beta\, f\beta|$, write down all of the possible VB determinants for the preceding six-electron system. How many of these will be needed to represent each Kekule or Dewar structure?

Solution

(A) We have six electrons, three of which must have α spin functions. The number of ways this can be accomplished is the number of combinations of six objects taken three at a time, viz.,

$$C(6,3) = \frac{6!}{3!\,3!} = \frac{720}{36} = \underline{20}. \quad \textbf{(1)}$$

Consequently, we expect 20 Slater determinants that satisfy the perfect pairing approximation.

(B) Let us start with the determinant

$$D_1 = (\alpha\alpha\alpha\beta\beta\beta) \quad \textbf{(2)}$$

and execute single exchanges. There are three α spins that can be exchanged with three β spins. Therefore, there are nine possible single exchanges. These exchanges give the determinants

(3) $D_2 = (\alpha\alpha\beta\alpha\beta\beta)$, **(4)** $D_3 = (\alpha\alpha\beta\beta\alpha\beta)$, **(5)** $D_4 = (\alpha\alpha\beta\beta\beta\alpha)$,
(6) $D_5 = (\alpha\beta\alpha\alpha\beta\beta)$, **(7)** $D_6 = (\alpha\beta\alpha\beta\alpha\beta)$, **(8)** $D_7 = (\alpha\beta\alpha\beta\beta\alpha)$,
(9) $D_8 = (\beta\alpha\alpha\alpha\beta\beta)$, **(10)** $D_9 = (\beta\alpha\alpha\beta\alpha\beta)$, and **(11)** $D_{10} = (\beta\alpha\alpha\beta\beta\alpha)$.

We now execute the double exchanges. There are three possible doubles of the α spins that can be exchanged with three possible doubles of the β spins. This gives another nine determinants:

(12) $D_{11} = (\alpha\beta\beta\alpha\alpha\beta)$; **(13)** $D_{12} = (\alpha\beta\beta\alpha\beta\alpha)$; **(14)** $D_{13} = (\alpha\beta\beta\beta\alpha\alpha)$;
(15) $D_{14} = (\beta\alpha\beta\alpha\alpha\beta)$; **(16)** $D_{15} = (\beta\alpha\beta\alpha\beta\alpha)$; **(17)** $D_{16} = (\beta\alpha\beta\beta\alpha\alpha)$;
(18) $D_{17} = (\beta\beta\alpha\alpha\alpha\beta)$; **(19)** $D_{18} = (\beta\beta\alpha\alpha\beta\alpha)$; **(20)** $D_{19} = (\beta\beta\alpha\beta\alpha\alpha)$.

Finally, there is one triple exchange that produces

$$D_{20} = (\beta\beta\beta\alpha\alpha\alpha). \quad \textbf{(21)}$$

These are the 20 possible Slater determinants that obey the perfect-pairing approximation.

Each Kekulé or Dewar structure contains three valence bonds. We will, therefore, need $\boxed{2^3 = 8 \text{ Slater determinants}}$ to represent the *VB* eigenfunction for each of these structures.

14.37 Let us consider only the six π electrons in benzene. These six electrons are on six atoms arranged in a regular hexagon as shown in the following diagram:

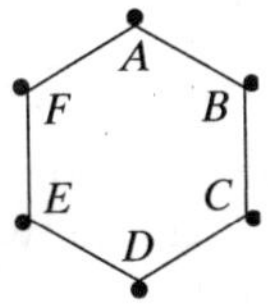

The black dots represent the six π electrons. In the perfect-pairing approximation used in VB theory, three of these electrons must have α spin functions and three β spin functions. This yields a total of 20 Slater determinants that satisfy the perfect-pairing requirement. These determinants are given in Problem 14.36. In terms of these determinants, construct VB wave functions for the three Dewar structures of benzene shown in Figure 14.25.

Solution

The three Dewar bonding structures are

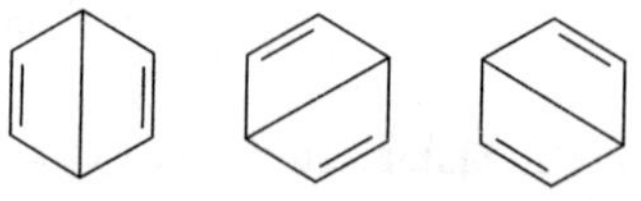

III IV V

For structure III, the *B–C*, *A–D*, and *E–F* spins must be paired. The Slater determinants that have these spins paired are D_3, D_4, D_6, D_7, D_{14}, D_{15}, D_{17}, and D_{18}. Referencing the signs to D_3, we see that determinants D_4, D_6, and D_{14} have one pair of spins exchanged relative to D_3. Determinants D_7, D_{15}, and D_{17} have two pairs of spins exchanged relative to D_3, and D_{18} has all three pairs exchanged. The *VB* wave function for structure III is, therefore,

$$\boxed{\psi_{\text{III}}^{VB} = [D_3 - D_4 - D_6 - D_{14} + D_7 + D_{15} + D_{17} - D_{18}]}. \quad \textbf{(1)}$$

For structure IV, the *A–F*, *B–E*, and *C–D* spins must be paired. The Slater determinants that have these spins paired are D_1, D_2, D_6, D_{10}, D_{11}, D_{15}, D_{19}, and D_{20}. Referencing the signs to D_1, we see that determinants D_2, D_6, and D_{10} have one pair of spins exchanged relative to D_1. Determinants D_{11}, D_{15}, and D_{19} have two pairs of spins exchanged relative to D_1, and D_{20} has all three pairs exchanged. The *VB* wave function for structure IV is, therefore,

$$\boxed{\psi_{\text{IV}}^{VB} = [D_1 - D_2 - D_6 - D_{10} + D_{11} + D_{15} + D_{19} - D_{20}]}. \quad \textbf{(2)}$$

For structure V, the *A–B*, *C–F*, and *D–E* spins must be paired. The Slater determinants that have these spins paired are D_5, D_6, D_8, D_9, D_{12}, D_{13}, D_{15}, and D_{16}. Referencing the signs to D_5, we see that determinants D_6, D_8, and D_{12} have one pair of spins exchanged relative to D_5. Determinants D_9, D_{13}, and D_{15} have two pairs of spins exchanged relative to D_5, and D_{16} has all three pairs exchanged. The *VB* wave function for structure V is, therefore,

$$\boxed{\psi_{\text{V}}^{VB} = [D_5 - D_6 - D_8 - D_{12} + D_9 + D_{13} + D_{15} - D_{16}]}. \quad \textbf{(3)}$$

14.39 A variational calculation on benzene is carried out using the trial wave function

$$\psi_{\text{total}}^{\text{VB}} = C_{\text{I}}\psi_{\text{I}}^{\text{VB}} + C_{\text{II}}\psi_{\text{II}}^{\text{VB}} + C_{\text{III}}\psi_{\text{III}}^{\text{VB}} + C_{\text{IV}}\psi_{\text{IV}}^{\text{VB}} + C_{\text{V}}\psi_{\text{V}}^{\text{VB}},$$

where $\psi_{\text{I}}^{\text{VB}}$ and $\psi_{\text{II}}^{\text{VB}}$ are the VB wave functions for the two Kekule structure of benzene and $\psi_{\text{III}}^{\text{VB}}$, $\psi_{\text{IV}}^{\text{VB}}$, and $\psi_{\text{V}}^{\text{VB}}$ are the wave functions for the three Dewar structures. After completion of the calculation, what relationship would be observed between C_{I}, C_{II}, C_{III}, C_{IV}, and C_{V}? Are any structures in resonance?

Solution

The three short π bonds present in the Kekulé structures are more stable than the two short and one long π bond in the Dewar structures. Hence, the Kekulé structures make a larger contribution to the overall wave function for the ground state of benzene than do the Dewar structures. This will be reflected by that fact that $|C_I|^2$ and $|C_{II}|^2$ will both exceed $|C_{III}|^2$, $|C_{IV}|^2$, and $|C_V|^2$. Since the two Kekulé structures and the three Dewar structures are equivalent, we will also find that

$$\boxed{|C_I|^2 = |C_{II}|^2} \qquad (1)$$

and

$$\boxed{|C_{III}|^2 = |C_{IV}|^2 = |C_V|^2}. \qquad (2)$$

Therefore, *VB* structures I and II are resonance forms, as are III, IV, and V, but I and II are not in resonance with III, IV, and V.

CHAPTER 15

Rotational, Vibrational and Electronic Spectra

15.1 The spacing between two energy levels is 0.2380 kJ mol^{-1}. What are the wavelength and wave number of the electromagnetic radiation needed to produce the resonance condition? Refer to Table 15.1 and state what type of energy levels we might reasonably expect to be associated with that wavelength and wave number?

Solution

Using Eq. 15.1, we can write the resonance condition as

$$h\nu = \frac{hc}{\lambda} = \Delta E. \quad (1)$$

Solving for the wavelength, we obtain

$$\lambda = \frac{hc}{\Delta E} = \frac{(6.6261 \times 10^{-34}\ \mathrm{J\,s})(2.9979 \times 10^{8}\ \mathrm{m\,s^{-1}})(6.0221 \times 10^{23}\ \mathrm{mol^{-1}})}{(238.0\ \mathrm{J\,mol^{-1}})}$$
$$= \underline{0.0005026\ \mathrm{m}} = \underline{0.05026\ \mathrm{cm}}. \quad (2)$$

The wave number, $\overline{\upsilon}$, is the reciprocal of the wavelength. Therefore,

$$\overline{\upsilon} = \frac{1}{\lambda} = \frac{1}{0.05026\ \mathrm{cm}} = \underline{19.90\ \mathrm{cm}^{-1}}. \quad (3)$$

Since the wavelength required to produce resonance is on the order of 10^{-1} cm, we are in the microwave region of the spectrum, and the energy levels are probably rotational-energy states of some molecule.

15.3 The selection rule for microwave rotational spectra is $\Delta J = \pm 1$. In this problem, you will show that this rule holds in two specific cases. The normalized eigenfunctions for the ($J = 1, M = 0$), ($J = 2, M = 0$), and ($J = 3, M = 0$) rotational states are

$$Y_1^0(\theta, \phi) = [3/4\pi]^{1/2} \cos\theta,$$
$$Y_2^0(\theta, \phi) = [5/16\pi]^{1/2}[3\cos^2\theta - 1],$$

and

$$Y_3^0(\theta, \phi) = [7/16\pi]^{1/2}[2\cos\theta - 5\cos\theta\sin^2\theta],$$

respectively. Using these data,

(A) show that the ($J = 1, M = 0$) $\longrightarrow$ ($J = 2, M = 0$) rotational transition is allowed, and

(B) show that the ($J = 1, M = 0$) $\longrightarrow$ ($J = 3, M = 0$) rotational transition is forbidden.

Solution

(A) For the ($J = 1, M = 0$) $\longrightarrow$ ($J = 2, M = 0$) transition, the transition matrix element is $\langle [Y_1^0(\theta, \phi)]^* | x + y + z | Y_2^0(\theta, \phi)\rangle$. The x integral is

$$\int_{\phi=0}^{2\pi}\int_{\theta=0}^{\pi}\left[\frac{3}{4\pi}\right]^{1/2}\cos\theta[R\sin\theta\cos\phi]\left[\frac{5}{16\pi}\right]^{1/2}[3\cos^2\theta - 1]\sin\theta\, d\theta\, d\phi$$
$$= R\left[\frac{3}{4\pi}\right]^{1/2}\left[\frac{5}{16\pi}\right]^{1/2}\int_{\theta=0}^{\pi}\cos\theta\sin^2\theta[3\cos^2\theta - 1]\,d\theta\int_{\phi=0}^{2\pi}\cos\phi\, d\phi = 0, \quad (1)$$

since the integral over ϕ is zero. The y integral in the transition matrix element is

$$\int_{\phi=0}^{2\pi}\int_{\theta=0}^{\pi}\left[\frac{3}{4\pi}\right]^{1/2}\cos\theta[R\sin\theta\sin\phi]\left[\frac{5}{16\pi}\right]^{1/2}[3\cos^2\theta-1]\sin\theta d\theta d\phi$$

$$= R\left[\frac{3}{4\pi}\right]^{1/2}\left[\frac{5}{16\pi}\right]^{1/2}\int_{\theta=0}^{\pi}\cos\theta\sin^2\theta[3\cos^2\theta-1]d\theta\int_{\phi=0}^{2\pi}\sin\phi d\phi = 0, \quad (2)$$

since the integral over ϕ again vanishes. The z integral in the transition matrix element is

$$\int_{\phi=0}^{2\pi}\int_{\theta=0}^{\pi}\left[\frac{3}{4\pi}\right]^{1/2}\cos\theta[R\cos\theta]\left[\frac{5}{16\pi}\right]^{1/2}[3\cos^2\theta-1]\sin\theta d\theta d\phi$$

$$= R\left[\frac{3}{4\pi}\right]^{1/2}\left[\frac{5}{16\pi}\right]^{1/2}\int_{\theta=0}^{\pi}\cos^2\theta\sin\theta[3\cos^2\theta-1]d\theta\int_{\phi=0}^{2\pi}d\phi$$

$$= R2\pi\left[\frac{3}{4\pi}\right]^{1/2}\left[\frac{5}{16\pi}\right]^{1/2}\int_{\theta=0}^{\pi}\cos^2\theta\sin\theta[3\cos^2\theta-1]d\theta$$

$$= R2\pi\left[\frac{3}{4\pi}\right]^{1/2}\left[\frac{5}{16\pi}\right]^{1/2}\int_{\theta=0}^{\pi}[3\cos^4\theta-\cos^2\theta]\sin\theta d\theta$$

$$= 2\pi R\left[\frac{3}{4\pi}\right]^{1/2}\left[\frac{5}{16\pi}\right]^{1/2}\left[-\frac{3\cos^5\theta}{5}+\frac{\cos^3\theta}{3}\right]_0^{\pi}$$

$$= 2\pi R\left[\frac{3}{4\pi}\right]^{1/2}\left[\frac{5}{16\pi}\right]^{1/2}\left[\frac{3}{5}-\frac{1}{3}+\frac{3}{5}-\frac{1}{3}\right] \neq 0. \quad (3)$$

Since the transition matrix is not zero, the $(J = 1, M = 0) \longrightarrow (J = 2, M = 0)$ rotational transition is allowed.

(B) For the $(J = 1,\ M = 0) \longrightarrow (J = 3,\ M = 0)$ transition, the transition matrix element is $\langle[Y_1^0(\theta,\phi)]^*|x + y + z|Y_3^0(\theta,\phi)\rangle$. The x integral is

$$\int_{\phi=0}^{2\pi}\int_{\theta=0}^{\pi}\left[\frac{3}{4\pi}\right]^{1/2}\cos\theta[R\sin\theta\cos\phi]\left[\frac{7}{16\pi}\right]^{1/2}[2\cos\theta-5\cos\theta\sin^2\theta]\sin\theta d\theta d\phi$$

$$= R\left[\frac{3}{4\pi}\right]^{1/2}\left[\frac{7}{16\pi}\right]^{1/2}\int_{\theta=0}^{\pi}\cos\theta\sin^2\theta\,[2\cos\theta-5\cos\theta\sin^2\theta]d\theta\int_{\phi=0}^{2\pi}\cos\phi d\phi = 0, \quad (4)$$

since the integral over ϕ is zero. The y integral in the transition matrix element is

$$\int_{\phi=0}^{2\pi}\int_{\theta=0}^{\pi}\left[\frac{3}{4\pi}\right]^{1/2}\cos\theta\,[R\sin\theta\sin\phi]\left[\frac{7}{16\pi}\right]^{1/2}[2\cos\theta-5\cos\theta\sin^2\theta]\sin\theta d\theta d\phi$$

$$= R\left[\frac{3}{4\pi}\right]^{1/2}\left[\frac{7}{16\pi}\right]^{1/2}\int_{\theta=0}^{\pi}\cos\theta\sin^2\theta[2\cos\theta-5\cos\theta\sin^2\theta]d\theta\int_{\phi=0}^{2\pi}\sin\phi d\phi$$

$$= 0, \quad (5)$$

since the integral over ϕ again vanishes. The z integral in the transition matrix element is

$$\int_{\phi=0}^{2\pi}\int_{\theta=0}^{\pi}\left[\frac{3}{4\pi}\right]^{1/2}\cos\theta[R\cos\theta]\left[\frac{7}{16\pi}\right]^{1/2}[2\cos\theta-5\cos\theta\sin^2\theta]\sin\theta\,d\theta\,d\phi$$

$$=R\left[\frac{3}{4\pi}\right]^{1/2}\left[\frac{7}{16\pi}\right]^{1/2}\int_{\theta=0}^{\pi}\cos^2\theta[2\cos\theta-5\cos\theta\sin^2\theta]\sin\theta\,d\theta\int_{\phi=0}^{2\pi}d\phi$$

$$=2\pi R\left[\frac{3}{4\pi}\right]^{1/2}\left[\frac{7}{16\pi}\right]^{1/2}\int_{\theta=0}^{\pi}\cos^2\theta[2\cos\theta-5\cos\theta\sin^2\theta]\sin\theta\,d\theta. \quad (6)$$

Let us now examine the integral over θ that appears in Eq. 6. We have

$$\int_{\theta=0}^{\pi}\cos^2\theta[2\cos\theta-5\cos\theta\sin^2\theta]\sin\theta\,d\theta$$

$$=\int_{\theta=0}^{\pi}[2\cos^3\theta-5\cos^3\theta\{1-\cos^2\theta\}]\sin\theta\,d\theta$$

$$=\left[-\frac{2\cos^4\theta}{4}+\frac{5\cos^4\theta}{4}-\frac{5\cos^6\theta}{6}\right]_0^{\pi}=0, \quad (7)$$

since the upper and lower limits give identical values. Therefore,

$$\langle[Y_1^0(\theta,\phi)]^*[x+y+z]Y_3^0(\theta,\phi)\rangle=0, \quad (8)$$

and the $(J=1, M=0)\longrightarrow(J=3, M=0)$ rotational transition is forbidden.

15.5 To the extent that the Born–Oppenheimer approximation holds, the Hamiltonian is independent of nuclear mass. This means that molecular structures, bond lengths, and bond angles will not depend upon the isotopic mass of the atoms that make up the molecule. Suppose we mix BeH and BeD (Be^2H) and take the microwave rotational spectrum of the mixture. Using the data obtained in Example 15.2, determine the wave numbers for the first three rotational absorption bands for both BeH and BeD.

Solution

The band spacing for BeH given in Example 15.2 is 20.62 cm^{-1}. Equation 15.29 shows that this spacing is equal to 2Bh. Since the rotational absorption bands occur at 2Bh, 4Bh, 6Bh, ..., the first three absorption bands for BeH will occur at wave numbers 20.62, 41.24, and 61.86 cm^{-1}.

The reduced mass for Be^2H is greater than that for BeH. Therefore, the band spacings and the location of the absorption bands will be different. Since we have

$$B=\frac{\hbar^2}{2\mu R^2 h}, \quad (1)$$

and since R is the same for both BeH and Be^2H, the ratio of the rotation constants for BeH and Be^2H is

$$\frac{B_{BeH}}{B_{BeD}}=\frac{\mu_{BeD}}{\mu_{BeH}}=\frac{m_{Be}m_D/(m_{Be}+m_D)}{m_{Be}m_H/(m_{Be}+m_H)}=\frac{m_D(m_{Be}+m_H)}{m_H(m_{Be}+m_D)}$$

$$=\frac{(2.0141)(9.012+1.0079)}{(1.0079)(9.012+2.0141)}=1.8160. \quad (2)$$

The rotational absorption band spacing is proportional to the rotational constant. Therefore,

$$\Delta\varepsilon_{\text{BeD}} = \frac{B_{\text{BeD}}}{B_{\text{BeH}}}\Delta\varepsilon_{\text{BeH}} = \frac{20.62\ \text{cm}^{-1}}{1.8160} = 11.35\ \text{cm}^{-1}. \tag{3}$$

The first three absorption bands for BeD will be observed at 11.35, 22.70, and 34.05 cm^{-1}.

15.7 Since chlorine has isotopes of nominal mass 35 and 37, HCl is a mixture of $H^{35}Cl$ and $H^{37}Cl$. How much spectral resolution in terms of cm^{-1} must our spectrometer have to permit us to separate the lowest energy rotational absorption band of $H^{35}Cl$ from that of $H^{37}Cl$? The atomic masses of ^{35}Cl and ^{37}Cl are 34.9688 and 36.9659, respectively, and the HCl equilibrium bond length is 1.3152×10^{-10} m. To the extent that the Born–Oppenheimer approximation holds, the equilibrium bond lengths of $H^{35}Cl$ and $H^{37}Cl$ are the same. What is the separation between the $H^{35}Cl$ and $H^{37}Cl$ absorption bands for the $J = 7$ to $J = 8$ transition?

Solution

The band spacings are given by Eq. 15.29:

$$\Delta\varepsilon = 2Bh = \frac{h^2}{4\pi^2 I} = \frac{h^2}{4\pi^2\mu R^2} = \frac{\hbar^2}{\mu R^2}. \tag{1}$$

Therefore, we must compute the reduced mass for $H^{35}Cl$ and $H^{37}Cl$, viz.,

$$\mu_{35} = \frac{(34.9688)(1.0079)}{(34.9688 + 1.0079)}\text{g mol}^{-1} \times \frac{1}{6.02214 \times 10^{23}\ \text{mol}^{-1}} \times \frac{1\ \text{kg}}{1{,}000\ \text{g}}$$
$$= 1.6268 \times 10^{-27}\ \text{kg} \tag{2}$$

and

$$\mu_{37} = \frac{(36.9659)(1.0079)}{(36.9659 + 1.0079)}\text{g mol}^{-1} \times \frac{1}{6.02214 \times 10^{23}\ \text{mol}^{-1}} \times \frac{1\ \text{kg}}{1{,}000\ \text{g}}$$
$$= 1.6292 \times 10^{-27}\ \text{kg}. \tag{3}$$

In atomic mass units, these values are

$$\mu_{35} = 1.6268 \times 10^{-27}\ \text{kg} \times \frac{1}{9.10939 \times 10^{-31}\ \text{kg}} = 1{,}785.8 \tag{4}$$

and

$$\mu_{37} = 1.6292 \times 10^{-27}\ \text{kg} \times \frac{1}{9.10939 \times 10^{-31}\ \text{kg}} = 1{,}788.5. \tag{5}$$

The HCl bond length, in bohr, is

$$R = 1.3152 \times 10^{-10}\ \text{m} \times \frac{1\ \text{bohr}}{0.529177 \times 10^{-10}\ \text{m}} = 2.4854\ \text{bohr}. \tag{6}$$

The corresponding rotational band spacings are

$$[\Delta\varepsilon]_{35} = \frac{(1)^2}{(1{,}785.8)(2.4854)^2}\ \text{hartree} = 9.0651 \times 10^{-5}\ \text{hartree} \tag{7}$$

and

$$[\Delta\varepsilon]_{37} = \frac{(1)^2}{(1{,}788.5)(2.4854)^2}\ \text{hartree} = 9.0515 \times 10^{-5}\ \text{hartree} \tag{8}$$

In wave numbers, the corresponding spacings are

$$[\Delta\varepsilon]_{35} = 9.0651 \times 10^{-5} \text{ hartree} \times \frac{219{,}475 \text{ cm}^{-1}}{\text{hartree}} = 19.896 \text{ cm}^{-1} \qquad \textbf{(9)}$$

and

$$[\Delta\varepsilon]_{37} = 9.0515 \times 10^{-5} \text{ hartree} \times \frac{219{,}475 \text{ cm}^{-1}}{\text{hartree}} = 19.866 \text{ cm}^{-1}. \qquad \textbf{(10)}$$

The first absorption band for the $J = 0$ to the $J = 1$ state, therefore, occurs at 19.866 cm^{-1} for $H^{37}Cl$ and at 19.896 cm^{-1} for $H^{35}Cl$. To distinguish these bands, our spectrometer must have a resolution sufficient to resolve differences of 0.030 cm^{-1}.

The energy difference between the absorption bands of $H^{35}Cl$ and $H^{37}Cl$ increases as J increases. Equation 15.26 shows that the $J = 7 \longrightarrow J = 8$ transition occurs at $16Bh$ instead of $2Bh$. Therefore, for $H^{35}Cl$, it occurs at 8(19.896) cm^{-1} = 159.17 cm^{-1}. For $H^{37}Cl$, the result is 8(19.866) cm^{-1} = 158.93 cm^{-1}. Hence, the band separation is now 0.24 cm^{-1}.

15.9 An investigator finds that if he increases the path length of the electromagnetic radiation beam through the sample cell in an absorption measurement by 50%, the ratio of the transmitted radiation intensity to the incident intensity drops by a factor of two, provided that the measurements are made at the same wavelength and concentration. Is this sufficient information to determine the fraction of the radiation transmitted in the original experiment? If so, compute that fraction. If not, state what additional information is required to do the calculation.

Solution

The ratio of the transmitted intensity to the incident intensity is given by Eq. 15.34:

$$\left[\frac{I}{I_o}\right] = \exp[-\varepsilon[C]\ell]. \qquad \textbf{(1)}$$

The problem tells us that if we increase ℓ to 1.5ℓ, the ratio of I to I_o decreases by a factor of two. If the transmitted intensity at the increased path length is I', then

$$\frac{[I'/I_o]}{[I/I_o]} = \frac{I'}{I} = \frac{\exp[-1.5\varepsilon[C]\ell]}{\exp[-\varepsilon[C]\ell]} = \exp[-0.5\varepsilon[C]\ell] = 0.500. \qquad \textbf{(2)}$$

Taking logarithms of both sides gives

$$-0.5\varepsilon[C]\ell = \ln(0.5). \qquad \textbf{(3)}$$

Solving for $\varepsilon[C]\ell$, we obtain

$$\varepsilon[C]\ell = -\frac{\ln(0.5)}{0.5} = 1.38629. \qquad \textbf{(4)}$$

Therefore, the fraction of the radiation transmitted in the original experiment is

$$\left[\frac{I}{I_o}\right] = \exp[-\varepsilon[C]\ell] = \exp[-1.38629] = \underline{0.25}. \qquad \textbf{(5)}$$

The problem contains sufficient data to perform the calculation.

15.11 (A) Compute the ratio of spontaneous emission to stimulated emission for a rotational transition whose transition frequency is 2×10^{11} s^{-1}, assuming that the incident radiation density per unit frequency is that for a blackbody radiator at a temperature of 500 K.

(B) At what temperature would spontaneous emission be 20% that of stimulated emission?

Solution

(A) The rates of stimulated and spontaneous emission are given by Eqs. 15.38 and 15.39, respectively:

$$w_J' = B_J\rho; \tag{1}$$

$$A_J = \left(\frac{8\pi h\nu^3}{c^3}\right)B_J. \tag{2}$$

The A_J–w_J' ratio is, therefore,

$$\frac{A_J}{w_J'} = \frac{\frac{8\pi h\nu^3}{c^3}}{\rho}. \tag{3}$$

Example 15.4 shows that the energy density per unit frequency for a black-body radiator is given by

$$\rho = \frac{8\pi h\nu^3}{c^3}\left\{\exp\left[\frac{h\nu}{kT}\right] - 1\right\}^{-1}. \tag{4}$$

Combining Eqs. 3 and 4 produces

$$\frac{A_J}{w_J'} = \left\{\exp\left[\frac{h\nu}{kT}\right] - 1\right\}. \tag{5}$$

With $\nu = 2 \times 10^{11}\ \text{s}^{-1}$ and $T = 500$ K, this ratio is

$$\frac{A_J}{w_J'} = \exp\left[\frac{(6.6261 \times 10^{-34})(2 \times 10^{11})}{(1.381 \times 10^{-23})(500)}\right] - 1 = \underline{0.0194}. \tag{6}$$

Slightly less than 2% of the total emission will be spontaneous.

(B) We seek the temperature for which we have

$$\frac{A_J}{w_J'} = \left\{\exp\left[\frac{h\nu}{kT}\right] - 1\right\} = 0.2000. \tag{7}$$

Rearranging and then taking logarithms of both sides produces

$$\frac{h\nu}{kT} = \ln(1.2), \tag{8}$$

so that the required temperature is

$$T = \frac{h\nu}{k\ln(1.2)} = \frac{(6.6261 \times 10^{-34})(2. \times 10^{11})}{(1.381 \times 10^{-23})\ln(1.2)} = \underline{52.63\ \text{K}}. \tag{9}$$

15.13 Consider a hypothetical situation in which the pure rotational selection rule is

$$\Delta J = \pm 2 \quad \text{instead of} \quad \Delta J = \pm 1.$$

(A) Derive a general formula giving the observed absorption frequencies for the allowed transitions from rotational state J for the rigid rotor.

(B) Describe, by means of an appropriate sketch, the expected appearance of a rotational spectrum of a rigid rotor for this selection rule.

Solution

(A) The energy levels for the rigid rotor are

$$E_J = J(J + 1)Bh, \qquad (1)$$

where B = rotational constant = $h/(8\pi^2 I)$. If we must have $\Delta J = \pm 2$, then the transition energies are

$$E_{J+2} - E_J = (J + 2)(J + 3)Bh - J(J + 1)Bh$$
$$= (J^2 + 5J + 6)Bh - (J^2 + J)Bh = (4J + 6)Bh. \qquad (2)$$

Absorptions will occur whenever the energy of the electromagnetic radiation, $h\nu$, is equal to the energy spacing given by Eq. 2. Thus, the absorption frequencies can be obtained from

$$h\nu = E_{J+2} - E_J = (4J + 6)Bh, \qquad (3)$$

which gives

$$\nu = (4J + 6)B. \qquad (4)$$

(B) The following transitions are expected in the table at the bottom left of the page. The absorption bands are now spaced by a frequency difference of $4B$. The spectrum will, therefore, have the appearance of the following sketch at the bottom right:

Rotational Transition	Frequency (s^{-1})
$J = 0$ to $J = 2$	6B
$J = 1$ to $J = 3$	10B
$J = 2$ to $J = 4$	14B
$J = 3$ to $J = 5$	18B
$J = 4$ to $J = 6$	22B

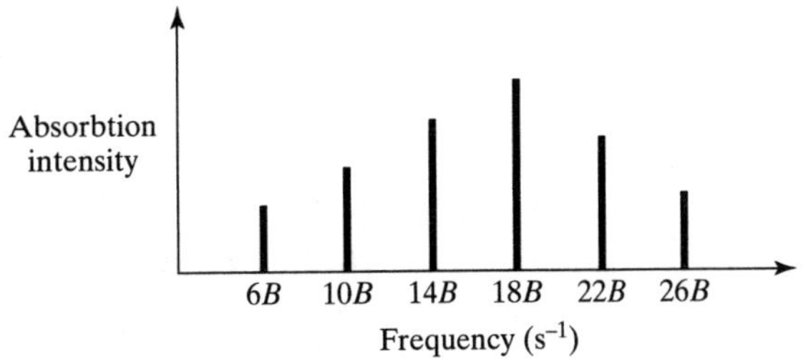

The absorption intensities are hypothetical; they are simply meant to show that not all absorptions have equal intensity and that some line will exhibit a maximum intensity. In order to predict this result, we must know the actual molecule and the temperature at which the spectrum is taken.

15.15[#] The absorption band for the pure rotational transition from $J = 0$ to $J = 1$ in a diatomic molecule possessing a permanent electric dipole moment is observed at an energy corresponding to 18.00 cm^{-1}.

(A) If we arbitrarily assign a value of 1.000 to the intensity of this band at 298 K, compute and plot the normalized relative intensities of the 18 absorption bands with lowest energy for this molecule at 298 K.

(B) Use the same procedure to compute and plot the relative intensities of the same 18 absorption bands at 150 K. (Assume that the absorption intensity is determined solely by the population of the initial state and that the spectrometer is capable of spanning the necessary range of frequencies.) Make separate plots of the results of (A) and (B) by plotting the normalized relative intensities versus the wave number at which the transition occurs. What happens to the intensities as the temperature drops?

Solution

(A) We first need to determine the value of Bh for our hypothetical molecule. The $J = 0$ to $J = 1$ transition occurs at an energy equal to $2Bh$. Using the conversion factors in Table 13.2, we obtain

$$Bh = \frac{18.00\ \text{cm}^{-1}}{2} \times \frac{2{,}625{,}500\ \text{J mol}^{-1}}{219{,}475\ \text{cm}^{-1}} \times \frac{1}{6.02214 \times 10^{23}\ \text{mol}^{-1}}$$

$$= 1.788 \times 10^{-22}\ \text{J}. \qquad \textbf{(1)}$$

The energy of state J is given by

$$E_J = J(J + 1)Bh = 1.788 \times 10^{-22}\, J(J + 1)\ \text{joules}. \qquad \textbf{(2)}$$

The population of state J can be obtained using Eq. 15.37:

$$N_J = Cg_J \exp\left[-\frac{E_J}{kT}\right] = C(2J + 1)\exp\left[-\frac{1.788 \times 10^{-22}\, J(J + 1)}{(1.381 \times 10^{-23})(298)}\right]$$

$$= C(2J + 1)\exp[-0.04344\, J(J + 1)]. \qquad \textbf{(3)}$$

For $J = 0$, we have $N_J = C$. Therefore, if we arbitrarily set the intensity of the $J = 0$ to $J = 1$ transition to 1.000, the intensities of the bands relative to this standard will be given by

$$\text{Relative intensity} = \frac{N_J}{C} = (2J + 1)\exp[-0.04344\, J(J + 1)]. \qquad \textbf{(4)}$$

Using Eq. 4 to compute the relative intensities of each of the spectral lines, we obtain the table on the following page.

(B) At $T = 150$ K, the relative intensities are given by Eq. 4, with 298 K replaced by 150 K. This result is

$$\text{Relative intensity} = \frac{N_J}{C} = (2J + 1)\exp[-0.08630\, J(J + 1)]. \qquad \textbf{(5)}$$

The results at 150 K are also shown in the table on the top of the next page.

The normalized relative intensities are obtained by dividing the relative intensities given by Eqs. 4 and 5 by the sum of all the computed relative intensities. For the transitions at 298 K and 150 K, these sums are 23.361 and 11.926, respectively. Therefore,

$$\text{Normalized relative intensity at 298} = \frac{(2J + 1)\exp[-0.04344J(J + 1)]}{23.361}. \qquad \textbf{(6)}$$

At 150 K, the corresponding result is

$$\text{Normalized relative intensity at 150} = \frac{(2J + 1)\exp[-0.08630\, J(J + 1)]}{11.926}. \qquad \textbf{(7)}$$

Initial State	Final State	Wave Number (cm^{-1})	Relative Intensity	
			298 K	150 K
0	1	18.00	1.000	1.000
1	2	36.00	2.750	2.524
2	3	54.00	3.853	2.979
3	4	72.00	4.156	2.485
4	5	90.00	3.775	1.602
5	6	108.00	2.988	0.826
6	7	126.00	2.097	0.347
7	8	144.00	1.317	0.119
8	9	162.00	0.745	0.034
9	10	180.00	0.381	0.0081
10	11	198.00	0.177	0.0016
11	12	216.00	0.074	0.0003
12	13	234.00	0.0285	
13	14	252.00	0.0099	
14	15	270.00	0.0082	
15	16	288.00	0.0009	
16	17	306.00	0.0002	
17	18	324.00	0.0001	

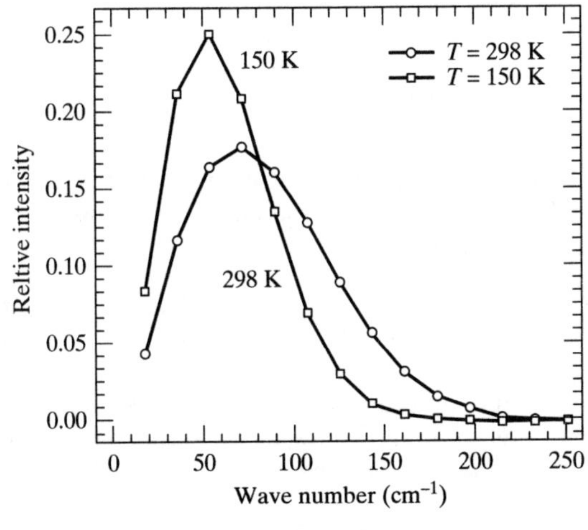

Plots of the normalized relative intensities at 298 K and 150 K are shown in the above graph. The plots demonstrate that the relative intensities of the lower energy transitions are enhanced as the temperature decreases, while those at higher energies show decreased relative intensities.

15.17 An investigator wishes to use the intensity of one band in the microwave rotational spectrum of the molecule described in Problem 15.15 as a means of detecting changes in temperature. The investigator states that the temperatures will range from a low of 298 K to a high of 400 K. The most intense spectral band at 298 K is the band at 72.00 cm^{-1} that corresponds to the $J = 3$ to $J = 4$ transition, while the least intense observable band is the one corresponding to the $J = 12$ to $J = 13$ transition at 234.00 cm^{-1}. (See Problem 15.15.) The investigator would like to have the advice of the physical chemistry class as to which rotational band should be selected so as to provide the maximum sensitivity to temperature changes. One member of the class suggests that the band at 72.00 cm^{-1} be used. Sam, on the other hand, believes that the 234.00 cm^{-1} band is the best choice. What do you think? Why? (Assume that the absorption intensity is determined solely by the population of the initial state and that the spectrometer is capable of spanning the necessary range of frequencies.)

Solution

The best choice is the band whose intensity shows the maximum percent change for a given percent change in the temperature. To determine which band this is, we first need to find the value of Bh for the hypothetical molecule in Problem 15.15. The $J = 0$ to $J = 1$ transition occurs at an energy equal to $2Bh$. Using the conversion factors in Table 13.2, we obtain

$$Bh = \frac{18.00 \text{ cm}^{-1}}{2} \times \frac{2{,}625{,}500 \text{ J mol}^{-1}}{219{,}475 \text{ cm}^{-1}} \times \frac{1}{6.02214 \times 10^{23} \text{ mol}^{-1}}$$
$$= 1.788 \times 10^{-22} \text{ J.} \qquad (1)$$

The energy of state J is given by

$$E_J = J(J+1)Bh = 1.788 \times 10^{-22}\, J(J+1) \text{ joules.} \qquad (2)$$

The population of state J can be obtained using Eq. 15.37:

$$N_J = Cg_J \exp\left[-\frac{E_J}{kT}\right] = C(2J+1)\exp\left[-\frac{1.788 \times 10^{-22}\, J(J+1)}{(1.381 \times 10^{-23})T}\right]$$
$$= C(2J+1)\exp\left[-\frac{12.95\, J(J+1)}{T}\right]. \qquad (3)$$

For $J = 0$, we have $N_J = C$. Therefore, if we arbitrarily set the intensity of the $J = 0$ to $J = 1$ transition to 1.000, the intensities of the bands relative to this standard will be given by

$$R = \text{Relative intensity} = \frac{N_J}{C} = (2J+1)\exp\left[-\frac{12.95\, J(J+1)}{T}\right]. \qquad (4)$$

The important point is the percent change in R for a given percent change in T. That is, we wish to know how $dR/R \times 100$ depends upon $dT/T \times 100$. Taking the differential of R for a change in T, we obtain

$$dR = \frac{12.95\, J(J+1)(2J+1)}{T^2}\exp\left[-\frac{12.95\, J(J+1)}{T}\right] dT. \qquad (5)$$

Dividing both sides of Eq. 5 by the corresponding sides of Eq. 4 produces

$$\boxed{\frac{dR}{R} = \frac{12.95\, J(J+1)}{T^2} dT = \frac{12.95\, J(J+1)}{T}\left(\frac{dT}{T}\right)}. \qquad (6)$$

For a given percent change in T at some fixed temperature, the percent change in the relative intensity is maximal when J is as large as possible, subject only to the constraints that the absorption band have sufficient intensity to be detected and that the spectrometer span the frequency range necessary to observe the transition. Hence, the most sensitive detection will be obtained if the investigator monitors the band at 234.00 cm^{-1} or the highest one observable in his or her spectrometer.

So Sam is correct. This fellow has real promise!

15.19 Determine the transition energies for all of the $J = 0$ to $J = 1$ and $J = 1$ to $J = 2$ transitions, with $\Delta J = \pm 1$ and $\Delta M = 0, \pm 1$ for a polar rigid rotor in an electric field $\mathcal{E}$ in terms of the Bh and the constant $C = I\mu_e^2\mathcal{E}^2/\hbar^2$. Obtain an expression for the electric dipole moment in terms of $\mathcal{E}$, I, $\hbar$, and B and the largest energy spacing for the stated transitions.

Solution

The energy eigenvalues in an electric field are given by Eqs. 15.47 and 15.48; thus,

$$E_{JM} = J(J+1)Bh - \frac{I\mu_e^2\mathcal{E}^2}{\hbar^2}\left[\frac{3M^2 - J(J+1)}{J(J+1)(2J-1)(2J+3)}\right] \quad \textbf{(1)}$$

for $J \neq 0$ and

$$E_{00} = -\frac{I\mu_e^2\mathcal{E}^2}{3\hbar^2}. \quad \textbf{(2)}$$

Inserting $C = I\mu_e^2\mathcal{E}^2/\hbar^2$ into these two equations gives

$$E_{JM} = J(J+1)Bh - C\left[\frac{3M^2 - J(J+1)}{J(J+1)(2J-1)(2J+3)}\right] \quad \textbf{(3)}$$

for $J \neq 0$ and

$$E_{00} = -\frac{C}{3}. \quad \textbf{(4)}$$

Since the J states are no longer degenerate, the possible transitions, which have $J = 0$ going to $J = 1$ or $J = 1$ going to $J = 2$, are as shown in the following table:

Transition		Energy of Upper State	Energy of Lower State	ΔE
$J\ M$	$J'\ M'$			
0 0	1 0	$2Bh + \frac{2C}{10}$	$-\frac{C}{3}$	$2Bh + \frac{8C}{15}$
0 0	1 ±1	$2Bh - \frac{C}{10}$	$-\frac{C}{3}$	$2Bh + \frac{7C}{30}$
1 0	2 0	$6Bh + \frac{C}{21}$	$2Bh + \frac{2C}{10}$	$4Bh - \frac{16C}{105}$
1 0	2 ±1	$6Bh + \frac{C}{42}$	$2Bh + \frac{2C}{10}$	$4Bh - \frac{37C}{210}$
1 ±1	2 0	$6Bh + \frac{C}{21}$	$2Bh - \frac{C}{10}$	$4Bh + \frac{31C}{210}$
1 ±1	2 ±1	$6Bh + \frac{C}{42}$	$2Bh - \frac{C}{10}$	$4Bh + \frac{13C}{105}$
1 ±1	2 ±2	$6Bh - \frac{C}{21}$	$2Bh - \frac{C}{10}$	$4Bh + \frac{11C}{210}$

The largest energy spacing for these transitions is that for the $(1,\pm 1) \longrightarrow (2,0)$ transition, for which

$$\Delta E_{(1\pm 1,\,20)} = 4Bh + \frac{31C}{210}. \tag{5}$$

Solving Eq. 5 for C, we obtain

$$C = \frac{210[\Delta E_{(1\pm 1,\,20)} - 4Bh]}{31}. \tag{6}$$

Inserting the definition of C produces

$$\frac{I\mu_e^2\mathcal{E}^2}{\hbar^2} = \frac{210[\Delta E_{(1\pm 1,\,20)} - 4Bh]}{31}. \tag{7}$$

The electric dipole moment is, therefore, given by

$$\boxed{\mu_e = \frac{\hbar}{\mathcal{E}}\left[\frac{210[\Delta E_{(1\pm 1,\,20)} - 4Bh]}{31I}\right]^{1/2}}. \tag{8}$$

15.21 Derive an expression in terms of B, D_J, and J giving the spacing between adjacent rotational states for a diatomic rotor whose rotational energy levels are given by Eq. 15.51. Does the deviation from the rigid-rotor result increase or decrease as J increases? Explain.

Solution

The rotational-energy eigenvalue for state J with centrifugal distortion included is given by

$$E_J = J(J+1)Bh - D_J J^2(J+1)^2 = (J^2 + J)Bh - D_J(J^4 + 2J^3 + J^2). \tag{1}$$

The corresponding energy for state $J+1$ is obtained by substituting $J+1$ for J in Eq. 1:

$$\begin{aligned} E_{J+1} &= (J+1)(J+2)Bh - D_J(J+1)^2(J+2)^2 \\ &= (J^2 + 3J + 2)Bh - D_J(J^2 + 2J + 1)(J^2 + 4J + 4). \end{aligned} \tag{2}$$

Expanding the second term on the right-hand side of Eq. 2 produces

$$\begin{aligned} E_{J+1} &= (J^2 + 3J + 2)Bh - D_J(J^4 + 4J^3 + 4J^2 + 2J^3 + 8J^2 + 8J + J^2 + 4J + 4) \\ &= (J^2 + 3J + 2)Bh - D_J(J^4 + 6J^3 + 13J^2 + 12J + 4). \end{aligned} \tag{3}$$

The energy spacing between rotational states $J+1$ and J is

$$\begin{aligned} \Delta E_{J,J+1} &= E_{J+1} - E_J \\ &= (J^2 + 3J + 2)Bh - D_J(J^4 + 6J^3 + 13J^2 + 12J + 4) \\ &\quad - (J^2 + J)Bh - D_J(J^4 + 2J^3 + J^2) \\ &= 2(J+1)Bh - D_J(4J^3 + 12J^2 + 12J + 4) \\ &= 2(J+1)Bh - 4D_J(J^3 + 3J^2 + 3J + 1). \end{aligned} \tag{4}$$

Since the term $J^3 + 3J^2 + 3J + 1 = (J+1)^3$, we may use the more compact form

$$\boxed{\Delta E_{J,J+1} = E_{J+1} - E_J = 2(J+1)Bh - 4D_J(J+1)^3}. \tag{5}$$

The rigid-rotor result is given by Eq. 15.26:

$$[\Delta E_{J,J+1}]_{\text{rigid rotor}} = 2(J+1)Bh. \qquad \textbf{(6)}$$

The deviation of Eq. 4 from the rigid-rotor result is, therefore,

$$\boxed{\text{Difference} = \Delta E_{J,J+1} - [\Delta E_{J,J+1}]_{\text{rigid rotor}} = -4D_J(J+1)^3}. \qquad \textbf{(7)}$$

Equation 7 shows that the difference between a real molecule with centrifugal distortion and the rigid-rotor result decreases as J increases. Since the difference is negative, this means that the magnitude of the difference increases as J increases. The larger the value of J, the greater the molecular orbital angular momentum and the greater the centrifugal force. The larger centrifugal force at elevated J states increases the importance of the distortion term, and the magnitude of the difference increases.

15.23 The measured pure rotational spectrum of $^{12}C^{16}O$ shows that the five bands with the lowest energy are at 3.86337 cm^{-1}, 7.72659 cm^{-1}, 11.5895 cm^{-1}, 15.4520 cm^{-1}, and 19.3138 cm^{-1}.

(A) Demonstrate that this spectrum cannot be described as that of a rigid-rotor.

(B) If $^{12}C^{16}O$ is treated as a rigid rotor, the five lowest energy rotational bands are at 3.8634 cm^{-1}, 7.7268 cm^{-1}, 11.5902 cm^{-1}, 15.4536 cm^{-1}, and 19.317 cm^{-1}. (See Problem 15.22.) Plot the magnitude of the difference between the observed transition energies and those predicted by a rigid-rotor approximation as a function of the J state from which the transition originates.

(C) Use the data given in (B) to obtain the $^{12}C^{16}O$ rotational constant in units of s^{-1}.

(D) The measured fundamental vibrational frequency of $^{12}C^{16}O$ is 6.6381×10^{13} s^{-1}. Compute the value of D_J that would result if the interatomic potential for $^{12}C^{16}O$ were a Morse potential. Express the result in both J and cm^{-1}.

(E) The magnitude of the difference between the observed rotational bands and those predicted by a rigid-rotor approximation is given by

$$\text{Difference Magnitude} = 4D_J[J+1]^3.$$

(See Problem 15.21.) Use the value of D_J obtained in (D) to compute the expected differences, and compare them with the experimentally measured results.

Solution

(A) The spacings between the observed bands are given in the table below. As the table shows, the band spacings are not constant, as they must be if the molecule is described by a rigid-rotor quantization. Instead, the spacings decrease slowly with increasing J because of the effect of centrifugal distortion.

Upper Band (cm^{-1})	Lower Band (cm^{-1})	Spacing (cm^{-1})
7.72659	3.86337	3.86322
11.5895	7.72659	3.86291
15.4520	11.5895	3.86250
19.3138	15.4520	3.86180

(B) The following table gives the magnitude of the differences in the band positions for a rigid rotor with the $^{12}C^{16}O$ masses and equilibrium distance and those observed for $^{12}C^{16}O$:

Transition	Rigid Rotor (cm^{-1})	Observed (cm^{-1})	|Difference| (cm^{-1})
$0 \longrightarrow 1$	3.8634	3.86337	0.00003
$1 \longrightarrow 2$	7.7268	7.72659	0.00021
$2 \longrightarrow 3$	11.5902	11.5895	0.00070
$3 \longrightarrow 4$	15.4536	15.4520	0.00160
$4 \longrightarrow 5$	19.317	19.3138	0.00320

The plot at the end of the problem shows the variation of |Difference| with the initial J state. As expected, centrifugal effects become more important as J increases.

(C) The $0 \longrightarrow 1$ rotational transition for the rigid rotor occurs at an energy of $2Bh$. This transition was computed to lie at 3.8634 cm^{-1}. The equivalent energy is $3.8634hc$. Therefore,

$$2Bh = 3.8634hc, \tag{1}$$

so that

$$B = \frac{3.8634\ cm^{-1}(2.9979 \times 10^{10}\ cm\ s^{-1})}{2} = \underline{5.7910 \times 10^{10}\ s^{-1}}. \tag{2}$$

(D) The distortion constant for a Morse potential is given by Eq. 15.50:

$$D_J = \frac{4B^3h}{\nu_o^2}. \tag{3}$$

Direct substitution into Eq. 3 gives

$$D_J = \frac{4(5.7910 \times 10^{10}\ s^{-1})^3(6.62608 \times 10^{-34}\ J\ s)}{(6.6381 \times 10^{13}\ s^{-1})^2} = \underline{1.168 \times 10^{-28}\ J}. \tag{4}$$

In terms of cm^{-1},

$$D_J(cm^{-1}) = \frac{D_J}{hc} = \frac{1.168 \times 10^{-28}\ J}{(6.62608 \times 10^{-34}\ J\ s)(2.9979 \times 10^{10}\ cm\ s^{-1})}$$

$$= \underline{5.88 \times 10^{-6}\ cm^{-1}}. \tag{5}$$

(E) With the foregoing value of D_J, the computed differences are too small. Using the result obtained in Problem 15.21, viz.,

$$|\text{Difference}| = 4D_J(J + 1)^3, \tag{6}$$

we compute differences of 0.0000235 cm^{-1}, 0.000188 cm^{-1}, 0.000635 cm^{-1}, 0.00150 cm^{-1}, and 0.00294 cm^{-1} for the five transitions. These results are shown as open circles on the plot on top of the next page. As can be seen, the result obtained from the Morse potential lies below the observed values, but is reasonably accurate.

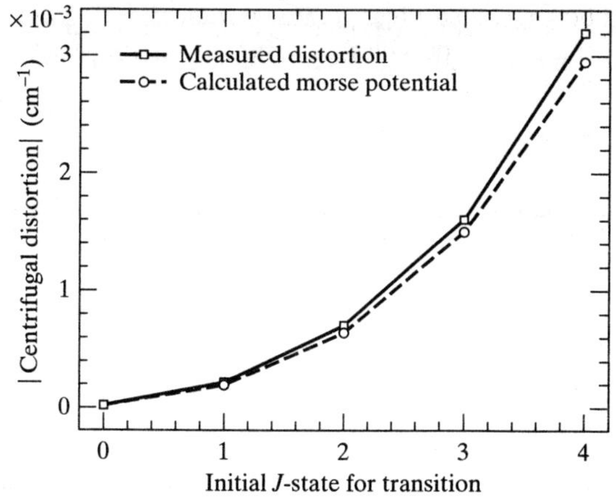

15.25 Which of the following molecules do we expect to exhibit pure rotational absorption spectra: N_2, H_2O, NF_3, C_6H_6, $H_2C{=}CH_2$, CO_2, H_2, O_2, CH_4, NO, HCN?

Solution

The molecule must possess a permanent dipole moment in order to exhibit a pure rotational spectrum. The only ones in the preceding list that do are H_2O, NF_3, NO, and HCN. The remaining molecules are nonpolar. Therefore, they will not undergo electric dipole transitions, and hence, they will not exhibit a rotational spectrum.

15.27 Prove that if the ratio of the path length difference Δp to the wavelength between two waves with the same wavelength and amplitude equals k, where k is an integer, constructive interference will be observed. Also, demonstrate that if $\Delta p/\lambda = k + \frac{1}{2}$, destructive interference will result.

Solution

Let the first wave be represented by

$$\Psi_1 = A \sin\left[\frac{2\pi x}{\lambda}\right]. \quad \textbf{(1)}$$

If the path length difference between Ψ_1 and Ψ_2 is Δp, then

$$\Psi_2 = A \sin\left[\frac{2\pi(x + \Delta p)}{\lambda}\right]. \quad \textbf{(2)}$$

The sum of the two waves is

$$S = \Psi_1 + \Psi_2 = A \sin\left[\frac{2\pi x}{\lambda}\right] + A \sin\left[\frac{2\pi(x + \Delta p)}{\lambda}\right]$$

$$= A\left\{\sin\left[\frac{2\pi x}{\lambda}\right] + \sin\left[\frac{2\pi x}{\lambda} + \frac{2\pi \Delta p}{\lambda}\right]\right\}. \quad \textbf{(3)}$$

Using the fact that $\sin(A + B) = \sin A \cos B + \cos A \sin B$, we can write Eq. 3 in the form

$$S = A\left\{\sin\left[\frac{2\pi x}{\lambda}\right] + \sin\left[\frac{2\pi x}{\lambda}\right]\cos\left[\frac{2\pi \Delta p}{\lambda}\right] + \cos\left[\frac{2\pi x}{\lambda}\right]\sin\left[\frac{2\pi \Delta p}{\lambda}\right]\right\}. \quad \textbf{(4)}$$

Let us now assume that the path length difference is such that $\Delta p/\lambda = k$, with k an integer. Substituting this condition into Eq. 4 produces

$$S = A\left\{\sin\left[\frac{2\pi x}{\lambda}\right] + \sin\left[\frac{2\pi x}{\lambda}\right]\cos[2\pi k] + \cos\left[\frac{2\pi x}{\lambda}\right]\sin[2\pi k]\right\}. \quad (5)$$

If k is integer, then $\cos(2k\pi) = 1$ and $\sin(2\pi k) = 0$. This gives

$$\boxed{S = 2A\sin\left[\frac{2\pi x}{\lambda}\right]}, \quad (6)$$

which is the condition for constructive interference. If, on the other hand, we have $\Delta p/\lambda = k + \frac{1}{2}$, then Eq. 4 becomes

$$S = A\left\{\sin\left[\frac{2\pi x}{\lambda}\right] + \sin\left[\frac{2\pi x}{\lambda}\right]\cos[(2k+1)\pi] + \cos\left[\frac{2\pi x}{\lambda}\right]\sin[(2k+1)\pi]\right\}. \quad (7)$$

We now have $\sin[(2k+1)\pi] = 0$ and $\cos[(2k+1)\pi] = -1$. Substituting into Eq. 7 yields

$$\boxed{S = A\left\{\sin\left[\frac{2\pi x}{\lambda}\right] - \sin\left[\frac{2\pi x}{\lambda}\right] + 0\right\} = 0}, \quad (8)$$

and we have destructive interference between the two waves.

15.29 Show that the electric dipole transition from $v = 0$ to $v = 2$ for a harmonic oscillator is forbidden.

Solution

For the transition to be forbidden, the transition matrix element between the $v = 0$ and $v = 2$ harmonic oscillator states must be zero. That is, we must have

$$\langle\psi_0(x)|x|\psi_2(x)\rangle = 0, \quad (1)$$

where the eigenfunctions are

$$\psi_0(x) = \left[\frac{\alpha}{\pi}\right]^{1/4}\exp\left[-\frac{\alpha x^2}{2}\right] \quad (2)$$

and

$$\psi_2(x) = \left[\frac{\alpha}{4\pi}\right]^{1/4}(1 - 2\alpha x^2)\exp\left[-\frac{\alpha x^2}{2}\right]. \quad (3)$$

The transition matrix element is, therefore,

$$\mathrm{H}_{01} = \left[\frac{\alpha^2}{4\pi^2}\right]^{1/4}\int_{x=-\infty}^{\infty}\exp\left[-\frac{\alpha x^2}{2}\right]x(1 - 2\alpha x^2)\exp\left[-\frac{\alpha x^2}{2}\right]dx$$

$$= \left[\frac{\alpha^2}{4\pi^2}\right]^{1/4}\int_{x=-\infty}^{\infty}(x - 2\alpha x^3)\exp[-ax^2]dx. \quad (4)$$

Since $F(x) = x - 2\alpha x^3$ is an odd function for which $F(x) = -F(-x)$, whereas $G(x) = \exp[-\alpha x^2]$ is an even function for which $G(x) = G(-x)$, H_{01} is zero. We may easily see this by writing

$$\mathrm{H}_{01} = \left[\frac{\alpha^2}{4\pi^2}\right]^{1/4}\int_{-\infty}^{\infty}F(x)G(x)dx = \left[\frac{\alpha^2}{4\pi^2}\right]^{1/4}\left\{\int_{x=-\infty}^{0}F(x)G(x)dx + \int_{0}^{\infty}F(x)G(x)dx\right\}, \quad (5)$$

where $F(x)$ and $G(x)$ are as just defined. In the first integral in Eq. 5, let us make the substitution $z = -x$, so that $dz = -dx$. This produces

$$\int_{x=-\infty}^{0} F(x)G(x)dx = -\int_{\infty}^{0} F(-z)G(-z)dz. \quad \textbf{(6)}$$

But $F(-z) = -F(z)$ and $G(-z) = G(z)$. Substituting these into Eq. 6 yields

$$\int_{x=-\infty}^{0} F(x)G(x)dx = \int_{\infty}^{0} F(z)G(z)dz = -\int_{0}^{\infty} F(z)G(z)dz = -\int_{0}^{\infty} F(x)G(x)dx, \quad \textbf{(7)}$$

since the name of the variable of integration does not affect the value of the definite integral. Combining Eqs. 5 and 7 produces

$$\mathrm{H}_{01} = \left[\frac{\alpha^2}{4\pi^2}\right]^{1/4}\left\{-\int_{0}^{\infty} F(x)G(x)dx + \int_{0}^{\infty} F(x)G(x)dx\right\} = 0. \quad \textbf{(8)}$$

Hence, the $v = 0$ to $v = 2$ transition is forbidden. In general, if we do not have $\Delta v = \pm 1$, the transition will be forbidden.

15.31 The FTIR spectrum of gaseous $H^{127}I$ exhibits absorption bands at the following wave numbers: 2,257.1, 2,270.2, 2,283.3, 2,296.4, 2,322.6, 2,335.7, and 2,348.8 cm^{-1}.

(A) What is the vibrational frequency of HI, expressed in cm^{-1}? Compute the harmonic vibrational force constant for $H^{127}I$.

(B) Compute the moment of inertia of $H^{127}I$ from the given data.

(C) Calculate the equilibrium H–I bond distance using the given data. Assume that Eq. 15.62 describes the vibrational–rotational quantum states of $H^{127}I$ with sufficient accuracy. The atomic mass of ^{127}I is 126.9045.

Solution

(A) The key to this problem is the spacings between the observed absorption bands, which are

$$2{,}270.2 - 2{,}257.1 = 13.1 \text{ cm}^{-1},$$

$$2{,}283.3 - 2{,}270.2 = 13.1 \text{ cm}^{-1},$$

$$2{,}296.4 - 2{,}283.3 = 13.1 \text{ cm}^{-1},$$

$$2{,}322.6 - 2{,}296.4 = 26.2 \text{ cm}^{-1},$$

$$2{,}335.7 - 2{,}322.6 = 13.1 \text{ cm}^{-1},$$

and

$$2{,}348.8 - 2{,}335.7 = 13.1 \text{ cm}^{-1}.$$

The spacing between the bands at 2,322.6 and 2,296.4 cm^{-1} is twice that of the other spacings. Therefore, this gap is the one that separates the P and R branches. We know that the band origin lies at the center of the gap. Consequently, we can compute

$$\bar{\nu}_{\text{origin}} = 2{,}296.4 + \frac{26.2}{2}\text{ cm}^{-1} = \underline{2{,}309.5 \text{ cm}^{-1}}. \quad \textbf{(1)}$$

This is the vibrational wave number for HI. The corresponding fundamental frequency is

$$\nu_{\text{origin}} = 2{,}309.5 \text{ cm}^{-1}(2.9979 \times 10^{10} \text{ cm s}^{-1}) = \underline{6.9236 \times 10^{13} \text{ s}^{-1}}. \quad \textbf{(2)}$$

The harmonic force constant is given by

$$k = 4\pi^2\mu[\nu_{\text{origin}}]^2. \quad \textbf{(3)}$$

The reduced mass of $H^{127}I$ is

$$\begin{aligned}\mu &= \frac{m_H m_I}{m_H + m_I}\\ &= \frac{(1.0078)(126.9045)}{1.0078 + 126.9045}\ \text{g mol}^{-1} \times \frac{1\ \text{kg}}{1{,}000\ \text{g}} \times \frac{1}{6.02214 \times 10^{23}\ \text{mol}^{-1}}\\ &= 1.6603 \times 10^{-27}\ \text{kg}.\end{aligned} \quad \textbf{(4)}$$

Substituting into Eq. 3 gives

$$k = 4(3.14159)^2(1.6603 \times 10^{-27}\ \text{kg})(6.9236 \times 10^{13}\ \text{s}^{-1})^2 = \underline{314.20\ \text{kg s}^{-2}}. \quad \textbf{(5)}$$

(B) The rotation band spacings are $2Bh$. Therefore, we have

$$\begin{aligned}2Bh &= \frac{h^2}{4\pi^2 I} = (13.1\ \text{cm}^{-1})hc\\ &= (13.1\ \text{cm}^{-1})(6.62608 \times 10^{-34}\ \text{J s})(2.9979 \times 10^{10}\ \text{cm s}^{-1})\\ &= 2.6022 \times 10^{-22}\ J.\end{aligned} \quad \textbf{(6)}$$

Solving Eq. 2 for I, we obtain

$$\begin{aligned}I &= \frac{h^2}{4\pi^2(2.6022 \times 10^{-22}\ \text{J})} = \frac{(6.62608 \times 10^{-34}\ \text{J s})^2}{4(3.1415927)^2(2.6022 \times 10^{-22}\ \text{J})}\\ &= \underline{4.2738 \times 10^{-47}\ \text{kg m}^2}.\end{aligned} \quad \textbf{(7)}$$

(C) The moment of inertia is given by

$$I = \mu R_e^2, \quad \textbf{(8)}$$

so that

$$R_e = \left[\frac{I}{\mu}\right]^{1/2}. \quad \textbf{(9)}$$

Substituting Eqs. 4 and 7 into Eq. 9 yields

$$R_e = \left[\frac{4.2738 \times 10^{-47}\ \text{kg m}^2}{1.6603 \times 10^{-27}\ \text{kg}}\right]^{1/2} = \underline{1.604 \times 10^{-10}\ \text{m}} = \underline{1.604\ \text{Å}}. \quad \textbf{(10)}$$

15.33 The equilibrium bond length for $H^{79}Br$ is 1.413 Å. Assume that the atomic weights are H = 1.0078 and ^{79}Br = 78.9183.

(A) Calculate the value of the rotational constant for HBr. Express the answer in (a) s^{-1} and (b) cm^{-1}.

(B) A microwave spectrum of HBr is taken. At what wavelength will the $J = 2 \rightarrow J = 3$ rotational transition be observed? (Assume that HBr behaves as a rigid rotor.)

(C) Determine the most populated rotational state of HBr at a temperature of 800 K.

(D) The band origin for HBr is at 2,649.67 cm^{-1}. If the vibrational–rotational energy levels of HBr are given by a harmonic oscillator–rigid-rotor expression, calculate the energy in joules for the ($v = 0, J = 2$) vibrational–rotational state. At what wave number will the absorption band for the ($v = 0, J = 2$)→($v = 1$, $J = 3$) transition be observed?

Solution

(A) The rotational constant is given by

$$B = \frac{h}{8\pi^2\mu R^2}. \quad (1)$$

The reduced mass for HBr is

$$\mu = \frac{m_H m_{Br}}{m_H + m_{Br}}$$

$$= \frac{(1.0078)(78.9183)}{1.0078 + 78.9183} \times \frac{1}{6.02214 \times 10^{23}} \times \frac{1\text{ kg}}{1{,}000\text{ g}} = 1.6524 \times 10^{-27}\text{ kg}. \quad (2)$$

Therefore, the rotational constant is

$$B = \frac{6.62608 \times 10^{-34}\text{ J s}}{8(3.1415927)^2(1.6524 \times 10^{-27}\text{ kg})(1.413 \times 10^{-10}\text{ m})^2}$$

$$= \underline{2.5437 \times 10^{11}\text{ s}^{-1}}. \quad (3)$$

In cm^{-1},

$$B = \frac{2.5437 \times 10^{11}\text{ s}^{-1}}{2.9979 \times 10^{10}\text{ cm s}^{-1}} = \underline{8.485\text{ cm}^{-1}}. \quad (4)$$

(B) Since the rigid-rotor energy is given by $E_J = J(J + 1)Bh$, the energy associated with the $J = 2 \longrightarrow J = 3$ transition is

$$\Delta E = E_3 - E_2 = (3)(4)Bh - 2(3)Bh = 6Bh. \quad (5)$$

The electromagnetic radiation energy is given by hc/λ; therefore, at the resonance point, we have

$$\lambda = \frac{hc}{6Bh} = \frac{c}{6B} = \frac{2.9979 \times 10^{10}\text{ cm s}^{-1}}{6(2.5437 \times 10^{11}\text{ s}^{-1})} = \underline{0.01964\text{ cm}} = \underline{1.964 \times 10^{-4}\text{m}}. \quad (6)$$

(C) The relative number of $H^{79}Br$ molecules in each state is given by

$$N_J = Cg_J \exp\left[-\frac{E_J}{kT}\right]. \quad (7)$$

At 800 K,

$$N_J = C(2J + 1)\exp\left[-J(J + 1)\frac{Bh}{800k}\right]. \quad (8)$$

The most populated rotational state is the rotational state that has the largest value of $(2J + 1)\exp[-J(J + 1)Bh/800k]$. Using the result obtained in Part (A), we may compute

$$\frac{Bh}{800k} = \frac{(2.5437 \times 10^{11}\text{ s}^{-1})(6.62608 \times 10^{-34}\text{ J s})}{800\text{ K}(1.38066 \times 10^{-23}\text{ J K}^{-1})} = 0.015260. \quad (9)$$

Therefore, we need the value of J for which $(2J + 1)\exp[-J(J + 1)(0.01526)]$ is a maximum. Direct calculations yield the following results:

J	$(2J + 1)\exp[-J(J + 1)(0.01526)]$
0	1
1	2.910
2	4.563
3	5.829
4	6.633
5	6.959
6	6.848
7	6.382
8	5.666

We see that the most populated rotational state is $\underline{J = 5}$.

(D) If the rotational–vibrational energy is given by

$$E_{vJ} = (v + 0.5)h\nu_o + J(J + 1)Bh, \tag{10}$$

then the energy of the $v = 0, J = 2$ state is

$$\begin{aligned} E_{v=0,J=2} &= 0.5h\nu_o + 2(3)Bh = 0.5h\nu_o + 6Bh \\ &= 0.5(6.62608 \times 10^{-34}\text{ J s})(2649.67\text{ cm}^{-1})(2.9979 \times 10^{10}\text{ cm s}^{-1}) \\ &\quad + 6(2.5437 \times 10^{11}\text{ s}^{-1})(6.62608 \times 10^{-34}\text{ J s}) \\ &= 2.6317 \times 10^{-20}\text{ J} + 1.0113 \times 10^{-21}\text{ J} = \underline{2.7328 \times 10^{-20}\text{ J}}. \end{aligned} \tag{11}$$

Using Eq. 10, we see that the energy of the $v = 1, J = 3$ state is

$$E_{v=1,J=3} = 1.5h\nu_o + 12Bh. \tag{12}$$

The transition energy for the $(v = 0, J = 2) \longrightarrow (v = 1, J = 3)$ transition is, therefore,

$$\begin{aligned} \Delta E &= E_{v=1,J=3} - E_{v=0,J=2} = 1.5h\nu_o + 12Bh - [0.5h\nu_o + 6Bh] = h\nu_o + 6Bh \\ &= (6.62608 \times 10^{-34}\text{ J s})(2{,}649.67\text{ cm}^{-1})(2.9979 \times 10^{10}\text{ cm s}^{-1}) \\ &\quad + 6(2.5437 \times 10^{11}\text{ s}^{-1})(6.62608 \times 10^{-34}\text{ J s}) = 5.3645 \times 10^{-20}\text{ J}. \end{aligned} \tag{13}$$

In wave numbers, this energy is equivalent to

$$\begin{aligned} \text{wave number} &= \frac{\Delta E}{hc} = \frac{5.3645 \times 10^{-20}\text{ J}}{(6.62608 \times 10^{-34}\text{ J s})(2.9979 \times 10^{10}\text{ cm s}^{-1})} \\ &= \underline{2{,}700.6\text{ cm}^{-1}}. \end{aligned} \tag{14}$$

15.35 Using the data in Table 15.4, predict the wave numbers for the first, second, and third vibrational overtones for $^{14}N^{14}N$.

Solution

The vibrational-energy levels are empirically described by Eq. 15.63:

$$E_v^e = \frac{E_v}{hc} = (v + 0.5)\omega_e - x_e\omega_e(v + 0.5)^2 + y_e\omega_e(v + 0.5)^3. \tag{1}$$

Therefore, the energy for the first overtone is

$$\Delta E_{02} = E_2^e - E_0^e = 2.5\omega_e - x_e\omega_e(2.5)^2 + y_e\omega_e(2.5)^3 - [0.5\omega_e - x_e\omega_e(0.5)^2 + y_e\omega_e(0.5)^3]$$

$$= 2\omega_e - 6.0x_e\omega_e + 15.5y_e\omega_e. \quad (2)$$

For the second overtone, the result is

$$\Delta E_{03} = E_3^e - E_0^e$$

$$= 3.5\omega_e - x_e\omega_e(3.5)^2 + y_e\omega_e(3.5)^3 - [0.5\omega_e - x_e\omega_e(0.5)^2 + y_e\omega_e(0.5)^3]$$

$$= 3\omega_e - 12.0x_e\omega_e + 42.75y_e\omega_e. \quad (3)$$

The third overtone has energy

$$\Delta E_{04} = E_4^e - E_0^e$$

$$= 4.5\omega_e - x_e\omega_e(4.5)^2 + y_e\omega_e(4.5)^3 - [0.5\omega_e - x_e\omega_e(0.5)^2 + y_e\omega_e(0.5)^3]$$

$$= 4\omega_e - 20.0x_e\omega_e + 91.0y_e\omega_e. \quad (4)$$

From Table 15.4, we obtain $\omega_e = 2{,}359.61\ \text{cm}^{-1}$, $x_e\omega_e = 14.456\ \text{cm}^{-1}$, and $y_e\omega_e = 0.00751\ \text{cm}^{-1}$. Substituting these values into Eqs. 2, 3, and 4 gives the desired results:

$$\Delta E_{02} = 2(2{,}359.61) - 6.0(14.456) + 15.5(0.00751)\ \text{cm}^{-1} = \underline{4{,}632.6\ \text{cm}^{-1}}; \quad (5)$$

$$\Delta E_{03} = 3(2{,}359.61) - 12.0(14.456) + 42.75(0.00751)\ \text{cm}^{-1} = \underline{6{,}905.7\ \text{cm}^{-1}}; \quad (6)$$

$$\Delta E_{04} = 4(2{,}359.61) - 20.0(14.456) + 91.0(0.00751)\ \text{cm}^{-1} = \underline{9{,}150.0\ \text{cm}^{-1}}. \quad (7)$$

15.37 (A) Assuming that we can ignore centrifugal distortion effects on rotational-energy levels, use the data in Table 15.4 to predict the wave number at which the hot band for the transition $(v = 1,\ J = 2) \rightarrow (v = 2,\ J = 3)$ would be observed in ^{7}LiH.

(B) Compute the ratio of the intensity of the transition in (A) to that for the transition $(v = 0, J = 2) \rightarrow (v = 1, J = 3)$ at 300 K. What is the intensity ratio at 600 K? Assume that the intensities are determined solely for the population of the initial states involved in the transitions.

Solution

(A) If we combine Eq. 15.63 with the rigid-rotor expression for the rotational energy, we obtain

$$E_{vJ}^e = (v + 0.5)\omega_e - x_e\omega_e(v + 0.5)^2 + y_e\omega_e(v + 0.5)^3 + J(J + 1)\frac{Bh}{hc}$$

$$= (v + 0.5)\omega_e - x_e\omega_e(v + 0.5)^2 + y_e\omega_e(v + 0.5)^3 + J(J + 1)\frac{B}{c}\ \text{cm}^{-1}. \quad (1)$$

Therefore, the wave number for the $v = 1, J = 2$ state is

$$E_{12}^e = 1.5\omega_e - 2.25x_e\omega_e + 3.375y_e\omega_e + \frac{6B}{c}. \quad (2)$$

For the $v = 2, J = 3$ state, the result is

$$E^e_{23} = 2.5\omega_e - 6.25x_e\omega_e + 15.625y_e\omega_e + \frac{12B}{c}. \quad \textbf{(3)}$$

The wave number for the transition between these two states is, therefore,

$$\Delta E^e = E^e_{23} - E^e_{12} = \omega_e - 4x_e\omega_e + 12.25y_e\omega_e + \frac{6B}{c}. \quad \textbf{(4)}$$

The data in Table 15.4 for ^{7}LiH give $\omega_e = 1{,}405.1\ \text{cm}^{-1}$, $x_e\omega_e = 13.228$, $y_e\omega_e = 0.1633\ \text{cm}^{-1}$, and $B/c = 7.5131\ \text{cm}^{-1}$. Substituting these values into Eq. 4 produces

$$\Delta E = 1{,}405.1 - 4(13.228) + 12.25(0.1633) + 6(7.5131)\ \text{cm}^{-1} = \underline{1{,}399.27\ \text{cm}^{-1}}. \quad \textbf{(5)}$$

(B) The wave number for the state ($v = 0, J = 2$) is

$$E^e_{02} = 0.5\omega_e - 0.25x_e\omega_e + 0.125y_e\omega_e + \frac{6B}{c}$$

$$= 0.5(1{,}405.1) - 0.25(13.228) + 0.125(0.1633) + 6(7.5131)\ \text{cm}^{-1}$$

$$= 744.34\ \text{cm}^{-1}. \quad \textbf{(6)}$$

If the band intensity is dependent only upon the population in the initial states for the transition, the ratio of the band intensity for the ($v = 1, J = 2$) $\longrightarrow$ ($v = 2$, $J = 3$) transition to that for the ($v = 0, J = 2$) $\longrightarrow$ ($v = 1, J = 3$) transition will be given by

$$R = \frac{N_{12}}{N_{02}} = \frac{g_2 \exp[-E_{12}/kT]}{g_2 \exp[-E_{02}/kT]} = \exp\left[-\left(\frac{E^e_{12} - E^e_{02}}{kT}\right)\right]. \quad \textbf{(7)}$$

The energy difference $E^e_{12} - E^e_{02}$ is given by combining Eqs. 2 and 6:

$$E^e_{12} - E^e_{02} = \left\{\left[1.5\omega_e - 2.25x_e\omega_e + 3.375y_e\omega_e + \frac{6B}{c}\right]\right.$$

$$\left. - \left[0.5\omega_e - 0.25x_e\omega_e + 0.125y_e\omega_e + \frac{6B}{c}\right]\right\}hc$$

$$= [\omega_e - 2x_e\omega_e + 3.25y_e\omega_e]hc. \quad \textbf{(8)}$$

Inserting the data produces

$$E_{12} - E_{02} = [1{,}405.1 - 2(13.228) + 3.25(0.1633)]\ \text{cm}^{-1}$$

$$\times (6.62608 \times 10^{-34}\ \text{J s})(2.9979 \times 10^{10}\ \text{cm s}^{-1}) = 2.7394 \times 10^{-20}\ \text{J}.$$

Substituting Eq. 8 into Eq. 7 with $T = 300$ K gives

$$R = \exp\left[-\frac{2.7394 \times 10^{-20}\ \text{J}}{(1.38066 \times 10^{-23}\ \text{J K}^{-1})(300\ \text{K})}\right] = \underline{0.001342}. \quad \textbf{(9)}$$

At $T = 600$ K, the result is

$$R = \exp\left[-\frac{2.7394 \times 10^{-20}\ \text{J}}{(1.38066 \times 10^{-23}\ \text{J K}^{-1})(600\ \text{K})}\right] = \underline{0.03663}. \quad \textbf{(10)}$$

15.39 The overtone spectra of H_2^+ give the following data:

v	$\Delta\omega_v(\text{cm}^{-1})$	v	$\Delta\omega_v(\text{cm}^{-1})$
0	2,191	8	1,257
1	2,064	9	1,145
2	1,941	10	1,033
3	1,821	11	918
4	1,705	12	800
5	1,591	13	677
6	1,479	14	548
7	1,368		

Use the Birge–Spooner extrapolation method to obtain the dissociation energy of H_2^+. The measured dissociation energy is 2.648 eV. Compute the percent error in your result.

Solution

The dissociation energy is given by Eq. 15.69:

$$D_e = \Delta\omega_0 + \Delta\omega_1 + \Delta\omega_2 + \cdots + \Delta\omega_{v_m} + \varepsilon = \sum_{v=0}^{v=v_m} \Delta\omega_v + \varepsilon. \quad (1)$$

Over the range where data are available, we may execute the sum in Eq. 15.69 in a straightforward fashion. This partial sum is

$$X = \sum_{v=0}^{14} \Delta\omega_v$$

$$= 2{,}191 + 2{,}064 + 1{,}941 + 1{,}821 + 1{,}705 + 1{,}591 + 1{,}479 + 1{,}368$$

$$+ 1{,}257 + 1{,}145 + 1{,}033 + 918 + 800 + 677 + 548 \text{ cm}^{-1}$$

$$= 20{,}538 \text{ cm}^{-1}. \quad (2)$$

We now use the last two points, at $v = 13$ and $v = 14$, to fit a straight line to permit extrapolation to higher values of v. The linear fit to the points at $v = 13$ and 14 is

$$\Delta\omega_v = 2{,}354 - 129v \quad (3)$$

for $v \geq 13$. This gives the additional data in the table at the right. Adding these results to the partial sum X in Eq. 2, we obtain

v	$\Delta\omega_v$ (cm^{-1})
15	419
16	290
17	161
18	32

$$X' = X + 419 + 290 + 161 + 32 = 20{,}538 + 419 + 290 + 161 + 32 \text{ cm}^{-1}$$

$$= 21{,}440 \text{ cm}^{-1}. \quad (4)$$

We can now estimate the remaining energy gap ε between the $v = 19$ vibrational state and the dissociation limit by solving Eq. 3 for the value of v at which $\Delta\omega_v = 0$. This result is

$$v_{\max} = v_m = \frac{2{,}354}{129} = 18.248. \quad (5)$$

A linear extrapolation from the point ($v = 18, \Delta\omega_v = 32$) to the point ($v = 18.248, \Delta\omega_v = 0$) produces a triangular area whose value is an estimate of the energy gap between $v = 19$ and the dissociation limit. This area is

$$\text{Area} = \frac{(\text{height})(\text{base})}{2} = \frac{(32\ \text{cm}^{-1})(0.248)}{2} = 3.97\ \text{cm}^{-1} \approx \varepsilon. \quad \textbf{(6)}$$

Addition of this quantity to X' in Eq. 3 gives the best estimate for the dissociation energy:

$$D_e = 21{,}440 + 3.97\ \text{cm}^{-1} = \underline{21{,}444\ \text{cm}^{-1}}. \quad \textbf{(7)}$$

Using the fact that there are 8,065.7 cm^{-1} per eV, we calculate the dissociation energy in eV:

$$D_e = 21{,}444\ \text{cm}^{-1} \times \frac{1\ \text{eV}}{8{,}065.7\ \text{cm}^{-1}} = \underline{2.659\ \text{eV}}. \quad \textbf{(8)}$$

The percent error is

$$\%\ \text{error} = 100 \times \frac{2.659 - 2.648}{2.648} = \underline{0.42\%}. \quad \textbf{(9)}$$

15.41* (A) Use Eq. 15.64 to write $\Delta\omega_v$ as a quadratic function of v.

(B) Using the data given in Example 15.10 for H_2, execute a linear least-squares fitting of the function

$$\Delta\omega_v = a_o + a_1 v + a_2 v^2$$

to the data and determine the best values for a_o, a_1, and a_2. Plot the results of the fit versus v, and compare them with the data.

(C) By combining the results of (A) and (B), determine the values of ω_e, $x_e\omega_e$, and $y_e\omega_e$ for H_2.

Solution

(A) The adjacent vibrational spacing, in wave numbers, is given by Eq. 15.64:

$$\Delta\omega_v = \omega_e - 2x_e\omega_e[v + 1] + y_e\omega_e[3v^2 + 6v + 3.25]. \quad \textbf{(1)}$$

Collecting coefficients of the different powers of v, we obtain

$$\boxed{\Delta\omega_v = 3y_e\omega_e v^2 + [6y_e\omega_e - 2x_e\omega_e]v + [\omega_e - 2x_e\omega_e + 3.25y_e\omega_e]}, \quad \textbf{(2)}$$

which is the desired result.

(B) A least-squares fit to the data yields

$$\Delta\omega_v = -6.364v^2 - 178.31v + 4{,}100.78\ \text{cm}^{-1}. \quad \textbf{(3)}$$

The figure at the end of the problem shows a plot of Eq. 3 compared to the measured vibrational spacings for H_2.

(C) Comparing Eqs. 2 and 3 shows that we must have

$$3y_e\omega_e = -6.364\ \text{cm}^{-1}. \quad \textbf{(4)}$$

Therefore,

$$y_e\omega_e = \underline{-2.12\ \text{cm}^{-1}}. \quad \textbf{(5)}$$

Equating the coefficients of the linear term, we obtain

$$[6y_e\omega_e - 2x_e\omega_e] = -178.31 \text{ cm}^{-1}. \quad (6)$$

Solving for $x_e\omega_e$ yields

$$x_e\omega_e = \frac{6y_e\omega_e + 178.31}{2} = \frac{6(-2.12) + 178.31}{2} = \underline{82.80 \text{ cm}^{-1}}. \quad (7)$$

Finally, equating the constant terms in Eqs. 2 and 3, we get

$$[\omega_e - 2x_e\omega_e + 3.25y_e\omega_e] = 4{,}100.78 \text{ cm}^{-1}, \quad (8)$$

so that

$$\omega_e = 4{,}100.78 - 3.25(y_e\omega_e) + 2x_e\omega_e = 4{,}100.78 - 3.25(-2.12) + 2(82.80)$$
$$= \underline{4{,}273.3 \text{ cm}^{-1}}. \quad (9)$$

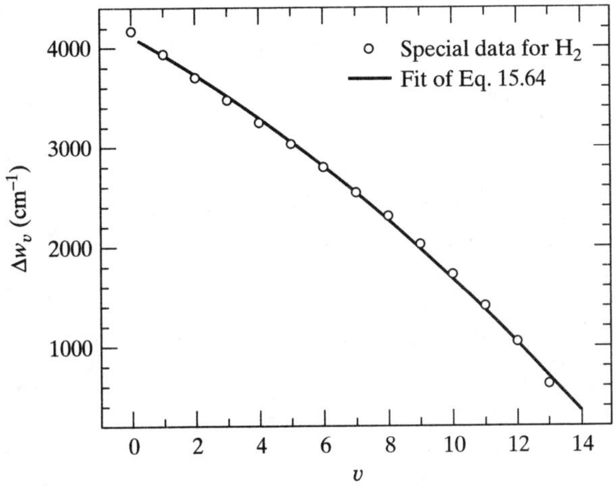

15.43 (A) Show that the minimum Raman shift for a line in the S branch of a Raman spectrum for a diatomic molecule is $h\nu_o + 6Bh$, provided that Eq. 15.62 accurately describes the vibrational–rotational energies.

(B) Show that the maximum Raman shift for a line in the O branch is $h\nu_o - 6Bh$.

(C) Show that the spacings between the lines in the S and O branches are both $4Bh$.

(D) Show that the spacings between the Q branch Raman line and the closest S and O branch lines are both $6Bh$.

Solution

(A) If Eq. 15.62 accurately describes the vibrational–rotational energy states, we have

$$E_{vJ} = (v + 0.5)h\nu_o + J(J + 1)Bh. \quad (1)$$

In the S branch, the selection rules for Raman scattering are $\Delta v = \pm 1$ and $\Delta J = +2$. Therefore, the energy spacings for the allowed transitions are

$$\Delta E_J^v(\text{S branch}) = E_{v+1,J+2} - E_{vJ}$$
$$= (v + 1.5)h\nu_o + (J + 2)(J + 3)Bh - (v + 0.5)h\nu_o - J(J + 1)Bh$$
$$= h\nu_o + [J^2 + 5J + 6]Bh - [J^2 + J]Bh = h\nu_o + [4J + 6]Bh. \quad (2)$$

Equation 2 shows that the energies of the allowed transitions are as shown in the following table:

Transition	ΔE_J^v (*S* branch)
$v = 0, J = 0 \longrightarrow v = 1, J = 2$	$h\nu_o + 6Bh$
$v = 0, J = 1 \longrightarrow v = 1, J = 3$	$h\nu_o + 10Bh$
$v = 0, J = 2 \longrightarrow v = 1, J = 4$	$h\nu_o + 14Bh$
$v = 0, J = 3 \longrightarrow v = 1, J = 5$	$h\nu_o + 18Bh$
$v = 0, J = 4 \longrightarrow v = 1, J = 6$	$h\nu_o + 22Bh$
$v = 0, J = 5 \longrightarrow v = 1, J = 7$	$h\nu_o + 26Bh$
and so on	

We see that the minimum Raman shift is obtained for the transition ($v = 0$, $J = 0 \longrightarrow v = 1, J = 2$), which occurs at $h\nu_o + 6Bh$.

(B) In the *O* branch, the selection rules for Raman scattering are $\Delta v = \pm 1$ and $\Delta J = -2$. Therefore, the energy spacings for the allowed transitions are

$$\begin{aligned}\Delta E_J^v(O\text{ branch}) &= E_{v+1,J-2} - E_{vJ}\\ &= (v + 1.5)h\nu_o + (J-2)(J-1)Bh - (v + 0.5)h\nu_o - J(J+1)Bh\\ &= h\nu_o + [J^2 - 3J + 2]Bh - [J^2 + J]Bh = h\nu_o + [-4J + 2]Bh. \qquad \textbf{(3)}\end{aligned}$$

Equation 3 shows that the energies of the allowed transitions are in the table below. We now see that the maximum Raman shift is obtained for the transition ($v = 0, J = 2 \longrightarrow v = 1, J = 0$), which occurs at $h\nu_o - 6Bh$.

Transition	ΔE_J^v (*O* branch)
$v = 0, J = 2 \longrightarrow v = 1, J = 0$	$h\nu_o - 6Bh$
$v = 0, J = 3 \longrightarrow v = 1, J = 1$	$h\nu_o - 10Bh$
$v = 0, J = 4 \longrightarrow v = 1, J = 2$	$h\nu_o - 14Bh$
$v = 0, J = 5 \longrightarrow v = 1, J = 3$	$h\nu_o - 18Bh$
$v = 0, J = 6 \longrightarrow v = 1, J = 4$	$h\nu_o - 22Bh$
$v = 0, J = 7 \longrightarrow v = 1, J = 5$	$h\nu_o - 26Bh$
and so on	

(C) The spacings between the energies shown in the tables in Parts (A) and (B) are all equal to $4Bh$ between adjacent states. In general, we have, for the *S* branch,

$$\begin{aligned}\text{Band Spacing} &= \Delta E_{J+1}^v(S\text{ branch}) - \Delta E_J^v(S\text{ branch})\\ &= [4(J+1) + 6]Bh - [4J + 6]Bh\\ &= [4J + 10 - 4J - 6]Bh = \underline{4Bh}. \qquad \textbf{(4)}\end{aligned}$$

For the *O* branch, the result is the same:

$$\begin{aligned}\text{Band Spacing} &= \Delta E_J^v(O\text{ branch}) - \Delta E_{J+1}^v(O\text{ branch})\\ &= [-4J + 2]Bh - [-4(J+1) + 2]Bh\\ &= [-4J + 2 + 4J + 2]Bh = \underline{4Bh}. \qquad \textbf{(5)}\end{aligned}$$

(D) The Q-branch line occurs at

$$\Delta E_J^v(Q \text{ branch}) = E_{v+1,J} - E_{vJ}$$
$$= (v + 1.5)h\nu_o + J(J+1)Bh - (v + 0.5)h\nu_o - J(J+1)Bh$$
$$= h\nu_o. \tag{6}$$

Therefore, the spacing between this line and the first S- and Q-branch lines, which occur at $h\nu_o + 6Bh$ and $h\nu_o - 6Bh$, respectively, is $\boxed{6Bh}$ in both cases.

15.45 A Raman sample cell contains a gaseous mixture of two diatomic molecules that are listed in Table 15.4. A Raman spectrum of the mixture shows a series of vibrational–rotational lines whose Raman shifts are 3,300.64, 3,543.88, 3,761.62, 3,787.11, 3,845.38, 3,929.13, 4,012.89, 4,030.35, 4,138.52, 4,264.15, 4,347.91, 4,395.20, 4,431.67, 4,515.42, 4,760.05, 5,003.29, 5,246.53, and 5,489.76 cm^{-1}.

(A) Identify the two molecules that are in the sample cell.

(B) Which Raman lines correspond to the Q branch for each molecule?

(C) Identify the S and O branch lines for each molecule.

Solution

(A) The only molecules listed in Table 15.4 that have vibrational frequencies sufficiently large to produce Raman shifts greater than 3,300.64 cm^{-1} are H_2, $H^{19}F$, and ^{16}OH. We know that the S- and O-branch line spacings are each $4Bh$. Thus, for H_2, the data in Table 15.4 show that there should be a set of lines whose spacings are $4(60.809\ cm^{-1}) = 243.24\ cm^{-1}$. If $H^{19}F$ is present, lines whose spacing is $4(20.939\ cm^{-1}) = 83.756\ cm^{-1}$ should appear in the combined spectrum. ^{16}OH should produce lines with a spacing $4(18.871) = 75.484\ cm^{-1}$. Thus, we need only examine the spacings between adjacent lines, shown in the following table:

Lower Line (cm^{-1})	Upper line (cm^{-1})	Spacing (cm^{-1})
3,300.64	3,543.88	243.24
3,543.88	3,761.62	217.74
3,761.62	3,787.11	25.49
3,787.11	3,845.38	58.27
3,845.38	3,929.13	83.75
3,929.13	4,012.89	83.76
4,012.89	4,030.35	17.46
4,030.35	4,138.52	108.17
4,138.52	4,264.15	125.63
4,264.15	4,347.91	83.76
4,347.91	4,395.20	47.29
4,395.20	4,760.05	364.85
4,760.05	5,003.29	243.24
5,003.29	5,246.53	243.24
5,246.53	5,489.76	243.23

There are no lines separated by 75.484 cm^{-1}. Therefore, ^{16}OH is not present. There are lines whose spacings are 83.76 and 243.24 cm^{-1}. Consequently, we know that H_2 and $H^{19}F$ are in the sample cell.

(B) The Raman shifts for the *Q*-branch lines will appear at the vibrational wave numbers of the two molecules: 4,395.2 cm^{-1} for H_2 and 4,138.52 cm^{-1} for $H^{19}F$. Both of these lines are present. This confirms our analysis of the contents of the sample cell.

(C) We can now pick out the *S*- and *O*-branch lines for both H_2 and $H^{19}F$. The *S* branch starts $6Bh$ below the *Q* branch, while the *O* branch begins $6Bh$ above the *Q* branch. Thereafter, the successive lines appear at separations of $4Bh$ in both branches. Thus, we have the results below:

Molecule: H_2	*Q Branch:* 4,395.2 cm^{-1}
S-Branch Lines (cm^{-1})	**O-Branch Lines (cm^{-1})**
4,030.35	4,760.05
3,787.11	5,003.29
3,543.88	5,246.53
3,300.64	5,489.76
Molecule: $H^{19}F$	*Q Branch:* 4,138.52 cm^{-1}
S-Branch Lines (cm^{-1})	**O-Branch Lines (cm^{-1})**
4,012.89	4,264.15
3,929.13	4,347.91
3,848.38	4,431.67
3,761.62	4,515.42

15.47 Show that the combination of Eqs. 15.87A, 15.87B, 15.87C, and 15.88 leads to Eq. 15.89, as long as θ is kept fixed at the equilibrium H_2O angle.

Solution

The derivatives of Eqs. 15.87A, 15.87B, and 15.87C with respect to time are as follows:

$$\frac{\partial \Delta x_o}{\partial t} = \frac{2m_H}{m_o}\sin\theta\frac{\partial S_3}{\partial t} = \frac{2m_H}{m_o}\sin\theta\, v_{S_3}; \quad (1)$$

$$\frac{\partial \Delta y_o}{\partial t} = \frac{2m_H}{m_o}\frac{\partial S_1}{\partial t} = \frac{2m_H}{m_o}v_{S_1}; \quad (2)$$

$$\frac{\partial \Delta z_o}{\partial t} = 0; \quad (3)$$

$$\frac{\partial \Delta x_{H2}}{\partial t} = -\frac{\partial S_2}{\partial t} - \frac{\partial S_3}{\partial t}\sin\theta = -v_{S_2} - v_{S_3}\sin\theta; \quad (4)$$

$$\frac{\partial \Delta y_{H2}}{\partial t} = -\frac{\partial S_1}{\partial t} + \frac{\partial S_3}{\partial t}\cos\theta = -v_{S_1} + v_{S_3}\cos\theta; \quad (5)$$

$$\frac{\partial \Delta z_{H2}}{\partial t} = 0; \quad (6)$$

$$\frac{\partial \Delta x_{H3}}{\partial t} = \frac{\partial S_2}{\partial t} - \frac{\partial S_3}{\partial t}\sin\theta = v_{S_2} - v_{S_3}\sin\theta; \quad (7)$$

$$\frac{\Delta y_{H3}}{\partial t} = -\frac{\partial S_1}{\partial t} - \frac{\partial S_3}{\partial t}\cos\theta = -v_{S_1} - v_{S_3}\cos\theta; \quad (8)$$

$$\frac{\partial \Delta z_{H3}}{\partial t} = 0. \tag{9}$$

The kinetic energy is given by Eq. 15.88:

$$T = \sum_{i=1}^{3} 0.5 m_i \left[\left(\frac{\partial \Delta x_i}{\partial t} \right)^2 + \left(\frac{\partial \Delta y_i}{\partial t} \right)^2 + \left(\frac{\partial \Delta z_i}{\partial t} \right)^2 \right]. \tag{10}$$

Combining Eqs. 1–10 produces

$$T = \frac{2m_H^2}{m_o}[\sin^2\theta\, v_{S_3}^2 + v_{S_1}^2]$$

$$+ \frac{m_H}{2}[v_{S_2}^2 + v_{S_3}^2 \sin^2\theta + 2v_{S_2}v_{S_3}\sin\theta + v_{S_1}^2 + v_{S_3}^2\cos^2\theta - 2v_{S_1}v_{S_3}\cos\theta]$$

$$+ \frac{m_H}{2}[v_{S_2}^2 + v_{S_3}^2 \sin^2\theta - 2v_{S_2}v_{S_3}\sin\theta + v_{S_1}^2 + v_{S_3}^2\cos^2\theta + 2v_{S_1}v_{S_3}\cos\theta]. \tag{11}$$

Collecting terms, using the fact that $(\sin^2\theta + \cos^2\theta) = 1$, and adding out the cross terms gives

$$\boxed{T = \left[m_H + \frac{2m_H^2}{m_o}\right] v_{S_1}^2 + m_H v_{S_2}^2 + \left[m_H + \frac{2m_H^2}{m_o}\sin^2\theta\right] v_{S_3}^2,} \tag{12}$$

which is Eq. 15.89, as required by the problem.

15.49 The figure below shows the FTIR spectrum of 4-bromo-*N*-isopropylaniline.

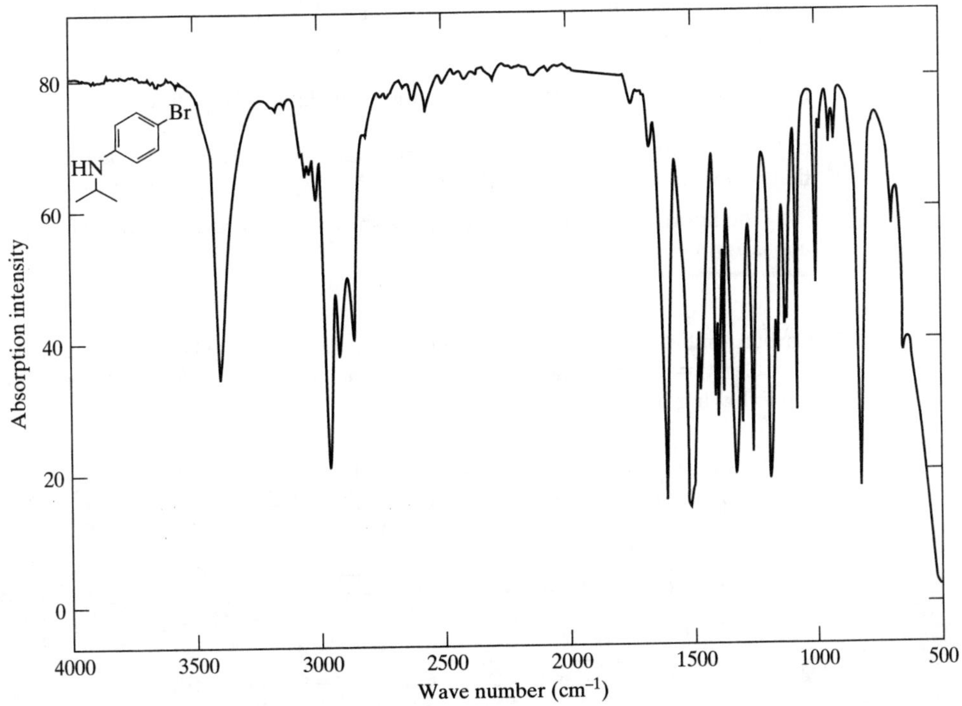

FTIR spectrum of 4-bromo-*N*-isopropylaniline. The author thanks Professor K. D. Berlin and Dr. M. M. Madler for providing this spectrum.

(A) Identify the absorption band due to the —N—H stretching mode.

(B) Which bands are due to the C—H stretching modes in the phenyl ring?

(C) What absorption bands are the result of C—H stretching motions in the $—CH_3$ groups?

(D) What strong absorption is the likely result of the —C=C— stretching modes in the phenyl ring?

(E) Which bands are associated with the —C=C—H and —C—C—H bending modes?

(F) There is a strong absorption around 1,450 cm^{-1}. What modes produce this absorption?

(G) Which absorption band is due to the —C—Br stretching mode?

Solution

(A) Table 15.5 indicates that the –N–H stretching mode should lie in the range from 3,100 to 3500 cm^{-1}. It is, therefore, clear that the band at 3,407 cm^{-1} reflects absorption from this mode.

(B) Using Table 15.5, we expect the =C–*H* stretching modes in the phenyl ring to appear slightly above 3,000 cm^{-1}. The small absorption bands in this region of the spectrum are the result of these modes.

(C) The –C–H stretching modes appear between 2,850 and 2,960 cm^{-1}. The three bands in this region are produced by absorptions in these modes.

(D) The –C=C– double-bond stretching modes generally absorb at around 1,620 to 1,680 cm^{-1}. The band around 1,650 cm^{-1} is the result of absorptions from these vibrational modes.

(E) The –C–C–H and –C=C–*H* bending modes absorb at around 1,000 and 1,100 cm^{-1}, respectively. There is a band at each of these wave numbers. The absorptions are probably due to these bending modes.

(F) The 1,450-cm^{-1} band is probably due to the H–C–H bending modes.

(G) The –C–Br stretch should appear at around 500 to 600 cm^{-1}. The bands in that region are due to this stretch.

The figure at the top of the next page shows the spectrum with the forego- ing assignments.

15.51 Carbon suboxide is a linear symmetric molecule whose structure is O=C=C=C=O. Eight molecular motions of this molecule are shown in the following diagram:

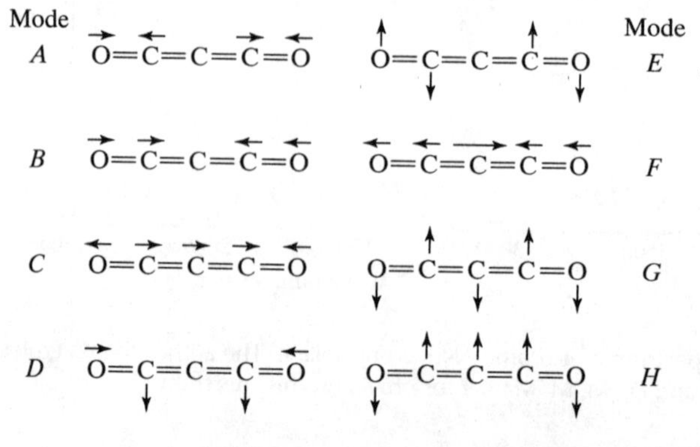

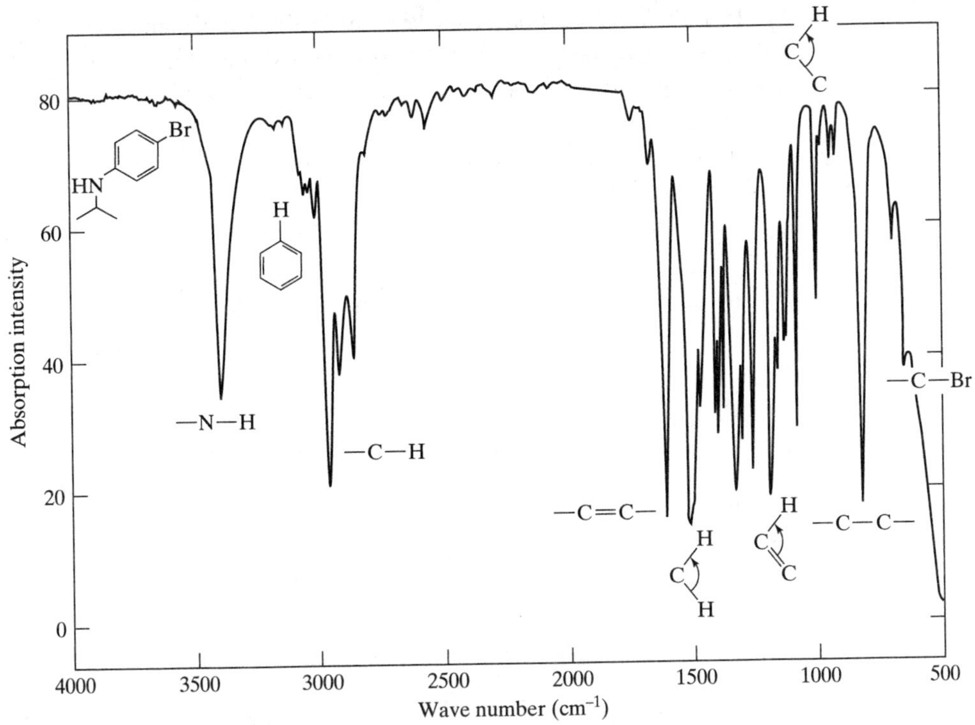

(A) How many vibrational modes does carbon suboxide have?

(B) Which of the molecular motions shown does not correspond to one of the vibrational modes of carbon suboxide?

(C) Which of the stretching modes shown above will be IR active? Which will be Raman active?

(D) Which of the molecular motions will be doubly degenerate vibrational modes of carbon suboxide? (The arrows show the normal-mode displacements in each mode.)

Solution

(A) In the preceding model, C_3O_2 is linear. The number of vibrational modes is, therefore,

$$3N - 5 = 3(5) - 5 = \underline{10 \text{ vibrational modes}}. \quad (1)$$

(B) Mode D is not a proper normal-mode vibrational coordinate, since the center of mass of the molecule is moving downward. It should be stationary.

(C) The stretching modes are A, B, C, and F. Modes A and B do not distort the symmetry of the molecule. Consequently, C_3O_2 never attains a dipole moment in these modes. They are, therefore, IR inactive. Modes C and F distort the symmetry and produce a momentary dipole moment. Consequently, these modes are IR active. Modes A and B significantly change the effective volume of the molecule during the vibration. The polarizability, therefore, is expected to vary with the vibrational motion such that $(\partial\alpha/\partial Q)_e \neq 0$. This makes these two modes Raman active.

(D) The bending motions can occur either in the plane of the paper or perpendicular to it. Each bending vibration—Modes E, G, and H—therefore represents two modes, one in the plane of the paper, the second perpendicular to that plane. These two motions will have the same frequency, so that they are doubly degenerate modes. In all, we have four stretching modes—*A*, *B*, *C*, and *F*—and three doubly degenerate bending modes—*E*, *G*, and *H*. This gives a total of 10 vibrational modes, which is the expected number.

15.53 Consider the ground state of the Li_2 molecule.

(A) What is the appropriate spectroscopic designation for the ground electronic state of this molecule?

(B) Suppose we excite Li_2 into the configuration $(1s\sigma_g)^2(1s\sigma_u^*)^2(2s\sigma_g)^1(2s\sigma_u^*)^1$. What are appropriate spectroscopic designations for the term states arising from this configuration?

Solution

(A) The electronic structure of Li_2 in the ground state is $(1s\sigma_g)^2(1s\sigma_u^*)^2(2s\sigma_g)^2$. We therefore have $|M| = 0$, since all electrons have a magnetic quantum number equal to zero. The spins are all paired, so $S = 0$. There is an even number of electrons in ungerade molecular orbitals, so the overall inversion symmetry is g. Finally, the reflection symmetry through a plane containing the internuclear axis must be +, since we have only sigma orbitals. Thus, the spectroscopic term is $\boxed{X^1\Sigma_g^+}$.

(B) In the configuration $(1s\sigma_g)^2(1s\sigma_u^*)^2(2s\sigma_g)^1(2s\sigma_u^*)^1$, all electrons still have magnetic quantum numbers of zero, so we can have only sigma states. However, the total spin can be either 0 or 1, since we need not worry about the Pauli principle, which is already satisfied by the fact that the two outer electrons are in different spatial orbitals. The overall inversion symmetry must be u, since we have an odd number of electrons in ungerade molecular orbitals. Again, the reflection symmetry is +, because we have only sigma orbitals. Thus, the two possible term states are $^1\Sigma_u^+$ and $^3\Sigma_u^+$. The first of these has the same multiplicity as the ground state. Therefore, we use a capital letter to designate it. The triplet term has a different multiplicity from the ground state, so we use a lower case letter for it.

The two states are $\boxed{A^1\Sigma_u^+}$ and $\boxed{a^3\Sigma_u^+}$.

15.55 A spectroscopist measures the vibrational bands between the $X^1\Sigma_g^+$ and the $B^1\Sigma_u^+$ states of H_2. The (02), (03), and (04) bands are found at 11.5043 eV, 11.6577 eV, and 11.8062 eV, respectively. The vibrational wave number and anharmonicity factor for the $X^1\Sigma_g^+$ ground state are known to be 4,395.2 cm^{-1} and 117.99 cm^{-1}, respectively. Use these data to determine the vibrational wave number and anharmonicity factor for the $B^1\Sigma_u^+$ state and the energy spacing between the potential minima of the two states.

Solution

The energy spacing between the vibrational bands is given by Eq. 15.93, viz.,

$$\Delta\omega_e = \omega_e + (v' + 0.5)\omega_e' - (x_e\omega_e)'(v' + 0.5)^2$$
$$- (v'' + 0.5)\omega_e'' + (x_e\omega_e)''(v'' + 0.5)^2, \qquad \textbf{(1)}$$

where ω_e is the energy spacing between the potential minima and ω_e' and $(x_e\omega)_e'$ are the vibrational wave number and anharmonicity factor respectively, for the excited $B^1\Sigma_u^+$ state. Substituting the data given in the problem produces three equations:

$$\Delta\omega_e(02) = 11.5043 \text{ eV } (8{,}065.7 \text{ cm}^{-1} \text{ eV}^{-1}) = 92{,}790.2$$
$$= \omega_e + 2.5\omega_e' - (2.5)^2(x_e\omega_e)' - 0.5(4{,}395.2) + 117.99(0.5)^2, \qquad \textbf{(2)}$$

$$\Delta\omega_e(03) = 11.6577 \text{ eV } (8{,}065.7 \text{ cm}^{-1} \text{ eV}^{-1}) = 94{,}027.5$$
$$= \omega_e + 3.5\omega_e' - (3.5)^2(x_e\omega_e)' - 0.5(4{,}395.2) + 117.99(0.5)^2, \quad \textbf{(3)}$$

and

$$\Delta\omega_e(04) = 11.8062 \text{ eV } (8{,}065.7 \text{ cm}^{-1} \text{ eV}^{-1}) = 95{,}225.3$$
$$= \omega_e + 4.5\omega_e' - (4.5)^2(x_e\omega_e)' - 0.5(4{,}395.2) + 117.99(0.5)^2. \quad \textbf{(4)}$$

We now have three linear equations and three unknowns. Subtracting Eq. 2 from 3 gives

$$94{,}027.5 - 92{,}790.2 = 1{,}237.3 = \omega_e' - 6(x_e\omega_e)'. \quad \textbf{(5)}$$

Subtracting Eq. 3 from Eq. 4 yields

$$95{,}225.3 - 94{,}027.5 = 1{,}197.8 = \omega_e' - 8(x_e\omega_e)'. \quad \textbf{(6)}$$

We can now subtract Eq. 6 from 5 to obtain

$$1{,}237.3 - 1{,}197.8 = 39.5 = 2(x_e\omega_e)', \quad \textbf{(7)}$$

so that

$$(x_e\omega_e)' = \underline{19.75 \text{ cm}^{-1}}. \quad \textbf{(8)}$$

Substituting Eq. 8 into Eq. 6 gives

$$\omega_e' = 1{,}197.8 + 8(19.75) \text{ cm}^{-1} = \underline{1{,}356 \text{ cm}^{-1}}. \quad \textbf{(9)}$$

Finally, we may substitute the results from Eqs. 8 and 9 into Eq. 2 to obtain

$$\omega_e = 92{,}790.2 - 2.5(1{,}356) + (2.5)^2(19.75) + 0.5(4{,}395.2) - 117.99(0.5)^2 \text{ cm}^{-1}$$
$$= \underline{91{,}692 \text{ cm}^{-1}} = \underline{11.368 \text{ eV}}. \quad \textbf{(10)}$$

15.57 The $A^3\Sigma_u^+$ state of O_2 has a rotational constant $B/c = B_e' = 1.05 \text{ cm}^{-1}$. The $O_2\ X^3\Sigma_g^-$ ground state has $B_e'' = 1.446 \text{ cm}^{-1}$. Thus, in this case, we have $B_e' < B_e''$. At what rotational quantum number J'' will the band head in the $X^3\Sigma_g^- \rightarrow A^3\Sigma_u^+$ vibrational bands appear? Plot the difference $(\Delta\omega_e^r - \Delta\omega_e)$ versus J'' to show that a Fortrat parabola exists in the R branch of the spectrum. Is the band head red-shifted or blue-shifted from $\Delta\omega_e$? What does this tell us about the equilibrium O_2 bond length in the two electronic states? The $X^3\Sigma_g^- \rightarrow A^3\Sigma_u^+$ vibrational bands are called Herzberg bands.

Solution

The difference $\Delta\omega_e^r - \Delta\omega_e$ is given by Eq. 15.95, derived in Problem 15.56. The result is

$$\Delta\omega_e^r - \Delta\omega_e = [B_e' - B_e''](J'')^2 + [3B_e' - B_e'']J'' + 2B_e'. \quad \textbf{(1)}$$

Inserting the data produces

$$\Delta\omega_e^r - \Delta\omega_e = -0.396(J'')^2 + 1.704J'' + 2.10 \text{ cm}^{-1}. \quad \textbf{(2)}$$

If J'' were a continuous variable, we could simply differentiate $\Delta\omega_e^r - \Delta\omega_e$ with respect to J'' and require that the derivative vanish to locate the point of maximum shift, which is the band head. This operation produces

$$\frac{d(\Delta\omega_e^r - \Delta\omega_e)}{dJ''} = -0.792J'' + 1.704 = 0. \quad \textbf{(3)}$$

The solution of Eq. 3 is

$$(J'')_{\max} = \frac{1.704}{0.792} = 2.15 \approx 2. \tag{4}$$

Therefore, we expect the band head to be at $J'' = 2$, but this is not absolutely certain, because J'' is not a continuous variable. The accompanying graph shows the Fortrat parabola, which is simply a plot of the left-hand side of Eq. 2 versus J''. Clearly, the maximum shift from $\Delta\omega_e$ occurs at $J'' = 2$. This is the band head, and it lies in the R branch, as we expected, since $B'_e < B''_e$.

Since the band head lies at an energy greater than $\Delta\omega_e$, it is blueshifted. We obtain a blueshifted band head whenever $B'_e < B''_e$. For this to be the case, we must have $R'_e > R''_e$; that is, the equilibrium O_2 bond distance must be greater in the upper than in the lower energy state.

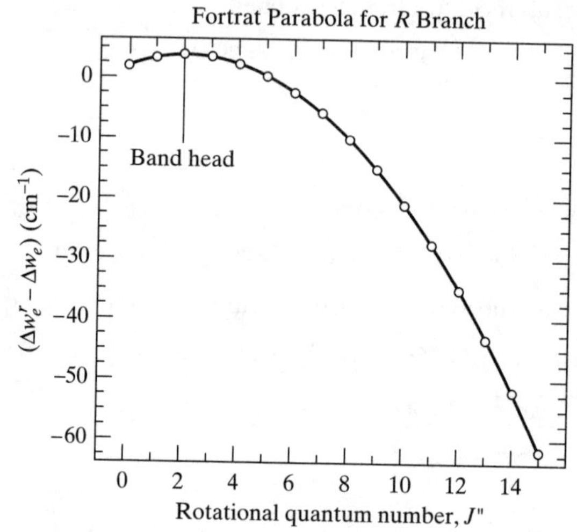

15.59 I'm ready to move onto nuclear magnetic resonance, but as usual, Sam has his hand up. It seems that he has a few questions about the James Bond (Agent 007) movie *Goldfinger.* For those members of the class who have not seen this movie, the scene is as follows: The archvillian Goldfinger has secret agent James Bond strapped to a platform. A large laser is moved into position with the beam directed at the foot of the platform between Bond's feet. When Goldfinger activates the laser, we see a brilliant golden-colored beam emerge and strike the platform. Immediately, the platform begins to melt under the action of the laser beam, which moves upward, threatening to cut 007 in half. Goldfinger announces that he expects Bond to die. Sam has the following questions:

(A) What is the approximate frequency of the electromagnetic radiation whose energy is in resonance with the transition responsible for Goldfinger's laser?

(B) What, if anything, do we know about the composition and color of the platform to which Bond is tied?

(C) What do we know about vibrational relaxation processes occurring in the platform under the action of the laser?

(D) What does this scene tell us, if anything, about the air quality in Goldfinger's laboratory? Explain.

Solution

(A) Since the color of the beam is golden or yellow, the wavelength of the radiation must be in the visible region of the electromagnetic spectrum. The radiation will appear to be yellow if its wavelength is about 580 nm. Therefore, the frequency of Goldfinger's laser is about

$$\nu = \frac{c}{\lambda} \approx \frac{2.9979 \times 10^{8}\ \mathrm{m\ s^{-1}}}{580 \times 10^{-9}\ \mathrm{m}} = \underline{5.17 \times 10^{14}\ \mathrm{s^{-1}}}. \tag{1}$$

(B) Since the platform is melting under the action of the laser, it must be absorbing the radiation. Therefore, the material is something that has an electronic transition in which the energy spacing corresponds to a frequency of about $5.17 \times 10^{14}\ \mathrm{s^{-1}}$. Furthermore, we know that the material absorbs strongly in this spectral region. With the yellow light of the visible spectrum being absorbed, we expect the platform to be green to blue in color.

(C) The platform is melting. Therefore, the radiant energy is being converted into thermal energy rather than being emitted as fluorescence or phosphorescence. This occurs by internal conversion or intersystem crossing, followed by vibrational relaxation to the lattice modes, where the energy will appear as thermal energy. Consequently, we know that a substantial amount of vibrational relaxation is taking place.

(D) The viewer easily sees the laser beam as it emerges from the optical cavity. This can occur only if the radiation in the beam is being scattered by particulate matter in the air. But then, Goldfinger's laboratory must be very dusty, as befits his archvillain status. If the EPA hears about this, Secret Agent 007, James Bond, will be the least of Goldfinger's problems.

CHAPTER 16

Magnetic and Diffraction Spectroscopy

16.1 (A) Compute the value the Z component of the spin magnetic moment of ^{19}F would have if the bare nucleus behaved as a circular conducting loop. The mass of ^{19}F is 3.1531×10^{-26} kg.
(B) Use the data in Table 16.1 to obtain the measured value for μ_z. You may express your answer in terms of m_I. What is the ratio of the results obtained in (A) and (B)?

Solution

(A) If the magnetic moment of the ^{19}F nucleus could be computed using a circular-conducting loop model, we could use Eq. 6.14 to obtain μ_z:

$$\mu_z = \frac{q}{2m} S_z = \frac{q\hbar}{2m} m_I. \quad (1)$$

The charge on a bare ^{19}F nucleus is $9 \times 1.602177 \times 10^{-19}$ C $= 1.44196 \times 10^{-18}$ C. Substituting this value and the mass into Eq. 1 produces

$$\mu_z = \frac{(1.44196 \times 10^{-18}\ \text{C})(6.62608 \times 10^{-34}\ \text{J s})}{4(3.1415927)(3.1531 \times 10^{-26}\ \text{kg})} m_I = \underline{2.4114 \times 10^{-27} m_I\ \text{J T}^{-1}}. \quad (2)$$

The measured value for μ_z is given by Eq. 16.16:

$$\mu_z = g_N \frac{e}{2m_p} S_z = g_N \frac{e\hbar}{2m_p} m_I = g_N \mu_N m_I. \quad (3)$$

Substituting the values of g_N from Table 16.1 and the nuclear magneton gives

$$\mu_z = 5.257736(5.0508 \times 10^{-27}\ \text{J T}^{-1}) m_I = \underline{26.556 \times 10^{-27} m_I\ \text{J T}^{-1}}. \quad (4)$$

The ratio of the two results is

$$\frac{[\mu_z]_{\text{loop}}}{[\mu z]_{\text{measured}}} = \frac{2.4114 \times 10^{-27} m_I\ \text{J T}^{-1}}{26.556 \times 10^{-27} \text{m}_\text{I}\ \text{J T}^{-1}} = \underline{0.0908}. \quad (5)$$

The circular-conducting loop model produces an answer that is one-eleventh that of the measured Z component of the nuclear spin magnetic moment. Obviously, it is very inaccurate. In this case, it does not even give the right order of magnitude.

16.3 Many NMR spectrometers use fixed magnetic fields such that ^1H resonance occurs at a frequency of 300 MHz, 400 MHz, 500 MHz, 600 MHz, or 750 MHz. Instruments using even higher frequencies will certainly be developed in the near future.
(A) Determine the magnetic fields required to produce proton resonance at the given frequencies.
(B) Determine the frequencies at which we would observe resonance for ^{13}C and ^{19}F bare nuclei in each of the indicated spectrometers.

Solution

(A) The condition for resonance is

$$h\nu = g_N \mu_N B. \quad (1)$$

Therefore, at a fixed frequency, the required magnetic field is

$$B = \frac{h\nu}{g_N \mu_N}. \quad (2)$$

Thus, for ^{1}H, the B field needed is

$$B(^1\text{H}) = \frac{6.62608 \times 10^{-34}\ \text{J s}}{(5.5856948)(5.0508 \times 10^{-31}\ \text{J G}^{-1})}\nu = 2.3487 \times 10^{-4}\nu\ \text{G s}. \quad \textbf{(3)}$$

Using Eqs. 3, we obtain the results below.

(B) At these magnetic fields, we will observe ^{13}C and ^{19}F resonance whenever we have

$$\nu_\text{C} = \frac{g_\text{C}\mu_N B_o}{h} = \frac{(1.40482)(5.0508 \times 10^{-31}\ \text{J G}^{-1})B_o}{6.62608 \times 10^{-34}\ \text{J s}} = 1{,}070.84 B_o\ s^{-1} \quad \textbf{(4)}$$

Nucleus	ν (MHz)	Magnetic Field Required for Resonance (G)
^{1}H	300	70,461
	400	93,948
	500	1.1744×10^5
	600	1.4092×10^5
	750	1.7615×10^5

and

$$\nu_\text{F} = \frac{g_\text{F}\mu_N B_o}{h} = \frac{(5.257736)(5.0508 \times 10^{-31}\ \text{J G}^{-1})B_o}{6.62608 \times 10^{-34}\ \text{J s}} = 4{,}007.77 B_o\ s^{-1}. \quad \textbf{(5)}$$

Inserting the B_o fields computed in Part (A) gives the results in the following table:

Nucleus	B_o	Frequency Required for Resonance (MHz)
^{13}C	70,461	75.452
	93,948	100.60
	1.1744×10^5	125.75
	1.4092×10^5	150.90
	1.7615×10^5	188.63
^{19}F	70,461	282.39
	93,948	376.52
	1.1744×10^5	470.65
	1.4092×10^5	564.78
	1.7615×10^5	705.98

16.5 An unknown nuclear particle is placed in an NMR spectrometer. When the applied field is 23,486 G, the particle is found to undergo nuclear spin transitions when the electromagnetic radiation has a frequency of 9.809 MHz. What is the nuclear particle whose spin transitions are being observed in the spectrometer?

Solution

The NMR resonant frequency is given by Eq. 16.19. Solving this equation for the nuclear g-factor, we obtain

$$g_N = \frac{h\nu}{\mu_N B}. \quad \textbf{(1)}$$

Substituting the data from the problem allows us to compute the g-factor for our unknown nucleus.

$$g_N = \frac{(6.62608 \times 10^{-34}\ \mathrm{J\ s})(9.809 \times 10^{6}\ \mathrm{s}^{-1})}{(5.0508 \times 10^{-31}\ \mathrm{J\ G}^{-1})(23{,}486\ \mathrm{G})} = 0.5479. \quad \textbf{(2)}$$

Inspecting the data given in Table 16.1 shows that the unknown nucleus must be $\boxed{^{35}\mathrm{Cl}}$, since that is the only one with this value for g_N.

16.7 An NMR ^{13}C spectrum of *p*-methylbenzoic acid (see Problem 16.6) is taken with a 400-MHz spectrometer.

(A) If there were no chemical shift between carbon nuclei, what frequency would be required to satisfy the NMR resonance condition?

(B) Estimate the frequency differences for the CH_3, ring, and acidic ^{13}C nuclei when the chemical shift is present.

Solution

(A) The external field employed in a 400-MHz NMR spectrometer is that required to produce ^{1}H resonance at 400 MHz. This field can be computed using Eq. 16.19:

$$B_o = \frac{h\nu}{g_H \mu_N} = \frac{(6.62608 \times 10^{-34}\ \mathrm{J\ s})(4 \times 10^{8}\ \mathrm{s}^{-1})}{(5.58569)(5.0508 \times 10^{-31}\ \mathrm{J\ G}^{-1})} = 93{,}946\ \mathrm{G}. \quad \textbf{(1)}$$

At this magnetic field, the frequency required for ^{13}C resonance is

$$\nu = \frac{g_C \mu_N B_o}{h} = \frac{(1.40482)(5.0508 \times 10^{-31}\ \mathrm{J\ G}^{-1})(93{,}946\ \mathrm{G})}{6.62608 \times 10^{-34}\ \mathrm{J\ s}} = 1.0060 \times 10^{8}\ \mathrm{s}^{-1}$$

$$= \underline{100.60\ \mathrm{MHz}}. \quad \textbf{(2)}$$

(B) We now need a reasonable estimate for the chemical shifts of the ^{13}C nuclei in the –COOH, $-CH_3$, and Ph groups of the molecule. Table 16.3 tells us that ^{13}C in CH_3– has a chemical shift in the range from −2.3 to 25.2. The ring ^{13}C's have a chemical shift of 128.5, and the acidic ^{13}C's exhibit shifts in the range from 166.0 to 178.1. If we take intermediate values as estimates, we obtain

$$\delta_{CH_3} \approx 11.4, \quad \textbf{(3)}$$

$$\delta_{Ph} \approx 128.5, \quad \textbf{(4)}$$

and

$$\delta_{COOH} \approx 172.1. \quad \textbf{(5)}$$

We can now use Eq. 16.30 to compute the difference in resonant frequencies:

$$\nu_{Ph} - \nu_{CH3} = \frac{g_C \mu_N}{h} B_o[\delta_{Ph} - \delta_{CH3}]10^{-6} = 100.60 \times 10^{6}\ \mathrm{Hz}\ [\delta_{Ph} - \delta_{CH3}]10^{-6}$$

$$= 100.60[\delta_{Ph} - \delta_{CH3}]\ \mathrm{Hz} = 100.60(128.5 - 11.4) = \underline{1.18 \times 10^{4}\ \mathrm{Hz}}. \quad \textbf{(6)}$$

For the –COOH and Ph groups, the result is

$$\nu_{COOH} - \nu_{Ph} = 100.60[\delta_{COOH} - \delta_{Ph}]\ \text{Hz} = 100.60(172.1 - 128.5) = \underline{4.39 \times 10^3\ \text{Hz}}. \quad \textbf{(7)}$$

If you have worked the previous problem, you will notice that the frequency separations are much larger for ^{13}C than for ^{1}H.

16.9 The propane molecule is $H_3C–CH_2–CH_3$.

(A) Qualitatively speaking, what is the nature of the spin–spin splitting of the $–CH_3$ and $–CH_2–$ protons that we would expect to observe in a ^{1}H NMR spectrum taken with a 400-MHz spectrometer.

(B) What would be the approximate ratio of the absorption intensities for each of these groups of protons? Explain.

(C) What are the approximate intensity ratios for each of the spin–spin bands within each individual proton group?

Solution

(A and C) The six $–CH_3$ protons are chemically equivalent, so that their chemical shifts are identical. They will, therefore, exhibit the same NMR spectrum. Each $–CH_3$ is adjacent to the $–CH_2–$ group with two equivalent protons. The two $–CH_2–$ protons can have three different total spin arrangements, which produce $M_I = 1, 0$, and -1. Thus, the $–CH_3$ proton resonance is split into three different bands. The intensity ratios of these three bands are 1 : 2 : 1, since the $M_I = 1$ and -1 spin states can be obtained in only $2!/[2!(2-2)!] = 1$ way, whereas the $M_I = 0$ state can be obtained in $2!/[1!(2-1)!] = 2$ ways. Therefore, the qualitative form of the $–CH_3$ NMR bands is as shown in the following diagram:

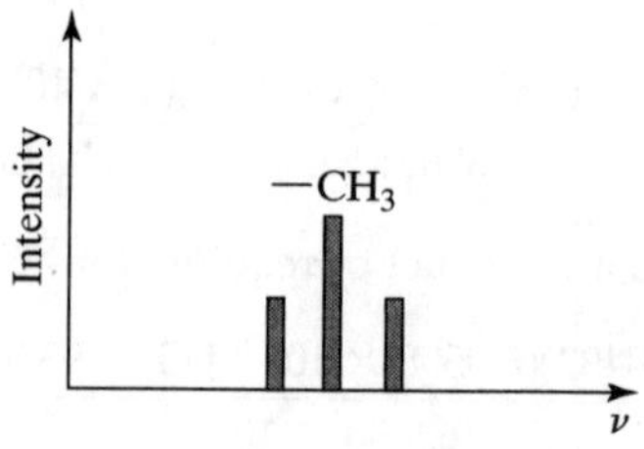

The two equivalent $–CH_2–$ protons are adjacent to six equivalent $–CH_3$ protons, which can have 0, 1, 2, 3, 4, 5, or 6 spin magnetic quantum numbers equal to $\frac{1}{2}$. Thus, the total spin magnetic quantum number can have values of $M_I = 3, 2, 1, 0, -1, -2$, or -3. As a result, the spin–spin coupling splits the $–CH_2–$ NMR resonance band into seven different bands. The intensity ratios are

$$I(M_I = 3) = I(M_I = -3) = \frac{6!}{6!(6-6)!} = 1, \quad \textbf{(1)}$$

$$I(M_I = 2) = I(M_I = -2) = \frac{6!}{5!(6-5)!} = 6, \quad \textbf{(2)}$$

$$I(M_I = 1) = I(M_I = -1) = \frac{6!}{4!(6-4)!} = 15, \quad \textbf{(3)}$$

and

$$I(M_I = 0) = \frac{6!}{3!(6-3)!} = 20. \quad \textbf{(4)}$$

Therefore, the qualitative form of the $–CH_3$ NMR band is at the top of the next page.

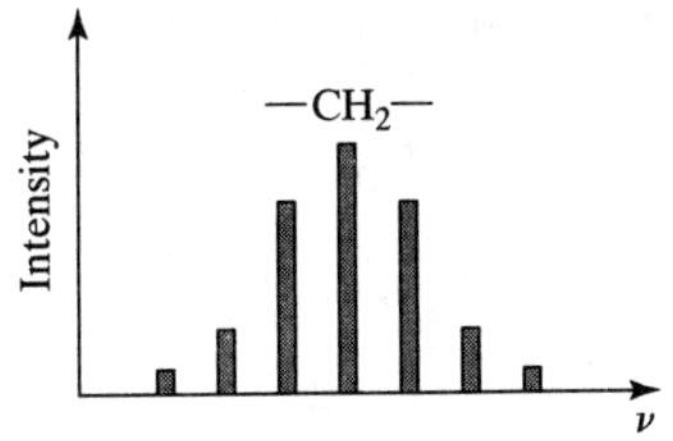

(B) The total absorption intensity for the $-CH_3$ protons will be about three times that for the $-CH_2-$ protons, since we have six methyl group protons and only two on the $-CH_2-$ group.

16.11 Use Eq. 16.43 to obtain the sensitivity of 200-, 300-, 400-, 500-, 600-, and 750-MHz NMR spectrometers relative to that of a spectrometer operating at 100 MHz at a temperature of 298 K. How does the sensitivity vary with frequency?

Solution

The sensitivity of a 100-MHz spectrometer was shown in the text to be

$$\text{Intensity(100-MHz spectrometer)} = I_o(100) = 0.0000161\ C'. \tag{1}$$

For a spectrometer operating at frequency ν, the sensitivity is given by Eq. 16.43:

$$I_o(\nu) = C'\left[1 - \exp\left\{-\frac{h\nu}{kT}\right\}\right]. \tag{2}$$

Substituting values gives

$$I_o(\nu) = C'\left[1 - \exp\left\{-\frac{(6.626 \times 10^{-34}\ \text{J s})\nu}{(1.381 \times 10^{-23}\ \text{J K}^{-1})(298\ \text{K})}\right\}\right]$$

$$= C'[1 - \exp\{-1.6101 \times 10^{-13}\nu\}]. \tag{3}$$

Using Eq. 3, we obtain the following table:

ν (MHz)	$I_o(\nu)$	$I_o(\nu) / I_o(100)$
200	0.0000322 C′	2.000
300	0.0000483 C′	3.000
400	0.0000644 C′	4.000
500	0.0000805 C′	5.000
600	0.0000966 C′	6.000
750	0.000121 C′	7.500

We see that the sensitivity is a near linear function of the spectrometer frequency. This is proven in Problem 16.12.

16.13 Prove that the commutator $[t, i\hbar\,(\partial/\partial t)] = -i\hbar$.

Solution

Consider the operation of the commutator on a function F, $[t, i\hbar\,\partial/\partial t]F$. Expanding the commutator gives

$$\left[t, i\hbar\frac{\partial}{\partial t}\right]F = t\,i\hbar\frac{\partial F}{\partial t} - i\hbar\frac{\partial}{\partial t}[tF]. \tag{1}$$

Expanding the derivative of the product in the second term of Eq. 1, we obtain

$$\left[t, i\hbar\frac{\partial}{\partial t}\right]F = t\,i\hbar\frac{\partial F}{\partial t} - i\hbar F - i\hbar t\frac{\partial F}{\partial t} = -i\hbar F. \quad \textbf{(2)}$$

For Eq. 2 to hold, we must have

$$\left[t, i\hbar\frac{\partial}{\partial t}\right] = -i\hbar. \quad \textbf{(3)}$$

16.15* The propanoic acid molecule is CH_3–CH_2–COOH.

(A) Qualitatively describe the expected appearance of the ^{1}H NMR spectrum of this molecule if the spectrum is first order. Give the expected line intensities.

(B) Using Tables 16.2 and 16.4, estimate the ^{1}H chemical shifts and spin–spin coupling constants.

(C) Use intermediate values of the data estimated in (B) to compute the expected resonance frequencies for all lines in the NMR spectrum if a 60-MHz spectrometer is used.

(D) Assuming a spin relaxation time constant of 500 ms, obtain a mathematical function describing the FID that would be expected if the NMR spectrum of propanoic acid were taken on a 60-MHz NMR spectrometer using the resonant frequency for $Si(CH_3)_4$ as ν_{ref}. Plot this function for times from zero to 1,200 ms.

Solution

(A) Three sets of absorption bands will appear in the ^{1}H NMR spectrum. The acidic proton will appear as a singlet, since there are no other spins within three chemical bonds to split the line. The $-CH_2-$ protons will appear as a quartet with the intensity ratio 1 : 3 : 3 : 1, for the reasons discussed in the text related to Figures 16.12 and 16.13. The total intensity of the $-CH_2-$ group should be about twice that of the acidic proton, because there are now two protons present. The $-CH_3$ group will appear as a triplet with an intensity ratio 1 : 2 : 1, because there are three ways for the two adjacent $-CH_2-$ spins to be aligned. The qualitative appearance of the spectrum will be as follows:

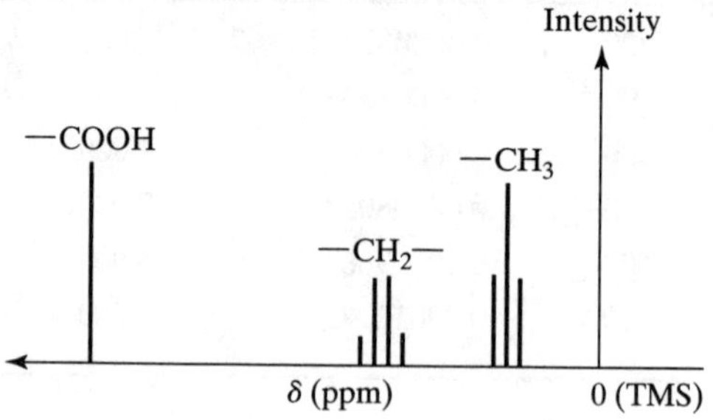

(B) The estimated chemical shifts, obtained from Table 16.2, are $\delta(CH_3) \approx$ 1.2–1.3, $\delta(CH_2) \approx$ 2.5–2.8, and $\delta(COOH) \approx$ 10.2–11.4. The spin–spin coupling constant, estimated using Table 16.4, is J_{CH-CH} = 5–9 Hz.

(C) Taking intermediate values for the chemical shifts and J, we have $\delta(CH_3) \approx$ 1.25, $\delta(CH_2) \approx$ 2.65, $\delta(COOH) \approx$ 10.8, and J = 7 Hz. The four resonant frequencies for the $-CH_2-$ protons are given by Eqs. 16.41 A–D if we replace J_{CHO-CH} with the CH–CH spin–spin coupling constant and $B_o(1 - \sigma_{CHO})$ with the appropriate resonant magnetic field for $-CH_2-$ protons given by Eq. 16.28:

$$\nu_1^{CH_2} = \frac{g_H \mu_N B_o(1 - \sigma_{ref} + 10^{-6}\delta_{CH_2})}{h} + \frac{3J_{CH-CH}}{2}, \quad (1)$$

$$\nu_2^{CH_2} = \frac{g_H \mu_N B_o(1 - \sigma_{ref} + 10^{-6}\delta_{CH_2})}{h} + \frac{J_{CH-CH}}{2}, \quad (2)$$

$$\nu_3^{CH_2} = \frac{g_H \mu_N B_o(1 - \sigma_{ref} + 10^{-6}\delta_{CH_2})}{h} - \frac{J_{CH-CH}}{2}, \quad (3)$$

and

$$\nu_4^{CH_2} = \frac{g_H \mu_N B_o(1 - \sigma_{ref} + 10^{-6}\delta_{CH_2})}{h} - \frac{3J_{CH-CH}}{2}. \quad (4)$$

The $-CH_3$ protons will appear as a triplet, since there are three possible orientations of the three spins on the adjacent $-CH_2-$ group. The orientations give total magnetic quantum numbers of $M_I = -1, 0,$ and 1. Consequently, the resonant $-CH_3$ frequencies are

$$\nu_1^{CH_3} = \frac{g_H \mu_N B_o(1 - \sigma_{ref} + 10^{-6}\delta_{CH_3})}{h} + J_{CH-CH}, \quad (5)$$

$$\nu_2^{CH_3} = \frac{g_H \mu_N B_o(1 - \sigma_{ref} + 10^{-6}\delta_{CH_3})}{h} \quad (6)$$

and

$$\nu_3^{CH_3} = \frac{g_H \mu_N B_o(1 - \sigma_{ref} + 10^{-6}\delta_{CH_3})}{h} - J_{CH-CH}. \quad (7)$$

The –COOH appears as a single band, since there is no spin–spin splitting. The resonance frequency is given by an equation similar to Eqs. 1–7, but without the spin–spin splitting term:

$$\nu^{COOH} = \frac{g_H \mu_N B_o(1 - \sigma_{ref} + 10^{-6}\delta_{COOH})}{h}. \quad (8)$$

In a 60-MHz spectrometer,

$$B_o = \frac{(60 \times 10^6\ s^{-1})h}{g_H \mu_N}. \quad (9)$$

Substituting this result into Eqs. 1–8 produces

$$\nu_1^{CH_2} = 60 \times 10^6 - 60 \times 10^6 \sigma_{ref} + 60\delta_{CH_2} + \frac{3J_{CH-CH}}{2}$$

$$= 60 \times 10^6 - 60 \times 10^6 \sigma_{ref} + 60(2.65) + \frac{3(7)}{2}$$

$$= 60 \times 10^6 - 60 \times 10^6 \sigma_{ref} + 169.5\ Hz, \quad (10)$$

$$\nu_2^{CH_2} = 60 \times 10^6 - 60 \times 10^6 \sigma_{ref} + 60\delta_{CH_2} + \frac{J_{CH-CH}}{2}$$

$$= 60 \times 10^6 - 60 \times 10^6 \sigma_{ref} + 60(2.65) + \frac{7}{2}$$

$$= 60 \times 10^6 - 60 \times 10^6 \sigma_{ref} + 162.5\ Hz, \quad (11)$$

$$\nu_3^{CH_2} = 60 \times 10^6 - 60 \times 10^6 \sigma_{ref} + 60\delta_{CH_2} - \frac{J_{CH-CH}}{2}$$

$$= 60 \times 10^6 - 60 \times 10^6 \sigma_{ref} + 60(2.65) - \frac{7}{2}$$

$$= 60 \times 10^6 - 60 \times 10^6 \sigma_{ref} + 155.5 \text{ Hz}, \quad (12)$$

$$\nu_4^{CH_2} = 60 \times 10^6 - 60 \times 10^6 \sigma_{ref} + 60\delta_{CH_2} - \frac{3J_{CH-CH}}{2}$$

$$= 60 \times 10^6 - 60 \times 10^6 \sigma_{ref} + 60(2.65) - \frac{3(7)}{2}$$

$$= 60 \times 10^6 - 60 \times 10^6 \sigma_{ref} + 148.5 \text{ Hz}, \quad (13)$$

$$\nu_1^{CH_3} = 60 \times 10^6 - 60 \times 10^6 \sigma_{ref} + 60\delta_{CH_3} + J_{CH-CH}$$

$$= 60 \times 10^6 - 60 \times 10^6 \sigma_{ref} + 60(1.25) + 7$$

$$= 60 \times 10^6 - 60 \times 10^6 \sigma_{ref} + 82 \text{ Hz}, \quad (14)$$

$$\nu_2^{CH_3} = 60 \times 10^6 - 60 \times 10^6 \sigma_{ref} + 60\delta_{CH_3}$$

$$= 60 \times 10^6 - 60 \times 10^6 \sigma_{ref} + 60(1.25)$$

$$= 60 \times 10^6 - 60 \times 10^6 \sigma_{ref} + 75 \text{ Hz}, \quad (15)$$

$$\nu_3^{CH_3} = 60 \times 10^6 - 60 \times 10^6 \sigma_{ref} + 60\delta_{CH_3} - J_{CH-CH}$$

$$= 60 \times 10^6 - 60 \times 10^6 \sigma_{ref} + 60(1.25) - 7$$

$$= 60 \times 10^6 - 60 \times 10^6 \sigma_{ref} + 68 \text{ Hz}, \quad (16)$$

and

$$\nu^{COOH} = 60 \times 10^6 - 60 \times 10^6 \sigma_{ref} + 60\delta_{COOH}$$

$$= 60 \times 10^6 - 60 \times 10^6 \sigma_{ref} + 60(10.8)$$

$$= 60 \times 10^6 - 60 \times 10^6 \sigma_{ref} + 648 \text{ Hz}. \quad (17)$$

(D) The resonant frequency for the reference TMS is

$$\nu_{TMS} = \nu_{ref} = \frac{g_H \mu_N B_o (1 - \sigma_{ref} + 10^{-6}\delta_{TMS})}{h} = \frac{g_H \mu_N B_o}{h}(1 - \sigma_{ref}), \quad (18)$$

since $\delta_{TMS} = 0$. Inserting Eq. 9 for B_o converts Eq. 18 to

$$\nu_{ref} = 60 \times 10^6 (1 - \sigma_{ref}) = 60 \times 10^6 - 60 \times 10^6 \sigma_{ref}. \quad (19)$$

The detector measures a set of frequency differences between the resonance lines for propanoic acid and that for the reference. These differences are

$$\nu_1^{CH_2} - \nu_{ref} = 169.5 \text{ Hz} = 0.1695 \text{ ms}^{-1}, \quad (20)$$

$$\nu_2^{CH_2} - \nu_{ref} = 162.5 \text{ Hz} = 0.1625 \text{ ms}^{-1}, \quad (21)$$

$$\nu_3^{CH_2} - \nu_{ref} = 155.5 \text{ Hz} = 0.1555 \text{ ms}^{-1}, \quad (22)$$

$$\nu_4^{CH_2} - \nu_{ref} = 148.5 \text{ Hz} = 0.1485 \text{ ms}^{-1}, \quad (23)$$

$$\nu_1^{CH_3} - \nu_{ref} = 82.0 \text{ Hz} = 0.0820 \text{ ms}^{-1}, \quad (24)$$

$$\nu_2^{CH_3} - \nu_{ref} = 75.0 \text{ Hz} = 0.0750 \text{ ms}^{-1}, \quad (25)$$

$$\nu_3^{CH_3} - \nu_{ref} = 68.0 \text{ Hz} = 0.0680 \text{ ms}^{-1}, \quad (26)$$

and

$$\nu^{\text{COOH}} - \nu_{\text{ref}} = 648 \text{ Hz} = 0.648 \text{ ms}^{-1}. \quad \textbf{(27)}$$

The signal for each of these excitations has the form given by Eq. 16.48:

$$s_i = n_i \exp\left(-\frac{t}{\tau}\right) \cos[2\pi(\nu_i - \nu_{\text{ref}})t]. \quad \textbf{(28)}$$

Thus, the signal due to the band at frequency ν_1^{CH3} is

$$s_1^{\text{CH3}} = n_1^{\text{CH3}} \exp\left[-\frac{t}{500}\right] \cos[2\pi(0.082)t], \quad \textbf{(29)}$$

provided that we measure t in ms. Similar forms give the other seven components of the total signal.

We can obtain the expected intensity ratios by adjusting the n_i appropriately. In doing this, we must remember that the intensity of a wave is proportional to the absolute square of its amplitude. The $-CH_3$ resonance appears as a triplet with line intensity ratio of 1:2:1. Therefore, we take $n_1^{\text{CH3}} = n_3^{\text{CH3}} = (\frac{1}{4})^{1/2}$ and $n_2^{\text{CH3}} = (\frac{1}{2})^{1/2}$. The $-CH_2$ resonance will be a quartet, with ν_2 and ν_3 having $\frac{3}{8}$ of the total intensity and ν_1 and ν_4 having $\frac{1}{8}$. We can obtain this result by taking $n_1^{\text{CH2}} = n_4^{\text{CH2}} = (\frac{1}{8})^{1/2}$ and $n_2^{\text{CH2}} = n_3^{\text{CH2}} = [\frac{3}{8}]^{1/2}$. Finally, the total CH_3 intensity will be three times that of the COOH resonance, while the $-CH_2-$ lines have two times the total intensity of the acidic proton line. These results can be gotten by multiplying the sum of the three CH_3 bands by $(3)^{1/2}$ and the sum of the four $-CH_2-$ lines by $(2)^{1/2}$. With these points in mind, we have, for the total FID signal,

$$\begin{aligned} S_{\text{total}} &= (3)^{1/2}\left[\left(\frac{1}{4}\right)^{1/2}\{s_1^{\text{CH3}} + s_3^{\text{CH3}}\} + \left(\frac{1}{2}\right)^{1/2} s_2^{\text{CH3}}\right] \\ &\quad + (2)^{1/2}\left[\left(\frac{1}{8}\right)^{1/2}\{s_1^{\text{CH2}} + s_4^{\text{CH2}}\} + \left(\frac{3}{8}\right)^{1/2}\{s_2^{\text{CH2}} + s_3^{\text{CH2}}\}\right] \\ &\quad + s^{\text{COOH}}, \end{aligned} \quad \textbf{(30)}$$

or

$$\begin{aligned} s_{\text{total}} &= (3)^{1/2} \exp\left[-\frac{t}{500}\right]\left[\left(\frac{1}{4}\right)^{1/2}\{\cos[2\pi(0.082)t] + \cos[2\pi(0.068)t]\}\right. \\ &\quad \left. + \left(\frac{1}{2}\right)^{1/2} \cos[2\pi(0.075)t]\right] \\ &\quad + (2)^{1/2} \exp\left[-\frac{t}{500}\right]\left[\left(\frac{1}{8}\right)^{1/2}\{\cos[2\pi(0.1695)t] + \cos[2\pi(0.1485)t]\}\right] \\ &\quad + (2)^{1/2} \exp\left[-\frac{t}{500}\right]\left[\left(\frac{3}{8}\right)^{1/2}\{\cos[2\pi(0.1625)t] + \cos[2\pi(0.1555)t]\}\right] \\ &\quad + \exp\left[-\frac{t}{500}\right] \cos[2\pi(0.648)t]. \end{aligned} \quad \textbf{(31)}$$

The figures on the following page show plots of this FID over the range 0 to 1,200 ms and the associated power spectrum of the FID.

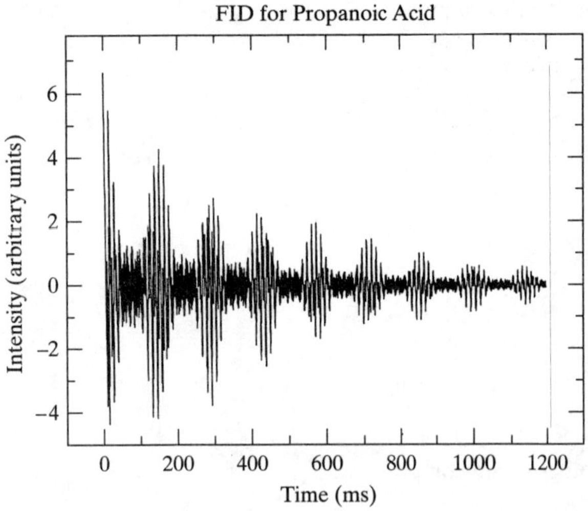

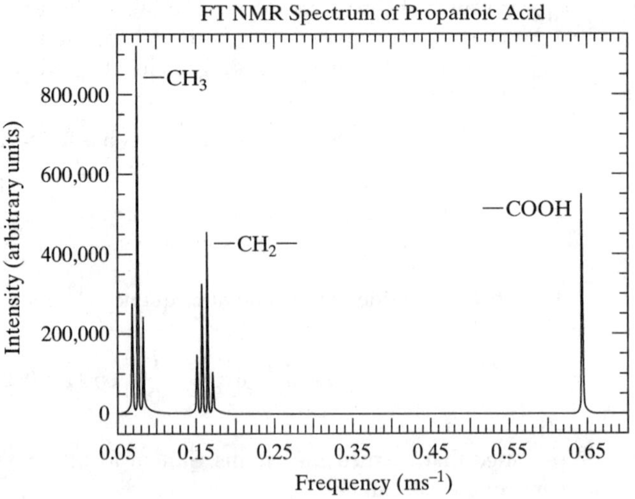

16.17 An investigator needs a ^1H NMR spectrum of a biological sample. The amount of material available is so small that it is estimated that 15,000 spectral scans will be required to obtain an accurate spectrum with sufficient resolution. Estimate the amount of spectrometer time required to obtain the data if a CW spectrometer is used. Estimate the time required for an FT NMR spectrometer. Which would you use?

Solution

The CW instrument will require about 10 to 15 minutes per scan. If we require 15,000 scans, the time needed is

$$15{,}000 \text{ scans} \times \left[\frac{10 \text{ to } 15 \text{ min}}{\text{scan}}\right] = 150{,}000 \text{ min to } 225{,}000 \text{ min}$$

$$= \underline{104.2 \text{ to } 156.2 \text{ days.}} \qquad \textbf{(1)}$$

Using an FT NMR spectrometer, the result is

$$15{,}000 \text{ scans} \times \left[\frac{1.5 \text{ to } 2.0 \text{ s}}{\text{scan}}\right] = 22{,}500 \text{ s to } 30{,}000 \text{ s}$$

$$= \underline{6.25 \text{ to } 8.33 \text{ hours.}} \qquad \textbf{(2)}$$

The last part of the question is known in physical-chemistry circles as a "no-brainer."

16.19 The figure below shows the ^1H NMR spectrum of ethyl 2,4-dimethylbenzoate. The spectrum was recorded on a 300-MHz FT NMR spectrometer with TMS as the reference. The positions of the bands are given in terms of δ (parts per million, or ppm). The conversion between δ and the frequency required for resonance is given by Eq. 16.28. (See Shankar Subramanian, *Modified Heteroarotinoids: Potential Anticancer Agents, Ph.D. dissertation,* Oklahoma State University, 1993.) The spectrum shows that the chemical shifts are much larger than the spin–spin splittings for many, but not all, of the bands. Identify the proton groups giving rise to each of the spectral lines.

Solution

The $-CH_3$ protons in the aliphatic chain will exhibit resonance around $\delta = 1.1$–1.2, according to Table 16.2. The line will appear as a triplet, due to the spin–spin coupling of the adjacent $-CH_2-$ protons. This triplet is seen clearly in the spectrum in the expected range. The two $-CH_3$ groups on the ring should exhibit resonance around $\delta = 2.3$. Each will appear as a singlet, since there are no other spins within three chemical bonds of it. The two singlets at $\delta = 2.31$ and 2.58 must be due to these protons. The protons in the $-CH_2-$ group should appear around $\delta = 3.8$ to 5.1, according to Table 16.2. These proton resonances will be split into a quartet by the adjacent $-CH_3$ protons. The quartet appearing around $\delta \approx 4.3$–4.4 is due to resonance of the protons in the $-CH_2-$ group. The absorptions due to the ring protons should appear around $\delta = 7.2$ to 7.3. H_a and H_b are nonequivalent protons that split each other into doublets. H_c is not equivalent to either H_a or H_b; hence, it will appear as a singlet. The two doublets are seen at $\delta = 7.05$ and $\delta = 7.8$–7.9, respectively. The singlet is superimposed on the doublet at $\delta = 7.05$. All assignments are shown in the following diagram:

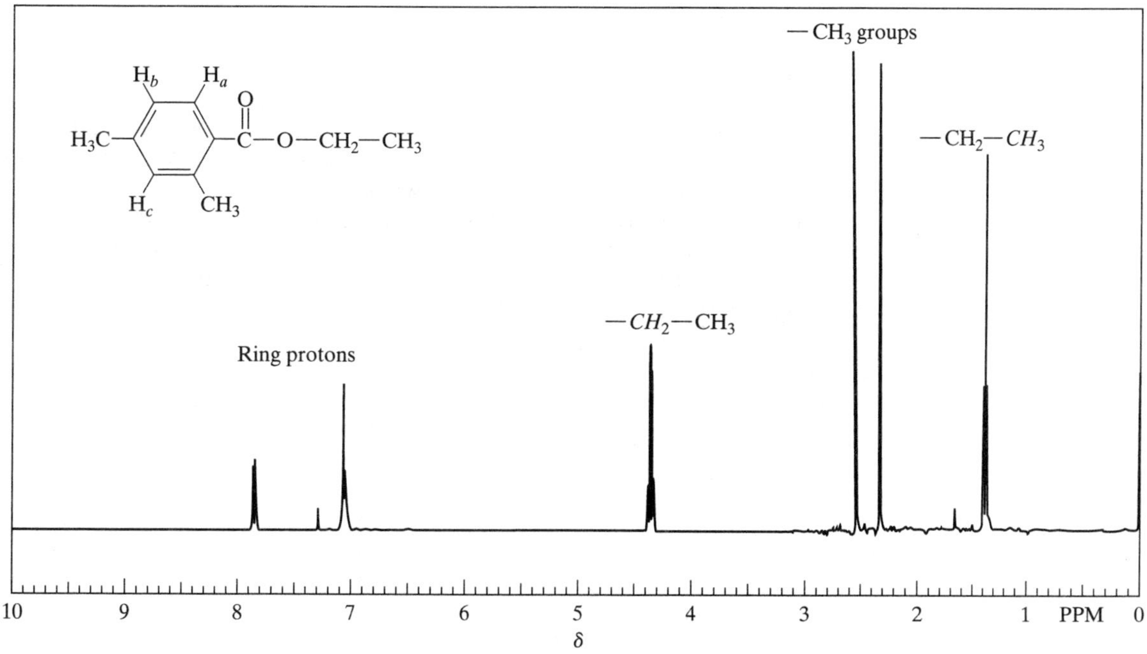

16.21 (A) A spin-decoupled, natural-abundance ^{13}C NMR spectrum of propanone (acetone) $[CH_3\text{–}CO\text{–}CH_3]$ is taken on a 400-MHz spectrometer. Describe qualitatively the expected appearance of the spectrum.

(B) A spin-decoupled ^{13}C NMR spectrum of ^{13}C-enriched propanone is taken on a 400-MHz spectro-meter. Describe qualitatively the expected appearance of the spectrum. Assume that the ^{13}C enrichment is in excess of 70%.

(C) A natural-abundance ^{13}C NMR spectrum of propanone is taken on a 400-MHz spectrometer without spin decoupling. Describe qualitatively the expected appearance of the spectrum.

(D) A ^{13}C NMR spectrum of ^{13}C-enriched propanone is taken on a 400-MHz spectrometer without spin decoupling. How many absorption bands might be seen? Explain.

Solution

(A) With spin decoupling, there will be no spin–spin splitting from the ^1H nuclei. The low-percentage natural abundance (1.1%) means that ^{13}C nuclei will be adjacent only rarely. Consequently, ^{13}C spin–spin splitting may also be ignored. Each chemically different carbon atom will exhibit exactly one absorption band. Table 16.3 indicates that the $-CH_3$ carbons should appear around $\delta = -2.3$ to 25.2. The ketone carbon resonance will be in the range $\delta = 206–212$. The $-CH_3$ ^{13}C resonance will be twice as intense, since there are two $-CH_3$ carbon atoms. Qualitatively, we expect the spectrum to resemble the following figure:

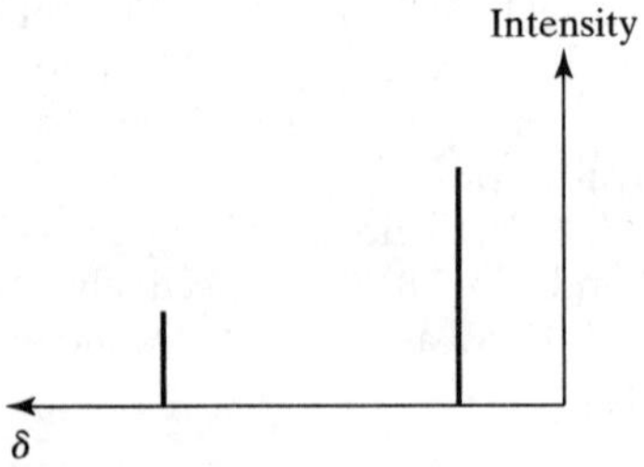

(B) With spin decoupling, there will be no spin–spin splitting from the ^1H nuclei. However, with 70% or more ^{13}C, we will observe spin–spin splitting from the adjacent ^{13}C nuclei. Therefore, the $-CH_3$ resonance will be a doublet, since the C=O carbon with spin $\frac{1}{2}$ has two possible spin orientations, which will split the $-CH_3$ absorption into two bands. The C=O resonance will be split by two equivalent ^{13}C nuclei, which can have $M_I = 1, 0$ or -1, depending upon the alignment of the two spins. The ketone carbon will therefore appear as a triplet with a 1:2:1 ratio. The spectrum will resemble the following diagram:

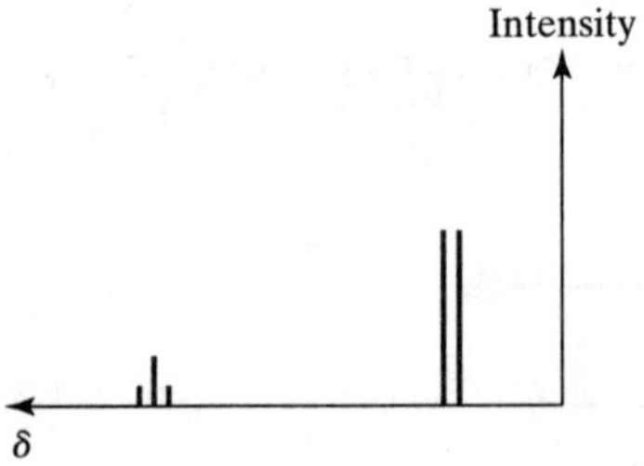

(C) The low-percentage natural abundance (1.1%) means that ^{13}C nuclei will be adjacent only rarely. Consequently, ^{13}C spin–spin splitting may be ignored, but without spin decoupling, the ^{13}C absorption bands will be split by the ^1H spins. The two equivalent $-CH_3$ carbon nuclei will each be split by three equivalent protons that can have $M_{IH} = \frac{3}{2}, \frac{1}{2}, -\frac{1}{2}$, or $-\frac{3}{2}$. The numbers of ways these values can be obtained are 1, 3, 3, and 1, respectively. Therefore, the $-CH_3$ resonance line will appear as a quartet with an intensity ratio of about 1:3:3:1. The C=O absorption band will be split by the six equivalent protons on the adjacent $-CH_3$ groups. Six protons can have $M_{IH} = 3, 2, 1, 0, -1, -2$, or -3. Consequently, the C=O band will appear as a septet (seven lines). The numbers of ways these proton spin states can be achieved are as follows:

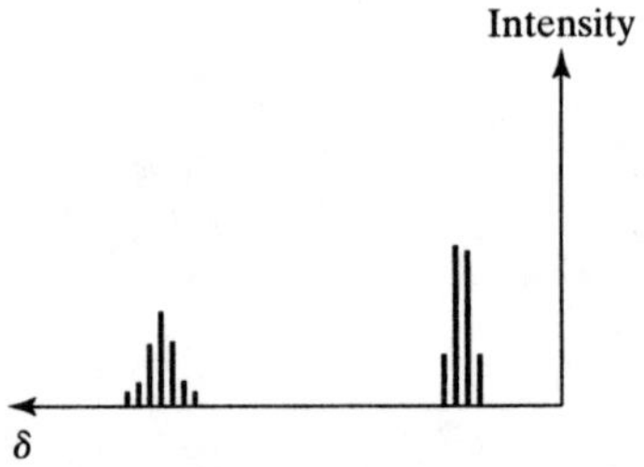

$$M_{IH} = 3 \text{ or } -3: \quad \frac{6!}{6!(6-6)!} = 1; \qquad M_{IH} = 2 \text{ or } -2: \quad \frac{6!}{5!(6-5)!} = 6;$$

$$M_{IH} = 1 \text{ or } -1: \quad \frac{6!}{2!(6-2)!} = 15; \qquad M_{IH} = 0: \quad \frac{6!}{3!(6-3)!} = 20.$$

Therefore, the seven bands for the C=O absorption will have an intensity ratio of about 1:6:15:20:15:6:1. The spectrum is something like the diagram to the left.

(D) With no spin decoupling and ^{13}C enrichment, we will observe spin–spin splitting from both the ^{1}H and ^{13}C spins. Therefore, the ^{13}C resonance band from the methyl carbons will be split into a doublet by the adjacent ^{13}C spin on the C=O group. Each of these lines will be further split into a quartet by the three ^{1}H spins. A total of eight lines might, therefore, be observed. In practice, a complex multiplet would probably be observed. The appearance of the C=O ^{13}C absorption will be even more complex. The two equivalent ^{13}C spins on the methyl groups will split the –C=O line into a triplet as described in Part (B). Each of these three lines will be further split into seven bands by the six equivalent adjacent ^{1}H spins. (See Part (C).) This gives a theoretical total of 21 lines. The result will almost surely be a complex multiplet.

16.23 The figure below shows a spin-decoupled ^{13}C NMR spectrum of thiochroman-6-ethanol in $DCCl_3$ solvent.

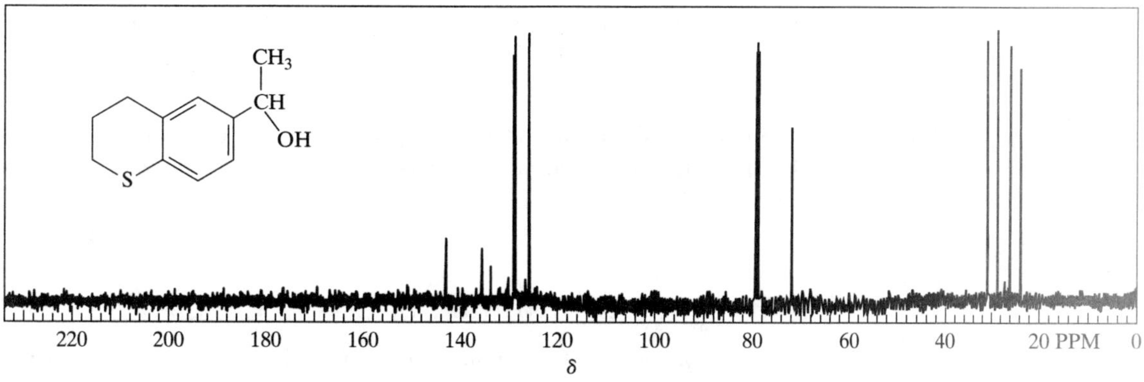

▲ Spin-decoupled ^{13}C NMR spectrum of thiochroman-6-ethanol

The spectrum was recorded on a 300-MHz FT NMR spectrometer with TMS as the reference and $DCCl_3$ as the solvent. The positions of the bands are given in terms of δ, as described in Example 16.7. (See Githarangi Mahika Weerasekare, *Chiral α-Substituted Sulfoxides: Liquid Crystals and Potential Anticancer Agents*, Ph.D. dissertation, Oklahoma State University, 1994.) Identify the ^{13}C atoms giving rise to each of the spectral lines.

Solution

Table 16.3 indicates that ^{13}C resonance for alkanes and cycloalkanes appears in the range $\delta = -2.3$ to 25.2. The four bands seen between 22.87 and 29.77 are due to the three ^{13}C spins in the heterocyclic ring and the aliphatic $-CH_3$ group. Without additional information, such as that which might be obtained from isotopic enrichment studies, we cannot assign these four bands to individual carbon atoms. The ^{13}C spin in the –CH– group should appear in the range $\delta = 49.0–57.0$, according to the data in Table 16.3, since this carbon is attached to an alcohol group. The band at $\delta = 70.3$ results from this ^{13}C spin. The solvent resonance line appears at $\delta \approx 78$ as a triplet. This identification is easy because we know that deuterium with $I = 1$ has three possible m_{ID} values: 1, 0, and −1. Therefore, the ^{13}C resonance line for $DCCl_3$ will appear as a triplet. The attached Cl nuclei have $I = \frac{3}{2}$ and a large quadrupole moment, which suppresses spin–spin splitting. The aromatic ^{13}C spins exhibit NMR resonance around $\delta = 128.5$. There should be six such lines, and six lines do appear in the spectrum on the following page.

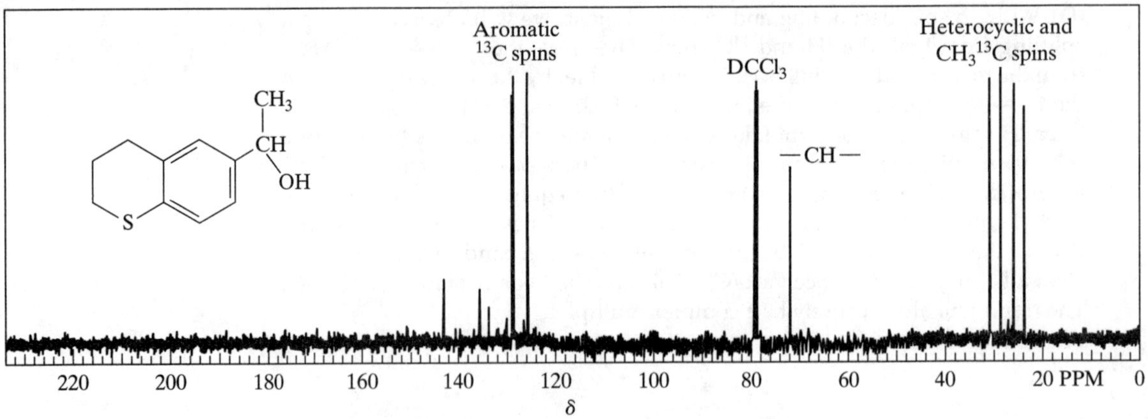

16.25 Two ESR sample tubes are resting on the laboratory bench. One contains N_2, the other O_2. Explain how we might use ESR spectral measurements to determine which sample tube holds the O_2.

Solution

N_2 and O_2 contain 14 and 16 electrons, respectively. In Chapter 14, we found that the simple MO electronic configurations for these two molecules are

$$(1s\sigma_g)^2(1s\sigma_u^*)^2(2s\sigma_g)^2(2s\sigma_u^*)^2(2p\sigma_g)^2(2p_{+1}\pi_u)^2(2p_{-1}\pi_u)^2$$

and

$$(1s\sigma_g)^2(1s\sigma_u^*)^2(2s\sigma_g)^2(2s\sigma_u^*)^2(2p\sigma_g)^2(2p_{+1}\pi_u)^2(2p_{-1}\pi_u)^2(2p_{+1}\pi_g^*)^1(2p_{-1}\pi_g^*)^1,$$

respectively. For N_2, all the orbitals contain two electrons. Therefore, all spins are paired, and there is no net electron spin angular momentum or spin magnetic moment. Consequently, N_2 does not exhibit an ESR spectrum. O_2, on the other hand, has two unpaired electrons in $2p_{+1}\pi_g^*$ and $2p_{-1}\pi_g^*$ orbitals. Hund's rule tells us that the ground state is triplet. Therefore, the molecule possesses spin angular momentum and a spin magnetic moment. As a result, O_2 has an ESR spectrum, which consists of a single unsplit band. The sample tube giving an ESR resonance signal contains O_2. For this reason, we say that O_2 is paramagnetic, whereas N_2 is diamagnetic.

16.27# Consider the ethyl radical (CH_3–$CH_2\cdot$). The measured hyperfine splitting constant for the ^1H nuclei on the –CH_2 group is −22.4 G. That for the ^1H nuclei on the –CH_3 group is 26.9 G.

(A) How many bands will be seen in the ESR spectrum of this radical?

(B) Assume that in the absence of hyperfine splitting, the ESR resonance line for the ethyl radical is seen at a field of G_o gauss. Determine the positions of all lines, in terms of G_o, when hyperfine splitting is present.

(C) What are the expected relative intensities of the bands? Prepare a plot of the spectrum, showing the positions and intensities of the bands.

Solution

(A) The hyperfine splitting caused by the two ^1H spins on the –CH_2 groups will split the ESR band into a triplet, since there are three possible orientations of these spins, leading to M_{IH} values of 1, 0, and −1. Each of these three bands will be further split into a quartet by the three ^1H spins on the –CH_3 group, since these spins can have four orientations, producing $M_{IH} = \frac{3}{2}, \frac{1}{2}, -\frac{1}{2}$, and $-\frac{3}{2}$. We will, therefore, see a total of $3 \times 4 = $ 12 bands in the spectrum.

(B) By assumption, the unsplit ESR resonance line appears at a magnetic field G_o. The hyperfine splitting by the ^1H nuclei on the $-CH_2$ group splits this band into a triplet with a hyperfine coupling constant $a = -22.4$ G. Equation 16.55 shows that the energy change due to the hyperfine splitting is

$$\Delta E_{\text{spin-spin}} = ahm_s m_{IH}. \tag{1}$$

Therefore, the three M_{IH} spin states produce the following splitting and transitions:

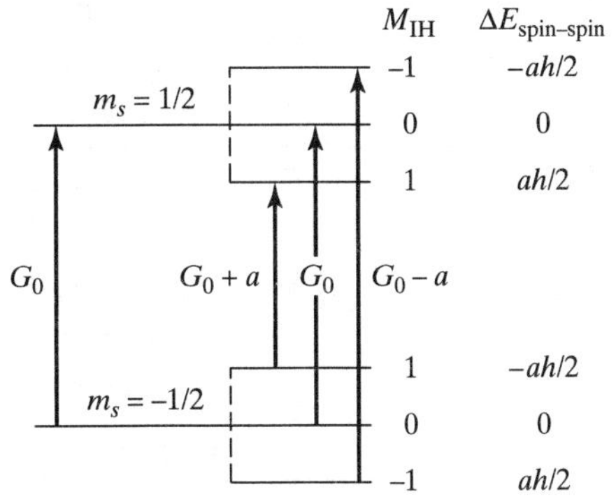

The triplet bands appear at frequencies of G_o, $G_o + a$, and $G_o - a$, with relative intensities in the ratio of 2 : 1 : 1, respectively.

The three ^1H nuclei on the $-CH_3$ group can have $M_{IH} = \frac{3}{2}, \frac{1}{2}, -\frac{1}{2}$, and $-\frac{3}{2}$. The hyperfine splitting produced by these spin states will be that shown in Figure 16.13. Each band will be split into four bands. If the original position of the band prior to the hyperfine splitting was E_o, Eqs. 16.41A–D show that the hyperfine interaction will produce absorption bands at $(E_o + a'/2)$, $(E_o + 3a'/2)$, $(E_o - a'/2)$ and $(E_o - 3a'/2)$, where $a' = 26.9$ G. The relative intensities of these bands are in the ratio of 3 : 1 : 3 : 1, respectively. Consequently, the three bands resulting from the $-CH_2$ hyperfine splitting will be further split as shown in the table on the top of the following page.

(C) The relative intensities of the 12 bands are the products of the 1 : 2 : 1 relative intensities of the hyperfine splitting resulting from coupling to the ^1H spins in the $-CH_2$ group and the 1 : 3 : 3 : 1 relative intensities from the $-CH_3$ hyperfine splittings. These products are given in the last column of the table. A plot of the ESR spectrum follows:

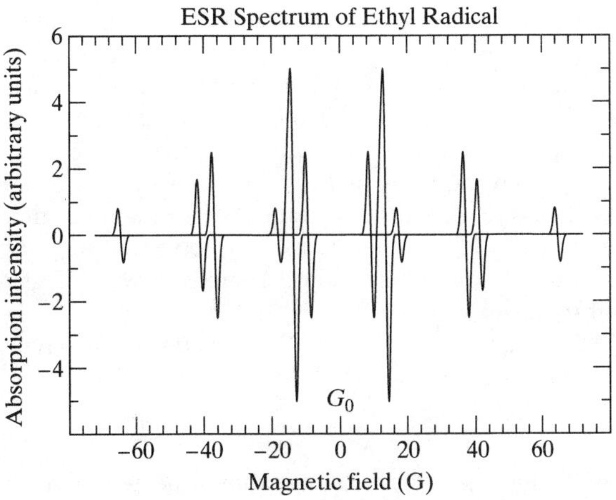

$-CH_2$ Band	Relative Intensity	New band $-CH_3$ Splitting	Relative Intensity	Position in Gauss	Total Intensity
$G_o + a$	1	$G_o + a + a'/2$	3	$G_o - 22.4 + 13.45 = G_o - 8.95$	$1 \times 3 = 3$
	1	$G_o + a + 3a'/2$	1	$G_o - 22.4 + 40.35 = G_o + 17.95$	$1 \times 1 = 1$
	1	$G_o + a - a'/2$	3	$G_o - 22.4 - 13.45 = G_o - 35.85$	$1 \times 3 = 3$
	1	$G_o + a - 3a'/2$	1	$G_o - 22.4 - 40.35 = G_o - 62.75$	$1 \times 1 = 1$
$G_o - a$	1	$G_o - a + a'/2$	3	$G_o + 22.4 + 13.45 = G_o + 35.85$	$1 \times 3 = 3$
	1	$G_o - a + 3a'/2$	1	$G_o + 22.4 + 40.35 = G_o + 62.75$	$1 \times 1 = 1$
	1	$G_o - a - a'/2$	3	$G_o + 22.4 - 13.45 = G_o + 8.95$	$1 \times 3 = 3$
	1	$G_o - a - 3a'/2$	1	$G_o + 22.4 - 40.35 = G_o - 17.95$	$1 \times 1 = 1$
G_o	2	$G_o + a'/2$	3	$G_o + 13.45$	$2 \times 3 = 6$
	2	$G_o + 3a'/2$	1	$G_o + 40.35$	$2 \times 1 = 2$
	2	$G_o - a'/2$	3	$G_o - 13.45$	$2 \times 3 = 6$
	2	$G_o - 3a'/2$	1	$G_o - 40.35$	$2 \times 1 = 2$

16.29 (A) Qualitatively describe the expected appearance of the ESR spectrum of a ^{14}N atom.

(B) What spacing, expressed in gauss, would be observed between the ESR bands?

Solution

A nitrogen atom has the electronic structure $1s^22s^22p^1_{-1}2p^1_02p^1_1$. There are three unpaired equivalent electron spins that will provide an ESR spectrum. The ^{14}N nucleus has spin quantum number $I = 1$. (See Table 16.1.) As a result, it has three possible spin orientations with $m_{IN} = 1, 0$, and -1. Each of these has equal probability. Consequently, the single ESR absorption line will be split into a triplet with all components having equal intensity. The lines will be broader than usual because of the presence of the nuclear quadrupole moment on the ^{14}N nucleus. This causes the nuclear spin to undergo rapid transitions between the three possible spin states. As a result, the unpaired electrons will tend to react to an average nuclear magnetic moment. The transition rate with $I = 1$, however, is usually too small to completely average out the individual spin orientations. Therefore, we expect a triplet, but a somewhat broadened one.

The hyperfine splitting for an ^{14}N nucleus is that given in Table 16.5: 0.0552 T, or 552 G.

16.31 The diagram on the top of the next page shows the X–Y plane of a cubic crystal:

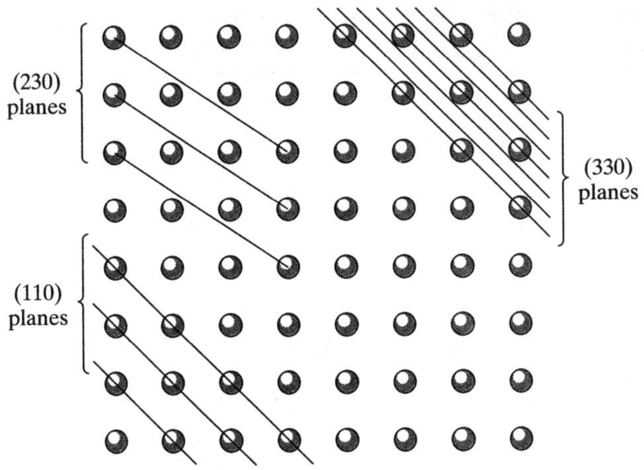

Use this diagram to illustrate sets of crystal planes that are described by the following Miller indices:

(A) (330);
(B) (110);
(C) (230).

Solution

The planes requested in the problem are shown in the preceding figure. The (110) planes are equivalent to the Weiss $[1, 1, \infty]$ planes, so that the distances from the origin to the intersection with the plane along the X- and Y-axes are both one unit cell. The lines show the edges of the planes, which are perpendicular to the plane of the paper and parallel to the Z-axis. The (330) Miller planes are parallel to the (110) planes, but they have one-third the spacing of the (110) planes. The Miller (230) planes are the same as the Weiss $[3, 2, \infty]$ planes, so that the distances from the origin to the intersection with the plane along the X- and Y-axes are 3 and 2 unit cell distances, respectively.

16.33 An orthorhombic crystal has unit cell distances $a = 5.5$ Å, $b = 6.7$ Å, and $c = 7.1$ Å. What is the spacing between the following planes?

(A) $[1, 4, \infty]$;
(B) (321).

Solution

(A) For Weiss indices, Eq. 16.73 gives the spacing:

$$\frac{1}{\Delta^2} = \frac{h^2}{b^2} + \frac{k^2}{a^2} \quad (1)$$

(for planes parallel to Z).

Substituting values produces

$$\frac{1}{\Delta^2} = \frac{1^2}{6.7^2} + \frac{4^2}{5.5^2} = 0.5512 \text{ Å}^{-2}. \quad (2)$$

Solving for Δ, we obtain

$$\underline{\Delta = 1.35 \text{ Å}.} \quad (3)$$

(B) For Miller indices, the general result is given in Eq. 16.74:

$$\frac{1}{\Delta^2} = \frac{h^2}{a^2} + \frac{k^2}{b^2} + \frac{\ell^2}{c^2}. \quad (4)$$

For the (321) plane, we obtain

$$\frac{1}{\Delta^2} = \frac{3^2}{5.5^2} + \frac{2^2}{6.7^2} + \frac{1^2}{7.1^2} = 0.4065\ \text{Å}^{-2}. \tag{5}$$

This gives

$$\underline{\Delta = 1.57\ \text{Å}.} \tag{6}$$

16.35 Determine the condition for complete destructive interference (complete cancellation of waves) for coherent X rays of wavelength λ scattering from planes with a separation d.

Solution

Equation 16.76 shows that the path length difference between waves reflected from successive planes is

$$\Delta X = 2a = 2d \sin \Theta. \tag{1}$$

Therefore, the summation of Ψ_1 and Ψ_2 at the detector will be given by

$$\Psi_{\text{total}} = \Psi_1 + \Psi_2 = A \sin\left[\frac{2\pi X_o}{\lambda}\right] + A \sin\left[\frac{2\pi(X_o + \Delta X)}{\lambda}\right], \tag{2}$$

where A and λ are, respectively, the amplitude and wavelength of the coherent radiation. Using the trigonometric identity $\sin(\alpha + \beta) = \sin \alpha \cos \beta + \sin \beta \cos \alpha$, we may write Eq. 2 in the form

$$\Psi_{\text{total}} = A \sin\left[\frac{2\pi X_o}{\lambda}\right] + A \sin\left[\frac{2\pi X_o}{\lambda}\right] \cos\left[\frac{2\pi \Delta X}{\lambda}\right] + A \cos\left[\frac{2\pi X_o}{\lambda}\right] \sin\left[\frac{2\pi \Delta X}{\lambda}\right]. \tag{3}$$

In order for Ψ_1 and Ψ_2 to add to zero (producing complete destructive interference), the third term in Eq. 16.78 must vanish and the second term must be equal to $-A \sin[2\pi X_o/\lambda]$. For this to occur, we must have $[2\pi\Delta X/\lambda]$ equal to an odd integral multiple of π radians. That is, we must have

$$\left[\frac{2\pi \Delta X}{\lambda}\right] = (2n + 1)\pi \tag{4}$$

for integer n. This condition requires that

$$\boxed{\Delta X = \left[n + \frac{1}{2}\right]\lambda \qquad \text{with } n \text{ integer}}. \tag{5}$$

Equation 5 is the condition for complete destructive interference. In terms of the separation, we know that $\Delta X = 2d \sin \Theta$. Therefore, for destructive interference, we must have

$$\boxed{2d \sin \Theta = \left[n + \frac{1}{2}\right]\lambda \quad \text{with } n \text{ integer}}. \tag{6}$$

16.37 The unit cell spacings of an orthorhombic crystal are $a = 4.7$ Å, $b = 5.2$ Å, and $c = 6.1$ Å. An X-ray diffraction pattern is taken of this crystal, using radiation whose wavelength is 1.54 Å. Reflections from the $(hk\ell)$ Miller planes exhibits a first-order maximum at 30.53°. From what Miller planes did the reflections occur?

Solution

The crystal plane spacing can be determined by means of the Bragg equation:

$$d = \frac{\lambda}{2 \sin \Theta} = \frac{1.54 \text{ Å}}{2 \sin(30.53°)} = 1.516 \text{ Å}. \qquad (1)$$

Equation 16.74 gives the spacing between the Miller $(hk\ell)$ planes:

$$\frac{1}{\Delta^2} = \frac{h^2}{a^2} + \frac{k^2}{b^2} + \frac{\ell^2}{c^2}. \qquad (2)$$

For the orthorhombic crystal in the problem, we have

$$\frac{1}{d^2} = \frac{h^2}{4.7^2} + \frac{k^2}{5.2^2} + \frac{\ell^2}{6.1^2} = \frac{1}{1.516^2} = 0.4351. \qquad (3)$$

Consequently, we seek a set of small integers that satisfies Eq. 3. This set can be found by trial and error. We obtain the table at the right.

h	k	ℓ	$\frac{h^2}{4.7^2} + \frac{k^2}{5.2^2} + \frac{\ell}{6.1^2}$
2	0	0	0.1811
3	0	0	0.4074
4	0	0	0.7243
0	2	0	0.1479
0	3	0	0.3328
0	4	0	0.5917
0	0	3	0.2419
0	0	4	0.4300
0	0	5	0.6719
1	1	1	0.1091
1	2	2	0.3007
1	3	2	0.4856
1	2	3	0.4351

Examining the results shows that the set of planes we seek are the (123) Miller planes.

16.39 A powder diffraction spectrum of a cubic lattice of a crystal having only one type of atom exhibits intensity maxima at 14.15°, 20.22°, 25.04°, 29.26°, 33.13°, 36.77°, 43.73°, 47.16°, 50.61°, 54.15°, 57.85°, 61.79°, and 66.13°. No other maxima are observed.

(A) If the crystal structure is known to be either primitive or face-centered cubic, which is it? Prove that your answer is correct.

(B) What is the unit cell spacing of the crystal if the wavelength of the X ray is 1.54 Å?

Solution

(A) Let us assume that the crystal is primitive cubic. If this is the case, the intensity maximum at 14.15° must be due to first-order reflections from the (100), (010), or (001) Miller planes. This permits the value of $\lambda/(2a)$ to be computed, using Eq. 16.80:

$$\sin \Theta = \sin(14.15°) = \frac{\lambda}{2a} = 0.2445. \qquad (1)$$

Indexing of the planes responsible for the intensity maxima can, therefore, be accomplished with the use of Eq. 16.82:

$$(h^2 + k^2 + \ell^2) = \left[\frac{2a}{\lambda}\right]^2 \sin^2 \Theta = 16.73 \sin^2 \Theta. \qquad (2)$$

The resulting indexing is shown in the table on the following page.

So far, everything fits. However, where is the intensity maximum for reflections from the (400) Miller planes, which should be observed at 77.94°? The fact that this maximum is missing suggests that the lattice is not a primitive cubic lattice.

If we assume that the structure is face-centered cubic, then the first reflection will occur from the (200), (020), or (002) planes. Using Eq. 16.80, we have

$$\sin \Theta = \sin(14.15°) = \frac{2\lambda}{2a} = \frac{\lambda}{a} = 0.2445. \qquad (3)$$

Θ_{max} (deg)	$16.73 \sin^2 \Theta_{max} = (h^2 + k^2 + \ell^2)$	Miller Planes
14.15	1.00	(100), (010), (001)
20.22	2.00	(110), (101), (011)
25.04	3.00	(111)
29.26	4.00	(200), (020), (002)
33.13	5.00	(210), (201), (120), (102), (021), (012)
36.77	6.00	(211), (121), (112)
43.73	8.00	(220), (202), (022)
47.16	9.00	(221), (212), (122), (300), (030), (003)
50.61	10.00	(310), (301), (031), (013), (130), (103)
54.15	11.00	(311), (131), (113)
57.85	12.00	(222)
61.79	13.00	(320), (302), (032), (023) (230), (203)
66.13	14.00	(321), (312), (213), (231), (132), (123)

Equation 16.82 then yields

$$(h^2 + k^2 + \ell^2) = \left[\frac{2a}{\lambda}\right]^2 \sin^2 \Theta = 66.91 \sin^2 \Theta. \quad \textbf{(4)}$$

If the lattice is face-centered cubic, only planes with even Miller indices can appear. Therefore, the indexing is on the top of the next page.
The next plane would be (644), for which we would have to have $\sin \Theta = 1.008$, which is impossible. Therefore, we have exactly the right number of total reflections, and the lattice fits a face-centered cubic assignment.

(B) With the assignment of a face-centered cubic lattice, the unit cell spacing can be computed using Eq. 3:

$$a = \frac{\lambda}{0.2445} = \frac{1.54 \text{ Å}}{0.2445} = 6.30 \text{ Å}. \quad \textbf{(6)}$$

16.41 Powder X-ray diffraction patterns are obtained for three samples, each containing a single type of atom. The first sample contains microcrystals of a primitive cubic crystal, the second sample is an ensemble of microcrystals for a face-centered cubic crystal, and the third sample holds body-centered cubic crystals. All samples have reflections from the following Miller planes: (100), (110), (111), (200), (210), (211), (220), (221), (300), (310), (311), (222), (320), (321), (400), (420), and (422). Which of these reflections might produce intensity maxima in the diffraction patterns for each sample? What condition on the spacing between the planes must be satisfied to observe intensity maxima.

Θ_{max} (deg)	$66.91 \sin^2\Theta_{max} = (h^2 + k^2 + \ell^2)$	Miller Planes
14.15	4.00	(200), (020), (002)
20.22	8.00	(220), (202), (022)
25.04	12.00	(222)
29.26	16.00	(400), (040), (004)
33.13	20.00	(420), (402), (240), (204), (042), (024)
36.77	24.00	(422), (242), (224)
43.73	32.00	(440), (404), (044)
47.16	36.00	(442), (424), (244), (600), (060), (006)
50.61	40.00	(620), (602), (062), (026), (260), (206)
54.15	44.00	(622), (262), (226)
57.85	48.00	(444)
61.79	52.00	(640), (604), (064), (046) (460), (406)
66.13	56.00	(642), (624), (426), (462), (264), (246)

Solution

Since there are no systematic absences in the X-ray diffraction spectrum of a primitive crystal, all of the listed reflections will produce intensity maxima for Sample 1 at the angle that satisfies the Bragg equation, provided that the spacing between the planes has a value that permits this to occur. That is, for a maximum, we must have

$$\boxed{\sin \Theta = \frac{\lambda}{2\Delta_{hk\ell}},} \quad (1)$$

with the right-hand side of Eq. 1 less than or equal to unity. If the spacing between the $(hk\ell)$ planes is such that $\lambda/2\Delta_{hk\ell} > 1$, no intensity maximum will be observed.

For the face-centered cubic microcrystals present in Sample 2, only the reflections with h, k, and / all even will produce intensity maxima. The other reflections will undergo destructive interference that produces systematic absences. Therefore, we might expect to observe maxima for the (200), (220), (222), (400), (420), and (422) reflections. The same conditions with respect to the plane spacings as discussed previously must also hold.

The body-centered microcrystals in Sample 3 exhibit maxima only for reflections from planes in which the sum of the Miller indices is even. Therefore, maxima might be observed for the (110), (200), (211), (220), (310), (222), (321), (400), (420), and (422) reflections, provided that the spacings between these planes meet the conditions discussed earlier.

CHAPTER 17

Molecular Energy Distributions—Kinetic Theory of Gases

17.1 Twelve coins are randomly tossed onto a table. What is the probability that we will obtain three heads and nine tails?

Solution

The requested probability is given by Eq. 17.1 in the text:

$$P(H^3T^9) = \frac{\text{number of ways to obtain three heads and nine tails}}{\text{total possible arrangements}}. \quad (1)$$

The number of possible arrangements is

$$s^N = 2^{12} = 4{,}096. \quad (2)$$

The number of ways to get three heads and nine tails is just the binomial expansion coefficient

$$C(12, 3, 9) = \frac{12!}{3!9!} = \frac{(12)(11)(10)}{3!} = \frac{1{,}320}{6} = 220 \text{ ways}. \quad (3)$$

Thus, the probability we seek is

$$P(H^3T^9) = \frac{220}{4{,}096} = \underline{0.05371}\ldots. \quad (4)$$

17.3 A particular board game uses eight-sided dice, with the probability of all faces being in the totally concealed position ("down") being equal. A player throws three such dice. What is the probability of his obtaining two fives and a seven in the down positions of the dice?

Solution

The total number of arrangements of the three dice is

$$s^N = 8^3 = 512. \quad (1)$$

The number of ways to obtain any given result is the value of the octanomial expansion coefficient:

$$C(3, n_1, n_2, n_3, n_4, n_5, n_6, n_7, n_8) = \frac{3!}{n_1!n_2!n_3!n_4!n_5!n_6!n_7!n_8!}. \quad (2)$$

Therefore, the number of ways to get two fives and one seven is

$$C(3, 0, 0, 0, 0, 2, 0, 1, 0) = \frac{3!}{0!0!0!0!2!0!1!0!} = 3. \quad (3)$$

The probability of this result is thus

$$P(f_5^2 f_7^1) = \frac{\text{number of ways to get two fives and a seven}}{\text{total arrangements}} = \frac{3}{512} = \underline{0.005859}\ldots. \quad (4)$$

17.5 Consider N coins randomly tossed onto a table, where N is very large.

(A) Obtain a general equation giving the ratio R of the probability of observing a p% deviation from the most probable distribution of heads and tails to the probability of the most probable distribution.

(B) Use Stirling's approximation to put the result obtained in (A) into a more convenient form. In particular, show that

$$R = \exp[2n\,\ell n(n) - (n + c)\ell n(n + c) - (n - c)\ell n(n - c)]$$

where

$$n = \frac{N}{2} \text{ and } c = \frac{np}{100}.$$

(C) Compute the ratio for the case $p = 0.01\%$ and $N = 2 \times 10^7$ coins. (Assume that the most probable distribution corresponds to having an equal number of heads and tails.)

(D) Compute the ratio for case $p = 0.01\%$ and $N = 6.0220 \times 10^{23}$.

Solution

(A) The most probable distribution contains $N/2$ heads and tails. Let us define

$$n = \frac{N}{2}. \quad \textbf{(1)}$$

If N is odd, we will agree to round n to the nearest even integer. A $p\%$ deviation corresponds to changing $np/100$ coins from tails to heads or vice versa. Let us define

$$c = \frac{np}{100}. \quad \textbf{(2)}$$

Again, if c is not an integer, we will round to the nearest integer or to the nearest even integer if c is half-integral. When N is large, these round-off procedures are of no practical consequence. The probability of the most probable distribution is

$$P(H^nT^n) = \frac{(2n)!}{n!n!2^{2n}}. \quad \textbf{(3)}$$

For the distribution corresponding to a $p\%$ deviation from the most probable distribution, the probability is

$$P(H^{n+c}T^{n-c}) = \frac{(2n)!}{(n+c)!(n-c)!2^{2n}}. \quad \textbf{(4)}$$

The ratio of these probabilities is

$$\boxed{R = \frac{P(H^{n+c}T^{n-c})}{P(H^nT^n)} = \frac{\dfrac{(2n)!}{(n+c)!(n-c)!2^{2n}}}{\dfrac{(2n)!}{n!n!2^{2n}}} = \frac{n!n!}{(n+c)!(n-c)!} = \frac{(n!)^2}{(n+c)!(n-c)!}}. \quad \textbf{(5)}$$

Equation 5 is the exact result, except for the round-off problems mentioned.

(B) Stirling's approximation for large N is

$$\ln(N!) = N\ln N - N = N[\ln N - 1]. \quad \textbf{(6)}$$

Taking logarithms of both sides of Eq. 5 produces

$$\ln R = \ln(n!)^2 - \ln(n+c)! - \ln(n-c)! = 2\ln n! - \ln(n+c)! - \ln(n-c)!. \quad \textbf{(7)}$$

Applying Stirling's approximation for the logarithms of factorial numbers, we obtain

$$\ln R = 2n[\ln n - 1] - (n + c)[\ln(n + c) - 1] - (n - c)[\ln(n - c) - 1]$$
$$= 2n \ln n - (n + c) \ln(n + c) - (n - c) \ln(n - c) - 2n + n + c + n - c$$
$$= 2n \ln n - (n + c) \ln(n + c) - (n - c) \ln(n - c). \quad \textbf{(8)}$$

Therefore,

$$\boxed{R = \exp[2n \ln(n) - (n + c) \ln(n + c) - (n - c) \ln(n - c)]}, \quad \textbf{(9)}$$

which is the equation we were asked to obtain.

(C) When $p = 0.01$ and $N = 2 \times 10^7$, we have

$$n = 1 \times 10^7 \quad \text{and } c = \frac{(0.01)(1 \times 10^7)}{100} = 10^3. \quad \textbf{(10)}$$

Therefore, $(n + c) = 10{,}000{,}000 + 1{,}000 = 10{,}001{,}000$ and $(n - c) = 9{,}999{,}000$. This gives

$$2n \ln(n) - (n + c) \ln(n + c) - (n - c) \ln(n - c)$$
$$= (20{,}000{,}000) \ln(10{,}000{,}000) - 10{,}001{,}000 \ln(10{,}001{,}000)$$
$$- 9{,}999{,}000 \ln(9{,}999{,}000) = -0.100000. \quad \textbf{(11)}$$

Therefore,

$$R = \exp(-0.1000) = \underline{0.9048}. \quad \textbf{(12)}$$

(D) If $p = 0.01\%$ and $N = 6.0220 \times 10^{23}$, we have $n = 3.0110 \times 10^{23}$ and $c = 3.0110 \times 10^{19}$. This produces

$$n + c = 3.0113 \times 10^{23} \quad \text{and} \quad n - c = 3.0107 \times 10^{23}. \quad \textbf{(13)}$$

Using Eq. 9, we obtain

$$2n \ln(n) - (n + c) \ln(n + c) - (n - c) \ln(n - c)$$
$$= 6.022 \times 10^{23} \ln(3.011 \times 10^{23}) - (3.0113 \times 10^{23}) \ln(3.0113 \times 10^{23})$$
$$- (3.0107 \times 10^{23}) \ln(3.0107 \times 10^{23}) = -3 \times 10^{15}. \quad \textbf{(14)}$$

Therefore, the ratio is

$$\boxed{R = \exp[-3.0 \times 10^{15}]}. \quad \textbf{(15)}$$

Oh, well, ... let's just call it zero.

17.7* Consider a weightless lever arm of length $3L$ that rotates about an axis perpendicular to the paper on which this problem is written at $X = 0$. (See the accompanying figure.)

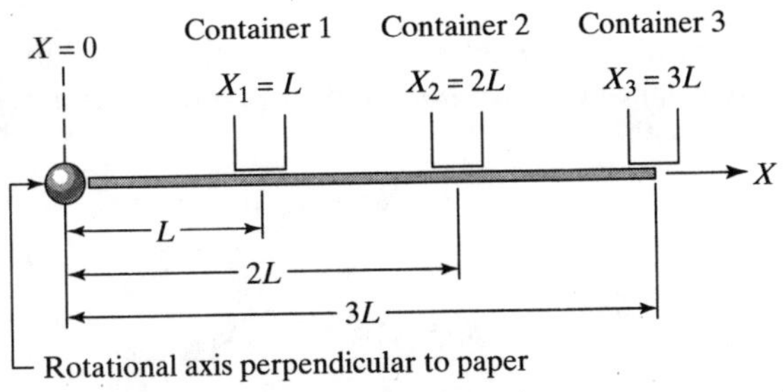

N identical particles, each of mass m_o, are distributed into three containers located at $X_1 = L$, $X_2 = 2L$, and $X_3 = 3L$. Use Lagrangian multipliers to determine the distribution of particles among the three containers that maximizes W subject to the constraints that mass be conserved and that $I = (7/3)Nm_oL^2$, where I is the moment of inertia about the rotation axis. That is, determine the fraction of the particles in each container that maximizes W subject to the said constraints. Assume that N is large.

Solution

As usual, we can maximize W by maximizing $\ln W$. The latter is obtained directly from Eq. 17.8 and Stirling's approximation. We obtain

$$\ln W = \ln N! - \ln \prod_{i=1}^{3} n_i! = \ln N! - \sum_{i=1}^{3} \ln n_i!$$

$$= N \ln N - N - \sum_{i=1}^{3} (n_i \ln n_i - n_i), \qquad (1)$$

since the number of aspects in this case is three. The mass constraint is

$$\sum_{i=1}^{3} n_i = N. \qquad (2)$$

The moment of inertia of the beam about its axis of rotation is

$$I = \sum_{i=1}^{3} m_o n_i X_i^2 = m_o \sum_{i=1}^{3} n_i X_i^2. \qquad (3)$$

The constraint on I requires that we have

$$I = m_o \sum_{i=1}^{3} n_i X_i^2 = \frac{7Nm_oL^2}{3}. \qquad (4)$$

We now multiply each constraint equation by a Lagrangian multiplier and then add the two constraint equations. Subtracting Eq. 1 from this sum yields

$$F(n_i; \alpha, \beta) = \alpha N + \beta I - \ln W$$

$$= \sum_{i=1}^{3} [\alpha n_i + \beta m_o X_i^2 n_i + n_i \ln n_i - n_i] + N - N \ln N. \qquad (5)$$

The requirement that $\ln W$ be a maximum requires that we have $dF(n_i; \alpha, \beta) = d \ln W = 0$. Taking differentials of Eq. 5 while recognizing that $[N - N \ln N]$ is a constant produces

$$dF(n_i; \alpha, \beta) = d \ln W = \sum_{i=1}^{3} [\alpha + \beta m_o X_i^2 + \ln n_i + 1 - 1] dn_i = 0, \qquad (6)$$

which can be written as

$$\sum_{i=1}^{3} [\alpha + \beta m_o X_i^2 + \ln n_i] dn_i = 0. \qquad (7)$$

Equation 7 contains five variables with two constraints. The three independent variables can be taken to be dn_1, dn_2, and dn_3. With this choice, the coefficients of the dn_i must vanish. This gives

$$[\alpha + \beta m_o X_i^2 + \ln n_i] = 0, \qquad (8)$$

so that

$$\ln n_i = -[\alpha + \beta m_o X_i^2] \qquad (9)$$

and

$$n_i = \exp(-\alpha)\exp(-\beta m_o X_i^2). \tag{10}$$

Substituting Eq. 10 into Eq. 3 yields

$$\exp(-\alpha)\sum_{i=1}^{3}\exp(-\beta m_o X_i^2) = N. \tag{11}$$

Thus,

$$\exp(-\alpha) = \frac{N}{\sum_{i=1}^{3}\exp(-\beta m_o X_i^2)}. \tag{12}$$

The remaining Lagrangian multiplier can be obtained by substituting into Eq. 4:

$$m_o\sum_{i=1}^{3} n_i X_i^2 = m_o\exp(-\alpha)\sum_{i=1}^{3} X_i^2\exp(-\beta m_o X_i^2) = I = \frac{7m_oNL^2}{3}. \tag{13}$$

This requires that we have

$$\begin{aligned} &m_o\exp(-\alpha)\sum_{i=1}^{3} X_i^2\exp(-\beta m_o X_i^2)\\ &= m_o\exp(-\alpha)[L^2\exp(-\beta m_o L^2) + 4L^2\exp(-4\beta m_o L^2)\\ &+ 9L^2\exp(-9\beta m_o L^2)] = \frac{7m_oNL^2}{3}. \end{aligned} \tag{14}$$

Canceling m_o and L^2 on both sides of Eq. 14 produces

$$\exp(-\alpha)[\exp(-\beta m_o L^2) + 4\exp(-4\beta m_o L^2) + 9\exp(-9\beta m_o L^2)] = \frac{7N}{3}. \tag{15}$$

Substituting Eq. 12 for $\exp(-\alpha)$ and canceling N gives

$$\frac{[\exp(-\beta m_o L^2) + 4\exp(-4\beta m_o L^2) + 9\exp(-9\beta m_o L^2)]}{[\exp(-\beta m_o L^2) + \exp(-4\beta m_o L^2) + \exp(-9\beta m_o L^2)]} = \frac{7}{3}. \tag{16}$$

If we now substitute $y = \exp(-\beta m_o L^2)$, we obtain, from Eq. 16,

$$\frac{(y + 4y^4 + 9y^9)}{(y + y^4 + y^9)} = \frac{7}{3}. \tag{17}$$

This equation can easily be solved by one-dimensional grid methods. A plot of the function appears at the end of the problem. As can be seen, the solution occurs at $y = 0.7462$. Therefore,

$$y = \exp(-\beta m_o L^2) = 0.7462. \tag{18}$$

Substituting this result into Eq. 12 gives us the value of $\exp(-\alpha)$:

$$\exp(-\alpha) = \frac{N}{(.7462) + (0.7462)^4 + (0.7462)^9} = \frac{N}{1.128}. \tag{19}$$

Equations 18 and 19 can now be substituted into Eq. 10 to obtain the fractions in each container that maximize the thermodynamic probability. We get

$$n_1 = \frac{N}{1.128}\exp(-\beta m_o L^2) = \frac{N}{1.128}(0.7462) = \underline{0.6615N}, \tag{20}$$

$$n_2 = \frac{N}{1.128}\exp(-4\beta m_o L^2) = \frac{N}{1.128}(0.7462)^4 = \underline{0.2749N}, \tag{21}$$

and

$$n_3 = \frac{N}{1.128}\exp(-9\beta m_o L^2) = \frac{N}{1.128}(0.7462)^9 = \underline{0.0636N}. \tag{22}$$

These fractions add to 1.000 N, so mass is conserved. We could also check the total moment of inertia:

$$m_o N(0.6615)(L^2) + m_o N(0.2749)(4L^2) + m_o N(0.0636)(9L^2)$$

$$= m_o NL^2[0.6615 + 1.0996 + 0.5724] = 2.3335 m_o NL^2 \approx \frac{7m_o NL^2}{3}, \tag{23}$$

as required.

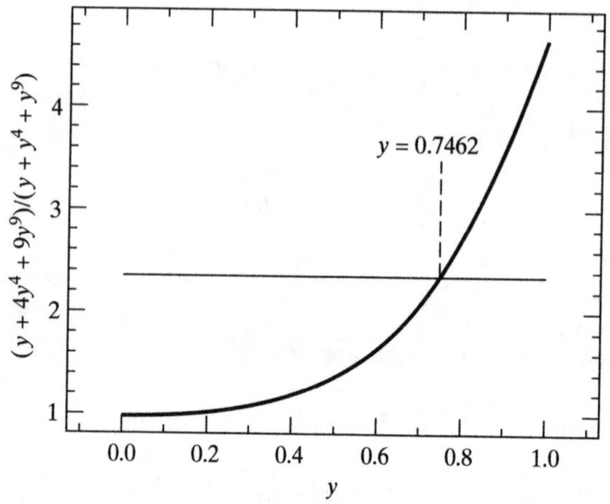

17.9 Four normal cubic dice are thrown. Compute the probability that the sum of the numbers thrown will add to 15.

Solution

All the possible combinations that yield a sum equal to 15 are in the table on top of the following page.

The number of ways the combinations with three numbers the same can occur in the throw of four dice is

$$N(3\text{ same}) = \frac{4!}{1!3!0!0!0!0!} = 4\text{ ways}. \tag{1}$$

For example, the combination of three fours and a three can be obtained from 3–4–4–4, 4–3–4–4, 4–4–3–4, or 4–4–4–3.

The number of ways to have two numbers the same and two different is

$$N(2\text{ same}) = \frac{4!}{2!1!1!0!0!0!} = 12\text{ ways}. \tag{2}$$

For example, for the case with two twos, one six, and one five, we could have 2–2–5–6, 2–2–6–5, 2–5–2–6, 2–6–2–5, 2–5–6–2, 2–6–5–2, 5–2–2–6, 6–2–2–5, 5–2–6–2, 6–2–5–2, 5–6–2–2, or 6–5–2–2, a total of 12 possible arrangements.

Finally, the number of ways to get a particular combination with all numbers different is

$$N(\text{all different}) = \frac{4!}{1!1!1!1!0!0!} = 24\text{ ways}. \tag{3}$$

	Number of Dice Showing					
	1	2	3	4	5	6
All four numbers the same			none			
Three numbers the same	0	0	1	3	0	0
	0	0	3	0	0	1
Two numbers the same	0	2	0	0	1	1
	0	0	2	1	1	0
	0	1	0	2	1	0
	1	0	0	2	0	1
	1	0	0	1	2	0
	0	1	1	0	2	0
	1	1	0	0	0	2
All numbers different	1	0	1	0	1	1
	0	1	1	1	0	1

Since we have two combinations with three numbers the same, this gives 8 ways to get 15. The table shows that there are seven combinations with two numbers the same. This adds another 7×12, or 84, ways to get 15, which brings us to 92 ways total. Finally, there are two combinations with all the numbers different. This contributes $2 \times 24 = 48$ additional arrangements. Therefore, we have a total of 140 different arrangements that yield a total of 15 on the four dice. The probability of obtaining a 15 total is thus

$$P(15) = \frac{\text{number of ways to get 15}}{\text{total number of arrangements}} = \frac{140}{6^4} = \frac{140}{1{,}296} = \underline{0.1080\ldots}. \quad \textbf{(4)}$$

17.11 A hypothetical system has four different energy levels with energies and degeneracies as given in the table at the right.

Level (n)	E(J mol^{-1})	g_n
1	0	1
2	200	2
3	400	3
4	600	1

At $T = 300$ K, four molecules are picked out of the system at random. Compute the probability that one of the molecules will be in the $n = 1$ state, one in the $n = 2$ state, one in the $n = 3$ state, and one in the $n = 4$ state.

Solution

We need the value of the molecular partition function for the system at 300 K. This is given by

$$z = \sum_{i=1}^{4} g_i \exp\left[-\frac{E_i}{RT}\right], \quad \textbf{(1)}$$

where the exponent is now written in terms of energy per mole. Substituting $T = 300$ K and data given in the statement of the problem produces

$$z_{300} = 1\exp(0) + 2\exp\left[-\frac{200}{(8.314)(300)}\right] + 3\exp\left[-\frac{400}{(8.314)(300)}\right]$$

$$+ \exp\left[-\frac{600}{(8.314)(300)}\right] = 1 + 1.846 + 2.555 + 0.786 = 6.187. \quad \textbf{(2)}$$

The probability that we will find a molecule in quantum state i is

$$P(i) = \frac{\text{number of molecules in quantum state } i}{\text{total number of molecules}} = \frac{n_i}{N} = \frac{1}{z} g_i \exp\left[-\frac{E_i}{RT}\right]. \quad \textbf{(3)}$$

Thus, we obtain

$$P(1) = \frac{1}{z}g_1 \exp\left[-\frac{E_1}{RT}\right] = \frac{1}{6.187}(1)\exp(0) = \frac{1}{6.187} = 0.162, \quad \textbf{(4)}$$

$$P(2) = \frac{1}{z}g_2 \exp\left[-\frac{E_2}{RT}\right] = \frac{1}{6.187}(2)\exp\left[-\frac{200}{(8.314)(300)}\right] = 0.298, \quad \textbf{(5)}$$

$$P(3) = \frac{1}{z}g_3 \exp\left[-\frac{E_3}{RT}\right] = \frac{1}{6.187}(3)\exp\left[-\frac{400}{(8.314)(300)}\right] = 0.413, \quad \textbf{(6)}$$

and

$$P(4) = \frac{1}{z}g_4 \exp\left[-\frac{E_4}{RT}\right] = \frac{1}{6.187}\exp\left[-\frac{600}{(8.314)(300)}\right] = 0.127. \quad \textbf{(7)}$$

The probability that the first molecule will be in state $n = 1$, the second in state $n = 2$, the third in state $n = 3$, and the fourth in state $n = 4$ is

$$P = P(1)P(2)P(3)P(4) = (0.162)(0.298)(0.413)(0.127) = 0.00253, \quad \textbf{(8)}$$

since these probabilities are independent. However, there are 4! different arrangements that yield one molecule in each of the four quantum states. The probabilities of each of these arrangements is that given in Eq. 8. Therefore, the total probability that we will observe one molecule in each quantum state if we choose four random molecules is

$$P_{\text{total}}(1 \text{ molecule in each quantum state}) = 4!(0.00253) = \underline{0.0608}. \quad \textbf{(9)}$$

17.13 Obtain the translational molecular partition function for a quantum "particle" of mass m in a three-dimensional potential well whose shape is that of a rectangular parallelepiped whose side lengths are a, b, and c. Assume that the translational quantum states can be treated as being continuous. Express your result in terms of the volume of the well.

Solution

The energy eigenvalues for this system were obtained in Chapter 12. They are

$$\varepsilon(n_x, n_y, n_z) = \frac{h^2}{8m}\left[\frac{n_x^2}{a^2} + \frac{n_y^2}{b^2} + \frac{n_z^2}{c^2}\right], \quad \textbf{(1)}$$

where n_x, n_y, and n_z are the three translational quantum numbers, each of which can take integer values between 1 and ∞. For this system, the molecular partition function is

$$\begin{aligned} z &= \sum_{n_x=1}^{\infty}\sum_{n_y=1}^{\infty}\sum_{n_z=1}^{\infty} \exp\left\{-\frac{h^2}{8mkT}\left[\frac{n_x^2}{a^2} + \frac{n_y^2}{b^2} + \frac{n_z^2}{c^2}\right]\right\} \\ &= \sum_{n_x=1}^{\infty}\exp\left\{-\frac{h^2 n_x^2}{8ma^2kT}\right\} \sum_{n_y=1}^{\infty}\exp\left\{-\frac{h^2 n_y^2}{8mb^2kT}\right\} \sum_{n_z=1}^{\infty}\exp\left\{-\frac{h^2 n_z^2}{8mc^2kT}\right\}. \quad \textbf{(2)} \end{aligned}$$

Since the translational energy levels are very closely spaced, we may replace the summations with integrals over the limits 0 to ∞. This replacement produces

$$z = \int_0^{\infty}\exp\left\{-\frac{h^2 n_x^2}{8ma^2kT}\right\} dn_x \int_0^{\infty}\exp\left\{-\frac{h^2 n_y^2}{8mb^2kT}\right\} dn_y \int_0^{\infty}\exp\left\{-\frac{h^2 n_z^2}{8mc^2kT}\right\} dn_z. \quad \textbf{(3)}$$

J	2J + 1	0.01105J(J + 1)	(2J + 1) exp [−0.01105J(J + 1)]
0	1	0	1
1	3	0.0221	2.93442725
2	5	0.0663	4.67925034
3	7	0.1326	6.13070744
4	9	0.221	7.21545012
5	11	0.3315	7.89630772
6	13	0.4641	8.17310886
7	15	0.6188	8.07885538
8	17	0.7956	7.67227625
9	19	0.9945	7.0282587
10	21	1.2155	6.22779562
11	23	1.4586	5.34891756
12	25	1.7238	4.45967469
13	27	2.0111	3.61371694
14	29	2.3205	2.84850937
15	31	2.652	2.18581161
16	33	3.0056	1.63379832
17	35	3.3813	1.19011276
18	37	3.7791	0.84519992
19	39	4.199	0.58541252
20	41	4.641	0.39556984
21	43	5.1051	0.26082649
22	45	5.5913	0.1678579
23	47	6.0996	0.10545696
24	49	6.63	0.06468799
25	51	7.1825	0.03874807
26	53	7.7571	0.02266784
27	55	8.3538	0.0129525
28	57	8.9726	0.00722977
29	59	9.6135	0.00394241
30	61	10.2765	0.0021004
31	63	10.9616	0.0010934
32	65	11.6688	0.00055618
33	67	12.3981	0.00027647
34	69	13.1495	0.00013431
35	71	13.923	6.3764E-05
36	73	14.7186	2.9588E-05
37	75	15.5363	1.3419E-05
		Sum =	90.8317987

Each of these integrals is a standard form. Integrating yields

$$z = \frac{1}{2}\left[\frac{8\pi m a^2 kT}{h^2}\right]^{1/2} \frac{1}{2}\left[\frac{8\pi m b^2 kT}{h^2}\right]^{1/2} \frac{1}{2}\left[\frac{8\pi m c^2 kT}{h^2}\right]^{1/2}$$

$$= \frac{1}{h^3}[2\pi m kT]^{3/2} abc. \quad \textbf{(4)}$$

The volume of a rectangular parallelepiped is the product of the lengths of the sides:

$$V = abc. \quad \textbf{(5)}$$

Thus, in terms of the volume, the translational molecular partition function is

$$\boxed{z = \frac{V}{h^3}[2\pi m kT]^{3/2}}. \quad \textbf{(6)}$$

17.15* The diatomic rigid-rotor rotational molecular partition function has the form $z = kT/(Bh)$ if we assume that the rotational quantum number J may be treated as a classical continuous variable. (See Problem 17.14.) Let us assess the accuracy of this approximation for two systems: $H^{35}Cl$ and $^{35}Cl^{35}Cl$. Use the data given in Table 15.4 to compute the correct rotational molecular partition function for both of these systems at 200 K, assuming that they act as rigid rotors. Compare your results with the value of the classical rotational molecular partition function that is obtained if we assume J to be a continuous variable. Compute the percent error in the classical approximation for each molecule. For which molecule is the approximation more accurate? Explain.

Solution

The molecular partition function for the rigid-rotor system with two quantum numbers J and m is

$$z = \sum_{J=0}^{\infty} \sum_{m=-J}^{+J} \exp\left[-\frac{J(J+1)Bh}{kT}\right]. \quad \textbf{(1)}$$

Since the energy is independent of m, Eq. 1 becomes

$$z = \sum_{J=0}^{\infty} (2J+1) \exp\left[-\frac{J(J+1)Bh}{kT}\right], \quad \textbf{(2)}$$

because there are $2J + 1$ different values of m that have the same energy. We see that Eq. 2 is just the molecular partition function written with the degeneracy factor $2J + 1$. The rotational constants for HCl and Cl_2 with ^{35}Cl in each case are given in Table 15.4. They are

$$B(\text{HCl}) = 10.5909c \text{ s}^{-1} = 10.5909(2.998 \times 10^{10}) \text{ s}^{-1} = 3.175 \times 10^{11} \text{ s}^{-1} \quad \textbf{(3)}$$

and

$$B(\text{Cl}_2) = 0.2438c \text{ s}^{-1} = 0.2438(2.998 \times 10^{10}) \text{ s}^{-1} = 7.309 \times 10^{9} \text{ s}^{-1}. \quad \textbf{(4)}$$

At 200 K, we have

$$\left(\frac{Bh}{kT}\right)_{\text{HCl}} = \frac{(3.175 \times 10^{11} \text{ s}^{-1})(6.626 \times 10^{-34} \text{ J s})}{(1.381 \times 10^{-23} \text{ J K}^{-1})(200 \text{ K})} = 0.07617 \quad \textbf{(5)}$$

and

$$\left(\frac{Bh}{kT}\right)_{\text{Cl2}} = \frac{(7.309 \times 10^{9} \text{ s}^{-1})(6.626 \times 10^{-34} \text{ J s})}{(1.381 \times 10^{-23} \text{ J K}^{-1})(200 \text{ K})} = 0.001753. \quad \textbf{(6)}$$

Substituting these results into Eq. 2 gives

$$z(\text{HCl}) = \sum_{J=0}^{\infty}(2J + 1)\exp[-(0.07617)J(J + 1)] \quad (7)$$

and

$$z(\text{Cl}_2) = \sum_{J=0}^{\infty}(2J + 1)\exp[-(0.001753)J(J + 1)]. \quad (8)$$

The sums in Eqs. 7 and 8 can be executed by using an Excel file or a C or FORTRAN program. (See Example 17.4.) The results are

$$z(\text{HCl}) = \underline{13.47} \quad (9)$$

and

$$z(\text{Cl}_2) = \underline{570.78}. \quad (10)$$

The classical molecular partition functions obtained by using the assumption of continuous J are

$$z_c(\text{HCl}) = \frac{kT}{Bh} = \frac{(1.381 \times 10^{-23}\ \text{J K}^{-1})(200\ \text{K})}{(3.175 \times 10^{11}\ \text{s}^{-1})(6.626 \times 10^{-34}\ \text{J s})} = \underline{13.13} \quad (11)$$

and

$$z_c(\text{Cl}_2) = \frac{kT}{Bh} = \frac{(1.381 \times 10^{-23}\ \text{J K}^{-1})(200\ \text{K})}{(7.309 \times 10^{9}\ \text{s}^{-1})(6.626 \times 10^{-34}\ \text{J s})} = \underline{570.3}. \quad (12)$$

The percent errors in the classical approximation are

$$\%\text{ error for HCl} = \frac{100 \times (13.13 - 13.47)}{13.47} = \underline{-2.52\%} \quad (13)$$

and

$$\%\text{ error for Cl}_2 = \frac{100 \times (570.3 - 570.78)}{570.78} = \underline{-0.175\%}. \quad (14)$$

The percent error for HCl is a factor of 14.4 greater than that for Cl_2. The classical approximation is more accurate for Cl_2 because, with the much smaller rotational constant resulting from the much larger moment of inertia of Cl_2, the rotational energy levels are much closer together than is the case for HCl, which has a significantly larger rotational constant.

17.17 The energy eigenvalues for a one-dimensional harmonic oscillator are $E_v = (v + 0.5)h\nu_o$, where v, the vibrational quantum number, can take the integral values 1, 2, 3, ..., ∞. Make the classical approximation that v may be treated as a continuous variable, and derive the form of the vibrational molecular partition function for a diatomic harmonic oscillator.

Solution

The correct quantum form for the vibrational molecular partition function for a harmonic oscillator is

$$z = \sum_{v=0}^{\infty}\exp\left[-\frac{(v + 0.5)h\nu_o}{kT}\right], \quad (1)$$

since all one-dimensional systems are nondegenerate. If v can be treated as a continuous variable, the summation in Eq. 1 can be replaced with an integral, to obtain

$$z \approx z_c = \int_0^\infty \exp\left[-\frac{(v+0.5)h\nu_o}{kT}\right]dv = \exp\left[-\frac{h\nu_o}{2kT}\right]\int_0^\infty \exp\left[-\frac{vh\nu_o}{kT}\right]dv$$

$$= -\frac{kT}{h\nu_o}\exp\left[-\frac{h\nu_o}{2kT}\right]\exp\left[-\frac{vh\nu_o}{kT}\right]\Bigg|_{v=0}^{v=\infty} = \frac{kT}{h\nu_o}\exp\left[-\frac{h\nu_o}{2kT}\right]. \quad (2)$$

The right-hand side of Eq. 2 provides the classical form for the harmonic oscillator vibrational partition function. The quantity $h\nu_o/(kT)$ is often represented by X. With this notation, the result is

$$\boxed{z_c = \frac{\exp\left(-\frac{X}{2}\right)}{X}}. \quad (3)$$

17.19 Show that, in general, the reciprocal of the molecular partition function plays the role of a normalization constant for the probability distribution describing the system.

Solution

The Boltzmann distribution giving the number of molecules in quantum state i at temperature T is

$$n_i = \frac{N}{z}\exp\left[-\frac{\varepsilon_i}{kT}\right], \quad (1)$$

where the partition function is

$$z = \sum_{i=1}^{\infty}\exp\left[-\frac{\varepsilon_i}{kT}\right] \quad (2)$$

and the limits on the sum indicate summation over all possible quantum states. The probability that we will observe a molecule in quantum state i is given by

$$P_i = \frac{\text{number of molecules in state } i}{\text{total number of molecules}} = \frac{n_i}{N}. \quad (3)$$

Combining Eqs. 1 and 3 gives

$$P_i = \frac{1}{z}\exp\left[-\frac{\varepsilon_i}{kT}\right]. \quad (4)$$

If the probability distribution is normalized, we will have $\sum_{i=1}^{\infty} P_i = 1$. Summing both sides of Eq. 4 over all quantum states, we obtain

$$\sum_{i=1}^{\infty} P_i = \sum_{i=1}^{\infty}\frac{1}{z}\exp\left[-\frac{\varepsilon_i}{kT}\right] = \frac{1}{z}\sum_{i=1}^{\infty}\exp\left[-\frac{\varepsilon_i}{kT}\right]. \quad (5)$$

The summation on the right-hand side of Eq. 5 is the partition function; therefore, Eq. 5 becomes

$$\sum_{i=1}^{\infty} P_i = \frac{1}{z}z = 1, \quad (6)$$

the distribution is normalized, and $1/z$ acts as the normalization constant.

17.21 In Example 17.5, we found that a fraction equal to 0.01409 of a beam of argon atoms would successfully traverse a slotted-disk speed selector if the entering distribution is Maxwell–Boltzmann at 300 K and the selector transmits all atoms with speeds in the range from 347 m s^{-1} to 353 m s^{-1}. If the rotation speed of the disks is changed so that speeds in the range $(450 - \varepsilon)\ \mathrm{m\ s^{-1}} \le C \le (450 + \varepsilon)\ \mathrm{m\ s^{-1}}$ are permitted to traverse the selector, what must ε be in order for the fraction of the total atoms transmitted to be the same as that in Example 17.5?

Solution

The Maxwell–Boltzmann probability for speed C to $C + dC$ is given by Eq. 17.62:

$$P(C)dC = 4\pi\left[\frac{m}{2\pi kT}\right]^{3/2} \exp\left[-\frac{mC^2}{2kT}\right] C^2 dC. \quad \textbf{(1)}$$

Replacing k with R/N converts Eq. 1 into

$$P(C)dC = 4\pi\left[\frac{M}{2\pi RT}\right]^{3/2} \exp\left[-\frac{MC^2}{2RT}\right] C^2 dC. \quad \textbf{(2)}$$

For argon, $M = 0.039948\ \mathrm{kg\ mol^{-1}}$. At 300 K, Eq. 2 becomes

$$P(C)dC = 4(3.14159)\left[\frac{(0.039948\ \mathrm{kg\ mol^{-1}})}{2(3.14159)(8.314\ \mathrm{J\ mol^{-1}\ K^{-1}})(300\ \mathrm{K})}\right]^{3/2}$$
$$\times \exp\left[-\frac{(0.039948\ \mathrm{kg\ mol^{-1}})C^2}{2(8.314\ \mathrm{J\ mol^{-1}\ K^{-1}})(300\ \mathrm{K})}\right], \quad \textbf{(3)}$$

which gives

$$P(C)dC = 5.114 \times 10^{-8} \exp[-8.008 \times 10^{-6} C^2] C^2 dC. \quad \textbf{(4)}$$

The probability of finding a speed in the range $(450 - \varepsilon)\ \mathrm{m\ s^{-1}} \le C \le (450 + \varepsilon)\ \mathrm{m\ s^{-1}}$ is, therefore,

$$P((450 - \varepsilon)\ \mathrm{m\ s^{-1}} \le C \le (450 + \varepsilon)\ \mathrm{m\ s^{-1}}) = \int_{450-\varepsilon}^{450+\varepsilon} P(C)dC. \quad \textbf{(5)}$$

If ε is small, we may assume that $P(C)$ is constant over the range of integration at the value $P(450)$. This assumption produces

$$P((450 - \varepsilon)\ \mathrm{m\ s^{-1}} \le C \le (450 + \varepsilon)\ \mathrm{m\ s^{-1}}) = P(450)\int_{450-\varepsilon}^{450+\varepsilon} dC = 2\varepsilon P(450). \quad \textbf{(6)}$$

Solving for ε, we obtain

$$\varepsilon = \frac{P((450 - \varepsilon)\ \mathrm{m\ s^{-1}} \le C \le (450 + \varepsilon)\ \mathrm{m\ s^{-1}})}{2P(450)}. \quad \textbf{(7)}$$

If we wish the probability to be 0.01409, Eq. 7 becomes

$$\varepsilon = \frac{0.01409}{2P(450)}. \quad \textbf{(8)}$$

Using Eq. 4, we obtain

$$P(450) = 5.114 \times 10^{-8} \exp[-8.008 \times 10^{-6} C^2] C^2$$
$$= 5.114 \times 10^{-8} \exp[-8.008 \times 10^{-6}(450)^2](450)^2 = 0.002046. \quad \textbf{(9)}$$

Combining Eqs. 8 and 9 produces

$$\varepsilon = \frac{0.01409}{2(0.002046)} = \underline{3.443 \text{ m s}^{-1}}. \tag{10}$$

We can check the accuracy of our approximation by actually executing the integral. This gives

$$\int_{450-\varepsilon}^{450+\varepsilon} P(C)dC = 5.114 \times 10^{-8} \int_{446.56}^{453.44} \exp[-8.008 \times 10^{-6}C^2]C^2 dC$$
$$= (5.114 \times 10^{-8})(2.753 \times 10^{-5}) = 0.01408. \tag{11}$$

Thus, we see that our approximation produces an error of one unit in the fourth significant digit.

17.23 Classical particles with translational energy ε_t are incident upon a potential barrier whose height is $V_o = 30$ kJ mol^{-1}.

(A) Compute the fraction F of molecules that will have sufficient energy to overcome this barrier at 300 K.

(B) Repeat (A) at temperatures of 330 K, 360 K, 400 K, and 500 K.

(C) Use your results to plot $\ln(F)$ versus T^{-1}. What type of behavior do you notice? Explain.

Solution

(A) and (B) The fraction of molecules with energy ε_t to $\varepsilon_t + d\varepsilon_t$ is given by Eq. 17.64:

$$P(\varepsilon_t)d\varepsilon_t = 4(2)^{1/2}\pi\left[\frac{1}{2\pi kT}\right]^{3/2} \exp\left[-\frac{\varepsilon_t}{kT}\right]\varepsilon_t^{1/2}d\varepsilon_t. \tag{1}$$

Replacing k with R/N produces

$$P(\varepsilon_t)d\varepsilon_t = 4(2)^{1/2}\pi\left[\frac{1}{2\pi RT}\right]^{3/2} \exp\left[-\frac{E_t}{RT}\right]E_t^{1/2}dE_t, \tag{2}$$

where E_t is the translational energy per mole. Inserting the values of the constants yields

$$P(\varepsilon_t)d\varepsilon_t = 17.7715\left[\frac{0.01914}{T}\right]^{3/2} \exp\left[-\frac{E_t}{(8.314)T}\right]E_t^{1/2}dE_t. \tag{3}$$

The fraction of molecules that have E_t greater than or equal to 30,000 J mol^{-1} is

$$F(E_t \geq 30{,}000 \text{ J mol}^{-1}) = \int_{300}^{\infty} P(\varepsilon_t)d\varepsilon_t$$
$$= 17.7715\left[\frac{0.01914}{T}\right]^{3/2} \int_{30{,}000}^{\infty} \exp\left[-\frac{E_t}{(8.314)T}\right]E_t^{1/2}dE_t. \tag{4}$$

Equation 4 can be integrated numerically. The results at the temperatures requested in Parts (A) and (B) are given in the table on top of the next page.

(C) The data required for the plot are given in the second table on the next page.

T (K)	$17.7715\left[\frac{0.01914}{T}\right]^{3/2}$	$\int_{30,000}^{\infty} \exp\left[-\frac{E_t}{(8.314)T}\right] E_t^{1/2}\,dE_t$	$F(E_t \geq 30,000 \text{ J mol}^{-1})$
300	9.056×10^{-6}	2.685	0.00002432
330	7.850×10^{-6}	8.846	0.00006944
360	6.889×10^{-6}	24.09	0.0001660
400	5.882×10^{-6}	73.29	0.0004311
500	4.209×10^{-6}	563.12	0.002370

T (K)	T^{-1} (K^{-1})	ln F
300	0.0033333	−10.624
330	0.0030303	−9.575
360	0.0027777	−8.704
400	0.0025000	−7.749
500	0.0020000	−6.045

The accompanying plot, shown at the bottom of the page, is nearly linear. A least-squares fit to the data gives the line appearing in the plot. The equation of this line is

$$\ln(F) = 0.8322 - \frac{3{,}435}{T}. \tag{5}$$

The discussion in the text and Eq. 17.70 show that we expect to have

$$\ln(F) = \text{constant} - \frac{30{,}000}{R}\left(\frac{1}{T}\right) \tag{6}$$

if the temperature range is narrow and if $RT/(2E_o) \ll 1$. At the highest temperature considered in the problem (500 K), we have

$$\left(\frac{RT}{2E_o}\right) = \frac{(8.314 \text{ J mol}^{-1} \text{ K}^{-1})(500 \text{ K})}{2(30{,}000 \text{ J mol}^{-1})} = 0.06928\ldots, \tag{7}$$

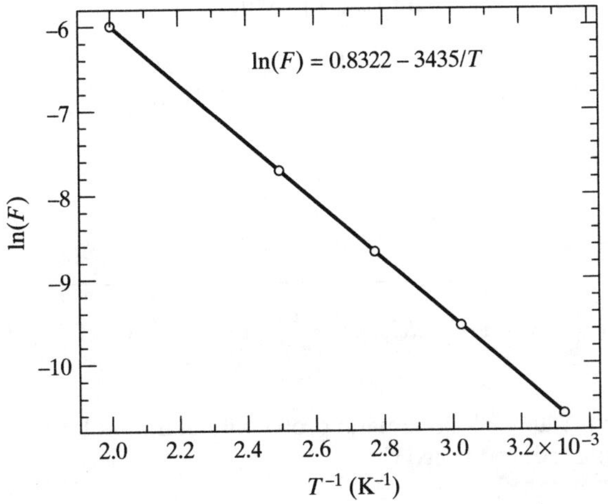

which is much less than unity. Therefore, we expect linear behavior. If this analysis were exact, then, according to Eq. 6, the slope of the straight line would be $-30{,}000/R = -3{,}608$ K. The actual result from the least-squares fit is $-3{,}435$ K.

17.25 Let us define a reduced speed as $C_R = C/C_{mp}$, where C_{mp} is the most probable speed, given by Eq. 17.72. Transform the molecular speed probability distribution for molecular speeds given by Eq. 17.62 into a probability distribution of reduced speeds.

Solution

The Maxwell–Boltzmann probability distribution for molecular speeds is

$$P(C)dC = 4\pi\left[\frac{m}{2\pi kT}\right]^{3/2}\exp\left[-\frac{mC^2}{2kT}\right]C^2dC. \qquad \textbf{(1)}$$

The most probable speed is

$$C_{mp} = \left(\frac{2kT}{m}\right)^{1/2}. \qquad \textbf{(2)}$$

Therefore, the reduced speed is

$$C_R = \left(\frac{m}{2kT}\right)^{1/2}C, \qquad \textbf{(3)}$$

and the speed is

$$C = \left(\frac{2kT}{m}\right)^{1/2}C_R. \qquad \textbf{(4)}$$

Squaring both sides, we obtain

$$C^2 = \left(\frac{2kT}{m}\right)C_R^2, \qquad \textbf{(5)}$$

and it follows that

$$dC = \left(\frac{2kT}{m}\right)^{1/2}dC_R. \qquad \textbf{(6)}$$

Substituting Eqs. 5 and 6 into Eq. 1 produces

$$P(C_R)dC_R = 4\pi\left[\frac{m}{2\pi kT}\right]^{3/2}\exp[-C_R^2]\left(\frac{2kT}{m}\right)C_R^2\left(\frac{2kT}{m}\right)^{1/2}dC_R.$$

$$= 4\pi\,\pi^{-3/2}\exp[-C_R^2]C_R^2dC_R^2, \qquad \textbf{(7)}$$

so that

$$\boxed{P(C_R)dC_R = \left(\frac{16}{\pi}\right)^{1/2}\exp[-C_R^2]C_R^2dC_R^2}, \qquad \textbf{(8)}$$

which is the Maxwell–Boltzmann probability of molecular speeds expressed in terms of the reduced speed C_R.

17.27 Obtain an expression in terms of temperature and mass for the uncertainty of speeds when the probability distribution is Maxwell–Boltzmann.

Solution

In Chapter 10, we found that the definition of the uncertainty in the measurement of the variable X is given by

$$\langle \Delta X^2 \rangle^{1/2} = [\langle X^2 \rangle - \langle X \rangle^2]^{1/2}. \tag{1}$$

That is, the squared uncertainty is the average square minus the square of the average of X. For speeds in a Maxwell–Boltzmann distribution, we have

$$\langle C^2 \rangle = \frac{3RT}{M} \tag{2}$$

and

$$\langle C \rangle = \left(\frac{8RT}{\pi M}\right)^{1/2}. \tag{3}$$

Substituting Eqs. 2 and 3 into Eq. 1 produces

$$\langle \Delta C^2 \rangle^{1/2} = \left[\frac{3RT}{M} - \left(\frac{8RT}{\pi M}\right)\right]^{1/2} = \left(\frac{RT}{M}\right)^{1/2}\left[3 - \frac{8}{\pi}\right]^{1/2}$$

$$= \boxed{0.6734\ldots\left(\frac{RT}{M}\right)^{1/2}}, \tag{4}$$

which is the desired expression.

Equation 4 demonstrates that the uncertainty of measurement increases as T increases. An examination of Figure 17.8 illustrates this point. The distributions become much broader and, hence, more uncertain as T increases.

17.29 A physical chemistry professor stands in front of the class and opens a container of $N_2O(g)$, a common anesthetic called "laughing gas" used in dental work. At the same time, the professor's assistant opens a container of benzoyl chloride in the back of the classroom. Benzoyl chloride (C_6H_5COCl) is a lacrimator (a substance that produces tears). If the classroom contains 25 rows of students, in which row do the students begin to laugh and cry at the same time? [The author is indebted to Fredrick L. Minn, M.D. Ph.D. for providing this problem and the associated solution.]

Solution

Let D_N and D_B represent the diffusion rates in units of rows s^{-1} of $N_2O(g)$ and benzoyl chloride, respectively. Let t_o be the moment of time at which students in some row start to laugh and cry at the same time. We then have

$$\text{No. rows traversed by benzoyl chloride} = D_B t_o \tag{1}$$

and

$$\text{No. rows traversed by } N_2O(g) = D_N t_o. \tag{2}$$

The conditions of the problem require that we have

$$\text{No. rows traversed by benzoyl chloride} + \text{No. rows traversed by } N_2O(g) = 25. \tag{3}$$

Thus,

$$[D_B + D_N]t_o = 25. \tag{4}$$

Graham's law implies that

$$\frac{D_N}{D_B} = \left[\frac{M_B}{M_N}\right]^{1/2} = \left[\frac{140.45}{44.0}\right]^{1/2} = 1.787, \tag{5}$$

since the molar masses of N_2O and benzoyl chloride are 44.0 and 140.45 g mol^{-1}, respectively. Using Eq. 5, we have

$$D_N = 1.787 D_B. \tag{6}$$

Substituting into Eq. 4 produces

$$[1 + 1.787]D_B t_o = 2.787 D_B t_o = 25. \tag{7}$$

This gives

$$D_B t_o = \frac{25}{2.787} = 8.97 \approx \underline{9 \text{ rows}}. \tag{8}$$

Thus, the benzoyl chloride has diffused 9 rows toward the front, while the $N_2O(g)$ has diffused

$$D_N t_o = 1.787 D_B t_o = 1.787(8.97) = 16.02 \approx \underline{16 \text{ rows}} \tag{9}$$

toward the back of the room. Students in the 16th row from the front (9 rows from the back) will begin to laugh and cry at the same time, as both N_2O and benzoyl chloride reach them simultaneously.

17.31 One of the most powerful experimental techniques for the study of chemical reaction dynamics is crossed molecular beam measurement. In this method, a well-collimated beam of molecules is directed into a high-vacuum chamber, where it is caused to intersect a second well-collimated molecular beam. The two beams interact, and suitable detectors measure what occurs during the collisions that take place when the beams intersect. Assume that a velocity-selected beam of tritium atoms (^3H) intersects a velocity-selected beam of H_2 molecules at an angle of 90°.

(A) If the tritium atoms each have a speed of 2,500 m s^{-1}, while the H_2 molecules are each moving with a speed of 1,750 m s^{-1}, what fraction of the total kinetic energy is in relative motion and what fraction is in center-of-mass motion?

(B) What fraction of the total kinetic energy would you expect to be effective in promoting chemical reactions between the tritium atoms and H_2 molecules? Explain. The atomic mass of tritium is 3.016029 amu.

Solution

(A) A reference to Figure 17.15 for the case where the two laboratory velocity vectors intersect at an angle of 90° shows that

$$V^2 = v_T^2 + v_{H2}^2 = (2{,}500)^2 + (1{,}750)^2 \text{ m}^2 \text{ s}^{-2}. \tag{1}$$

This gives

$$V = 3{,}051.64 \text{ m s}^{-1}. \tag{2}$$

The relative kinetic energy is given by

$$T_{\text{rel}} = 0.5\,\mu\,V^2 \tag{3}$$

where

$$\mu = \frac{m_{H2}m_T}{m_{H2} + m_T} = \frac{(2.0158)(3.016029)}{2.0158 + 3.016029} \times \frac{1}{6.02214 \times 10^{23}} \times \frac{1\text{ kg}}{1{,}000\text{ g}}$$

$$= 2.0063 \times 10^{-27}\text{ kg} \tag{4}$$

is the tritium–H_2 reduced mass. The relative kinetic energy is, therefore,

$$T_{\text{rel}} = 0.5(2.0063 \times 10^{-27}\text{ kg})(3{,}051.64)^2\text{ m}^2\text{ s}^{-2} = 9.342 \times 10^{-21}\,J. \tag{5}$$

The total kinetic energy is

$$T_{\text{total}} = 0.5 m_T v_T^2 + 0.5 m_{H2} v_{H2}^2 = (0.5)\frac{0.003016029}{6.02214 \times 10^{23}}\text{kg}(2{,}500\text{ m s}^{-1})^2$$

$$+ (0.5)\frac{0.0020158}{6.02214 \times 10^{23}}\text{kg}(1{,}750\text{ m s}^{-1})^2 = 2.078 \times 10^{-20}\text{ J}. \tag{6}$$

The fraction of this total that is in the form of relative motion is

$$f_{\text{rel}} = \frac{T_{\text{rel}}}{T_{\text{total}}} = \frac{9.342 \times 10^{-21}\text{ J}}{2.078 \times 10^{-20}\text{ J}} = \underline{0.4496}. \tag{7}$$

The fraction of the kinetic energy in the form of motion of the center of mass is, therefore,

$$f_{\text{CM}} = 1 - 0.4496 = \underline{0.5504}. \tag{8}$$

(B) The energy present in the motion of the center of mass does nothing to increase the collision frequency or provide the energy needed to produce chemical reactions, since this energy must be conserved by Newton's first law. The only translational energy that is of any consequence in influencing chemical reactions is that in relative motion.

17.33 The 78th edition of the CRC *Handbook of Chemistry and Physics* reports that, at an altitude of 10^6 m (621.4 miles), the atmospheric pressure is 7.514×10^{-9} Pa and the number of molecules per m^3 is 5.442×10^{11}. The temperature of the gas at that altitude is estimated to be about 1,000 K. These figures assume the air is dry and that it obeys the ideal-gas equation of state. Let us further assume that the air is pure N_2 gas with a hard-sphere collision radius σ_{AA} of 5×10^{-10} m.

(A) Compute the collision frequency of each N_2 molecule.

(B) Compute the mean free path of N_2 at 10^6 m altitude. You may use the average N_2 molecular mass (i.e., the molecular weight), for the mass of N_2.

(C) If you were an astronaut taking a space walk 10^6m above the earth's surface, would you expect to be incinerated by the 1000K gas present at that altitude? Explain. (The author thanks Fredrick L. Minn, M.D., Ph.D., for this part of the question.)

Solution

(A) The collision frequency of each N_2 molecule is given by Eq. 17.103, provided that we divide this result by a factor of two to avoid counting the N_2–N_2 collisions twice:

$$z_{AA} = \frac{N_A}{2}(\pi\sigma_{AA}^2)\left[\frac{8kT}{\pi\mu}\right]^{1/2}. \quad (1)$$

The reduced mass of the $N_2 - N_2$ system is

$$\mu = \frac{m_{N2}m_{N2}}{m_{N2} + m_{N2}} = \frac{m_{N2}}{2} = \frac{0.0140067 \text{ kg mol}^{-1}}{6.02214 \times 10^{23} \text{ mol}^{-1}} = 2.326 \times 10^{-26} \text{ kg}. \quad (2)$$

Substituting the given data yields

$$z_{AA} = \frac{5.442 \times 10^{11} \text{ m}^{-3}}{2}(3.14159)(5 \times 10^{-10} \text{ m})^2 \times$$

$$\left[\frac{8(1.38066 \times 10^{-23} \text{ J K}^{-1})(1{,}000 \text{ K})}{(3.14159)(2.326 \times 10^{-26} \text{ kg})}\right]^{1/2}$$

$$= \underline{0.000263 \text{ s}^{-1}}. \quad (3)$$

(B) The mean free path is given by Eq. 17.104, provided that we note that A and B are both N_2 molecules and replace N_B with $N_A/2$. The mass of N_2 is twice the reduced mass of N_2–N_2. Therefore, the mean free path of N_2 is

$$\langle L_{AA}\rangle = \frac{2(\mu/m_A)^{1/2}}{N_A(\pi\sigma_{AB}^2)} = \frac{(2)^{1/2}}{N_A(\pi\sigma_{AB}^2)} = \frac{(2)^{1/2}}{5.442 \times 10^{11} \text{ m}^{-3}(3.14159)(5 \times 10^{-10} \text{ m})^2}$$

$$= \underline{3.309 \times 10^6 \text{ m}} \approx \underline{2{,}056 \text{ miles}}. \quad (4)$$

Not many collisions occur at this altitude.

(C) Since the collision frequency is 1 collision ever 2.63×10^{-4} s, the astronaut would be struck by about $1/2.63 \times 10^{-4} \approx 3{,}802$ N_2 molecules each second. In one minute, about 22,814 N_2 molecules would strike the astronaut. The average knetic energy of each N_2 molecule is

$$<T> = \frac{3kT}{2} = \frac{3(1.38 \times 10^{-23} \text{ J K}^{-1})(1000 \text{ K})}{2} = 2.07 \times 10^{-19} \text{ J}. \quad (5)$$

If every N_2 molecule were to transfer all its kinetic energy to the astronaut, a total of

$$E_{total} = (22{,}814 \text{ molecules min}^{-1})(2.07 \times 10^{-19} \text{ J molecule}^{-1}) = 4.72 \times 10^{-15} \text{ J min}^{-1}$$

would be imparted to the astronaut. This amount of energy would be too small to detect, much less incinerate our astronaut.

17.35 In this problem, you will develop the probability distribution of free paths in terms of a reduced free path. Let us define the reduced free path as $X_R = X/\langle L\rangle$, where $\langle L\rangle$ is the mean free path.

(A) Obtain the normalized probability distribution function for observing a reduced free path in the range X_R to $X_R + dX_R$.

(B) Use the expression obtained in (A) to determine the median free path in terms of $\langle L \rangle$.

Solution

(A) The free-path distribution is

$$P(X)dX = \frac{1}{\langle L \rangle} \exp\left[-\frac{X}{\langle L \rangle} \right] dX. \tag{1}$$

If we define $X_R = X/\langle L \rangle$, then

$$dX = \langle L \rangle dX_R. \tag{2}$$

Direct substitution into Eq. 1 produces

$$\boxed{P(X_R)dX_R = \frac{1}{\langle L \rangle} \exp(-X_R)\langle L \rangle dX_R = \exp(-X_R)dX_R}, \tag{3}$$

which is the desired result.

(B) The median free path X_m is that free path for which half of the observed free paths are greater than X_m and half are less. Therefore, the probability of observing a free path less than X_m is exactly $\frac{1}{2}$. This means we must have

$$\int_{X_R=0}^{X_{Rm}} \exp(-X_R)dX_R = \frac{1}{2}. \tag{4}$$

Integrating gives

$$-\exp(-X_R)\Big|_0^{X_{Rm}} = -\exp(-X_{Rm}) + 1 = \frac{1}{2}, \tag{5}$$

so that

$$\exp(-X_{Rm}) = 1 - \frac{1}{2} = \frac{1}{2}. \tag{6}$$

Taking logarithms of both sides produces

$$-X_{Rm} = \ln(1/2) = -\ln(2), \tag{7}$$

so that

$$X_{Rm} = \ln(2). \tag{8}$$

But $X_{Rm} = X_m/\langle L \rangle$; therefore,

$$\boxed{X_m = \langle L \rangle \ln(2)}. \tag{9}$$

This result is illustrated in Figure 17.19.

17.37 Consider the system described in Example 17.9. Assume that the total path length to the detector in the experimental apparatus is 0.20 m, that the molecules may be treated as hard spheres with a collision radius of 6×10^{-10} m, and that the gases are ideal. Determine the pressure the investigator should use inside the chamber to maximize the probability of observing exactly one collision of an entering A molecule with a B molecule inside the chamber. Let A be a potassium atom and B a Cl_2 molecule. Assume that the temperature is 298 K and that the masses are the average atomic and molecular masses (i.e., the atomic and molecular weights) of the elements.

Solution

In Example 17.9, we found that the probability that molecule A undergoes exactly one collision in a total path length of X_o is

$$P(1) = \frac{X_o}{\langle L \rangle} \exp\left[-\frac{X_o}{\langle L \rangle}\right]. \tag{1}$$

Our first task is to find the value of X_o at which $P(1)$ attains a maximum. The condition for a maximum is that $dP(1)/dX_o = 0$. Thus, we require that

$$\frac{dP(1)}{dX_o} = \frac{1}{\langle L \rangle} \exp\left[-\frac{X_o}{\langle L \rangle}\right] - \frac{X_o}{\langle L \rangle^2} \exp\left[-\frac{X_o}{\langle L \rangle}\right] = 0. \tag{2}$$

Rearranging terms in Eq. 2 gives

$$\frac{1}{\langle L \rangle} \exp\left[-\frac{X_o}{\langle L \rangle}\right]\left[1 - \frac{X_o}{\langle L \rangle}\right] = 0. \tag{3}$$

This equation is satisfied for

$$X_o = \langle L \rangle. \tag{4}$$

Consequently, if $\langle L \rangle = 0.20$ m, the path length in the apparatus will be the mean free path for which $P(1)$ is a maximum. The mean free path for hard spheres is given by Eq. 17.104:

$$\langle L_{AB} \rangle = \frac{(\mu/m_A)^{1/2}}{N_B(\pi\sigma_{AB}^2)}. \tag{5}$$

The mass ratio appearing in Eq. 5 is

$$\frac{\mu}{m_A} = \frac{m_A m_B}{m_A(m_A + m_B)} = \frac{m_B}{m_A + m_B} = \frac{70.90}{39.098 + 70.90} = 0.6446. \tag{6}$$

The hard-sphere cross section is

$$\pi\sigma_{AB}^2 = (3.14159)(6 \times 10^{-10}\ \text{m})^2 = 1.131 \times 10^{-18}\ \text{m}^2. \tag{7}$$

The number of Cl_2 molecules per m^3 can be obtained from the ideal gas law:

$$PV = n_B RT = \frac{W_B RT}{N}, \tag{8}$$

where N is Avogadro's constant and W_B is the number of B molecules in the chamber. Therefore,

$$N_B = \frac{W_B}{V} = \frac{PN}{RT}$$

$$= \frac{(6.02214 \times 10^{23}\ \text{mol}^{-1})P}{(0.08206\ \text{L atm mol}^{-1}\ \text{K}^{-1})(298\ \text{K})} \times \frac{1\ \text{L}}{1{,}000\ \text{cm}^3} \times \frac{10^6\ \text{cm}^3}{\text{m}^3}$$

$$= 2.463 \times 10^{25}\ \text{P atm}^{-1}\ \text{m}^{-3}. \qquad \textbf{(9)}$$

Substituting the data from Eqs. 6–9 into Eq. 5 gives

$$\langle L_{AB}\rangle = \frac{(0.6446)^{1/2}}{(2.463 \times 10^{25}\ \text{atm}^{-1}\ \text{m}^{-3})(1.131 \times 10^{-18}\ \text{m}^2)P} = \frac{2.882 \times 10^{-8}\ \text{atm m}}{P}. \qquad \textbf{(10)}$$

To achieve a maximum probability for $P(1)$, we require that $\langle L_{AB}\rangle$ be the average path length in the apparatus, which is 0.20 m. Thus, $P(1)$ is a maximum if

$$\frac{2.882 \times 10^{-8}\ \text{atm m}}{P} = 0.20\ \text{m}. \qquad \textbf{(11)}$$

The pressure required to achieve this condition is

$$P = \frac{2.882 \times 10^{-8}\ \text{atm m}}{0.20\ \text{m}} = 1.441 \times 10^{-7}\ \text{atm} = \underline{1.095 \times 10^{-4}\ \text{torr}}. \qquad \textbf{(12)}$$

17.39 For the system described in Example 17.9, obtain an expression in terms of X_o and $\langle L\rangle$ giving the probability that an entering A molecule will undergo exactly two collisions with a B molecule as it moves through a total path length of X_o inside the chamber. In terms of $\langle L\rangle$, what value of X_o makes this probability a maximum?

Solution

To obtain two collisions within a total path of length X_o, molecule A must move a distance X_1 and then collide. Then it must move an additional distance X_2, at which point it collides again. After that, molecule A must move the remaining distance $X_o - X_1 - X_2$ without colliding. The probability that two collisions will occur is the sum of the probability of the event just described over all possible collision points X_1 and X_2. The following diagram summarizes the situation:

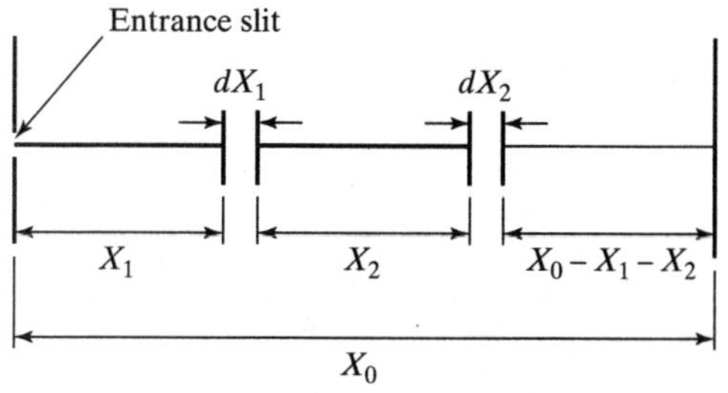

The probability of collision after a path of length X_1 is

$$P(X_1)dX_1 = \frac{1}{\langle L\rangle}\exp\left[-\frac{X_1}{\langle L\rangle}\right]dX_1. \qquad \textbf{(1)}$$

The probability of a second collision after a path of length X_2 is

$$P(X_2)dX_2 = \frac{1}{\langle L\rangle}\exp\left[-\frac{X_2}{\langle L\rangle}\right]dX_2. \tag{2}$$

Finally, the molecule must move through the remaining distance, $X_o - X_1 - X_2$, without colliding. The probability that this will occur is given by Eq. 17.107 with a set equal to $1/\langle L\rangle$:

$$P(0) = \exp\left[-\frac{X_o - X_1 - X_2}{\langle L\rangle}\right]. \tag{3}$$

The combined probability that all three of these independent events will occur is the product of the individual probabilities:

$$\begin{aligned} P_2(X_1, X_2)dX_1dX_2 &= \text{probability of two collisions at points } X_1 \text{ and } X_2 \\ &= P(X_1)dX_1P(X_2)dX_2P(0). \end{aligned} \tag{4}$$

Substituting Eqs. 1–3 into Eq. 4 produces

$$\begin{aligned} &P_2(X_1, X_2)dX_1dX_2 \\ &= \frac{1}{\langle L\rangle}\exp\left[-\frac{X_1}{\langle L\rangle}\right]dX_1\frac{1}{\langle L\rangle}\exp\left[-\frac{X_2}{\langle L\rangle}\right]dX_2\exp\left[-\frac{X_o - X_1 - X_2}{\langle L\rangle}\right] \\ &= \frac{1}{\langle L\rangle^2}\exp\left[-\frac{X_o}{\langle L\rangle}\right]dX_1dX_2. \end{aligned} \tag{5}$$

The total probability of two collisions is obtained by summing $P_2(X_1, X_2)dX_1dX_2$ over all possible collision points. The point X_2 can be anywhere in the range $0 \le X_2 \le X_o - X_1$; the point X_1 can be anywhere in the range $0 \le X_1 \le X_o$. Therefore, the required sum is

$$\begin{aligned} P(2) &= \int_{X_1=0}^{X_o}\int_{X_2=0}^{X_o-X_1} P_2(X_1, X_2)dX_2dX_1 \\ &= \frac{1}{\langle L\rangle^2}\exp\left[-\frac{X_o}{\langle L\rangle}\right]\int_{X_1=0}^{X_o}\int_{X_2=0}^{X_o-X_1} dX_2dX_1. \end{aligned} \tag{6}$$

We first perform the integration over X_2 to obtain

$$P(2) = \frac{1}{\langle L\rangle^2}\exp\left[-\frac{X_o}{\langle L\rangle}\right]\int_{X_1=0}^{X_o}[X_o - X_1]dX_1. \tag{7}$$

Integrating a second time yields

$$\begin{aligned} P(2) &= \frac{1}{\langle L\rangle^2}\exp\left[-\frac{X_o}{\langle L\rangle}\right]\int_{X_1=0}^{X_o}[X_o - X_1]dX_1 \\ &= \frac{1}{\langle L\rangle^2}\exp\left[-\frac{X_o}{\langle L\rangle}\right]\left[X_oX_1 - \frac{X_1^2}{2}\right]_0^{X_o} = \frac{1}{\langle L\rangle^2}\exp\left[-\frac{X_o}{\langle L\rangle}\right]\left[X_o^2 - \frac{X_o^2}{2}\right] \\ &= \boxed{\frac{X_o^2}{2\langle L\rangle^2}\exp\left[-\frac{X_o}{\langle L\rangle}\right]}, \end{aligned} \tag{8}$$

which is the desired expression. $P(2)$ attains a maximum value whenever

$$\frac{dP(2)}{dX_o} = \frac{X_o}{\langle L\rangle^2}\exp\left[-\frac{X_o}{\langle L\rangle}\right] - \frac{X_o^2}{2\langle L\rangle^3}\exp\left[-\frac{X_o}{\langle L\rangle}\right] = 0. \tag{9}$$

Rearranging terms in Eq. 8 produces

$$\frac{X_o}{\langle L\rangle^2}\exp\left[-\frac{X_o}{\langle L\rangle}\right]\left[1-\frac{X_o}{2\langle L\rangle}\right]=0. \qquad \textbf{(10)}$$

Equation 9 is satisfied if

$$\boxed{X_o = 2\langle L\rangle}, \qquad \textbf{(11)}$$

which is the requirement for a maximum probability of exactly two collisions.

CHAPTER 18

Statistical Thermodynamics

18.1 (A) Compute the effective number of available translational states for one argon atom in a three-dimensional potential well with a potential $V(x, y, z) = 0$ whose shape is that of a rectangular parallelepiped, the lengths of whose sides are $a = b = c = 0.200$ m. Assume that the temperature is 298 K.

(B) Let us now assume that we have 1 mole of noninteracting argon atoms in the potential well. Suppose that we can examine each possible translational quantum state to see whether it is occupied by any of the argon atoms. Suppose further that we can examine two such states per second and that we continue our examination 10 hours per day, every day for one year. Approximately how many times will we find that a translational quantum state is occupied by one or more argon atoms?

Solution

(A) The effective number of translational states is given by Eq.18.5 in the text:

$$z = s_{\text{eff}} = \frac{V}{h^3}[2\pi mkT]^{3/2}. \tag{1}$$

The mass of an argon atom is

$$m = \frac{0.039948 \text{ kg mol}^{-1}}{6.02214 \times 10^{23} \text{ mol}^{-1}} = 6.634 \times 10^{-26} \text{ kg}. \tag{2}$$

The volume of the system is

$$V = abc = (0.20 \text{ m})^3 = 0.0080 \text{ m}^3. \tag{3}$$

Therefore,

$$z = s_{\text{eff}} = \frac{0.0080 \text{ m}^3}{(6.62608 \times 10^{-34} \text{ J s})^3}$$
$$\times [2(3.14159)(6.634 \times 10^{-26} \text{ kg})(1.3807 \times 10^{-23} \text{ J K}^{-1})(298 \text{ K})]^{3/2}$$
$$= \underline{1.953 \times 10^{30} \text{ effective states.}} \tag{4}$$

(B) If we examine two quantum states per second 10 hours per day for 365 days, we will have examined

$$(2 \text{ states s}^{-1}) \times \frac{3{,}600 \text{ s}}{\text{hour}} \times \frac{10 \text{ hours}}{\text{day}} \times 365 \text{ days}$$
$$= 26{,}280{,}000 \text{ states}. \tag{5}$$

The average number of states per argon atom is

$$\frac{1.953 \times 10^{30} \text{ states}}{6.02214 \times 10^{23} \text{ argon atoms}} = 3.243 \times 10^6 \text{ states atom}^{-1}. \tag{6}$$

Therefore, we expect to find that

$$\text{number of filled quantum states} \approx \frac{\text{no. states examined}}{\text{no. states atom}^{-1}}$$
$$= \frac{26{,}280{,}000 \text{ states}}{3.243 \times 10^6 \text{ states atom}^{-1}} = 8.10 \text{ atoms}. \tag{7}$$

Thus, we have eight, maybe nine, atoms for a year's work—not much of a return on our investment.

18.3 (A) Obtain an expression for the ratio of $s_{\text{eff}}^4/4!$ to the number of distinguishable microstates obtained when four indistinguishable objects are distributed among s_{eff} different states.

(B) Compute this ratio when s_{eff}/N is 10, 10^2, 10^3, 10^4, 10^5, and 10^6, where N is the number of objects to be distributed into the s_{eff} quantum states.

Solution

(A) The number of distinguishable microstates for four indistinguishable objects distributed among 12 states is given in Table 18.4:

$$\text{no. distinguishable microstates} = \frac{s_{\text{eff}}^4}{4!} + \frac{s_{\text{eff}}^3}{4} + \frac{11 s_{\text{eff}}^2}{4!} + \frac{s_{\text{eff}}}{4}. \quad (1)$$

The requested ratio is, therefore, given by

$$\boxed{R = \frac{\dfrac{s_{\text{eff}}^4}{4!}}{\dfrac{s_{\text{eff}}^4}{4!} + \dfrac{s_{\text{eff}}^3}{4} + \dfrac{11 s_{\text{eff}}^2}{4!} + \dfrac{s_{\text{eff}}}{4}}}. \quad (2)$$

(B) Since $N = 4$, the ratio $s_{\text{eff}}/N = 10^n$ whenever $s_{\text{eff}} = 4 \times 10^n$. We can, therefore, compute R as a function of n:

$$R(n) = \frac{\dfrac{(4 \times 10^n)^4}{4!}}{\dfrac{(4 \times 10^n)^4}{4!} + \dfrac{(4 \times 10^n)^3}{4} + \dfrac{11(4 \times 10^n)^2}{4!} + \dfrac{(4 \times 10^n)}{4!}}. \quad (3)$$

Direct computation yields the results to the right. These results show that the leading term becomes the only term of importance when the ratio of s_{eff} to N becomes large.

18.5 Compute the entropy of argon at pressures of 1, 10, and 100 bar for a temperature of 300 K. The experimentally determined argon entropies at 300 K listed in the 78th edition of the CRC *Handbook of Chemistry and Physics* at these pressures at 300 K are 155.0, 135.6, and 114.7 J mol^{-1} K^{-1}, respectively. Compute the percent error in your result as a function of pressure. Why does the error increase as the pressure rises?

Solution

If we assume that the molecules do not interact, the entropy will be given by the Sackur-Tetrode equation, since Ar atoms have only translational energy. Using Eq. 18.64, we have

$$S = S_{tr} = \frac{3R}{2}\ln(M/\text{g mol}^{-1}) + \frac{5R}{2}\ln(T/K) - R\ln(p/\text{bar}) - 1.15169R. \quad (1)$$

For Ar at $T = 300$ K, this gives

$$S = R[1.5\ln(39.948) + 2.5\ln(300) - \ln(p/\text{bar}) - 1.15169]$$
$$= R[18.639 - \ln(p/\text{bar})]. \quad (2)$$

Accordingly, we have the table at the top of the next page.

Argon becomes less ideal at higher pressures. The molecules are in closer proximity; consequently, they exhibit more interaction, and the molecular partition function computed by assuming zero interaction is less accurate at higher pressures.

p (bar)	S [Eq.(2)] (J mol^{-1} K^{-1})	% Error
1	155.0	0.000
10	135.8	0.147
100	116.7	1.174

18.7 Show that the assumption that we can obtain the translational molecular partition function by using the eigenvalues for independent particles in a potential well with zero potential energy leads directly to the ideal-gas equation of state. (*Hint*: Compute the pressure from the translational molecular partition function.)

Solution

The pressure is given by

$$p = RT\left(\frac{\partial \ln z}{\partial V}\right)_T. \quad (1)$$

For translation of the center of mass, the molecular partition function obtained from the particle in a zero potential well is

$$z_{tr} = \frac{V}{h^3}[2\pi mkT]^{3/2}. \quad (2)$$

The natural logarithm of z_{tr} can be written in the form

$$\ln z_{tr} = \ln V + \ln\left[\frac{[2\pi mkT]^{3/2}}{h^3}\right]. \quad (3)$$

When the temperature is held constant, the second term in Eq. 3 is also a constant, whose derivative is zero. Therefore,

$$\left(\frac{\partial \ln z}{\partial V}\right)_T = \left(\frac{\partial \ln V}{\partial V}\right)_T = \frac{1}{V}. \quad (4)$$

Substituting Eq. 4 into Eq. 1 produces

$$\boxed{p = \frac{RT}{V} \text{ or } pV = RT}. \quad (5)$$

It was obvious that we were going to obtain this equation of state when we employed the model of noninteracting molecules in a zero-potential well to compute the translational molecular partition function.

18.9 Starting with Eq. 18.82, derive Eq. 18.83. Show all details.

Solution

Equation 18.82 for the rotational contribution to the internal energy is

$$U_{\text{rot}} = RT^2\left(\frac{\partial \ln z_{\text{rot}}}{\partial T}\right)_V = \frac{RT^2}{z_{\text{rot}}}\left(\frac{\partial z_{\text{rot}}}{\partial T}\right)_V. \quad (1)$$

The heat capacity is given by

$$[C_v]_{\text{rot}} = \left(\frac{\partial U_{\text{rot}}}{\partial T}\right)_V. \quad (2)$$

Differentiating Eq. 1 directly with respect to T at constant V gives

$$\left(\frac{\partial U_{\text{rot}}}{\partial T}\right)_V = \frac{2RT}{z_{\text{rot}}}\left(\frac{\partial z_{\text{rot}}}{\partial T}\right)_V - \frac{RT^2}{(z_{\text{rot}})^2}\left(\frac{\partial z_{\text{rot}}}{\partial T}\right)_V\left(\frac{\partial z_{\text{rot}}}{\partial T}\right)_V + \frac{RT^2}{z_{\text{rot}}}\left(\frac{\partial^2 z_{\text{rot}}}{\partial T^2}\right)_V. \quad (3)$$

Multiplying the first term in the numerator and denominator by T and the second term in the numerator and denominator by RT^2 produces

$$\left(\frac{\partial U_{\text{rot}}}{\partial T}\right)_V = \frac{2}{T}\left[\frac{RT^2}{z_{\text{rot}}}\left(\frac{\partial z_{\text{rot}}}{\partial T}\right)_V\right] - \frac{1}{RT^2}\left[\frac{RT^2}{z_{\text{rot}}}\left(\frac{\partial z_{\text{rot}}}{\partial T}\right)_V\right]^2 + \frac{RT^2}{z_{\text{rot}}}\left(\frac{\partial^2 z_{\text{rot}}}{\partial T^2}\right)_V. \quad (4)$$

The factors within brackets on the right-hand side of Eq. 4 can be seen from Eq. 1 to be U_{rot}. Therefore,

$$\left(\frac{\partial U_{\text{rot}}}{\partial T}\right)_V = \frac{2U_{\text{rot}}}{T} - \frac{(U_{\text{rot}})^2}{RT^2} + \frac{RT^2}{z_{\text{rot}}}\left(\frac{\partial^2 z_{\text{rot}}}{\partial T^2}\right)_V, \quad (5)$$

and

$$z_{\text{rot}} = \sum_{J=0}^{\infty}(2J+1)\exp\left[-\frac{J(J+1)Bh}{kT}\right]. \quad (6)$$

The first derivative of z_{rot} with respect to T at constant V can be obtained by differentiating within the summation, since the derivative and summation operators commute. Thus,

$$\left(\frac{\partial z_{\text{rot}}}{\partial T}\right)_V = \sum_{J=0}^{\infty}(2J+1)J(J+1)\frac{Bh}{kT^2}\exp\left[-\frac{J(J+1)Bh}{kT}\right]. \quad (7)$$

The second derivative is the derivative of the first derivative. Therefore,

$$\left(\frac{\partial^2 z_{\text{rot}}}{\partial T^2}\right)_V = \sum_{J=0}^{\infty}\left\{-2(2J+1)J(J+1)\frac{Bh}{kT^3}\exp\left[-\frac{J(J+1)Bh}{kT}\right]\right.$$
$$\left. + (2J+1)J^2(J+1)^2\left(\frac{Bh}{kT^2}\right)^2\exp\left[-\frac{J(J+1)Bh}{kT}\right]\right\}. \quad (8)$$

Factoring out the common terms, we can express Eq. 8 in the form

$$\left(\frac{\partial^2 z_{\text{rot}}}{\partial T^2}\right)_V = \frac{1}{T^2}\sum_{J=0}^{\infty}\left\{(2J+1)J(J+1)\frac{Bh}{kT}\exp\left[-\frac{J(J+1)Bh}{kT}\right]\right\}$$
$$\left\{\frac{J(J+1)}{T}\left(\frac{Bh}{k}\right) - 2\right\}$$
$$= -\frac{2Bh}{kT^3}\sum_{J=0}^{\infty}(2J+1)J(J+1)\exp\left[-\frac{J(J+1)Bh}{kT}\right]$$
$$+ \frac{1}{T^2}\left(\frac{Bh}{kT}\right)^2\sum_{J=0}^{\infty}(2J+1)J^2(J+1)^2\exp\left[-\frac{J(J+1)Bh}{kT}\right]. \quad (9)$$

The first term on the right-hand side of Eq. 9 can be expressed as

$$-\frac{2Bh}{kT^3}\sum_{J=0}^{\infty}(2J+1)J(J+1)\exp\left[-\frac{J(J+1)Bh}{kT}\right]$$
$$= -\frac{2}{T}\left(\frac{Bh}{kT^2}\right)\sum_{J=0}^{\infty}(2J+1)J(J+1)\exp\left[-\frac{J(J+1)Bh}{kT}\right]. \quad (10)$$

With the use of Eq. 7, Eq. 10 becomes

$$-\frac{2Bh}{kT^3}\sum_{J=0}^{\infty}(2J+1)J(J+1)\exp\left[-\frac{J(J+1)Bh}{kT}\right] = -\frac{2}{T}\left(\frac{\partial z_{rot}}{\partial T}\right)_V. \quad (11)$$

Therefore, the second derivative in Eq. 9 can be written as

$$\left(\frac{\partial^2 z_{rot}}{\partial T^2}\right)_V = -\frac{2}{T}\left(\frac{\partial z_{rot}}{\partial T}\right)_V + \frac{1}{T^2}\left(\frac{Bh}{kT}\right)^2 \times \sum_{J-0}^{\infty}(2J+1)J^2(J+1)^2\exp\left[-\frac{J(J+1)Bh}{kT}\right]. \quad (12)$$

Combining Eqs. 5 and 12 yields

$$[C_v]_{rot} = \left(\frac{\partial U_{rot}}{\partial T}\right)_V = \frac{2U_{rot}}{T} - \frac{(U_{rot})^2}{RT^2} - \frac{2}{T}\left[\frac{RT^2}{z_{rot}}\left(\frac{\partial z_{rot}}{\partial T}\right)_V\right] + \frac{R}{z_{rot}}\left(\frac{Bh}{kT}\right)^2\sum_{J=0}^{\infty}(2J+1)J^2(J+1)^2\exp\left[-\frac{J(J+1)Bh}{kT}\right]. \quad (13)$$

Since $[(RT^2/z_{rot})(\partial z_{rot}/\partial T)_V] = U_{rot}$, we have

$$[C_v]_{rot} = \left(\frac{\partial U_{rot}}{\partial T}\right)_V = \frac{2U_{rot}}{T} - \frac{(U_{rot})^2}{RT^2} - \frac{2U_{rot}}{T} + \frac{R}{z_{rot}}\left(\frac{Bh}{kT}\right)^2\sum_{J=0}^{\infty}(2J+1)J^2(J+1)^2\exp\left[-\frac{J(J+1)Bh}{kT}\right]. \quad (14)$$

The first and third terms add to zero to give

$$\boxed{[C_v]_{rot} = \left(\frac{\partial U_{rot}}{\partial T}\right)_V = -\frac{(U_{rot})^2}{RT^2} + \frac{R}{z_{rot}}\left(\frac{Bh}{kT}\right)^2\sum_{J=0}^{\infty}(2J+1)J^2(J+1)^2\exp\left[-\frac{J(J+1)Bh}{kT}\right]}, \quad (15)$$

which is Eq. 18.83.

18.11* Use the data in Table 15.4 to compute the rotational heat capacity, $[C_v]_{rot}$, for HF at temperatures between 5 K and 40 K at 5-K intervals and then at 50 K, 100 K, 200 K, and 300 K. Compute the ratio $[C_v]_{rot}/R$ at each temperature, and plot the results as a function of temperature. At approximately what temperature does the error in the classical approximation for $[C_v]_{rot}$ for HF approach 1.0%? Assume that the rotational-energy levels are given by a rigid-rotor expression.

Solution

The rotational contribution to C_v is given by Eq. 18.83, viz.,

$$[C_v]_{rot} = \left(\frac{\partial U_{rot}}{\partial T}\right)_V = -\frac{(U_{rot})^2}{RT^2} + \frac{R}{z_{rot}}\left(\frac{Bh}{kT}\right)^2 \times \sum_{J=0}^{\infty}(2J+1)J^2(J+1)^2\exp\left[-\frac{J(J+1)Bh}{kT}\right], \quad (1)$$

where

$$z_{rot} = \sum_{J=0}^{\infty}(2J+1)\exp\left[-\frac{J(J+1)Bh}{kT}\right] \quad (2)$$

and the rotational internal energy is given by Eq. 18.82:

$$U_{\text{rot}} = \frac{R}{z_{\text{rot}}} \sum_{J=0}^{\infty} (2J+1)J(J+1)\frac{Bh}{k} \exp\left[-\frac{J(J+1)Bh}{kT}\right]. \quad \textbf{(3)}$$

The quantity required inside the sums is

$$\frac{Bh}{k} = \frac{(20.939)(2.9979 \times 10^{10})\ \text{s}^{-1}(6.62608 \times 10^{-34}\ \text{J s})}{1.38066 \times 10^{-23}\ \text{J K}^{-1}} = 30.126\ \text{K}. \quad \textbf{(4)}$$

The summations are, therefore,

$$z_{\text{rot}} = \sum_{J=0}^{\infty} (2J+1) \exp\left[-\frac{30.126J(J+1)}{T}\right], \quad \textbf{(5)}$$

$$U_{\text{rot}} = \frac{8.314\ \text{J mol}^{-1}}{z_{\text{rot}}} \sum_{J=0}^{\infty} 30.126(2J+1)J(J+1) \exp\left[-\frac{30.126J(J+1)}{T}\right], \quad \textbf{(6)}$$

and

$$\frac{R}{z_{\text{rot}}}\left(\frac{Bh}{kT}\right)^2 \sum_{J=0}^{\infty} (2J+1)J^2(J+1)^2 \exp\left[-\frac{J(J+1)Bh}{kT}\right] = \frac{(30.126)^2(8.314)}{T^2 z_{\text{rot}}}$$

$$\times \sum_{J=0}^{\infty} (2J+1)J^2(J+1)^2 \exp\left[-\frac{30.126J(J+1)}{T}\right]. \quad \textbf{(7)}$$

These sums can easily be obtained using an Excel spreadsheet. The results are shown in the table below. The percent error can be computed from the ratio $r = [C_v]_{\text{rot}}/R$, using

$$\%\ \text{error} = 100 \times \frac{R - [C_v]_{\text{rot}}}{[C_v]_{\text{rot}}} = 100[r^{-1} - 1]. \quad \textbf{(8)}$$

A plot of r versus temperature is shown on the next page. The presence of the maximum in the ratio of the quantum mechanical to the classical result is interesting and unexpected. The error in the classical value for $[C_v]_{\text{rot}}$ approaches 1% somewhere between 50 and 100 K. A linear interpolation suggests a temperature of 71 K.

T (K)	$[C_v]_v$ (J mol^{-1} K^{-1})	$r = \frac{[C_v]_{\text{rot}}}{R}$	% Error
5	0.0212	0.00254	39,270.
10	2.157	0.2595	285.4
15	6.556	0.7886	26.8
20	8.769	1.055	−5.2
25	9.120	1.097	−8.8
30	8.926	1.074	−6.9
35	8.713	1.048	−4.6
40	8.573	1.031	−3.0
50	8.442	1.015	−1.5
100	8.336	1.003	−0.3
200	8.319	1.0006	−0.06
300	8.316	1.000	−0.00

$[C_v]_{rot}/R$ for HF as a Function of T

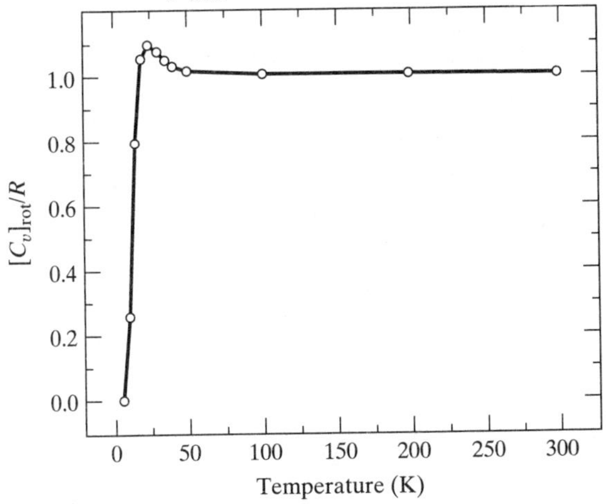

18.13 Table 12.1 lists the $\Theta_\ell^m(\theta)$ portion of the $Y_\ell^m(\theta, \phi)$ spherical harmonic for $\ell = 0$, 1, and 2. The $\Theta_\ell^m(\theta)$ functions for $\ell = 3$ are as follows,

$$\Theta_3^0(\theta) = \frac{3(14)^{1/2}}{4}\left[\frac{5}{3}\cos^3\theta - \cos\theta\right],$$

$$\Theta_3^{\pm 1}(\theta) = \frac{(42)^{1/2}}{8}\sin\theta[5\cos^2\theta - 1],$$

$$\Theta_3^{\pm 2}(\theta) = \frac{(105)^{1/2}}{4}\sin^2\theta\cos\theta,$$

and

$$\Theta_3^{\pm 3}(\theta) = \frac{(70)^{1/2}}{8}\sin^3\theta.$$

For the spherical harmonics with $\ell \le 3$, show that $\Theta_\ell^m(\theta)$ is symmetric to nuclear exchange in a homonuclear rotor if ℓ is even and antisymmetric if ℓ is odd. Discuss the significance of this result with respect to the rotational molecular partition function.

Solution

If $\Theta_\ell^m(\theta)$ is symmetric to the exchange of nuclei in a homonuclear diatomic rotor, we must have $\Theta_\ell^m(\theta) = \Theta_\ell^m(\theta + \pi)$, since a rotation of π radians about θ is equivalent to nuclear exchange. For $\ell = 0$,

$$\Theta_0^0(\theta) = \left(\frac{1}{2}\right)^{1/2}, \quad (1)$$

so we have

$$\Theta_0^0(\theta) = \Theta_0^0(\theta + \pi) = \left(\frac{1}{2}\right)^{1/2}, \quad (2)$$

and the function is symmetric. For $\ell = 2$, we have

$$\Theta_2^0(\theta) = \left(\frac{5}{8}\right)^{1/2}[3\cos^2\theta - 1], \quad (3)$$

so that

$$\Theta_2^0(\theta + \pi) = \left(\frac{5}{8}\right)^{1/2}[3\cos^2(\theta + \pi) - 1] = \left(\frac{5}{8}\right)^{1/2}[3(-\cos\theta)^2 - 1]$$

$$= \left(\frac{5}{8}\right)^{1/2}[3\cos^2\theta - 1] = \Theta_2^0(\theta), \quad \textbf{(4)}$$

and hence, $\Theta_2^0(\theta)$ is symmetric with respect to exchange. Another possibility is

$$\Theta_2^{\pm 1}(\theta) = \left(\frac{15}{4}\right)^{1/2}\cos\theta\sin\theta. \quad \textbf{(5)}$$

Rotation by π radians produces

$$\Theta_2^{\pm 1}(\theta) = \left(\frac{15}{4}\right)^{1/2}\cos(\theta + \pi)\sin(\theta + \pi) = \left(\frac{15}{4}\right)^{1/2}[-\cos\theta][-\sin\theta]$$

$$= \left(\frac{15}{4}\right)^{1/2}\cos\theta\sin\theta = \Theta_2^{\pm 1}(\theta). \quad \textbf{(6)}$$

Thus, $\Theta_2^{\pm 1}(\theta)$ is also symmetric.

The last possibility with $\ell = 2$ is

$$\Theta_2^{\pm 2}(\theta) = \left(\frac{15}{4}\right)^{1/2}\sin^2\theta. \quad \textbf{(7)}$$

Rotation by π radians produces

$$\Theta_2^{\pm 2}(\theta + \pi) = \left(\frac{15}{4}\right)^{1/2}[\sin(\theta + \pi)]^2 = \left(\frac{15}{4}\right)^{1/2}[-\sin\theta]^2$$

$$= \left(\frac{15}{4}\right)^{1/2}\sin^2\theta = \Theta_2^{\pm 2}(\theta), \quad \textbf{(8)}$$

so that the function is symmetric to exchange. Therefore, all $\ell = 2$ functions are symmetric.

For the odd value $\ell = 1$, we have

$$\Theta_1^0(\theta) = \left(\frac{3}{2}\right)^{1/2}\cos\theta. \quad \textbf{(9)}$$

Since $\cos(\theta + \pi) = -\cos\theta$, this function is antisymmetric to nuclear exchange. The other possibility is

$$\Theta_1^{\pm 1}(\theta) = \left(\frac{3}{4}\right)^{1/2}\sin\theta. \quad \textbf{(10)}$$

Since $\sin(\theta + \pi) = -\sin\theta$, this function is antisymmetric to nuclear exchange as well. Therefore, all the spherical harmonics with $\ell = 1$ are antisymmetric with respect to nuclear exchange. For $\ell = 3$, there are the four possibilities listed in the problem. First,

$$\Theta_3^0(\theta) = \frac{3(14)^{1/2}}{4}\left[\frac{5}{3}\cos^3\theta - \cos\theta\right] \quad \textbf{(11)}$$

will be antisymmetric, because $\cos(\theta + \pi) = -\cos\theta$. Also,

$$\Theta_3^{\pm 1}(\theta) = \frac{(42)^{1/2}}{8}\sin\theta[5\cos^2\theta - 1] \quad \textbf{(12)}$$

is antisymmetric, because the $\sin\theta$ factor changes sign upon rotation by π radians, since $\sin(\theta + \pi) = -\sin\theta$. However, the second factor, $[5\cos^2\theta - 1]$, does not change sign, because $\cos^2(\theta + \pi) = [-\cos\theta]^2 = \cos^2\theta$. Third,

$$\Theta_3^{\pm 2}(\theta) = \frac{(105)^{1/2}}{4}\sin^2\theta\cos\theta \tag{13}$$

is antisymmetric, because the cosine factor changes sign when we rotate by π radians, but the $\sin^2\theta$ factor does not. Finally, the function

$$\Theta_3^{\pm 3}(\theta) = \frac{(70)^{1/2}}{8}\sin^3\theta \tag{14}$$

is antisymmetric to nuclear exchange, because $\sin^3(\theta + \pi) = [-\sin\theta]^3 = -\sin\theta$. Therefore, all of the spherical harmonics with either $\ell = 1$ or $\ell = 3$ are antisymmetric to nuclear exchange.

The previous results mean that we must eliminate all the even rotational eigenstates whenever the spherical harmonic must be antisymmetric, because the nuclear spin function is symmetric, and we must eliminate all the odd rotational eigenstates whenever the spherical harmonic must be symmetric, because the nuclear spin function is antisymmetric. The result is an elimination of half of the quantum states. We know that z is the effective number of quantum states available to the molecules. If half are eliminated by exchange symmetry requirements, we must divide z_{rot} by two to get the correct result. This factor is called the "symmetry factor."

18.15 F. Hynne [*Am. J. Phys.*, **49**, 125 (1981)] and H. Kroemer [*Am. J. Phys.* **48**, 962 (1980)] have pointed out that omitting the $c_{N-1}s_{eff}^{N-1}$ term in Eq. 18.44 that includes the contribution of doubly occupied states produces a large error in the value of Z given by Eq. 18.45. Their analysis indicates that the canonical partition function should have the form $Z = fz^N/N!$, where f can be very large.

(A) Show that the presence of the factor f in the expression for Z makes no difference in the computed values of U and C_v.

(B) Show that the only effect of the factor f on the entropy is to alter the value of the constant in Eq. 18.38.

Solution

(A) From Eq. 18.32, the internal energy in terms of Z is given by

$$U = kT^2\left(\frac{\partial \ln Z}{\partial T}\right)_V. \tag{1}$$

With Z given by $fz^N/N!$, Eq. 1 becomes

$$U = kT^2\frac{\partial}{\partial T}\left[\ln\left(\frac{fz^N}{N!}\right)\right]_V = kT^2\frac{\partial}{\partial T}[\ln(f) + N\ln(z) - \ln(N!)]_V. \tag{2}$$

Now, f and N are independent of T, so that their derivatives are both zero. This gives

$$U = NkT^2\left(\frac{\partial \ln z}{\partial T}\right)_V = RT^2\left(\frac{\partial \ln z}{\partial T}\right)_V, \tag{3}$$

which is the same result we have without the factor of f. Since $C_v = (\partial U/\partial T)_V$ and U is the same with or without the factor of f, no change in C_v is produced by inserting f. The same analysis shows that the expression for the pressure is also unaltered by including the f factor. The only difference is that the derivative is now with respect to volume rather than temperature.

(B) From Eq. 18.38, the entropy is

$$S = k \ln Z + \frac{U}{T} + \text{constant.} \tag{4}$$

We already know from Part (A) that there is no change in U produced by the factor f. Therefore, substituting $Z = fZ'$ in Eq. 4, where Z' is the canonical partition function without f, produces

$$S = k \ln(fZ') + \frac{U}{T} + \text{constant} = k[\ln(f) + \ln Z'] + \frac{U}{T} + \text{constant}$$

$$= k \ln Z' + \frac{U}{T} + [k \ln(f) + \text{constant}]. \tag{5}$$

The third law requires that the entropy approach a constant for all perfect crystals at $T = 0$ K. We know that Z' and U approach zero as $T \longrightarrow 0$ K. Therefore, we have

$$\lim_{T \longrightarrow 0\,\text{K}} [S] = k \ln(f) + \text{constant.} \tag{6}$$

If we wish to take the value of S at 0 K to be zero, then the constant is chosen to be $-k \ln(f)$. This produces

$$S = k \ln Z' + \frac{U}{T}, \tag{7}$$

which is the same result as that listed in Table 18.1.

This analysis demonstrates that, because thermodynamic quantities depend upon $\ln(Z)$ rather than upon Z itself, even if Z is wrong by a large factor f, the thermodynamic quantities will still be correct.

18.17 In this problem, you will derive a result for the classical rotation of an asymmetric polyatomic molecule that is very close to the exact result given in Eq. 18.91. The various parts of the problem serve as a procedure guide.

(A) Suppose we have a particle of mass m rotating in a plane about the origin at a fixed distance R with the potential energy equal to zero. Since R is fixed and the motion is planar, this is a single-variable problem. Let the angle between the radial vector to the particle and the X-axis be ϕ. Figure 11.5 illustrates the model if we replace θ with ϕ. The classical Hamiltonian for this system is given by Eq. 11.22:

$$\mathbf{H} = \frac{P_R^2}{2m} + \frac{P_\phi^2}{2mR^2} + V(R, \phi).$$

In the case under examination, P_R and $V(R, \phi)$ are both zero, so that

$$\mathbf{H} = \frac{P_\phi^2}{2mR^2}.$$

The quantum mechanical operator corresponding to P_ϕ is $(\hbar/i)(\partial/\partial\phi)$. Therefore, the Schrödinger equation for the system is

$$-\frac{\hbar^2}{2mR^2}\frac{\partial^2\psi(\phi)}{\partial\phi^2} = E\psi(\phi).$$

Solve the Schrödinger equation and obtain the eigenfunctions and eigenvalues for this two-dimensional rotational motion. What quantum restrictions are there on the rotational quantum number? Explain.

(B) Write down the quantum mechanical expression for the rotational molecular partition function for the system described in (A). By assuming the rotational levels to be continuous, derive an appropriate expression for the molecular partition function.

(C) Now assume that the Hamiltonian for rotation in three-dimensional space about the three principal axes with moments of inertia I_1, I_2, and I_3 is the sum of Hamiltonians that look exactly like the one in (A). (This is not correct.) Then $E^c_{rot} = E_1 + E_2 + E_3$, where E_i is the energy for rotation about the principal axis whose moment of inertia is I_i. With this assumption, show that total molecular partition function for classical rotation of the polyatomic molecule is the product of molecular partition functions for each of the independent rotations. Determine the form that the total molecular partition function for classical rotation would have. Compare your result with Eq. 18.91.

Solution

(A) Mutiplying the Schrödinger equation by $2mR^2/\hbar^2$ and then rearranging terms gives

$$\frac{\partial^2\psi(\phi)}{\partial\phi^2} + \frac{2mER^2}{\hbar^2}\psi(\phi) = \frac{\partial^2\psi(\phi)}{\partial\phi^2} + n^2\psi(\phi) = 0, \tag{1}$$

where $n^2 = 2mER^2/\hbar^2$. In Chapter 12, we found that the solution of this type of equation is

$$\psi(\phi) = A\exp[in\phi], \tag{2}$$

where A is a constant. Since the eigenfunction must be single valued, we must have

$$\psi(\phi) = \psi(\phi + 2\pi). \tag{3}$$

This requires that we have

$$A\exp[in\phi] = A\exp[in(\phi + 2\pi)] = Ae^{2in\pi}\exp[in\phi]. \tag{4}$$

Consequently,

$$e^{2in\pi} = \exp[2in\pi] = \cos(2n\pi) + i\sin(2n\pi) = 1. \tag{5}$$

Equation 5 can be true only if n is zero or a positive or negative integer. The eigenfunctions are, therefore,

$$\psi(\phi) = A\exp[in\phi] \quad \text{for } n = 0, \pm1, \pm2, \pm3, \ldots. \tag{6}$$

Normalizing gives us the value of A:

$$\langle\psi(\phi)|\psi(\phi)\rangle = \int_{\phi=0}^{2\pi} A^*\exp[-in\phi]A\exp[in\phi]\,d\phi = 2\pi A^2. \tag{7}$$

Thus, $A = (2\pi)^{-1/2}$, and

$$\boxed{\psi(\phi) = (2\pi)^{-1/2}\exp[in\phi]} \tag{8}$$

for $n = 0, \pm1, \pm2, \pm3, \ldots$.

Since $n^2 = 2mER^2/\hbar^2$, the rotational eigenvalues are

$$E_m = \frac{n^2\hbar^2}{2mR^2} = \frac{n^2\hbar^2}{2I_1}, \tag{9}$$

where I_1 is the moment of inertia.

(B) The quantum mechanical expression for the rotational-molecular-partition function for the system described in Part (A) is

$$z_{1-\mathrm{rot}} = \sum_{n=-\infty}^{\infty} \exp\left[-\frac{n^2\hbar^2}{2I_1kT}\right]. \tag{10}$$

If we assume that n is a classical continuous variable, the summation in Eq. 10 can be replaced by an integral. This gives

$$z^c_{1-\mathrm{rot}} = \int_{-\infty}^{\infty} \exp\left[-\frac{n^2\hbar^2}{2I_1kT}\right] \mathrm{dn}. \tag{11}$$

This integral is a standard form that can be found in any table of integrals. The result is

$$\boxed{z^c_{1-\mathrm{rot}} = \left[\frac{2\pi I_1kT}{\hbar^2}\right]^{1/2} = \left[\frac{8\pi^3I_1kT}{h^2}\right]^{1/2}}. \tag{12}$$

(C) Under the assumptions in Part (C), we have

$$E_{\mathrm{rot}} = E_1 + E_2 + E_3, \tag{13}$$

where

$$E_i = \frac{n_i^2\hbar^2}{2I_i} \quad (i = 1, 2, \text{or } 3). \tag{14}$$

The partition function is, therefore,

$$\begin{aligned} z^c_{\mathrm{rot}} &= \sum_{n_1=-\infty}^{\infty}\sum_{n_2=-\infty}^{\infty}\sum_{n_3=-\infty}^{\infty} \exp\left[-\left\{\frac{n_1^2\hbar^2}{2I_1kT} + \frac{n_2^2\hbar^2}{2I_2kT} + \frac{n_3^2\hbar^2}{2I_3kT}\right\}\right] \\ &= \sum_{n_1=-\infty}^{\infty} \exp\left[-\frac{n_1^2\hbar^2}{2I_1kT}\right] \sum_{n_2=-\infty}^{\infty} \exp\left[-\frac{n_2^2\hbar^2}{2I_2kT}\right] \sum_{n_3=\infty} \exp\left[-\frac{n_3^2\hbar^2}{2I_3kT}\right] \\ &= z^c_{1-\mathrm{rot}} z^c_{2-\mathrm{rot}} z^c_{3-\mathrm{rot}}. \end{aligned} \tag{15}$$

That is, the total molecular-partition function for classical rotation of the polyatomic molecule is the product of molecular-partition functions for each of the independent rotations. Each of the classical partition functions in Eq. 15 has the form of Eq. 12. Therefore, the result is

$$z^c_{\mathrm{rot}} = \left[\frac{8\pi^3I_1kT}{h^2}\right]^{1/2}\left[\frac{8\pi^3I_2kT}{h^2}\right]^{1/2}\left[\frac{8\pi^3I_3kT}{h^2}\right]^{1/2}. \tag{16}$$

If we now divide Eq. 16 by the symmetry number to account for indistinguishable orientations upon rotation, we obtain

$$\boxed{z^c_{\mathrm{rot}} = \frac{1}{\sigma h^3}[8\pi^3kT]^{3/2}[I_1I_2I_3]^{1/2} = \frac{\pi^{3/2}}{\sigma h^3}[8\pi^2kT]^{3/2}[I_1I_2I_3]^{1/2}}. \tag{17}$$

Equation 17 is almost identical to Eq. 18.91. The difference is that the factor of $\sqrt{\pi}$ in Eq. 18.91 appears as $\pi^{3/2}$ in Eq. 17. Although the results are close, the "derivation" used in this problem is not correct, because we have assumed an incorrect form for the quantum mechanical Hamiltonian for rotation. Nevertheless, the result does provide insight as to why Eq. 18.91 has the form it does.

18.19 The moments of inertia of NH_3 are $I_1 = I_2 = 2.816 \times 10^{-47}$ kg m^2 and $I_3 = 4.43 \times 10^{-47}$ kg m^2. Compute the contribution of the rotational motion of NH_3 to S, A, and G at 298 K.

Solution

Using Eq. 18.93, we have

$$S_{rot} = \frac{3R}{2} \ln T + R \ln\left[\frac{[I_1 I_2 I_3]^{1/2}}{\sigma}\right] + 158.9R. \tag{1}$$

The symmetry number for NH_3 is 3, since there are three orientations about the figure axis that produce indistinguishable configurations of the four atoms.

Substituting the data gives

$$S_{rot} = 1.5(8.314\text{ J mol}^{-1}\text{ K}^{-1}) \ln(298.15) + (8.314\text{ J mol}^{-1}\text{ K}^{-1}) \ln\left[\frac{[(2.816 \times 10^{-47})^2(4.43 \times 10^{-47})]^{1/2}}{3}\right] + 158.9(8.314\text{ J mol}^{-1}\text{ K}^{-1}), \tag{2}$$

or

$$S_{rot} = 8.314\text{ J mol}^{-1}\text{ K}^{-1}[8.546 - 161.65 + 158.9] = \underline{48.19\text{ J mol}^{-1}\text{ K}^{-1}}. \tag{3}$$

The Helmholtz free energy of rotation for NH_3 at 298.15 K is (see Eq. 18.94)

$$A_{rot} = -\left[\frac{3RT}{2} \ln T + RT \ln\left[\frac{[I_1 I_2 I_3]^{1/2}}{\sigma}\right] + 157.4RT\right] = -RT\left[\frac{3}{2} \ln T + \ln\left[\frac{[I_1 I_2 I_3]^{1/2}}{\sigma}\right] + 157.4\right]. \tag{4}$$

Substituting the data produces

$$A_{rot} = -(8.314\text{ J mol}^{-1}\text{ K}^{-1})(298.15\text{ K}) \times \left[\frac{3}{2} \ln(298.15) + \ln\left[\frac{[(2.816 \times 10^{-47})^2(4.43 \times 10^{-47})]^{1/2}}{3}\right] + 157.4\right] = -2{,}478.8\text{ J mol}^{-1}[8.546 - 161.65 + 157.4] = \underline{-1.065 \times 10^4\text{ J mol}^{-1}}. \tag{5}$$

Since we have already added the pV term for translation,

$$G_{rot} = A_{rot} = \underline{-1.065 \times 10^4\text{ J mol}^{-1}}. \tag{6}$$

18.21 Deuterium (^2H) has a nuclear spin quantum number $I = 1$.

(A) Does ortho–para deuterium exist? Explain.

(B) What is the high-temperature ratio of ortho to para D_2? Explain. (*Hint*: Is deuterium a fermion or a boson? How many spin states can each deuterium atom have? How many symmetric and antisymmetric spin functions can be constructed?)

Solution

(A) Yes, ortho–para D_2 exists. Since deuterium has integral spin, it is a boson. Hence, its total rotational wave function must be symmetric with respect to nuclear exchange. Therefore, if the nuclear spin wave function is symmetric, the spatial part of the wave function must also be symmetric. That is, $Y_J^M(\theta, \phi)$ must have only even values of J. If, on the other hand, the nuclear spin function is antisymmetric to exchange, $Y_J^M(\theta, \phi)$ must also be antisymmetric, which requires that J be odd. Therefore, two forms of D_2 exist. These forms are D_2 with only even values of J (para-D_2) and D_2 with only odd values of J (ortho-D_2).

(B) To determine the weighting of para-D_2 to ortho-D_2, we need to find the number of possible symmetric and antisymmetric nuclear spin wave functions. The nuclear spin magnetic quantum number, m_I, must take values in the range $-I \leq m_I \leq +I$, and it must go by integer steps from 2I to 1I. Since we have I 5 1, m_I can have one of three values: 21, 0, or 11. Consequently, there are now three possible spin states for each deuterium nucleus. Let us label these b, g, and a for $m_I = -1$, 0, and 11, respectively. If we label the two deuterium nuclei as A and B, we can form six possible symmetric spin functions: $\beta_A\beta_B$, $\gamma_A\gamma_B$, $\alpha_A\alpha_B$, $\beta_A\gamma_B + \gamma_A\beta_B$, $\beta_A\alpha_B + \alpha_A\beta_B$, and $\gamma_A\alpha_B + \alpha_A\gamma_B$. In contrast, there are only three possible antisymmetric nuclear spin functions: $\beta_A\gamma_B - \gamma_A\beta_B$, $\beta_A\alpha_B - \alpha_A\beta_B$, and $\gamma_A\alpha_B - \alpha_A\gamma_B$. The six symmetric nuclear spin functions must have symmetric spatial wave functions to give a total wave function that is symmetric to nuclear exchange. This leads to even values of J and para-D_2. The three antisymmetric nuclear spin functions require antisymmetric spatial parts and odd values of J, so that we have ortho-D_2. As a result, the high-temperature ratio of para-D_2 to ortho-D_2 must reflect the two-to-one ratio of symmetric to antisymmetric spin functions. Therefore the high-temperature para-D_2 to ortho-D_2 ratio must be two to one.

18.23 Figure 18.9 indicates that the rotational contribution to the constant-volume heat capacity of ortho-H_2 at 175 K is 3.173 J mol^{-1} K^{-1}, while that for para-H_2 is 12.178 J mol^{-1} K^{-1} at this temperature. Use an appropriately modified form of Eq. 18.83 to verify the correctness of these two results. The rotational constant for H_2 can be found in Table 15.4.

Solution

The rotational contribution to C_v is given by Eq. 18.83, modified so that the summations run only over even J states for para-H_2 and only over odd J states for ortho-H_2. Thus, $[C_v]_{\text{rot}}$ is

$$[C_v]_{\text{rot}}^{\text{para}} = -\frac{U_{\text{rot}}^2}{RT^2} + \frac{R}{z_{\text{rot}}}\left(\frac{Bh}{kT}\right)^2 \sum_{\text{even } J}^{\infty} (2J+1)J^2(J+1)^2 \exp\left[-\frac{J(J+1)Bh}{kT}\right] \quad \textbf{(1)}$$

and

$$[C_v]_{\text{rot}}^{\text{ortho}} = -\frac{U_{\text{rot}}^2}{RT^2} + \frac{R}{z_{\text{rot}}}\left(\frac{Bh}{kT}\right)^2 \sum_{\text{odd } J}^{\infty} (2J+1)J^2(J+1)^2 \exp\left[-\frac{J(J+1)Bh}{kT}\right]. \quad \textbf{(2)}$$

U_{rot} is given by Eq. 18.82, again modified to include only even or odd J states:

$$U_{\text{rot}}^{\text{para}} = \frac{R}{z_{\text{rot}}} \sum_{\text{even } J}^{\infty} (2J+1)J(J+1)\frac{Bh}{k} \exp\left[-\frac{J(J+1)Bh}{kT}\right] \quad \textbf{(3)}$$

and

$$U_{\text{rot}}^{\text{ortho}} = \frac{R}{z_{\text{rot}}} \sum_{\text{odd } J}^{\infty} (2J+1)J(J+1)\frac{Bh}{k} \exp\left[-\frac{J(J+1)Bh}{kT}\right]. \quad \textbf{(4)}$$

The molecular-rotational-partition functions for para and ortho hydrogen are

$$z_{\text{rot}}^{\text{para}} = \sum_{\text{even } J}^{\infty} (2J+1) \exp\left[-\frac{J(J+1)Bh}{kT}\right] \quad \textbf{(5)}$$

and

$$z_{\text{rot}}^{\text{ortho}} = \sum_{\text{odd } J}^{\infty} (2J+1) \exp\left[-\frac{J(J+1)Bh}{kT}\right], \quad \textbf{(6)}$$

respectively. The quantities required inside the summations are

$$\frac{Bh}{k} = \frac{(60.809\ \text{cm}^{-1})(2.9979 \times 10^{10}\ \text{cm s}^{-1})(6.62608 \times 10^{-34}\ \text{J s})}{1.38066 \times 10^{-23}\ \text{J K}^{-1}} = 87.489\ \text{K} \quad (7)$$

and

$$\frac{Bh}{kT} = \frac{87.489\ \text{K}}{175\ \text{K}} = 0.49994. \quad (8)$$

We could use an Excel spreadsheet or simple FORTRAN code to execute the summations, but in this case, convergence is so rapid that we can do the computation by brute force:

$$z_{\text{rot}}^{\text{ortho}} = \sum_{\text{odd } J}^{\infty} (2J+1)\exp[-0.49994J(J+1)] = 3\exp(-0.99988)$$

$$+ 7\exp(-5.99928) + 11\exp(-14.9982) + \cdots$$

$$= 1.1038 + 0.01736 + 0.000003371 + \cdots = 1.1212. \quad (9)$$

For para-H_2, the result is

$$z_{\text{rot}}^{\text{para}} = \sum_{\text{even } J}^{\infty} (2J+1)\exp[-0.49994J(J+1)] = 1 + 5\exp(-2.99964)$$

$$+ 9\exp(-9.9988) = 1 + 0.24902 + 0.00041 = 1.2494. \quad (10)$$

The rotational energy for ortho-H_2 is

$$U_{\text{rot}}^{\text{ortho}} = \frac{(8.314\ \text{J mol}^{-1}\ \text{K}^{-1})(87.489\ \text{K})}{1.1212} \sum_{\text{odd } J}^{\infty} (2J+1)J(J+1)\exp[-0.49994J(J+1)]$$

$$= (648.75\ \text{J mol}^{-1})[6\exp(-0.99988)$$

$$+ 84\exp(-5.99928) + 330\exp(-14.9982) = 1{,}567.4\ \text{J mol}^{-1}. \quad (11)$$

The para-H_2 rotational energy is

$$U_{\text{rot}}^{\text{para}} = \frac{(8.314\ \text{J mol}^{-1}\ \text{K}^{-1})(87.489\ \text{K})}{1.2494} \sum_{\text{even } J}^{\infty} (2J+1)J(J+1)\exp[-0.49994J(J+1)]$$

$$= (582.19\ \text{J mol}^{-1})[0 + 30\exp(-2.99964) + 180\exp(-9.9988) + 546\exp(-20.9975)]$$

$$+ \cdots = 582.19\ \text{J mol}^{-1})(1.4941 + 0.0081818 + 0.000000415)]$$

$$= 874.6\ \text{J mol}^{-1}. \quad (12)$$

Using Eq. 2, we can now compute

$$[C_v]_{\text{rot}}^{\text{ortho}} = -\frac{U_{\text{rot}}^2}{RT^2} + \frac{R}{z_{\text{rot}}}\left(\frac{Bh}{kT}\right)^2 \sum_{\text{odd } J}^{\infty} (2J+1)J^2(J+1)^2 \exp\left[-\frac{J(J+1)Bh}{kT}\right]$$

$$= -\frac{(1{,}567.4)^2}{(8.314)(175)^2} + \frac{(8.314)(0.49994)^2}{1.1212} \sum_{\text{odd } J}^{\infty} (2J+1)J^2$$

$$(J+1)^2\exp[-0.49994J(J+1)] = -9.6488 + 1.8534$$

$$[12\exp(-0.99988) + 1{,}008\exp(-5.99928) + 9{,}900\exp(-14.9982)]$$

$$+ 47{,}040\exp(-27.9966) = -9.6488 + 1.8534$$

$$\times [4.4151 + 2.5004 + 0.003034 + 0.00000003] = \underline{3.174\ \text{J mol}^{-1}\ \text{K}^{-1}},$$

(13)

which is essentially the result given in the problem, except for a small round-off error in the fourth significant digit. For para-H_2, the result is

$$[C_v]_{rot}^{para} = -\frac{U_{rot}^2}{RT^2} + \frac{R}{z_{rot}}\left(\frac{Bh}{kT}\right)^2 \sum_{\text{even } J}^{\infty} (2J+1)J^2(J+1)^2 \exp\left[-\frac{J(J+1)Bh}{kT}\right]$$

$$= -\frac{(874.6)^2}{(8.314)(175)^2} + \frac{(8.314)(0.49994)^2}{1.2494} \sum_{\text{even } J}^{\infty} (2J+1)J^2$$

$$(J+1)^2 \exp[-0.49994J(J+1)]$$

$$= -3.0042 + 1.6632\,[0 + 180\exp(-2.99964) + 3{,}600\exp(-9.9988)$$

$$+ 22{,}932\exp(-20.9975)] = -3.0042 + 1.6632(8.96490 + 0.16364$$

$$+ 0.0000174) = \underline{12.178 \text{ J mol}^{-1}\text{ K}^{-1}}. \quad (14)$$

This result is in accord with the data given in Figure 18.9.

18.25 Starting with the fact that $U = RT^2(\partial \ln z/\partial T)_V$, show that ortho-$H_2$ has a rotational zero-point energy equal to $2NBh$, where N is Avogadro's constant and B is the H_2 rotational constant. Evaluate the expected rotational zero-point energy for ortho-H_2. Is the result consistent with Figure 18.8? (*Hint*: Write down the appropriate form for z_{ortho}^{rot}. Expand the summation by writing down the first few terms, factor out anything you can, and then take the natural logarithm of the result. Finally, take the derivative of the natural logarithm of the result with respect to temperature at constant volume, and then examine the limit of the result as $T \to 0$ K.)

Solution

The rotational-partition function for ortho-H_2 is given by Eq. 18.96:

$$z_{rot}^{ortho} = \frac{3}{4}\sum_{\text{odd } J}^{\infty} (2J+1)\exp\left[-\frac{J(J+1)Bh}{kT}\right]. \quad (1)$$

Following the hints given in the problem, let us write out the summation in expanded form:

$$z_{rot}^{ortho} = \frac{3}{4}\left\{3\exp\left[-\frac{2Bh}{kT}\right] + 7\exp\left[-\frac{12Bh}{kT}\right]\right.$$

$$\left. + 11\exp\left[-\frac{30Bh}{kT}\right] + 15\exp\left[-\frac{56Bh}{kT}\right] + \cdots\right\}. \quad (2)$$

We can factor out $\exp[-2Bh/(kT)]$ from this sum and write the result as

$$z_{rot}^{ortho} = \frac{3}{4}\exp\left[-\frac{2Bh}{kT}\right]\left\{3 + 7\exp\left[-\frac{10Bh}{kT}\right]\right.$$

$$\left. + 11\exp\left[-\frac{28Bh}{kT}\right] + 15\exp\left[-\frac{54Bh}{kT}\right] + \cdots\right\}. \quad (3)$$

Let us represent the summation in Eq. 3 by X. That is,

$$X = \left\{3 + 7\exp\left[-\frac{10Bh}{kT}\right] + 11\exp\left[-\frac{28Bh}{kT}\right] + 15\exp\left[-\frac{54Bh}{kT}\right] + \cdots\right\}. \quad (4)$$

In this notation, we have

$$z_{rot}^{ortho} = \frac{3}{4}\exp\left[-\frac{2Bh}{kT}\right]X. \quad (5)$$

The natural logarithm of this expression is

$$\ln z_{\text{rot}}^{\text{ortho}} = \ln(0.75) + \ln X - \frac{2Bh}{kT}. \tag{6}$$

The derivative of $\ln z_{\text{rot}}^{\text{ortho}}$ with respect to temperature at constant V is

$$\left(\frac{\partial \ln z_{\text{rot}}^{\text{ortho}}}{\partial T}\right)_V = \frac{1}{X}\left(\frac{\partial X}{\partial T}\right)_V + \frac{2Bh}{kT^2}. \tag{7}$$

Using Eq. 4, we have

$$\left(\frac{\partial X}{\partial T}\right)_V = \left\{\frac{70Bh}{kT^2}\exp\left[-\frac{10Bh}{kT}\right] + \frac{308Bh}{kT^2}\exp\left[-\frac{28Bh}{kT}\right] + \frac{810Bh}{kT^2}\exp\left[-\frac{54Bh}{kT}\right] + \cdots\right\}. \tag{8}$$

Multiplying Eq. 7 by RT^2 and then replacing R/k with $N = 6.02214 \times 10^{23}\ \text{mol}^{-1}$ gives

$$U_{\text{rot}}^{\text{ortho}} = RT^2\left(\frac{\partial \ln z_{\text{rot}}^{\text{ortho}}}{\partial T}\right)_V$$

$$= RT^2 \frac{\left\{70Bh\exp\left[-\frac{10Bh}{kT}\right] + 308Bh\exp\left[-\frac{28Bh}{kT}\right] + 810Bh\exp\left[-\frac{54Bh}{kT}\right] + \cdots\right\}}{X} + 2NBh. \tag{9}$$

We now take the limit of $U_{\text{rot}}^{\text{ortho}}$ as $T \longrightarrow 0$ K. When this is done, all the exponential functions become $\exp(-\infty) = 0$. Therefore, these terms approach zero. Equation 4 shows that X will then approach 3. Consequently, we obtain

$$\lim_{T\longrightarrow 0\,\text{K}} U_{\text{rot}}^{\text{ortho}} = \frac{RT^2(0)}{3} + 2NBh = 2NBh. \tag{10}$$

Thus, the internal energy of ortho-H_2 is $2NBh$ at $T = 0$ K. It is, therefore, appropriate to call this quantity the rotational zero-point energy of ortho-H_2.

For ortho-H_2, $B = 60.809\ \text{cm}^{-1}(2.9979 \times 10^{10}\ \text{cm s}^{-1}) = 1.8230 \times 10^{12}\ \text{s}^{-1}$. The zero-point rotational energy of ortho-H_2 is

$$2NBh = 2(6.02214 \times 10^{23}\ \text{mol}^{-1})(1.8230 \times 10^{12}\ \text{s}^{-1})(6.62608 \times 10^{-34}\ \text{J s})$$

$$= \underline{1{,}454.9\ \text{J mol}^{-1}}. \tag{11}$$

An examination of Figure 18.9 shows that the result obtained in Eq. 11 is exactly the intercept of $U_{\text{rot}}^{\text{ortho}}$ with energy axis at $T = 0$ K.

18.27 Consider the series $y = 1 + 2 + 2^2 + 2^3 + 2^4 + 2^5 + \ldots$. An investigator attempts to sum this series as follows:

$$y = 1 + 2 + 2^2 + 2^3 + 2^4 + 2^5 + \cdots$$

$$= 1 + 2[1 + 2 + 2^2 + 2^3 + 2^4 + 2^5 + \cdots]$$

$$= 1 + 2y.$$

Solving for y, he obtains

$$y - 2y = 1 = -y,$$

so that the sum is -1. The investigator has the nagging feeling that something is wrong. Can you help him out?

Solution

The problem is that the summation in this case is not convergent. Clearly, we have

$$y = 1 + 2 + 2^2 + 2^3 + 2^4 + 2^5 + \cdots = \infty. \tag{1}$$

Therefore, the equation

$$y = 1 + 2y \tag{2}$$

tells us that

$$\infty = 1 + 2(\infty), \tag{3}$$

which simply says that an infinite quantity is equal to an infinite quantity. With infinities present in the equation, the simple rules of algebra do not hold, and the solution yielding $y = -1$ is invalid. For this type of summation procedure to be valid, the series must be convergent. Therefore, if we have

$$y = 1 + \left(\frac{1}{2}\right) + \left(\frac{1}{2}\right)^2 + \left(\frac{1}{2}\right)^3 + \left(\frac{1}{2}\right)^4 + \left(\frac{1}{2}\right)^5 + \ldots, \tag{4}$$

we can write

$$y = 1 + \frac{1}{2}\left[1 + \left(\frac{1}{2}\right) + \left(\frac{1}{2}\right)^2 + \left(\frac{1}{2}\right)^3 + \left(\frac{1}{2}\right)^4 + \left(\frac{1}{2}\right)^5 + \cdots\right] = 1 + \frac{y}{2}, \tag{5}$$

so that

$$y - \frac{y}{2} = \frac{y}{2} = 1, \tag{6}$$

and therefore,

$$y = 2, \tag{7}$$

which is the correct sum for the series. The difference is that this sum is convergent, whereas the first one is not. Likewise, the sum we executed to obtain z^h_{vib} is convergent.

18.29* (A) Compute the rotational contribution to the entropy of para-H_2 and ortho-H_2 at 298.15 K.

(B) Assuming that the rotation can be treated classically at 298.15 K, use a symmetry number in lieu of explicit consideration of ortho–para forms to compute the rotational contribution to the entropy of 1 mole of H_2 at that temperature. Refer to Table 15.4 for any data you may need.

(C) Use the results of (A) to compute the expected ratio of ortho to para hydrogen at 298.15 K.

(D) Calculate the total entropy of 1 mole of H_2 that has the ratio of ortho to para forms computed in (C). Compare your result with that obtained in (B). How much error is made by using a classical approximation with a symmetry number instead of actually considering the two forms of H_2?

Solution

(A) The rotational-molecular-partition functions for para- and ortho-H_2 are

$$z_{\text{rot}}^{\text{para}} = \sum_{\text{even } J}^{\infty} (2J + 1) \exp\left[-\frac{J(J + 1)Bh}{kT}\right] \tag{1}$$

and

$$z_{\text{rot}}^{\text{ortho}} = \sum_{\text{odd } J}^{\infty} (2J + 1) \exp\left[-\frac{J(J + 1)Bh}{kT} \right]. \tag{2}$$

The corresponding internal energies are

$$U_{\text{rot}}^{\text{para}} = \frac{R}{z_{\text{rot}}} \sum_{\text{even } J}^{\infty} (2J + 1)J(J + 1)\frac{Bh}{k} \exp\left[-\frac{J(J + 1)Bh}{kT} \right] \tag{3}$$

and

$$U_{\text{rot}}^{\text{ortho}} = \frac{R}{z_{\text{rot}}} \sum_{\text{odd } J}^{\infty} (2J + 1)J(J + 1)\frac{Bh}{k} \exp\left[-\frac{J(J + 1)Bh}{kT} \right]. \tag{4}$$

Using the rotational constant for H_2 given in Table 15.4, we have

$$\frac{Bh}{k} = \frac{(60.809)(2.9979 \times 10^{10})\ \text{s}^{-1}(6.62608 \times 10^{-34}\ \text{J s})}{1.38066 \times 10^{-23}\ \text{J K}^{-1}} = 87.489\ \text{K}. \tag{5}$$

We now need to execute the four summations contained in Eqs. 1–4, using a FORTRAN or C program or a spreadsheet. The results of this operation for $T = 298.15$ K are

$$z_{\text{rot}}^{\text{para}} = 1.8851, \quad z_{\text{rot}}^{\text{ortho}} = 1.8767, \quad U_{\text{rot}}^{\text{para}} = 2{,}187.4\ \text{J mol}^{-1}, \tag{6}$$

and

$$U_{\text{rot}}^{\text{ortho}} = 2{,}274.7\ \text{J mol}^{-1}.$$

The rotational contribution to the entropy is given by

$$S_{\text{rot}} = R \ln z_{\text{rot}} + \frac{U_{\text{rot}}}{T}. \tag{7}$$

Using Eq. 7 and the data given in Eq. 6, we obtain

$$S_{\text{rot}}^{\text{para}} = R \ln(1.88511) + \frac{2{,}187.4}{298.15}\ \text{J mol}^{-1}\ \text{K}^{-1} = \underline{12.608\ \text{J mol}^{-1}\ \text{K}^{-1}} \tag{8}$$

and

$$S_{\text{rot}}^{\text{ortho}} = R \ln(1.8767) + \frac{2{,}274.7}{298.15}\ \text{J mol}^{-1}\ \text{K}^{-1} = \underline{12.863\ \text{J mol}^{-1}\ \text{K}^{-1}}. \tag{9}$$

(B) If H_2 rotation is classical, the energy levels are continuous, and it makes no difference whether we sum over even or odd J. That is, the rotational entropy of ortho and para hydrogen will be the same. The classical rotational entropy is given by

$$S_{\text{rot}} = S_{\text{rot}}^{c} = R \ln\left[\frac{kT}{\sigma Bh} \right] + R. \tag{10}$$

The symmetry number for hydrogen is 2; therefore,

$$S_{\text{rot}}^{c} = R \ln\left[\frac{298.15}{(2)(87.489)} \right] + \text{R} = \underline{12.745\ \text{J mol}^{-1}\ \text{K}^{-1}}. \tag{11}$$

(C) The ratio of ortho to para hydrogen is given by Eq. 18.97:

$$\frac{[\text{ortho-H}_2]}{[\text{para-H}_2]} = 3 \exp\left[\frac{(U_{\text{rot}}^{\text{para}} - U_{\text{rot}}^{\text{ortho}})}{RT} \right]. \tag{12}$$

Using the results of Part (A), Eq. 6 and Eq. 12 gives

$$\frac{[\text{ortho-H}_2]}{[\text{para-H}_2]} = 3\exp\left[\frac{(2{,}187.4 - 2{,}274.7)}{RT}\right] = \underline{2.8962.} \quad \textbf{(13)}$$

(D) If we have n_p moles of para-H_2 and n_o moles of ortho-H_2, then with 1 mole of total hydrogen,

$$n_p + n_o = 1. \quad \textbf{(14)}$$

From Part (C), we also know that

$$\frac{n_o}{n_p} = 2.8962. \quad \textbf{(15)}$$

Solving for n_o and n_p, we obtain

$$n_p + n_o = n_p + 2.8962n_p = 3.8962n_p = 1, \quad \textbf{(16)}$$

so

$$n_p = 0.25667 \quad \text{and } n_o = 0.74334. \quad \textbf{(17)}$$

Therefore, the total entropy of n_p moles of para-H_2 plus n_o moles of ortho-H_2 can be obtained by using the results of Part (A). We have

$$S_{\text{total}} = n_p S_{\text{rot}}^{\text{para}} + n_o S_{\text{rot}}^{\text{ortho}} = 0.25667(12.608) + 0.74334(12.863)\ \text{J mol}^{-1}\,\text{K}^{-1}$$

$$= \underline{12.798\ \text{J mol}^{-1}\,\text{K}^{-1}}. \quad \textbf{(18)}$$

The percent error in the classical approximation is therefore

$$\%\ \text{error} = 100 \times \frac{12.745 - 12.798}{12.798} = \underline{-0.414\%}. \quad \textbf{(19)}$$

This is not bad, and it's a lot easier to use the classical approximation.

18.31* If the temperature is not too high, the vibrational energy will be only slightly greater than the zero-point energy. Under these conditions, we can regard the vibrational modes as being uncoupled. In this problem, you will examine how the total vibrational energy of a diatomic molecule whose vibrational mode is treated harmonically varies with the ratio of ν_o to T. For a diatomic molecule whose vibrational mode is assumed to be harmonic, compute the ratio of U_{vib} to the vibrational zero-point energy as a function of $x = h\nu_o/(kT)$ over the range $1.0 \le x \le 6$. Plot the ratio as a function of x. What is the value of the ratio for H_2, O_2, and $^{35}Cl^{35}Cl$ at 300 K?

Solution

The vibrational energy is given by Eq. 18.107:

$$U_{\text{vib}} = \frac{Nh\nu_o}{2} + \frac{Nh\nu_o}{e^x - 1} = \frac{RTx}{2} + \frac{RTx}{e^x - 1}. \quad \textbf{(1)}$$

The second expression is obtained directly from the first by multiplying and dividing by kT and noting that $NkT = RT$. The zero-point energy (ZPE) per mole is $RTx/2$. Therefore,

$$r = \frac{U_{\text{vib}}}{\text{ZPE}} = 1 + \frac{2}{e^x - 1}. \quad \textbf{(2)}$$

As $x \longrightarrow \infty$, $r \longrightarrow 1$, and as $x \longrightarrow 0$, $r \longrightarrow \infty$. The plot at the end of the problem shows this function over the range $1.0 \le x \le 6.0$.

The values of x for H_2, O_2, and Cl_2 at 300 K are, respectively,

$$x(H_2) = \frac{h\nu_o}{kT} = \frac{(6.62608 \times 10^{-34}\ \text{J s})(4{,}395.2\ \text{cm}^{-1})(2.9979 \times 10^{10}\ \text{cm s}^{-1})}{(1.38066 \times 10^{-23}\ \text{J K}^{-1})(300\ \text{K})}$$

$$= 21.08, \tag{3}$$

$$x(O_2) = \frac{(6.62608 \times 10^{-34}\ \text{J s})(1{,}580.36\ \text{cm}^{-1})(2.9979 \times 10^{10}\ \text{cm s}^{-1})}{(1.38066 \times 10^{-23}\ \text{J K}^{-1})(300\ \text{K})}$$

$$= 7.579, \tag{4}$$

and

$$x(Cl_2) = \frac{(6.62608 \times 10^{-34}\ \text{J s})(564.9\ \text{cm}^{-1})(2.9979 \times 10^{10}\ \text{cm s}^{-1})}{(1.38066 \times 10^{-23}\ \text{J K}^{-1})(300\ \text{K})}$$

$$= 2.709. \tag{5}$$

Thus, even for very low vibrational frequencies, U_{vib} is only about 20% greater than the zero-point energy at 300 K. For this reason, the assumption of separable vibrational modes is usually reasonably accurate.

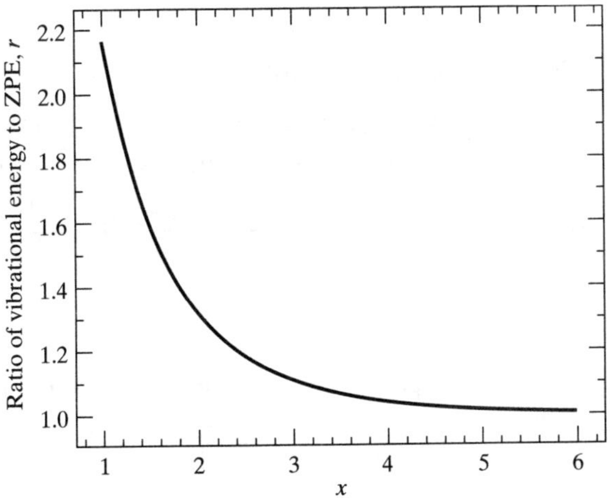

18.33 (A) Obtain an expression for K_p for the equilibrium reaction $H_2(g) + D_2(g) = 2\ HD(g)$ in terms of fundamental frequencies, moments of inertia, masses, and temperature.

(B) The atomic mass of deuterium is 2.0141 amu. The vibrational frequencies of H_2, D_2, and HD, expressed in wave numbers, are 4,395.2 cm^{-1}, 3,118.4 cm^{-1}, and 3,817.09 cm^{-1}, respectively. Obtain K_p for the reaction in (A) as a function of temperature alone.

(C) Plot K_p versus T and determine the temperature at which $K_p = 1$.

Solution

(A) K_p is given by Eq. 18.124:

$$K_p = \frac{(z_{HD}')^2}{(z_{H2}')(z_{D2}')}. \tag{1}$$

The ratio of translational-molecular-partition functions is

$$\frac{(z_{tr}')^2_{HD}}{(z_{tr}')_{H2}(z_{tr}')_{D2}} = \frac{m^3_{HD}}{(m_{H2}m_{D2})^{3/2}}, \tag{2}$$

because the remaining constants cancel.

If we assume that we may treat rotation classically, the ratio of the rotational-molecular-partition functions is

$$\frac{(z_{rot})^2_{HD}}{(z_{rot})_{H2}(z_{rot})_{D2}} = \frac{(\sigma B)_{H2}(\sigma B)_{D2}}{(\sigma B)^2_{HD}}, \tag{3}$$

since $z_{rot} = kT/\sigma Bh$.

The symmetry number for HD is unity. For H_2 and D_2, it is two. The rotational constant is $B = h/(8\pi^2 \mu R_e^2)$. Since the equilibrium bond lengths of H_2, D_2, and HD are the same, Eq. 3 becomes

$$\frac{(z_{rot})^2_{HD}}{(z_{rot})_{H2}(z_{rot})_{D2}} = \frac{4\mu^2_{HD}}{\mu_{H2}\mu_{D2}}. \tag{4}$$

The ratio of the reduced masses is

$$\frac{\mu^2_{HD}}{\mu_{H2}\mu_{D2}} = \frac{\dfrac{(m_H m_D)^2}{(m_H + m_{HD})^2}}{(m_H/2)(m_D/2)} = \frac{4m_H m_D}{(m_H + m_D)^2}. \tag{5}$$

Thus,

$$\frac{(z_{rot})^2_{HD}}{(z_{rot})_{H2}(z_{rot})_{D2}} = \frac{16 m_H m_D}{(m_H + m_D)^2}. \tag{6}$$

The ratio of vibrational-molecular-partition functions is

$$\frac{(z_{vib})^2_{HD}}{(z_{vib})_{H2}(z_{vib})_{D2}} = \left[\frac{\exp(-x_{HD}/2)}{1-\exp(-x_{HD})}\right]^2 \left[\frac{\exp(-x_{H2}/2)}{1-\exp(-x_{H2})}\right]^{-1} \left[\frac{\exp(-x_{D2}/2)}{1-\exp(-x_{D2})}\right]^{-1}$$

$$= \exp\left[-\left(x_{HD} - \frac{x_{H2}}{2} - \frac{x_{D2}}{2}\right)\right]\left[\frac{[1-\exp(-x_{H2})][1-\exp(-x_{D2})]}{[1-\exp(-x_{HD})]^2}\right]$$

$$= \exp\left[-\frac{h(2\nu_{oHD} - \nu_{oH2} - \nu_{oD2})}{2kT}\right]\left[\frac{[1-\exp(-x_{H2})][1-\exp(-x_{D2})]}{[1-\exp(-x_{HD})]^2}\right]. \tag{7}$$

The electronic-molecular-partition function for each molecule is

$$z_{el} = g_o \exp\left[-\frac{\varepsilon_o}{kT}\right]. \tag{8}$$

The ground-state electronic energy for H_2, D_2, and HD are all the same, and $g_o = 1$ in each case. Therefore, the ratio of electronic partition functions is unity. The overall result is

$$K_p = \frac{m^3_{HD}}{(m_{H2}m_{D2})^{3/2}} \frac{16 m_H m_D}{(m_H + m_D)^2} \times \exp\left[-\frac{h(2\nu_{oHD} - \nu_{oH2} - \nu_{oD2})}{2kT}\right] \left[\frac{[1-\exp(-x_{H2})][1-\exp(-x_{D2})]}{[1-\exp(-x_{HD})]^2}\right]. \tag{9}$$

The product of mass factors simplifies to $2m_{HD}/(m_H m_D)^{1/2}$. If we write the difference $2\nu_{oHD} - \nu_{oH2} - \nu_{oD2}$ as $\Delta\nu_o$, then Eq. 9 has the form

$$\boxed{K_p = \frac{2m_{HD}}{(m_H m_D)^{1/2}}\left[\frac{[1-\exp(-x_{H2})][1-\exp(-x_{D2})]}{[1-\exp(-x_{HD})]^2}\right]\exp\left[-\frac{h(\Delta\nu_o)}{2kT}\right]}. \tag{10}$$

Equation 10 is the requested result.

(B) The individual terms are

$$\frac{2m_{HD}}{(m_H m_D)^{1/2}} = \frac{2(1.0078 + 2.0141)}{[(1.0078)(2.0141)]^{1/2}} = 4.2421. \tag{11}$$

All of the factors in the second term are essentially unity. The last term is

$$\Delta\nu_o = [2(3817.09) - 4{,}395.2 - 3{,}118.4](2.9979 \times 10^{10}\ \text{s}^{-1}) = 3.6149 \times 10^{12}\ \text{s}^{-1}. \quad \textbf{(12)}$$

Hence,

$$\exp\left[-\frac{h(\Delta\nu_o)}{2kT}\right] = \exp\left[-\frac{(6.62608 \times 10^{-34}\ \text{J s})(3.6149 \times 10^{12}\ \text{s}^{-1})}{2(1.38066 \times 10^{-23}\ \text{J K}^{-1})T}\right]$$

$$= \exp\left[-\frac{86.743}{T}\right]. \quad \textbf{(13)}$$

Therefore, we obtain

$$\boxed{K_p = 4.2421 \exp\left[-\frac{86.743}{T}\right]}. \quad \textbf{(14)}$$

(C) A plot of K_p follows. The temperature at which $K_p = 1$ is given by

$$-\frac{86.743}{T} = \ln\left[\frac{1}{4.2421}\right] = -1.4451, \quad \textbf{(15)}$$

so that

$$T = \underline{60.03\ \text{K}}. \quad \textbf{(16)}$$

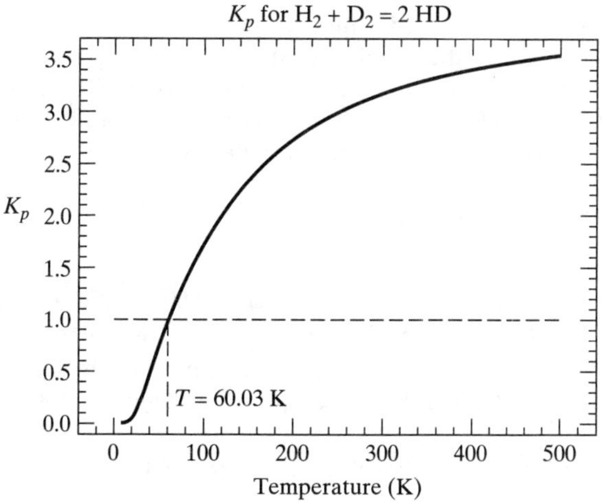

18.35 An investigator has 1 mole of $H_2(g)$ molecules at 300 K inside a rectangular parallelepiped, the lengths of whose sides are all 0.2000 m. If the molecules do not interact, the ground-state translational energy is given by $E_{gs} = 3h^2/(8ma^2)$, where m is the mass of H_2 and a is the length of the sides of the cubical container.

(A) Use the Boltzmann distribution to compute the number of $H_2(g)$ molecules in the translational ground state.

(B) Is the calculation meaningful? Discuss this point.

Solution

(A) The Boltzmann distribution tells us that the number of molecules in quantum state i is

$$n_i = \frac{N}{z}\exp\left[-\frac{\varepsilon_i}{kT}\right]. \quad \textbf{(1)}$$

For this system, the mass of H_2 is

$$m = \frac{(2.0156\ \text{g mol}^{-1})}{(1{,}000\ \text{g kg}^{-1})(6.02214 \times 10^{23}\ \text{mol}^{-1})} = 3.3470 \times 10^{-27}\ \text{kg}. \quad (2)$$

The ground-state translational energy is, therefore,

$$E_{gs} = \frac{3(6.62608 \times 10^{-34}\ \text{J s})^2}{8(3.3470 \times 10^{-27}\ \text{kg})(0.0200\ \text{m}^2)} = 1.230 \times 10^{-39}\ \text{J}. \quad (3)$$

The translational-partition function is given by Eq. 18.4:

$$z = \frac{V}{h^3}[2\pi mkT]^{3/2} = \frac{(0.200\ \text{m})^3}{(6.62609 \times 10^{-34}\ \text{J s})^3}$$

$$\times [(2)(3.14159)(3.3470 \times 10^{-27}\ \text{kg})(1.38066 \times 10^{-23}\ \text{J K}^{-1})(300\ \text{K})]^{3/2}$$

$$= 2.2356 \times 10^{28}. \quad (4)$$

Therefore, the number of molecules in the ground translational state is

$$n_o = \frac{6.02214 \times 10^{23}}{2.2356 \times 10^{28}} \exp\left[-\frac{1.230 \times 10^{-39}\ \text{J}}{(1.38066 \times 10^{-23}\ \text{J K}^{-1})(300\ \text{K})}\right]$$

$$= \underline{0.0000269\ \text{molecule}}. \quad (5)$$

(B) No, the calculation is not meaningful. The Boltzmann distribution law does not hold when the number of molecules in a given quantum state is extremely small. Its derivation in Chapter 17 assumed that n_i was large, in that we replaced $\ln(n_i!)$ with Stirling's approximation. When n_i turns out to be less than one molecule per energy state, the Boltzmann distribution does not hold.

18.37 One mole of helium atoms at a temperature of 500 K is distributed into four containers at heights of zero, 2,000, 3,000, and 4,000 m above the earth's surface.

(A) If the gravitational energy is assumed to be given by $E_{\text{gravitation}} = mgh$ and the helium atoms do not interact, compute the number of helium atoms in each container.

(B) Compute the gravitational contribution to the entropy for the system described in (A).

Solution

(A) The mass of helium is

$$m = \frac{(4.0026\ \text{g mol}^{-1})}{(1{,}000\ \text{g kg}^{-1})(6.02214 \times 10^{23}\ \text{mol}^{-1})} = 6.6465 \times 10^{-27}\ \text{kg}. \quad (1)$$

The gravitation energies of the containers are

$$E_{\text{gravitation}}(h = 0) = 0.0\ \text{J}, \quad (2)$$

$$E_{\text{gravitation}}(h = 2{,}000\ \text{m}) = (6.6465 \times 10^{-27}\ \text{kg})(9.80665\ \text{m s}^{-2})(2{,}000\ \text{m})$$

$$= 1.3036 \times 10^{-22}\ \text{J}, \quad (3)$$

$$E_{\text{gravitation}}(h = 3{,}000\ \text{m}) = 1.9554 \times 10^{-22}\ \text{J}, \quad (4)$$

and

$$E_{\text{gravitation}}(h = 4{,}000\ \text{m}) = 2.6072 \times 10^{-22}\ \text{J}. \quad (5)$$

The gravitational-molecular-partition function is

$$z_g = \sum_{i=1}^{4} \exp\left[-\frac{\varepsilon_i}{kT}\right] = e^0 + \exp\left[-\frac{1.3036 \times 10^{-22}\,\text{J}}{(1.38066 \times 10^{-23}\,\text{J K}^{-1})(500\,\text{K})}\right]$$

$$+ \exp\left[-\frac{1.9554 \times 10^{-22}\,\text{J}}{(1.38066 \times 10^{-23}\,\text{J K}^{-1})(500\,\text{K})}\right]$$

$$+ \exp\left[-\frac{2.6072 \times 10^{-22}\,\text{J}}{(1.38066 \times 10^{-23}\,\text{J K}^{-1})(500\,\text{K})}\right]$$

$$= 1 + 0.9813 + 0.9721 + 0.9629 = 3.9163. \qquad \textbf{(6)}$$

The number of helium atoms in container i is given by the Boltzmann distribution

$$n_i = \frac{N}{z_g} \exp\left[-\frac{\varepsilon_i}{kT}\right]. \qquad \textbf{(7)}$$

Thus,

$$n_1 = \frac{6.02214 \times 10^{23}(1)}{3.9163} = \underline{1.5377 \times 10^{23}\text{ helium atoms}}, \qquad \textbf{(8)}$$

$$n_2 = \frac{6.02214 \times 10^{23}(0.9813)}{3.9163} = \underline{1.5090 \times 10^{23}\text{ helium atoms}}, \qquad \textbf{(9)}$$

$$n_3 = \frac{6.02214 \times 10^{23}(0.9721)}{3.9163} = \underline{1.4948 \times 10^{23}\text{ helium atoms}}, \qquad \textbf{(10)}$$

and

$$n_4 = \frac{6.02214 \times 10^{23}(0.9629)}{3.9163} = \underline{1.4807 \times 10^{23}\text{ helium atoms}}. \qquad \textbf{(11)}$$

(B) The gravitational contribution to the entropy of the system in Part (A) is given by Eq. 18.131, viz.,

$$S_{\text{grav}} = R \ln z + \frac{U_{\text{grav}}}{T}, \qquad \textbf{(12)}$$

since the number of He atoms is large relative to the number of gravitational energy states. The energy of the system is

$$U_{\text{grav}} = 1.5377 \times 10^{23}\,\text{mol}^{-1}(0.0\,\text{J}) + 1.5090 \times 10^{23}\,\text{mol}^{-1}(1.3036 \times 10^{-22}\,\text{J})$$

$$+ 1.4948 \times 10^{23}\,\text{mol}^{-1}(1.9554 \times 10^{-22}\,\text{J})$$

$$+ 1.4807 \times 10^{23}\,\text{mol}^{-1}(2.6072 \times 10^{-22}\,\text{J}) = 87.505\,\text{J mol}^{-1}. \qquad \textbf{(13)}$$

The entropy is, therefore,

$$S_{\text{grav}} = R \ln z + \frac{U_{\text{grav}}}{T} = R \ln(3.9163) + \frac{87.505\,\text{J mol}^{-1}}{500\,\text{K}}$$

$$= \underline{11.52\,\text{J mol}^{-1}\,\text{K}^{-1}}. \qquad \textbf{(14)}$$

18.39 "Sir, I have a question."

"What is it, Sam?"

"It concerns the electronic partition function for the hydrogen atom that you discussed."

"Yes, what about it, Sam?"

"Well, sir, I really don't know how to say this, but I don't think your result is correct."

"Why is that, Sam," I ask.

"Well, sir, if we take the ground state of the hydrogen atom as our zero for energy, as you did, the electronic partition function is given by

$$z_{el} = \sum_{n=1}^{\infty} n^2 \exp\left[-\frac{1.313 \times 10^6[1 - n^{-2}]}{RT}\right].$$

You then expanded this summation at 298 K and wrote

$$z_{el}(T = 298\text{ K}) = \sum_{n=1}^{\infty} n^2 \exp[-530.0(1 - n^{-2})]$$
$$= e^o + 4\exp[-397.5]$$
$$+ 9\exp[-471.1] + \cdots = 1.$$

"Yes. That's correct, Sam. So what's your point?"

"Well, sir, the exponential factor in the series approaches a limiting value for large n that is *not* zero. In fact, its limiting value at $n = \infty$ is $\exp(-530.0) = 6.667 \times 10^{-231}$. Therefore, the exponential factor approaches a constant, while the n^2 factor continues to increase without bound. Consequently, $z_{el}(T = 298.15\text{ K}) = \infty$, not unity. Not only is the first term not the only important one; it is, in fact, negligible compared to the term with $n = 10^{300}$. The electronic partition function appears to be divergent. I hate to say this, sir, but, I think we've been bamboozled."

(*Author's note*: To bamboozle—to deceive by underhand methods; to dupe, to hoodwink.)

What do you think of Sam's comment? Have *you* been bamboozled? If z_{el} is indeed infinite, what is the significance of such a result? If z_{el} is not infinite, what is wrong with Sam's analysis?

Solution

As usual, Sam's point is excellent. In fact, it raises an issue that arises with some frequency in various applications of statistical mechanics. The electronic-molecular-partition function for the hydrogen atom is indeed infinite if the upper limit on the summation is allowed to be infinite. However, if we are going to employ statistical models to represent experimental systems, we must be certain that our model conforms to the experimental system.

Suppose we have the hydrogen atoms in a rectangular parallelepiped container, the length of whose sides is 0.3 meter. Let us estimate the average distance of the hydrogen-atom electron from the nucleus. If we use a Slater orbital, the wave function for the nth eigenstate is

$$\psi(r, \theta, \phi) = Nr^{n-1}\exp\left[-\frac{r}{na_o}\right],$$

where a_o is the Bohr radius. The average radial distance of the electron from the nucleus is, therefore, about

$$\langle r \rangle = \frac{\langle\psi(r,\theta,\phi)|r|\psi(r,\theta,\phi)\rangle}{\langle\psi(r,\theta,\phi)|\psi(r,\theta,\phi)\rangle} \approx \frac{4\pi N^2 \int_0^{\infty} r^{2n+1}\exp\left[-\frac{2r}{na_o}\right]dr}{4\pi N^2 \int_0^{\infty} r^{2n}\exp\left[-\frac{2r}{na_o}\right]dr} = \left[\frac{2n^2 + n}{2}\right]a_o.$$

Now, if our model is to conform to the experimental situation, the electron must remain inside the container. Therefore, the maximum value n can have is that value for which

$$\left(\frac{n^2 + n}{2}\right)a_o = 0.3\text{ m}. \qquad (1)$$

This gives $n_{max} \approx 75{,}294$. If we use this value as our upper limit for the summation in the electronic-partition function, the result is

$$z_{el}(T = 298 \text{ K}) = \sum_{n=1}^{75{,}294} n^2 \exp[-530.0(1 - n^{-2})]. \quad \textbf{(2)}$$

The last term in this series expansion is $(75{,}294)^2 \exp(-530) \approx 3.78 \times 10^{-221}$, which we can safely say is zero, so that $z_{el}(T)$ is equal to the first term in the expansion.

The point of the foregoing analysis and the conclusion we reach from Sam's question is that in statistical mechanical applications, infinite upper limits sometimes must be replaced with finite limits to ensure that our statistical mechanical model conforms to the experimental situation.

Now let us address the second part of the question. Suppose we had 1 mole of hydrogen atoms isolated in a volume equal to that of the universe rather than in a rectangular parallelepiped container. In this case, the upper limit of the summation for z_{el} will be infinite, provided that we assume a universe without limits. The result will be, as Sam pointed out, an infinite value for z_{el}. In effect, this means that all the hydrogen atoms will spontaneously ionize into protons and electrons, since it is the terms with huge values of n that make $z_{el} = \infty$. The huge values of n are associated with huge distances of the electron from the proton. Therefore, the system is, in effect, $H^+ + e^-$, an ionized hydrogen atom. (Remember, thermodynamics doesn't say how fast this ionization will occur; it only says that the process will be spontaneous.) These protons and electrons will now have translational energy only, and the translational-molecular-partition function, which has the form

$$z_{tr} = \frac{V}{h^3}[2\pi mkT]^{3/2},$$

will also be infinite, since the volume of the system is, by assumption, infinite. Therefore, everything is internally consistent. Of course, 1 mole of hydrogen atoms in an isolated, infinite universe is not a system that can ever be realized. Therefore, this infinite result has no real significance.

So have you been bamboozled? Maybe a little bit, but only a little bit. Note that I did not write the upper limit on the summation of the electronic-molecular partition as being infinite. However, it is true that I glossed over this point. It's hard for professors to get away with anything when there are students like Sam in the class.

CHAPTER 19

Phenomenological Kinetics

19.1 (A) If we measure concentrations in mol L^{-1}, what are the units on the rate coefficient for a concerted bimolecular reaction $A + B \longrightarrow$ products?

(B) What are the units on $k(T)$ for the third-order reaction discussed in Example 19.1?

Solution

(A) The concerted bimolecular rate expression is

$$\mathcal{R} = -\frac{d[A]}{dt} = k(T)[A][B]. \tag{1}$$

When the concentrations are given in mol L^{-1}, the left side of Eq. 1 has the units mol L^{-1} s^{-1} or mol L^{-1} min^{-1}. Of course, any time unit might be used. Therefore, the right side of the equation must have these same units. The unit equivalence is

$$\text{mol L}^{-1}\,\text{s}^{-1} = k(T)\,\text{mol}^2\,\text{L}^{-2}. \tag{2}$$

Thus, $k(T)$ must have the units mol L^{-1} s^{-1}/(mol^2 L^{-2}) = $\boxed{\text{L mol}^{-1}\,\text{s}^{-1}}$.

In general, the time unit on the left must be same as the time unit on $k(T)$.

(B) The termolecular reaction in Example 19.1 has the rate expression

$$\mathcal{R} = -\frac{1}{2}\frac{d[A]}{dt} = k(T)[A]^2[B]. \tag{3}$$

Now the unit equivalence equation is

$$\text{mol L}^{-1}\,\text{s}^{-1} = k(T)\,\text{mol}^3\,\text{L}^{-3}, \tag{4}$$

and $k(T)$ has the units mol L^{-1} s^{-1}/(mol^3 L^{-3}) = $\boxed{\text{L}^2\,\text{mol}^{-2}\,\text{s}^{-1}}$.

Again, the time unit on $k(T)$ must match the time unit on the rate.

19.3 (A) Show that Eq. 19.11 is correct.

(B) Starting with Eq. 19.12, derive Eqs. 19.13 and 19.14.

Solution

(A) Equation 19.11 is

$$\frac{1}{\{[A]_o - x\}\{[B]_o - x\}} = \frac{1}{[A]_o - [B]_o}\left[\frac{1}{[B]_o - x} - \frac{1}{[A]_o - x}\right]. \tag{1}$$

We can verify the accuracy of this equation by taking a common denominator on the right-hand side. Doing so gives

$$\frac{1}{[A]_o - [B]_o}\left[\frac{1}{[B]_o - x} - \frac{1}{[A]_o - x}\right] = \frac{1}{[A]_o - [B]_o}\left[\frac{[A]_o - x - ([B]_o - x)}{\{[B]_o - x\}\{[A]_o - x\}}\right]$$

$$= \frac{1}{[A]_o - [B]_o}\left[\frac{[A]_o - [B]_o}{\{[B]_o - x\}\{[A]_o - x\}}\right] = \frac{1}{\{[A]_o - x\}\{[B]_o - x\}}, \tag{2}$$

so that the two expressions on both sides of Eq. 1 are the same.

(B) Starting with Eqs. 19.10 and 19.11, we have

$$\int_{x=0}^{x}\frac{dx}{\{[A]_o - x\}\{[B]_o - x\}} = \frac{1}{[A]_o - [B]_o}\left[\int_0^x \frac{dx}{[B]_o - x} - \int_0^x \frac{dx}{[A]_o - x}\right] = k(T)t. \tag{3}$$

The two integrals both give natural logarithm functions, so that

$$\frac{1}{[A]_o - [B]_o}\left[-\ln\{[B]_o - x\}\Big|_0^x + \ln\{[A]_o - x\}\Big|_0^x\right]$$
$$= \frac{1}{[A]_o - [B]_o}\left[\ln\left\{\frac{[B]_o}{[B]_o - x}\right\} + \ln\left\{\frac{[A]_o - x}{[A]_o}\right\}\right] = k(T)t. \quad \textbf{(4)}$$

Combining the logarithms in Eq. 4 and multiplying by $\{[A]_o - [B]_o\}$, we obtain

$$\ln\left[\frac{[A]_o - x}{[B]_o - x}\right] + \ln\left[\frac{[B]_o}{[A]_o}\right] = ([A]_o - [B]_o)k(T)t. \quad \textbf{(5)}$$

Using the fact that $\ln[[B]_o/[A]_o] = -\ln[[A]_o/[B]_o]$, we may rearrange Eq. 5 to obtain

$$\ln\left[\frac{[A]_o - x}{[B]_o - x}\right] = \ln\left[\frac{[A]_o}{[B]_o}\right] + ([A]_o - [B]_o)k(T)t, \quad \textbf{(6)}$$

which is Eq. 19.13. If we now substitute Eqs. 19.8A and 19.8B, which give $[A] = [A]_o - x$ and $[B]_o - x = [B]$, we obtain

$$\ln\left[\frac{[A]}{[B]}\right] = \ln\left[\frac{[A]_o}{[B]_o}\right] + ([A]_o - [B]_o)k(T)t, \quad \textbf{(7)}$$

which is Eq. 19.14.

19.5 For the reaction $2A \longrightarrow B + C$, the following data are obtained for $[A]$ as a function of time at 310 K.

Time (min)	$[A]$ (mol L^{-1})
0	0.800
8	0.659
24	0.487
40	0.387
60	0.302
100	0.218

(A) By suitable means, establish the order of the reaction.

(B) What is the value of the rate coefficient?

(C) Calculate the rate of formation of B at $t = 30$ min.

Solution

(A and B) We can establish the order of the reaction either by plotting the data in the appropriate manner or by computating the rate coefficient directly. Since there is only one reactant involved, we expect the kinetics to be described by Eq. 19.16:

$$\frac{1}{[A]} = \frac{1}{[A]_o} + 2k(T)t. \quad \textbf{(1)}$$

If this is the case, a plot of $[A]^{-1}$ against time should be linear within the accuracy of the experimental data. The plot appears at the end of the problem; as can be seen, the linearity is very good. The line is a least-squares fit to the data. The average square deviation from the line σ^2 is 0.0026. The least-squares fit gives

$$\frac{1}{[A]} = 1.253 \text{ L mol}^{-1} + 0.03355(\text{L mol}^{-1}\text{ min}^{-1})\text{t}. \qquad \textbf{(2)}$$

Therefore, $2k(310 \text{ K})$ is

$$2k(T) = 2k(310 \text{ K}) = 0.03355 \text{ L mol}^{-1}\text{ min}^{-1}, \qquad \textbf{(3)}$$

so that

$$k(310 \text{ K}) = \underline{0.01677 \text{ L mol}^{-1}\text{ min}^{-1}}. \qquad \textbf{(4)}$$

Alternatively, we can compute $k(T)$ at each data point. Solving Eq. 1 for $k(T)$, we obtain

$$k(T) = \frac{[A]^{-1} - [A]_o^{-1}}{2t}. \qquad \textbf{(5)}$$

Computating this quantity at each time greater than zero gives the following table of values:

Time (min)	$\frac{[A]^{-1} - [A]_o^{-1}}{2t}$ (L mol^{-1} min^{-1})
8.00	0.01672
24.00	0.01674
40.00	0.01667
60.00	0.01718
100.00	0.01669

Except for the point at $t = 60.0$ min, these results for $k(T)$ are quite constant. Examining the plot of $1/[A]$ versus time at the end of the problem also shows that the data point at 60.0 min. falls off the line. This point is, therefore, suspect. The investigator might well repeat the experiment. The average of the preceding results for $k(T)$ gives

$$\langle k(310 \text{ K})\rangle = \underline{0.01680 \text{ L mol}^{-1}\text{ min}^{-1}} \qquad \textbf{(6)}$$

for the rate coefficient. This is very close to the result obtained from the least-squares fitting of the plot. Taking the average of the two methods, we might report a measured rate coefficient of

$$k(310 \text{ K}) = \underline{0.01678 \text{ L mol}^{-1}\text{ min}^{-1}}. \qquad \textbf{(7)}$$

(C) At $t = 30$ min, the concentration of compound A can be computed by using Eq. 1 with our computed value for $k(310 \text{ K})$. This gives

$$\frac{1}{[A]_{30\text{ min}}} = \frac{1}{0.80} + 2(0.01678)(30) \text{ L mol}^{-1} = 2.257 \text{ L mol}^{-1}, \qquad \textbf{(8)}$$

so that

$$[A]_{30\text{ min}} = \frac{1}{2.257} \text{ mol L}^{-1} = 0.4431 \text{ mol L}^{-1}. \qquad \textbf{(9)}$$

The rate of the reaction is

$$\mathcal{R} = k(310\text{ K})[A]^2 = 0.01678 \text{ L mol}^{-1} \text{ min}^{-1}[A]^2. \quad \textbf{(10)}$$

At $t = 30$ min, this rate is

$$\mathcal{R} = (0.01678)(0.4431)^2 \text{ mol L}^{-1} \text{ min}^{-1} = \underline{0.003294 \text{ mol L}^{-1} \text{ min}^{-1}}. \quad \textbf{(11)}$$

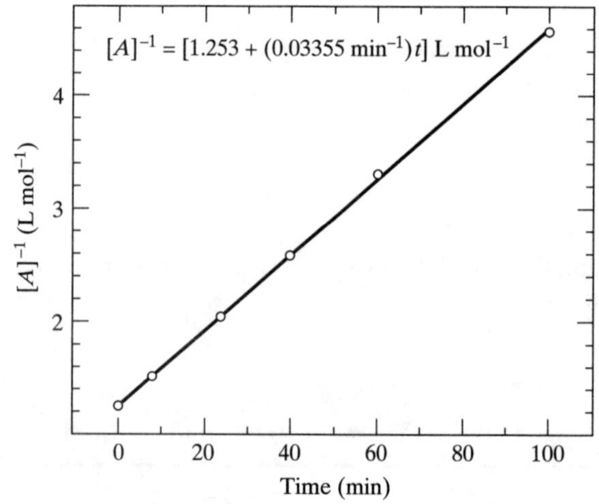

19.7 An investigator is measuring kinetic data for a rather slow concerted bimolecular reaction between compounds A and B at $T = 320$ K. The initial concentrations of A and B are $[A]_o = 0.0400 \text{ mol L}^{-1}$ and $[B]_o = 0.0300 \text{ mol L}^{-1}$. While the experiment is in progress, the investigator leaves the laboratory for a few moments, during which a jealous rival enters, raises the temperature of the reaction vessel to 335 K, and then leaves. The investigator returns and completes the measurements of the temporal variation of $[A]$. The data are as follows:

Time (min)	$[A]$ (mol L^{-1})	Time(min)	$[A]$ (mol L^{-1})
0.00	0.0400	160.00	0.0306
10.00	0.0392	180.00	0.0298
20.00	0.0384	200.00	0.0290
40.00	0.0370	220.00	0.0271
60.00	0.0357	240.00	0.0256
80.00	0.0346	260.00	0.0242
100.00	0.0333	280.00	0.0230
120.00	0.0324	300.00	0.0220
140.00	0.0315		

(A) How will the investigator know that the data have been corrupted by the trick played by the jealous rival?

(B) Can the data be salvaged so that the investigator will be able to obtain the value of the rate coefficient at 320 K?

(C) Is it also possible to obtain the rate coefficient at 335 K? If so, how? If not, what additional information would be needed?

Solution

(A) Our investigator is unaware of the trick played by the jealous rival. Expecting the data to be described by Eq. 19.14 for a concerted bimolecular reaction involving two reactants, the investigator needs to manipulate the data so as to permit plotting of $\ln[A]/[B]$ against time. This should yield a straight line whose slope is $[[A]_o - [B]_o]k(T)$. Since 1 mole of B reacts each time 1 mole of A reacts, we have

$$x = \text{number of moles of } A \text{ and } B \text{ that have reacted at time } t = [A]_o - [A]. \quad \textbf{(1)}$$

At time t, therefore,

$$[B] = [B]_o - x = [B]_o - [A]_o + [A] = 0.0300 - 0.400 + [A] = [A] - 0.0100. \quad \textbf{(2)}$$

Hence, we have

$$\ln\left[\frac{[A]}{[B]}\right] = \ln\left[\frac{[A]}{[A] - 0.0100}\right]. \quad \textbf{(3)}$$

Consequently, the data the investigator needs to plot are those given in the following table:

Time (min)	$\left[\frac{[A]}{[A] - 0.0100}\right]$	$\ln\left[\frac{[A]}{[A] - 0.0100}\right]$
0.00	1.333	0.288
10.00	1.342	0.295
20.00	1.352	0.302
40.00	1.370	0.315
60.00	1.389	0.329
80.00	1.407	0.341
100.00	1.429	0.357
120.00	1.446	0.369
140.00	1.465	0.382
160.00	1.485	0.396
180.00	1.505	0.409
200.00	1.526	0.423
220.00	1.585	0.460
240.00	1.641	0.495
260.00	1.704	0.533
280.00	1.769	0.571
300.00	1.833	0.606

Our investigator now plots $\ln[[A]/([A] - 0.0100)]$ against time and obtains the result shown in the figure at the end of the problem. As can be seen, the plot is linear up to $t = 200$ min, at which point there is a sharp break in the curve. Beyond $t = 200$ min, the result is still linear, but with a different slope.

Our investigator easily deduces that someone has altered the temperature of the constant-temperature bath. Naturally, the experiment will be repeated, but we already know what the new results will be.

(B and C) A least-squares fit to the data between $0 \leq t \leq 200$ min shows that in this range,

$$\ln\left[\frac{[A]}{[A] - 0.0100}\right] = 0.2883 + 0.0006727t, \tag{4}$$

and in the range 200 min $\leq t \leq$ 300 min, a second least-squares fit gives

$$\ln\left[\frac{[A]}{[A] - 0.0100}\right] = 0.05538 + 0.001837t. \tag{5}$$

Therefore, at $T = 320$ K, the rate coefficient is

$$k(320\text{ K}) = \frac{\text{slope}}{[A]_o - [B]_o} = \frac{0.0006727\text{ min}^{-1}}{0.0100\text{ mol L}^{-1}} = \underline{0.0673\text{ L mol}^{-1}\text{ min}^{-1}}. \tag{6}$$

At $T = 335$ K, the rate coefficient is

$$k(335\text{ K}) = \frac{\text{slope}}{[A]_o - [B]_o} = \frac{0.001837\text{ min}^{-1}}{0.0100\text{ mol L}^{-1}} = \underline{0.184\text{ L mol}^{-1}\text{ min}^{-1}}. \tag{7}$$

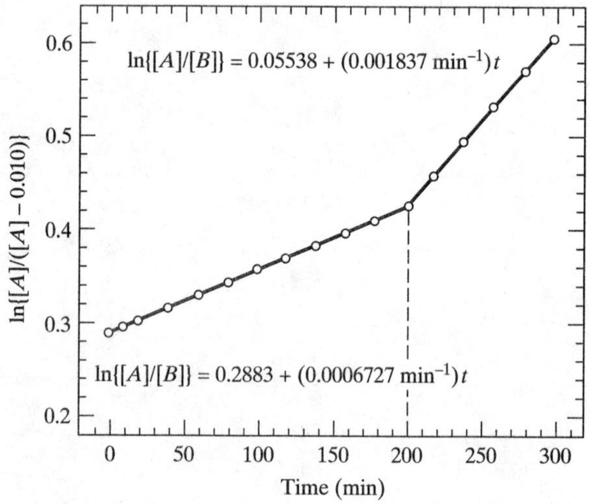

19.9 Show that, under the conditions that $[B]$ is constant at a value $[B]_o$ with $[B]_o \gg [A]$, Eq. 19.14 reduces to Eq. 19.22 for a pseudo first-order reaction.

Solution

For a concerted, bimolecular second-order reaction, the integrated rate expression is

$$\ln\left[\frac{[A]}{[B]}\right] = \ln\left[\frac{[A]_o}{[B]_o}\right] + ([A]_o - [B]_o)k(T)t. \tag{1}$$

If $[B]$ is constant at the value $[B]_o$, Eq. 1 becomes

$$\ln\left[\frac{[A]}{[B]_o}\right] = \ln[A] - \ln[B]_o = \ln\left[\frac{[A]_o}{[B]_o}\right] + ([A]_o - [B]_o)k(T)t$$

$$= \ln[A]_o - \ln[B]_o + ([A]_o - [B]_o)k(T)t. \tag{2}$$

Canceling the $-\ln[B]_o$ terms on both sides, we obtain

$$\ln[A] = \ln[A]_o + ([A]_o - [B]_o)k(T)t. \tag{3}$$

If we now have $[B]_o \gg [A]$, then

$$[A]_o - [B]_o \approx -[B]_o, \tag{4}$$

so that Eq. 3 becomes

$$\ln[A] = \ln[A]_o - [B]_o k(T)t. \tag{5}$$

If we now define the pseudo-first-order reaction rate coefficient as

$$k_p(T) = [B]_o k(T), \tag{6}$$

Eq. 5 may be expressed as

$$\ln[A] = \ln[A]_o - k_p(T)t. \tag{7}$$

Exponentiation of both sides of Eq. 7 gives

$$\boxed{[A] = [A]_o \exp[-k_p(T)t]}, \tag{8}$$

which is Eq. 19.22.

19.11. Derive Eqs. 19.26A and 19.26B, which respectively give the half-life and relaxation time for a second-order concerted reaction involving a single reactant.

Solution

The integrated rate expression for the concerted second-order reaction with a single reactant is given by Eq. 19.16:

$$\frac{1}{[A]} = \frac{1}{[A]_o} + 2k(T)t. \tag{1}$$

At time $t = 0$, $[A] = [A]_o$. At the point $t = \tau$, by definition, we will have $[A] = [A]_o/e$. Substituting of this requirement into Eq. 1 produces

$$\frac{e}{[A]_o} = \frac{1}{[A]_o} + 2k(T)\tau. \tag{2}$$

Solving for τ, we obtain

$$\boxed{\tau = \frac{e-1}{2[A]_o k(T)} = \frac{0.8591409\ldots}{[A]_o k(T)}}, \tag{3}$$

which is Eq. 19.26B.

At the point $t = t_{1/2}$, by definition, we will have $[A] = [A]_o/2$. Substituting this requirement into Eq. 1 produces

$$\frac{2}{[A]_o} = \frac{1}{[A]_o} + 2k(T)t_{1/2}. \tag{4}$$

Solving for $t_{1/2}$, we obtain

$$\boxed{t_{1/2} = \frac{2-1}{2[A]_o k(T)} = \frac{1}{2[A]_o k(T)}}, \tag{5}$$

which is Eq. 19.26A.

19.13 A compound A undergoes a first-order decomposition reaction at 320 K. An investigator conducts experiments that yield the following concentration data for A as a function of time:

Time (min)	Concentration (mmol L^{-1})	Time (min)	Concentration (mmol L^{-1})
0.00000	10.00000	37.89474	7.38482
4.21053	9.66877	42.10526	7.14022
8.42105	9.34851	46.31579	6.90371
12.63158	9.03885	50.52632	6.67504
16.84211	8.73946	54.73684	6.45394
21.05263	8.44998	58.94737	6.24016
25.26316	8.17009	63.15789	6.03347
29.47368	7.89947	67.36842	5.83362
33.68421	7.63781	71.57895	5.64039
75.78947	5.45356	140.00000	1.67831
80.00000	5.27292	144.00000	1.48829
84.00000	4.99374	148.00000	1.31978
88.00000	4.72933	152.00000	1.17035
92.00000	4.47893	156.00000	1.03784
96.00000	4.24179	160.00000	0.92034
100.00000	4.01720	164.00000	0.81614
104.00000	3.80450	168.00000	0.72373
108.00000	3.60306	172.00000	0.64179
112.00000	3.41229	176.00000	0.56913
116.00000	3.23162	180.00000	0.50469
120.00000	3.06052	184.00000	0.44755
124.00000	2.71400	188.00000	0.39687
128.00000	2.40672	192.00000	0.35194
132.00000	2.13423	196.00000	0.31209
136.00000	1.89259	200.00000	0.27676

Without his knowledge, a practical joker alters the system temperature to 340 K partway into the experiment. At a later time, this same practical joker raises the temperature again. The investigator cannot understand his data and concludes that his apparatus must have a serious design flaw. He needs your assistance.

(A) Determine the time at which the practical joker altered the temperature to 340 K and the time at which she raised the temperature a second time.

(B) Determine the specific reaction rate coefficient for the decomposition at 320 K.

(C) Determine the activation energy for the reaction.

(D) Determine the final temperature to which the practical joker raised the system.

Solution

(A) Since we know that the reaction is first order, we should see linear behavior when $\ln[A]$ is plotted against time. This plot is shown at the end of the problem for the given data. As can be seen, there are clear breaks in the plot at $t = 80$ min and at $t = 120$ min. These must be times at which our practical joker raised the temperature and thereby increased the rate coefficient, which produces the increase in the magnitude of the slopes seen in the plot.

(B) We can determine the specific reaction rate coefficient for the reaction by analyzing the data between 0 and 80 minutes, a time during which we know the temperature was 320 K. This plot is shown at the end of the problem. A least-squares linear fit gives a slope of $-0.00800\ \text{min}^{-1}$. Therefore, the specific reaction rate coefficient, which is the negative of this slope, is $k(320\ \text{K}) = 0.00800\ \text{min}^{-1}$.

(C) We can determine the specific reaction rate coefficient for the reaction at 340 K by analyzing the data between 80 and 120 minutes, a time during which we know the temperature was 340 K. This plot is shown at the end of the problem. A least-squares linear fit gives a slope of $-0.0136\ \text{min}^{-1}$. Therefore, the specific reaction rate coefficient, which is the negative of this slope, is $k(340\ \text{K}) = 0.0136\ \text{min}^{-1}$. The variation of k with T is given by the Arrhenius equation:

$$k(T) = A \exp\left[-\frac{E_a}{RT}\right]. \qquad \textbf{(1)}$$

Therefore, the ratio of rate coefficients at temperatures T_1 and T_2 is

$$\frac{k_1(T_1)}{k_2(T_2)} = \exp\left[-\frac{E_a}{R}\left\{\frac{1}{T_1} - \frac{1}{T_2}\right\}\right]. \qquad \textbf{(2)}$$

Solving for E_a, we obtain

$$E_a = -\frac{R \ln\left[\dfrac{k_1(T_1)}{k_2(T_2)}\right]}{\left\{\dfrac{1}{T_1} - \dfrac{1}{T_2}\right\}} = -\frac{8.314\ \text{J mol}^{-1}\ \text{K}^{-1} \ln(0.00800/0.0136)}{(1/320 - 1/340)\ \text{K}^{-1}} = \underline{24{,}000\ \text{J mol}^{-1}}. \qquad \textbf{(3)}$$

(D) To obtain the final temperature, we must first compute the value of $k(T)$ at that temperature by analyzing the data between 120 and 200 min. This analysis is shown in one of the plots at the end of the problem. The slope of the line in this range is $-0.0300\ \text{min}^{-1}$. Therefore, the rate coefficient at the final temperature T_3 is $k(T_3) = 0.0300\ \text{min}^{-1}$. Using this result and the data at $T = 320$ K in Eq. 2, we obtain

$$\frac{0.00800}{0.0300} = \exp\left[-\frac{24{,}000}{8.314}\left(\frac{1}{320} - \frac{1}{T_3}\right)\right]. \qquad \textbf{(4)}$$

Solving for $1/T_3$, we get

$$\frac{1}{T_3} = \frac{8.314\ \text{J mol}^{-1}\ \text{K}^{-1}}{24{,}000\ \text{J mol}^{-1}} \ln\left[\frac{0.00800}{0.0300}\right] + \frac{1}{320}$$
$$= [-0.0004579 + 0.003125]\ \text{K}^{-1} = 0.002667\ \text{K}^{-1}. \qquad \textbf{(5)}$$

The final temperature is

$$T_3 = \frac{1}{0.002667}\ \text{K} = \underline{374.9\ \text{K}}. \qquad \textbf{(6)}$$

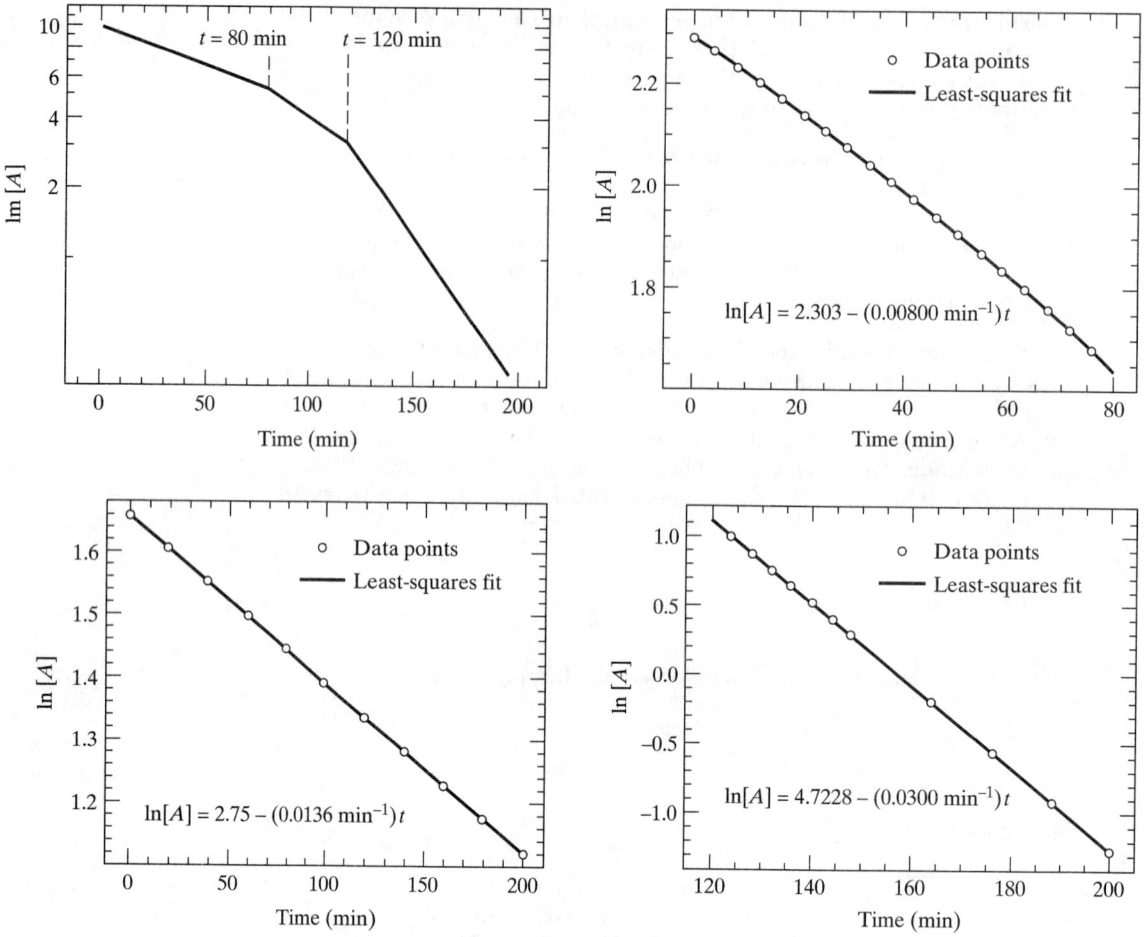

19.15 A series of rate measurements is conducted on a pseudo first-order reaction. The data obtained at 300 K, 320 K, 340 K, and 360 K are given in the table at the top of the next page (time is given in minutes and concentration in moles L^{-1}). Determine the activation energy and frequency factor for the reaction.

Solution

The integrated rate expression for a pseudo-first-order reaction is

$$\ln[A] = \ln[A]_o - k_p(T)t, \tag{1}$$

so that a plot of $\ln[A]$ against t will be linear with a slope $= -k_p(T)$. Such plots are shown at the end of the problem. Each data set produces a linear result. The lines are least-squares fits to the data. The equations of the lines are

$$\ln[A] = -3.912 - 0.00100t \text{ at } T = 300 \text{ K}, \tag{2}$$

$$\ln[A] = -3.916 - 0.00472t \text{ at } T = 320 \text{ K}, \tag{3}$$

$$\ln[A] = -3.923 - 0.0185t \text{ at } T = 340 \text{ K}, \tag{4}$$

and

$$\ln[A] = -3.912 - 0.0630t \text{ at } T = 360 \text{ K}. \tag{5}$$

	T = 300 K	T = 320 K	T = 340 K		T = 360 K	
Time	$[A]$	$[A]$	Time	$[A]$	Time	$[A]$
0.00	0.02000	0.02000	0.00	0.02000	0.00	0.02000
200.00	0.01638	0.00777	50.00	0.00789	10.00	0.01065
400.00	0.01341	0.00302	100.00	0.00311	20.00	0.00567
600.00	0.01098	0.00117	150.00	0.00123	30.00	0.00302
800.00	0.00899	0.00046	200.00	0.00048	40.00	0.00161
1000.0	0.00736	0.00018	250.00	0.00019	50.00	0.00086
1200.0	0.00602	0.00007	300.00	0.00008	60.00	0.00046
1400.0	0.00493	0.00003	350.00	0.00003	70.00	0.00024
1600.0	0.00404				80.00	0.00013
1800.0	0.00331				90.00	0.00007
2000.0	0.00271					
2200.0	0.00222					
2400.0	0.00181					
2600.0	0.00148					
2800.0	0.00122					
3000.0	0.00100					

The pseudo-first-order rate coefficients are, therefore, 0.00100 min^{-1}, 0.00472 min^{-1}, 0.0185 min^{-1}, and 0.0630 min^{-1} at 300 K, 320, 340 K, and 360 K, respectively. In logarithmic form, the Arrhenius equation is

$$\ln k_p(T) = \ln A - \frac{E_a}{RT}, \tag{6}$$

so that a plot of $\ln k_p(T)$ against T^{-1} should be linear with a slope equal to $-E_a/R$.

The plot at the end of the problem is an Arrhenius plot of the data. The line is a least-squares fit to the four data points. The equation of this line is

$$\ln k_p(T) = 17.95 - \frac{7{,}458\ \text{K}}{T}. \tag{7}$$

Since the intercept must be $\ln A$, where A is the frequency factor, we have

$$\ln A = 17.95, \tag{8}$$

so that

$$A = \exp(17.95) = 6.246 \times 10^7\ \text{min}^{-1}. \tag{9}$$

The slope of the line is equal to $-E_a/R$. Thus,

$$-\frac{E_a}{R} = -7{,}458\ \text{K}, \tag{10}$$

which gives

$$E_a = 7{,}458\ \text{K}\ (8.314\ \text{J mol}^{-1}\ \text{K}^{-1}) = \underline{6.201 \times 10^5\ \text{J mol}^{-1}}. \tag{11}$$

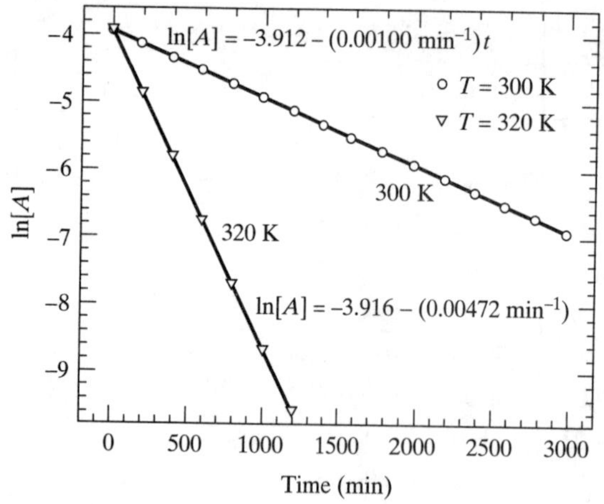

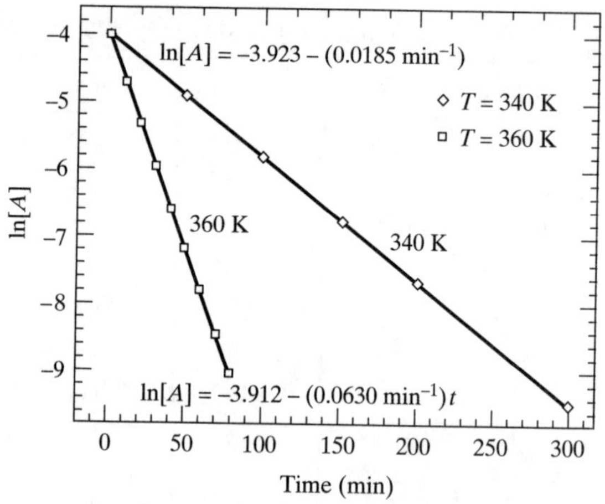

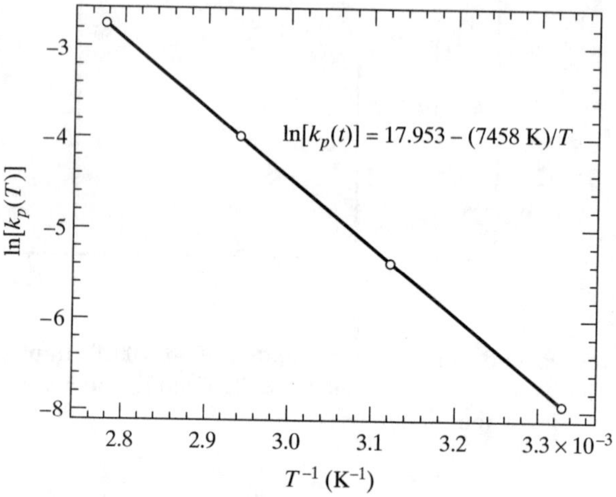

19.17 Consider two different radioactive isotopes A and B that each decay via ${}_{-1}^{0}\beta$ emission, but with different half-lives. A scientist prepares a 2-gram sample containing isotope A and a second 2-gram sample containing isotope B. Suppose it is known that each of the samples initially contains the same number of atoms of radioactive isotope and that the half-life of the first isotope is 49.90 min. The scientist now mixes the two samples together and measures the total ${}_{-1}^{0}\beta$ emission rate of the combined samples as a function of time. The data are obtained at the top of the next page.

Use the data to determine the half-life of the second radioactive isotope and the total number of radioactive atoms of each type that were present in the original 2-gram mixtures.

Solution

The total decay rate is the sum of the rates for each of the two isotopes that we denote as A and B. That is,

$$\text{Rate} = \mathcal{R} = k_1 A + k_2 B. \tag{1}$$

Since each rate is first order, we have

$$A = A_o e^{-k_1 t} \tag{2}$$

Time (min)	Emissions (counts) per minute
0	2,464
10	2,174
30	1,694
45	1,406
60	1,168
100	713
150	387
200	212

and

$$B = B_o e^{-k_2 t}. \tag{3}$$

With $A_o = B_o = C$ as stipulated in the problem, we obtain [by combining Eqs. 1, 2, and 3]

$$\text{Rate} = k_1 C e^{-k_1 t} + k_2 C e^{-k_2 t}. \tag{4}$$

At time $t = 0$, Eq. 4 gives

$$2{,}464 = C[k_1 + k_2]. \tag{5}$$

At time $t = 60$ min, at which the total rate is 1,168 counts min^{-1}, we obtain

$$1{,}168 = C[k_1 e^{-60k_1} + k_2 e^{-60k_2}]. \tag{6}$$

Dividing Eq. 5 by Eq. 6 yields

$$\frac{2{,}464}{1{,}168} = \frac{k_1 + k_2}{k_1 e^{-60k_1} + k_2 e^{-60k_2}} = 2.1096. \tag{7}$$

Since the half-life of radioactive isotope A is 49.90 min,

$$k_1 = \frac{\ln 2}{49.90} = 0.01389 \text{ min}^{-1}. \tag{8}$$

Substituting Eq. 8 into Eq. 7 gives

$$2.1096 = \frac{0.01389 + k_2}{0.01389\, e^{-0.83344} + k_2 e^{-60k_2}} = \frac{0.01389 + k_2}{0.006036 + k_2 e^{-60k_2}}. \tag{9}$$

Rearranging terms in of Eq. 9, gives

$$0.012734 + 2.1096 k_2 e^{-60k_2} - k_2 - 0.01389 = 0 \tag{10}$$

or

$$2.1096 k_2 e^{-60k_2} - k_2 - 0.001156 = 0. \tag{11}$$

Equation 11 is a transcendental equation that must be solved numerically. A one-dimensional grid search of k_2 values shows that a value $k_2 = 0.01074$ min^{-1} makes the left side nearly zero. This gives a half-life for B of

$$t_{1/2} = \frac{\ln 2}{k_2} = \frac{\ln 2}{0.01074} \text{ min} = \underline{64.54 \text{ min}}. \tag{12}$$

Using Eq. 5, we may now find the number of atoms initially in each sample:

$$C = \frac{2{,}464}{k_1 + k_2} = \frac{2{,}464}{0.01389 + 0.01074} = \underline{1.000 \times 10^5}. \quad \textbf{(13)}$$

As a check, we may compute the number of counts per minute expected at $t = 150$ min. This is given by Eq. 4 with the results inserted for the variables:

$$\text{Rate at } t = 150 \text{ min} = 1.000 \times 10^5[0.01389e^{-150(0.01389)} + 0.01074e^{-150(0.01074)}]$$

$$= 1.000 \times 10^5[0.001729 + 0.002145] = 387. \quad \textbf{(14)}$$

Equation 14 is in excellent accord with the data. Therefore,

$$k_2 = 0.01074 \text{ min}^{-1},$$

$$t_{1/2} \text{ for isotope } B = \underline{64.54 \text{ min}},$$

and

$$C = \underline{1.000 \times 10^5} \text{ atoms initially in each sample.} \quad \textbf{(15)}$$

19.19 What is the probability that a randomly selected molecule in a system undergoing a first-order reaction will survive for at least 2τ, where τ is the relaxation time?

Solution

The distribution of lifetimes is given by Eq. 19.39:

$$P(t_L) = k \exp[-kt_L]dt_L. \quad \textbf{(1)}$$

The probability that we will observe $t_L \geq 2\tau$ is the sum of the probabilities for all such lifetimes. Since we sum a continuous distribution via integration, the desired result is

$$P(t_L \geq 2\tau) = \int_{t_L=2\tau}^{\infty} P(t_L) = k\int_{2\tau}^{\infty} \exp[-kt_L]dt_L$$

$$= -\exp[-kt_L]\Big|_{2\tau}^{\infty} = \exp[-2k\tau]. \quad \textbf{(2)}$$

Because $\tau = 1/k$, the probability is

$$P(t_L \geq 2\tau) = \exp[-2] = \underline{0.1353\ldots}. \quad \textbf{(3)}$$

Thus, about 13.53 % of the molecules will live for a time 2τ or longer.

19.21 Starting with Eq. 19.42A, derive Eq. 19.43.

Solution

Equation 19.42A is

$$\left(\frac{d[A]}{dt}\right)_{\text{total}} = -k_1[A] + k_2[B]. \quad \textbf{(1)}$$

Separating the concentration and time variables produces

$$\frac{d[A]}{-k_1[A] + k_2[B]} = dt. \quad \textbf{(2)}$$

As suggested in the text, we now let

$$[A] = [A]_o - x, \quad [B] = [B]_o + x, \quad \text{and } d[A] = -dx. \tag{3}$$

Substituting the relationships in Eq. 3 into Eq. 2 produces

$$\frac{dx}{(k_1[A]_o - k_2[B]_o) - x(k_1 + k_2)} = dt. \tag{4}$$

Integrating Eq. 4 between corresponding limits gives

$$\int_{x=0}^{x} \frac{dx}{(k_1[A]_o - k_2[B]_o) - x(k_1 + k_2)} = \int_{t=0}^{t} dt = t. \tag{5}$$

The integral of the left side of Eq. 5 is a simple ln function:

$$-\frac{1}{k_1 + k_2} \ln[(k_1[A]_o - k_2[B]_o) - x(k_1 + k_2)]\Big|_0^x = t. \tag{6}$$

Evaluating between limits produces

$$-\frac{1}{k_1 + k_2}\{\ln[(k_1[A]_o - k_2[B]_o) - x(k_1 + k_2)] - \ln[(k_1[A]_o - k_2[B]_o)]\} = t. \tag{7}$$

The argument of the first ln function is

$$(k_1[A]_o - k_2[B]_o) - x(k_1 + k_2) = k_1\{[A]_o - x\} - k_2\{[B]_o + x\} = k_1[A] - k_2[B]. \tag{8}$$

Substituting Eq. 8 into Eq. 7 and then rearranging terms gives

$$\boxed{\ln[k_1[A] - k_2[B]] = \ln[k_1[A]_o - k_2[B]_o] - (k_1 + k_2)t}, \tag{9}$$

which is Eq. 19.43.

19.23 (A) Determine the median lifetime in terms of the specific reaction rate coefficient for molecules undergoing a first-order reaction.
(B) Compute the probability of observing a lifetime at or between the median and the average lifetime for molecules undergoing a first-order reaction?

Solution

(A) The distribution of lifetimes is given by Eq. 19.39:

$$P(t_L) = k \exp[-kt_L]dt_L. \tag{1}$$

Exactly half of the lifetimes are greater than the median lifetime t_m, and half are less. Therefore,

$$\int_{t_L=0}^{t_m} P(t_L) = k \int_0^{t_m} \exp[-kt_L]dt_L = 0.5. \tag{2}$$

Integrating yields

$$-\exp[-kt_L]\Big|_0^{t_m} = -\exp[-kt_m] + 1 = 0.5. \tag{3}$$

Rearranging produces

$$\exp[-kt_m] = 0.5. \tag{4}$$

Taking logarithms of both sides, we obtain

$$-kt_m = \ln(0.5) = \ln\left(\frac{1}{2}\right) = -\ln(2). \quad (5)$$

The median lifetime is, therefore,

$$\boxed{t_m = \frac{\ln(2)}{k}}, \quad (6)$$

and we see that t_m is identical to the half-life, $t_{1/2}$.

(B) The probability of observing a lifetime in the range $t_m \le t_L \le \langle t_L \rangle$ is

$$P(t_m \le t_L \le \langle t_L \rangle) = \int_{t_m}^{\langle t_L \rangle} P(t_L) = k \int_{t_m}^{\langle t_L \rangle} \exp[-kt_L]\,dt_L = -\exp[-kt_L]\Big|_{t_m}^{\langle t_L \rangle}$$

$$= -\exp[-k\langle t_L \rangle] + \exp[-kt_m]. \quad (7)$$

Equation 6 shows that $t_m = \ln(2)/k$, and Example 19.7 demonstrates that $\langle t_L \rangle = \tau = 1/k$. Substituting these results into Eq. 7 gives

$$P(t_m \le t_L \le \langle t_L \rangle) = -\exp[-1] + \exp[-\ln(2)] = -0.367879\cdots + 0.5000$$

$$= \underline{0.1321\ldots}. \quad (8)$$

19.25 Compound A simultaneously undergoes a pseudo first-order reaction and a second-order reaction, so that the mechanism is

$$A \xrightarrow{k_1} B \qquad \text{and} \qquad A + A \xrightarrow{k_2} C$$

Reaction 1 Reaction 2

Obtain the integrated rate expression for $[A]$ as a function of time.

Solution

The differential rate expressions for these reactions are

$$\left(\frac{d[A]}{dt}\right)_1 = -k_1[A] \quad (1)$$

and

$$-\frac{1}{2}\left(\frac{d[A]}{dt}\right)_2 = k_2[A]^2, \quad (2)$$

so that

$$\left(\frac{d[A]}{dt}\right)_2 = -2k_2[A]^2. \quad (3)$$

The total rate for both reactions is, therefore,

$$\left(\frac{d[A]}{dt}\right) = \left(\frac{d[A]}{dt}\right)_1 + \left(\frac{d[A]}{dt}\right)_2 = -k_1[A] - 2k_2[A]^2. \quad (4)$$

Separating the variables yields

$$\frac{d[A]}{k_1[A] + 2k_2[A]^2} = \frac{d[A]}{[A](k_1 + 2k_2[A])} = -dt. \quad (5)$$

This expression may be integrated easily if we express the denominator of the left-hand side as the sum of two fractions. That is, we write

$$\frac{1}{[A](k_1 + 2k_2[A])} = \frac{a}{[A]} + \frac{b}{k_1 + 2k_2[A]} \tag{6}$$

and ask what values a and b must have to make Eq. 6 an identity. Taking a common denominator on the right-hand side of Eq. 6 produces

$$\frac{1}{[A](k_1 + 2k_2[A])} = \frac{a(k_1 + 2k_2[A]) + b[A]}{[A](k_1 + 2k_2[A])} = \frac{ak_1 + [A](2ak_2 + b)}{[A](k_1 + 2k_2[A])}. \tag{7}$$

In order for Eq. 7 to be an identity, we must have

$$ak_1 = 1$$

and

$$[A](2ak_2 + b) = 0. \tag{8}$$

Solving for a and b, we obtain

$$a = \frac{1}{k_1}$$

and

$$b = -2ak_2 = -\frac{2k_2}{k_1}. \tag{9}$$

Equation 9 allows Eq. 5 to be written in the form

$$\frac{d[A]}{k_1[A]} - \frac{2k_2 d[A]}{k_1(k_1 + 2k_2[A])} = -dt. \tag{10}$$

Integrating between corresponding limits gives

$$\int_{[A]_o}^{[A]} \frac{d[A]}{k_1[A]} - \frac{1}{k_1}\int_{[A]_o}^{[A]} \frac{2k_2 d[A]}{(k_1 + 2k_2[A])} = -\int_0^t dt = -t. \tag{11}$$

Both integrals give logarithmic functions, so that we have

$$\frac{1}{k_1}\ln\left[\frac{[A]}{[A]_o}\right] - \frac{1}{k_1}\ln\left[\frac{k_1 + 2k_2[A]}{k_1 + 2k_2[A]_o}\right] = -t. \tag{12}$$

Combining the logarithmic terms and multiplying by $-k_1$ produces

$$\boxed{\ln\left[\frac{[A]_o\{k_1 + 2k_2[A]\}}{[A]\{k_1 + 2k_2[A]_o\}}\right] = k_1 t}, \tag{13}$$

which is the integrated rate law for the process.

19.27 An investigator carries out a chemical process involving a single reactant A. He obtains the data on top of the next page.

It is quickly found that these data do not fit any simple reaction rate law. After much effort, the investigator learns an interesting fact: A plot of $\ln[(1/[A]) + 1]$ vs. time is linear. (See figure on the bottom of the next page.) Being unable to determine the reason behind this behavior, he submits the problem to you for solution.

Time (min)	$[A]$ (mol L^{-1})	$1/[A]$ (mol L^{-1})
0	1.0	1.0
10	0.82621	1.2103
20	0.69309	1.4428
30	0.58833	1.6997
40	0.50412	1.9836
50	0.43527	2.2974
60	0.37818	2.6442
70	0.33030	3.0275
80	0.28976	3.4511
90	0.25515	3.9192

(A) Show that the reaction mechanism

$$A + A \xrightarrow{k_1} P_1 \quad \text{and} \quad A \xrightarrow{k_2} P_2$$

obeys the integrated rate law

$$\frac{1}{[A]} = a \exp[k_2 t] - b,$$

where

$$a = \frac{2k_1}{\beta k_2} \quad \text{and} \quad b = \frac{2k_1}{k_2},$$

in which

$$\beta = \frac{2k_1[A]_o}{(2k_1[A]_o + k_2)}.$$

(B) In view of the result in (A) and the data given in the problem, determine k_1 and k_2 for the reaction mechanism.
(P.S. If you solved the problem, the investigator extends his profound thanks. In the future, remember to ask him for a consultant's fee.)

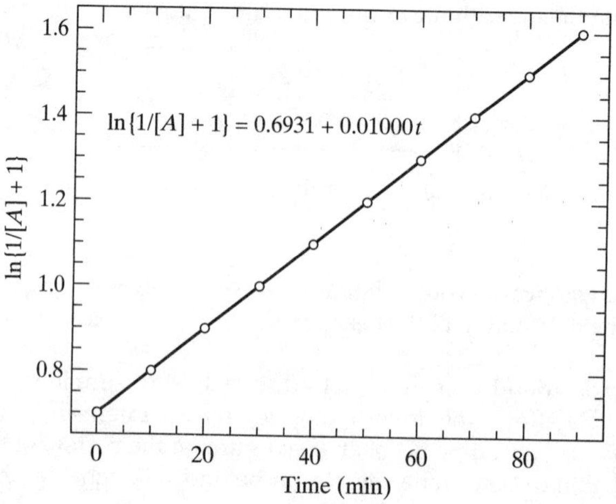

Solution

(A) The differential rate expressions for these reactions are

$$-\frac{1}{2}\left(\frac{d[A]}{dt}\right)_1 = k_1[A]^2, \tag{1}$$

so that

$$\left(\frac{d[A]}{dt}\right)_1 = -2k_1[A]^2 \tag{2}$$

and

$$\left(\frac{d[A]}{dt}\right)_2 = -k_2[A]. \tag{3}$$

The total rate for both reactions is, therefore,

$$\left(\frac{d[A]}{dt}\right) = \left(\frac{d[A]}{dt}\right)_1 + \left(\frac{d[A]}{dt}\right)_2 = -2k_1[A]^2 - k_2[A]. \tag{4}$$

Separating the variables yields

$$\frac{d[A]}{k_2[A] + 2k_1[A]^2} = \frac{d[A]}{[A](k_2 + 2k_1[A])} = -dt. \tag{5}$$

This expression may be integrated easily if we express the denominator of the left-hand side as the sum of two fractions. That is, we write

$$\frac{1}{[A](k_2 + 2k_1[A])} = \frac{a}{[A]} + \frac{b}{k_2 + 2k_1[A]} \tag{6}$$

and ask what values a and b must have to make Eq. 6 an identity. Taking a common denominator on the right-hand side of Eq. 6, produces

$$\frac{1}{[A](k_2 + 2k_1[A])} = \frac{a(k_2 + 2k_1[A]) + b[A]}{[A](k_2 + 2k_1[A])} = \frac{ak_2 + [A](2ak_1 + b)}{[A](k_2 + 2k_1[A])}. \tag{7}$$

In order for Eq. 7 to be an identity, we must have

$$ak_2 = 1$$

and

$$[A](2ak_1 + b) = 0. \tag{8}$$

Solving for a and b, we obtain

$$a = \frac{1}{k_2}$$

and

$$b = -2ak_1 = -\frac{2k_1}{k_2}. \tag{9}$$

Equation 9 allows Eq. 5 to be written in the form

$$\frac{d[A]}{k_2[A]} - \frac{2k_1 d[A]}{k_2(k_2 + 2k_1[A])} = -dt. \tag{10}$$

Integrating between corresponding limits gives

$$\int_{[A]_o}^{[A]} \frac{d[A]}{k_2[A]} - \frac{1}{k_2}\int_{[A]_o}^{[A]} \frac{2k_1 d[A]}{(k_2 + 2k_1[A])} = -\int_0^t dt = -t. \tag{11}$$

Both integrals give logarithmic functions, so that we have

$$\frac{1}{k_2}\ln\left[\frac{[A]}{[A]_o}\right] - \frac{1}{k_2}\ln\left[\frac{k_2 + 2k_1[A]}{k_2 + 2k_1[A]_o}\right] = -t. \tag{12}$$

Combining the logarithmic terms and multiplying by $-k_2$ produces

$$\ln\left[\frac{[A]_o\{k_2 + 2k_1[A]\}}{[A]\{k_2 + 2k_1[A]_o\}}\right] = k_2 t, \tag{13}$$

which is the integrated rate law for the process. Taking the exponentials of both sides, we obtain

$$\frac{[A]_o\{k_2 + 2k_1[A]\}}{[A]\{k_2 + 2k_1[A]_o\}} = \exp(k_2 t). \tag{14}$$

Rearranging terms in Eq. 14 gives

$$\frac{k_2[A]_o}{[A]} + 2k_1[A]_o = \{k_2 + 2k_1[A]_o\}\exp(k_2 t). \tag{15}$$

Putting $2k_1[A]_o$ on the right-hand side of Eq. 15 and then dividing by $k_2[A]_o$ produces

$$\frac{1}{[A]} = \frac{k_2 + 2k_1[A]_o}{k_2[A]_o}\exp(k_2 t) - \frac{2k_1}{k_2}. \tag{16}$$

In terms of the variables defined in the problem, we have

$$\frac{k_2 + 2k_1[A]_o}{k_2[A]_o} = \frac{2k_1}{\beta k_2} = a \text{ and } b = \frac{2k_1}{k_2}. \tag{17}$$

Substituting into Eq. 16 gives

$$\boxed{\frac{1}{[A]} = \frac{2k_1}{\beta k_2}\exp(k_2 t) - b = a\exp(k_2 t) - b}, \tag{18}$$

which is the equation that we are supposed to derive.

(B) Rearranging terms in Eq. 18 gives

$$\frac{1}{[A]} + b = \frac{2k_1}{\beta k_2}\exp(k_2 t). \tag{19}$$

Taking logarithms of both sides produces

$$\ln\left[\frac{1}{[A]} + b\right] = k_2 t + \ln\left[\frac{2k_1}{\beta k_2}\right]. \tag{20}$$

Equation 2 is the equation of a straight line. Our investigator found that a plot of $\ln[(1/[A]) + 1]$ is linear. This means that the value of b in his system must be unity. Therefore, we know that the magnitudes of k_1 and k_2 are such that $k_2 = 2k_1$. We can also see from Eq. 20 that the slope of line must be the value of k_2. A least-squares fit to the measured data gives the result

$$\ln\left[\frac{1}{[A]} + b\right] = 0.6931 + 0.01000t. \tag{21}$$

Therefore, the slope is 0.01000 min^{-1}, and we have

$$k_2 = \underline{0.01000\ \text{min}^{-1}} \quad \text{and } k_1 = \underline{0.005000\ \text{L mol}^{-1}\ \text{min}^{-1}}. \tag{22}$$

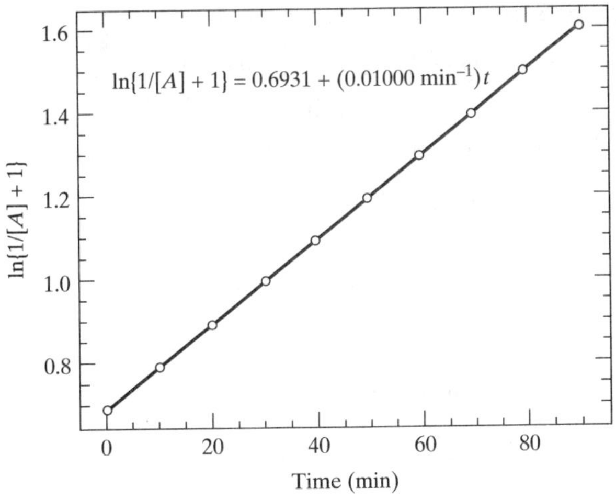

19.29 Show that the lifetime probability distribution given by Eq. 19.39 is properly normalized.

Solution

The lifetime distribution is

$$P(t_L) = k\exp[-kt_L]\,dt_L. \tag{1}$$

If the distribution is normalized, the probabilities will sum to unity. The summation is executed by integrating between $t_L = 0$ and $t_L = \infty$:

$$\int_{tL=0}^{\infty} P(t_L) = k\int_0^{\infty} \exp[-kt_L]\,dt_L = -\exp[-kt_L]\Big|_0^{\infty} = 0 - (-1) = 1. \tag{2}$$

Therefore, $P(t_L)$ is properly normalized.

19.31 Compound A is forming Compounds B and C by concurrent reactions. The following data are obtained for the reactions:

Time (min)	B (mol L^{-1})	C (mol L^{-1})
5.00	2.00a	4.00a
10.00	b	2.00b
15.00	c	2.00c
20.00	d	2.00d
25.00	e	2.00e

where a, b, c, d, and e represent some set of concentrations. What can be said about the orders and the rate coefficients for the two concurrent reactions? Justify your answer.

Solution

Let us assume that the reaction forming B is nth order, so that we have

$$\frac{d[B]}{dt} = k_1[A]^n, \quad (1)$$

while the reaction forming C is mth order, which gives

$$\frac{d[C]}{dt} = k_2[A]^m. \quad (2)$$

Dividing Eq. 1 by Eq. 2 produces

$$\frac{\frac{d[B]}{dt}}{\frac{d[C]}{dt}} = \frac{d[B]}{d[C]} = \frac{k_1[A]^n}{k_2[A]^m} = \frac{k_1[A]^{n-m}}{k_2}. \quad (3)$$

But the data show clearly that we have

$$\frac{d[B]}{d[C]} = \frac{1}{2}, \quad (4)$$

so that

$$2[B] = [C], \quad (5)$$

which is exactly the form of the experimental data.

The only way we can make Eq. 3 compatible with Eq. 4 is to have

$$\boxed{n = m \text{ and } \frac{k_1}{k_2} = \frac{1}{2}, \text{ or } k_2 = 2k_1}. \quad (6)$$

This is as much as we can determine without knowing the values of a, b, c, d, and e.

19.33 Consider two consecutive first-order nuclear decay reactions with rate coefficients k_1 and k_2 in which

$$A \xrightarrow{k_1} B \quad \text{and} \quad B \xrightarrow{k_2} C.$$

Show that if the initial numbers of atoms of A and B are $[A]_o$ and $[B]_o$, respectively, with $[B]_o \neq 0$, the temporal dependence of the number of atoms of B is given by Eq. 19.65.

Solution

The analysis is exactly that given in the text down to Eq. 19.63, where we integrated the left side of the differential equation between the limits $(t = 0, [B] = 0)$ and $(t = t, [B] = [B])$. We now wish the lower limit to be $(t = 0, [B] = [B]_o)$. Thus, Eq. 19.63 becomes

$$[B]\exp[k_2 t]\Big|_{0,[B]_o}^{t,[B]} = [B]\exp[k_2 t] - [B]_o = \frac{k_1[A]_o}{k_2 - k_1}\exp[(k_2 - k_1)t]\Big|_0^t$$
$$= \frac{k_1[A]_o}{k_2 - k_1}\{\exp[(k_2 - k_1)t] - 1\}. \quad (1)$$

Dividing by $\exp[k_2t]$ now produces

$$[B] - [B]_o \exp[-k_2t] = \frac{k_1[A]_o}{k_2 - k_1}\exp[-k_1t] - \frac{k_1[A]_o}{k_2 - k_1}\exp[-k_2t]. \qquad (2)$$

Rearranging terms in Eq. 2 gives

$$\boxed{[B] = \frac{k_1[A]_o}{k_2 - k_1}\exp[-k_1t] + \left[[B]_o - \frac{k_1[A]_o}{k_2 - k_1}\right]\exp[-k_2t]}, \qquad (3)$$

which is Eq. 19.65 in the text.

19.35 Consider two consecutive first-order nuclear decay reactions with rate coefficients k_1 and k_2 in which

$$A \xrightarrow{k_1} B \quad \text{and} \quad B \xrightarrow{k_2} C.$$

In the general case when $[B]_o \neq 0$, does the number of atoms or the concentration of the intermediate B always exhibit a maximum when plotted against time? If so, prove that this is the case. If not, determine the condition that must hold for a maximum to exist.

Solution

The temporal variation in the concentration or the number of atoms of B is given by Eq. 19.65:

$$[B] = \frac{k_1[A]_o}{k_2 - k_1}\exp[-k_1t] + \left[[B]_o - \frac{k_1[A]_o}{k_2 - k_1}\right]\exp[-k_2t]. \qquad (1)$$

For a maximum in a plot of $[B]$ versus time to exist, we must have $d[B]/dt = 0$. Using Eq. 1, we find that

$$\frac{d[B]}{dt} = -\frac{k_1^2[A]_o}{k_2 - k_1}\exp[-k_1t] - k_2\left[[B]_o - \frac{k_1[A]_o}{k_2 - k_1}\right]\exp[-k_2t]. \qquad (2)$$

Let us assume that when $t = t_m$, we have $d[B]/dt = 0$ and a maximum in $[B]$. This says that we have

$$-\frac{k_1^2[A]_o}{k_2 - k_1}\exp[-k_1t_m] - k_2\left[[B]_o - \frac{k_1[A]_o}{k_2 - k_1}\right]\exp[-k_2t_m] = 0. \qquad (3)$$

Rearranging terms in Eq. 3, we get

$$k_2\left[\frac{k_1[A]_o}{k_2 - k_1} - [B]_o\right]\exp[-k_2t_m] = \frac{k_1^2[A]_o}{k_2 - k_1}\exp[-k_1t_m]. \qquad (4)$$

Isolating the exponentials, we obtain

$$\frac{\exp[-k_1t_m]}{\exp[-k_2t_m]} = \exp[(k_2 - k_1)t_m] = \frac{k_2\left[\dfrac{k_1[A]_o}{k_2 - k_1} - [B]_o\right]}{\dfrac{k_1^2[A]_o}{k_2 - k_1}}$$

$$= \frac{(k_2 - k_1)k_2}{k_1^2[A]_o}\left[\frac{k_1[A]_o}{k_2 - k_1} - [B]_o\right] = \frac{k_2}{k_1} - \frac{(k_2 - k_1)k_2[B]_o}{k_1^2[A]_o}. \qquad (5)$$

Taking logarithms of both sides and solving for t_m, we obtain

$$\boxed{t_m = \frac{1}{(k_2 - k_1)} \ln\left[\frac{k_2}{k_1} - \frac{(k_2 - k_1)k_2[B]_o}{k_1^2[A]_o}\right]}. \quad \textbf{(6)}$$

For a real solution of t_m to exist, t_m must be positive and real. This means that if $k_2 - k_1 > 0$, the argument of the logarithm function must be greater than unity. If $k_2 - k_1 < 0$, the argument of the logarithm function must lie in the range $0 <$ argument < 1. We consider these two cases separately.

If $k_2 - k_1 > 0$, a maximum in B will occur if

$$\frac{k_2}{k_1} - \frac{(k_2 - k_1)k_2[B]_o}{k_1^2[A]_o} > 1. \quad \textbf{(7)}$$

Rearranging terms in Eq. 7, we get

$$\frac{k_2}{k_1} - 1 > \frac{(k_2 - k_1)k_2[B]_o}{k_1^2[A]_o}. \quad \textbf{(8)}$$

Since $k_2 - k_1$ is positive, we can divide by this quantity without reversing the direction of the inequality. Multiplying both sides by $k_1^2[A]_o/[(k_2 - k_1)k_2]$ gives

$$\frac{k_1^2[A]_o}{(k_2 - k_1)k_2}\left[\frac{k_2}{k_1} - 1\right] > [B]_o. \quad \textbf{(9)}$$

A little algebra on the left side of Eq. 9 produces

$$\frac{k_1[A]_o}{(k_2 - k_1)} - \frac{k_1^2[A]_o}{(k_2 - k_1)k_2} = \boxed{\frac{k_1[A]_o}{(k_2 - k_1)}\left[1 - \frac{k_1}{k_2}\right] > [B]_o}. \quad \textbf{(10)}$$

If the inequality in Eq. 10 is satisfied when $k_2 - k_1 > 0$, a maximum will occur in a plot of $[B]$ versus time.

If $k_2 - k_1 < 0$, a maximum in B will occur if

$$0 < \frac{k_2}{k_1} - \frac{(k_2 - k_1)k_2[B]_o}{k_1^2[A]_o} < 1. \quad \textbf{(11)}$$

Rearranging terms in Eq. 11, we obtain

$$-1 < \frac{k_2}{k_1} - 1 < \frac{(k_2 - k_1)k_2[B]_o}{k_1^2[A]_o}. \quad \textbf{(12)}$$

Since $k_2 - k_1$ is negative, dividing by this quantity reverses the directions of the inequalities. Therefore, multiplying Eq. 12 by $k_1^2[A]_o/[(k_2 - k_1)k_2]$ gives

$$-\frac{k_1^2[A]_o}{(k_2 - k_1)k_2} > \frac{k_1^2[A]_o}{(k_2 - k_1)k_2}\left[\frac{k_2}{k_1} - 1\right] > [B]_o. \quad \textbf{(13)}$$

The same algebraic manipulation as done in Eq. 10 produces

$$-\frac{k_1^2[A]_o}{(k_2 - k_1)k_2} > \frac{k_1[A]_o}{(k_2 - k_1)}\left[1 - \frac{k_1}{k_2}\right] > [B]_o. \quad \textbf{(14)}$$

Thus, the requirement for a maximum in a plot of $[B]$ versus time is

$$\boxed{\frac{k_1[A]_o}{(k_2 - k_1)}\left[1 - \frac{k_1}{k_2}\right] > [B]_o},$$

regardless of the sign of $k_2 - k_1$.

19.37 For the sequence of three first-order consecutive reactions

$$A \xrightarrow{k_1} B \qquad B \xrightarrow{k_2} C \quad \text{and} \quad C \xrightarrow{k_3} D$$

show that the temporal variation of $[C]$ is given by Eq. 19.69, provided that $[B]_o = [C]_o = 0$. You may use the results obtained in the text for the first two reactions in the sequence.

Solution

The differential rate equation for C is

$$\frac{d[C]}{dt} = k_2[B] - k_3[C]. \tag{1}$$

Multiplying by dt and rearranging terms gives

$$d[C] + k_3[C]dt = k_2[B]dt. \tag{2}$$

The temporal variation in B is given by Eq. 19.64:

$$[B] = -\frac{k_1[A]_o}{k_2 - k_1}e^{-k_2t} + \frac{k_1[A]_o}{k_2 - k_1}e^{-k_1t} = \frac{k_1[A]_o}{k_2 - k_1}[e^{-k_1t} - e^{-k_2t}] \tag{3}$$

if $[B]_o = 0$, as stipulated in the problem. Substituting Eq. 3 into Eq. 2 and then rearranging terms gives

$$d[C] + k_3[C]dt = \frac{k_1k_2[A]_o}{k_2 - k_1}[e^{-k_1t} - e^{-k_2t}]dt. \tag{4}$$

In the text, it is shown that equations of this type may be integrated by using an integrating factor given by

$$F = e^{\int k_3 dt} = e^{k_3t}. \tag{5}$$

Multiplying both sides of Eq. 4 by F yields

$$e^{k_3t}d[C] + k_3e^{k_3t}[C]dt = d\{[C]e^{k_3t}\} = \frac{k_1k_2[A]_o}{k_2 - k_1}[e^{-(k_1-k_3)t} - e^{-(k_2-k_3)t}]dt. \tag{6}$$

Integrating between the limits $(t = 0, [C] = 0)$ and $(t = t, [C] = [C])$ gives

$$[C]e^{k_3t} = \frac{k_1k_2A_o}{k_2 - k_1}\int_0^t [e^{-(k_1-k_3)t} - e^{-(k_2-k_3)t}]dt$$

$$= \frac{k_1k_2A_o}{k_2 - k_1}\left[\frac{1}{k_3 - k_1}e^{-(k_1-k_3)t} + \frac{1}{k_2 - k_3}e^{-(k_2-k_3)t}\right]_0^t$$

$$= \frac{k_1k_2A_o}{k_2 - k_1}\left[\frac{1}{k_3 - k_1}e^{-(k_1-k_3)t} + \frac{1}{k_2 - k_3}e^{-(k_2-k_3)t} - \frac{1}{k_3 - k_1} - \frac{1}{k_2 - k_3}\right]. \tag{7}$$

Dividing both sides by e^{k_3t} yields

$$[C] = \frac{k_1k_2A_o}{(k_2 - k_1)(k_3 - k_1)}e^{-k_1t} + \frac{k_1k_2A_o}{(k_2 - k_1)(k_2 - k_3)}e^{-k_2t}$$

$$- \frac{k_1k_2A_o}{k_2 - k_1}\left[\frac{1}{k_3 - k_1} + \frac{1}{k_2 - k_3}\right]e^{-k_3t}$$

$$= \frac{k_1k_2A_o}{(k_2 - k_1)(k_3 - k_1)}e^{-k_1t} + \frac{k_1k_2A_o}{(k_2 - k_1)(k_2 - k_3)}e^{-k_2t}$$

$$-\frac{k_1k_2A_o}{k_2-k_1}\left[\frac{k_2-k_3+k_3-k_1}{(k_3-k_1)(k_2-k_3)}\right]e^{-k_3t}$$

$$=\frac{k_1k_2A_o}{(k_2-k_1)(k_3-k_1)}e^{-k_1t}+\frac{k_1k_2A_o}{(k_2-k_1)(k_2-k_3)}e^{-k_2t}-\frac{k_1k_2A_o}{(k_3-k_1)(k_2-k_3)}e^{-k_3t}. \quad (8)$$

Rewriting Eq. 8 so that all terms are positive, we obtain

$$\boxed{[C]=\frac{k_1k_2A_o}{(k_2-k_1)(k_3-k_1)}e^{-k_1t}+\frac{k_1k_2A_o}{(k_2-k_1)(k_2-k_3)}e^{-k_2t}+\frac{k_1k_2A_o}{(k_3-k_1)(k_3-k_2)}e^{-k_3t}}, \quad (9)$$

which is the desired solution.

19.39 Consider the series of consecutive first-order reactions

$$A_1 \xrightarrow{k} A_2,\quad A_2 \xrightarrow{k} A_3,\quad A_3 \xrightarrow{k} A_4,\ \ldots,$$

$$A_i \xrightarrow{k} A_{i+1},\ldots,$$

in which all rate coefficients are equal and all initial concentrations are zero save for $[A_1]_o$.

(A) Obtain $[A_1]$ as a function of time.

(B) Use the result of (A) and an integrating factor as described in the text to obtain $[A_2]$ as a function of time.

(C) Use the result of (B) and an integrating factor as described in the text to obtain $[A_3]$ as a function of time.

(D) Using the results obtained in (A), (B), and (C), infer the form of the equation that gives $[A_i]$ as a function of time. Prove by induction that your inference is correct.

Solution

(A) Compound A_1 decays in the simple first-order reaction

$$\frac{d[A_1]}{dt}=-k[A_1]. \quad (1)$$

Separating the variables and integrating between corresponding limits gives

$$\int_{[A_1]=[A_1]_o}^{[A_1]}\frac{d[A_1]}{[A_1]}=\ln\left[\frac{[A_1]}{[A_1]_o}\right]=-k\int_0^t dt=-kt. \quad (2)$$

Exponentiation of both sides produces

$$\boxed{[A_1]=[A_1]_oe^{-kt}}. \quad (3)$$

(B) The rate of change of A_2 is

$$\frac{d[A_2]}{dt}=k[A_1]-k[A_2]. \quad (4)$$

Substituting Eq. 3 for $[A_1]$ in Eq. 4 and then rearranging terms yields

$$d[A_2]+k[A_2]dt=k[A_1]_oe^{-kt}dt. \quad (5)$$

Equation 5 has the same form as Eq. 19.59, so it can be integrated by using an integrating factor given by

$$I = \exp\left[\int k\,dt\right] = e^{kt}. \tag{6}$$

Multiplying both sides of Eq. 5 by I produces

$$e^{kt}d[A_2] + ke^{kt}[A_2]dt = d\{[A_2]e^{kt}\} = k[A_1]_o dt. \tag{7}$$

Integrating between corresponding limits gives

$$\int_{t=[A_2]=0}^{t[A_2]} d\{[A_2]e^{kt}\} = [A_2]e^{kt} = \int_0^t k[A_1]_o dt = k[A_1]_o t. \tag{8}$$

Dividing by e^{kt} yields the final result:

$$\boxed{[A_2] = [A_1]_o(kt)e^{-kt}}. \tag{9}$$

(C) The rate of change of A_3 is

$$\frac{d[A_3]}{dt} = k[A_2] - k[A_3]. \tag{10}$$

Substituting Eq. 9 for $[A_2]$ in Eq. 10 and then rearranging terms yields

$$d[A_3] + k[A_3]dt = [A_1]_o k^2 t e^{-kt} dt. \tag{11}$$

Equation 11 has the same form as Eq. 19.59 and Eq. 5, so it can be integrated by using an integrating factor given by

$$I = \exp\left[\int k\,dt\right] = e^{kt}. \tag{12}$$

Multiplying both sides of Eq. 11 by I produces

$$e^{kt}d[A_3] + ke^{kt}[A_3]dt = d\{[A_3]e^{kt}\} = [A_1]_o k^2 t\,dt. \tag{13}$$

Integrating between corresponding limits gives

$$\int_{t=[A_3]=0}^{t[A_3]} d\{[A_3]e^{kt}\} = [A_3]e^{kt} = k^2[A_1]_o \int_0^t t\,dt = \frac{k^2[A_1]_o t^2}{2}. \tag{14}$$

Dividing by e^{kt} yields the final result:

$$\boxed{[A_3] = \frac{[A_1]_o(kt)^2 e^{-kt}}{2} = \frac{[A_1]_o(kt)^2 e^{-kt}}{2!}}. \tag{15}$$

(D) The results for $[A_1]$, $[A_2]$, and $[A_3]$ can respectively be written in the forms

$$[A_1] = [A_1]_o e^{-kt} = \frac{[A_1]_o(kt)^0 e^{-kt}}{(1-1)!}, \tag{16}$$

$$[A_2] = [A_1]_o(kt)e^{-kt} = \frac{[A_1]_o(kt)^1 e^{-kt}}{(2-1)!}, \tag{17}$$

and

$$[A_3] = \frac{[A_1]_o(kt)^2 e^{-kt}}{(3-1)!}. \tag{18}$$

The result for $[A_i]$ appears to be

$$\boxed{[A_i] = \frac{[A_1]_o(kt)^{i-1}e^{-kt}}{(i-1)!}}. \tag{19}$$

This can be simply, but rigorously, proven by induction. The first step is to assume that the form deduced in Eq. 19 is correct for some value of i. We know that this assumption is valid, since we have rigorously shown Eq. 19 to be true if $i = 1, 2$, or 3. The rate of production of A_{i+1} is

$$\frac{d[A_{i+1}]}{dt} = k[A_i] - k[A_{i+1}]. \tag{20}$$

Rearranging terms in Eq. 20 and then substituting Eq. 19 for $[A_i]$ gives

$$d[A_{i+1}] + k[A_{i+1}] = k[A_i]dt = \frac{[A_1]_o k^i t^{i-1} e^{-kt}}{(i-1)!} dt. \tag{21}$$

Again, we employ an integrating factor of the form $I = e^{kt}$. This yields

$$e^{kt}[d[A_{i+1}] + k[A_{i+1}]] = d[e^{kt}[A_{i+1}]] = \frac{[A_1]_o k^i t^{i-1}}{(i-1)!} dt. \tag{22}$$

Integrating both sides between the limits $t = [A_{i+1}] = 0$ and arbitrary upper limits t and $[A_{i+1}]$ produces

$$\int_{t=[A_{i+1}]=0}^{t[A_{i+1}]} d[e^{kt}[A_{i+1}]] = e^{kt}[A_{i+1}] = \frac{[A_1]_o k^i}{(i-1)!}\int_{t=0}^{t} t^{i-1} dt = \frac{[A_1]_o k^i t^i}{i(i-1)!}$$

$$= \frac{[A_1]_o (kt)^i}{(i)!}. \tag{23}$$

Dividing by e^{kt} gives the final result:

$$\boxed{[A_{i+1}] = \frac{[A_1]_o (kt)^i}{(i)!} e^{-kt}}. \tag{24}$$

Thus, we have proven that if Eq. 19 holds for $[A_i]$, then Eq. 24 must hold for $[A_{i+1}]$. Hence, an equation of the form of Eq. 24 holds for all $[A_i]$ for any value of i. Equations 19 and 24 are identical to Eq. 19.70 in the text, which was stated without proof.

19.41 Monomer M_1 is polymerized inside the reaction vessel shown in the diagram at the top of the next page.

The overall process is illustrated in the above figure. Monomer is continuously added to the chamber at a rate sufficient to keep the rate of monomer reaction, $d[M_1]/dt$, equal to a constant that we denote by R mol L^{-1} min^{-1}. During the reaction, the temperature is held constant, and the concentration of M_1 is sufficiently large to make all the polymerization steps behave as pseudo first-order reactions with a common pseudo first-order rate coefficient k. That is, we have $k = k_{\text{true}}[M_1] = k'[M_1]$ for all polymerization steps, where $k_{\text{true}} = k'$ is the actual second-order rate coefficient. Subsequent to the addition of monomer, polymerization begins. The first two steps in the process are

$$M_1 + M_1 \xrightarrow{k_2} M_2$$

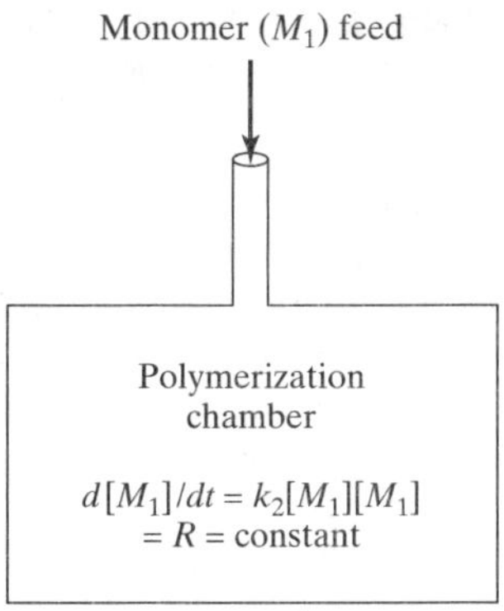

and

$$M_1 + M_2 \xrightarrow{k=k_{\text{true}}[M_1]=k'[M_1]} M_3.$$

(A) Obtain an expression giving $[M_2]$ as a function of time, k, and R. Use an integrating factor to solve the differential equation.
(B) Does $[M_2]$ attain a steady state? If not, why not? If it does, what is the steady-state $[M_2]$, and does the result agree with the prediction of the stationary-state approximation?

Solution

(A) The rate of change of M_2 is given by

$$\frac{d[M_2]}{dt} = k_2[M_1]^2 - k'[M_1][M_2] = R - k[M_2], \quad \textbf{(1)}$$

since the conditions of the problem tell us that $k_2[M_1]^2 = R =$ constant and the second step can be treated as a pseudo-first-order reaction with $k = k'[M_1]$. Rearranging terms in Eq. 1 yields

$$d[M_2] + k[M_2]dt = Rdt. \quad \textbf{(2)}$$

The left side of Eq. 2 can be made into an exact differential by multiplying by the integrating factor

$$I = \exp\left[\int k dt\right] = e^{kt}. \quad \textbf{(3)}$$

Doing so gives

$$e^{kt}d[M_2] + ke^{kt}[M_2]dt = d\{e^{kt}[M_2]\} = Re^{kt}dt. \quad \textbf{(4)}$$

Integrating both sides of Eq. 4 between the corresponding limits ($t = 0$, $[M_2] = 0$) and ($t = t$, $[M_2] = [M_2]$) produces

$$\int_{t=0[M_2]=0}^{t[M_2]} d\{e^{kt}[M_2]\} = [M_2]e^{kt} = \int_0^t Re^{kt}dt = \frac{R}{k}e^{kt}\Big|_0^t = \frac{R}{k}[e^{kt} - 1]. \quad \textbf{(5)}$$

Dividing by e^{kt} gives the desired result:

$$\boxed{[M_2] = \frac{R}{k} - \frac{Re^{-kt}}{k}}. \quad \textbf{(6)}$$

(B) Equation 6 shows that as $t \longrightarrow \infty$, $[M_2]$ attains a steady-state concentration equal to R/k. If we employ the stationary-state approximation for M_2, we have, from Eq. 1,

$$\frac{d[M_2]}{dt} = R - k[M_2] \approx 0, \tag{7}$$

so that

$$\boxed{[M_2]_{ss} = \frac{R}{k}}, \tag{8}$$

which is the same result as that predicted by Eq. 6.

19.43 The concentration of the nth-mer, $[M_n]$, in the polymerization process described in Problems 19.41 and 19.42 is

$$[M_n] = \frac{R}{k} - \frac{R}{k}\left[1 + kt + \frac{(kt)^2}{2!} + \frac{(kt)^3}{3!} + \cdots + \frac{(kt)^{n-2}}{(n-2)!}\right]e^{-kt}.$$

(A) Does $[M_n]$ attain a steady state? If so, what is its value in the steady state?
(B) Show that at any finite time the concentration of M_∞ is zero.

Solution

(A) Yes, $[M_n]$ attains a steady state as t becomes large. The limiting value of $[M_n]$ as $t \longrightarrow \infty$ can be seen to be R/k, which is the steady-state concentration.

(B) The concentration of M_∞ is given by

$$[M_\infty] = \frac{R}{k} - \frac{Re^{-kt}}{k} \lim_{n \longrightarrow \infty}\left[1 + kt + \frac{(kt)^2}{2!} + \frac{(kt)^3}{3!} + \cdots + \frac{(kt)^{n-2}}{(n-2)!}\right]. \tag{1}$$

However, the infinite series produced by the limit expression in Eq. 1 is e^{kt}. That is, the series expansion for e^{kt} is

$$e^{kt} = 1 + kt + \frac{(kt)^2}{2!} + \frac{(kt)^3}{3!} + \cdots + \frac{(kt)^{n-2}}{(n-2)!} + \cdots + \frac{(kt)^\infty}{\infty!}. \tag{2}$$

Therefore,

$$[M_\infty] = \frac{R}{k} - \frac{Re^{-kt}}{k}e^{kt} = \frac{R}{k} - \frac{R}{k} = 0, \tag{3}$$

which is what we are asked to prove.

19.45 The N_2O_5-catalyzed gas-phase decomposition of O_3 to produce O_2 is presumed to occur via the following mechanism:

$$N_2O_5 \xrightarrow{k_1} NO_2 + NO_3 \qquad \text{(Reaction 1);}$$

$$NO_2 + NO_3 \xrightarrow{k_2} N_2O_5 \qquad \text{(Reaction 2);}$$

$$NO_2 + O_3 \xrightarrow{k_3} NO_3 + O_2 \qquad \text{(Reaction 3);}$$

$$2\,NO_3 \xrightarrow{k_4} 2\,NO_2 + O_2 \qquad \text{(Reaction 4).}$$

For the first two reactions, the equilibrium constant $K_{12} \ll 1$, with both k_1 and k_2 large relative to k_3 and k_4.

(A) Assuming that Reaction 1 is first order and that all other reactions are second order, write down the differential rate equations for all the species.

(B) By employing the stationary-state approximation, obtain expressions for the steady-state concentrations, and express the rate of reaction of O_3 in terms of $[O_3]$ and $[N_2O_5]$.

Solution

(A) The differential rate equations are

$$\frac{d[N_2O_5]}{dt} = -k_1[N_2O_5] + k_2[NO_2][NO_3], \quad (1)$$

$$\frac{d[NO_2]}{dt} = k_1[N_2O_5] - k_2[NO_2][NO_3] - k_3[NO_2][O_3] + 2k_4[NO_3]^2, \quad (2)$$

$$\frac{d[NO_3]}{dt} = k_1[N_2O_5] - k_2[NO_2][NO_3] + k_3[NO_2][O_3] - 2k_4[NO_3]^2, \quad (3)$$

$$\frac{d[O_3]}{dt} = -k_3[NO_2][O_3], \quad (4)$$

and

$$\frac{d[O_2]}{dt} = k_3[NO_2][O_3] + k_4[NO_3]^2, \quad (5)$$

where the factors of two appear because of the stoichiometric coefficients in reaction (4).

(B) Since the problem tells us that k_1 and k_2 are both large relative to k_3 and k_4, we can be assured that N_2O_5, NO_2, and NO_3 are in a state of equilibrium throughout the reaction. Therefore, the rates of reactions (1) and (2) are equal, which gives

$$k_1[N_2O_5] = k_2[NO_2][NO_3], \quad (6)$$

provided that we can use concentrations rather than activities. Solving for $[NO_2]$, we obtain

$$[NO_2] = \frac{k_1[N_2O_5]}{k_2[NO_3]}. \quad (7)$$

Combining Eqs. 6 and 3 shows that we have

$$\frac{d[NO_3]}{dt} = k_1[N_2O_5] - k_2[NO_2][NO_3] + k_3[NO_2][O_3] - 2k_4[NO_3]^2$$

$$= k_3[NO_2][O_3] - 2k_4[NO_3]^2. \quad (8)$$

If we use the steady-state approximation and set $d[NO_3]/dt \approx 0$, Eq. 8 gives

$$\frac{d[NO_3]}{dt} = k_3[NO_2][O_3] - 2k_4[NO_3]^2 = 0, \quad (9)$$

so that

$$[NO_3]^2 = \frac{k_3[NO_2][O_3]}{2k_4}. \quad (10)$$

Substituting Eq. 7 into Eq. 10 for $[NO_2]$ produces

$$[NO_3]^2 = \left[\frac{k_3k_1[N_2O_5][O_3]}{2k_2k_4[NO_3]}\right]. \tag{11}$$

Multiplying by $[NO_3]$ gives

$$[NO_3]^3 = \left[\frac{k_3k_1[N_2O_5][O_3]}{2k_2k_4}\right]. \tag{12}$$

Therefore,

$$[NO_3] = \left[\frac{k_3k_1[N_2O_5][O_3]}{2k_2k_4}\right]^{1/3}. \tag{13}$$

Equations 7 and 13 give us the stationary-state concentrations of NO_2 and NO_3, respectively. The differential rate of reaction of $[O_3]$ is given by Eq. 4. If we substitute Eq. 7 for $[NO_2]$, the result is

$$\frac{d[O_3]}{dt} = -\frac{k_3k_1[N_2O_5][O_3]}{k_2[NO_3]}. \tag{14}$$

Inserting Eq. 13 for $[NO_3]$ into Eq. 14 produces

$$\boxed{\frac{d[O_3]}{dt} = -\frac{\dfrac{k_3k_1[N_2O_5][O_3]}{k_2}}{\left[\dfrac{k_3k_1[N_2O_5][O_3]}{2k_2k_4}\right]^{1/3}} = -(2k_4)^{1/3}\left[\frac{k_3k_1[N_2O_5][O_3]}{k_2}\right]^{2/3}}, \tag{15}$$

so that $d[O_3]/dt$ is expressed in terms of $[N_2O_5]$ and $[O_3]$ alone. We see, then, that the rate of reaction of O_3 will be of two-thirds order in both $[N_2O_5]$ and $[O_3]$. When we employ the stationary-state approximation, such fractional orders are often the result.

19.47 Compounds A and B are in equilibrium with compound C, which also reacts via a pseudo first-order process to form product P. The reaction mechanism is

$$A + B \xrightarrow{k_1} C;$$
$$C \xrightarrow{k_2} A + B;$$
$$C \xrightarrow{k_3} P.$$

(A) Show that this mechanism will appear to be one for which the rate of change of A is first order with respect to both A and B and second-order overall, provided that a stationary-state approximation is accurate for compound C.
(B) What will be the apparent second-order rate coefficient in terms of k_1, k_2, and k_3?

Solution

If a stationary-state approximation can be made for compound C, then we have

$$\frac{d[C]}{dt} = k_1[A][B] - k_2[C] - k_3[C] \approx 0. \tag{1}$$

The steady-state concentration of C is, therefore,

$$[C]_{ss} = \frac{k_1[A][B]}{k_2 + k_3}. \tag{2}$$

The rate of change of compound A is given by

$$\frac{d[A]}{dt} = -k_1[A][B] + k_2[C]. \tag{3}$$

Inserting $[C]_{ss}$ for $[C]$ in Eq. 3 produces

$$\frac{d[A]}{dt} = -k_1[A][B] + \frac{k_2k_1[A][B]}{k_2 + k_3} = -\left[k_1 - \frac{k_1k_2}{k_2 + k_3}\right][A][B], \quad \textbf{(4)}$$

and the rate will appear to be first order with respect to both A and B and second order overall. The apparent rate coefficient for the process will be

$$\boxed{k_a = \left[k_1 - \frac{k_1k_2}{k_2 + k_3}\right] = \left[\frac{k_1k_3}{k_2 + k_3}\right].} \quad \textbf{(5)}$$

19.49 The initial rate of an enzyme-catalyzed reaction is measured as a function of the concentration of the substrate by using an initial enzyme concentration of 1.75×10^{-9} mol L^{-1}. The data obtained are shown in the table to the right. Use these data to determine k_2 and the Michaelis constant for the reaction at the temperature of the experiment. (For a definition of k_2, see the discussion preceding Eq. 19.80 in the text.)

$[S]_o$ (mol L^{-1})	Initial Rate $\mathcal{R}_o$ (mol L^{-1} s^{-1})
0.0059	0.000048
0.0108	0.000066
0.0157	0.000077
0.0206	0.000085
0.0255	0.000091
0.0304	0.000095
0.0353	0.000098
0.0402	0.000100
0.0451	0.000103

Solution

Using the preceding data, we may easily produce a Lineweaver plot in which $1/\mathcal{R}_o$ is plotted against $1/[S]_o$. According to Eq. 19.87, we should observe

$$\frac{1}{\mathcal{R}_o} = \frac{1}{k_2[E]_o} + \frac{K_M}{k_2[S]_o[E]_o}, \quad \textbf{(1)}$$

so that the intercept is $1/(k_2[E]_o)$ and the slope is $K_M/(k_2[E]_o)$. The plot is shown at the end of the problem. A least-squares fit to the data points yields

$$\frac{1}{\mathcal{R}_o} = 8{,}060 + \frac{75.7}{[S]_o}\,(\text{L s mol}^{-1}). \quad \textbf{(2)}$$

Using the intercept, we obtain

$$k_2 = \frac{1}{(\text{intercept})[E]_o} = \frac{1}{(8{,}060 \text{ L s mol}^{-1})(1.75 \times 10^{-9} \text{ mol L}^{-1})} = \underline{7.09 \times 10^4 \text{ s}^{-1}}. \quad \textbf{(3)}$$

From the slope, we have

$$K_M = k_2[E]_o(\text{slope}) = (7.09 \times 10^4 \text{ s}^{-1})(1.75 \times 10^{-9} \text{ mol L}^{-1})(75.7 \text{ s})$$
$$= \underline{9.39 \times 10^{-3} \text{ mol L}^{-1}}. \quad \textbf{(4)}$$

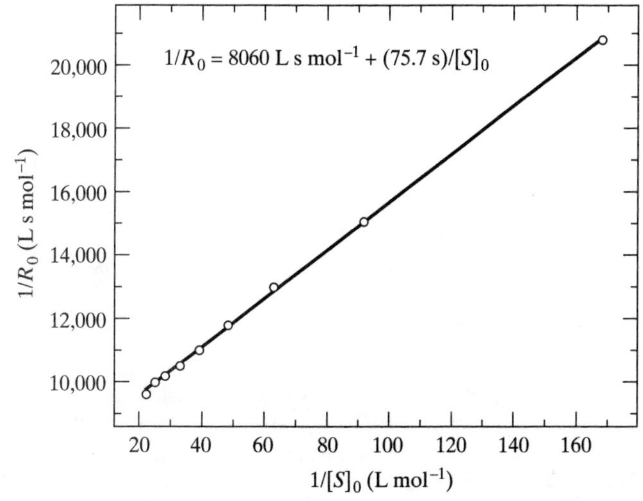

19.51 The extent of surface coverage for an absorbate $A(g)$ on a particular surface is measured as a function of the pressure of $A(g)$ at 273 K. The data obtained are as follows:

p_A (torr)	θ_e	p_A (torr)	θ_e
100.0	0.2593	600.0	0.6774
200.0	0.4118	700.0	0.7101
300.0	0.5122	800.0	0.7368
400.0	0.5833	900.0	0.7590
500.0	0.6364		

From these data, determine K, assuming that the absorption process is accurately described by a Langmuir isotherm. (For a definition of K, see Eq. 19.91 in the text.)

Solution

Equation 19.92,

$$\frac{1}{\theta_e} = 1 + \frac{1}{Kp_A} \tag{1}$$

shows that a plot of $1/\theta_e$ versus $1/p_A$ should be linear. Such a plot is shown at the end of the problem. The best straight-line fit to the data is

$$\frac{1}{\theta_e} = 1.00 + \frac{285.6}{p_A}. \tag{2}$$

Therefore,

$$\frac{1}{K} = \text{slope} = 285.6 \text{ torr}, \tag{3}$$

so that

$$K = \frac{1}{285.6 \text{ torr}} = \underline{0.00350 \text{ torr}^{-1}}. \tag{4}$$

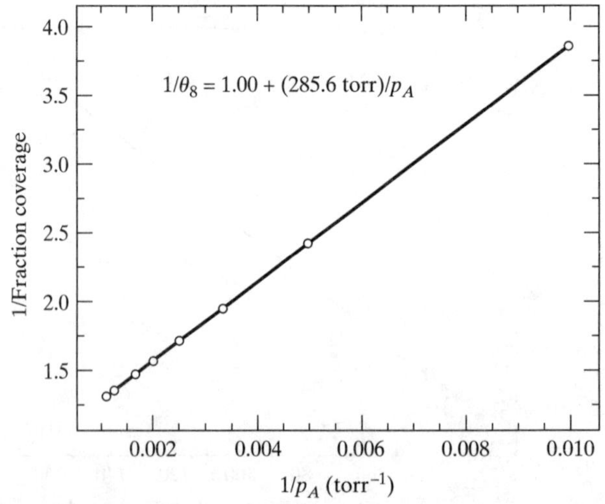

19.53 Starting with Eq. 19.98 and a similar equation for Compound B, derive Eq. 19.99. (*Hint*: Remember that the fractional coverage at steady state or equilibrium, θ_e, can be written as the sum of the fractions covered by each of the absorbing compounds. That is, $\theta_e = \theta_e^A + \theta_e^B + \theta_e^C + \ldots$.)

Solution

Equation 19.98, written for both compounds A and B, is

$$\theta_e^A = \frac{k_a p_A(1-\theta_e)}{k_{da}} = K_A p_A(1-\theta_e) = K_A p_A(1-\theta_e^A-\theta_e^B) \quad (1)$$

and

$$\theta_e^B = \frac{k_b p_B(1-\theta_e)}{k_{db}} = K_B p_B(1-\theta_e) = K_B p_B(1-\theta_e^A-\theta_e^B). \quad (2)$$

Solving Eq. 2 for θ_e^B, we obtain

$$\theta_e^B[1+K_B p_B] = K_B p_B(1-\theta_e^A), \quad (3)$$

so that

$$\theta_e^B = \frac{K_B p_B(1-\theta_e^A)}{1+K_B p_B}. \quad (4)$$

Substituting Eq. 4 into Eq. 1 gives

$$\theta_e^A = K_A P_A\left(1-\theta_e^A-\frac{K_B p_B(1-\theta_e^A)}{1+K_B p_B}\right). \quad (5)$$

Taking a common denominator on the right-hand side of Eq. 5, we obtain

$$\theta_e^A = \frac{K_A p_A(1+K_B p_B-\theta_e^A-\theta_e^A K_B p_B-K_B p_B+\theta_e^A K_B p_B)}{1+k_B p_B}$$

$$= \frac{K_A p_A(1-\theta_e^A)}{1+K_B p_B}. \quad (6)$$

Isolating θ_e^A, we get

$$\theta_e^A + \frac{K_A p_A \theta_e^A}{1+K_B p_B} = \frac{\theta_e^A(1+K_B p_B+K_A p_A)}{1+K_B p_B} = \frac{K_A p_A}{1+K_B p_B}. \quad (7)$$

Canceling $1+K_B p_B$ in both denominators in Eq. 7 and solving for θ_e^A gives

$$\boxed{\theta_e^A = \frac{K_A p_A}{1+K_A p_A+K_B p_B}}. \quad (8)$$

Substituting Eq. 8 into Eq. 4 produces

$$\theta_e^B = [1+K_B p_B]^{-1} K_B p_B\left[1-\frac{K_A p_A}{1+K_A p_A+K_B p_B}\right]$$

$$= [1+K_B p_B]^{-1} K_B p_B\left[\frac{1+K_A p_A+K_B p_B-K_A p_A}{1+K_A p_A+K_B p_B}\right]$$

$$= [1+K_B p_B]^{-1} K_B p_B\left[\frac{1+K_B p_B}{1+K_A p_A+K_B p_B}\right] = \boxed{\frac{K_B p_B}{1+K_A p_A+K_B p_B}}. \quad (9)$$

Equations 8 and 9 are identical to Eq. 19.99 in the text.

Time (s)	$\ln(p_A)$ (torr)
0.0	2.303
100.0	2.196
200.0	2.090
300.0	1.983
400.0	1.877
500.0	1.771
600.0	1.664
700.0	1.558
800.0	1.451
900.0	1.345
1,000.0	1.239

Time (s)	$[F]$ (mol cm^{-3})
0.0	0.00
1000.0	5.69×10^{-6}
5000.0	2.84×10^{-5}
10,000.0	5.69×10^{-5}
30,000.0	1.71×10^{-4}

19.55[#] An investigator is examining the surface-catalyzed unimolecular decomposition of compound $A(g)$ at 300 K. After cleaning the catalyst, he follows the gas-phase pressure of $A(g)$ as a function of time, starting with an initial pressure of 10.00 torr. His data are in the top left table. In a second experiment, the investigator starts with p_A equal to 76,000 torr. He then measures the concentration of decomposition product F as a function of time. His results are shown in the bottom left table.

(A) Express the low-pressure limiting rate for the surface-catalyzed unimolecular decomposition of compound $A(g)$ at 300 K in terms of the rate of change of the gas-phase pressure of A.

(B) Use the low-pressure data to determine a value for the $k_r NK$ product, where the symbols are as defined in Eq. 19.101.

(C) Use the high-pressure data to obtain the value for the $k_r N$ product. What is the value of K?

(D) Plot the surface-catalyzed rate as a function of p_A at 300 K over the range $0 \leq p_A \leq 2{,}000$ torr.

Solution

(A) The rate of the surface-catalyzed, unimolecular decomposition of compound $A(g)$ at 300 K is given by Eq. 19.101:

$$\mathcal{R} = \frac{d[F]}{dt} = -\frac{d[A(g)]}{dt} = \frac{k_r NK p_A}{1 + Kp_A}. \quad (1)$$

At low pressure, we have $Kp_A \ll 1$, so that

$$-\frac{d[A(g)]}{dt} = k_r NK p_A. \quad (2)$$

Using the ideal-gas law, we can write $[A(g)] = n/V = p_A/RT$. Substituting this expression into Eq. 2 gives

$$\boxed{\mathcal{R} = \frac{dp_A}{dt} = -RTk_r NK p_A}, \quad (3)$$

which is the desired equation.

(B) Separating the variables in Eq. 3 produces

$$\frac{dp_A}{p_A} = -RTk_r NK\,dt = -C\,dt. \quad (4)$$

Integrating Eq. 4 between corresponding limits yields

$$\int_{p_A=10}^{p_A} \frac{dp_A}{p_A} = \ln(p_A) - \ln(10) = -C\int_{t=0}^{t} dt = -Ct. \quad (5)$$

Thus,

$$\ln(p_A) = \ln(10) - Ct. \quad (6)$$

If we plot $\ln(p_A)$ against time, the slope will yield the value of C. Such a plot is shown at the top of the next page on the left. The least-squares linear-fit equation is

$$\ln(p_A) = 2.303 - 0.001064t. \quad (7)$$

Therefore, $C = 0.001064\ \text{s}^{-1}$.

Using the fact that $C = RTk_r NK$, we have

$$k_r NK = \frac{C}{RT} = \frac{(0.001064\ \text{s}^{-1})}{(62{,}370\ \text{cm}^3\ \text{torr mol}^{-1}\ \text{K}^{-1})(300\ \text{K})}$$
$$= \underline{5.69 \times 10^{-11}\ \text{mol cm}^{-3}\ \text{torr}^{-1}\ \text{s}^{-1}}. \quad (8)$$

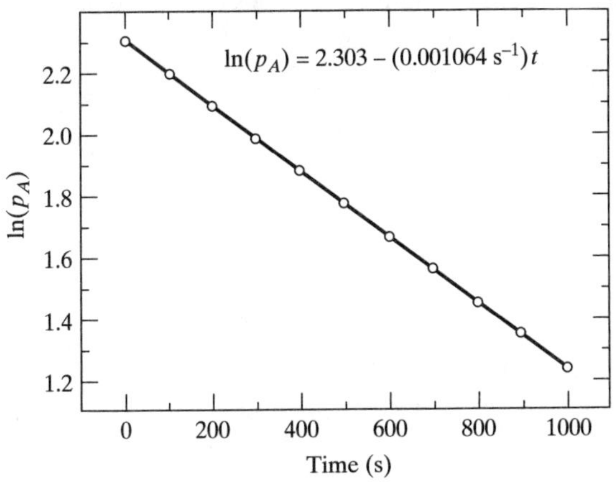

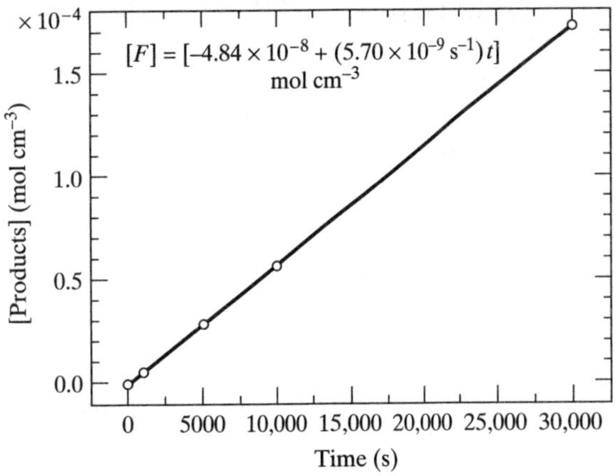

(C) At high pressure, we have $Kp_A \gg 1$, so that the high-pressure limiting rate expression, from Eq. 1, is

$$\mathcal{R} = \frac{d[F]}{dt} = k_r N. \quad \textbf{(9)}$$

Consequently, the rate is zero order, and we have

$$\int_{[F]=0}^{[F]} d[F] = [F] = \int_{t=0}^{t} k_r N\,dt = k_r N t. \quad \textbf{(10)}$$

Therefore, a plot of $[F]$ versus time should be linear with a slope equal to $k_r N$. Such a plot is shown above at the right. The slope is 5.70×10^{-9} mol cm^{-3} s^{-1}. Consequently,

$$k_r N = \underline{5.70 \times 10^{-9}\ \text{mol cm}^{-3}\ \text{s}^{-1}}. \quad \textbf{(11)}$$

Dividing Eq. 8 by Eq. 11 gives

$$\frac{k_r N K}{k_r N} = K = \frac{5.69 \times 10^{-11}\ \text{mol cm}^{-3}\ \text{torr}^{-1}\ \text{s}^{-1}}{5.70 \times 10^{-9}\ \text{mol cm}^{-3}\ \text{s}^{-1}} = \underline{9.98 \times 10^{-3}\ \text{torr}^{-1}}. \quad \textbf{(12)}$$

(D) The requested plot is shown below.

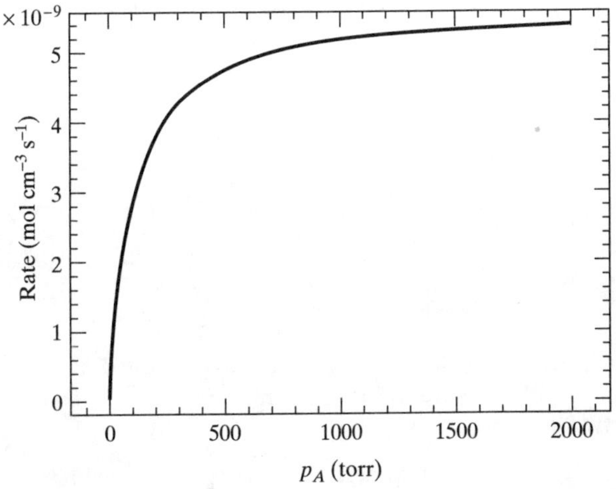

19.57 An investigator suspects that a single reactant is simultaneously undergoing first- and second-order reactions via the mechanism

$$A \xrightarrow{k_1} B \quad \text{and} \quad A + A \xrightarrow{k_1} C.$$

She measures the concentration of A as a function of time during the initial stages of the reaction for two different concentrations. Her data are shown in the following pair of tables:

Data Set 1	
Time (min)	**[A] (mol L^{-1})**
0.00	0.800
1.00	0.784
2.00	0.768
3.00	0.752
4.00	0.737
5.00	0.723
6.00	0.709

Data Set 2	
Time (min)	**[A] (mol L^{-1})**
0.00	0.400
1.00	0.395
2.00	0.390
3.00	0.385
4.00	0.380
5.00	0.375
6.00	0.371

Using Eqs. 9.173 and 9.174 to obtain $d[A]/dt$ at $t = 3.00$ min for each data set, determine k_1 and k_2.

Solution

Equations 9.173 and 9.174, which are derived in Problem 9.52, tell us that the derivative at $t = 3.00$ min is

$$\frac{d[A]}{dt} = \frac{(0.75([A]_{4.0} - [A]_{2.0}) - 0.15([A]_{5.0} - [A]_{1.0}) + ([A]_{6.0} - [A]_{0.0})/60)}{1.00}, \tag{1}$$

since the step size h is 1.00 min for the data. Substituting the data into Eq. 1 gives

$$\left(\frac{d[A]}{dt}\right)_{\text{Set 1}} = \frac{0.75[0.737 - 0.768] - 0.15[0.723 - 0.784] + [0.709 - 0.800]/60}{1.00}$$

$$= -0.01562 \text{ mol L}^{-1} \text{ min}^{-1} \tag{2}$$

and

$$\left(\frac{d[A]}{dt}\right)_{\text{Set 2}} = \frac{0.75[0.380 - 0.390] - 0.15[0.375 - 0.395] + [0.371 - 0.4000]/60}{1.00}$$

$$= -0.004983 \text{ mol L}^{-1} \text{ min}^{-1}. \tag{3}$$

With the two rates in hand, we can now employ the methods used in Example 19.13. Substituting the two rates into the differential rate equation yields two linear equations:

$$-0.01562 = -k_1(0.752) - 2k_2(0.752)^2 = -0.752k_1 - 1.131k_2 \tag{4}$$

and

$$0.004983 = 0.385k_1 + 2(0.385)^2k_2 = 0.385k_1 + 0.2964k_2. \quad (5)$$

Solving Eq. 4 for k_1 in terms of k_2 gives

$$k_1 = \frac{0.01562 - 1.131k_2}{0.752} = 0.02077 - 1.504k_2. \quad (6)$$

Substituting Eq. 6 into Eq. 5 produces

$$0.004983 = 0.385[0.02077 - 1.504k_2] + 0.2964k_2 = 0.007996 - 0.2826k_2. \quad (7)$$

Thus,

$$k_2 = \frac{0.007996 - 0.004983}{0.2826}\ \mathrm{L\ mol^{-1}\ min^{-1}} = \underline{0.0107\ \mathrm{L\ mol^{-1}\ min^{-1}}}, \quad (8)$$

where the units are those for a second-order reaction. Substituting k_2 into Eq. 6 gives

$$k_1 = 0.02077 - 1.504(0.0107) = \underline{0.00468\ \mathrm{min^{-1}}}, \quad (9)$$

which are the units for a first-order reaction.

19.59 Temperature-jump measurements are taken on an equilibrium system exhibiting reactions of the form

$$A + B \xrightarrow{k_1} \mathrm{C} - \Delta\overline{\mathrm{H}}$$

and

$$-\Delta\overline{\mathrm{H}} + \mathrm{C} \xrightarrow{k_2} A + B.$$

The investigator uses different initial concentrations of A and B in three different kinetic measurements. In each case, he determines the relaxation time of the equilibrium system after a small jump in the temperature. The results are given in the following table:

$[A]_e$	$[B]_e$ (mmol L^{-1})	τ (s)
22.36	22.36	2.14×10^{-7}
10.00	10.00	4.55×10^{-7}
7.07	7.07	6.20×10^{-7}

Use these data to determine k_1 and k_2 for the equilibrium reaction.

Solution

Equation 19.120 describes the dependence of the relaxation time for this type of equilibrium upon the equilibrium concentrations of A and B:

$$\frac{1}{\tau} = k_1\{[A]_e + [B]_e\} + k_2. \quad (1)$$

If we plot τ^{-1} against $[A]_e + [B]_e$, the result should be linear with a slope equal to k_1 and an intercept equal to k_2. Such a plot is shown in the figure on the next page. A linear least-squares fit of a straight line to the data yields the result

$$\frac{1}{\tau} = 1.97 \times 10^5 + 1.00 \times 10^5\{[A]_e + [B]_e\}\ \text{s}^{-1}. \quad \textbf{(2)}$$

Therefore, we must have

$$k_1 = \underline{1.00 \times 10^5\ \text{L mmol}^{-1}\ \text{s}^{-1}} \quad \textbf{(3)}$$

and

$$k_2 = \underline{1.97 \times 10^5\ \text{s}^{-1}}. \quad \textbf{(4)}$$

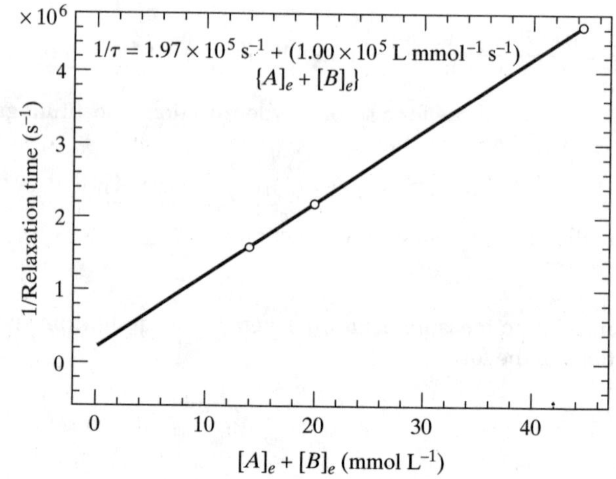

19.61 Sam has located some kinetic data that puzzle him. He would like your comments. The data, for the gas-phase decomposition of a single reactant that we shall label $A(g)$, are as follows:

Time (min)	$[A(g)]$ (mol L^{-1})
0.0	0.1000
4.0	0.0961
8.0	0.0923
12.0	0.0887
16.0	0.0852
20.0	0.0819

In an effort to determine the order of the reaction, Sam has plotted $\ln[A]$ versus time to test for first-order kinetics and also $[A]^{-1}$ versus time to test for second-order kinetics. His plots are shown at the top of the next page, where the lines are least-square fits to the data.

As can be seen, both plots are essentially linear. The plot of $\ln[A]$ versus time has a slope of $-0.00998\ \text{min}^{-1}$, which suggests that the reaction is first order with a rate coefficient $k = 0.00998\ \text{min}^{-1}$. The plot of $[A]^{-1}$ versus time has a slope of $0.110\ \text{L mol}^{-1}\ \text{min}^{-1}$. This suggests that the kinetics are second order

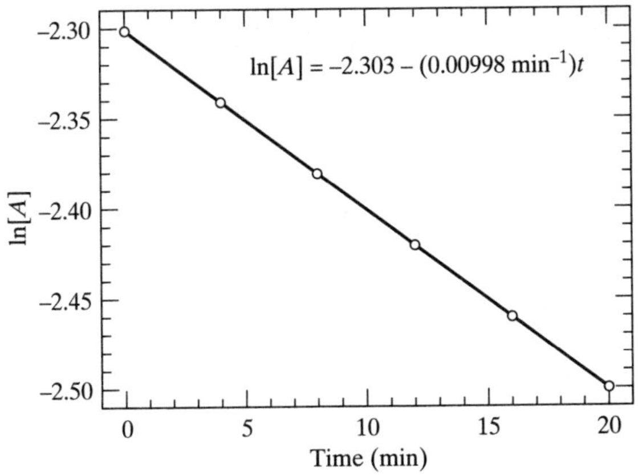

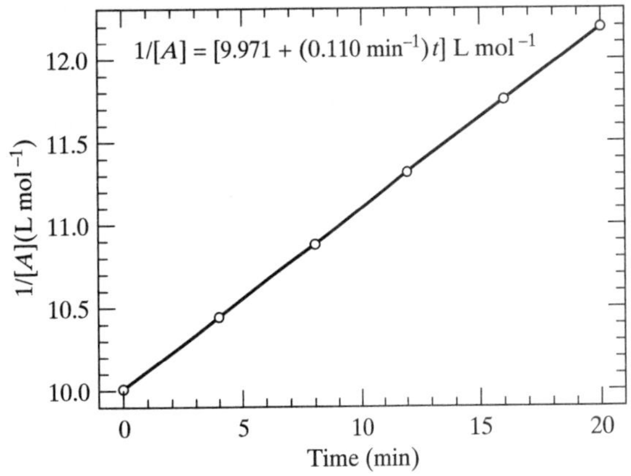

with a rate coefficient such that $2k = 0.110 \text{ L mol}^{-1} \text{ min}^{-1}$. What is wrong? Can you assist Sam? Here are some suggestions:

(A) The integrated rate expression for a first-order reaction of Compound A is $[A] = [A]_o e^{-kt}$. Write this expression in the form of a power series expansion in time about the point $t = 0$.

(B) The integrated rate law for a second-order reaction with a single component is given by Eq. 19.16. If we combine the factor of two with the rate coefficient, this expression is $[A]^{-1} = [A]_o^{-1} + k_2 t$. Solve the preceding equation for $[A]$, and expand the result in a power series in time about the point $t = 0$.

(C) Let us define the quantity $k_2[A]_o = k$. In terms of k and time, what is the power series expansion for $[A]$ in a second-order reaction with a single component? Compare this expression with the result obtained in (A). Comment on the similarities and the differences.

(D) Compute $[A]$ for both a first-order and a second-order process with $[A]_o = 0.100 \text{ mol L}^{-1}$ and $k = k_2[A]_o = 0.0100 \text{ min}^{-1}$. Plot $\ln[A]$ versus time for both cases over the interval $0 \leq t \leq 30$ minutes. Comment on the result. Having done (A)–(D), can you help Sam?

Solution

(A) The power series expansion for e^{-ax} is

$$e^{-ax} = 1 - ax + \frac{(ax)^2}{2!} - \frac{(ax)^3}{3!} + \frac{(ax)^4}{4!} - \ldots. \tag{1}$$

Therefore, the integrated first-order rate law can be written in the form

$$\boxed{[A] = [A]_o e^{-kt} = [A]_o\left\{1 - kt + \frac{(kt)^2}{2!} - \frac{(kt)^3}{3!} + \frac{(kt)^4}{4!} - \cdots\right\}}. \tag{2}$$

(B) The integrated second-order expression is

$$\frac{1}{[A]} = \frac{1}{[A]_o} + k_2 t = \frac{1 + k_2[A]_o t}{[A]_o}. \tag{3}$$

Inverting this result, we obtain

$$[A] = \frac{[A]_o}{1 + k_2[A]_o t}. \tag{4}$$

We now wish to expand the function $[1 + k_2[A]_o t]^{-1}$ in a power series in time about the point $t = 0$. The technique for executing this expansion was described in Chapter 1. If we write

$$f(t) = [1 + k_2[A]_o t]^{-1}, \tag{5}$$

the nth expansion coefficient is

$$a_n = \frac{1}{n!}\frac{d^n f(t)}{dt^n}\bigg|_{t=0}. \tag{6}$$

Thus,

$$a_0 = \frac{1}{0!}f(0) = 1, \tag{7}$$

$$a_1 = \frac{1}{1!}\frac{df(t)}{dt}\bigg|_{t=0} = -[1 + k_2[A]_o t]^{-2}k_2[A]_o\bigg|_{t=0} = -k_2[A]_o, \tag{8}$$

$$a_2 = \frac{1}{2!}\frac{d^2 f(t)}{dt^2}\bigg|_{t=0} = \frac{2}{2!}[1 + k_2[A]_o t]^{-3}\{k[A]_o\}^2\bigg|_{t=0} = \{k_2[A]_o\}^2, \tag{9}$$

$$a_3 = \frac{1}{3!}\frac{d^3 f(t)}{dt^3}\bigg|_{t=0} = -\frac{6}{3!}[1 + k_2[A]_o t]^{-4}\{k[A]_o\}^3\bigg|_{t=0} = -\{k_2[A]_o\}^3, \tag{10}$$

and

$$a_4 = \frac{1}{4!}\frac{d^4 f(t)}{dt^4}\bigg|_{t=0} = \frac{24}{4!}[1 + k_2[A]_o t]^{-5}\{k[A]_o\}^4\bigg|_{t=0} = \{k_2[A]_o\}^4. \tag{11}$$

Therefore, the expansion is

$$[1 + k_2[A]_o t]^{-1} = 1 - k_2[A]_o t + \{k_2[A]_o\}^2 t^2 - \{k_2[A]_o\}^3 t^3 + \{k_2[A]_o\}^4 t^4 - \cdots, \tag{12}$$

so that the integrated rate law, from Eq. 4, is

$$\boxed{[A] = [A]_o\{1 - k_2[A]_o t + \{k_2[A]_o\}^2 t^2 - \{k_2[A]_o\}^3 t^3 + \{k_2[A]_o\}^4 t^4 - \cdots\}}, \tag{13}$$

(C) If we define $k_2[A]_o = k$, Eq. 13 becomes the second-order equation

$$[A] = [A]_o\{1 - kt + (kt)^2 - (kt)^3 + (kt)^4 - \cdots\}. \tag{14}$$

The result for a first-order reaction is given in Eq. 2. Repeating this for convenience, we have

$$[A] = [A]_o\left\{1 - kt + \frac{(kt)^2}{2!} - \frac{(kt)^3}{3!} + \frac{(kt)^4}{4!} - \cdots\right\}. \tag{15}$$

We see that Eqs. 14 and 15 are identical through the linear term. In order to distinguish the two rate laws, we will have to examine $[A]$ at times for which the quadratic and higher terms become important. This is the reason experimental kineticists usually take data for several half-lives. If that is not done, it becomes very difficult to distinguish one reaction order from the next.

(D) The plot is shown in the figure on the next page. The two reactions appear to be nearly the same. That is, the second-order reaction is nearly the same as the first-order process. The problem is that t = 30 minutes is only 43% of one half-life. Data need to be taken for a much longer period of time if a clear distinction between the reaction orders is to be obtained.

With regard to Sam's problem, we can see that with $k = 0.01$ min^{-1}, the second-order reaction looks almost identical to the first-order reaction over the range $0 \le t \le 20$ min, which is the range for the kinetic data. The problem is that the half-life for the reaction is $\ln 2/k = 0.693/0.010\ \text{min}^{-1} = 69.3$ min. The person who took the data quit too soon! He took data only over approximately one-third of one half-life. It was probably time for his favorite TV program. His research director is not going to be a happy camper!

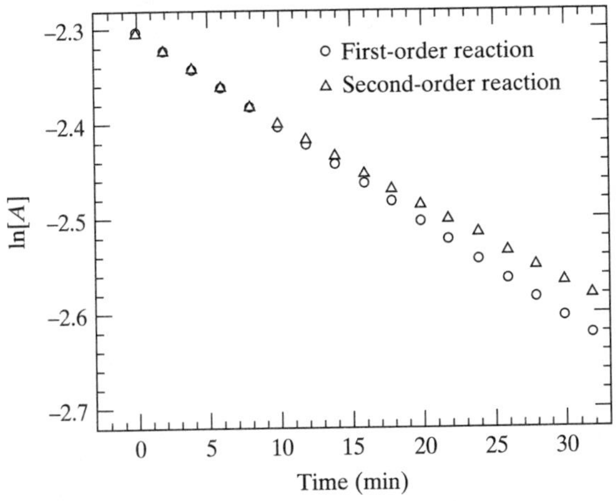

CHAPTER 20

Theoretical Kinetics and Reaction Dynamics

20.1 Starting with Eq. 20.13, obtain Eq. 20.14.

Solution

Equation 20.13 is

$$\mathcal{R} = 4\pi p N_A N_B \left[\frac{\mu}{2\pi kT}\right]^{3/2} \int_{b=0}^{\sigma_{AB}} (2\pi b)\,db \int_{\left(\frac{2E_a}{\mu}\right)^{1/2}}^{\infty} \exp\left[-\frac{\mu V^2}{2kT}\right] V^3\,dV. \quad \textbf{(1)}$$

The integral over the impact parameter is given by

$$\int_{b=0}^{\sigma_{AB}} (2\pi b)\,db = 2\frac{\pi b^2}{2}\bigg|_0^{\sigma_{AB}} = \pi\sigma_{AB}^2. \quad \textbf{(2)}$$

We need to integrate by parts over relative speed. Let us first transform variables. If we define $x = (\mu/2kT)^{1/2}V$, we have

$$dV = \left(\frac{\mu}{2kT}\right)^{-1/2} dx, \quad \textbf{(3)}$$

$$V^3 = \left(\frac{\mu}{2kT}\right)^{-3/2} x^3, \quad \textbf{(4)}$$

and

$$\exp\left[-\frac{\mu V^2}{2kT}\right] = \exp(-x^2). \quad \textbf{(5)}$$

The upper limit will still be infinite. The lower limit becomes

$$x_{\text{lower}} = \left(\frac{\mu}{2kT}\right)^{1/2}\left(\frac{2E_a}{\mu}\right)^{1/2} = \left(\frac{E_a}{kT}\right)^{1/2}. \quad \textbf{(6)}$$

With this substitution, the integral is

$$\int_{\left(\frac{2E_a}{\mu}\right)^{1/2}}^{\infty} \exp\left[-\frac{\mu V^2}{2kT}\right] V^3\,dV = \left(\frac{2kT}{\mu}\right)^2 \int_{x=\left(\frac{E_a}{kT}\right)^{1/2}}^{0} x^3 \exp(-x^2)\,dx. \quad \textbf{(7)}$$

To integrate by parts the first time, we take $dv = x\exp(-x^2)$ and $u = x^2$. This gives $du = 2x\,dx$ and $v = -\frac{1}{2}\exp(-x^2)$. Thus, the integral becomes

$$\int_{x=\left(\frac{E_a}{kT}\right)^{1/2}}^{0} x^3 \exp(-x^2)\,dx = -\frac{x^2}{2}\exp(-x^2)\bigg|_{\left(\frac{E_a}{kT}\right)^{1/2}}^{\infty}$$

$$+ \int_{x=\left(\frac{E_a}{kT}\right)^{1/2}}^{0} x\exp(-x^2)\,dx = \frac{E_a}{2kT}\exp\left(-\frac{E_a}{kT}\right) - \frac{1}{2}\exp(-x^2)\bigg|_{\left(\frac{E_a}{kT}\right)^{1/2}}^{\infty}$$

$$= \frac{E_a}{2kT}\exp\left(-\frac{E_a}{kT}\right) + \frac{1}{2}\exp\left(-\frac{E_a}{kT}\right) = \frac{1}{2}\exp\left(-\frac{E_a}{kT}\right)\left[1 + \frac{E_a}{kT}\right]. \quad \textbf{(8)}$$

Substituting the result in Eq. 8 into Eq. 7 gives

$$\int_{\left(\frac{2E_a}{\mu}\right)^{1/2}}^{\infty} \exp\left[-\frac{\mu V^2}{2kT}\right] V^3\,dV = \left(\frac{2kT}{\mu}\right)^2 \frac{1}{2}\exp\left(-\frac{E_a}{kT}\right)\left[1 + \frac{E_a}{kT}\right]. \quad \textbf{(9)}$$

Combining the results contained in Eqs. 2 and 9 with Eq. 1 produces

$$\mathcal{R} = 4\pi p N_A N_B \left[\frac{\mu}{2\pi kT}\right]^{3/2} (\pi\sigma_{AB}^2)\left(\frac{2kT}{\mu}\right)^2 \left(\frac{1}{2}\right)\exp\left(-\frac{E_a}{kT}\right)\left[1 + \frac{E_a}{kT}\right]$$

$$= 2\pi p N_A N_B \pi^{-3/2}\left(\frac{2kT}{\mu}\right)^{1/2} (\pi\sigma_{AB}^2)\exp\left(-\frac{E_a}{kT}\right)\left[1 + \frac{E_a}{kT}\right]$$

$$= 2pN_AN_B\left(\frac{2kT}{\pi\mu}\right)^{1/2}(\pi\sigma_{AB}^2)\exp\left(-\frac{E_a}{kT}\right)\left[1+\frac{E_a}{kT}\right]$$

$$= \boxed{pN_AN_B\left(\frac{8kT}{\pi\mu}\right)^{1/2}(\pi\sigma_{AB}^2)\exp\left(-\frac{E_a}{kT}\right)\left[1+\frac{E_a}{kT}\right]}. \quad (10)$$

The last expression is identical to Eq. 20.14 in the text.

20.3 The rate coefficient for the bimolecular gas-phase reaction $NO_2 + NO_2 \longrightarrow 2\,NO + O_2$ at 300 K has been found to be 1.90×10^{-10} L mol^{-1} s^{-1}. The measured activation energy for the reaction is 111.3 kJ mol^{-1}. Using the hard-sphere model developed in the text, along with a hard-sphere collision radius of 8.0×10^{-10} m (8 Å), determine the steric factor for this reaction. Discuss the reasons the steric factor should be less than unity.

Solution

The hard-sphere rate coefficient is given by Eq. 20.15:

$$k(T) = p[\pi\sigma_{AB}^2]\left[\frac{8kT}{\pi\mu}\right]^{1/2}\left[1+\frac{E_a}{kT}\right]\exp\left[-\frac{E_a}{kT}\right]. \quad (1)$$

The NO_2–NO_2 reduced mass is

$$\mu = \frac{m_{NO_2}}{2} \times \frac{1}{N} = \frac{46.00}{2}\text{ g mol}^{-1} \times \frac{1}{6.022 \times 10^{23}\text{ mol}^{-1}}$$

$$\times \frac{1\text{ kg}}{1{,}000\text{ g}} = 3.82 \times 10^{-26}\text{ kg}. \quad (2)$$

The average relative speed at 300 K is, therefore,

$$\left[\frac{8kT}{\pi\mu}\right]^{1/2} = \left[\frac{(8)(1.381 \times 10^{-23}\text{ J K}^{-1})(300\text{ K})}{(3.14159)(3.82 \times 10^{-26}\text{ kg})}\right]^{1/2} = 525.5\text{ m s}^{-1}. \quad (3)$$

Substituting these results and the data given in the problem into Eq. 1 gives

$$1.90 \times 10^{-10}\text{ L mol}^{-1}\text{ s}^{-1} = p(3.14159)(8.0 \times 10^{-10}\text{ m})^2(525.5\text{ m s}^{-1})$$

$$\times\left[1+\frac{111{,}300}{(8.314)(300)}\right]\exp\left[-\frac{111{,}300}{(8.314)(300)}\right] \times \left(\frac{100\text{ cm}}{\text{m}}\right)^3$$

$$\times \frac{1\text{ L}}{1{,}000\text{ cm}^3} \times 6.022 \times 10^{23}\text{ mol}^{-1}. \quad (4)$$

The units on both sides now agree, and we have

$$1.90 \times 10^{-10} = 1.21 \times 10^{-6}p. \quad (5)$$

Solving for the steric factor, we obtain

$$p = \frac{1.90 \times 10^{-10}}{1.21 \times 10^{-6}} = 0.00016. \quad (6)$$

The steric factor suggests that it will be difficult to arrange the two NO_2 molecules in a configuration suitable for the production of two NO molecules and an O_2 molecule. The steric factor computed in Problem 20.2 for the H_2–I_2 gas-phase reaction was 0.00279. The smaller value of p in this problem is not surprising, considering the more complex structure of the reactants.

20.5* The total reaction cross section for a particular reaction is computed at 300 K. The results obtained at different relative speeds are given in the following table:

V (m s^{-1})	σ^2 (m^2)	V (m s^{-1})	σ^2 (m^2)
2,000.0	0.00	2,600.0	2.16×10^{-20}
2,100.0	1.01×10^{-22}	2,700.0	3.43×10^{-20}
2,200.0	8.09×10^{-22}	2,800.0	5.13×10^{-20}
2,300.0	2.71×10^{-21}	2,900.0	7.30×10^{-20}
2,400.0	6.42×10^{-21}	3,000.0	1.00×10^{-19}
2,500.0	1.25×10^{-20}		

The reduced mass of the reactants is 6.75×10^{-27} kg.

(A) Using a linear least-squares method, fit the computed total reaction cross section to a cubic equation in V. That is, writing $\sigma^2 = a_0 + a_1 V + a_2 V^2 + a_3 V^3$, determine the best values for a_0, a_1, a_2, and a_3.

(B) Compute the threshold energy.

(C) Using any convenient numerical integration method, determine the value of the specific reaction rate coefficient at 300 K, 330 K, 375 K, and 400 K. Assume that σ^2 is independent of temperature over this range.

(D) Determine the activation energy for the reaction. Compare your result for E_a with the threshold energy. Comment on the result.

Solution

(A) A standard linear least-squares fitting program in the software package "Passage" was employed to execute the fit. The results are

$$a_0 = -7.99 \times 10^{-19}\ \text{m}^2, \quad a_1 = 1.20 \times 10^{-21}\ \text{m s}, \quad a_2 = -6.00 \times 10^{-25}\ \text{s}^2,$$

and

$$a_3 = 1.00 \times 10^{-28}\ \text{m}^{-1}\ \text{s}^3. \quad \textbf{(1)}$$

The reaction cross section is, therefore, best fit by

$$\sigma^2 = -7.99 \times 10^{-19} + 1.20 \times 10^{-21} V - 6.00 \times 10^{-25} V^2 + 1.00 \times 10^{-28} V^3. \quad \textbf{(2)}$$

The following graph shows the computed data points and the fit to the data given in Eq. 2.

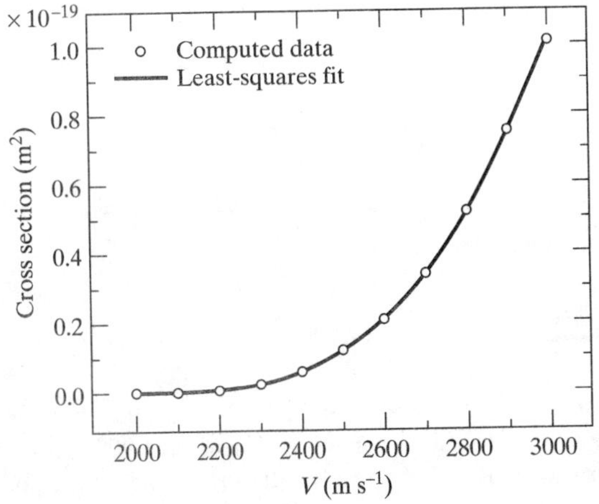

(B) The threshold is at or around 2,000 m s^{-1}. Therefore, the threshold energy is

$$E_o = 0.5\mu V_o^2 = 0.5(6.75 \times 10^{-27}\ \text{kg})(2{,}000\ \text{m s}^{-1})^2$$
$$= \underline{1.35 \times 10^{-20}\ \text{J}} = \underline{8{,}130\ \text{J mol}^{-1}}. \quad (3)$$

(C and D) The rate coefficient is given by Eq. 20.20:

$$k(T) = 4\pi\left[\frac{\mu}{2\pi kT}\right]^{3/2}\int_{V=0}^{\infty}\sigma^2V^3\exp\left[-\frac{\mu V^2}{2kT}\right]dV. \quad (4)$$

Inserting the data permits us to compute the necessary constants:

$$4\pi\left[\frac{\mu}{2\pi kT}\right]^{3/2} = 4(3.14159)\left[\frac{6.75 \times 10^{-27}}{2(3.14159)(1.381 \times 10^{-23})}\right]^{3/2}T^{-3/2}$$
$$= 8.622 \times 10^{-6}T^{-3/2}; \quad (5)$$

$$\frac{\mu}{2k} = \frac{6.75 \times 10^{-27}}{2(1.381 \times 10^{-23})} = 0.0002444. \quad (6)$$

Inserting these results into Eq. 4 gives

$$k(T) = 8.622 \times 10^{-6}T^{-3/2}\int_{2{,}000}^{\infty}\sigma^2V^3\exp\left[\frac{-0.0002444\,V^2}{T}\right]dV, \quad (7)$$

T (K)	$k(T)$ m^3 s^{-1}
300	3.30×10^{-18}
330	5.57×10^{-18}
375	1.08×10^{-17}
400	1.50×10^{-17}

where we can replace the lower limit of $V = 0$ with $V = 2{,}000$ m s^{-1}, since σ^2 is zero for $V < 2{,}000$ m s^{-1}. If we now utilize Eq. 2 for σ^2, we have an integral that can be evaluated by using Simpson's rule, a trapezoid rule, or any other numerical method. When this is done with the temperature either 300, 330, 375, or 400 K, the results are shown in the table to the left. If we now plot $\ln[k(T)]$ versus T^{-1}, the slope of the line will be equal to $-E_a/R$. The plot follows. The equation of the best linear fit is $\ln[k(T)] = -34.202 - 1{,}820\ \text{K}/T$. Therefore,

$$E_a = 1{,}820R = 1{,}820\ \text{K}(8.314\ \text{J mol}^{-1}\,\text{K}^{-1}) = \underline{15{,}131\ \text{J mol}^{-1}}. \quad (8)$$

As usual, we have $E_a > E_o$. The activation energy always exceeds the threshold energy (except for recombination reactions, when E_a can be negative). In this case, the difference is $E_a - E_o = 7{,}000$ J mol^{-1}.

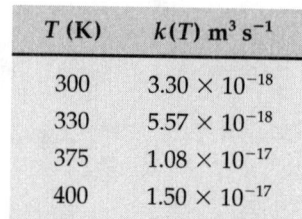

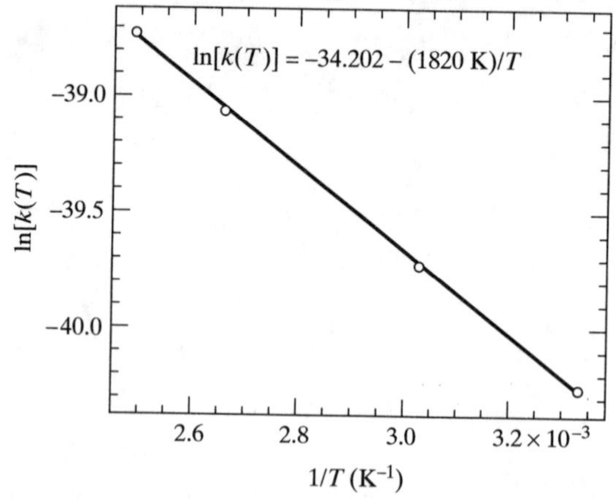

20.7 Using Figures 20.6, 20.7, and 20.9, plot, as accurately as possible, the potential energy for the D_2H system for the collinear and 90° configurations for structures where $R_1 = R_2$. Show the results on the same graph. Assume that the saddle point for the 90° configuration lies at 1.3 eV and that the dissociated-atom energy is 4.74 eV.

Solution

This is just a matter of reading the values of $R = R_1 = R_2$ along the figure diagonal of the figure at the intersection points with the contour lines. By enlarging the figures and measuring, I obtained the following results:

90° Configurations	
R (bohr)	*E* (eV)
1.20	4.0
1.28	3.0
1.41	2.0
1.76	1.3
2.39	2.0
3.03	3.0
3.95	4.0
5.00	≈4.5

180° Linear Configurations	
R (bohr)	*E* (eV)
1.11	4.0
1.17	3.0
1.25	2.0
1.39	1.0
1.70	0.396
1.87	0.5
2.15	1.0
2.61	2.0
3.13	3.0
3.98	4.0
5.00	≈4.5

The accompanying plot shows these data. The smooth curves are nonlinear least-squares fits to the data of the Morse-like function

$$V = a_o + a_1[\exp\{-2a_2(R - a_3)\} - 2\exp\{-a_2(R - a_3)\}]. \quad (1)$$

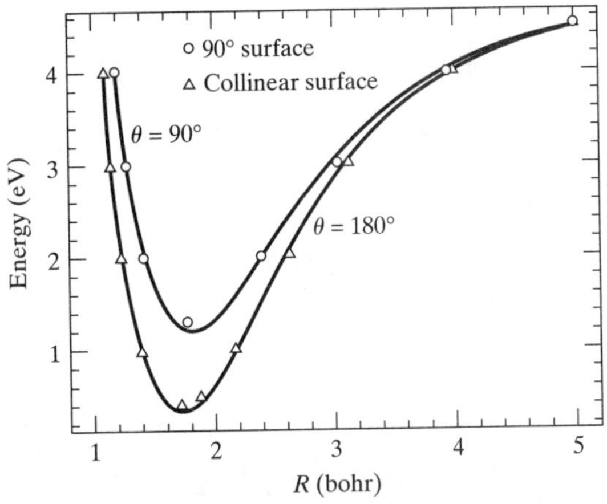

The fitting parameters for the two curves are as follows:

Configuration	a_o	a_1	a_2	a_3
90°	4.67	3.47	1.0695	1.79
180°	4.69	3.368	1.092	1.70

We see that the saddle point is a minimum along the diagonal, which is perpendicular to the reaction coordinate at the saddle point.

20.9 Figure 20.11 represents a possible potential-energy contour map for a linear, symmetric reaction of two iodine atoms with H_2 to form two HI molecules. Sketch this contour map roughly, and show the following features on your sketch:

(A) the $H_2 + 2I$ reactant valley,

(B) the 2HI product valley,

(C) the saddle point for reaction along a symmetric, linear reaction coordinate, and

(D) the reaction coordinate. What is the approximate barrier height along the reaction coordinate? What is the structure of the system at the saddle point?

Solution

(A) The reactant valley occurs when the H_2 separation is in the near vicinity of the equilibrium distance, 1.40 bohr, and the iodine atoms are very far apart so that R_{II} is large.

(B) The product valley occurs when the separation between both the hydrogen and iodine atoms is very large, but with the I–I separation being twice the equilibrium H–I distance larger than is the H–H separation. Since the HI equilibrium distance is 3.032 bohr, the product valley occurs in the region where $R_{II} - R_{HH} = 2(3.032) = 6.064$ bohr and both R_{II} and R_{HH} are large.

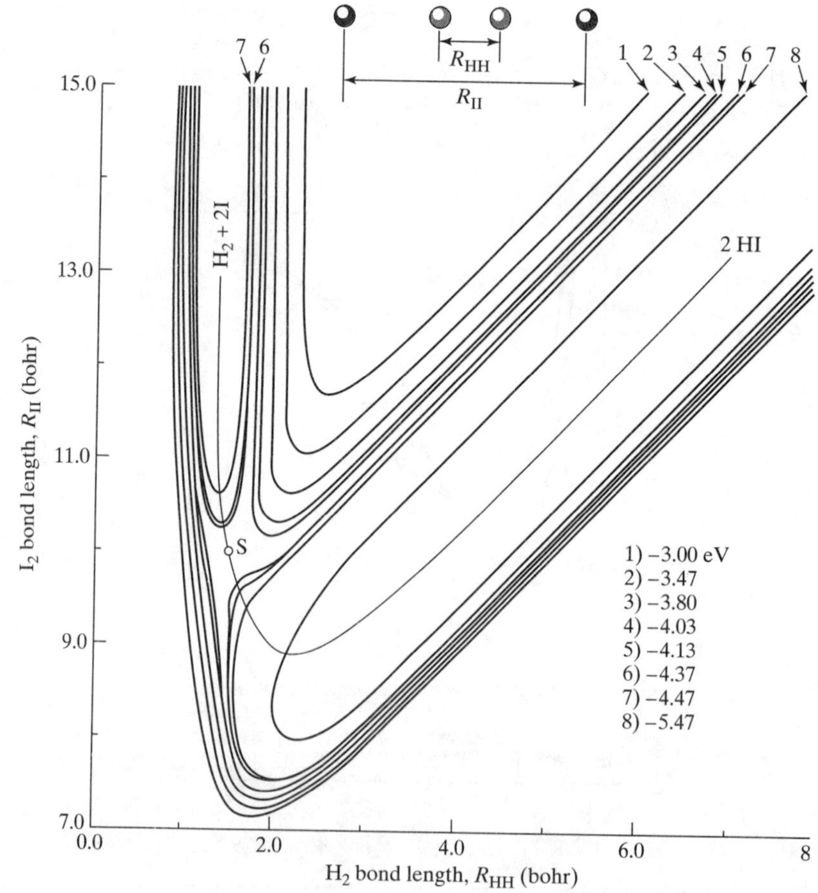

(C) The saddle point occurs at the point marked with an "S." All of these features, as well as the approximate reaction coordinate, are shown on the contour map on the previous page.

As can be seen, to reach the saddle point from the reactant valley, we must cross the contour line whose energy is -4.37 eV, but we do not have to cross the -4.13 eV contour. Therefore, the saddle-point energy lies between these two limits. We might estimate a value midway between the two, -4.25 eV. Since the energy of the reactants, $H_2 + 2I$, was given in the text as -4.746 eV, the barrier height for the reaction $H_2 + 2I \longrightarrow 2\,HI$ on this contour map is

$$E_b \approx -4.25 \text{ eV} - (-4.746 \text{ eV}) = 0.496 \approx 0.50 \text{ eV}, \quad \textbf{(1)}$$

which is about 48,000 J mol^{-1}.

(D) The R_{HH} and R_{II} distances at the saddle point are approximately 1.6 and 10.0 bohr, respectively. Therefore, the saddle structure is

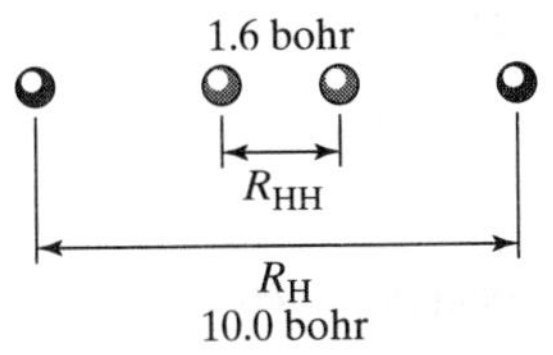

20.11 Consider a hypothetical three-body reaction $A + BC \longrightarrow$ products. Suppose we have determined that the reaction probability, is given by

$$P(V, b, \theta, \phi, \gamma, R, v, j, m) = K \sin\theta \exp[-\alpha b]\, F(V, v),$$

where K and α are constants and where

$$F(V, v) = a_v V^3 + b_v V^2 + c_v V + d_v \qquad \text{for } V \geq V_o$$

and

$$F(V, v) = 0 \qquad \text{for } V < V_o.$$

The parameters a_v, b_v, c_v, and d_v are constants whose values depend upon the BC vibrational state. In general,

$$a_v = a_n,\ b_v = b_n,\ c_v = c_n, \text{ and } d_v = d_n$$

if the BC molecule is in the nth vibrational state. Therefore, in this model, the reaction probability is independent of ϕ, γ, R, j, and m. Use Eq. 20.18 and other appropriate equations derived in the text to determine the proper form for the reaction cross section for the vibrational–rotational state (v, j). That is, derive $\sigma^2(V, v, j)$ in terms of the constants and parameters of the problem.

Solution

The reaction cross section for vibrational-rotational state (v, j) is given by Eq. 20.18:

$$\sigma^2(V, v, j) = \int_{b=0}^{\infty}\int_{\theta=0}^{\pi}\int_{\phi=0}^{2\pi}\int_{\gamma=0}^{2\pi}\int_{R=R_{min}}^{R_{max}} \sum_{m=-j}^{+j} P(V, b, \theta, \phi, \gamma, R, v, j, m)$$
$$\times\, (2\pi b\, db) P_1(\theta, \phi) P_2(R) P_3(\gamma). \quad \textbf{(1)}$$

Substituting $P(V, b, \theta, \phi, \gamma, R, v, j, m)$ from the problem and $P_1(\theta, \phi)P_2(R)P_3(\gamma)$ from the text gives

$$\sigma^2(V, v, j) = KF(V, v) \int_{b=0}^{\infty} (2\pi b) \exp[-\alpha b]\, db \int_{\theta=0}^{\pi} \int_{\phi=0}^{2\pi} \frac{\sin^2\theta\, d\phi\, d\theta}{4\pi}$$
$$\times \sum_{m=-j}^{+j} \int_{\gamma=0}^{2\pi} \left[\frac{d\gamma}{2\pi}\right] \int_{R=R_{min}}^{R_{max}} \left[\frac{dR}{\pi[\alpha^{-1} - (R - R_{eq})^2]^{1/2}}\right]. \quad (2)$$

The integrals over γ and R are both unity, since the probability distributions are normalized. The same result is obtained if we use a quantum mechanical expression for $P_2(R)$, because the reaction probability in this problem is independent of R. Therefore, in the summation over m, we are simply adding unity to itself $(2j + 1)$ times. Thus, the overall result is

$$\sum_{m=-j}^{+j} \int_{\gamma=0}^{2\pi} \left[\frac{d\gamma}{2\pi}\right] \int_{R=R_{min}}^{R_{max}} \left[\frac{dR}{\pi[\alpha^{-1} - (R - R_{eq})^2]^{1/2}}\right] = (2j + 1). \quad (3)$$

The integrals over orientation angles are

$$\int_{\theta=0}^{\pi} \int_{\phi=0}^{2\pi} \frac{\sin^2\theta\, d\phi\, d\theta}{4\pi} = \frac{2\pi}{4\pi}\left[\frac{\theta}{2} - \frac{\sin(2\theta)}{4}\right]_0^{\pi} = \frac{\pi}{4}. \quad (4)$$

The integral over the impact parameter is

$$\int_{b=0}^{\infty} (2\pi b) \exp[-\alpha b]\, db = 2\pi\left[\frac{1!}{\alpha^2}\right] = \frac{2\pi}{\alpha^2}. \quad (5)$$

Combining the results from all integrations with Eq. 2, we obtain

$$\sigma^2(V, v, j) = K(2j + 1)F(V, v)\frac{2\pi}{\alpha^2}\frac{\pi}{4} = \frac{K(2j + 1)\pi^2 F(V, v)}{2\alpha^2}. \quad (6)$$

Inserting the definition for $F(V, v)$ gives

$$\boxed{\sigma^2(V, v, j) = \frac{K(2j + 1)\pi^2}{2\alpha^2}[a_v V^3 + b_v V^2 + c_v V + d_v]} \quad (7)$$

for $V \geq V_o$ and

$$\boxed{\sigma^2(V, v, j) = 0}$$

for $V < V_o$, which is the desired expression.

20.13 Suppose we have means of computing the total reaction cross section in a certain reaction. The reaction rate coefficient is then given by Eq. 20.20. Using a cumulative distribution function, express this integral in the form of a Monte Carlo summation. (*Hint*: You will first need to normalize the probability distribution function for the relative speed.)

Solution

The reaction rate coefficient is given by Eq. 20.20:

$$k(T) = \int_{V=0}^{\infty} \sigma^2 Z_{AB} = 4\pi\left[\frac{\mu}{2\pi kT}\right]^{3/2} \int_{V=0}^{\infty} \sigma^2 V^3 \exp\left[-\frac{\mu V^2}{2kT}\right] dV. \quad (1)$$

To reduce the complexity of the notation, let us define a new variable. Let

$$z = \left(\frac{\mu}{2kT}\right)^{1/2} V. \tag{2}$$

In terms of z and dz, we have

$$dV = \left(\frac{2kT}{\mu}\right)^{1/2} dz, \tag{3}$$

$$V = \left(\frac{2kT}{\mu}\right)^{1/2} z, \tag{4}$$

and

$$\frac{\mu V^2}{2kT} = z^2. \tag{5}$$

Substituting into Eq. 1 gives

$$\begin{aligned} k(T) &= 4\pi\left[\frac{\mu}{2\pi kT}\right]^{3/2}\left[\frac{2kT}{\mu}\right]^2 \int_0^\infty z^3 \exp(-z^2)\sigma^2(z)\,dz \\ &= 4\left[\frac{2kT}{\pi\mu}\right]^{1/2} \int_0^\infty z^3 \exp(-z^2)\sigma^2(z)\,dz. \end{aligned} \tag{6}$$

The distribution function for z is, therefore, $w(z) = z^3 \exp(-z^2)dz$. As the hint in the problem suggests, we must normalize this distribution function before proceeding. Thus, we require that

$$I = N \int_0^\infty z^3 \exp(-z^2)\,dz = 1. \tag{7}$$

The integration in Eq. 7 can be done by parts. In the first step, we let $u = z^2$ and $dv = z\exp(-z^2)dz$, so that $du = 2z\,dz$ and $v = -\frac{1}{2}\exp(-z^2)$. Therefore, the integral is

$$I = N\left[\left[-\frac{z^2\exp(-z^2)}{2}\right]_0^\infty + \int_0^\infty z\exp(-z^2)\,dz\right] = N\int_0^\infty z\exp(-z^2)\,dz, \tag{8}$$

since the first term is zero. The new expression is easily integrated. The result is

$$I = -\frac{N\exp(-z^2)}{2}\bigg|_0^\infty = \frac{N}{2}. \tag{9}$$

To satisfy Eq. 7, we must have $N = 2$. Therefore, we write Eq. 6 in the form

$$k(T) = 2\left[\frac{2kT}{\pi\mu}\right]^{1/2} \int_0^\infty 2z^3 \exp(-z^2)\sigma^2(z)\,dz, \tag{10}$$

so that the distribution function for z, $w(z) = 2z^3\exp(-z^2)dz$, is properly normalized. Now that this is done, we can obtain the cumulative distribution function for $w(z)$:

$$G(z) = \int_0^z w(z) = 2\int_0^z z^3\exp(-z^2)\,dz. \tag{11}$$

Integrating by parts exactly as before, we obtain

$$\begin{aligned} G(z) &= 2\left[\left[-\frac{z^2\exp(-z^2)}{2}\right]_0^z + \int_0^z z\exp(-z^2)\,dz\right] \\ &= -z^2\exp(-z^2) + 2\int_0^z z\exp(-z^2)\,dz = -z^2\exp(-z^2) - \exp(-z^2)\bigg|_0^z \\ &= -z^2\exp(-z^2) - \exp(-z^2) + 1 = 1 - [z^2+1]\exp(-z^2). \end{aligned} \tag{12}$$

With this cumulative distribution function, we have

$$G(0) = 1 - 1 = 0 \quad \text{and } G(\infty) = 1 - [\infty^2 + 1]\exp(-\infty^2) = 1, \tag{13}$$

as expected.

We also have

$$dG = [-2z\exp(-z^2) + 2z[z^2+1]\exp(-z^2)]dz = 2z^3\exp(-z^2)dz. \tag{14}$$

Substituting Eqs. 13 and 14 into Eq. 10 gives

$$k(T) = 2\left[\frac{2kT}{\pi\mu}\right]^{1/2}\int_0^1 \sigma^2(G)dG = \left[\frac{8kT}{\pi\mu}\right]^{1/2}\int_0^1 \sigma^2(G)dG. \tag{15}$$

Evaluating the integral in Eq. 15 with the use of Monte Carlo methods produces

$$\boxed{k(T) = \left[\frac{8kT}{\pi\mu}\right]^{1/2}\frac{1}{N}\sum_{i=1}^{N}\sigma^2(G_i)}. \tag{16}$$

To perform the Monte Carlo summation, we select $G(z)$ randomly. This selection gives

$$G_i(z) = \xi_i = 1 - [z_i^2 + 1]\exp(-z_i^2), \tag{17}$$

where ξ_i is our usual random number uniform over the interval from zero to unity. To obtain the value of z_i corresponding to this selection, we must solve the transcendental equation

$$(1 - \xi_i) - [z_i^2 + 1]\exp(-z_i^2) = 0 \tag{18}$$

numerically for z_i. Having obtained the solution, we then compute $\sigma^2(z_i) = \sigma^2(G_i)$. This procedure is repeated N times, and the Monte Carlo sum in Eq. 16 is evaluated. After a simple multiplication, we have the specific reaction rate coefficient.

20.15* An engineer wishes to evaluate the two-dimensional integral

$$I = \int_{x=0}^{10}\int_{y=0}^{10} xy^2 \sin\left[\frac{\pi x}{10}\right] dy\,dx.$$

(A) Perform the integration analytically, and obtain the correct answer to eight significant digits.

(B) Write a FORTRAN or C computer code to evaluate the integral via Monte Carlo methods.

(C) Run your computer code using 10^2, 10^3, 10^4, 10^5, and 10^6 points for the evaluation. Compute the percent error in the result for each evaluation.

(D) Plot $\ln[|\%\ \text{error}|]$ from (C) versus $\ln[N^{-1/2}]$ where N is the number of points used in the Monte Carlo evaluation of the integral. Comment on the results.

Solution

(A) The integral over x may be integrated by parts. If we let $u = x$ and $dv = \sin(\pi x/10)dx$, we obtain $du = dx$ and $v = -10\cos(\pi x/10)/\pi$. Thus,

$$\int x\sin\left(\frac{\pi x}{10}\right)dx = -\frac{10x\cos(\pi x/10)}{\pi} + \frac{10}{\pi}\int\cos\left(\frac{\pi x}{10}\right)dx$$

$$= -\frac{10x\cos(\pi x/10)}{\pi} + \frac{100}{\pi^2}\sin\left(\frac{\pi x}{10}\right) = \frac{100}{\pi^2}\left[\sin\left[\frac{\pi x}{10}\right] - \frac{\pi x\cos\left[\frac{\pi x}{10}\right]}{10}\right]. \tag{1}$$

Hence, the exact value of the integral is given by

$$\int_{x=0}^{10}\int_{y=0}^{10} xy^2 \sin\left[\frac{\pi x}{10}\right] dy\,dx = \frac{y^3}{3}\Bigg|_0^{10} \frac{100}{\pi^2}\left[\sin\left[\frac{\pi x}{10}\right] - \frac{\pi x \cos\left[\frac{\pi x}{10}\right]}{10}\right]\Bigg|_0^{10}$$

$$= \frac{10^3}{3} \times \frac{100}{\pi^2}[\sin(\pi) - \pi\cos(\pi) - 0 + 0] = \frac{10^5}{3\pi} = \underline{10{,}610.329}. \qquad (2)$$

(B) The Monte Carlo expression for I can be obtained by combining Eqs. 20.28 and 20.31. If we choose x and y randomly over the range from 0 to 10 for each variable,

$$I = \frac{(10-0)(10-0)}{N}\sum_{i=1}^{N} x_i y_i^2 \sin\left[\frac{\pi x_i}{10}\right] = \frac{100}{N}\sum_{i=1}^{N} x_i y_i^2 \sin\left[\frac{\pi x_i}{10}\right], \qquad (3)$$

where N is the number of points computed and (x_i, y_i) are the random points selected for each point, using

$$x_i = 10\xi_{1i} \quad \text{and} \quad y_i = 10\xi_{2i}. \qquad (4)$$

There are many FORTRAN codes that might be written. The following is a simple one:

```
      N=100
      do 2 j=1, 5
      sum=0.0
      do 1 i=1, N
      r1=ran(1)
      r2=ran(2)
      x=10.0 * r1
      y=10.0 * r2
      z=x*y*y*sin(3.1415927*x/10)
      sum=sum+z
1     continue
      sum=100*sum/N
      write(6, 100) N, sum
100   format(2x, 'Number of points=',I8, 2x,
      'Integral=', f14.4)
      N=10*N
2     continue
      stop
      end
```

The subroutine ran(1) or ran(2) is the call for the random-number generator.

(C) The foregoing computer code was run. The results obtained are given in the following table:

N	Monte Carlo Integral	|% Error|
100	9,947.199	6.25
1,000	10,758.55	1.40
10,000	10,601.20	0.086
100,000	10,628.93	0.175
1,000,000	10,609.20	0.011

The percent error is given by

$$\% \text{ error} = \left| \frac{100(I_{mc} - 10{,}610.329)}{10{,}610.329} \right|, \tag{5}$$

where I_{mc} is the Monte Carlo value.

(D) The requested plot appears at the end of the problem. Combining Eqs. 20.34 and 20.35 in the text, we obtain

$$\% \text{ error} = \frac{100N^{-1/2}[\langle f^2\rangle' - (\langle f\rangle')^2]^{1/2}}{\langle f\rangle}. \tag{6}$$

Taking absolute values and natural logarithms of both sides gives

$$\ln[|\% \text{ error}|] = \ln[N^{-1/2}] + \ln\left[\frac{100[\langle f^2\rangle' - (\langle f\rangle')^2]^{1/2}}{\langle f\rangle}\right] = \ln[N^{-1/2}] + \text{constant}. \tag{7}$$

Thus, we expect nearly linear behavior with a slope of unity when we plot $\ln[|\% \text{ error}|]$ versus $\ln[N^{-1/2}]$. The plot shows that the result is approximately linear. Since the calculations are statistical, perfect linearity cannot be expected. A linear least-squares fit to the data gives

$$\ln[|\% \text{ error}|] = 4.596 + 1.282 \ln[N^{-1/2}], \tag{8}$$

so that the slope is near unity, as expected.

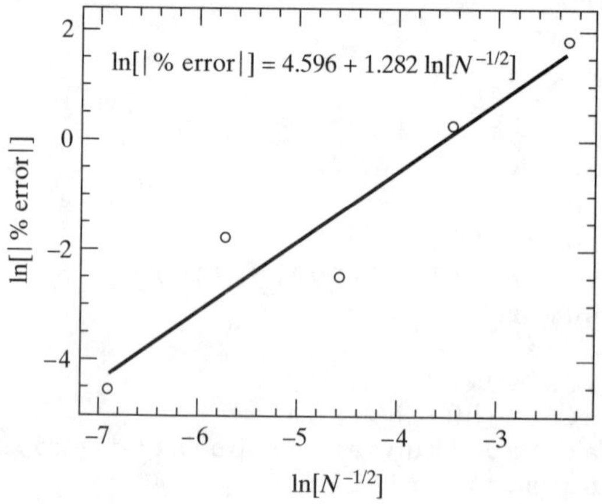

20.17 Karplus, Porter, and Sharma [*J. Chem. Phys.* **43**, 3259 (1965)] computed cross sections for the $H + H_2$ exchange reaction as a function of relative speed for particular vibrational–rotational states. When $v = j = 0$ and $V = 1.76 \times 10^4$ m s^{-1}, they obtained a reaction cross section of 1.15×10^{-20} m^2.

(A) If 400 total trajectories were computed with a maximum impact parameter $b_m = 1.058 \times 10^{-10}$ m, how many trajectories resulted in a reaction?

(B) What is the expected absolute percent error in the result?

Solution

(A) The reaction cross section is given by Eq. 20.45:

$$\sigma^2(V, v, j) = \frac{N_R(2j + 1)\pi b_m^2}{N}. \tag{1}$$

Solving Eq. 1 for the number of reactive trajectories, N_R, we obtain

$$N_R = \frac{N\sigma^2(V, v, j)}{(2j + 1)\pi b_m^2}. \tag{2}$$

Substituting the data given in the problem yields

$$N_R = \frac{400(1.15 \times 10^{-20}\ \text{m}^2)}{(1)(3.14159)(1.058 \times 10^{-10}\ \text{m})^2} = 130.8 = \underline{131\ \text{reactions}}. \quad \textbf{(3)}$$

(B) The expected absolute percent error is given by Eq. 20.46:

$$|\%\ \text{error}| = 100 \times \left[\frac{N - N_R}{NN_R}\right]^{1/2}. \quad \textbf{(4)}$$

Using the result obtained in Part (A), we obtain

$$|\%\ \text{error}| = 100 \times \left[\frac{400 - 131}{(400)(131)}\right]^{1/2} = \underline{7.16\%}. \quad \textbf{(5)}$$

20.19 Figure 20.13 shows the details of a reactive hydrogen exchange trajectory in which $H_a + H_b\text{–}H_c \longrightarrow H_a\text{–}H_b + H_c$. Initially, H_b–H_c was in the $j = 0$ rotational state, so there was no molecular rotational angular momentum present. How can we tell from the figure that the newly formed H_a–H_b molecule is not in the $j = 0$ rotational state? Explain your reasoning.

Solution

If the H_a–H_b molecule were in the $j = 0$ rotational state with no angular momentum, the trajectory details would reflect motion such as that illustrated in the following diagram:

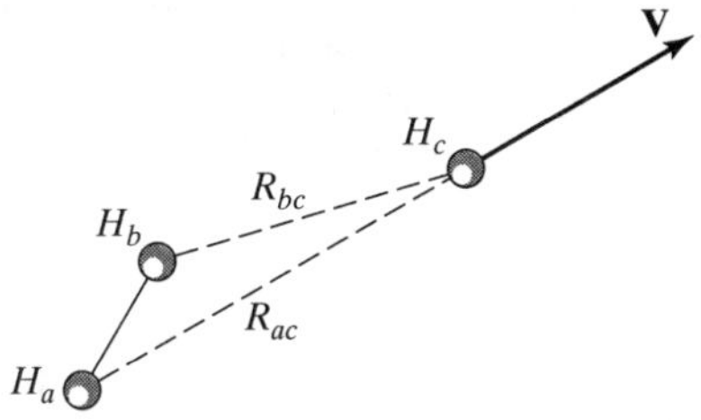

If the relative-velocity vector between H_c and the center of mass of H_a–H_b is **V**, as shown in the diagram, and if H_a–H_b is in the $j = 0$ state, so that it is not rotating, we will have $R_{ac} > R_{bc}$ at all times. Referring to Figure 20.13, we see that we do have this inequality from the moment of formation of H_a–H_b around $t \approx 5.0$ time units up to $t \approx 8.0$ time units. However, at $t \approx 8.0$ time units, the inequality reverses, so that $R_{ac} < R_{bc}$. Since the direction of **V** cannot change once H_c is out of interaction range, this reversal must reflect the fact that H_a–H_b is rotating, so that $t \approx 8.0$ time units and we have something like the following situation:

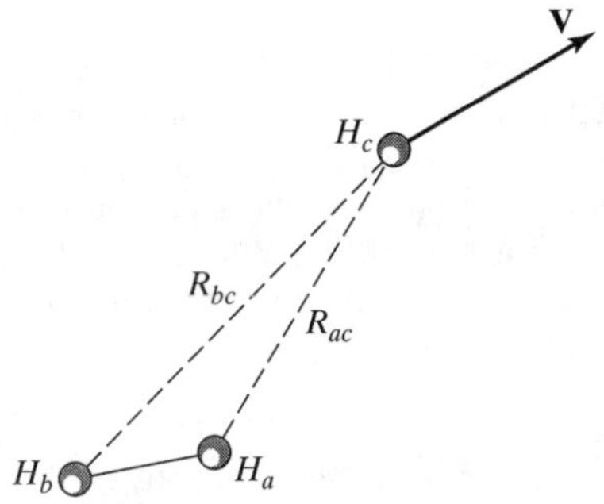

If H_a–H_b is rotating, we cannot have $j = 0$. Therefore, the dynamics of the collision have imparted rotational angular momentum to the new H_2 molecule, whereas the original one was in the $j = 0$ state. Since angular momentum must be conserved, it is appropriate to ask where the new angular momentum came from. The answer is that it came from the angular momentum associated with the relative motion of the incoming H_a atom and the H_b–H_c molecule about the H_3 center of mass. This type of angular momentum is called "orbital angular momentum," as opposed to "molecular rotational angular momentum," which is associated with rotation of a molecule about its own center of mass. But as Gertrude Stein might have said, "Angular momentum is angular momentum is angular momentum is angular momentum."

20.21 Figure 20.17 indicates that the vibrational motion of the D–F bond is much more erratic between $t \approx 60$ tu and $t \approx 170$ tu than it is beyond 170 tu. Explain qualitatively why this is true. (1 tu $= 1.019 \times 10^{-14}$ s.)

Solution

In the range between 60 and 170 tu, the D–F atoms are still bonded to the rest of the molecule. As a result, there is energy exchange between the D–F bond and the other bonds and vibrational modes of the eight-atom molecule. This energy exchange is reflected in the time-varying amplitude of the D–F vibrational motion. As energy enters the bond, the amplitude increases; as it leaves, the amplitude decreases.

For $t > 170$ tu, the C–F bond has ruptured, and DF is no longed bonded to the rest of the molecular system. It moves off into the distance. This means that energy transfer is no longer possible. Consequently, the amplitude must become a constant. The trajectory details show that this is indeed the case. In fact, one can use the moment at which the D–F vibrational amplitude becomes constant as a marker for the rupture of the C–F bond.

20.23 Use simple transition-state theory to compute the reaction rate at 300 K for the hydrogen exchange reaction $H + H_2 \longrightarrow H_2 + H$. Required data can be found in Example 20.8, Figures 20.6 and 20.7, the related discussion in the text, and Table 15.4. You may use a classical expression for the rotational partition functions and a harmonic quantization for the vibrational partition functions. Express the result in units of $m^3\ s^{-1}$ and $cm^3\ mol^{-1}\ s^{-1}$.

Solution

Using Eq. 20.66 for the rate, along with the assumption that the flux is sufficiently close to the true rate that we can use the equals sign, we have

$$k_H(T) = \frac{kTV}{h}\frac{g_{el}^{\ddagger}}{g_{el}^{H} g_{el}^{H_2}} \exp\left[-\frac{E_b}{kT}\right]\frac{[z_{tr}^{\ddagger} z_{rot}^{\ddagger} z_{vib}^{\ddagger}]_{H_3}}{z_{tr}^{H} z_{rot}^{H} z_{vib}^{H} z_{tr}^{H_2} z_{rot}^{H_2} z_{vib}^{H_2}}. \quad \textbf{(1)}$$

The electronic degeneracy of H_3 is 2, since we have an unpaired electron and, therefore, a doublet electronic state. The same is true for the H atom. The degeneracy of H_2 is unity. Thus, the ratio of degeneracy factors is $2/[2(1)] = 1$. Using the Porter–Karplus potential whose contours are plotted in Figures 20.6 and 20.7, we find that the barrier height for the hydrogen exchange reaction in the transition state is 38,200 J mol^{-1}, as stated in the text. At 300 K, the exponential in Eq. 1, therefore, has the value

$$\exp\left[-\frac{E_b}{kT}\right] = \exp\left[-\frac{38{,}200}{(8.314)(300)}\right] = 2.231 \times 10^{-7}. \quad \textbf{(2)}$$

Also, 300 K,

$$\frac{kT}{h} = \frac{(1.381 \times 10^{-23}\,\mathrm{J\,K^{-1}})(300\,\mathrm{K})}{6.626 \times 10^{-34}\,\mathrm{J\,s}} = 6.253 \times 10^{12}\,\mathrm{s^{-1}}. \quad \textbf{(3)}$$

Let us now compute the ratio of translational partitions in Eq. 1 and include the factor V in this calculation. If we use the energy levels for a particle in an infinite well, the molecular-translational partition is given by $z_{tr} = V/h^3[2\pi mkT]^{3/2}$. Thus, the required ratio is

$$\frac{Vz_{tr}^{\ddagger}}{z_{tr}^{H} z_{tr}^{H_2}} = \frac{\frac{V^2}{h^3}[2\pi m_{H_3}kT]^{3/2}}{\frac{V}{h^3}[2\pi m_{H}kT]^{3/2}\frac{V}{h^3}[2\pi m_{H_2}kT]^{3/2}} = \frac{h^3}{(2\pi kT)^{3/2}}\left[\frac{m_{H_3}}{m_{H}m_{H_2}}\right]^{3/2}$$

$$= \frac{(6.626 \times 10^{-34}\,\mathrm{J\,s})^3}{[2(3.14159)(1.381 \times 10^{-23}\,\mathrm{J\,K^{-1}})(300\,\mathrm{K})]^{3/2}}\left[\frac{3(1.674 \times 10^{-27})\,\mathrm{kg}}{2(1.674 \times 10^{-27})^2\,\mathrm{kg}^2}\right]^{3/2}$$

$$= 1.858 \times 10^{-30}\,\mathrm{m^3}. \quad \textbf{(4)}$$

Notice that the volume cancels in Eq. 4. This is not unexpected, since we know that the rate coefficient for a chemical reaction cannot depend upon the size of the container in which we run the reaction.

There is no rotational or vibrational energy for a hydrogen atom. Therefore, the ratio of molecular-rotational partition functions required by Eq. 1 is just that for H_3 divided by that for H_2. If a classical rigid-rotor expression is used for these partition functions, we have

$$z_{rot} = \frac{kT}{\sigma Bh} = \frac{8\pi^2 IkT}{\sigma h^2}. \quad \textbf{(5)}$$

The symmetry numbers for H_2 and H_3 are both 2. Therefore, the ratio is

$$\frac{z_{rot}^{\ddagger}}{z_{rot}^{H_2}} = \frac{I_{H_3}}{I_{H_2}}. \quad \textbf{(6)}$$

The H_3 moment of inertia was computed in Example 20.8. The result is $I_{H_3} = 2.71 \times 10^{-47}\,\mathrm{kg\,m^2}$. Table 15.4 lists the H_2 equilibrium distance as 0.7416×10^{-10} m. The moment of inertia for a diatomic rigid rotor is

$$I_{H_2} = \mu R_{eq}^2 = \frac{m_H R_{eq}^2}{2} = \frac{(1.674 \times 10^{-27}\,\mathrm{kg})(0.7416 \times 10^{-10}\,\mathrm{m})^2}{2}$$

$$= 4.603 \times 10^{-48}\,\mathrm{kg\,m^2}. \quad \textbf{(7)}$$

The ratio of molecular-rotational partition functions is

$$\frac{z_{rot}^{\ddagger}}{z_{rot}^{H_2}} = \frac{I_{H_3}}{I_{H_2}} = \frac{2.71 \times 10^{-47}\,\mathrm{kg\,m^2}}{4.603 \times 10^{-48}\,\mathrm{kg\,m^2}} = 5.887. \quad \textbf{(8)}$$

The vibrational partition function for each mode is given by Eq. 18.105 if we assume a harmonic quantization. This equation is

$$z_{vib}^{h} = \frac{e^{-x/2}}{1 - e^{-x}}, \quad \textbf{(9)}$$

where $x = h\nu_o/(kT)$. At $T = 300$ K, we obtain, for H_3,

$$x(\text{symmetric stretch}) = \frac{(6.626 \times 10^{-34}\,\mathrm{J\,s})(2{,}184\,\mathrm{cm^{-1}})(2.998 \times 10^{10}\,\mathrm{cm\,s^{-1}})}{(1.381 \times 10^{-23}\,\mathrm{J\,K^{-1}})(300\,\mathrm{K})}$$

$$= 10.471 \quad \textbf{(10)}$$

and

$$x(\text{bending}) = \frac{(6.626 \times 10^{-34}\,\text{J s})(979\,\text{cm}^{-1})(2.998 \times 10^{10}\,\text{cm s}^{-1})}{(1.381 \times 10^{-23}\,\text{J K}^{-1})(300\,\text{K})} = 4.694. \quad \textbf{(11)}$$

The vibrational partition function for H_3 is, therefore,

$$(z^{\ddagger}_{\text{vib}})_{H_3} = \frac{\exp(-10.471/2)}{1 - \exp(-10.471)} \times 2 \times \frac{\exp(-4.694/2)}{1 - \exp(-4.694)}$$

$$= (0.005324)(2)(0.09654) = 0.001028. \quad \textbf{(12)}$$

H_2 has only one vibrational mode. Table 15.4 lists its vibrational wave number as 4,395.2 cm^{-1}. Thus,

$$x_{H_2} = \frac{(6.626 \times 10^{-34}\,\text{J s})(4{,}395.2\,\text{cm}^{-1})(2.998 \times 10^{10}\,\text{cm s}^{-1})}{(1.381 \times 10^{-23}\,\text{J K}^{-1})(300\,\text{K})} = 21.072. \quad \textbf{(13)}$$

The H_2 vibrational partition function is

$$z^{H_2}_{\text{vib}} = \frac{\exp(-21.072/2)}{1 - \exp(-21.072)} = 2.656 \times 10^{-5}. \quad \textbf{(14)}$$

The ratio of the molecular-vibrational partition functions required by Eq. 1 is

$$\frac{z^{\ddagger}_{\text{vib}}}{z^{H_2}_{\text{vib}}} = \frac{0.001028}{2.656 \times 10^{-5}} = 38.70. \quad \textbf{(15)}$$

Combining all factors, we obtain

$$k_H(300\,\text{K}) = (6.253 \times 10^{12}\,\text{s}^{-1})(1)(2.231 \times 10^{-7})(1.858 \times 10^{-30}\,\text{m}^3)$$

$$\times (5.887)(38.742) = \underline{5.91 \times 10^{-22}\,\text{m}^3\,\text{s}^{-1}}. \quad \textbf{(16)}$$

Converting this result to $\text{cm}^3\,\text{mol}^{-1}\,\text{s}^{-1}$, yields

$$5.91 \times 10^{-22}\,\text{m}^3\,\text{s}^{-1} \times \left[\frac{100\,\text{cm}}{\text{m}}\right]^3 \times \frac{6.022 \times 10^{23}}{\text{mol}} = \underline{3.56 \times 10^8\,\text{cm}^3\,\text{mol}^{-1}\,\text{s}^{-1}}.$$

(17)

20.25 The ratio of the rate coefficients for the reactions

$$H + H_2 \longrightarrow H_2 + H$$

and

$$D + H_2 \longrightarrow HD + H$$

at 300 K is $k_H(300\,\text{K})/k_D(300\,\text{K}) = 0.187$. (See Example 20.8.) The deviation of this result from unity in spite of the fact that the potential-energy hypersurfaces for both reactions are identical is called the isotope effect. Is the isotope effect temperature dependent? Explain. If it is, discuss the origin of the temperature dependence within the framework of simple transition-state theory.

Solution

In Example 20.8, it is shown that the isotope effect is given by

$$\frac{k_H(T)}{k_D(T)} = \frac{z^D_{\text{tr}}}{z^H_{\text{tr}}}\left[\frac{\{z^{\ddagger}_{\text{tr}} z^{\ddagger}_{\text{rot}} z^{\ddagger}_{\text{vib}}\}_{H_3}}{\{z^{\ddagger}_{\text{tr}} z^{\ddagger}_{\text{rot}} z^{\ddagger}_{\text{vib}}\}_{DH_2}}\right]. \quad \textbf{(1)}$$

The first factor is

$$\frac{z_{tr}^{D}}{z_{tr}^{H}} = \left[\frac{m_D}{m_H}\right]^{3/2}. \tag{2}$$

Consequently, there is no temperature dependence in this term. The ratio of the translational partition functions in the second factor has the same form, viz.,

$$\frac{(z_{tr}^{\ddagger})_{H_3}}{(z_{tr}^{\ddagger})_{DH_2}} = \left[\frac{m_{H_3}}{m_{DH_2}}\right]^{3/2}, \tag{3}$$

so that there is no temperature dependence. The ratio of the rotational partition functions in Eq.1 has the form

$$\frac{(z_{rot}^{\ddagger})_{H_3}}{(z_{rot}^{\ddagger})_{DH_2}} = \frac{I_{H_3}\sigma_D}{I_{DH_2}\sigma_H}, \tag{4}$$

and again, no temperature dependence is present. The ratio of the vibrational partition functions in Eq. 1 is given by

$$\frac{(z_{vib}^{\ddagger})_{H_3}}{(z_{vib}^{\ddagger})_{DH_2}} = \frac{\left[\dfrac{e^{-x/2}}{1-e^{-x}}\right]_{H_3(ss)}(2)\left[\dfrac{e^{-x/2}}{1-e^{-x}}\right]_{H_3(bend)}}{\left[\dfrac{e^{-x/2}}{1-e^{-x}}\right]_{DH_2(ss)}(2)\left[\dfrac{e^{-x/2}}{1-e^{-x}}\right]_{DH_2(bend)}}$$

$$= \frac{\left[\dfrac{e^{-x/2}}{1-e^{-x}}\right]_{H_3(ss)}\left[\dfrac{e^{-x/2}}{1-e^{-x}}\right]_{H_3(bend)}}{\left[\dfrac{e^{-x/2}}{1-e^{-x}}\right]_{DH_2(ss)}\left[\dfrac{e^{-x/2}}{1-e^{-x}}\right]_{DH_2(bend)}}. \tag{5}$$

In Eq. 5, the subscripts "ss" and "bend" refer to the symmetric stretch and bending vibrational modes of H_3 and DH_2, respectively.

Since $x = h\nu_o/(kT)$, Eq. 5 depends upon the temperature. Therefore, the isotope effect is temperature dependent, and the temperature dependence arises entirely from the ratio of vibrational partition functions that appear in the expression for the effect.

20.27 The simple transition-state theory rate coefficient for the $H + H_2$ exchange reaction at 300 K on a Porter–Karplus potential-energy hypersurface is 3.56×10^8 cm^3 mol^{-1} s^{-1}. (See Problem 20.23.) The rate coefficient obtained from trajectory calculations at 300 K is 1.84×10^8 cm^3 mol^{-1} s^{-1} [M. Karplus, R. N. Porter, and R. D. Sharma, *J. Chem. Phys.* **43**, 3,259 (1965)]. Using these data, estimate the value of the transmission coefficient in Eq. 20.67.

Solution

The trajectory calculations automatically correct for recrossings of the dividing surface, since each reactive trajectory is counted only once, no matter how many times it may cross a dividing surface. Therefore, we expect to have

$$k_{traj} < k_{TST}, \tag{1}$$

where k_{TST} is the simple transition-state–theory rate coefficient. Both calculations ignore tunneling, so the presence of such quantum effects should not reverse the inequality in Eq. 1. If we use Eq. 20.67 with the transmission coefficient present, this correction factor can be employed to bring k_{traj} and k_{TST} into agreement. In the present case, we have

$$k_{traj} = 1.84 \times 10^8 \text{ cm}^3 \text{ mol}^{-1} \text{ s}^{-1} = \kappa k_{TST} = 3.56 \times 10^8 \text{ cm}^3 \text{ mol}^{-1} \text{ s}^{-1}\kappa. \tag{2}$$

Solving for κ, we obtain

$$\kappa = \frac{1.84 \times 10^8 \text{ cm}^3 \text{ mol}^{-1} \text{ s}^{-1}}{3.56 \times 10^8 \text{ cm}^3 \text{ mol}^{-1} \text{ s}^{-1}} = \underline{0.517}. \quad (3)$$

This result suggests that there are about twice as many crossings of the dividing surface as there are reactions.

20.29 The largest bimolecular gas-phase rate coefficient that we can have is one for a reaction with zero activation energy and a steric factor of unity. Use the simple hard-sphere model to evaluate this rate coefficient at 300 K for a reaction in which the hard-sphere collision radius is 3.0×10^{-10} m and the reduced mass is 2.21×10^{-27} kg. What is the entropy of activation for this process?

Solution

The gas-kinetic rate coefficient is given by Eq. 20.16:

$$k_{\text{max}}(T) = [\pi\sigma_{AB}^2]\left[\frac{8kT}{\pi\mu}\right]^{1/2}. \quad (1)$$

For the reaction in this problem, the result is

$$k_{\text{max}}(300 \text{ K}) = (3.14159)(3.0 \times 10^{-10} \text{ m})^2\left[\frac{8(1.381 \times 10^{-23})(300)}{(3.14159)(2.21 \times 10^{-27})}\right]^{1/2}$$

$$= 6.178 \times 10^{-16} \text{ m}^3 \text{ s}^{-1}. \quad (2)$$

Converting to $\text{m}^3 \text{ mol}^{-1} \text{ s}^{-1}$, we obtain

$$k_{\text{max}}(300) = 6.178 \times 10^{-16} \text{ m}^3 \text{ s}^{-1} \times 6.022 \times 10^{23} \text{ mol}^{-1}$$

$$= \underline{3.72 \times 10^8 \text{ m}^3 \text{ mol}^{-1} \text{ s}^{-1}}. \quad (3)$$

The entropy of activation can be obtained from Eq. 20.80:

$$Nk(T) = \left(\frac{ekTV'}{h}\right)\exp\left[\frac{\Delta\bar{S}_o^\ddagger}{R}\right]\exp\left[-\frac{E_a}{RT}\right]. \quad (4)$$

Equating $Nk(T)$ to the result for the hard sphere at 300 K with $E_a = 0$, we obtain

$$\left(\frac{ekTV'}{h}\right)\exp\left[\frac{\Delta\bar{S}_o^\ddagger}{R}\right] = 3.72 \times 10^8, \quad (5)$$

where the units are already compatible.

Solving Eq. 5 for $\Delta\bar{S}_o^\ddagger$, we get

$$\Delta\bar{S}_o^\ddagger = R\ln\left[\frac{3.72 \times 10^8 h}{ekTV'}\right] = R\ln\left[\frac{(3.72 \times 10^8)(6.626 \times 10^{-34})}{(2.71828)(1.381 \times 10^{-23})(300)(1)}\right]$$

$$= \underline{-89.2 \text{ J mol}^{-1} \text{ K}^{-1}}. \quad (6)$$

Since we cannot have a bimolecular reaction with a rate coefficient substantially larger than k_{max}, this value of $\Delta\bar{S}_o^\ddagger$ represents the maximum value (smallest negative magnitude) we expect to observe for a bimolecular gas-phase reaction.

20.31 The data used to plot Figure 20.27 in the text are given in the table on the top of the next page. Since chemiluminescence cannot be observed from the ground vibrational state, the result for $v = 0$ is assumed. Use these data to verify the result given in the text that, on the average, 249 kJ mol^{-1} of vibrational excitation energy is present as HF(g) vibrational energy.

HF(g) Vibrational State	Excitation Energy (kJ mol^{-1})	Probability
0	0	0.000
1	47	0.017
2	93	0.026
3	136	0.046
4	178	0.086
5	217	0.205
6	255	0.291
7	291	0.137
8	326	0.103
9	359	0.089
10	390	0.000

Solution

The average value of any discrete quantity f is given by

$$\langle f \rangle = \sum_{\text{all } i} f_i P(f_i), \tag{1}$$

where $P(f_i)$ is the normalized probability of observing a value f_i in a measurement of f. The average vibrational excitation of HF(g) is, therefore,

$$\begin{aligned}\langle E_{\text{ex}} \rangle = \sum_{i=0}^{10} E_i P(E_i) &= (0)(0) + (47)(0.017) + (93)(0.026) + (136)(0.046) \\ &+ (178)(0.086) + (217)(0.205) + (255)(0.291) + (291)(0.137) \\ &+ (326)(0.103) + (359)(0.089) + (390)(0.00)\ \text{kJ mol}^{-1} = \underline{249\ \text{kJ mol}^{-1}},\end{aligned} \tag{2}$$

which is the result given in the text.

20.33 Sam seems to have one last question. "Last chance, Sam. What's the problem?"

"I know everyone is anxious to wrap it up and leave, but I'm curious. Collision theory shows that $k(T)$ for the reaction of two hard spheres A and B with a collision radius σ_{AB} is

$$k(T) = p[\pi\sigma_{AB}^2]\left[\frac{8kT}{\pi\mu}\right]^{1/2}\left[1 + \frac{E_a}{kT}\right]\exp\left[-\frac{E_a}{kT}\right],$$

provided that the reaction probability is given by Eq. 20.12. Suppose we evaluate this rate coefficient using Eq. 20.67, with the assumption that the transition state corresponds to the structure illustrated at the right. What relationship must exist between the transmission coefficient κ and the steric factor p in order for the two results to be identical?"

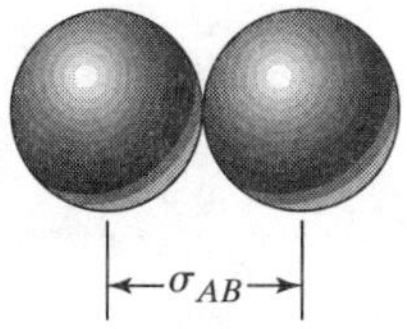

"Sam," I reply, "I have the feeling that you already know the answer to your question. Do you?"

With a sly grin, "I think so."

"I think so, too."

As one last problem for the road, can you answer Sam's question?

Solution

The transition-state-theory expression for the rate coefficient is given by Eq. 20.67:

$$k_{\text{TST}}(T) = \frac{\kappa kTV}{h}\frac{g_{\text{el}}^{\ddagger}}{g_{\text{el}}^{A}g_{\text{el}}^{B}}\exp\left[-\frac{E_b}{kT}\right]\frac{z_{\text{tr}}^{\ddagger}z_{\text{rot}}^{\ddagger}z_{\text{vib}}^{\ddagger}}{z_{\text{tr}}^{A}z_{\text{rot}}^{A}z_{\text{vib}}^{A}z_{\text{tr}}^{B}z_{\text{rot}}^{B}z_{\text{vib}}^{B}}. \tag{1}$$

When A and B are hard spheres, they possess only translational energy. Therefore, Eq. 1 can be written in the form

$$k_{\text{TST}}(T) = \frac{\kappa kTV}{h}\exp\left[-\frac{E_b}{kT}\right]\frac{z_{\text{tr}}^{\ddagger}z_{\text{rot}}^{\ddagger}z_{\text{vib}}^{\ddagger}}{z_{\text{tr}}^{A}z_{\text{tr}}^{B}} \tag{2}$$

where we have omitted the ratio of electronic degeneracies, since hard spheres do not have them. The diatomic transition state A–B has only one vibrational degree of freedom, so it must correspond to motion along the reaction coordinate, which has already been included in obtaining the factor kT/h in Eq. 2. Therefore, we can omit the factor $z_{\text{vib}}^{\ddagger}$. This gives

$$k(T) = \frac{\kappa kTV}{h}\exp\left[-\frac{E_b}{kT}\right]\frac{z_{\text{tr}}^{\ddagger}z_{\text{rot}}^{\ddagger}}{z_{\text{tr}}^{A}z_{\text{tr}}^{B}}. \tag{3}$$

If we use the energy levels for a particle in an infinite well, the molecular-translational partition is given by $z_{\text{tr}} = V/h^3[2\pi mkT]^{3/2}$. Thus, the ratio of the translational partition functions multiplied by V is

$$\frac{Vz_{\text{tr}}^{\ddagger}}{z_{\text{tr}}^{A}z_{\text{tr}}^{B}} = \frac{V\left[\frac{V}{h^3}[2\pi MkT]^{3/2}\right]}{\left[\frac{V}{h^3}[2\pi m_A kT]^{3/2}\right]\left[\frac{V}{h^3}[2\pi m_B kT]^{3/2}\right]} = \frac{h^3}{(2\pi kT)^{3/2}}\left[\frac{M}{m_A m_B}\right]^{3/2}, \tag{4}$$

where $M = m_A + m_B$. Recognizing the fact that the reduced mass μ is $m_A m_B/M$, we can write Eq. 4 in the form

$$\frac{Vz_{\text{tr}}^{\ddagger}}{z_{\text{tr}}^{A}z_{\text{tr}}^{B}} = \frac{h^3}{(2\pi\mu kT)^{3/2}}. \tag{5}$$

The rotational partition function for the diatomic transition state is

$$z_{\text{rot}}^{\ddagger} = \frac{kT}{\sigma Bh} = \frac{8\pi^2 kTI}{h^2} = \frac{8\pi^2 kT\mu\sigma_{AB}^2}{h^2}, \tag{6}$$

provided that we use a rigid-rotor expression for the rotational energy levels and assume that the temperature is sufficiently high for us to treat rotation classically. To obtain Eq. 6, we have made use of the fact that the symmetry number for the A–B transition state is unity. Combining Eqs. 3, 5, and 6, we obtain

$$\begin{aligned} k_{\text{TST}}(T) &= \frac{\kappa kT}{h}\frac{h^3}{(2\pi\mu kT)^{3/2}}\frac{8\pi^2 kT\mu\sigma_{AB}^2}{h^2}\exp\left[-\frac{E_b}{kT}\right] \\ &= \kappa\left[\frac{(kT)^2 8\pi\mu}{(2\pi)^{3/2}(kT)^{3/2}\mu^{3/2}}\right](\pi\sigma_{AB}^2)\exp\left[-\frac{E_b}{kT}\right]. \end{aligned} \tag{7}$$

A little algebra produces

$$k_{\text{TST}}(T) = 2\kappa\left[\frac{2kT}{\pi\mu}\right]^{1/2}(\pi\sigma_{AB}^2)\exp\left[-\frac{E_b}{kT}\right] = \kappa\left[\frac{8kT}{\pi\mu}\right]^{1/2}(\pi\sigma_{AB}^2)\exp\left[-\frac{E_b}{kT}\right]. \tag{8}$$

The hard-sphere rate coefficient derived in the text is

$$k_c(T) = p[\pi\sigma_{AB}^2]\left[\frac{8kT}{\pi\mu}\right]^{1/2}\left[1 + \frac{E_a}{kT}\right]\exp\left[-\frac{E_a}{kT}\right]. \tag{9}$$

Equations 8 and 9 will be identical if we have

$$\boxed{\kappa = p\left[1 + \frac{E_a}{kT}\right]}. \tag{10}$$